PROBABILITY AND STATISTICS WITH APPLICATIONS:

A PROBLEM SOLVING TEXT

SECOND EDITION

For Marilyn, - L.A.

ACTEX Learning
Greenland, NH

Requests for permission should be addressed to

ACTEX Learning
PO Box 69
Greenland, NH 03840

Manufactured in the United States of America

10 9 8 7 6 5 4 3 2

Cover design by Jeff Melaragno

Library of Congress Cataloging-in-Publication Data

Asimow, Leonard A., 1939-
Probability and statistics with applications : a problem solving text / Leonard A. Asimow, Mark M. Maxwell. -- Second Edition.
pages cm
Revised edition of the authors' Probability and statistics with applications, 2010.
Includes bibliographical references and index.
Summary: "Probability and Statistics with Applications is an introductory textbook designed to make the subject accessible to college freshmen and sophomores concurrent with their study of calculus"-- Provided by publisher.
ISBN 978-1-62542-472-3 (alk. paper)
1. Casualty Actuarial Society--Examinations--Study guides. 2. Mathematical statistics--Examinations, questions, etc. 3. Insurance--Statistics. I. Maxwell, Mark M., 1967- II. Title.
HG8045.A85 2015
519.5024'368--dc23
2015010222

ISBN: 978-1-62542-472-3

PREFACE

Preface to the Second Edition

This edition contains a number of new topics, primarily in the mathematical statistics portion of the text. We continue to cover all of the material necessary for the SOA Exam P (Probability) in Chapters 1-8. We have expanded the second part of the book, now Chapters 9-12, to provide full coverage of the syllabus for the Casualty Actuarial Society Exam ST, and the statistics portion of the new CAS Exam S.

New sections include explanations and worked examples on the following topics:

- Mixture distributions,
- Non-homogeneous Poisson processes,
- Sufficient statistical estimators and the linear exponential family,
- Bayesian analysis and conjugate prior distributions,
- Nonparametric statistical methods,
- Graphical methods.

As in the first edition, our aim is to provide the best qualities of both a standard textbook and an actuarial exam study manual. We attempt to provide full explanations and derivations or proofs wherever feasible for an introductory text. We also provide a multitude of examples and problems, including many from earlier actuarial exams. Space limitations prevent us from providing complete solutions to all of the exercises in the body of the text. However, a complete solutions manual is available as a separate volume.

We have benefited from much assistance from reviewers and editors for this edition. In particular we express our appreciation to Ali Ishaq, Tom Lonergan, Rajesh Sahasrabuddhe and Emiliano Valdez, who read all or portions of the manuscript and provided valuable suggestions.

We are indebted to David Hudak for his assistance with the new material on non-homogeneous Poisson processes.

We also gratefully acknowledge the detailed editorial review provided by Geoff Tims. Finally, we extend heartfelt thanks to Stephen Camilli and Garrett Doherty of ACTEX, the former for his editorial guidance and encouragement, the latter for his formidable technical skills in producing the final manuscript pages.

L.A.

Preface to the First Edition

A cursory search of Google Books reveals thousands of titles with the words *Probability and Statistics*, or *Mathematical Statistics*, thus prompting the inevitable question, why yet another? In view of this superabundance of choices we feel impelled to offer a few words of explanation in our defense for adding to this already crowded marketplace.

The idea for this textbook evolved out of a two-semester course in probability and statistics we have been offering for many years to sophomore and qualified freshman students, predominantly actuarial science majors. The goals of our course are twofold: to lay the foundations of calculus-based probability theory for our students, and to prepare them to pass the actuarial exam covering this material as early as possible in their collegiate careers.

The primary market segment we are targeting consists of students like ours – freshmen and sophomores – who are studying calculus-based probability while simultaneously learning the selfsame calculus it is based upon. We desire that our actuarial students be prepared to pass Exam P/1 (jointly offered by the Society of Actuaries of Actuaries (SOA) and the Casualty Actuarial Society (CAS)) no later than the end of their sophomore year. Consequently, the probability and statistics component of their education tends to overlap with the typical 3-semester calculus sequence they all take.

Our experience has been that the myriad existing textbooks in probability and statistics fall into two types. They are either designed to support a "calculus-free" environment suitable for general business school statistics courses, or they are intended for more advanced, calculus-based mathematical statistics courses for juniors and seniors with the requisite technical background. The first category does not provide the depth of understanding required for the actuarial exam, and the second type of book tends to be too formal and advanced for our students.

For these reasons we decided to produce an introductory text in calculus-based probability and statistics whose level is comparable to a modern-day calculus book, and that could reasonably be used by freshman or sophomore students studying the material concurrently with their calculus classes. We consciously strive to pace the material in a way that makes it accessible to a student whose background consists of just one semester of college-level calculus. This might be, for example, either entering freshmen with AP or high-school/college credit concurrently taking Calculus II, or sophomores with one or two semesters of calculus already under their belts.

Chapters 1-4 present the rudiments of probability theory for discrete distributions with little or no reference to calculus topics, save for the basic knowledge of infinite series required for understanding the geometric and the Poisson distributions. Chapter 5, entitled Calculus and Probability, introduces continuous random variables and is the first chapter heavily dependent on derivatives and integrals. The material on continuous, jointly distributed random variables comes in Chapter 7, by which time our students will have been introduced to double integrals in their Calculus III class. The second part of the book, comprising Chapters 9-11, covers all of the syllabus topics for the statistics portion of CAS Exam 3L – Life Contingencies and Statistics Segment. Taken as a whole, this book provides ample content to serve as the text for the standard two-semester introductory sequence in mathematical statistics and probability.

The text contains nearly 800 exercises, many with multiple parts. Numerical answers are given in the back of the book and a supplementary manual with complete solutions is available separately. Many of the exercises, and some examples, are based on previous actuarial exam questions released by the SOA and the CAS. All of the SOA/CAS Exam P/1 Sample Exam Questions (142 in number at press time in late 2009) have been incorporated into the text. In addition, we have used many statistics questions from CAS Exam 3 and 3L, as well as questions from the earlier Exam 110 (Probability and Statistics), in Chapters 8-11. We are grateful to the Society of Actuaries and the Casualty Actuarial Society for their permission to use these materials.

While designed primarily with the actuarial audience in mind, we hope this text will have appeal for a broader audience of mathematics and statistics students. We have sought to engage students with a light, informal style, emphasizing detailed explanations and providing a multitude of examples. At the same time, we have sought at each stage to present a sufficient glimpse into the theoretical underpinnings to make the text suitable for more advanced students. The overarching leitmotif, however, is problem solving, and we emphasize the requisite skills throughout. We hope to have a struck a balance that will allow students at all levels to benefit from a close reading of the text.

We have benefited from the many helpful comments and valuable insights provided by our reviewers: Carolyn Cuff, Westminster College; Thomas Herzog, Department of Housing and Urban Development; Thomas Lonergan, CIGNA; Jeffrey Mitchell, Robert Morris University; Emiliano Valdez, University of Connecticut; Charles Vinsonhaler, University of Connecticut. We would also like to acknowledge the invaluable editorial assistance provided by Gail Hall of ACTEX, whose firm but gentle hand managed to guide this project to completion. We are also indebted to Marilyn Baleshiski, whose patience and skill rendered a lumpy manuscript into a finished text. Our thanks to you all. Needless to say, for all remaining errors and indiscretions, the authors have only each other to blame.

Finally, we note that personages (both named and unnamed) who appear in various exercises and examples are completely fictitious and any resemblance to real people (living or departed) is purely coincidental.

Len Asimow, Moon Township, PA
Mark Maxwell, Austin, TX

TABLE OF CONTENTS

CHAPTER 3 DISCRETE RANDOM VARIABLES 89

CHAPTER 4 SOME DISCRETE DISTRIBUTIONS 133

CHAPTER 5 CALCULUS, PROBABILITY, AND CONTINUOUS DISTRIBUTIONS 175

CHAPTER 6 SOME CONTINUOUS DISTRIBUTIONS 233

CHAPTER 7 MULTIVARIATE DISTRIBUTIONS 305

CHAPTER 8 A PROBABILITY POTPOURRI 377

CHAPTER 11 THEORY OF ESTIMATION AND HYPOTHESIS TESTING 537

CHAPTER 12 A STATISTICS POTPOURRI 631

CHAPTER 1

COMBINATORIAL PROBABILITY

1.1 THE PROBABILITY MODEL

The fundamental object of interest in this text is the ***probability model***. Probability models are based on experiments, such as flipping a coin or tossing a pair of dice, for which there are multiple possible outcomes. It is impossible to say in advance which outcome will occur. In this sense a probability model constitutes an apt metaphor for life, where we rarely know the ultimate results, or future outcomes, of our actions. This is in contrast to *deterministic* models, such as a well-designed chemistry experiment, where one knows in advance the outcome with a high degree of certainty (a bad odor will arise, a small explosion will occur, a penny will completely dissolve in 60 seconds, etc.).

A probability model has three essential elements. These three elements can be defined as follows:

The Elements of a Probability Model

i. The description of the underlying probability experiment,
ii. A list, or a procedure for listing, all possible ***outcomes*** of the experiment,
iii. A rule for assigning numbers between 0 (impossible) and 1 (certain) to subsets of outcomes.

The set of all outcomes on the list in (ii) is called the ***sample space***. Subsets of the sample space are called ***events***. The numbers in (iii) are called ***probabilities***, and they measure the likelihood of an outcome along a scale from 0% to 100%. These numbers must satisfy one other property, denoted below as the ***additive property***, which we will consider in more detail as we continue.

The Additive Property

iii′. The "sum" of the probabilities of all the outcomes must equal one.

"Sum" is written in quotation marks because the summing process may be more complicated than simple addition, possibly involving an infinite series or a definite integral. As a consequence of the additive property, it seems sensible (and it is correct) to define the *probability of an **event*** as the "sum" of the probabilities of all the outcomes contained in the event. The additive property assures that the event consisting of the entire sample space has probability one. That is, it is certain that one of the outcomes listed in (ii) will occur.

Example 1.1-1 Elements of a Probability Model

The description of an experiment could be as simple as "*toss a coin once*." A complete list of outcomes would be the set consisting of *heads* or *tails*. A simple rule for assigning probabilities is $\frac{1}{2}$ for heads and $\frac{1}{2}$ for tails. Notation that summarizes these statements is

$$\text{sample space} = \{heads,\ tails\},\quad \Pr(heads) = \tfrac{1}{2},\ \text{and}\ \Pr(tails) = \tfrac{1}{2}.$$

In this case the experiment, the outcomes, and the probability assignments constitute an accurate model of the real-world activity of tossing a balanced (fair) coin. In life there is no such thing as a perfectly balanced coin, and there is always the remote possibility of a third outcome (e.g., *coin stands on edge*). But since we are constructing an idealized mathematical model of an experiment, we keep things as simple as possible while preserving the essential structure of the real world phenomena we are modeling.

Example 1.1-2 Sample Space

Suppose that a couple conducts the experiment of having three children. Their concern is the *gender* of the children.

a. List the sample space of all possible outcomes.
b. List the event that a boy is the first born child.

Solution

The sample space can be represented by the set $\{BBB, BBG, BGB, BGG, GBB, GBG, GGB, GGG\}$, where B denotes the birth of a boy and G denotes the birth of a girl. Therefore the *outcome* BGG denotes a boy being the first born, followed by the births of two girls. The *event* of a boy being the first born consists of the subset $\{BBB, BBG, BGB, BGG\}$.

Here, the set of outcomes was identified as a list of *words*. By *words* in this text we mean any systematic grouping of letters or numbers used to describe outcomes. In this example, the outcomes consist of all 3 letter *words* (not just dictionary words) consisting of the letters B and G exclusively. Note that the words are systematically listed in alphabetical order, which is not only convenient, but helps to assure that we don't inadvertently omit an outcome. ❐

If all experiments were as simple as tossing a single coin, or having three children (which is conceptually simple, although more complicated in actual practice[1]), then you wouldn't need this text. Needless to say, probability models can be quite complex, with a very large (or even infinite) number of outcomes.

---

[1] This comment was written by Len, father of two.

There are a number of different ways of classifying probability models, but the most basic is the distinction between *discrete* probability models and *continuous* probability models. In *discrete* probability models, the outcomes can be enumerated in a list, as *outcome*(1), *outcome*(2), etc. This list can be *finite* or *countably infinite*. An example of the latter would be the experiment of tossing a coin until the first head appeared. One way of listing this sample space would be,

$$\{H,TH,TTH,TTTH,...\}$$

Alternatively and equivalently, we could describe the sample space as $\{1,2,3,4,...\}$, where the natural number stands for the coin toss on which the first head occurs. If this were a fair coin, then an appropriate assignment of probabilities would be $\Pr(H) = \Pr(1) = .5$, $\Pr(TH) = \Pr(2) = .25$, $\Pr(TTH) = \Pr(3) = .125$, and so forth. In Chapter 4 we will verify that the infinite sum of these probabilities equals one.

A *continuous* model has outcomes along a *continuum*. A description of such an experiment might be, "*toss a dart at the unit interval* $[0,1]$." The set of outcomes consists of all real numbers between 0 and 1. For those readers impertinently protesting that one can't toss a dart at a one-dimensional line in the real world, we can reformulate the experiment as "*spin a wheel-of-fortune*" whose edge is calibrated with the unit interval. Here, the fixed pointer plays the role of the dart, and we are now "*throwing the unit interval at the dart*," so to speak. In either case, the mathematical model is the same. The real difficulty here is in assigning the probability numbers. Since the model is continuous, it should not be surprising that we require the tools of calculus to do this properly. Essentially, probabilities are assigned to intervals, not single points, and the summing of the probabilities referred to in *additive property* (iii') requires *integration*. In this example, the probability of the pointer landing in the interval $[a,b]$ is its length, $b-a$. We will leave the details until Chapter 5.

Finally, it must be emphasized that the *way* in which the list in (ii) is constructed can be extremely important.

Example 1.1-3 Sample Space and Assigning Probabilities

Consider the experiment of *rolling a pair of fair dice.*

(a) List all outcomes in the sample space.

(b) Assign probabilities to the outcomes.

(c) Find the probability that the sum of the dice is 8.

Solution

Generally, when we roll a pair of dice the outcome of real interest is the *sum*. In particular, we would like to model the real-life probability of the outcome *sum equals* 8. Our list of all outcomes could therefore be the eleven integers from 2 to 12. That leaves the issue of assigning probabilities. Since this is a mathematical model we can assign probabilities any way we like so long as the probabilities sum to one. In particular, we could arbitrarily assign *equally likely outcomes* so that the probability of a sum equal to 8 would be the same as the probability of a sum equal to 12 or any other outcome, namely 1/11. However, this model will not conform to our experience gained over the many years of playing with dice.

We will therefore take a different approach to listing the outcomes. We first imagine in our mathematical model that we can distinguish between the two dice (assume that one is green and the other is grey). We then list all outcomes as two-letter *words* with the *letters* being the *integers* from 1 to 6), with the green die result followed by the grey die:

For example, the word (14) represents the green die (left) showing one dot and the grey die (right) showing four dots.

11	21	31	41	51	61
12	22	32	42	52	62
13	23	33	43	53	63
14	24	34	44	54	64
15	25	35	45	55	65
16	26	36	46	56	66

In this second method of listing, it *is* reasonable to assume that the outcomes are all equally likely. This is because we assume that the individual die gives equally likely outcomes from 1 to 6 (such dice are called fair and balanced - like the ideal newscast), and therefore the pairs are also equally likely to occur. In this second method, there are a total of 5 outcomes out of 36 with a *sum equal to* 8. Thus, "*sum of 8*" is no longer an *outcome,* but an *event.* If the 36 words are all assigned the probability of 1/36, then the probability of the event *sum of* 8 is $\frac{5}{36}$, which conforms to actual experience. Even though there are more outcomes (36 versus 11) to deal with, the individual outcomes have the advantage of being naturally equally likely. This provides a mathematical model that conforms more readily to reality.

In general it is better to list outcomes in a way that preserves as much information as possible about the underlying experiment, and which takes advantage of an intuitively natural assignment of equally likely outcomes. □

Exercise 1-1 At a picnic, you select a main course from a hamburger or a chicken sandwich. You select a side from potato chips or coleslaw, and you select a beverage from soda, water, or milk. These 3 selections result in a **meal.**

(a) Write out the sample space of all **meals**.

(b) Write out the event (subset) consisting of all meals with soda as the beverage.

(c) Assuming all meals are equally likely, calculate the probability of getting soda with your meal.

Exercise 1-2 Roll a pair of fair six-sided dice.

(a) How many ways can *doubles* (both die have the same number of dots) be rolled?

(b) What is the probability that the *sum of the two dice is a prime number*?

Exercise 1-3 A convenience store has three packages of plain M&M'S®,[2] two packages of peanut M&M'S, one package of dark chocolate M&M'S, one package of peanut butter M&M'S, and one package of almond M&M'S. Select two packages of candy.

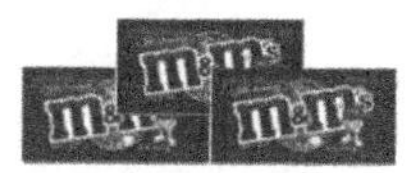

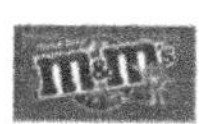

(a) Write out the sample space of all possible outcomes.

(b) Assume the outcomes are equally likely. Calculate the probability you select exactly one package of plain M&M'S, (and the other package is **not** plain – either peanut, dark chocolate, peanut butter, or almond).

(c) Calculate the probability of getting **at least** one package of plain M&M'S.

1.2 FINITE DISCRETE MODELS WITH EQUALLY LIKELY OUTCOMES

In this section we consider in more detail the wide class of probability models in which there is only a finite number of possible outcomes. Probabilities are assigned, as in the dice example in the previous section, so that all outcomes are equally likely. This situation of *equally likely outcomes* governs many of the typical probability problems that arise in modeling games (drawing cards, tossing dice, flipping coins, etc.).

In the case of equally likely outcomes, the assignment of probabilities is straightforward. Since the sum of the probabilities of outcomes must equal one, and the outcomes are equally likely, each outcome has probability $\frac{1}{n}$, where n is the total number of possible outcomes.

Probability with Equally Likely Outcomes

Let S be a sample space consisting of n equally likely outcomes. Let E be an event in S consisting of r outcomes. Then the probability of event E, denoted $\Pr(E)$, is

$$\Pr(E) = \frac{\text{number of outcomes in event } E}{\text{total number of possible outcomes}} = \frac{r}{n} = \frac{N(E)}{N(S)},$$

where $N(E)$ denotes the number of elements in the set E.

Determining the value of n or r is often the most complicated part. This leads us to a discussion of various counting techniques (called *combinatorics*) such as *permutations* and *combinations*.

[2] ®/™ M&M'S is a registered trademark of Mars, Incorporated and its affiliates. Mars, Incorporated is not associated with ACTEX Publications, Inc. The M&M'S® trademark is used with permission of Mars, Incorporated. ©Mars, Inc. 2010.

1.2.1 TREE DIAGRAMS

A tree diagram is a fundamental conceptual tool for constructing lists of outcomes. It is also a valuable approach for tracking probabilities in a multistage experiment.

Example 1.2-1 Tree Diagram

A family has two children. Construct a tree diagram showing all possible combinations of boys and girls.

Solution

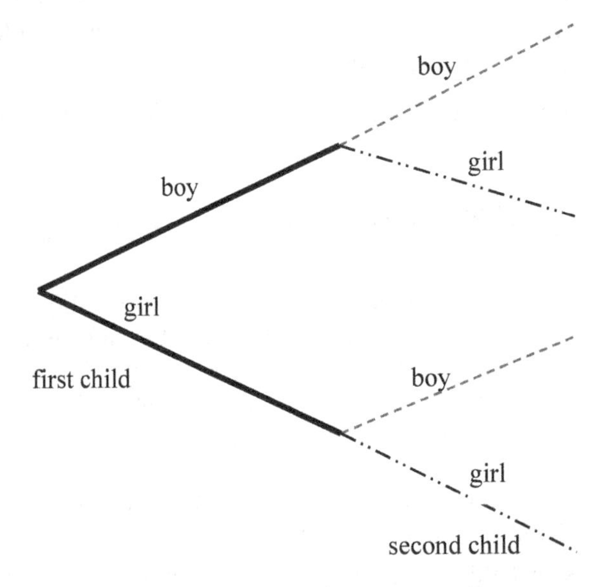

There are four distinct possible outcomes, each corresponding to exactly one path through the diagram. Traveling across the top path, for example, denotes the birth of two boys. Using the diagram, we can write the sample space as $\{BB, BG, GB, GG\}$ where B denotes the birth of a boy and G denotes the birth of a girl.

Example 1.2-2 Tree Diagram of Your Wardrobe

Suppose that your wardrobe consists of two types of shoes (Birkenstocks and Nikes), three different shirts (baseball jersey, polo, tee-shirt), and two types of pants (shorts and Levi button-fly 501 jeans).

(a) Construct a tree diagram showing all possible outfits that you could wear.

(b) Assuming all outfits are equally likely, calculate the probability of choosing an outfit that includes a polo shirt.

Solution

(a)

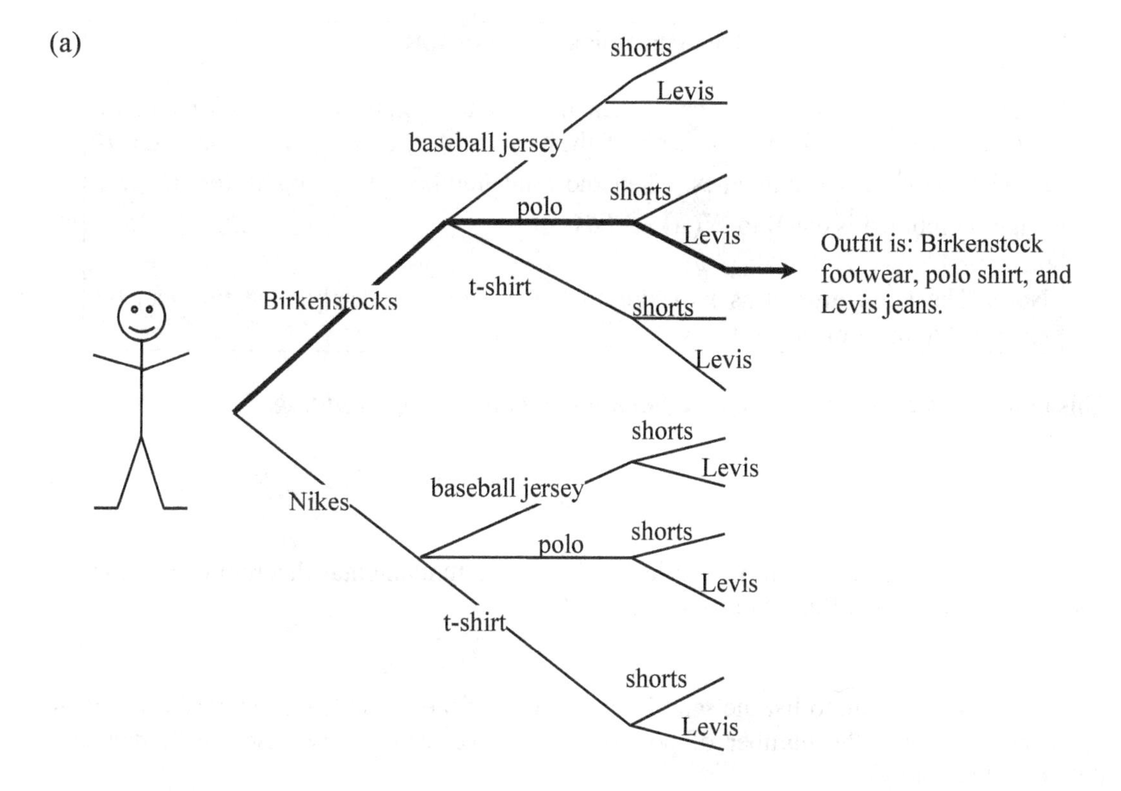

(b) The diagram consists of twelve distinct branches, each corresponding to a distinct outfit. Four of these branches include a polo shirt, so that the probability is 4/12 = 1/3. This is, of course, the same as the probability of choosing a polo shirt from among the 3 possible shirts available. ❐

In these examples, each *outcome* can be represented by a particular path through the tree. In the first example there are two possible results (*boy* or *girl*) for the gender of the first child, and each of these leads to two possible results for the gender of the second child. This leads to a total of $2 \times 2 = 4$ paths (outcomes) through the tree. In the second example we have 2 possible pairs of shoes, and for each of these there are 3 possible shirts. For each choice of shoes and shirt, there are 2 choices for pants. The total number of paths through the tree is $2 \times 3 \times 2 = 12$.

The general rule at work here is called the *multiplication principle*.

1.2.2 THE MULTIPLICATION PRINCIPLE

The Multiplication Principle

Suppose an experiment can be broken down into a first stage A consisting of $N(A)$ outcomes **and** that for each of these outcomes, there is a second stage B consisting of $N(B)$ outcomes. Then the total number of outcomes for the two stages combined is equal to $N(A)\times N(B)$.

Note: The word **and** is as indicator that one should multiply. The rule can be extended to three or more stages.

This principle is commonly called *the* ***fundamental theorem of counting***.

Example 1.2-3 The Multiplication Principle

For the wardrobe described in Example 1.2-2, use the fundamental theorem of counting to determine how many different outfits you could wear.

Solution

We used a tree diagram to list the sample space of all of the possible outfits. In this problem, our concern is only the number of possible outfits we have, so we use the fundamental theorem of counting.

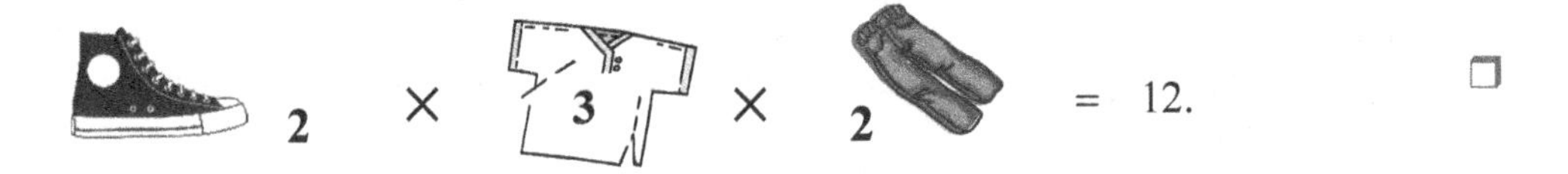

❐

Example 1.2-4 The Multiplication Principle

Assume standard license plates in the state of Oregon are comprised of three letters followed by three numbers. How many different license plates could theoretically be fabricated?

Solution

The total number of standard license plates is $26\cdot 26\cdot 26\cdot 10\cdot 10\cdot 10 = 26^3\cdot 10^3 = 17{,}576{,}000$. Using a tree diagram to list all possible outcomes from AAA000 through ZZZ999 would require more space than is available here.

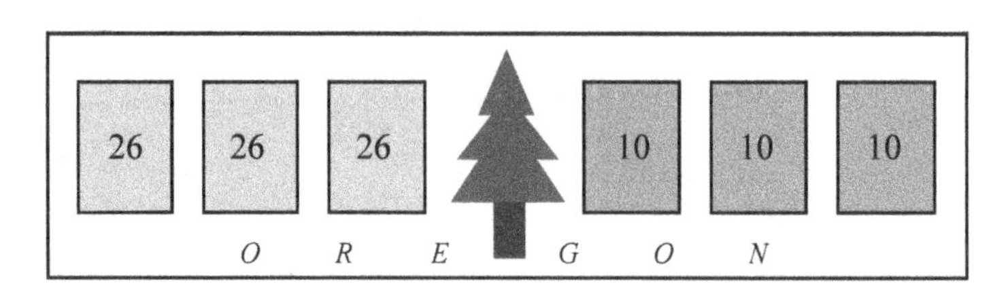

❐

Exercise 1-4 Your current wardrobe consists of eleven different items: three pairs of shoes, two pairs of pants, and six different shirts. Your mom offered to buy you a new shirt **or** a new pair of shoes. Which item (shoes or shirt) should your mom buy for you to maximize your total number of outfits (shoes-shirt-pant)?

Exercise 1-5 Roll three fair six-sided dice.

(a) Assuming the dice are distinguishable and always tossed in a particular order (e.g., red-white-blue), how many different outcomes are possible?

(b) How many outcomes are there in which the first two dice come up 3 and 4 respectively?

(c) What is the probability that the first two dice come up 3 and 4 respectively?

(d) Would you like to use a tree diagram to list all of the possible outcomes?

Exercise 1-6 Assuming there are no restrictions, how many different 9-digit social security numbers are possible?

Exercise 1-7 Assume that a standard California license plate consists of a digit, 3 letters, and then 3 more digits (e.g., 1SAM123), except that I, O, and Q are excluded from the first and third alpha positions. How many possible standard license plates are there in California?

Exercise 1-8 A thief plans to rob four banks. His town of Podunk, Idaho has six banks. He could rob a bank twice, but never consecutively. In how many ways may he select the four banks?

Exercise 1-9 How many ten-digit phone numbers are possible in the United States? **Note**: Creating a mathematical model that accurately reflects the real world is extremely important and frequently difficult. Rules that govern the problem should be understood and listed. In this case, we assume that the first digit cannot be 0 or 1 and the fourth digit cannot be 0.

1.2.3 PERMUTATIONS

Permutations

Given a set of n distinguishable objects, an ordered selection of r different elements of the set is called a ***permutation of n objects chosen r at a time.***

Example 1.2-5 Permutations

In May you travel to Louisville and make a trifecta wager on the Kentucky Derby. A trifecta bet consists of selecting the first three finishers in order. That is, the one that finishes first (win), the one that finishes second (place), and the one that finishes third (show).. How many different trifecta wagers are possible if 14 horses go to post? Rephrased mathematically, how many permutations are there of 14 objects (distinguishable horses) chosen 3 at a time?

Solution

First we note that any of the 14 horses could win. But the winning horse cannot also finish second, so there remain 13 horses that could be selected for second. Similarly, there remain 12 horses that can be selected for third. Using the multiplication principle directly we see the number of distinct trifecta wagers is,

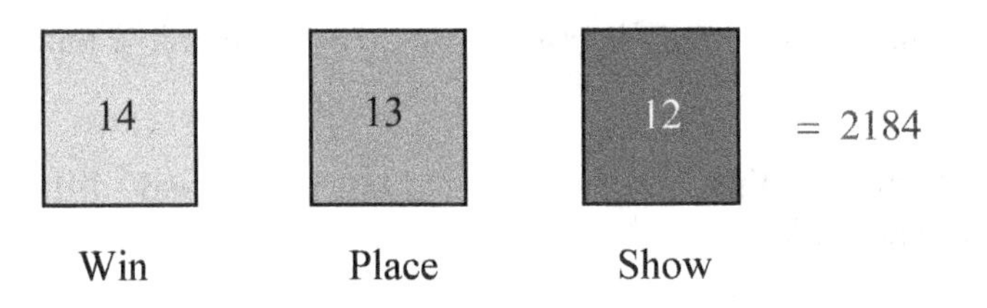

❒

Factorials

Let n be a whole number. Then $n!$ (read as "n factorial") is defined by

$$n! = n \cdot (n-1) \cdot (n-2) \cdot \cdots \cdot 3 \cdot 2 \cdot 1.$$

By convention, we define $0! = 1$.

Alternatively, we can express the solution to Example 1.2-5 using factorial notation.

Using factorial notation we can easily generalize from the horserace example to a formula for the number of permutations.

Permutation Formula

The number of permutations on n objects chosen r at a time is denoted by ${}_nP_r$.

$${}_nP_r = \underbrace{n(n-1)(n-2)\cdots(n-r+1)}_{\text{first } r \text{ factors of } n!} = \frac{n!}{(n-r)!}.$$

Note: The $(n-r)!$ in the denominator cancels out the "tail" of $n!$, leaving just the first r factors.

Note: Other common ways to denote permutations: ${}_nP_r = P(n,r) = P_r^n$.

This formula results from applying the multiplication principle to the r stage process of picking the r objects, one at a time and in order.

Note

${}_nP_n = \dfrac{n!}{(n-n)!} = \dfrac{n!}{0!} = n!$ is simply the number of distinct ways of rearranging (choosing a different ordering) all n objects.

The solution to the horserace problem can be rephrased as the number of permutations on 14 horses chosen 3 at a time,

$${}_{14}P_3 = \frac{14!}{(14-3)!} = \frac{14!}{11!} = \underbrace{14}_{\text{win}} \cdot \underbrace{13}_{\text{place}} \cdot \underbrace{12}_{\text{show}} = 2{,}184.$$

We use permutations when the ***order*** of selection is important.

Exercise 1-10 Find the numerical values of the following permutations:

(a) ${}_7P_6$ (b) ${}_{13}P_{12}$ (c) ${}_7P_4$
(d) ${}_5P_2$ (e) ${}_4P_4$ (f) ${}_4P_6$
(g) ${}_{23}P_3$ (h) ${}_3P_0$ (i) ${}_5P_5$
(j) ${}_{10}P_1$ (k) $\dfrac{{}_7P_4}{4!}$ (l) ${}_0P_0$

Exercise 1-11 An engineering faculty member shuffles the quizzes of his 17 students and redistributes them randomly for peer grading.

(a) In how many different ways may the quizzes be handed out?
(b) In how many ways may Kendra, one of the 17 students, be assigned her own quiz for grading?
(c) What is the probability that Kendra will receive her own quiz?

Exercise 1-12 Peter Pickemfast spends a day at the races and decides to play the trifecta in an eight horse field. Knowing nothing about any of the horses, he chooses horses randomly and buys one ticket. Calculate the probability he has the winning trifecta ticket.

Exercise 1-13 Henry Hoodlum speaks to two of the jockeys in an 8 horse field, politely requesting they not finish in the top 3. Assuming they comply, how many trifecta tickets must Henry buy to assure he holds a winning ticket?

Exercise 1-14 A psychology professor has twelve students in class. She insults one student, throws an eraser at a second student, lowers the grade of a third student, and sends a fourth student to the dean's office. In how many different ways can she perform these motivational techniques?

1.2.4 THE BIRTHDAY PROBLEM AND ITS GENRE

There is a classic problem in probability that asks, "What are the chances that two people in this room share the same birthday (month and day only)?" The answer, of course, depends on how many people are in the room, and how many possibilities there are for birthdays. To keep things simple we assume that there are exactly 365 equally likely birthdays. Thus, if you happen to have been born on February 29, you don't count. We could let you celebrate on February 28, but then we would no longer have equally likely outcomes. It is easier to just assume you do not exist in our simplified model. You still exist in the real world and our model aspires to be a simple approximation to reality. We are also aware that birthdays are not equally likely throughout the year, but dealing with this peculiarity would further complicate our model.

We will illustrate the use of the multiplication rule and permutations to answer the question in a general format. Assume there are n distinguishable objects ("birthdays") and we wish to assign them to r "people." We will calculate the probability of the event, E, that there is *no* duplication of birthdays. The event that there *exists* a duplication of birthdays is called the *complement* of E and is denoted E'. Its probability must equal one minus the probability of the event E.

The sample space, S, consists of all possible ways of distributing n "birthdays" to r "people." This is an r stage process with n ways to choose at each stage. Thus, the size of the sample space equals n^r (the multiplication principle in action). Now, the event E consists of all outcomes in which there are *no* duplications, that is, permutations of the n "birthdays" chosen r at a time. Therefore, the event, E, consists of ${}_nP_r$ outcomes. Thus, the probability of no duplications is given by

$$\Pr[E] = \Pr[\text{unique birthdays}] = \frac{{}_nP_r}{n^r},$$

and the probability of at least one duplication is given by,

$$\Pr[E'] = 1 - \frac{{}_nP_r}{n^r}.$$

Example 1.2-6 Birthday Problem Probability

Calculate the probability that in a class of 25 students at least one pair of students will have the same birthday.

Solution

From the second formula, the answer is

$$\Pr\left[\text{at least one shared birthday}\right] = 1 - \frac{{}_{365}P_{25}}{365^{25}} = 0.57.$$ ❐

Exercise 1-15 Find the smallest number of students for which the probability of a birthday duplication is at least ½.

Exercise 1-16 What is the smallest number of people in a room to assure that the probability that at least two were born on the same day of the week is at least 40%?

Exercise 1-17 Seven people enter the elevator on the first floor of a 12 story building. What is the probability that no two will get off on the same floor? You may assume that all floors are equally likely and no one gets off the elevator before it starts up.

Exercise 1-18 The church of Individual Self-Actualization has a hymnal consisting of 35 songs. Twelve members come in and open their hymnals to a random song and commence singing (all songs are equally likely). What is the probability that no two congregants are on the same song?

Exercise 1-19 Each individual in a group of n students is asked to pick an integer at random between 1 and 10 (inclusive). What is the smallest value of n that assures at least a 50% chance that at least two students select the same integer?

1.2.5 COMBINATIONS

We now turn to situations in which the *order* of selection doesn't matter. It only matters if an object is selected or not selected. Think of a poker hand. It does not matter the order that you are dealt all four aces and a fifth card; what matters is that you were dealt all of the aces. (Advice: Bet big, this happens infrequently).

Combinations

Given a set of n distinguishable objects, an unordered selection of r different elements of the set is called a ***combination of** n **objects chosen** r **at a time*** and is denoted by ${}_nC_r$ and read as n choose r.

Combinations apply when order does **not** matter. In determining whether to use combinations or permutations, it is helpful to ask, "Is this like the outcomes of a horserace where order matters?", or "Is this like choosing pizza toppings where order does not matter?"

Example 1.2-7 Combinations

Papa Your Mommas Pizza Parlor has the following four pizza toppings: pepperoni, mushroom, sausage, and olives.

(a) How many different pizzas with two distinct toppings could be ordered?

(b) List the sample space for pizzas with two distinct toppings.

(c) If all two-topping pizzas are equally likely to be ordered, find the probability that sausage is on the pizza.

Solution

The order in which *Papa* places toppings on your pie does not matter. That is, unless you are hopelessly obsessive-compulsive (same as compulsive-obsessive), a pepperoni and mushroom pizza is identical to a mushroom and pepperoni pizza.

(a) Using permutations, we calculate the number of distinct pizzas as ${}_4P_2$ and then divide by the 2! possible orderings of the 2 toppings. Thus, ${}_4C_2 = \frac{4\cdot 3}{2\cdot 1} = 6$.

(b) The six pizzas are pepperoni-mushroom, pepperoni-sausage, pepperoni-olive, mushroom-sausage, mushroom-olive, and sausage-olive. Shorthand notation for our sample space is $\text{Sample Space} = \{PM, PS, PO, MS, MO, SO\}$.

(c) Three pizzas have sausage: pepperoni-sausage, mushroom-sausage, and sausage-olive. Therefore $\Pr(\text{sausage on the 2 topping pizza}) = \frac{3}{6} = 0.5$. ❐

The formula for the number ${}_nC_r$ of combinations can be deduced from the multiplication principle, applied to a *two*-step process for listing the ${}_nP_r$ ordered outcomes.

(1) Choose r objects without regard to order (${}_nC_r$ ways),

(2) Choose a particular ordering (permutation) of the r objects (${}_rP_r = r!$ ways).

Then, ${}_nP_r$ is the product resulting from steps (1) and (2). That is, ${}_nP_r = \frac{n!}{(n-r)!} = {}_nC_r \times r!$.

Solving for ${}_nC_r$ leads to the following formula.

Combination Formula

The number of combinations of r objects chosen from a collection of n distinguishable objects is given by,

$$ {}_nC_r = \frac{{}_nP_r}{r!} = \frac{n!}{r!(n-r)!} = \frac{\overbrace{n\cdot(n-1)\cdot(n-2)\cdots(n-r+1)}^{\text{first } r \text{ factors of } n! \text{ on top}}}{\underbrace{r\cdot(r-1)\cdot(r-2)\cdots 3\cdot 2\cdot 1}_{\text{all } r \text{ factors of } r! \text{ on bottom}}} $$

Note: Other common ways to denote combinations: ${}_nC_r = \binom{n}{r} = C(n,r) = C_r^n$

The form ${}_nC_r = \binom{n}{r}$ is especially common and is referred to as the ***binomial coefficient***. The reason for this will be explained in Subsection 1.4.1.

Exercise 1-20 Find the numerical values of the following combinations:

(a) ${}_7C_6$ (b) ${}_{13}C_{12}$ (c) ${}_{13}C_1$

(d) ${}_5C_2$ (e) ${}_4C_4$ (f) ${}_2C_3$

(g) ${}_5C_3$ (h) ${}_4C_0$ (i) ${}_5C_5$

(j) ${}_{5867}C_1$ (k) $\frac{{}_7P_4}{4!}$ (l) ${}_0C_0$

Exercise 1-21 Verify that ${}_{12}C_3 = {}_{12}C_9$.

Exercise 1-22 Verify the relationship ${}_nC_r = {}_nC_{n-r}$. In terms of pizza toppings, explain this relationship.

Exercise 1-23 A father of 6 children comes home to a disheveled house. His slovenly children are to blame. Since he does not know who is responsible, he will select four of the six children for clean-up duty. In how many ways can he do this?

Note: ${}_6C_4 = 15 = {}_6C_2$, since selecting four children to clean up is equivalent to selecting two children to spare from chores.

Exercise 1-24 *Papa Your Mommas Pizza Parlor* has 6 meat toppings and 7 vegetable toppings from which to select. The parlor has three different sizes of pizza (individual, large, and giant) and two different types of crust (deep-dish and thin).

(a) How many different two-topping pizzas could be ordered?

(b) How many different two-topping pizzas could be ordered with exactly one meat topping and exactly one vegetable topping?

(c) How many different four-topping vegetarian pizzas could be ordered?

Exercise 1-25 Coach Cramer has 15 basketball players: four centers, five forwards, and six guards.

(a) How many ways can she select five players to form her starting line-up, regardless of position?

(b) How many ways can she select five players to form her starting line-up if she needs one center, two forwards, and two guards?

Exercise 1-26 Your local Ice Cream Shoppe has eleven flavors of ice cream, three types of cones (waffle, chocolate, and sugar), and five types of toppings (broken peanuts, whipped cream, hot fudge, maraschino cherries, and sprinkles). How many different two-scoop cones can be ordered that have two different flavors of ice cream, exactly one topping, and your choice of cone?

Exercise 1-27 A young man has a collection of fourteen earrings, three nose rings, seven necklaces, and five hemp bracelets. How many different groupings of jewelry may he wear if he selects three earrings, one nose ring, two necklaces, and one bracelet?

Exercise 1-28 A high school lottery uses two sets of numbered balls. One set consists of ten white balls numbered 1-10 and the second set contains twenty blue balls numbered 1-20. To play, you select two white balls and two blue balls.

(a) How many different outcomes are possible?

(b) Your lottery ticket consists of four numbers: two white numbers, each between 1 and 10 inclusive, and two blue numbers, each between 1 and 20, inclusive. What is the probability that your lottery ticket contains exactly one matching white number and two matching blue numbers?

Exercise 1-29 At a picnic, there was a bowl of chocolate candy that had 10 pieces each of Milky Way®, Almond Joy®, Butterfinger®, Nestle Crunch®, Snickers®, and Kit Kat®. Jen grabbed six pieces at random from this bowl of 60 chocolate candies.

(a) What is the probability that she got one of each variety?

(b) What is the probability that Jen grabs exactly five varieties?

1.2.6 PARTITIONS

Whenever we select a combination of r objects from n distinguishable objects, we are in effect dividing the n objects into two classes, those that are "in" and those that are "out." For example, in selecting 5 starters from 15 players the coach probably just reads off the names of the 5 who are "in." Alternatively though, she could convey the same information by reading off the names of the 10 "out" players, telling them to go sit on the bench (admittedly, an approach that emphasizes the negative and probably wouldn't do much for morale.)

Selecting a combination of objects in effect *partitions* the n objects in two sets, one of size r (the *in* set), the other of size $n-r$ (the *out* set). This is another explanation for the relationship ${}_nC_r = {}_nC_{n-r}$ (see Exercise 1-22).

Many combinatorial problems involve partitioning the population (the n distinguishable objects) into *more* than two subsets. The sizes of the subsets are always prescribed in advance.

Partitions

Let A be a set consisting of n distinguishable objects. Let whole numbers $\{r_1, r_2, \ldots, r_k\}$ be given such that $r_1 + r_2 + \cdots + r_k = n$. A ***partition of*** A ***into subsets of sizes*** $\{r_1, r_2, \ldots, r_k\}$ is a particular distribution of the n objects into disjoint subsets $A_1, A_2, \ldots, A_k$ of sizes $r_1, r_2, \ldots, r_k$ respectively.

Notes

(1) Since the A_i are disjoint and $r_1 + r_2 + \cdots + r_k = n$, it follows that $A = \bigcup_{i=1}^{k} A_i$.

(2) Once objects are assigned to one of the subsets, their order *within* the subset is irrelevant.

(3) As we saw above, a *combination* of r objects is a partition of A into subsets of size r (the *in* set) and size $n - r$ (the *out* set).

Example 1.2-8 Partitions

An intramural basketball team has five highly versatile starters (Amy, Brandy, Chrystal, Diane, and Erin) each capable of playing any one of the three positions of center, forward, and guard. How many different starting *lineups* consisting of one center, two forwards and two guards are possible? We do not distinguish here between *power forward* and *small forward*, or *point guard* and *shooting guard*.

Solution

A *lineup* consists of one center, two forwards and two guards. Thus, the solution consists of calculating the number of partitions of 5 objects (the starters) into 3 subsets (center, forwards, guards) of sizes 1, 2, and 2 respectively. Any of the five players could play at center, so there are ${}_5C_1 = 5$ choices for center. Of the four remaining players, we must choose two to play at forward – and there are ${}_4C_2 = 6$ ways to accomplish this. The remaining two players must be assigned to play guard – there are ${}_2C_2 = 1$ ways to do this (the one way is to have both remaining ladies play as guards). By the multiplication principle, the total number of line-ups is ${}_5C_1 \times {}_4C_2 \times {}_2C_2 = 5 \times 6 \times 1 = 30$.

Since the numbers are manageable, we can try listing all the partitions.

Position	Center	Forwards	Guards
Size	1	2	2
Set	#1	#2	#3
Lineup (1)	Amy	Brandy, Chrystal	Diane, Erin
Lineup (2)	Amy	Brandy, Diane	Chrystal, Erin
Lineup (3)	Amy	Brandy, Erin	Chrystal, Diane
Lineup (4)	Amy	Chrystal, Diane	Brandy, Erin
⋮	⋮	⋮	⋮
Lineup (30)	Erin	Chrystal, Diane	Amy, Brandy

The enterprising student should complete the list of all 30 partitions. The partial listing above is in alphabetical order. Note that, since the ordering of the forwards and guards doesn't matter, each pair is listed in alphabetical order within position for definiteness. ❒

Systematically listing all possible outcomes is an extremely useful tool. We have already used systematic listing to find the sample spaces in a variety of instances, including: families with three children, flipping coins until a head occurs, rolling a pair of fair dice, constructing wardrobes, and constructing basketball line-ups.

Example 1.2-9 Several Partitions of a Deck of Cards

Suppose the set under consideration is a ***standard deck*** of playing cards, consisting of 52 distinguishable objects (no jokers, missing cards, or wild cards). Dividing the deck into four ***suits*** (clubs, diamonds, hearts, and spades) is a particular partition into 4 subsets, all of size 13.

In the card game ***bridge***, the deck is dealt (divided randomly) into 4 hands (sets, often labeled North, East, South and West) each of size 13. Thus, a bridge hand is one of many possible partitions of the deck into 4 subsets of size 13.

Another special partition of a standard deck would be into the thirteen **denominations, or ranks** of cards (ace, king, queen, ..., 2), a partition into 13 subsets, each of size 4.

Partitions of cards need not be of the same size – like in Uno®, gin rummy, and Go Fish. If all 52 cards are dealt to three players, the sets might be partitioned into three hands of size 18, 17, and 17. ❐

Example 1.2-10 Partitions

How many different bridge games are possible?
In other words, how many distinct partitions of 52 objects into 4 subsets of size 13 are possible?

Sample calculations of this type will lead us to a general formula for counting the number of possible partitions. We can break the problem down into a four step process and employ the multiplication principle.

Solution

Step 1: How many ways to choose the cards for North?

Answer: $_{52}C_{13} = \frac{52!}{(13!)(39!)} = 635{,}013{,}559{,}600.$

Step 2: How many ways to choose the cards for South?
Answer (since there are just 39 cards left for South):

$$_{39}C_{13} = \frac{39!}{(13!)(26!)} = 8{,}122{,}425{,}444\ .$$

Step 3: How many ways to choose the cards for East?

Answer: $_{26}C_{13} = \frac{26!}{(13!)(13!)} = 10{,}400{,}600.$

Step 4: How many ways to choose the cards for West?
Answer (note that after Steps 1-3 there are only 13 cards left):

$$_{13}C_{13} = \frac{13!}{(13!)(0!)} = 1.$$

Thus, the total number of bridge hands is given by

$$\begin{aligned} {}_{52}C_{13} \cdot {}_{39}C_{13} \cdot {}_{26}C_{13} \cdot {}_{13}C_{13} &= \frac{52!}{(13!)(39!)} \cdot \frac{39!}{(13!)(26!)} \cdot \frac{26!}{(13!)(13!)} \cdot \frac{13!}{(13!)(0!)} \\ &= \frac{52!}{(13!)(13!)(13!)(13!)} \\ &= (635{,}013{,}559{,}600) \cdot (8{,}122{,}425{,}444) \cdot (10{,}400{,}600) \cdot (1) \\ &= 53{,}644{,}737{,}765{,}488{,}792{,}839{,}237{,}440{,}000, \end{aligned}$$

which is one heck of a lot of bridge hands. ❒

Example 1.2-11 Partitions of Students

There are a total of nine available seats in three different sections of MATH101. MATH101 section A (MATH101-A) has 3 available seats, MATH101-B has 2 available seats, and MATH101-C has 4 available seats. There are 9 students awaiting assignment to the remaining 9 open slots in MATH101. How many different assignments of students to sections are possible?

Solution
Since the order of the students selected for a given section does not matter, we can view this as the number of ways of partitioning 9 distinguishable objects (students) into sets of size 3, 2, and 4. Thus,

Total ways equals $ {}_9C_3 \times {}_6C_2 \times {}_4C_4 = \frac{9!}{(3!)(6!)} \times \frac{6!}{(4!)(2!)} \times \frac{4!}{(4!)(0!)} = \frac{9!}{(3!)(2!)(4!)} = 1260.$ ❒

In general the solution to a partitioning problem is called a ***multinomial coefficient***. The examples above can be generalized to the multinomial formula:

Multinomial Coefficients

The number of partitions of n distinct objects into k subsets of sizes $r_1, r_2, \cdots, r_k$, where $r_1 + r_2 + \cdots + r_k = n$ is called a ***multinomial coefficient***, denoted by $\binom{n}{r_1, r_2, \cdots, r_k}$. The resulting formula is:

$$\binom{n}{r_1, r_2, \cdots, r_k} = \frac{n!}{r_1! r_2! \cdots r_k!}.$$

Proof
First, using the multiplication principle as in the preceding examples, we have,

$$\binom{n}{r_1, r_2, \cdots, r_k} = \binom{n}{r_1}\binom{n - r_1}{r_2}\binom{n - r_1 - r_2}{r_3} \cdots \binom{r_k}{r_k}.$$

Thus,

$$\binom{n}{r_1, r_2, \cdots, r_k} = \left(\frac{n!}{r_1!(n-r_1)!}\right)\left(\frac{(n-r_1)!}{r_2!(n-r_1-r_2)!}\right)\left(\frac{(n-r_1-r_2)!}{r_3!(n-r_1-r_2-r_3)!}\right)$$

$$\cdots\left(\frac{(\overbrace{n-r_1-\cdots-r_{k-1}}^{\text{this equals } r_k})!}{r_k!0!}\right)$$

$$= \left(\frac{n!}{r_1!(n-r_1)!}\right)\left(\frac{(n-r_1)!}{r_2!(n-r_1-r_2)}\right)\left(\frac{(n-r_1-r_2!)}{r_3!(n-r_1-r_2-r_3)!}\right)\cdots\left(\frac{r_k!}{r_k!\cdot 0!}\right)$$

$$= \frac{n!}{r_1!r_2!r_3!\cdots r_k!},$$ where the first several cancellations are noted above.

Example 1.2-12 Multinomial Coefficients

The women's basketball team consists of 15 highly versatile players, each capable of playing all positions. How many different lineups consisting of one center, two forwards, and two guards are possible? **Note:** Chrystal playing guard results in a different lineup from Chrystal playing forward.

Solution 1

We can break this up into a two step process in which the first step is to choose the five starting players. The second step is to partition the five players into subsets of sizes 1, 2, and 2. Using the multiplication principle, the number of different lineups is

$$\underbrace{\binom{15}{5}}_{\text{ways to select 5 starters}} \times \underbrace{\binom{5}{1,2,2}}_{\text{among 5 starters, ways to assign positions}} = \frac{15!}{5!10!} \times \frac{5!}{1!2!2!} = (3003)(30) = 90{,}090.$$

Solution 2

Alternatively, we can select one student to start at center. There are ${}_{15}C_1 = \frac{15!}{14!\cdot 1!} = 15$ ways to do this. Then we select two of the remaining 14 players to play guard, ${}_{14}C_2 = \frac{14!}{12!\cdot 2!} = 91$ and two women to play forward, ${}_{12}C_2 = \frac{12!}{10!\cdot 2!} = 66$. By the multiplication principle, there are a total of ${}_{15}C_1 \cdot {}_{14}C_2 \cdot {}_{12}C_2 = 90{,}090$ ways.

Solution 3

Using the multinomial coefficient directly, we have that 15 players need to be partitioned into a group of 1 starting center, 2 starting guards, 2 starting forwards, and 10 players starting on the bench. There are

$$\binom{15}{1,2,2,10} = \frac{15!}{1!\cdot 2!\cdot 2!\cdot 10!} = \frac{15\cdot 14\cdot 13\cdot 12\cdot 11}{4} = 90{,}090$$

ways to do this. ❐

Note

Does the answer change if the coach selects two starting guards first, then 10 bench players, then two forwards, and finally the center?

Exercise 1-30 Coach Cramer has 15 basketball players – four centers, 5 forwards, and 6 guards. She starts one center, two guards, and two forwards. How many different groups of bench-warmers are possible?

Exercise 1-31 Due to budgetary constraints, Coach Cramer has just eight uniforms for her 15 girls. Conference rules require all players wear a uniform. In how many ways can the coach select five ladies to start, three ladies to dress as subs, and seven ladies to remain in street clothes (she no longer cares about filling guard/forward/center, just ways to start/sub/not dress)?

Using multinomial coefficients allows us to easily calculate the number of different arrangements possible when some of the objects are indistinguishable. For example, we might have a row of colored light-bulbs on a score board consisting of several indistinguishable red bulbs, several more indistinguishable blue bulbs, and so forth. The question would be, "How many distinct arrangements of the colored bulbs are possible?"

Example 1.2-13 Multinomial Coefficients and Different Words

How many different "words" of length 11 are possible using 4 I's, 1 M, 2 P's, and 4 S's?

Solution

The technique is to think of the eleven distinguishable objects in slots forming our "word."

I	I	I	I	M	P	P	S	S	S	S
1	2	3	4	5	6	7	8	9	10	11

Then the question reduces to partitioning the 11 slots into subsets of sizes 4,1,2,4, where the subsets are labeled (and populated) using the letters I, M, P, and S. One such partition is:

Subset Label	I	M	P	S
Subset Size	4	1	2	4
Slots Assigned	2,5,8,11	1	9,10	3,4,6,7

resulting in the particular partition,

Label	M	I	S	S	I	S	S	I	P	P	I
Slot	1	2	3	4	5	6	7	8	9	10	11

Thus, another formulation of the problem is, "how many different arrangements are possible using the letters MISSISSIPPI?"

Answer: $$\binom{11}{4,1,2,4} = \frac{11!}{4!1!2!4!} = 34,650.$$ ❐

Exercise 1-32 How many "words" can be spelled using two O's, an H and a P? What is the probability of choosing the word HOOP assuming arrangements are equally likely?

Exercise 1-33 How many words can be spelled using all of the letters in ABCBA? List them.

Exercise 1-34 How many words can be spelled using all of the letters in *Pennsylvania*?

Exercise 1-35 Six friends are playing poker. Each person is dealt five cards. In how many different ways can this be accomplished? That is, how many ways can five cards each be dealt to six different players in the initial deal of cards? Write your solution in terms of (a) the multiplication principle and (b) multinomial coefficients.

Exercise 1-36 Twenty-five girls are bored.

(a) In how many ways may five girls be chosen to go to a party and five different girls be chosen to volunteer at a soup kitchen (and the remaining 15 girls stay home)?

(b) How many ways may the girls form two teams of 5 to play basketball against each other?

Exercise 1-37 A fraternity consisting of 30 members wants to play "seven versus seven" flag football. How many different match-ups are possible?

Exercise 1-38 Nine workers are assigned to nine jobs. Two of the jobs are considered bad, four are considered average, and three of the jobs are considered good. The nine workers consist of seven men and two women. If workers are randomly assigned to jobs, calculate the probability that the two women are both assigned to bad jobs.

1.3 SAMPLING AND DISTRIBUTION

The best way to master combinatorial (counting) type problems is to completely understand a few of the basic building-blocks, and to learn how seemingly unrelated problems are in fact structurally the same. In this section we provide a structure to formally classify certain combinatorial problems by their particular type.

One common formulation for these types of problems – as we have seen in examples from earlier sections – involves choosing subsets from a set of n distinguishable objects. In statistics, these subsets (or events) are called ***samples***. A different way of formulating problems involves assigning, or ***distributing***, markers to the n objects. It turns out that these two types of problems, *sampling* and *distribution*, are closely related. In fact every sampling problem can be recast as a distribution problem, and vice-versa.

Mathematicians often couch sampling problems in terms of *removing balls from urns*. The related assignment problem would be posed as *distributing balls into urns*. In this sense, the distribution of balls into urns is frequently referred to as *occupancy*. We have never understood the fascination with balls and urns, but the terminology pervades in the classical textbooks and is convenient to use in designing examples.

1.3.1 SAMPLING

We begin with sampling problems. Consider a set consisting of n distinguishable objects (for example, numbered balls in an urn). The set of n objects is called the *population*. Assuming all outcomes are equally likely, the probability model boils down to the question "How many distinct samples of size r can be drawn from the n objects?"

Before we can answer this big question we need the answers to two related questions:

I. Are the samples taken with or without replacement (that is, can we pick an object at most once, or can we pick the same object more than once)?

II. Does the order in which we select the items in the sample matter?

Example 1.3-1 The Four Different Types of Sampling

For each of the four types of sampling described below, calculate the number of distinct outcomes.

(1) A club consisting of 26 members needs to select an executive board consisting of a president, a vice-president, a secretary and a treasurer. Each position has different duties and responsibilities. No individual can hold more than one office.

(2) A club consisting of 26 members needs to select a delegation of 4 (distinct) members to attend a convention. The delegation of 4 wears identical goofy hats.

(3) A club consisting of 26 members requires volunteers to complete 4 distinct chores; sweeping the clubhouse, printing off raffle tickets for the drawing at the party that night, picking up the prizes, and picking up the empties after the party. The same person can volunteer for more than one job, and each job requires only one volunteer.

(4) A club consisting of 26 members requires volunteers to make 4 identical recruiting phone calls, to be chosen later from a long list. A member can volunteer for one, two, three, or all four calls.

We classify these four situations with respect to replacement/without replacement, and order matters/order doesn't matter.

Example	n	r	Replacement?	Order Matters?
#1	26	4	No	Yes
#2	26	4	No	No
#3	26	4	Yes	Yes
#4	26	4	Yes	No

Or,

	Without replacement	With replacement
Order matters	Example 1	Example 3
Order doesn't matter	Example 2	Example 4

In (1) and (2) the context makes clear we want to select four different people, so in both cases we are sampling without replacement. In (1) the order matters since if Rajulio is selected it makes a difference as to which office he holds. In (2), order doesn't matter. We are only concerned with the members of the delegation, not the order in which they were selected. In (3) and (4), since members can volunteer for multiple chores, we are sampling with replacement. In (3) the chores are different, so we need to keep track of who is volunteering for which job. In (4) the calls are essentially identical, so we are only concerned with how many calls a volunteer makes.

Each member of the club can be uniquely identified (conveniently, with a single letter of the alphabet since there are exactly 26 members). The outcomes for each of these four sampling experiments can be put into one-to-one correspondence with certain types of four letter "words." "Words" (in quotes) means, as previously, any list of four letters, not just dictionary words.

In sampling without replacement, no letter can be repeated in a word. If order matters then ABCD is different from DCBA. If order does not matter, then we agree to list the four letters chosen for the sample just once, in alphabetical order. Thus, we can rephrase our four questions this way:

(1) Calculate the number of four letter words with no duplication of letters.

(2) Calculate the number of four letter words with no duplicate letters and the letters arranged in alphabetical order.

(3) Calculate the number of four letter words, duplicate letters allowed.

(4) Calculate the number of four letter words, duplicate letters allowed, and the letters arranged in alphabetical order.

Solution

The first three are straightforward to work out using permutations, combinations and the basic multiplication principle, respectively.

(1) Answer: $\underset{\text{president}}{\underline{26}} \cdot \underset{\text{vice-pres}}{\underline{25}} \cdot \underset{\text{secretary}}{\underline{24}} \cdot \underset{\text{treasurer}}{\underline{23}} = {}_{26}P_4 = 358{,}800.$

(2) Answer: ${}_{26}C_4 = \dfrac{{}_{26}P_4}{4!} = \dbinom{26}{4} = 14{,}950.$

(3) Answer: $\underset{\text{sweep}}{\underline{26}} \cdot \underset{\text{raffle}}{\underline{26}} \cdot \underset{\text{prizes}}{\underline{26}} \cdot \underset{\text{empties}}{\underline{26}} = 26^4 = 456{,}976.$

Example (4) is a little more subtle and we will defer the solution to the next section where it will be more easily understood as a distribution problem. ❒

1.3.2 DISTRIBUTIONS

Each of the four types of sampling experiments described above has a dual formulation as a *distribution*, or occupancy, experiment. For this purpose, we let n be the number of distinguishable (fixed and labeled) urns, and let r be the number of balls to be distributed into the urns. How many different distributions are possible? Again, before we can answer the question, we need to know:

I. Can an urn hold at most one ball (exclusive) or can it hold many balls (non-exclusive)?

II. Are the balls distinguishable (for example, bearing unique numbers) or are they indistinguishable (like plain white ping-pong balls)?

We can rephrase each of the four sampling problems above as a corresponding distribution problem:

(1) In how many ways can 4 executive board positions (distinguishable balls) be distributed among 26 members (urns) with exclusion (since no member can hold more than one position)?

(2) In how many ways can 4 delegation slots (indistinguishable balls) be distributed among 26 members (urns) with exclusion?

(3) In how many ways can 4 different jobs (distinguishable balls) be distributed among 26 members (urns) without exclusion (since one member can do multiple jobs)?

(4) In how many ways can 4 identical jobs (indistinguishable balls) be distributed among 26 members (urns) without exclusion (since one member can do multiple jobs)?

Statements (1)-(3) were solved in Section 1.3.1 formulated as equivalent sampling problems. Here is a paradigm for counting in sampling or distribution problems like (4).

Imagine the 26 urns lined up in a row sharing common partitions:

A	*B*	*C*	⋯	*Z*

We now distribute the 4 indistinguishable balls among the 26 urns. A sample outcome might be:

○	○○			○
A	*B*	*C*	⋯	*Z*

Now, observe that the top row in the above schematic can be thought of as a word consisting of (26 + 1) lines and 4 circles. However, the leftmost line and the rightmost line remain fixed in every word and are therefore superfluous. A moment's thought will show that there is a one-to-one correspondence between distinct distributions into the 26 slots, and distinct words consisting of $(26-1)$ lines and 4 circles. The distribution above appears as

$$\circ|\circ\circ|||\cdots|\circ.$$

Thus, the question is reduced to, "How many $(26-1+4)$ letter words are there consisting of four circles and $(26-1)$ vertical lines?" Therefore, the solution to example (4) is

$$\binom{26-1+4}{4} = \binom{29}{4} = 23{,}751.$$

The above discussion leads us to the following result:

Samples with Replacement When Order Does Not Matter

The number of unordered samples of r objects, with replacement, from n distinguishable objects is

$${}_{n+r-1}C_r = \binom{n+r-1}{r} = \binom{n+r-1}{n-1}.$$

This is equivalent to the number of ways to distribute r indistinguishable balls into n distinguishable urns without exclusion.

The same reasoning works in general, so that ${}_{n+r-1}C_r$ is the solution to the following sampling and distribution problems:

Sampling: How many distinct unordered samples of size r, with replacement, are there from n distinguishable objects?

Distribution: How many distinct ways are there to distribute r indistinguishable balls into n distinguishable urns, with multiple balls in an urn allowed?

We like to think of these types of problems in terms of distributing r dollar bills to n children. One needs $n-1$ partitions to separate the children, and then select the locations to place the r dollars.

Example 1.3-2 Samples with Replacement When Order Does Not Matter

Nicole wishes to select a dozen bagels from *Unleavened Bread Company.* Her choices include: Asiago cheese, plain, nine grain, cinnamon crunch, and very-very blueberry.

(a) How many different orders of a dozen bagels can she select?

(b) How many different orders of a dozen bagels can she select in which she has at least one of each kind?

Solution

Selecting a dozen bagels is equivalent to distributing 12 indistinguishable markers (ordering a bagel) into 5 (bagel) bins. Our answer to (a) is

$$_{(5-1+12)}C_{12} = {}_{16}C_{12} = 1,820.$$

For part (b), imagine that we first select one of each type of bagel. Our problem reduces to selecting the remaining seven bagels from any the five types. This can be done in

$$_{(5-1+7)}C_7 = {}_{11}C_7 = 330 \text{ ways.}$$ ❐

The "Boston Chicken" example, below, illustrates how identifying the particular type of sample can be very important. Boston Chicken has 16 distinct side-dishes (the population). Each dinner is served with the customer's choice of three side-dishes. Read the following article and explain the advertisement that there were "more than 3000 combinations" possible.

Thurs, Jan. 26, 1995. Rocky Mountain News, Retail & Marketing section

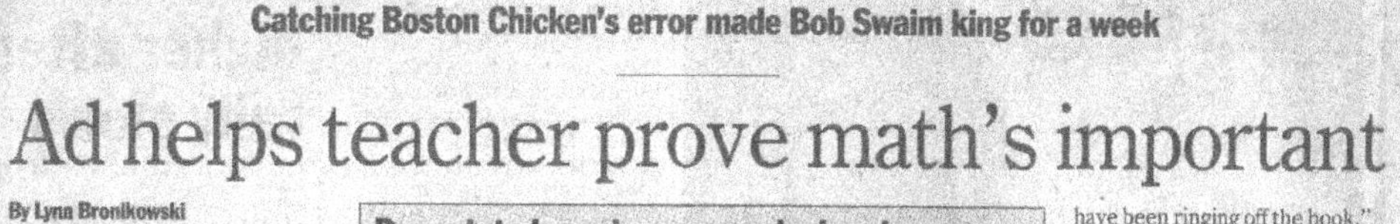

Catching Boston Chicken's error made Bob Swaim king for a week

Ad helps teacher prove math's important

By Lynn Bronikowski
Rocky Mountain News Staff Writer

Nearly every day, Pennsylvania high school teacher Bob Swaim harps about the importance of math skills.

Some students yawn; others chow down on chicken — the reward for Swaim's attention to detail in the world of television advertising.

The Golden-based Boston Chicken rotisserie chicken restaurant chain earlier this month aired an ad for its combo platter in which it said a customer could get more than 3,000 combinations by ordering a three-item combo.

But Swaim, who has taught math for 28 years, caught Boston Chicken with egg on its face after pointing out there are actually only 816 *combinations* but more than 3,000 *permutations.*

Boston Chicken, which corrected the ad by last weekend, won a flood of publicity by admitting the goof. Stories appeared locally and in *USA Today, The Toronto Star, Orange County Register, Chicago Sun Times, Cincinnati Post* and the Associated Press.

National Public Radio picked up the story and producers from TV network shows made calls.

Swaim, 48, became a small-town hero in Souderton, Pa. Boston Chicken donated $500 to the math department and treated 30 trigonometry students to lunch at a nearby Boston Chicken restaurant last Friday.

Swaim never actually saw the commercial starring Joe Montana, giving credit to his 15-year-old daughter, Joann, who first spotted the miscalculation.

"Everyday I'm telling my students that math is important," said Swaim, who teaches two trig classes, a geometry class and math for daily living. "I felt I had a moral responsibility to call about the mistake."

Swaim said at first he got a runaround before finally talking to the copywriter at the Integer Group in Golden.

He was surprised no one else had called. "Their phones should have been ringing off the hook," said Swaim. "There should have been more people than just me (questioning) it.

"It points out some of the weaknesses in our culture. It was such a simple problem and I'm no whiz kid."

He figures a second-year algebra student could have worked the formula.

"We're not used to this happening; Souderton is such a small town," said Jacey Stroback, a junior in one of Swaim's two trig classes. "He's always talking about how we're going to use math skills later on in life, no matter what you do."

Even before the math encounter, Swaim was a Boston Chicken fan, having eaten the restaurant's food at home. And after spotting a flier in a nearby restaurant, he invited a speaker to talk to two classes about career opportunities.

"I'm always looking for people to promote positive education and I liked their attitude," said Swaim.

After his fleeting moment of fame, life has returned to normal. But not before the producers of Tom Snyder's late-night talk show called about an appearance.

"I got bumped by a guy who collects stray birds."

Permutated veggies: a reworked recipe

Boston Chicken made its mistake by confusing combinations and permutations. One combination can count as several permutations.

Corn-corn-mashed potato and corn-mashed potato-corn are two different *permutations*, but you still wind up with the same *combination* of foods.

The chain provided this corrected formula for calculating the number of possible three-item samplers that can be made from 16 Boston Chicken side items.

Add up all of the following scenarios:

■ Selecting all three of the same items. For example, corn-corn-corn. Equals 16 ways.

■ Selecting two of the same items and a third different item. For example, corn-corn-mashed potato. The first one in 16 ways; second one in 15 ways; third one in two ways (as you selected just one of the two already selected); remove combinations that are the same. Equals 240 ways.

■ Selecting all three different items. For example, corn-stuffing-mashed potato. First one in 16 ways; second one in 15 ways; third one in 14 ways; remove combinations that are the same. Equals 560 ways.

■ Total all of the above number of ways. Equals 816 ways.

Exercise 1-39 Calculate the number of possibilities under ordered samples without replacement (customers eat their side dishes in the order of selection and must choose 3 different dishes).

Exercise 1-40 Calculate the number of possibilities using unordered samples without replacement (customers can eat their dishes in any order but still must choose 3 different dishes).

Exercise 1-41 Calculate the number of possibilities using ordered samples with replacement (customers eat their side dishes in a definite order, but can order, for example, corn-corn-mashed potato, which is different from corn-mashed potato-corn.

Exercise 1.42 Calculate the number of possibilities using unordered samples with replacement (customers eat their side dishes in any order, and corn-corn-mashed potato is possible and is identical to corn-mashed potato-corn.

Exercise 1-43 Which type of sampling gives an answer closest to "more than 3000?"

Exercise 1-44 Which type of sampling provides the most realistic result?
Hint: The correct answer of 816 is given in the side-bar of the article, although it is worked out differently using multiple steps.

Exercise 1-45 Assume that Boston Chicken customers choose three side-dishes from the 16 possible in such a way that all unordered samples with replacement are equally likely. What is the probability that a customer will choose all three side-dishes the same? That is, in poker parlance, what is the probability of three-of-a-kind?

Exercise 1-46
(a) How many ways can a parent distribute five one-dollar bills to her three children?
(b) How many ways can she accomplish this if each child gets at least one dollar?

Exercise 1-47
(a) How many ways can a witch distribute ten candy bars and seven packages of gum to four trick-or-treaters?
(b) How many ways can she do this if each child receives at least one candy bar and one package of gum?

Exercise 1-48 How many ways may a parent distribute ten identical pickled beets to his five children?

Exercise 1-49 How many different 13-card bridge hands are possible?

Exercise 1-50 How many 5-card poker hands are there?

Exercise 1-51 At a local fast-food restaurant in Oregon (no sales tax), fries, soda, hamburgers, cherry pie, and sundaes cost $1 each. Chicken sandwiches cost $2 each. You have five dollars. How many different meals can you order?

Exercise 1-52 I have fifteen certificates for a free pizza and 24 cans of Coca-Cola®. How many ways may I distribute the certificates and the cans of coke to 22 students?

1.3.3 SAMPLING AND OCCUPANCY UNITED

The following diagram shows the complete set of correspondences between sampling and distribution. Although the formulation of a sampling problem may appear to be quite dissimilar to the corresponding distribution problem, the two are in fact mathematically equivalent.

Samples of size r from n distinguishable objects	Without replacement	With replacement	
Order matters	${}_nP_r$ (Example 1)	n^r (Example 3)	Distinguishable balls
Order doesn't matter	$\binom{n}{r}$ (Example 2)	$\binom{n+r-1}{r}$ (Example 4)	Indistinguishable balls
	Exclusive	Non-exclusive	Distributions of r balls into n distinguishable urns

1.4 MORE APPLICATIONS

In this section we bring together a variety of special applications of the combinatorial methods previously discussed. We begin with an illustration of some basic results involving the algebra of polynomials.

1.4.1 THE BINOMIAL AND MULTINOMIAL THEOREMS

Consider the problem of expanding the binomial expression

$$(x+y)^n = \underbrace{(x+y)}_{\text{factor 1}}\underbrace{(x+y)}_{\text{factor 2}}\underbrace{(x+y)}_{\text{factor 3}}\cdots\underbrace{(x+y)}_{\text{factor } n}.$$

Algebraically, this amounts to adding together all possible products consisting of n letters, the first selected from "factor 1," the second from "factor 2," and so forth up to "factor n." In other words, the expansion consists of the sum of all n letter "words" consisting of the letters x and y.

To simplify our final result, we group together all words that contain the same number of x's and y's. Also, a stickler for algebra would insist on proper exponent notation (and the fact that multiplication is commutative), to express words like $xxyxyyx$ and $yxxxyxy$ (with 4 x's and $3y$'s) all as x^4y^3.

Now, for a given r $(r = 0,1,2,\ldots,n)$, the ***number*** of distinct n letter words with (r) x's and $(n-r)$ y's will be the coefficient of $x^r \cdot y^{n-r}$ in the expansion of $(x+y)^n$. As we have previously seen, this is just the combination ${}_nC_r = \binom{n}{r}$. This is the reason that the expression "binomial coefficient" is used to describe the coefficient $\binom{n}{r}$ of the term $x^r \cdot y^{n-r}$ of the binomial expansion for $(x+y)^n$.

The formula in the Binomial Theorem now follows from this observation.

The Binomial Theorem

For every non-negative integer n and real numbers x and y, we have

$$\begin{aligned}(x+y)^n &= \sum_{r=0}^{n} {}_nC_r \cdot x^r \cdot y^{n-r} \\ &= {}_nC_0 \cdot y^n + {}_nC_1 \cdot x^1 \cdot y^{n-1} + {}_nC_2 \cdot x^2 \cdot y^{n-2} + \cdots + {}_nC_n \cdot x^n,\end{aligned}$$

or equivalently,

$$\begin{aligned}(x+y)^n &= \sum_{r=0}^{n} {}_nC_r \cdot x^{n-r} \cdot y^r \\ &= {}_nC_0 \cdot x^n + {}_nC_1 \cdot x^{n-1} \cdot y^1 + {}_nC_2 \cdot x^{n-2} \cdot y^2 + \cdots + {}_nC_n \cdot y^n\end{aligned}$$

Note
The two forms of the theorem given are in fact equivalent since ${}_nC_r = {}_nC_{n-r}$ is the coefficient of both $x^r \cdot y^{n-r}$ and $x^{n-r} \cdot y^r$.

Example 1.4-1 The Binomial Theorem

Use the Binomial Theorem to expand $(x-2y)^3$.

Solution

To apply the binomial theorem, we rewrite $(x-2y)^3 = (x+\{-2y\})^3$.

$$\begin{aligned}(x-2y)^3 &= {}_3C_0 \cdot x^3 \cdot (-2y)^0 + {}_3C_1 \cdot x^2 \cdot (-2y)^1 + {}_3C_2 \cdot x^1 \cdot (-2y)^2 + {}_3C_3 \cdot x^0 \cdot (-2y)^3 \\ &= 1 \cdot x^3 \cdot 1 + 3 \cdot x^2(-2y) + 3 \cdot x \cdot (4y^2) + 1 \cdot 1 \cdot (-8y^3) = x^3 - 6x^2y + 12xy^2 - 8y^3\end{aligned}$$ ❒

Exercise 1-53 Use the Binomial Theorem to expand the following:

(a) $(x+3)^4$

(b) $(2x+y)^5$

(c) $(4x-5y)^3$

Exercise 1-54 Use the mathematical definition of combinations to verify the identity:

$$\binom{n}{r} = \binom{n-1}{r-1} + \binom{n-1}{r}.$$

Note 1

For a *non*-algebraic proof, imagine that you belong to a sorority with n members, exactly r of whom get to go to a party at the fraternity next door. How many ways to select the lucky r members? Break it down into those combinations that contain you, and those combinations that do not contain you. Are you one of the party-goers? If so, it remains to choose $r-1$ members from the $n-1$ members who are not you. If you stay home completing your probability homework, then we must count the ways r members are chosen from the $n-1$ members who are not you.

Note 2

This identity is what makes the generation of Pascal's[3] triangle possible. The nth row of Pascal's triangle contains the binomial coefficients for expanding $(x+y)^n$. Each new row is generated by adding the adjacent coefficients (as in the identity above) from the previous row:

Pascal's Triangle

$(x+y)^0$					1					0th row
$(x+y)^1$				1		1				1st row
$(x+y)^2$			1 ↘		↙ 2		1			2nd row
$(x+y)^3$		1		3		3		1		3rd row
$(x+y)^4$	1		4		6		4		1	4th row

[3] Pascal (1623-1662), considered to be, along with his contemporary, Pierre de Fermat (1601-1665), a progenitor of modern probability theory.

Exercise 1-55 Use the binomial theorem to verify the identity $\sum_{k=0}^{n} {}_nC_k = 2^n$. For example, if $n = 4$, then $\sum_{k=0}^{4} {}_4C_k = {}_4C_0 + {}_4C_1 + {}_4C_2 + {}_4C_3 + {}_4C_4 = 16 = 2^4$.

Note

${}_nC_k$ is the number of k – topping pizzas that can be made if there are n toppings from which to select. You may create a zero-topping pizza, or a one-topping pizza, up to an n-topping pizza. The left-hand side of the equation is the sum of different numbers of k-topping pizzas. On the other hand, each pizza topping may either be placed on the pizza or not. There are two ways to do this (put topping on pizza or do not put topping on pizza). There are n toppings from which to select. By the multiplication principle, there are $\underbrace{2 \cdot 2 \cdot \cdots \cdot 2}_{n \text{ times}} = 2^n$ ways to do this, which is the right-hand side of the identity.

Exercise 1-56 Explain the fundamental identity $\binom{n}{r} = \binom{n}{n-r}$ in terms of pizza toppings.

Exercise 1-57 In the expansion of $(x+y)^n$ the coefficient of $x^4 y^{n-4}$ is 3,876 and the coefficient of $x^5 y^{n-5}$ is 11,628. Find the coefficient of $x^5 y^{n-4}$ in the expansion of $(x+y)^{n+1}$. What is the value of n?

It is possible to generalize the arguments involving the binomial theorem in order to demonstrate the multinomial theorem. We wish to expand out the expression

$$(x_1 + x_2 + \cdots + x_r)^n = \underbrace{(x_1 + x_2 + \cdots + x_r)}_{\text{factor 1}}\underbrace{(x_1 + x_2 + \cdots + x_r)}_{\text{factor 2}}\cdots\underbrace{(x_1 + x_2 + \cdots + x_r)}_{\text{factor } n}.$$

The expansion now consists of all possible n-letter words consisting of the letters $x_1, x_2, \ldots, x_r$. We again group together all words with the same exponents on the letters. How many words are there with (n_1) x_1's, (n_2) x_2's, (n_3) x_3's, …, (n_r) x_r's (think MISSISSIPPI)? The answer is the number of partitions of n objects into subsets of sizes $n_1, n_2, n_3, \ldots, n_r$, that is $\binom{n}{n_1, n_2, \cdots, n_r}$. Therefore, the multinomial expansion can be expressed as follows:

The Multinomial Theorem

$$(x_1 + x_2 + \cdots + x_r)^n = \sum_{n_1 + \cdots + n_r = n} \binom{n}{n_1, n_2, \cdots, n_r} \cdot x_1^{n_1} \cdot x_2^{n_2} \cdot \cdots \cdot x_r^{n_r}.$$

The sum runs over all possible partitions $n_1, n_2, n_3, \ldots, n_r$ such that $n_1 + n_2 + n_3 + \cdots + n_r = n$. Finding all of these partitions is the challenging part.

Note

The number of such partitions is the number of ways of distributing n indistinguishable balls into r urns non-exclusively, that is, $\binom{n+r-1}{n}$.

Example 1.4-2 The Multinomial Theorem

Use the multinomial theorem to expand $(x+y+z)^4$.

Solution

You should take the time to find all of the ways to combine x, y, and z so that the total powers sum to 4.

$$\begin{aligned}(x+y+z)^4 &= \binom{4}{4,0,0}x^4y^0z^0 + \binom{4}{3,1,0}x^3y^1z^0 + \cdots + \binom{4}{0,0,4}x^0y^0z^4 \\ &= x^4 + 4x^3y + 4x^3z + 6x^2y^2 + 6x^2z^2 + 12x^2yz + 4xy^3 + 12xy^2z \\ &\quad + 12xyz^2 + 4xz^3 + y^4 + 4y^3z + 6y^2z^2 + 4yz^3 + z^4\end{aligned}$$

Note that the number of terms is 15, the same as $\binom{n+r-1}{n} = \binom{4+3-1}{4}$. ❒

Exercise 1-58 Use the multinomial theorem to expand the following:

(a) $(x-2y+5z)^3$

(b) $(w+x-y+2z)^2$.

1.4.2 POKER HANDS

Our simple version of ***poker*** is played with a standard four-suit, 52-card deck. We are serious players, so there are neither jokers nor wild cards in our deck. A poker hand consists of 5 cards dealt from a standard deck. In other words, a poker hand is an unordered random sample of size 5 chosen from a population of size 52, without replacement (you wouldn't want to be caught in Dodge City with 2 Queens of Hearts in your hand). The ace can be played as either high or low, as explained below. We present the definitions of the various types of poker hands.

Straight flush: Five cards of the same suit in sequence, such as 7♥6♥5♥4♥3♥. The Ace-King-Queen-Jack-Ten (A♣K♣Q♣J♣T♣) is called a ***royal flush***. The ace can also play low so that 5♣4♣3♣2♣A♣ is another straight flush.

***Four-of-a-kind*:** Four cards of the same denomination accompanied by another card, like 7♣7♦7♥7♠9♥.

***Full house* (*a.k.a.* a *boat*):** Three cards of one denomination accompanied by two of another, such as Q♣Q♦Q♥4♠4♥.

***Flush*:** Five cards of the same suit, such as K♠Q♠9♠6♠4♠. Straight flushes are excluded (they form their own category above).

***Straight*:** Five cards in sequence, such as J♥T♦9♣8♦7♠. The ace plays either high or low, but a collection like 32AKQ is not allowed. That is, there are no wrap-around straights. Again, straight flushes are excluded, since they have been counted separately.

***Three-of-a-kind*:** Three cards of the same denomination and two other cards of different denominations, such as 7♠7♣7♦K♣2♦.

***Two Pair*:** Two cards of one denomination, two cards of another denomination and a fifth card of a third denomination, such as K♠K♣8♦8♠7♥.

***One Pair*:** Two cards of one denomination accompanied by three cards of different denominations, such as T♠T♣Q♦8♠7♥.

***High Card* (*a.k.a.* Nothing):** Any hand that does not qualify as one of the hands above, such as A♠Q♣9♦8♠7♥.

Various combinatorial techniques are employed to calculate the probabilities of being dealt these hands on an initial deal from a standard deck of cards. We illustrate several of these calculations and provide a summary of the probabilities of all types of poker hands.

Since poker hands consist of five cards (any order) selected without replacement from a population of size 52 cards, the size of the sample space is $\binom{52}{5} = 2{,}598{,}960$.

Example 1.4-3 Straight Flush

Compute the probability of being dealt a straight flush.

Solution

The highest ranked straight flush is AKQJT in one of the four suits (clubs, diamonds, hearts, and spades), the lowest ranked straight flush is 5432A. There are 10 of these rankings. There are four suit choices. By the multiplication rule, there are $10 \cdot 4 = 40$ possible straight flushes.

$$\Pr(\text{straight flush}) = \frac{\text{number of straight flushes}}{\text{number of five card poker hands}}$$

$$= \frac{10 \cdot 4}{\binom{52}{5}} = \frac{40}{2{,}598{,}960} = .00001539.$$ ❐

Example 1.4-4 Full House

Compute the probability of being dealt a full house on the initial deal of 5 cards.

Solution

We begin by selecting a card denomination (Ace, King,...,2) for the three-of-a-kind. One has 13 ways to do this. We then select three of the four cards from the selected denomination to create the three-of-a-kind. For the pair, there are only 12 remaining denominations to select, since the card value chosen for the three-of-a-kind is no longer available. We need to choose two cards with this second denomination.

$$\Pr(\text{full house}) = \frac{\text{number of full houses}}{\text{number of five card poker hands}}$$

$$= \frac{\binom{13}{1} \cdot \binom{4}{3} \cdot \binom{12}{1} \cdot \binom{4}{2}}{\binom{52}{5}} = \frac{13 \cdot 4 \cdot 12 \cdot 6}{2{,}598{,}960} = .00144058.$$ ❐

Exercise 1-59 Compute the probabilities for all nine poker hand types. Do this by yourself prior to looking at the following summary. It is an important exercise, even if you struggle.

Poker Probability Summary:

1. **Straight Flush**

$$\Pr(\text{Straight Flush}) = \frac{\text{number of straight flushes}}{\text{number of five card poker hands}}$$

$$= \frac{4 \text{ suits} \cdot 10 \text{ possible straights}}{\binom{52}{5}} = \frac{40}{2{,}598{,}960} = .00001538.$$

2. **Four-of-a-Kind**

$$\Pr(\text{Four-of-a-Kind}) = \frac{\binom{13}{1} \cdot \binom{4}{4} \cdot \binom{12}{1} \cdot \binom{4}{1}}{\binom{52}{5}} = \frac{13 \cdot 1 \cdot 12 \cdot 4}{2{,}598{,}960} = \frac{624}{2{,}598{,}960} = .00024010.$$

3. **Full House**

$$\Pr(\text{Full House}) = \frac{\binom{13}{1}\cdot\binom{4}{3}\cdot\binom{12}{1}\cdot\binom{4}{2}}{\binom{52}{5}} = \frac{13\cdot 4\cdot 12\cdot 6}{2{,}598{,}960} = \frac{3{,}744}{2{,}598{,}960} = .00144058.$$

4. **Flush**

$$\Pr(\text{Flush}) = \frac{\overbrace{\binom{4}{1}}^{\text{suits}}\cdot\overbrace{\binom{13}{5}}^{\text{5 cards selected from the suit}} - \overbrace{40}^{\text{straight flushes already accounted}}}{\binom{52}{5}}$$

$$= \frac{4\cdot 1{,}287-40}{2{,}598{,}960} = \frac{5{,}108}{2{,}598{,}960} = .00196540.$$

5. **Straight**

$$\Pr(\text{Straight}) = \frac{10\cdot\binom{4}{1}\cdot\binom{4}{1}\cdot\binom{4}{1}\cdot\binom{4}{1}\cdot\binom{4}{1}-40}{\binom{52}{5}}$$

$$= \frac{10\cdot 4^5-40}{2{,}598{,}960} = \frac{10{,}200}{2{,}598{,}960} = .00392465.$$

6. **Three-of-a-Kind**

$$\Pr(\text{Three-of-a-Kind}) = \frac{\binom{13}{1}\cdot\binom{4}{3}\cdot\binom{12}{2}\cdot\binom{4}{1}\cdot\binom{4}{1}}{\binom{52}{5}}$$

$$= \frac{13\cdot 4\cdot 66\cdot 4\cdot 4}{2{,}598{,}960} = \frac{54{,}912}{2{,}598{,}960} = .02112846.$$

7. **Two Pair**

$$\Pr(\text{Two Pair}) = \frac{\binom{13}{2}\cdot\binom{4}{2}\cdot\binom{4}{2}\cdot\binom{11}{1}\cdot\binom{4}{1}}{\binom{52}{5}}$$

$$= \frac{78\cdot 6\cdot 6\cdot 11\cdot 4}{2{,}598{,}960} = \frac{123{,}552}{2{,}598{,}960} = .04753902.$$

8. **One Pair**

$$\Pr(\text{One Pair}) = \frac{\binom{13}{1}\cdot\binom{4}{2}\cdot\binom{12}{3}\cdot\binom{4}{1}\cdot\binom{4}{1}\cdot\binom{4}{1}}{\binom{52}{5}}$$

$$= \frac{13\cdot 6\cdot 220\cdot 4^3}{2{,}598{,}960} = \frac{1{,}098{,}240}{2{,}598{,}960} = .42256903$$

9. **Nothing**

$$\Pr(\text{Nothing}) = 1 - \Pr(\text{other possibilities})$$

$$= \frac{\binom{52}{5} - (40 + 624 + 3{,}744 + 5{,}108 + 10{,}200 + 54{,}912 + 123{,}552 + 1{,}098{,}240)}{\binom{52}{5}}$$

$$= \frac{2{,}598{,}960 - 1{,}296{,}420}{2{,}598{,}960} = \frac{1{,}302{,}540}{2{,}598{,}960} = .50117739.$$

Exercise 1-60 (Poker Dice): Play poker using 5 fair dice rather than a deck of cards. Roll the five dice onto the table. The possible *hands* are: five-of-a-kind, four-of-a-kind, a full house, three-of-a-kind, two pair, one pair, a straight, and nothing. Find the probabilities of these events on a single roll of the dice. Where should the straight ***rank***? The less likely (lower probability) an event is to happen, the higher it should rank.

1.4.3 THE POWERBALL® LOTTERY

The following rules, prizes, and odds were once found at www.musl.com. To play the game, we draw five balls out of a drum with 53 numbered white balls, and one power ball out of a drum with 42 numbered green balls.

Powerball® Prizes and Odds

Match	Prize	Odds
+	Grand Prize	1 in 120,526,770.00
	$100,000	1 in 2,939,677.32
+	$5,000	1 in 502,194.88
	$100	1 in 12,248.66
+	$100	1 in 10,685.00
	$7	1 in 260.61
+	$7	1 in 1696.85
+	$4	1 in 123.88
	$3	1 in 70.39

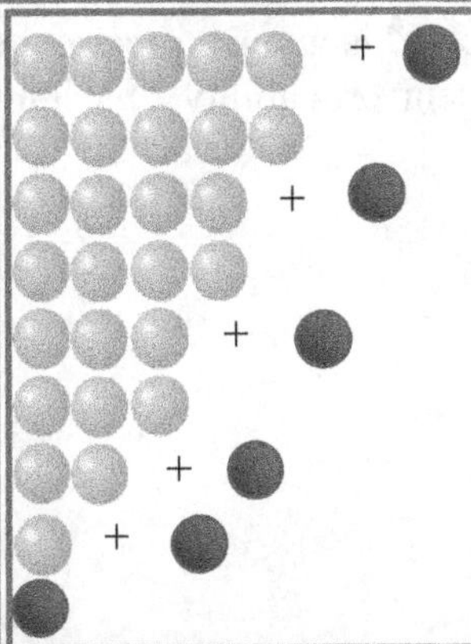

The overall odds of winning a prize are 1 in 36.06.
The odds presented here are based on a $1 play and are rounded to two decimal places.

Example 1.4-5 Probability of winning the $5,000 Powerball® lottery prize

Calculate the probability of winning the $5,000 prize (i.e., matching four of the five white balls and the green power ball).

Note
The amount that you win is $4,999 since it costs you a dollar to purchase the game ticket.

Solution
There are five *winning* white balls, of which we need to select four. There are forty-eight $(53-5)$ *losing* white balls, of which we need to select one. We also need to select the correct green power ball, of which there is only one.

$$\begin{aligned}
\Pr(\text{4 white and the power ball}) &= \frac{\binom{5}{4}\cdot\binom{48}{1}\cdot\binom{1}{1}}{\binom{53}{5}\cdot\binom{42}{1}} \\
&= \frac{5\cdot 48\cdot 1}{2{,}869{,}685\cdot 42} \\
&= \frac{240}{120{,}526{,}770} \\
&= \frac{1}{502{,}194.88} \\
&= 0.000001991.
\end{aligned}$$

The Powerball® graphic uses the expression "odds" in the third column. They should have used "probability" and written the expressions as $1/502{,}194.88$. There are a number of ways in which odds are stated, but all of them are variants of probability statements.

In gambling, it is common to give the ratio of expected losses to expected wins in a fixed number of plays. This ratio is referred to as "the odds against winning." For example, suppose that the betting public's favorite horse in the Santa Anita derby has odds of 2:1 (read as "two to one"). This implies that if three identical races were run, then our horse would be expected to *lose* twice and *win* once (the word "expect" is given a precise meaning in Chapter 3). We could translate these odds to the equivalent probability of winning the race, $\Pr(\text{win}) = \frac{\text{1 win}}{\text{1 win+2 losses}} = \frac{1}{3}$. ❐

Consider an event A whose probability of occurring is p. The ***complementary*** event (that A does ***not*** occur) is denoted A'.

Odds

The ***odds*** against the event A are quoted as the ratio,

$$\Pr(A \text{ does not occur}) : \Pr(A \text{ does occur}) = \Pr(A') : \Pr(A) = (1-p) : p.$$

Note: Odds are generally quoted using whole numbers. For example, if $P(A) = \frac{2}{5}$ then the odds against A are $\frac{3}{5} : \frac{2}{5} = 3:2.$

If you wanted to bet on A occurring, and the odds against A are 3:2, then you would put $2 in the pot and your opponent would put in $3. If the experiment could be repeated 5 times, then you would expect to win twice and lose three times. You would lose $6 on the three losses and win $6 on the two wins, breaking even. This would be considered a ***fair*** bet.

Converting Odds to Probability

If the odds against the event A are quoted as $b:a$, then $\Pr(A) = \dfrac{a}{a+b}$.

Example 1.4-6 Probability of winning the $4 Powerball® lottery prize

Calculate the probability of winning the $4 prize (matching one of the five white balls and the power ball).

Solution

We need to match one of the 5 winning white balls (therefore select 4 losing white balls) and match the one winning Powerball®.

$$\Pr(\text{one white and the power ball}) = \frac{\binom{5}{1}\cdot\binom{48}{4}\cdot\binom{1}{1}}{\binom{53}{5}\cdot\binom{42}{1}} = \frac{5\cdot 194{,}580\cdot 1}{2{,}869{,}685\cdot 42}$$

$$= \frac{972{,}900}{120{,}526{,}770} = \frac{1}{123.88} = .0081. \ \square$$

Example 1.4-7 Probability of winning nothing

Calculate the probability of losing one dollar (e.g., matching zero or one or two of the five white balls and not the power ball).

Solution

Losing options include matching 0, 1, or 2 winning white balls and not the power ball.

$$\Pr(\text{losing a dollar}) = \frac{\binom{5}{0}\cdot\binom{48}{5}\cdot\binom{41}{1}+\binom{5}{1}\cdot\binom{48}{4}\cdot\binom{41}{1}+\binom{5}{2}\cdot\binom{48}{3}\cdot\binom{41}{1}}{\binom{53}{5}\cdot\binom{42}{1}}$$

$$= \frac{1\cdot 1{,}712{,}304\cdot 41+5\cdot 194{,}580\cdot 41+10\cdot 17{,}296\cdot 41}{2{,}869{,}685\cdot 42} = \frac{117{,}184{,}724}{120{,}526{,}770}$$

$$= .9723$$

❐

Exercise 1-61 Check the remaining probabilities for the Powerball® game listed below. Verify that theses probabilities sum to 1.

Powerball® Probability Summary

Match	Prize	Odds
5 white + Powerball	Grand Prize	$\frac{1}{120{,}526{,}770}$
5 white	$100,000	$\frac{41}{120{,}526{,}770}$
4 white + Powerball	$5,000	$\frac{240}{120{,}526{,}770}$
4 white	$100	$\frac{9{,}840}{120{,}526{,}770}$
3 white + Powerball	$100	$\frac{11{,}280}{120{,}526{,}770}$
3 white	$7	$\frac{462{,}480}{120{,}526{,}770}$
2 white + Powerball	$7	$\frac{172{,}960}{120{,}526{,}770}$
1 white + Powerball	$4	$\frac{972{,}900}{120{,}526{,}770}$
Powerball	$3	$\frac{1{,}712{,}304}{120{,}526{,}770}$
Nothing	$0	$\frac{117{,}184{,}724}{120{,}526{,}770}$

Exercise 1-62 Suppose that the Grand Prize is 50 million dollars and paid in cash today and that there are no taxes. Find the *average* winning computed as the sum of the possible winning times its probability.

1.5 CHAPTER 1 SAMPLE EXAMINATION

1. A high school teacher has a rowdy home-room with 26 students. How many groups of five students can be sent to detention? How many ways may the teacher fail five students?

2. How many different telephone numbers can be formed from a 7-digit number if the first digit cannot be 0 or 1 and there are no 555-xxxx numbers?

3. There are 15 students in class. We give a Slinky® to one student, a root beer to a second student, and a bop-upon-the-nose to a third student. How many ways may we select the three students?

4. There are 15 students in this class. We give a Slinky® to two students and a bop-upon-the-nose to a third student. How many ways may we select the three students?

5. A special deck of Old Maid cards consists of 51 distinguishable cards, 25 pairs and a single old maid card. All 51 cards are dealt evenly between you and two other players – 17 cards per player.
   (a) How many different hands can be dealt to you?
   (b) What is the probability that your hand has exactly two pair (and 13 single cards)?

6. Five cards are drawn without replacement from a standard 52-card deck.
   (a) How many different hands are possible?
   (b) How many different hands have exactly four cards in the same suit?

7. Suppose that there are 7 soccer players and 6 football players on a bus.
   (a) How many ways can a group consisting of 3 soccer players and 2 football players be chosen?
   (b) Find the probability that a group of 5 players selected at random will consist of 3 soccer players and 2 football players.

8. List all possible meals that can be ordered at IN-N-OUT Burger of Southern California. A meal is defined as one sandwich, one side, and one drink.

Sandwich	Side	Drink
Hamburger	French Fries	Coke
Cheeseburger	Potato Chips	Diet Coke
Double-Double		Chocolate Shake

9. An old Powerball® lottery used two sets of balls. A set of white balls numbered 1-49 and a set of red balls (the power ball) numbered 1-42. To play, you select 5 white balls and one power ball. How many different lottery tickets were possible?

10. Use the Binomial Theorem to expand $(2x-y)^5$.

11. Use the multinomial theorem to expand $(x-2y+5z)^4$.

12. You are dealt three cards from a standard deck. Find the probability that all three cards have faces (Jacks, Queens, or Kings).

13. A drawer contains 4 black dress socks and 8 white athletic socks. Two socks are drawn at random. What is the probability that the socks are the same color?

14. Joe, an avid and properly licensed sportsman, is in his hunting blind when he locates 20 Canada geese, 25 Mallard ducks, 40 Bald Eagles, 10 Whooping Cranes, and 5 Flamingos. Joe randomly selects six birds to target. What is the probability that at least one of each species is targeted?

15. You estimate the odds against your passing an actuarial exam at 8:5. What is the corresponding probability that you will pass the exam?

16. **SOA Exam P Sample Exam Questions #141**
Thirty items are arranged in a 6-by-5 array as shown.

A_1	A_2	A_3	A_4	A_5
A_6	A_7	A_8	A_9	A_{10}
A_{11}	A_{12}	A_{13}	A_{14}	A_{15}
A_{16}	A_{17}	A_{18}	A_{19}	A_{20}
A_{21}	A_{22}	A_{23}	A_{24}	A_{25}
A_{26}	A_{27}	A_{28}	A_{29}	A_{30}

Calculate the number of ways to form a set of three distinct items such that no two of the selected items are in the same row or same column.

(A) 200 (B) 760 (C) 1200 (D) 4560 (E) 7200

CHAPTER 2

GENERAL RULES OF PROBABILITY

In Chapter 1 we introduced some of the basic concepts of probability, but confined our attention to the combinatorial issues surrounding sample spaces of equally likely outcomes. As we saw, this is a useful introduction to probability theory, with widespread applications. In this chapter we focus on the set theoretic foundations of probability, explaining the formal axioms and general rules applicable to *any* probability model. We will complete our discussion of sample spaces, and focus on two other techniques, namely Venn diagrams and probability trees.

In particular, we illustrate different methods to calculate the probability of an event, and we analyze the interaction of two or more events. This will lead us to a discussion of conditional probability and the concept of independent events.

2.1 SETS, SAMPLE SPACES, AND EVENTS

A set is like a grocery bag. Elements of the set (grocery bag) are items that one places in the bag. There may be a set containing "a coke, a candy bar, a roll of toilet paper, and a receipt" or "a pepperoni pizza, a block of ice, and a court summons". The grocery bag could even be empty. The elements in a set are presumed to be distinguishable. For example, if you have 12 beers and 6 cokes in your grocery bag the individual cans must be identifiable. Otherwise, the set would just consist of two elements, "beer and coke".

Sets and Elements

A ***set*** is a collection of objects. The objects belonging to the set are called ***elements*** or ***members***. We use the symbol $\in$ to denote that a particular object is an element of a set, and the symbol $\notin$ to denote that a particular object is not an element of a set. The ***universal*** set contains all possible objects, and every set is a ***subset*** of the universal set.

There are three common methods to define a particular set. Each method has its own strengths, depending upon context and desired information.

(a) *A description in words*: "The first six prime numbers,"

(b) *The listing method*: $\{2,3,5,7,11,13\}$,

(c) *Set-builder notation*: $\{x_n \mid x_n$ is the nth prime number, $n = 1, 2, ..., 6\}$. The vertical line following the first x_n can be read as "such that," or simply "where."

Example 2.1-1 Sets of Numbers

There are important sets of numbers throughout mathematics. Common sets include the natural numbers, the whole numbers, the integers, and the rational numbers. Use set theory notation to describe each of these.

Solution

Set of natural numbers: $\mathbf{N} = \{1, 2, 3, 4, ...\}$.

Whole numbers: $\mathbf{W} = \{0, 1, 2, 3, ...\}$,
which is just the set of natural numbers and zero.

Set of integers: $\mathbf{Z} = \{..., -3, -2, -1, 0, 1, 2, 3, ...\}$.

Set of rational numbers: $\mathbf{Q} = \left\{ \frac{r}{s} \middle| r \in \mathbf{Z},\ s \in \mathbf{N} \right\}$.

Rational numbers are those that can be expressed as a ratio of integers. For consistency, we write the ratio $\frac{r}{s}$ in reduced form. Rational numbers have decimal expansions with repeating cycles, including those with terminating decimal expansions, which can be thought of as a repeating cycle of zeros.

The set of irrational numbers, *I*, consists of all numbers whose decimal expansions are non-repeating. Some common examples of irrational numbers are π, $\sqrt{2}$, e, and $e + 4$. ❐

In probability vernacular, the universal set is referred to as the *sample space*, and subsets are called *events*. We began using these terms in Chapter 1 and we now place them in the context of basic set theory.

Sample Space

The ***sample space*** is the set (collection) of all possible outcomes of a probability experiment. An ***event*** is a subset of the sample space.

2.2 SET THEORY SURVIVAL KIT

In the course of solving problems involving two or more sets, we need to be able to combine sets in various ways. In this section we define the fundamental terminology and introduce the important concept of a Venn diagram.

Size of Finite Sets

For sets that contain a finite number of elements we denote the ***size***, or ***cardinality***, of the set by $N(A)$ = *number of elements in A*. A set containing zero elements is called the ***empty set*** or the ***null set*** and is denoted by $\varnothing$.

Set Operations

Suppose that A and B represent two sets. Suppose that U represents the universal set.

(1) ***Union*** $A \cup B = \{x \mid x \in A \text{ or } x \in B\}$.
This is read as "A union B" or as "A or B".

(2) ***Intersection*** $A \cap B = \{x \mid x \in A \text{ and } x \in B\}$.
This is read as "A intersect B" or as "A and B".

(3) ***Complement*** $A' = \{x \mid x \in U \text{ and } x \notin A\}$.
This is read as "A complement" or as "not A".

(4) ***Difference*** $A - B = \{x \mid x \in A \text{ and } x \notin B\}$.
This is read as "A takeaway B".

(5) ***Subset*** $A \subseteq B$ if for every $x \in A$, then $x \in B$.

Notes

1. There are many alternate ways to denote the complement[1] of a set. They include: A', A^C, $\sim A$, and $\overline{A}$.

2. Sets A and B are equal if every element in set A is also in set B and every element in set B is also in set A. That is, if $A \subseteq B$ and $B \subseteq A$, then $A = B$.

In probability theory we refer to a collection of events (subsets) as ***mutually exclusive*** if all of their pair-wise intersections consist of the empty set. For example, the sets A, B, and C are mutually exclusive if and only if $A \cap B = \varnothing$, $A \cap C = \varnothing$, and $B \cap C = \varnothing$.

Example 2.2-1 Set Theory

The Luisi family has the following pets (which constitute the universal set in this example): U = {dog, cat, chicken, steer, snake, rabbit, frog, rock}. The chicken, steer, and rabbit will eventually be eaten by the Luisi family. We denote this set as F = {chicken, steer, rabbit}. Some pets have names: Fenway the dog, Loosy the cat, T-Bone the steer, Roger the rabbit, and Francisco the frog. The set of named pets is denoted by N = {dog, cat, steer, rabbit, frog}. Using set operations, describe:

(a) $N \cap F$
(b) F'
(c) Is F a subset of N?

[1] Complement (*noun*) is something that completes. Compliment(*noun*) is an expression of praise.

Solution

(a) $N \cap F = \{\text{steer, rabbit}\}$ is the set of Luisi pets that have names and will eventually be eaten.

(b) $F' = \{\text{dog, cat, snake, frog, rock}\}$ is the set of pets that will not be eaten.

(c) $F \not\subseteq N$, because the pet chicken does not have a name. ❐

Exercise 2-1 Let O denote the set of odd natural numbers less than 10 and P denote the set of prime numbers less than 10. **Note:** The number 1 is not considered prime.

(a) List the sets O and P in set notation.

(b) Find the intersection of the odd natural numbers less than 10 and the prime numbers less than ten, $O \cap P$.

(c) Find the union of the odd natural numbers less than 10 and the prime integers less than ten, $O \cup P$.

(d) Find the difference of the odd natural numbers less than 10 and the prime integers less than ten, $O - P$.

Exercise 2-2 Suppose the universal set is $U = \{1, 2, \pi, \text{Jamaal, water, gum}\}$. Let $A = \{1, 2, \pi\}$ and $B = \{1, 2, \text{Jamaal, gum}\}$. Find the following.

(a) $A \cap B$.

(b) The number of elements in B complement, $N(B')$.

(c) $(A \cup B)'$.

2.2.1 THE VENN DIAGRAM

A graphical representation that is useful for illustrating various events is a ***Venn diagram***. The sample space (universal set) is represented by a rectangular box and events (subsets) are represented by overlapping ovals. These ovals partition the sample space into mutually exclusive events. We label each region with the event name and, if possible, the number of outcomes in the event.

Example 2.2-2 Venn Diagram

Here, the center area in the first Venn diagram represents $A \cap B$ and the shaded area in the second graph represents A'.

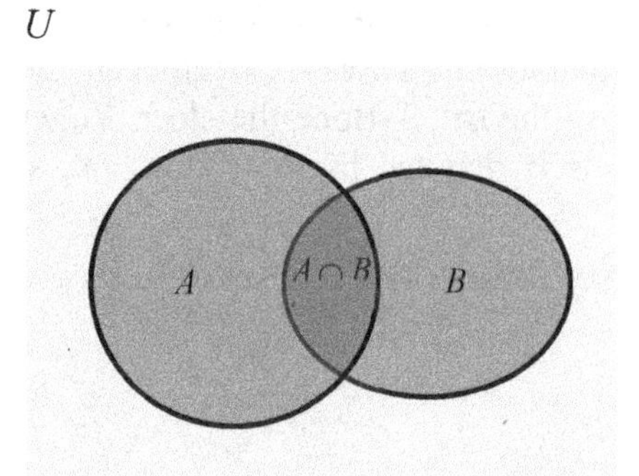

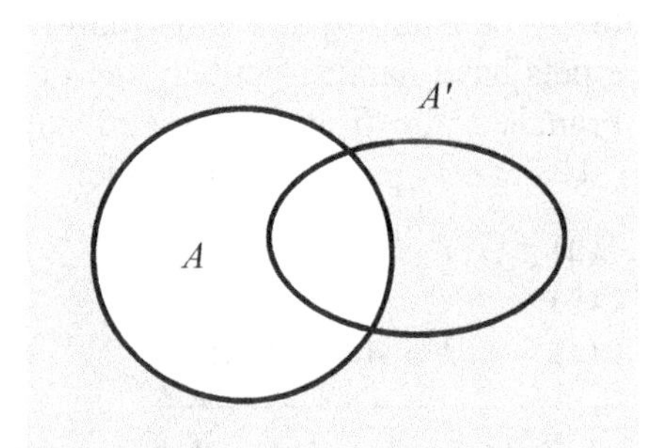

❐

Exercise 2-3 Refer to the Luisi family pet example in the previous section.

(a) Construct a Venn diagram based on the sets F and N that lists all of the cast members in their appropriate region.

(b) Construct a second Venn diagram that just lists the number of elements (cast members) in the appropriate region.

Exercise 2-4 Construct Venn diagrams for the operations: A, $A \cup B$, $A - B$, and B'.

2.2.2 BASIC RULES OF PROBABILITY

In Chapter 1 we described the three basic elements of a probability experiment, the last of which concerned the assignment of probabilities. Using the language of set theory we can state more precisely the properties of the probability assignments. Given a sample space (universal set), U, each event, E is assigned a probability number, $\Pr(E)$. This so-called probability function must satisfy three basic axioms:

Axioms of Probability Theory

(1) $0 \leq \Pr(E) \leq 1$ for any event E.

(2) $\Pr(U) = 1$, where U denotes the entire sample space.

(3) The probability of the union of mutually exclusive events is the sum of the individual probabilities of those disjoint sets:

$$\Pr\left(\bigcup_{\substack{\text{mutually} \\ \text{exclusive}}} E_i \right) = \sum_i \Pr(E_i).$$

The third axiom is the additivity property of the probability function, a term we first employed in Chapter 1. Recall that events are mutually exclusive if $E_i \cap E_j = \varnothing$ for all $i \neq j$.

Two of the most basic probability rules are the *negation rule* and the *inclusion-exclusion rule*.

Two Important Probability Rules

I. ***Negation Rule***: $\Pr(E') = 1 - \Pr(E)$.

II. ***Inclusion-Exclusion Rule***:

$$\Pr(E) + \Pr(F) = \Pr(E \cup F) + \Pr(E \cap F).$$

We can rewrite the inclusion-exclusion rule as,

$$\Pr(E \cup F) = \Pr(E) + \Pr(F) - \Pr(E \cap F).$$

On the left-hand side, to be in the union, an element is in set E, or set F, or possibly in both sets. The right-hand side counts the probability of being in set E plus the probability of being in set F. We must subtract the probability of the intersection because those elements were counted twice on the right-hand side of the equality.

Both rules follow immediately from the axioms:

For (I), since $\underbrace{U = E \cup E'}_{\text{disjoint}}$, then $1 \underbrace{=}_{\substack{\text{Axiom}\\(2.)}} \Pr(U) = \Pr(E) \underbrace{+}_{\substack{\text{Axiom}\\(3.)}} \Pr(E')$ which implies the negation rule $\Pr(E') = 1 - \Pr(E)$.

For (II) we partition the event $E \cup F$ into three mutually exclusive sets,

$$E \cup F = [E \cap F] \cup [E \cap F'] \cup [E' \cap F].$$

Refer to the Venn diagram for this identity.

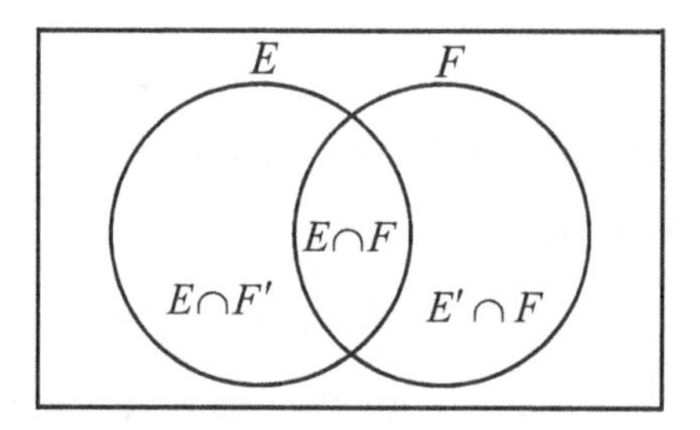

Then,
$$\begin{aligned}
\Pr(E \cup F) + \Pr(E \cap F) &= \Pr\{[E \cap F] \cup [E \cap F'] \cup [E' \cap F]\} + \Pr(E \cap F) \\
&= \{\Pr(E \cap F) + \Pr(E \cap F') + \Pr(E' \cap F)\} + \Pr(E \cap F) \\
&= [\Pr(E \cap F) + \Pr(E \cap F')] + [\Pr(E' \cap F) + \Pr(E \cap F)] \\
&= \Pr((E \cap F) \cup (E \cap F')) + \Pr((E' \cap F) \cup (E \cap F)) \\
&= \Pr(E) + \Pr(F).
\end{aligned}$$

Example 2.2-3 Negation Rule

You are dealt one card from a standard deck of 52 playing cards. Calculate the probability that your card is not the seven of diamonds.

Solution

$\Pr(\text{card is not the 7 of diamonds}) = 1 - \frac{1}{52} = \frac{51}{52}$. We hope that you would not count the 51 cards that are not the seven of diamonds. ❒

Example 2.2-4 Inclusion-Exclusion Rule

Suppose that we roll a pair of fair six-sided dice.

(a) Find the probability that the sum is equal to seven or eleven.

(b) Find the probability that either the sum is seven or exactly one of the dice is showing four.

Solution

In the margin of this book, try listing the 36 outcomes for this sample space (see Example 1.1-3). (Use a light pencil in case you intend to resell it.) The event that the sum of the two dice equals seven has six outcomes and is disjoint from the event that the sum is eleven, which happens twice. Hence,

(a) $\Pr(\text{sum equals 7 or 11}) = \frac{6}{36}+\frac{2}{36} = \frac{8}{36}$.

(b) Let the set $A=\{\text{sum equals 7}\}=\{(1,6),(2,5),(3,4),(4,3),(5,2),(6,1)\}$.

Let the set $B=\{\text{exactly one die shows 4} =$

$\{(4,1),(4,2),(4,3),(4,5),(4,6),(1,4),(2,4),(3,4),(5,4),(6,4)\}$

Then $\Pr(A)=\frac{6}{36}$, $\Pr(B)=\frac{10}{36}$, and $\Pr(A\cap B)=\frac{2}{36}$.

Using the inclusion-exclusion principle $\Pr(A\cup B)=\frac{6+10-2}{36}=\frac{14}{36}=\frac{7}{18}$. ❐

Exercise 2-5 Calculate the probability of losing one dollar in the Powerball® using the negation rule. Assume pre-2009 rules of selecting 5 of 53 white balls and 1 of 42 red balls.

Exercise 2-6 Suppose that we roll a pair of fair dice. Find the probability that the sum is divisible by three or that doubles are rolled.

Exercise 2-7 SOA Exam P Sample Exam Questions #134

A mattress store sells only king, queen, and twin-size mattresses. Sales records at the store indicate that one-fourth as many queen size mattresses are sold as king and twin-size mattresses combined. Records also indicate that three times as many king-size mattresses are sold as twin-size mattresses. Calculate the probability that the next mattress sold is either king or queen-size.

(A) 0.12 (B) 0.15 (C) 0.80 (D) 0.85 (E) 0.95

2.2.3 DE MORGAN'S LAWS

De Morgan's Laws give us a relationship between the complement of the union (and the complement of the intersection) of two sets or statements. For example, if the statements are "it is raining" and "it is cold," then we know what it means to be "raining *and* cold." But what is the opposite, or complement, of the statement "it is (raining and cold)?" The answer is "it is not raining, *or* it is not cold, or it is both not raining and not cold."

De Morgan's Laws

For any two sets A and B,

(1) $(A \cap B)' = A' \cup B'$ and (2) $(A \cup B)' = A' \cap B'$

Example 2.2-5 Set Theory

Suppose that $U = \{1, 2, 3, 4, 5, 6, 7, 8, 9, 10\}$, $A = \{1,3,5,7,9\}$, and $B = \{2,3,5,7\}$. Find $A - B$, $A \cap B'$, and $(A \cap B)'$.

Solution

First note that $A - B = A \cap B' = \{1,9\}$.
Second, De Morgan's law helps us see $(A \cap B)' = A' \cup B' = \{1,2,4,6,8,9,10\}$. ❒

You will be asked to verify De Morgan's Laws using Venn diagrams in one of the following exercises.

Exercise 2-8 Suppose that $U = \{1,2,3,4,5,6,7,8,9,10\}$, $A = \{1,3,5,7,9\}$, and $B = \{2,3,5,7\}$. Find the following: $A', B', A \cup B, A \cap B, A' \cup B', A' \cap B'$, and $(A \cup B)'$.

Exercise 2-9 Use a Venn diagram to verify that $A - B = A \cap B'$. That is, shade in the part of a Venn diagram that represents the left-hand side of the expression and one that represents the right-hand side. Observe that the shaded areas are the same. This implies that the concept of the difference between two sets is redundant. That is, we can write the difference operator using intersections and complements.

Exercise 2-10 Verify both of De Morgan's laws using Venn diagrams.

Exercise 2-11 Verify the distributive laws $A \cap (B \cup C) = (A \cap B) \cup (A \cap C)$

Complete the statement $A \cup (B \cap C) =$

Exercise 2-12 Use Venn diagrams to justify $N(A \cup B) = N(A) + N(B) - N(A \cap B)$, where N denotes the cardinality of a set.

Exercise 2-13 Construct a Venn diagram and shade in $(A \cup B \cup C)'$. Construct a second Venn diagram and shade in $A \cap (B \cup C)'$.

Exercise 2-14 Suppose $N(U) = 20$, $N(A') = 14$, $N(B') = 10$, and $N(A \cup B) = 12$. Find $N(A \cap B)$.

2.2.4 THE VENN BOX DIAGRAM

There is an alternative tabular form of the Venn diagram (hereafter referred to as the ***box diagram***) that is convenient for keeping track of cardinalities or probabilities for pairs of events and their complements. The idea is to represent A and A' as a vertical partition of the sample space (i.e., the universal set) U, and to consider B and B' as a horizontal partition of U:

(a)

A	A'	
$\Pr(A)$	$\Pr(A')$	1

(b)

B		$\Pr(B)$
B'		$\Pr(B')$
		1

(c)

	A	A'	
B	$\Pr(A \cap B)$	$\Pr(A' \cap B)$	$\Pr(B)$
B'	$\Pr(A \cap B')$	$\Pr(A' \cap B')$	$\Pr(B')$
	$\Pr(A)$	$\Pr(A')$	1

Diagram (a) shows the vertical partition, with the corresponding relationship $\Pr(A)+\Pr(A')=1$. Diagram (b) shows the horizontal partition, with $\Pr(B)+\Pr(B')=1$. Finally, (c) shows the overlay, with the sample space partitioned into 4 mutually exclusive events, as shown. The table forms a *magic square* in which the rows and columns sum to the numbers in the right and bottom margins, respectively. Finally, summing over the four mutually exclusive sets we obtain $\Pr(U)=1$, tabulated in the lower-right corner.

We note that in the case of equally likely outcomes we can use the cardinality of events, $N(A)$, in place of probabilities, by multiplying through by $N(U)$. This has the salutary effect of eliminating fractions and making the arithmetic easier.

Example 2.2-6 The Box Diagram

A small business with 120 workers provides two employee benefits, a matching retirement contribution and a health care subsidy. Seventy-one employees are signed up for the retirement benefit and 98 employees are signed up for the health care benefit. Fifteen employees have chosen neither benefit. Calculate the number of employees who have the retirement benefit but not the health care benefit.

Solution

Draw the Venn box diagram and begin by filling in all known data. We let A denote the set of employees with the retirement benefit and we let B denote the set of employees with the health care benefit. Then $N(S)=120, N(A)=71, N(B)=98,$ and $N(A'\cap B')=15$.

	Retirement Benefits A	No Retirement A'	
Health Care Benefits, B			98
No Health Care, B'		15	
	71		120

The rest of the numbers can be filled in (uniquely) to make the rows and columns sum correctly. Begin with the margins and work through the rest of the cells:

	A	A'	
B	64	34	98
B'	7	15	22
	71	49	120

Thus, the number with the retirement benefit, but not health-care equals seven. That is, $N(A\cap B')=7$.

Any question concerning the cardinality of A and B and their complements can be answered with ease from the completed table. For example, the number of employees with *at least one* benefit is $N(A\cup B) = 64+7+34 = 120-15 = 105$. Note that this is an instance of De Morgan's law (2) in the form $A\cup B=(A\cup B)''=[(A\cup B)']'=(A'\cap B')'$.

Any problem of this type involving two sets can be visualized using the box diagram. Enough information must be given to fill in at least one of the 4 inside cells. ❒

Example 2.2-7 The Box Diagram

Suppose that $\Pr(A\cup B)=.6,\ \Pr(A')=.6$ and $\Pr(A\cap B')=.3$. Find $\Pr(B)$.

Solution

The Venn box diagram follows:

	A	A'	
B			
B'	.3	$1-.6$	
		.6	1

(a)

	A	A'	
B	.1	.2	.3
B'	.3	.4	.7
	.4	.6	1

(b)

Box (a) shows the given information. The lower right hand corner, $A' \cap B'$, is the complement of the remaining three cells, whose "sum" is $A \cup B$ (De Morgan's law). Hence, the entry is shown as $1-.6$. Box (b) is easily obtained by filling in the margins and remaining cells to make everything add up correctly. It follows that $\Pr(B) = .3$. ❒

For questions involving three sets it is easier to revert to a conventional Venn drawing with the three sets all overlapping, unless one is prepared to construct a 3-dimensional box diagram. The resulting picture of A, B, and C is called the *Mickey Mouse diagram.*

It is most efficient to begin filling in the known cardinalities from the inside (greatest number of intersections) and then move outward (if you have the requisite information).

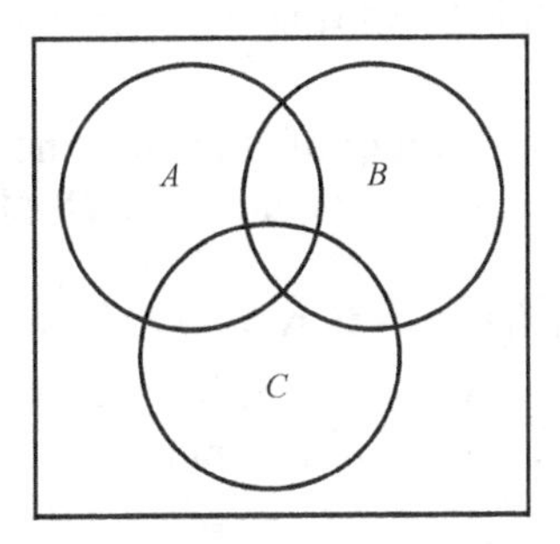

Example 2.2-8 The Mickey Mouse Diagram

A sophomore class of 36 mathematics majors was surveyed about their habits. Twelve students use aspirin, sixteen students consume beer, and twenty-two students use caffeine. Five students use all three, seven students use aspirin and beer, nine students use aspirin and caffeine, and eleven students use beer and caffeine.

(a) How many students practice exactly two habits?
(b) How many students practice exactly one habit?
(c) How many students practice none of these habits?

Solution

The first number inserted in our three set Venn diagram should be the 5 representing the intersection of all three sets (those students who use aspirin, drink beer, and consume coffee).

We know that exactly seven students use aspirin and drink beer, but we already know that five of those seven use all three. That leaves two students who use aspirin, drink beer, but do not use caffeine. We can place a 2 in the region aspirin $\cap$ beer $\cap$ (caffeine)$'$.

U is the set of mathematics majors.

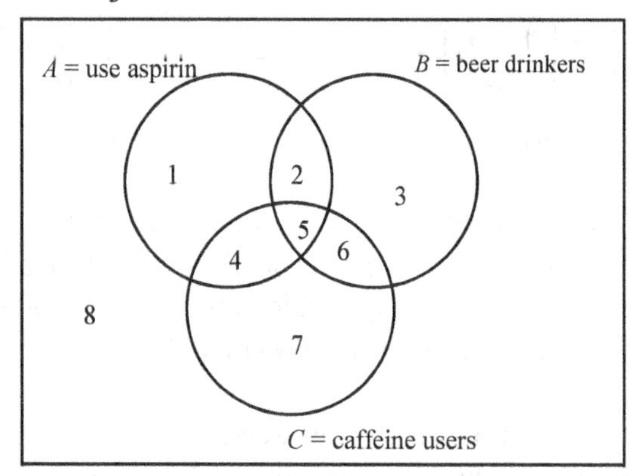

Since nine students use aspirin and caffeine, and we know 5 students use all three, that leaves $4 = 9 - 5$ students who are in the region aspirin $\cap$ (beer)$'$ $\cap$ caffeine. Similarly, $N[(\text{aspirin})' \cap \text{beer} \cap \text{caffeine}] = 6$.

We are told that a total of 12 students take aspirin. We have filled in regions with 2, 5, and 4. This leaves one student in the region aspirin $\cap$ (beer)$'$ $\cap$ (caffeine)$'$.

Finally, we need to determine the number of students who do not practice any of these habits. There are 36 math majors and we have accounted for 28 of the students.

There are twelve students who have exactly two habits $(2+4+6)$: 2 take aspirin and drink beer, but do not use caffeine, 4 who use aspirin, use caffeine but do not drink beer, and 6 students who drink beer and consume caffeine, but do not take aspirin.

Eleven students have exactly one habit $(1+3+7 = 11)$. Eight students are habit-free. ❐

Exercise 2-15 Verify the relationship

$$N(A \cup B \cup C) = N(A) + N(B) + N(C) - N(A \cap B) - N(A \cap C) - N(B \cap C) + N(A \cap B \cap C)$$

by drawing and labeling the Venn diagram with three intersecting sets.

Exercise 2-16 Generalize the inclusion-exclusion formula to derive a relationship for $N(A \cup B \cup C \cup D)$.

Exercise 2-17 Draw a Venn diagram with four intersecting regions.

Exercise 2-18 There are 28 people in your family consisting of 18 adults, 13 females, and 11 who have purple hair. There are eleven adult females, six of whom sport purple hair. There are ten adults with purple hair. No female children have purple hair. How many male children without purple hair are in the family?

Exercise 2-19 SOA/CAS Course 1 2000 Sample Examination #1

A marketing survey indicates that 60% of the population owns an automobile, 30% owns a house, and 20% owns both an automobile and a house. Calculate the probability that a person chosen at random owns an automobile or a house, but not both.

(A) 0.4 (B) 0.5 (C) 0.6 (D) 0.7 (E) 0.9

Exercise 2-20 SOA EXAM P Sample Exam Questions #5

An auto insurance company has 10,000 policyholders. Each policyholder is classified as

(i) young or old (ii) male or female (iii) married or single.

Of these policyholders, 3000 are young, 4600 are male, and 7000 are married. The policyholders can also be classified as 1320 young males, 3010 married males, and 1400 young married persons. Finally, 600 of the policyholders are young married males. How many of the company's policyholders are young, female, and single?

(A) 280 (B) 423 (C) 486 (D) 880 (E) 896

Exercise 2-21 SOA EXAM P Sample Exam Questions #1

A survey of a group's viewing habits over the last year revealed the following information:

(i) 28% watched gymnastics
(ii) 29% watched baseball
(iii) 19% watched soccer
(iv) 14% watched gymnastics and baseball
(v) 12% watched baseball and soccer
(vi) 10% watched gymnastics and soccer
(vii) 8% watched all three sports

Calculate the percentage of the group that watched none of the three sports during the last year.

(A) 24 (B) 36 (C) 41 (D) 52 (E) 60

Exercise 2-22 SOA EXAM P Sample Exam Questions #2

The probability that a visit to a primary care physician's (PCP) office results in neither lab work nor referral to a specialist is 35%. Of those coming to a PCP's office, 30% are referred to specialists and 40% require lab work. Determine the probability that a visit to a PCP's office results in both lab work and referral to a specialist.

(A) 0.05 (B) 0.12 (C) 0.18 (D) 0.25 (E) 0.35

Exercise 2-23 SOA EXAM P Sample Exam Questions #3

You are given $\Pr(A \cup B) = 0.7$ and $\Pr(A \cup B') = 0.9$. Determine $\Pr(A)$.

(A) 0.2 (B) 0.3 (C) 0.4 (D) 0.6 (E) 0.8

Exercise 2-24 SOA EXAM P Sample Exam Questions #8

Among a large group of patients recovering from shoulder injuries, it is found that 22% visit both a physical therapist and a chiropractor, whereas 12% visit neither of these. The probability that a patient visits a chiropractor exceeds by 0.14 the probability that a patient visits a physical therapist. Determine the probability that a randomly chosen member of this group visits a physical therapist.

(A) 0.26 (B) 0.38 (C) 0.40 (D) 0.48 (E) 0.62

Exercise 2-25 SOA EXAM P Sample Exam Questions #128

An insurance agent offers his clients auto insurance, homeowners insurance and renters insurance. The purchase of homeowners insurance and the purchase of renters insurance are mutually exclusive. The profile of the agent's clients is as follows:

(i) 17% of the clients have none of these three products.

(ii) 64% of the clients have auto insurance.

(iii) Twice as many of the clients have homeowners insurance as renters insurance.

(iv) 35% of the clients have two of these products.

(v) 11% of the clients have homeowners insurance, but not auto insurance.

Calculate the percentage of the agent's clients that have both auto and renters insurance.

(A) 7% (B) 10% (C) 16% (D) 25% (E) 28%

2.3 CONDITIONAL PROBABILITY

Often in a probability experiment we may have partial information about the outcome. This partial additional information is expressed through the concept of conditional probability. We want to know the probability of an event *given* that some other events have happened. For example, the probability that a person chosen at random from the general population will die in the next year would be revised if we also knew the person is 96 years old.

An insurance company would like to know the probability that a policyholder will be involved in an accident. There is an overall likelihood based upon the company's experience, but this probability changes if we know that the insured is a 16 year old male driving a red sports car and living in downtown Chicago. The probability of being dealt an ace is $\frac{4}{52}$ if the deck is freshly shuffled enough times[2], but it changes if we see some of the cards that have already been dealt. Gamblers who count cards can expect to benefit in Las Vegas if they vary their wagers based on the conditional probabilities of cards that remain in the deck(s). To combat this, casinos can shuffle the deck on each deal of the cards. But they find it easier to escort card counters from their casinos.

Conditional Probability

The conditional probability that event A occurs given that event B occurred is

$$\Pr(A \mid B) = \frac{\Pr(A \cap B)}{\Pr(B)}.$$

If the sample space consists of equally likely outcomes, then

$$\Pr(A \mid B) = \frac{N(A \cap B)}{N(B)}.$$

Note: Since event B has occurred it must have positive probability. This precludes division by zero in the definition.

Example 2.3-1 Conditional Probability

Consider Example 2.2-7, a small business with 120 workers and two employee benefits, A and B. This is the completed Venn box diagram:

	A	A'	
B	64	34	98
B'	7	15	22
	71	49	120

Calculate the probability that an employee chosen at random has the retirement benefit (A) given that she takes the health-care benefit (B).

[2] For most people, seven or more shuffles will sufficiently randomize the deck.

Solution

Choosing "an employee at random" means we are assuming a sample space of size 120 with equally likely outcomes. Thus, the probabilities of events are assigned according to their cardinality, as in Chapter 1.

Here we have $\dfrac{\Pr(A\cap B)}{\Pr(B)} = \dfrac{\frac{N(A\cap B)}{N(U)}}{\frac{N(B)}{N(U)}} = \dfrac{\frac{64}{120}}{\frac{98}{120}} = \dfrac{N(A\cap B)}{N(B)} = \dfrac{64}{98} = .6531.$ ❐

Intuitively, this says that if we are given an employee in set B, then we can just restrict our attention to the top row of the diagram and B becomes our modified sample space with cardinality 98. The only part of A that is relevant is the part in the top row, i.e., the part $A\cap B$ with cardinality 64.

2.3.1 CONDITIONAL PROBABILITY AND TREE DIAGRAMS

In introducing tree diagrams in Chapter 1 we mentioned they were helpful in analyzing probability models in which the outcomes develop in stages. They are most useful with conditional probability because outcomes of later stages can be conditioned on results from earlier stages. It is therefore not surprising that tree diagrams and conditional probability are intimately related.

The earlier conditional probability definition can be reformulated in multiplicative form:

$$\Pr(A\cap B) = \Pr(A\mid B)\cdot\Pr(B) = \Pr(B\mid A)\cdot\Pr(A).$$

This, in turn, can be generalized to any number of events. Thus, for example,

$$\Pr(A\cap B\cap C) = \Pr(C\cap B\cap A) = \Pr(C\mid B\cap A)\cdot\Pr(B\cap A) = \Pr(C\mid B\cap A)\cdot\Pr(B\mid A)\cdot\Pr(A).$$

In the following tree diagram analyses each possible path through the tree (from start to finish) represents exactly one outcome of the complete experiment (including all stages). There are two rules for listing probabilities in a tree diagram.

Tree Diagram Rules

1. The probabilities along a path are the conditional probabilities given that the prior events along the path have occurred.

2. The probability of an entire path (an outcome) is calculated as the product of the numbers along the path.

The second rule follows from the multiplicative form of conditional probability. In a two-stage experiment, for example, if A represents an outcome from the first stage, and B an outcome from the second stage given that A has occurred, then $A\cap B$ represents a complete outcome (path through the tree). By multiplying probabilities along a path we are calculating $\Pr(A\cap B) = \Pr(B\mid A)\cdot\Pr(A)$.

Example 2.3-2 Tree Diagrams

Suppose that you are dealt two cards sequentially (face down) from a standard deck. Let A be the event that the first card is an ace. Let B represent the event that the second card is an ace. We draw the associated tree diagram using the two rules. Observe that the probabilities in the second stage (drawing the second card and observing event B) are conditioned on the outcome in the first stage, that is, whether A or A' occurs. Also, the probabilities of the outcomes are the products of the probabilities along the path:

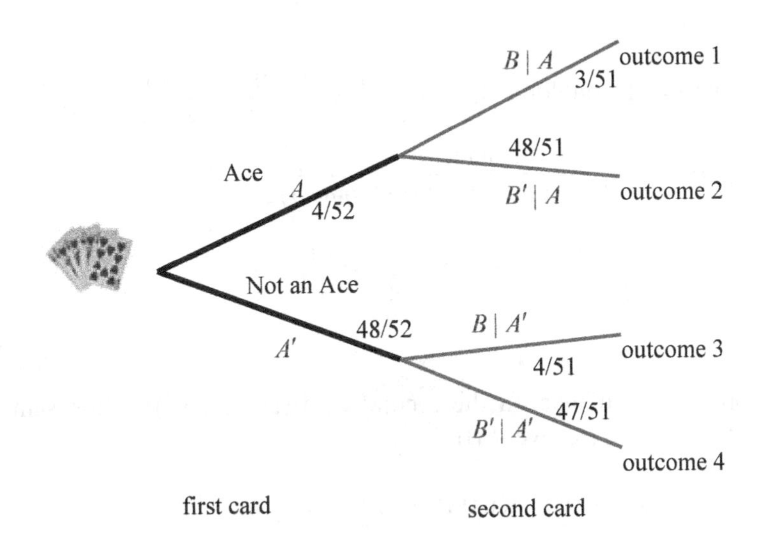

Notes

The sum of all the probabilities of the branches emanating from a branch point always equals one. The four paths through the tree correspond exactly to the four events in a Venn box diagram. They can be thought of as the four possible outcomes of the complete experiment of choosing two cards, assuming we are only concerned with "ace" or "not an ace." We exhibit the box below, together with the four probability calculations as products along the associated path:

	A	A'	
B	$\Pr(\text{outcome }1) = \Pr(A \cap B)$ $= \Pr(B \mid A)\Pr(A) = \frac{3}{51} \cdot \frac{4}{52}$	$\Pr(\text{outcome }3) = \Pr(A' \cap B)$ $= \Pr(B \mid A')\Pr(A') = \frac{4}{51} \cdot \frac{48}{52}$	$\Pr(B)$
B'	$\Pr(\text{outcome }2) = \Pr(A \cap B')$ $= \Pr(B' \mid A)\Pr(A) = \frac{48}{51} \cdot \frac{4}{52}$	$\Pr(\text{outcome }4) = \Pr(A' \cap B')$ $= \Pr(B' \mid A')\Pr(A') = \frac{47}{51} \cdot \frac{48}{52}$	$\Pr(B')$
	$\Pr(A)$	$\Pr(A')$	1

Here are some typical questions relating to this probability experiment:

(a) What is the probability that the first card is an ace?
(b) What is the probability that the second card is an ace given that the first card is not an ace?
(c) What is the probability that the second card is an ace given that the first card is an ace?

(d) What is the probability that the second card is an ace?
(e) What is the probability that the first card was an ace given that the second card is an ace?

Solution

(a) $\Pr(A) = \Pr(\text{outcome 1}) + \Pr(\text{outcome 2}) = \frac{3}{51}\cdot\frac{4}{52} + \frac{48}{51}\cdot\frac{4}{52} = \frac{204}{2{,}652} = \frac{4}{52} = \frac{1}{13}$.

(b) $\Pr(B \mid A') = \Pr(\text{second card is an ace} \mid \text{first card is not an ace}) = \frac{4}{51}$.

(c) $\Pr(B \mid A) = \frac{3}{51}$.

(d) $\Pr(B) = \Pr(\text{outcome 1}) + \Pr(\text{outcome 3}) = \frac{3}{51}\cdot\frac{4}{52} + \frac{4}{51}\cdot\frac{48}{52} = \frac{204}{2{,}652} = \frac{4}{52} = \frac{1}{13}$.

(e) $$\Pr(A \mid B) = \frac{\Pr(A \cap B)}{\Pr(B)} = \frac{\Pr(\text{outcome 1})}{\Pr(\text{outcome 1}) + \Pr(\text{outcome 3})}$$

$$= \frac{\frac{3}{51}\cdot\frac{4}{52}}{\frac{3}{51}\cdot\frac{4}{52} + \frac{4}{51}\cdot\frac{48}{52}} = \frac{12}{204} = \frac{1}{17}.$$

Note
The (unconditional) probability of ace on the second card (answer (d)) is the same as the probability of ace on the first card (answer (a)). ❐

2.3.2 BAYESIAN "CAUSE AND EFFECT" REASONING

In Example 2.3-2, part (e) requires us to speculate about what happened on the first stage after observing the result of the second stage. This can be thought of as an instance of inferring the *cause* by observing the *effect*. We cannot know for sure what happened in the first stage, but we can calculate conditional probabilities for the *causes* based on the observed *effect*. For example, an observer may pick up the second card and note that it is an ace, but she will not know whether this was *caused* by drawing an ace or a non-ace on the first card. She can however (with practice) calculate the conditional probabilities for the first card. This type of backward-looking reasoning is called Bayesian inference in mathematical statistics. We have further illustrations of this in some of the following examples.

Example 2.3-3 Bayesian Inference

A calculus class is composed of the following students:

	Male	Female
Engineering	12	10
Mathematics	8	4

Assume that all students are equally likely to be selected.
Calculate the following probabilities:

Pr(male), Pr(male|mathematics), Pr(engineering | female), and Pr(female|mathematics)

Solution

There are 34 students total. Construct the associated box diagram in terms of probabilities.

	Male	Female	
Engineering	12/34	10/34	22/34
Mathematics	8/34	4/34	12/34
	20/34	14/34	34/34

$$\Pr(\text{male}) = \frac{20}{34} = \frac{12}{34} + \frac{8}{34} = .5882.$$

$$\Pr(\text{male}|\text{mathematics}) = \frac{8/34}{12/34} = \frac{8}{12} = \frac{2}{3}.$$

$$\Pr(\text{engineering} \mid \text{female}) = \frac{10}{14} = .7143.$$

❐

In many problems we are able to calculate conditional probabilities directly from the data. For example, $\Pr(\text{female}|\text{mathematics})$ means the probability that a student is a female given that the student is a mathematics major. In the original table, we see that there are $8+4=12$ mathematics majors, four of whom are female. Hence we quickly calculate

$$\Pr(\text{female}|\text{mathematics}) = \frac{4}{12}.$$

Example 2.3-4 SOA EXAM P Sample Exam Questions #25

A blood test indicates the presence of a particular disease 95% of the time when the disease is actually present. The same test indicates the presence of the disease 0.5% of the time when the disease is not present. One percent of the population actually has the disease. Calculate the probability that a person has the disease given that the test indicates the presence of the disease.

(A) 0.324 (B) 0.657 (C) 0.945 (D) 0.950 (E) 0.995

Note

This problem is a classic example used to illustrate Bayesian reasoning. We first validate the blood test by administering it to people whose condition we already know by other means. Thus, we know before the test whether they are healthy (H) or have the disease (D). This is the first, or *causal*, stage. The second stage (the *effect*) is the outcome of the test, which can be positive (+) or negative (-).

Solution
The information given leads to the following tree:

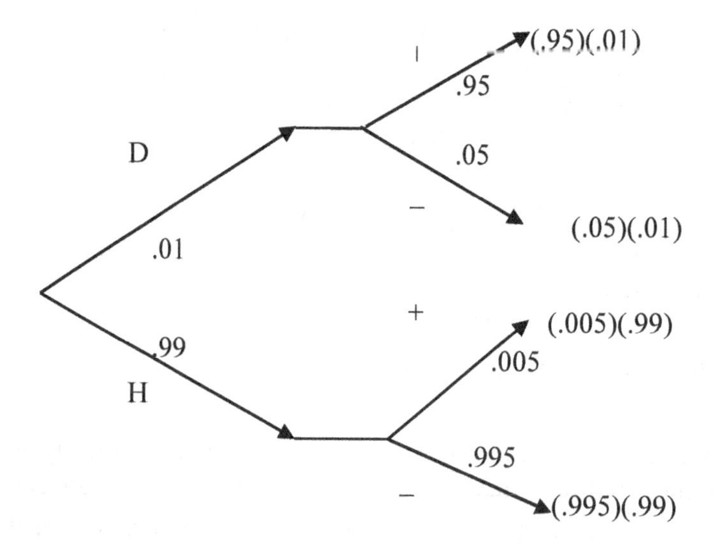

Now, the whole point of developing the blood test is to administer it to people whose condition we do not know in advance. Thus, we first observe the *effect* (positive test or negative test), and then *infer* the presence or absence of the disease. So how reliable is the test in making these inferences? In other words, given the effect of a positive (+) test, what is the probability it was *caused* by the person having the disease?

From the tree diagram we calculate that

$$\Pr(D|+) = \frac{\Pr(D \cap +)}{\Pr(+)} = \frac{(.95)(.01)}{(.95)(.01)+(.005)(.99)} = .6574,$$

which is answer B. ❐

In words, the *effect* of a positive test is *caused* by actually having the disease 65.74% of the time. This test seemed far more reliable from the original test results, 95% and 99.5% reliable for sick and healthy people, respectively. But when used as a diagnostic tool, it will cause a lot of unnecessary anxiety among the 34.26% of the people who have a positive test despite not having the disease. The high percentage of false-positives is due to the fact that most of the population was healthy and did not have the disease to begin with. This group vastly outnumbers the diseased group.

Example 2.3-5 Bayesian Inference

Your local fraternity, ΣΜΔ, is throwing a party. Three of its brothers are in charge of bringing beverages. Matt brings six root beers and six colas in his cooler. Chris brings twenty root beers and three colas in his cooler. And Greg brings twelve colas in his cooler. Suppose that Alicia is the first partygoer to select a beverage. She selects a cooler at random and a beverage from that cooler at random. Find the following:

(a) Pr(cola|Matt's cooler)

(b) Pr(cola|Chris' cooler)

(c) Pr(root beer|Chris' cooler)

(d) Pr(cola)

(e) Pr(root beer)

(f) Suppose that all beverages were combined into Chris' cooler. If Alicia selects a beverage at random, what is the probability that she selects a cola?

(g) The probability that Alicia selected Chris' cooler given that she is drinking a root beer.

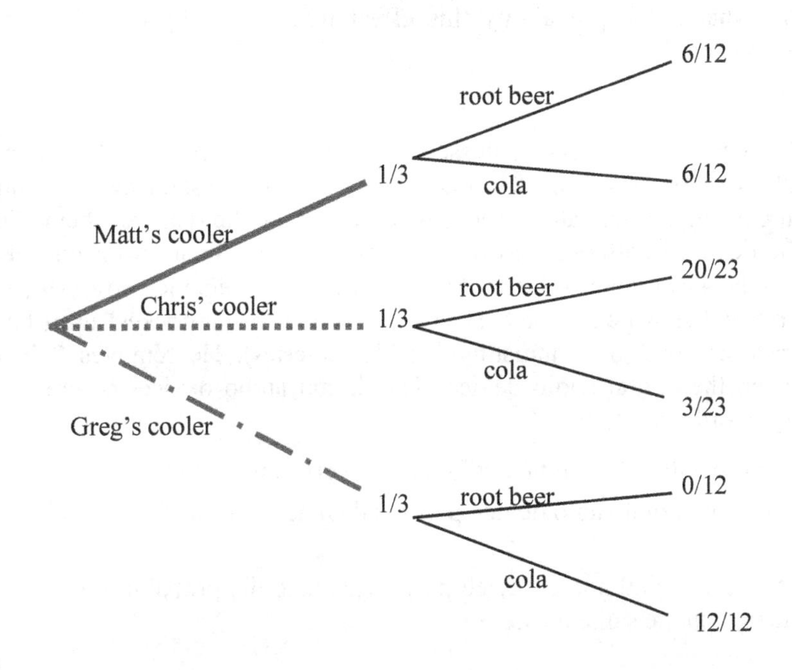

Solutions

(a) $\text{Pr(cola|Matt's cooler)} = \dfrac{\text{6 colas}}{\text{6 colas} + \text{6 root beers}} = \dfrac{\text{6 colas}}{\text{12 beverages}} = .5.$

(b) $\text{Pr(cola|Chris' cooler)} = \dfrac{3}{23}.$

(c) $\text{Pr(root beer|Chris' cooler)} = \dfrac{20}{23}.$

(d) $\text{Pr(cola)} = \underbrace{\dfrac{1}{3}}_{\text{Matt's cooler}} \cdot \underbrace{\dfrac{6}{12}}_{\text{cola given Matt's cooler}} + \underbrace{\dfrac{1}{3}}_{\text{Chris' cooler}} \cdot \underbrace{\dfrac{3}{23}}_{\text{cola given Chris' cooler}} + \underbrace{\dfrac{1}{3}}_{\text{Greg's cooler}} \cdot \underbrace{\dfrac{12}{12}}_{\text{cola given Greg's cooler}} = \dfrac{450}{828} = .5435.$

(e) $\text{Pr(root beer)} = \dfrac{1}{3}\cdot\dfrac{6}{12}+\dfrac{1}{3}\cdot\dfrac{20}{23}+\dfrac{1}{3}\cdot\dfrac{0}{12} = \dfrac{378}{828} = 1 - \text{Pr(coke)} = .4565.$

(f) $\Pr(\text{cola if combined cooler}) = \dfrac{\text{number of colas}}{\text{number of beverages}} = \dfrac{6+3+12}{6+6+20+3+12} = \dfrac{21}{47} = .4468.$

(g) $\Pr(\text{Chris' cooler}|\text{root beer}) = \dfrac{\Pr(\text{Chris' cooler} \cap \text{root beer})}{\Pr(\text{root beer})} = \dfrac{\frac{1}{3} \cdot \frac{20}{23}}{\frac{378}{828}} = .6349.$ ❐

Notes

The way that the beverages were located in the various coolers made it more likely that a cola was selected than if all of the beverages were grouped together. If we know nothing, each partygoer has a 1/3rd chance of selecting Chris' cooler. But if we see that Alicia has a root beer, then there is about a 63% chance that she got it from Chris' cooler. Certainly she could not have selected Greg's cooler, because he brought no root beer.

Part (g) is another instance of Bayesian inference: Given Alicia is holding a root beer (an observed *effect*), what is the probability this effect was *caused* by (that is, resulted from choosing) Chris' cooler?

Exercise 2-26 An absent-minded professor goes for a walk carrying a digital audio device using 2 batteries. He has 2 fresh replacement batteries stashed away in one of four pockets. Sure enough, both batteries lose their charge and he removes them. Not wanting to throw the depleted batteries into the woods, he places them into a pocket chosen at random from the 4 available. A little while later he remembers the two fresh batteries, but he cannot remember which pocket. He fishes around in his pockets until he finds one with batteries (either 2 or 4 indistinguishable batteries). He removes 2 batteries and inserts them in the digital audio device. The digital audio device requires at least one good battery in order to play.

(a) Find the probability the digital audio device works on the first try.

(b) Given that the digital audio device plays, calculate the probability he chose two good batteries.

(c) Given that the digital audio device plays, calculate the probability he had placed all four batteries in the same pocket.

Exercise 2-27 (The Monte Hall Problem, also known as the 3 door problem): On the game show *Let's Make a Deal*, there are three doors hiding three prizes. Two of the prizes are goats, and one prize is an expensive new automobile. Naturally you would like the car, but only Monte knows which door it is behind. You select one of the doors. Rather than opening the door you chose, Monte opens one of the two other doors to reveal a goat. You are now given the option to stick with your original selection or to switch to the other unopened door.

(a) What is the (conditional) probability that you win the car if you stick with your original choice of doors?

(b) What is the (conditional) probability that you win the car if you change your mind and select the other unopened door?

Exercise 2-28 SOA EXAM P Sample Exam Questions #12

A doctor is studying the relationship between blood pressure and heartbeat abnormalities in her patients. She tests a random sample of her patients and notes their blood pressures (high, low, or normal) and their heartbeats (regular or irregular). She finds that:

(i) 14% have high blood pressure

(ii) 22% have low blood pressure.

(iii) 15% have an irregular heartbeat

(iv) Of those with an irregular heartbeat, one-third have high blood pressure.

(v) Of those with normal blood pressure, one-eighth have an irregular heartbeat.

What portion of the patients selected have a regular heartbeat and low blood pressure?

(A) 2% (B) 5% (C) 8% (D) 9% (E) 20%

Exercise 2-29 SOA EXAM P Sample Exam Questions #26

The probability that a randomly chosen male has a circulation problem is 0.25. Males who have a circulation problem are twice as likely to be smokers as those who do not have a circulation problem. What is the conditional probability that a male has a circulation problem, given that he is a smoker?

(A) $\frac{1}{4}$ (B) $\frac{1}{3}$ (C) $\frac{2}{5}$ (D) $\frac{1}{2}$ (E) $\frac{2}{3}$

Exercise 2-30 SOA/CAS Course 1 2000 Sample Examination #3

Ten percent of a company's life insurance policyholders are smokers. The rest are non-smokers. For each non-smoker the probability of dying during the year is 0.01. For each smoker, the probability of dying during the year is 0.05. Given that a policyholder has died, what is the probability that the policyholder was a smoker?

(A) 0.05 (B) 0.20 (C) 0.36 (D) 0.56 (E) 0.90

Exercise 2-31 SOA EXAM P Sample Exam Questions #24

The number of injury claims per month is given by N where

$$P(N=n) = \frac{1}{(n+1)(n+2)},$$

where $n \geq 0$.

Determine the probability of at least one claim during a particular month, given that there have been at most four claims during that month.

(A) $\frac{1}{3}$ (B) $\frac{2}{5}$ (C) $\frac{1}{2}$ (D) $\frac{3}{5}$ (E) $\frac{5}{6}$

Exercise 2-32 The Pittsburgh Penguins win at home 55% of the time and on the road 35% of the time in games against the New Jersey Devils. Assume there are no ties. In a ***best-of three*** series (the first team to win 2 games wins the series), the Penguins play game 1 and game 3 (if necessary) at home.

(a) Determine the probability that the Penguins win the three game series.

(b) Determine the probability that the Devils win at least one game.

Exercise 2-33 SOA Exam P Sample Exam Questions #21
Upon arrival at a hospital's emergency room, patients are categorized according to their condition as critical, serious, or stable. In the past year:

(i) 10% of the emergency room patients were critical;

(ii) 30% of the emergency room patients were serious;

(iii) the rest of the emergency room patients were stable;

(iv) 40% of the critical patients died;

(v) 10% of the serious patients died; and

(vi) 1% of the stable patients died.

Given that the patient survived, what is the probability that the patient was categorized as serious upon arrival?

(A) 0.06 (B) 0.29 (C) 0.30 (D) 0.39 (E) 0.64

Exercise 2-34 Some people believe that baseball players get *hot*. One potential model is if a player had a hit in his last at-bat, then he has a .400 average in his next at-bat. If he failed to hit, then he has a .250 average in his next at-bat. Suppose that a player has not been hitting (so during his next at-bat we assume he hits .250) and has three at bats in a game. Find the probability that he gets exactly one hit in those 3 at-bats.

Exercise 2-35 In a small community, there is a polygamous man named Packy. He is married to one percent of the women in this community. If you were raised to call Packy "dad," then there is a 90% chance that he is in fact your father. If you call someone else "dad," there is still a 5% chance that Packy is your father. Given that Packy is your father, what is the probability that you call him dad?

Exercise 2-36 SOA Exam P Sample Exam Questions #10
An insurance company examines its pool of auto insurance customers and gathers the following information:

(i) All customers insure at least one car.
(ii) 64% of the customers insure more than one car.
(iii) 20% of the customers insure a sports car.
(iv) Of those customers who insure more than one car, 15% insure a sports car.

What is the probability that a randomly selected customer insures exactly one car, and that car is not a sports car?

(A) 0.16 (B) 0.19 (C) 0.26 (D) 0.29 (E) 0.31

2.4 INDEPENDENCE

In most of the examples we have seen involving multiple events, knowledge of whether one of the events has occurred affects our assessment of the likelihood of the other event. For example, if A represents the event "it is snowing in Pittsburgh" then we might assess the probability of A at 4%. But if we know that B has occurred, where B is the event "current temperature is above 95°F," then we know that the probability of A is smaller than 4%. Other examples of dependent events include drawing two cards from a deck consecutively, without replacement, or the events "Speeding," and "Receiving a speeding ticket."

Contrast these examples with the situation when two coins are tossed. Knowledge of the outcome of one toss should have no effect on our assessment of the outcome of the other toss. Similarly, suppose A is the event "outcome of the first spin on a roulette table is even" and B is the event "outcome of the second spin on a roulette table is red." All gamblers get to see what happened during the first spin, but the probabilities of the various outcomes involving the second spin do not change based on the outcome of the first spin. Other examples of independent events include drawing two cards from a deck with replacement or tossing multiple dice.

Independence

Let A and B be events with non-zero probabilities. We say A and B are ***independent*** if any (and hence all) of the following hold:

1. $\Pr(A|B) = \Pr(A)$,
2. $\Pr(B|A) = \Pr(B)$, or
3. $\Pr(A \cap B) = \Pr(A) \cdot \Pr(B)$. This is called the ***multiplicative rule***.

Otherwise the events are said to be ***dependent***.

Note
If $\Pr(B) = 0$ then the left-side of (1) is indeterminate. A similar remark applies to (2) if $\Pr(A) = 0$. However, (3) continues to make sense and be true if either one or both of A or B has zero probability. Therefore, statement (3) can be used as a stand-alone definition of independence that also applies to events with zero probability.

When analyzing two events A and B for independence we can make use of the Venn box diagram. Previously, we noted that in order to solve a problem we need to know one of the *inside* probabilities. But, if we know that A and B are independent, then (from (3) in the definition) we can calculate the probability of any[3] inside cell as the product of the probabilities on the margin.

Example 2.4-1 Independent Events

Suppose that the chance that a student earns a grade of "A" in a mathematics course is 40%, and the chance that the student is assigned an "A" in engineering is 70%. If these events are independent, then find the following:

(a) The probability that the student receives an A in both math and engineering.
(b) The probability that the student receives an A in neither.
(c) The probability that the student receives exactly one A.

Solution

Let M be the event "A in math" and let E be the event "A in engineering." In this problem the magic word is *independent*. This means that the margins of the box diagram can be completed and all of the inner cells are calculated as products.

	M	M'	
E	.28	.42	.70
E'	.12	.18	.30
	.40	.60	1.00

(a) $\Pr(M \cap E) = \Pr(M) \cdot \Pr(E) = (.40) \cdot (.70) = .28$.

(b) $\Pr(\text{no A's}) = \Pr(M' \cap E') = .18$.

(c) $\Pr(\text{exactly one A}) = \Pr(M \cap E') + \Pr(M' \cap E) = .42 + .12 = .54$.

The equivalent Venn diagram looks like this:

Grades a student receives

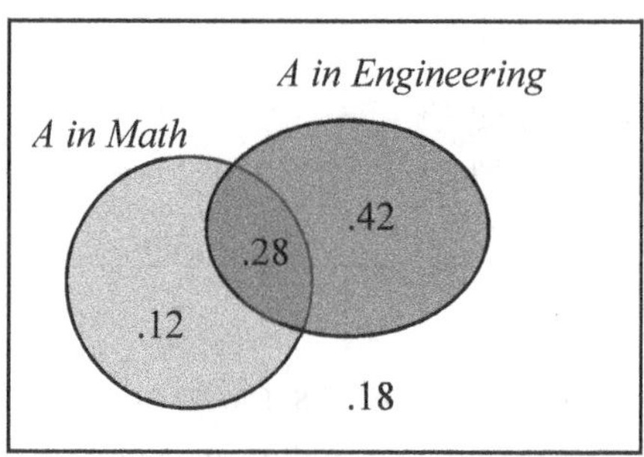

❐

[3] You will show that if two events A and B are independent, then so are A and B', A' and B, as well as A' and B'.

Example 2.4-2 Population Statistics

The table below was generated from the 2000 census and shows the population of the U.S. by gender.

Gender	Number	Percentage
Male	138,053,563	49.056
Female	143,368,343	50.944
Population	281,421,906	100.0

Two individuals are chosen at random in the U.S. Construct a tree diagram listing all possible outcomes with their associated probabilities.

Solution

The words "selected at random" typically imply independence. In all problems, we need to determine if there is replacement (the person selected first is replaced back into the population and can be selected again) or if we do not replace the person. In cases such as this where we have a large population, our rounding rules make this issue immaterial. That is,

$$\Pr(1^{st} \text{ person is male}) = \frac{138{,}053{,}563}{281{,}421{,}906} = .490557274.$$

The conditional probability that the second person is male, given the first person is male is

$$\Pr(2^{nd} \text{person is male} \mid 1^{st} \text{person is male}) = \frac{138{,}053{,}562}{281{,}421{,}905} = .490557272$$

These are almost exactly equal.

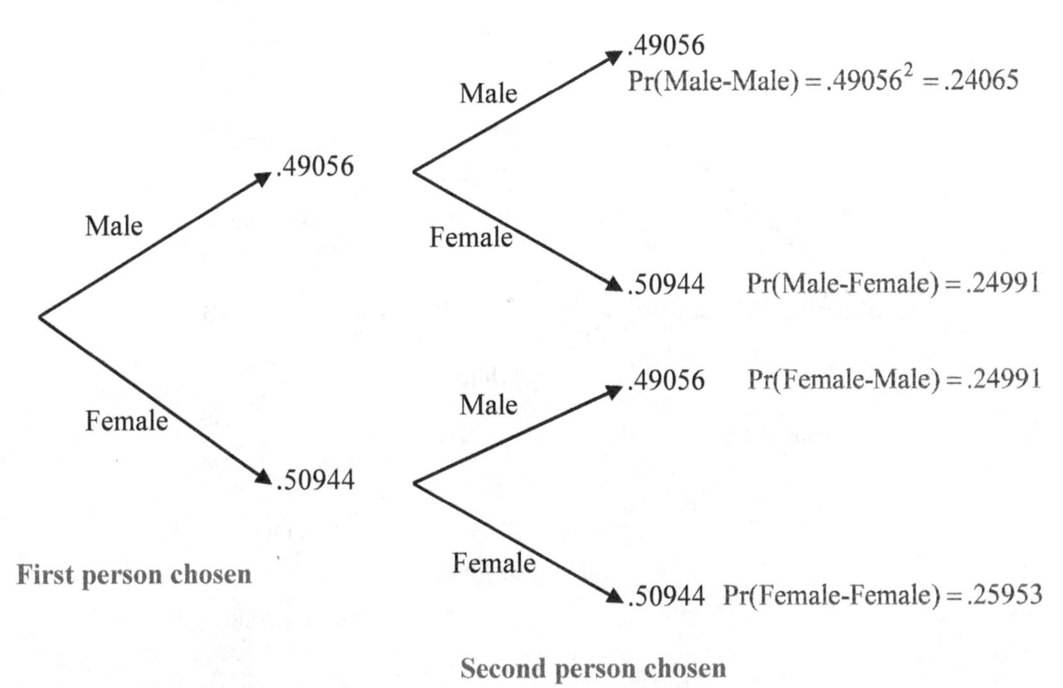

❐

Example 2.4-3 Tree Diagram Analysis

On a fireplace mantle sit three urns: a red urn that contains three red balls and two white balls, a white urn that contains one red ball and four white balls, and a blue urn that contains Aunt Mildred's cremation ashes. Pick the red or white urn randomly (this means either the red urn or the white urn is selected with probability one-half each). Randomly select a ball from the chosen urn. Do not replace this ball. Select a second ball at random from the urn whose color matches the color of the first ball.

(a) Find the probability that the second ball is white.
(b) Find the probability that the first ball is white.
(c) Find the probability that the first ball is white, given that the second ball is white.
(d) Are the events "first ball is white" and "second ball is white" independent?

Solution
Note that Aunt Mildred's blue urn is a red herring[4].

(a) $\Pr(2^{nd}\text{ ball is white}) = .5 \cdot .6 \cdot .5 + .5 \cdot .4 \cdot .8 + .5 \cdot .2 \cdot .4 + .5 \cdot .8 \cdot .75 = .65$

(b) $\Pr(1^{st}\text{ ball white}) = .5 \cdot .4 + .5 \cdot .8 = .6$

(c) $\Pr(1^{st}\text{ ball white} | 2^{nd}\text{ ball white}) = \dfrac{\Pr(1^{st}\text{ ball white} \cap 2^{nd}\text{ ball is white})}{\Pr(2^{nd}\text{ ball is white})}$

$$= \frac{(.50)(.40)(.80) + (.50)(.80)(.75)}{.65} = \frac{.46}{.65} = .7077$$

(d) No, since (b) and (c) are not equal.

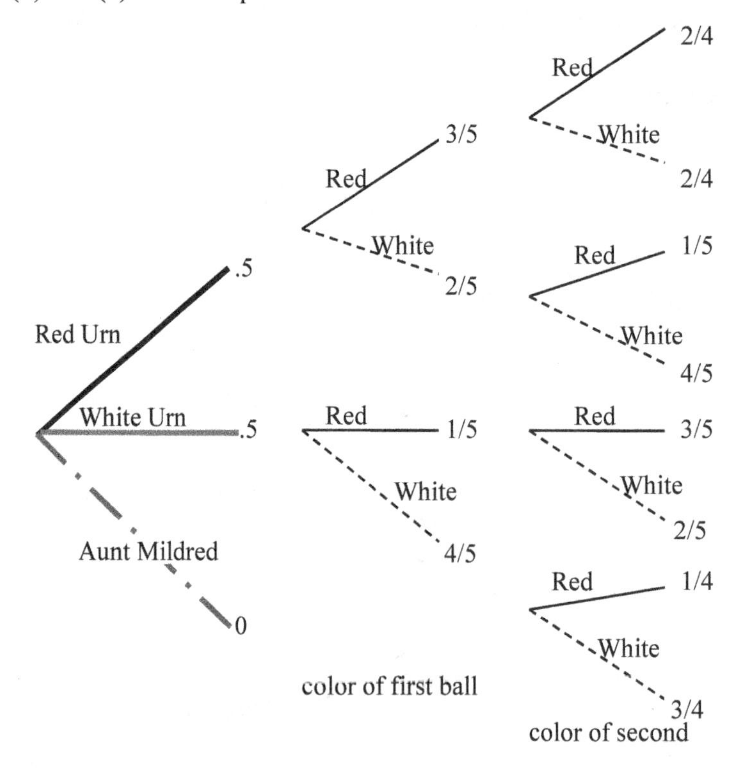

❐

4 "A diversion intended to draw attention away from the main issue." www.google.com/dictionary.

Exercise 2-37 The following table, taken from the National Center for Health Statistics, shows births for 2001, a typical year for newborns:

Gender	Number	Percentage
Male	2,057,922	51.1
Female	1,968,011	48.9
Population	4,025,933	100.0

For a family with two children, construct the tree diagram for child gender and order of birth. Calculate the probabilities for all outcomes, assuming the percentages for 2001 hold generally and that the genders of siblings are independent.

If the example and the exercise seem somewhat paradoxical (why are there more women in the U.S. if more men are born?) the answer is that males have higher mortality rates at all ages and therefore die sooner. The percentage of men and women under age 56 is about even, but the tide turns rapidly after that. Only 31.6% of the population age 85 or over is male.

Exercise 2-38 Suppose that we flip a loaded coin twice. The probability that an individual flip comes up heads is $\frac{2}{3}$. Use a tree diagram to calculate all probabilities.

Note
If you are having a difficult time imagining a ***loaded coin***, think of a six-sided fair die. On four of the sides, write an H with a black permanent Sharpie® marker. On the remaining two sides, write a T with a Sharpie® marker. This die will perform like a loaded coin with probability of flipping a head equal to $\frac{4}{6} = \frac{2}{3}$.

Exercise 2-39 John Stockton (number 12, retired, Utah Jazz, Gonzaga University) makes 38.4% of his three-point attempts. In a game against the New York Knickerbockers, he had 3 three-point attempts. Construct a tree diagram listing all possible outcomes and the associated probabilities. What is the probability that John Stockton makes exactly two of his three attempts?

Note
This exercise is an example of the binomial distribution which we will discuss in Chapter 4. At that time, we will derive formulas for probabilities of outcomes. For now, draw a small tree diagram and make some astute observations and reasonable assumptions.

Exercise 2-40 Two cards are dealt sequentially and without replacement from a standard deck. Suppose A is the event "the first card dealt is a spade" and B is the event "the second card dealt is a spade". Find $\Pr(B \mid A)$. Are the two events independent of one another?

Exercise 2-41 Suppose that events A and B are independent with $\Pr(A) = a$ and $\Pr(B) = b$. Make use of the Venn box diagram (and some algebra) to show that

(a) A and B' are independent.
(b) A' and B' are independent.
(c) A' and B are independent.

Exercise 2-42 Suppose that events A and B are independent and that events A and C are independent. Are the events B and C independent? Hint: What if B and C are the same event? The definition of independence can be extended to collections of more than two events.

Independence for Collections of Events

Let $\{E_1, E_2, \ldots, E_n\}$ be a collection of events. Then the collection is ***independent*** providing every sub-collection of these events satisfies the multiplicative rule. That is, every possible pair E_{i_1}, E_{i_2} satisfies $\Pr(E_{i_1} \cap E_{i_2}) = \Pr(E_{i_1})\Pr(E_{i_2})$, every possible triple $E_{i_1}, E_{i_2}, E_{i_3}$ satisfies $\Pr(E_{i_1} \cap E_{i_2} \cap E_{i_3}) = \Pr(E_{i_1})\Pr(E_{i_2})\Pr(E_{i_3})$, and so forth.

This is a complicated definition and it raises the question, why isn't pair-wise independence sufficient for three or more events to be independent? Here is an example based on the sample space for tossing three coins.

Exercise 2-43 Consider the experiment of tossing three distinguishable fair coins.

(a) List the sample space using 3-letter words with H representing heads and T representing tails.
(b) Let A be the event the first and second coins match, B the event the first and third coins match, and C the event that the second and third coins match. Calculate the probabilities of A, B, and C.
(c) Calculate the probabilities of the pair-wise intersections $A \cap B$, $A \cap C$, and $B \cap C$.
(d) Are these three events pair-wise independent?
(e) Calculate $\Pr(A \cap B \cap C)$.
(f) Are A, B, and C independent?

Exercise 2-44 Suppose that events A and B are independent with $\Pr(A \cap B') = .2$ and $\Pr(A' \cap B) = .3$. Find $\Pr(A \cup B)$. *Hint*: There are two correct answers.

Exercise 2-45 Flip a fair coin three times. Let A be the event "the first flip is tails" and B be the event "the third flip is heads."

(a) Create a tree diagram listing all possible outcomes of three flips of the coin.
(b) Find $\Pr(B \mid A)$.
(c) Are the events A and B independent of one another?

Exercise 2-46 Repeat Exercise 2-45 but change the coin so that it is loaded to come up heads 60% of the time.

Exercise 2-47 SOA/CAS Course 1 2000 Sample Examination #22
A dental insurance policy covers three procedures: orthodontics, fillings and extractions. During the life of the policy, the probability that the policyholder needs:

(i) orthodontic work is 1/2
(ii) orthodontic work or a filling 2/3
(iii) orthodontic work or an extraction is 3/4
(iv) a filling and an extraction is 1/8

The need for orthodontic work is independent of the need for a filling and is independent of the need for an extraction. Calculate the probability that the policyholder will need a filling or an extraction during the life of the policy.

(A) 7/24 (B) 3/8 (C) 2/3 (D) 17/24 (E) 5/6

Exercise 2-48 The Hulls were trying to get pregnant. Rhonda thought that she might be with child and so bought a home test that was advertised as 99% accurate. The results came back positive (indicating conception). Mark, her husband, came home and heard the news. Not being convinced, he went to a different store and purchased a different brand of early pregnancy test. It was advertised as 98% accurate. The second test also indicated a pregnancy. Mark was about to spend $20.00 on a third test, but Rhonda calculated the probability that both tests were incorrect. What was her result? What are you assuming about these two tests?

Exercise 2-49 Of 31 students in a course, 14 major in business, the rest in science. Assume there are no double majors. Grades assigned from a previous class showed the following results. Fill in the missing information.

Major	**Grades**				
	A	**B**	**C**	**D**	**F**
business majors	2	3	3		1
science majors	7		1	0	2

(a) What is the probability that a student received a grade of 'B'?
(b) What is the probability that a student is a science major?
(c) What is the probability that a student is a science major and received a grade of 'B'?
(d) Given that a student is a business major, what is the conditional probability that he received a grade of 'C'?
(e) Given that a student received a grade of 'A', what is the conditional probability that he is a business major?
(f) Are the events "business major" and "receiving a 'C'" independent?

Exercise 2-50 A flashlight has 5 batteries, two of which have lost their charge. Randomly select two batteries without replacement.

(a) Find the probability that both batteries have lost their charge.

(b) Find the probability that exactly one battery has lost its charge.

(c) Repeat (a) and (b), but assuming this time that the batteries are selected randomly ***with replacement.*** That is, you select a battery and test to see if it has lost its charge. That battery is then replaced and the second battery is chosen randomly from the original 5.

Exercise 2-51 SOA EXAM P Sample Exam Questions #11

An actuary studying the insurance preferences of automobile owners makes the following conclusions:

(i) An automobile owner is twice as likely to purchase collision coverage as disability coverage.

(ii) The event that an automobile owner purchases collision coverage is independent of the event that he or she purchases disability coverage.

(iii) The probability that an automobile owner purchases both collision and disability coverage is 0.15.

What is the probability that an automobile owner purchases neither collision nor disability coverage?

(A) 0.18 (B) 0.33 (C) 0.48 (D) 0.67 (E) 0.82

Exercise 2-52 Two cards are dealt from a standard deck consecutively and without replacement. Find the probability of being dealt two face cards (Kings, Queens, or Jacks).

Exercise 2-53 Suppose that you enter the Cal-Neva casino in Reno, Nevada and approach the roulette table. On the roulette wheel there are 18 red, 18 black and 2 green "house numbers." You watch for a while and observe seven consecutive black numbers.

(a) What is the probability that the next number is black?

(b) What is the probability that the next eight consecutive numbers are black?

Exercise 2-54 SOA/CAS Course 1 2000 Sample Examination #14

Workplace accidents are categorized into three groups: minor, moderate and severe. The probability that a given accident is minor is 0.5, that it is moderate is 0.4, and that it is severe is 0.1. Two accidents occur independently in one month.

Calculate the probability that neither accident is severe and at most one is moderate.

(A) 0.25 (B) 0.40 (C) 0.45 (D) 0.56 (E) 0.65

Exercise 2-55 SOA EXAM P Sample Exam Questions #4

An urn contains 10 balls: 4 red and 6 blue. A second urn contains 16 red balls and an unknown number of blue balls. A single ball is drawn from each urn. The probability that both balls are the same color is 0.44.

Calculate the number of blue balls in the second urn.

(A) 4 (B) 20 (C) 24 (D) 44 (E) 64

2.5 BAYES' THEOREM

Bayes' theorem looks difficult and intimidating as a formula, but the idea is quite simple and straightforward. In fact, we have already been applying Bayes' theorem without drawing much attention to it. For example, when we calculated such things as

$$\Pr(\text{Chris' cooler}|\text{drinking root beer}) \quad \text{and} \quad \Pr(1^{\text{st}}\text{ card an ace} \mid 2^{\text{nd}}\text{ card an ace}),$$

we were in fact using Bayes' theorem, without the fancy notation. Here is the general statement:

Bayes' Theorem

Suppose that the sample space S is partitioned into disjoint subsets $B_1, B_2, \ldots, B_n$. That is, $S = B_1 \cup B_2 \cup \cdots \cup B_n$, $\Pr(B_i) > 0$ for all $i = 1, 2, \ldots, n$, and $B_i \cap B_j = \varnothing$ for all $i \neq j$. Then for an event A,

$$\Pr(B_j \mid A) = \frac{\Pr(B_j \cap A)}{\Pr(A)} = \frac{\Pr(B_j) \cdot \Pr(A \mid B_j)}{\sum_{i=1}^{n} \Pr(B_i) \cdot \Pr(A \mid B_i)}.$$

Proof

Consider the following tree diagram.

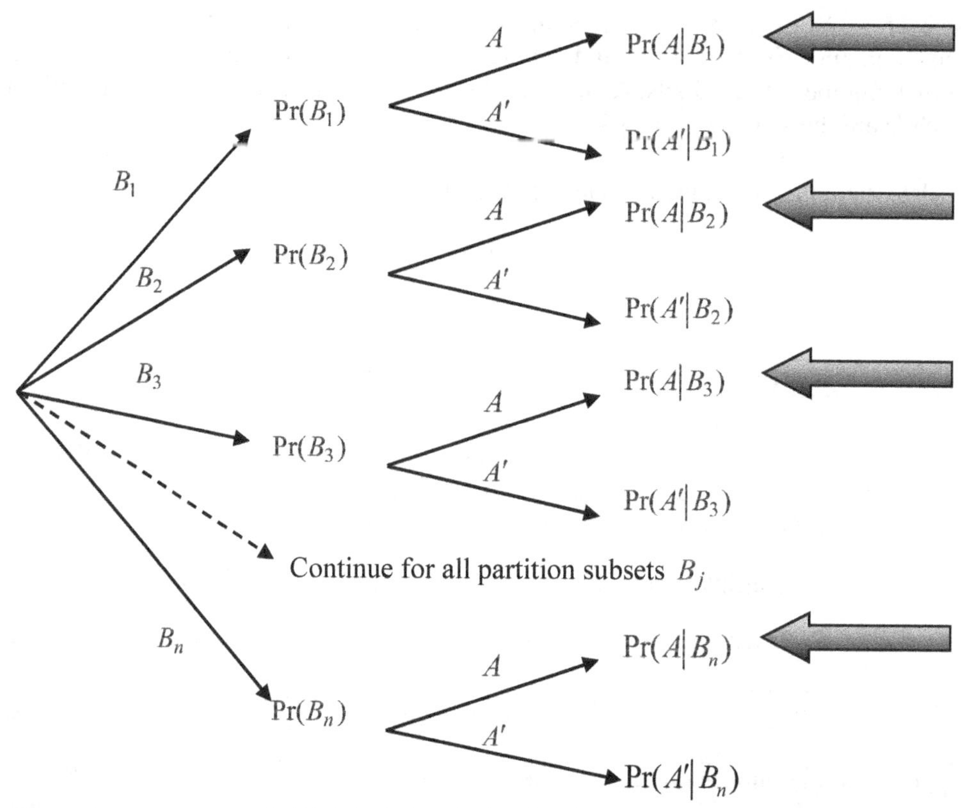

First, the numerator, $\Pr(B_j)\cdot\Pr(A\,|\,B_j)$, in the formula for $\Pr(B_j\,|\,A)$ follows from the definition of conditional probability in multiplicative form. Second, for the denominator of the formula, note that,

$$A = A\cap S = A\cap(B_1\cup B_2\cup\cdots\cup B_n) = (A\cap B_1)\cup(A\cap B_2)\cup\cdots\cup(A\cap B_n).$$

The sets $A\cap B_i$ are disjoint because the sets $B_1, B_2,\cdots, B_n$ form a partition. Therefore,

$$\begin{aligned}\Pr(A) &= \Pr(A\cap B_1)+\Pr(A\cap B_2)+\cdots+\Pr(A\cap B_n)\\ &= \Pr(B_1)\cdot\Pr(A\,|\,B_1)+\Pr(B_2)\cdot\Pr(A\,|\,B_2)+\cdots+\Pr(B_n)\cdot\Pr(A\,|\,B_n)\\ &= \sum_{i=1}^{n}\{\Pr(B_i)\cdot\Pr(A\,|\,B_i)\}.\end{aligned}$$

The Bayes' formula now follows.

Exercise 2-56 Suppose that you are dealt two cards face down from a standard deck.

(a) Find the probability that the first card is an ace.

(b) Find the probability that the second card is an ace, given that the first card is not.

(c) Bayes' Theorem: Find the probability that the first card is an ace, given that the second card is not.

Exercise 2-57 You are dealt three cards. The events of interest concern the number of face cards that you are dealt (0, 1, 2, or 3). Construct a tree diagram tracking face cards.

(a) What is the probability that you are dealt three face cards?

(b) What is the probability that you are dealt at least two face cards?

(c) What is the conditional probability that you are dealt three face cards given that the first two cards are face cards?

(d) What is the conditional probability that you are dealt three face cards given that you are dealt at least two face cards?

(e) What is the conditional probability that you are dealt at least two face cards given that the last card dealt to you was a face card?

2.6 CREDIBILITY

Bayesian analysis of the sort we have seen is regularly used by insurance companies in calculating premiums for automobile and other policies. Insurance actuaries, for example, are constantly evaluating the driving history of their policyholders in order to more accurately classify them according to various risk categories. This continual updating of probability estimates based on new experience is referred to as ***credibility theory***.

To illustrate the general principle we will take a simple example that can be analyzed using a tree diagram. Suppose the insurance company employs only unsophisticated actuaries and therefore uses just two risk categories, *good* driver and *bad* driver. Over the years the company has concluded that 80% of their customers are good drivers and 20% are bad drivers. A good driver has a probability of 0.10 of having exactly one accident in the next year, whereas a bad driver has a probability of 0.50 of having exactly one accident in the next year. Now, suppose you are the customer and we represent the insurance company (in a few years you may turn the tables on us).

We, as the actuaries, may be devilishly clever, but we are neither omniscient nor clairvoyant. Your risk category may be the *cause* for your accidents, but it is not directly *observable*. We can never know for certain, even after years of experience with you, whether you are a bad driver or simply a good driver who has had a run of bad fortune. (In the real world, insurance companies go to a great deal of trouble in their ***underwriting***, which is the industry term for attempting to ascertain your risk category in advance of issuing you a policy.) What we can do, however, is *draw inferences* about your risk category based on your number of accidents. While your risk category is an unobservable *cause*, your record is an *observable effect*, and we can use Bayes' Theorem to calculate the *probability* you belong to one risk category or another.

Example 2.6-1 Bayesian Credibility

We assume you are an average customer, so that there is an 80% chance you are good (G) and a 20% chance you are bad (B). Let A be the event you have an accident in the next year. For simplicity, we assume at most one accident per year. The tree diagram looks like this:

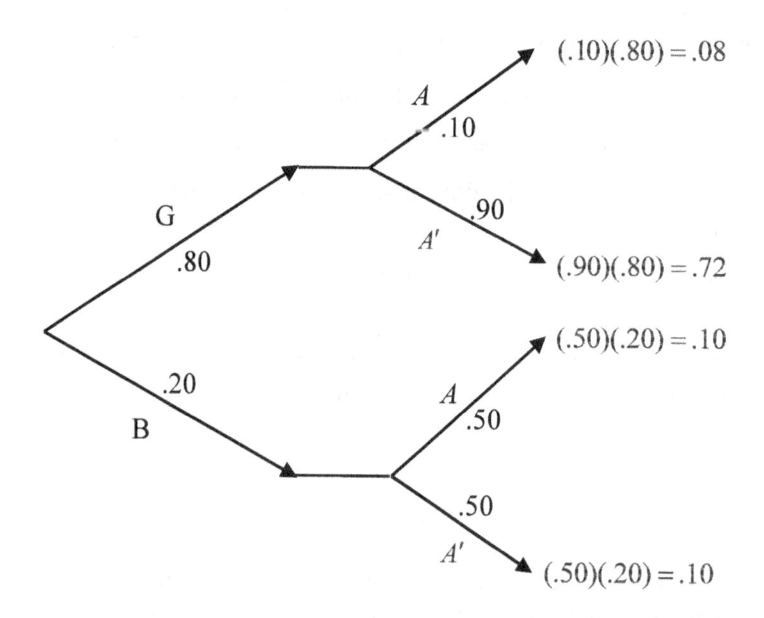

Suppose we observe an accident (an effect) in the first year. We are able to revise our estimate of your risk category based on your experience. Using standard Bayes' calculations, we see that the probability you have an accident is,

$$\Pr(A) = \Pr(A \mid G)P(G) + \Pr(A \mid B)P(B) = (.10)(.80) + (.50)(.20) = .18,$$

and

$$\Pr(G \mid A) = \frac{(.10)(.80)}{(.10)(.80) + (.50)(.20)} = \frac{4}{9},$$

with

$$\Pr(B \mid A) = 1 - \tfrac{4}{9} = \tfrac{5}{9}.$$

Thus, our assessment as to whether or not you are in the good category drops from 80% (allowing you the benefit of the doubt) to a far more suspicious 44.4%.

Example 2.6-2 Credibility Using 2 Years Experience

What if you have two consecutive years with an accident? Assuming accidents are independent from year to year, we can repeat the same tree diagram analysis for the second year, but now using our revised probabilities of good driver or bad driver:

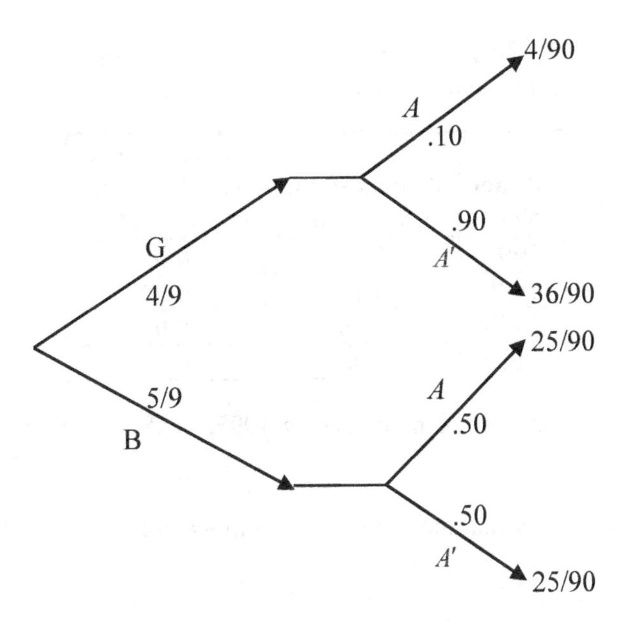

Now, given that you had an accident in the first year, the probability of an accident in the second year increases to 29/90, or almost 1/3. Moreover, our new assessment of your risk category – after two years of accidents - becomes,

$$\Pr(G \mid A) = \frac{4/90}{4/90 + 25/90} = \frac{4}{29} \text{ and } \Pr(B \mid A) = 1 - \frac{4}{29} = \frac{25}{29}.$$

Be assured, as the likelihood you are a good driver with bad luck plummets, your premiums will zoom in the opposite direction. That is, in essence, the logic of Bayesian credibility theory. Insurance companies have other approaches to assessing the reliability of experience data in making judgments about risk, but that is another story.

Exercise 2.58 Let A_1 be the event of an accident in the first year, and let A_2 be the event of an accident in the second year. Draw a three stage tree diagram, starting with risk categories (G) and (B) at the first stage, and then the complete results for years 1 and 2 in the following two stages. Use the same probabilities (.8 and .2) as in the example for the initial risk assignments, and the same conditional accident probabilities (.1 and .5, respectively).

(a) Calculate $\Pr(A_1 \cap A_2)$.

(b) Calculate $\Pr(A_2 \mid A_1)$ and compare to the above calculation.

(c) Calculate $\Pr(G \mid A_1 \cap A_2)$ and compare with the above.

(d) Find the probability that someone is a bad driver given he/she has at least one year with an accident in the two years.

(e) Find the probability that someone who has no accidents in each of the first two years is in fact a good driver.

Exercise 2-59 SOA EXAM P Sample Exam Questions #27

A study of automobile accidents produced the following data:

Model Year	Proportion of all vehicles	Probability of involvement in an accident
1997	0.16	0.05
1998	0.18	0.02
1999	0.20	0.03
Other	0.46	0.04

An automobile from one of the model years 1997, 1998, and 1999 was involved in an accident.

Determine the probability that the model year of this automobile is 1997.

(A) 0.22 (B) 0.30 (C) 0.33 (D) 0.45 (E) 0.50

Exercise 2-60 SOA EXAM P Sample Exam Questions #20

An insurance company issues life insurance policies in three separate categories: standard, preferred, and ultra-preferred. Of the company's policyholders, 50% are standard, 40% are preferred, and 10% are ultra-preferred. Each standard policyholder has a probability 0.010 of dying in the next year, each preferred policyholder has a probability 0.005 of dying in the next year, and each ultra-preferred policyholder has probability 0.001 of dying in the next year. A policyholder dies[5] in the next year.

What is the probability that the deceased policyholder was ultra-preferred?

(A) 0.0001 (B) 0.0010 (C) 0.0071 (D) 0.0141 (E) 0.2817

Exercise 2-61 SOA EXAM P Sample Exam Questions #23

An actuary studied the likelihood that different types of drivers would be involved in at least one collision during any one-year period. The results of the study are presented below.

Type of driver	Percentage of all drivers	Probability of at least one collision
Teen	8%	0.15
Young Adult	16%	0.08
Midlife	45%	0.04
Senior	31%	0.05

Given that a driver has been involved in at least one collision in the past year, what is the probability that the driver is a young adult driver?

(A) 0.06 (B) 0.16 (C) 0.19 (D) 0.22 (E) 0.25

[5] If you would like to live for a long time, you should be a right-handed woman who eats healthily, is wealthy, highly-educated, and lives in the United States.

Exercise 2-62 SOA EXAM P Sample Exam Questions #22

A health study tracked a group of persons for five years. At the beginning of the study, 20% were classified as heavy smokers, 30% as light smokers and 50% as nonsmokers. Results of the study showed that light smokers were twice as likely as nonsmokers to die during the five-year study, but only half as likely as heavy smokers. A randomly selected participant from the study died over the five-year period. Calculate the probability that the participant was a heavy smoker.

(A) 0.20 (B) 0.25 (C) 0.35 (D) 0.42 (E) 0.57

Exercise 2-63 SOA EXAM P Sample Exam Questions #19

An auto insurance company insures drivers of all ages. An actuary compiled the following statistics on the company's insured drivers:

Age of Driver	Probability Of Accident	Portion of Company's Insured Drivers
16-20	0.06	0.08
21-30	0.03	0.15
31-65	0.02	0.49
66-99	0.04	0.28

A randomly selected driver that the company insures has an accident. Calculate the probability that the driver was age 16-20.

(A) 0.13 (B) 0.16 (C) 0.19 (D) 0.23 (E) 0.40

2.7 CHAPTER 2 SAMPLE EXAMINATION

1. Draw a Venn diagram and shade the area $A \cap B'$ (A and not B).

2. What is the probability that a full house is dealt in poker?

3. An insurance agent sells two types of insurance, auto and dismemberment. Of her 82 clients, 62 have auto insurance and 37 have dismemberment insurance.
   (a) How many clients have both?
   (b) How many clients have exactly one type of insurance?
   (c) Are the events of having auto insurance and having dismemberment insurance independent?

4. A college class consists of 32 students of which 17 students are engineering majors, and 13 students are female. There are 8 male engineering students in this class.

(a) Make a Venn diagram that represents the aforementioned information. Construct a Venn box diagram as well.
(b) What is the probability that a person chosen at random from this class is a female engineering major?
(c) What is the probability that a person chosen at random from this class is female or an engineering major?
(d) What is the probability that a person chosen at random from this class is a female, given that the person is not an engineering major?
(e) Are the events "being a male" and "being an engineering major" independent in this class?

5. Suppose that you are dealt 3 cards from a standard deck of cards.
(a) Find the probability that you are dealt 3 Kings.
(b) Find the probability that you are not dealt 3 Kings (that is, zero, one or two Kings).
(c) Find the probability that you are dealt 3 Kings, given that the first two cards you were dealt are kings.
(d) Find the probability that you are dealt 3 Kings, given you were dealt at least 2 Kings.

6. Roll a pair of fair 4-sided dice. Each die is numbered with 2, 3, 4, and 5 respectively.
(a) List all possible outcomes (the sample space).
(b) Find the probability that the sum is nine.
(c) Find the probability that at least one die is a five.
(d) Find the probability that the sum is six given that at least one die is a five.
(e) Find the probability that at least one die is a five given that the sum is six.

7. There are two urns with numbered Ping-Pong balls in each (see diagram below). Draw one Ping-Pong ball from each urn.
(a) Construct a tree diagram listing all possible outcomes with associated probabilities.
(b) What is the probability that the sum of the numbers on the Ping-Pong balls is four?
(c) What is the probability that at least one Ping-Pong ball is numbered 2?
(d) What is the probability that the sum of the numbers on the Ping-Pong balls is four, given that the number on the Ping-Pong ball from Urn A is a 2?
(e) What is the probability that the number on the Ping-Pong ball from Urn B is a 2, given that the sum of the numbers on the Ping-Pong balls is four?

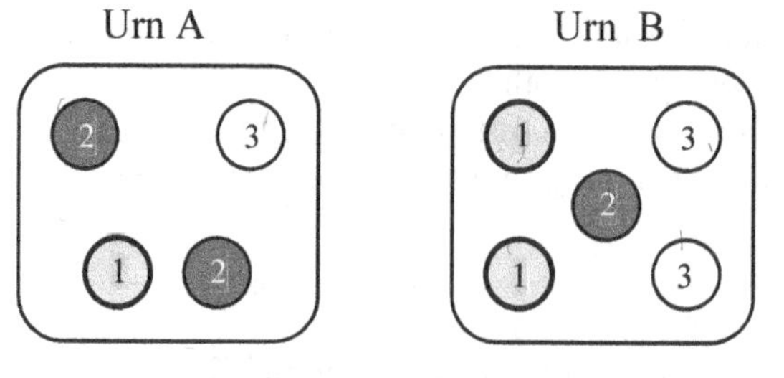

8. The favorite drinks of 30 students, 12 of whom are male, are summarized below.

Gender	Favorite Drink				
	Coke	Diet Pepsi	Milk	Water	Red Bull®
Female	2	3		4	1
Male		1	2	0	4

(a) Fill in the missing information.
(b) Calculate the probability that Red Bull® is the favorite drink of a student selected at random.
(c) Calculate the conditional probability that the student is female given that the student's favorite drink is milk.
(d) Calculate the probability that a student selected at random is either male, or the student's favorite drink is milk (or both).

9. Using information provided in the following Venn diagram, find the following probabilities:
(a) $\Pr(A)$
(b) $\Pr(A') = \Pr(\overline{A}) = \Pr(A^C) = \Pr(\text{not } A)$
(c) $\Pr(A \cap B) = \Pr(A \text{ and } B)$
(d) $\Pr(A \mid B)$
(e) $\Pr(A \cup B) = \Pr(A \text{ or } B)$
(f) $\Pr(B \mid A)$
(g) Are the events A and B independent?

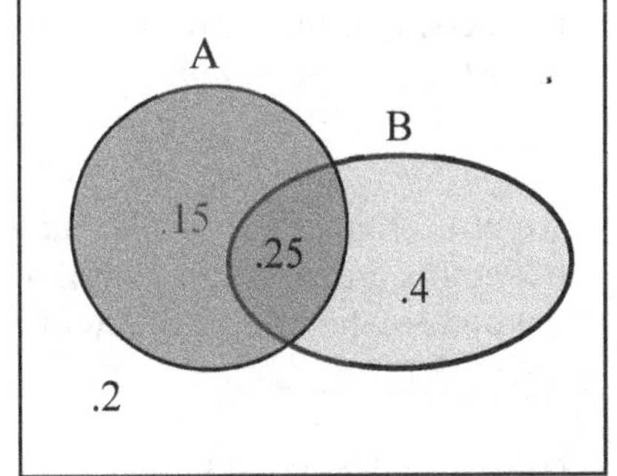

10. An ambidextrous mathematician with a very short attention span keeps two video game credit cards, one in each of his two front pockets. One game card has credit for 5 games. The other game card has credit for 4 games. The mathematician pays for a video game with a credit card selected from a random pocket and replaces the credit card once it is used to pay for the game. What is the probability that when the mathematician uses the last credit from one of her two cards, then the other contains 4 credits?

11. The ambidextrous mathematician in the previous problem now wants to know the probability that when she uses the last credit from one of her two cards, then the other contains 3 credits.

12. The NBA championship is won by the first team that wins 4 games in a best of seven series. Given two evenly matched teams, what is the probability that the team that has won three of the first four games will win the championship?

13. Rework the previous problem if the team that has won three of the first four games has a 60% chance of winning each of the subsequent games.

14. Suppose $\Pr(A) = \frac{2}{3}$, $\Pr(A|B) = \frac{7}{10}$, and $\Pr(B^C) = \frac{1}{3}$. Find $\Pr(A \cup B)$.

15. At a dance, there are $n = 3$ married couples: Ann and Andy Anderson, Betty and Boris Brown, and Danielle and Dan Daniels. The wives select a husband at random with whom to share a dance. What is the probability that each of the three men dances with a woman other than his spouse?

16. Repeat the problem with $n = 4$ couples.

17. There are 12 mathematics faculty at School University. Five are old, nine are bad teachers, and there are three old bad teachers. How many young good teachers are there?

18. An urn contains 7 red balls and 6 white balls. Players Beavis and Butthead alternately take turns drawing a ball until a player selects a red ball. What is the probability that the player who selects first will win?

19. (Wuss-*ian* Roulette) Suppose that there are 6 balls in an urn, 2 red and 4 white. There are two players. The first player draws a ball at random. If the ball is white then it is replaced and the other player draws, and so forth, continuing until a red ball is drawn. The player drawing the first red ball then buys a round of drinks.
    (a) What is the probability that the player who goes first gets a free drink?
    (b) What is the probability that the player who goes first gets a free drink if after each draw of a white ball it is replaced in the urn by a red ball?

20. **SOA EXAM P Sample Exam Questions #13**
An actuary is studying the prevalence of three health risk factors, denoted by A, B, and C, within a population of women. For each of the three factors, the probability is 0.1 that a woman in the population has only this risk factor (and no others). For any two of the three factors, the probability is 0.12 that she has exactly these two risk factors (but not the other). The probability that a woman has all three risk factors, given that she has A and B, is $\frac{1}{3}$. What is the probability that a woman has none of the three risk factors, given that she does not have risk factor A?

(A) 0.280 (B) 0.311 (C) 0.467 (D) 0.484 (E) 0.700

21. **SOA EXAM P Sample Exam Questions #15**
An insurer offers a health plan to the employees of a large company. As part of this plan, the individual employees may choose exactly two of the supplementary coverages A, B, and C, or they may choose no supplementary coverage. The proportions of the company's employees that choose coverages A, B, and C are $\frac{1}{4}$, $\frac{1}{3}$, and $\frac{5}{12}$, respectively. Determine the probability that a randomly chosen employee will choose no supplementary coverage.

(A) 0 (B) $\frac{47}{144}$ (C) $\frac{1}{2}$ (D) $\frac{97}{144}$ (E) $\frac{7}{9}$

22. **SOA EXAM P Sample Exam Questions #6**
A public health researcher examines the medical records of a group of 937 men who died in 1999 and discovers that 210 of the men died from causes related to heart disease. Moreover, 312 of the 937 men had at least one parent who suffered from heart disease, and, of these 312 men, 102 died from causes related to heart disease. Determine the probability that a man randomly selected from this group died of causes related to heart disease, given that neither of his parents suffered from heart disease.

(A) 0.115 (B) 0.171 (C) 0.224 (D) 0.327 (E) 0.514

23. **SOA EXAM P Sample Exam Questions #9**
An insurance company examines its pool of auto insurance customers and gathers the following information:

(i) All customers insure at least one car.
(ii) 70% of the customers insure more than one car.
(iii) 20% of the customers insure a sports car.
(iv) Of those customers who insure more than one car, 15% insure a sports car.

Calculate the probability that a randomly selected customer insures exactly one car and that car is not a sports car.

(A) 0.13 (B) 0.21 (C) 0.24 (D) 0.25 (E) 0.30

24. **SOA EXAM P Sample Exam Questions #7**
An insurance company estimates that 40% of policyholders who have only an auto policy will renew next year and 60% of policyholders who have only a homeowner's policy will renew next year. The company estimates that 80% of policyholders who have both an auto and a homeowner's policy will renew at least one of those policies next year.

Company records show that 65% of policyholders have an auto policy, 50% of policyholders have a homeowner's policy, and 15% of policyholders have both an auto and a homeowners policy. Using the company's estimates, calculate the percentage of policyholders that will renew at least one policy next year.

(A) 20 (B) 29 (C) 41 (D) 53 (E) 70

25. **SOA EXAM P Sample Exam Questions #17**
An insurance company pays hospital claims. The number of claims that include emergency room or operating room charges is 85% of the total number of claims. The number of claims that do not include emergency room charges is 25% of the total number of claims. The occurrence of emergency room charges is independent of the occurrence of operating room charges or hospital claims. Calculate the probability that a claim submitted to the insurance company includes operating room charges.

(A) 0.10 (B) 0.20 (C) 0.25 (D) 0.40 (E) 0.80

26. **SOA EXAM P Sample Exam Questions #16**
An insurance company determines that N, the number of claims received in a week, satisfies $\Pr[N=n]=\frac{1}{2^{n+1}}$, where $n \geq 0$. The company also determines that the number of claims received in a given week is independent of the number of claims received in any other week.

Determine the probability that exactly seven claims will be received during a given two-week period.

(A) $\frac{1}{256}$ (B) $\frac{1}{128}$ (C) $\frac{7}{512}$ (D) $\frac{1}{64}$ (E) $\frac{1}{32}$

27. **SOA EXAM P Sample Exam Questions #143**
The probability that a member of a certain class of homeowners with liability and property coverage will file a liability claim is 0.04, and the probability that a member of this class will file a property claim is 0.10. The probability that a member of this class will file a liability claim but not a property claim is 0.01.

Calculate the probability that a randomly selected member of this class of homeowners will not file a claim of either type.

(A) 0.850 (B) 0.860 (C) 0.864 (D) 0.870 (E) 0.890

28. **SOA EXAM P Sample Exam Questions #146**

A survey of 100 TV watchers revealed that over the last year:

(i) 34 watched CBS.
(ii) 15 watched NBC.
(iii) 10 watched ABC.
(iv) 7 watched CBS and NBC.
(v) 6 watched CBS and ABC.
(vi) 5 watched NBC and ABC.
(vii) 4 watched CBS, NBC, and ABC.
(viii) 18 watched HGTV and of these, none watched CBS, NBC, or ABC.

Calculate how many of the 100 TV watchers did not watch any of the four channels (CBS, NBC, ABC or HGTV).

(A) 1 (B) 37 (C) 45 (D) 55 (E) 82

CHAPTER 3

DISCRETE RANDOM VARIABLES

In this chapter we develop tools for studying discrete probability models whose outcomes need not be equally likely. The most important tool, and the foundation for all that follows, is the mathematical object we call a ***random variable***. We give a formal definition below, and then describe the resulting probability distribution of a random variable. We will continue with a discussion of basic properties of random variables and their probability distributions. These include standard measures of central tendency (averages) and of spread, two features that help to characterize random variables.

We will conclude the chapter previewing important concepts that also apply to non-discrete random variables, covered in later chapters. The first concerns *conditional* distributions and the second establishes some useful properties of *jointly distributed* random variables. The final section develops the important idea of *transforming* a discrete random variable into a power series, just like the ones encountered in calculus.

3.1 DEFINITION AND EXAMPLES OF DISCRETE RANDOM VARIABLES

Definition: Discrete Random Variable

We say X is a ***discrete random variable*** if X is a numerically valued function whose domain is the sample space of a probability experiment with a finite or countably infinite number of outcomes.

Notation: Random Variables

We use capital letters to denote random variables (e.g., X, Y, or Z). We use lower case letters to represent outcomes that the random variable may take on (e.g., x, y, or z).

Every random variable has a ***probability distribution*** associated with it. To illustrate, we continue with the 3 child example from earlier chapters.

Example 3.1-1 Discrete Random Variable

The probability experiment consists of having three children, with the assumption that a child is equally likely to be born male or female.[1] As we saw in Chapter 1, the 8 equally likely outcomes of this experiment can be represented as 3-letter words using the letters B and G. We let X represent the number of girls. Display in tabular form the probabilities $\Pr[X=x]$ for $x = 0,1,2,3$.

Solution

Each of the four values (0, 1, 2, 3) that X can take corresponds to an *event* whose probability can be calculated by inspecting the underlying sample space. For example, $\Pr(X=1)$ denotes the probability of the event $\{GBB, BGB, BBG\}$. Since the underlying outcomes are equally likely, we have $\Pr(X=1) = \frac{3}{8}$. The other values are calculated similarly.

The following table, listing all the values of X and the probabilities $\Pr[X=x]$ of the associated events, is called the ***probability distribution*** of X:

x	$\Pr[X=x]$
0	1/8 =.125
1	3/8 =.375
2	3/8 =.375
3	1/8 =.125

❒

As required for a probability model, the probabilities are all between 0 and 1 and the probabilities sum to 1. Although the underlying sample space has 8 equally likely outcomes, the random variable X has compressed the information into 4 outcomes, not all of which are equally likely. The general definition follows.

Probability Distribution and Probability Function

The tabulation of the probabilities for each possible value x of a discrete random variable X is called its ***probability distribution***. The probabilities must be positive and sum to one:

$$(1) \quad 0 < \Pr(X=x) = p(x) \le 1, \qquad (2) \sum_i \Pr(X=x_i) = \sum_i p(x_i) = 1.$$

The sum is over all of the values x_i that X takes with non-zero probability. The values x_i are typically displayed in the table in ascending order.

The function $p(x_i) = \Pr(X=x_i)$ on the values of the random variable X is called the ***probability function*** of X.

Here are some important observations. First, each random variable X has a unique associated probability distribution. Conversely, every probability distribution table satisfying the properties (1) and (2) in the box above can be thought of as *defining* a unique random variable X.

[1] In Chapter 2, we observed that the actual probability of being born a male in the U.S.A. is .511.

Second, this one-to-one correspondence between random variables and their probability distributions allows us to discuss probability models without reference to a particular underlying experiment. For example, the probability distribution for the number of girls in a family with 3 children is identical to the number of heads in 3 tosses of a fair coin, or the number of free-throws made by a 50% free-throw shooter in 3 attempts, or the number of correct answers on a 3 question true-false quiz in which the student randomly guesses. All of these experiments lead to the same random variable and the same probability distribution as the one that is displayed in Example 3.1-1.

Note
This is a particular instance of what is called a ***binomial distribution***, which we will study in detail in Section 4.2.

Employing the language of random variables and probability distributions allows us to disregard the underlying experiment and to focus entirely on the more abstract probability model that is summarized in the distribution table. Indeed, we can think of X itself as the experiment, with the possible values x_i comprising the outcomes in the sample space. In fact, we often call the values x_i ***observations*** (i.e., outcomes) of X. The probability function of X comprises the third element required for a complete probability model as described at the beginning of Chapter 1.

Exercise 3-1 Suppose a women's basketball team bus carries five starters and seven reserves. A random group of three players gets off of the bus. Let Y be the number of starters that get off the bus. Find the probability distribution for Y.

Note
This is a ***hyper-geometric*** distribution that we will study in Section 4.5.

Exercise 3-2 Pick an integer at random from 1 to 10 inclusively. Let Z denote the integer that is selected. Write the probability distribution for Z.

Note
This is a ***discrete uniform*** distribution that we will study in Section 4.1. Uniform distributions cover the case when outcomes are equally likely.

Exercise 3-3 Throw snowball(s) at your little brother until you hit him once. With winters of practice, you are able to hit him $\frac{18}{38}$ of the time. Let S denote the total number of snowballs that you throw at your brother.

(a) Find the probability that you hit him on your first attempt, $\Pr(S=1)$.
(b) Find the probability that you hit him on your third attempt, $\Pr(S=3)$.
(c) Complete the probability distribution for S.

Note
This is a ***geometric*** distribution that we will study in Section 4.3.

Exercise 3-4 You travel to the local riverboat casino. Your strategy is to bet on "black" during each spin of the roulette wheel until you win exactly once. Let S denote the total number of spins of the roulette wheel (wagers) until a ball lands on a black number. A ***Roulette wheel*** has 18 red numbers, 18 black numbers, and 2 green numbers – each equally likely.

(a) Find the probability distribution for S.

(b) Describe an entirely different situation that has the same probability distribution as S.

3.2 CUMULATIVE PROBABILITY DISTRIBUTION

Since a random variable X is numerically valued, it is possible to talk about the event that X takes a value less than or equal to a particular number x. As x increases, the probability that X is less than or equal to x also increases. This gives rise to a monotonically increasing (non-decreasing) function of x that is called the cumulative distribution function (CDF) for X.

Cumulative Distributive Function

Let X be a discrete random variable. For each real number x, let $F(x) = \Pr(X \le x)$. The function $F(x)$ is called the ***cumulative distribution function*** (CDF) for the random variable X and satisfies

(1) $0 \le F(x) = \Pr(X \le x)$ for all X.

(2) If $x_{i-1} < x_i$ are consecutive values in the probability distribution table of X, then $\Pr(X = x_i) = F(x_i) - F(x_{i-1}) = \Pr(X \le x_i) - \Pr(X \le x_{i-1}) = p(x_i)$.

(3) We define $F(\infty) = \Pr(X < \infty) = 1$.

The graph of the CDF for a discrete random variable X will be a step function. That is, it will consist of horizontal lines with discrete jumps at the points x_i for which $\Pr[X = x_i]$ is positive.

There is a variation on the CDF graph, which is called an ***ogive*** (pronounced ō-jive). The ogive graphs the discrete points $(x_i, F(x_i))$ and connects them with straight lines. We illustrate how these graphs compare in the next example.

Example 3.2-1 Cumulative Distribution and Ogive

The number of cases of coke purchased by customers of Fred G. Meyer has probability distribution given below. Find the cumulative probability distribution, graph the CDF, and construct an ogive.

X	0	1	2	3	4
$\Pr(X)$	0.65	0.20	0.07	0.03	0.05

Solution
First we compute the values of the cumulative distribution function. For example,

X	0	1	2	3	4
$F(x)$	0.65	0.85	0.92	0.95	1.00

$$F(2) = \Pr(X \leq 2) = 0.65 + 0.20 + 0.07 = 0.92.$$

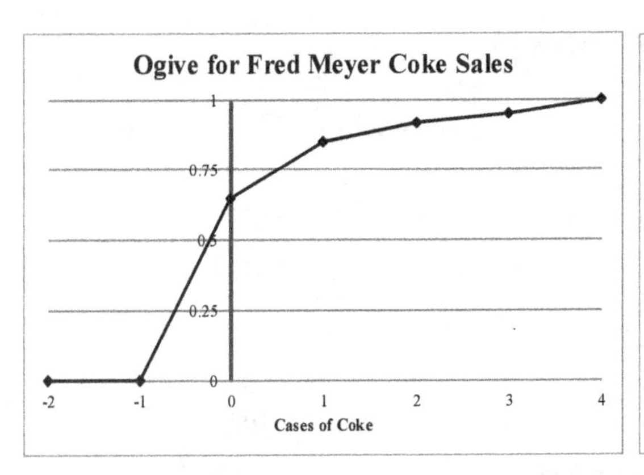

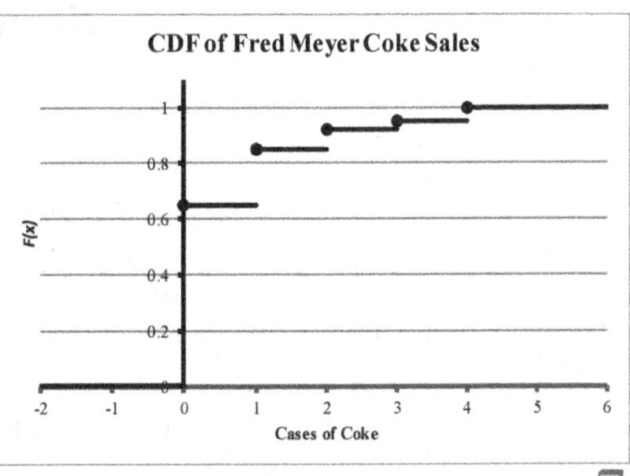

❐

Exercise 3-5 Roll a pair of fair six-sided dice. Let S denote the sum of the two dice.

(a) Find the probability distribution for S.

(b) Find the cumulative probability distribution for S.

(c) Graph the ogive and CDF for S.

Exercise 3-6 The ages of all college baseball players is denoted by the random variable A and has cumulative probability distribution given by:

a	17	18	19	20	21	22	23
$F(A=a)$	0	.23	.48	.89	.94	.99	1.00

(a) Find the probability distribution for A.

(b) Create an ogive for A.

Exercise 3-7 Flip a loaded coin four times. Assume that $\Pr(\text{head}) = 0.7$. Let the random variable H denote the number of heads that result.

(a) Find the cumulative probability distribution for H.

(b) Create an ogive for H.

Exercise 3-8 Flip a fair coin three times. Let T be the random variable that denotes the number of tails that occur *given that at least one head occurred.* Graph the cumulative probability distribution for T.

Exercise 3-9 Using the following ogive, estimate the probability that a house costs between ¼ and ½ million dollars.

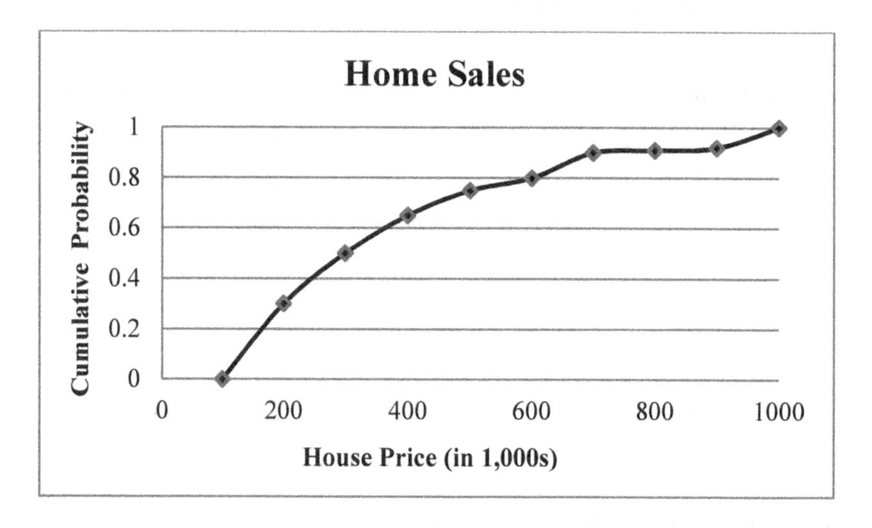

3.3 MEASURES OF CENTRAL TENDENCY

One learns in elementary statistics that when we have a large collection of data points, such as test scores or guinea pig weights, it is often more informative to summarize the data than it is to simply list the data points. If an employer were to ask the human resource (HR) personnel about the ages of her employees, she is probably not interested in pages of raw data consisting of individual birthdays. More likely, her preference would be for a summary of the data that would be easier to digest. The HR person may supply this in the form of summary statistics indicating the range and spread of the data, along with a graphical display that exhibits the overall age distribution.

The same desire for simplification holds true for a random variable and its associated probability table. Rather than just a list of the outcomes and their probabilities, some summary statistics and perhaps a picture would be valuable for improving the usefulness of the information being conveyed.

We will describe some of these summary measures for the probability distributions of random variables. An important special case arises when we are required to summarize the characteristics of a given data set. A simple set of n data points (i.e., numbers) may be considered as the sample space of a discrete random variable X whose values are the data points. The values are treated as equally likely, so that the associated probabilities are all equal to $1/n$. In this fashion the calculations for the average and the standard deviation (statistics to be defined shortly) of a data set can be considered as within the context of the calculations for a general random variable.

The term ***average*** is often used in casual conversation to represent a variety of things. This can cause confusion. The objective in speaking of an average is to provide a single number that best represents the entire distribution of the random variable. But there are several competing candidates when it comes to defining this *average* number.

Consider a group of fraternity brothers eating pizza. Suppose the numbers of slices that these seven fellows ate are: 2, 2, 2, 4, 6, 7, and 12, respectively. Is the average number of slices "2" because it is the most common number of slices consumed (the ***mode***)? Or is the average number of slices "4" because three brothers ate less than 4 slices and three brothers ate more than 4 slices (the ***median***)? Or is the average "7" since it is halfway between the least number of slices consumed and the maximum number (the ***midrange***)? Or is the average "5" because seven guys ate 35 total slices of pizza, which works out to an overall rate of 5 slices per person (the ***mean***)? Notice that while the mean number of slices is 5, no fraternity brother actually consumed exactly five slices.

The answer to all of the above is "yes." The word "average" is a generic term that in informal usage could represent the mode, the median, the midrange, the mean, or even some range about the mean (the *average* student) depending on the speaker and the context. In statistics, when the word "average" is used by itself, it is generally, but not always, the case that one is referring to the mean.

In addition to calculating the various measures of centrality or average, we can display the distribution graphically in a histogram, as illustrated below. A ***histogram*** is a bar-graph with the data values (in our example, slices consumed) along the x-axis and frequencies along the y-axis.

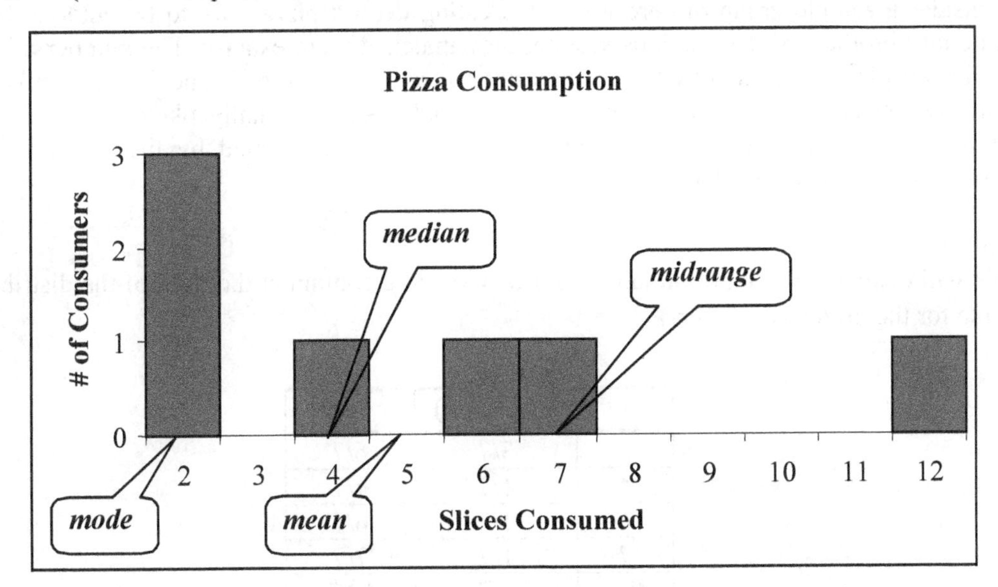

3.3.1 EXPECTED VALUE (MEAN)

The most common measure of the average, or central tendency, of a random variable is the ***expected value.*** This quantity is also frequently referred to as the ***mean*** of the random variable, or the mean of the associated probability distribution. The definition below calculates the mean as a weighted average of the values of the random variable, with the weights being the associated probabilities.

Note
The Greek letter μ (pronounced "mew") is commonly used to denote the mean of a random variable. The subscript refers to the random variable so that μ_X denotes the mean of the random variable X.

Expected Value (Mean) of a Discrete Random Variable

If X is a discrete random variable with probability function $\Pr(X=x_i) = p(x_i)$, then the ***expected value*** (**mean**) of the random variable X is given by,

$$\mu_X = E[X] = \sum_i x_i \cdot p(x_i).$$

In the case of n data points, treated as the equally likely values of a random variable, the calculation becomes the usual average, $\mu_X = \sum_{i=1}^{n} x_i \cdot \frac{1}{n} = \frac{x_1 + x_2 + \cdots + x_n}{n} = \frac{1}{n} \cdot \sum_{i=1}^{n} x_i$.

Example 3.3-1 Expected Value

Consider a sample group of sorority sisters eating dessert pizza. Not to be outdone by the fraternity brothers of the previous section, they matched them exactly. The number of slices of dessert pizza consumed by the seven sisters were: 2, 2, 2, 4, 6, 7, and 12, respectively. A sorority sister is selected at random (meaning each sister is equally likely to be selected). Define the random variable X as the number of slices consumed by the selected sister. Calculate the expected value of X.

Solution
We will display the solution in tabular form. We add a column at the right of the distribution table for the products $x_i \cdot p(x_i)$.

x	$\Pr(X = x)$	$x \cdot p(x)$
2	3/7	6/7
4	1/7	4/7
6	1/7	6/7
7	1/7	7/7
12	1/7	12/7
Total:	1	35/7

Observe that the $x \cdot p(x)$ column can be viewed as the total slices consumed, scaled by dividing by the total number (7) of sorority sisters in the sample space. This is equivalent to the usual calculation for the mean number of slices as total slices divided by number of consumers:

$$\mu_X = \frac{2+2+2+4+6+7+12}{7} = \frac{35}{7} = 5 \text{ slices.}$$ ❐

It is often the case that we will be creating new random variables from old ones. The technical term for this is ***transformation***, which just means forming a function of the random variable. For example, given the random variable X, some common transformations are X^2, $(X-\mu_X)^2$, and e^X. Each of these transformations of X results in a new random variable, with new values, and a new probability distribution. Calculating the expected values of the transformed random variables can be easily accomplished by referring back to the probability distribution of X.

Example 3.3-2 Expected Value of a Transformed Random Variable

Let X be the random variable of Example 3.3-1. Calculate $E[X^2]$. Again, we illustrate the procedure in tabular form:

x	$\Pr(X=x)$	$x^2 \cdot p(x)$
2	3/7	12/7
4	1/7	16/7
6	1/7	36/7
7	1/7	49/7
12	1/7	144/7
Total:	1	257/7

The $x^2 \cdot p(x)$ column calculates $E[X^2]$ as,

$$E[X^2] = 2^2\left(\frac{3}{7}\right)+4^2\left(\frac{1}{7}\right)+6^2\left(\frac{1}{7}\right)+7^2\left(\frac{1}{7}\right)+12^2\left(\frac{1}{7}\right) = \frac{257}{7}.$$

We observe that $E[X^2]=\frac{257}{7}$, which is **not** the same as $\left(E[X]\right)^2 = 5^2 = 25.$ ❐

This calculation obviates the need to write a whole new table for each transformed version of X. The procedure in general is sometimes disparagingly referred to by mathematicians as "the Law of the Unconscious Statistician."

Rule for Calculating the Mean of a Transformed Random Variable.

If X is a discrete random variable with probability function $\Pr(X=x_i)=p(x_i)$, and $Y=g(X)$ is a transformation of X, then

$$\mu_Y = E[Y] = E[g(X)] = \sum_i g(x_i)\cdot p(x_i).$$

An important special case is a linear transformation, $Y = aX + b$. Then

$$E[Y] = E[aX+b] = \sum_{x_i}(ax_i+b)p(x_i) = a\sum_{x_i} x_i p(x_i) + b\sum_{x_i} p(x_i) = a \cdot E[X] + b.$$

This can be easily extended to more general linear expressions. If, for example $Y = aX^2+bX+c$, then $E[Y] = aE[X^2]+bE[X]+c$. This can be expressed by saying that the expected value is a ***linear operator***.

Exercise 3-10 Suppose that there are 5 men and 6 women at a party. The task of going to a store for more food and libation is assigned to two party guests, chosen at random. Let W denote the number of women selected. Find $E[W]$ and $E[W^2]$.

Exercise 3-11 Suppose that the grand prize in the Powerball® lottery is 150 million dollars and that you purchase exactly one ticket. Calculate your expected winnings. Assume no taxes, no split/multiple winners, and that the grand prize is paid in cash. See Subsection 1.4.3 for a fuller discussion of the Powerball® lottery.

Exercise 3-12 The distribution of the ages of mathematics faculty is summarized below. Calculate the mean age.

x = age	32	39	45	57	62
$\Pr[X = x]$	.16	☺	.42	.10	.28

Exercise 3-13 Given the following cumulative probability distribution, calculate $E[X]$.

X	−2	0	1	3	5	6
$F(x)$	.12	.23	.48	.76	.94	1.00

Exercise 3-14 SOA EXAM P Sample Exam Questions #49

An insurance policy pays an individual 100 per day for up to 3 days of hospitalization and 25 per day for each day of hospitalization thereafter. The number of days of hospitalization, X, is a discrete random variable with probability function

$$P(X=k) = \begin{cases} \frac{6-k}{15} & \text{for } k = 1,2,3,4,5 \\ 0 & \text{otherwise} \end{cases}.$$

Calculate the expected payment for hospitalization under this policy.

(A) 85 (B) 163 (C) 168 (D) 213 (E) 255

Exercise 3-15 Consider a random variable M with mean $\mu_M = 2.3$ and cumulative probability distribution given below. Find ☺.

m	0	1	2	3	4
$F(M=m)$	0	.2	☺	.8	1.0

Exercise 3-16 **SOA EXAM P** **Sample Exam Questions** **#44**

An insurance policy pays an individual 100 per day for up to 3 days of hospitalization and 50 per day for each day of hospitalization thereafter. The number of days of hospitalization, X, is a discrete random variable with probability function

$$P(X=k) = \begin{cases} \frac{6-k}{15} & \text{for } k = 1,2,3,4,5 \\ 0 & \text{otherwise} \end{cases}.$$

Determine the expected payment for hospitalization under this policy.

(A) 123 (B) 210 (C) 220 (D) 270 (E) 367

3.3.2 MEDIAN OF A DATA SET

In general usage, the ***median*** of a data set is the middle number when the items are listed in ascending order of size (think of the median on a highway, where median means the middle). In the event that there is an even number of data points, the median is the average (mean) of the middle two data points.

Median

If $x_1, x_2, \cdots, x_n$ is a collection of n data points listed from smallest to largest, then the ***median*** of the data equals

(a) $x_{\frac{n+1}{2}}$ if n is odd. This is just the middle term in the sequence.

(b) $\dfrac{x_{\frac{n}{2}} + x_{\frac{n}{2}+1}}{2}$ if n is even. This is the mean of the two middle terms.

Example 3.3-3 Median of a Data Set

Calculate the median of the data: 3, 8, 1.5, 6, 3, 29, 14, and 2.

Solution

First we order the data from smallest to largest (ascending): 1.5, 2, 3, 3, 6, 8, 14, and 29. There are eight data points so the median is the mean of the forth and fifth data points. The median is equal to: $\frac{3+6}{2}=4.5$. ❐

Note

The median does not have to equal one of the data points, as demonstrated in this example.

3.3.3 MIDRANGE OF A DATA SET

The midrange is halfway between the minimum and maximum values of the set of data.

Midrange

If $\{x_1, x_2, \cdots, x_n\}$ is a collection of n data points listed from smallest to largest, then the ***midrange*** of the data is defined to be:

$$\frac{x_1 + x_n}{2} = \frac{\text{minimum} + \text{maximum}}{2}.$$

Example 3.3-4 Midrange of a Data Set

Calculate the midrange of the data: 3, 8, 1.5, 6, 3, 29, 14, and 2.

Solution

The midrange is equal to: $\frac{1.5+29}{2}=15.25$. **Note:** As this case shows, the midrange need not equal any particular data point. ❐

3.3.4 MODE OF A DATA SET

The mode represents the most frequently occurring value in a collection of data.

Mode

If $x_1, x_2, \ldots, x_n$ is a collection of n data points, then the ***mode*** of the data is defined as,

(a) The value x_i that occurs most frequently.

(b) The two values x_i and x_j if they occur the same number of times, and more frequently than the remaining points. In this case we say the data is ***bi-modal***.

(c) Otherwise, the mode does not exist.

Example 3.3-5 Mode

Calculate the mode of the data: 3, 8, 1.5, 6, 3, 29, 14, and 2.

Solution

The number 3 occurs twice, and all other data points occur once. Hence the mode equals 3. **Note**: For sets with more than one data point, if the mode exists, then it necessarily takes the value of at least two of the data points. ❒

3.3.5 QUARTILES AND PERCENTILES

Percentiles are used to rank items in relative order. You are familiar with your percentile score on standard tests (SAT, ACT, GRE). Parents receive percentile statistics of the height and weight of their children. Percentile scores are used for college admissions, for determining who qualifies for Medicaid or food stamps, and for measures of intelligence.

Percentiles

If $x_1, x_2, \cdots, x_n$ are n data points arranged in ascending order, then x_i corresponds to the

$$\left(100 \cdot \frac{i}{n+1}\right)^{th} \text{ percentile.}$$

If there were one data point, it would correspond to the 50th percentile. If there were 4 data points, they would correspond to the 20th, 40th, 60th, and 80th percentiles. Think of spreading out all of the data points evenly between the 0th and 100th percentiles.

Example 3.3-6 Percentiles

Consider the data points: 1.2, 5.0, 6.1, 7.8, 8.2, 9.3, and 13.7.

(a) What percentile corresponds to the data point 9.3?
(b) What is the 25th percentile?

Solution

There are $n = 7$ data points. The sixth data point is 9.3, so $x_6 = 9.3$. Therefore a score of 9.3 corresponds to the $100 \cdot \frac{6}{7+1} = 75^{th}$ percentile. For (b), $25 = 100 \cdot \frac{i}{8}$, which implies that $i = 2$. Therefore the 25th percentile corresponds to data point $x_2 = 5$. ❒

Often, a particular percentile does not correspond to a specific data point. That is, the 20th percentile would correspond to the 1.6th data point in Example 3.3-6. When deciding on the value of the fractional data point, we use ***linear interpolation***[2].

---

[2] In linear interpolation the procedure is based on the assumption that the three points having these values for the ordinates lie on a straight line. The formula is $f(x) = f(x_1) + \left[f(x_2) - f(x_1)\right]\frac{(x-x_1)}{(x_2-x_1)}$.

Example 3.3-7 Percentile Using Interpolation

Consider the data points: 1.2, 5.0, 6.1, 7.8, 8.2, 9.3, and 13.7. What is the 20th percentile?

Solution

$20 = 100 \cdot \frac{i}{8}$ implies that $i = 1.6$. To find the 20th percentile of this data set, we find the value of the "1.6^{th}" data point. That is, we go 60% of the way between the data points $x_1 = 1.2$ and $x_2 = 5$. Then, $x_{1.6} = x_1 + .6(x_2 - x_1) = 1.2 + .6 \cdot 3.8 = 3.48$. □

If there were many data points (like 9,999), then each percentile corresponds to a specific data point or very close to one. In this case, you will not have to worry about interpolating.

Exercise 3-17 Consider a set of starting salaries for engineers: \$42,000, \$43,412, \$45,500, and \$53,750.

(a) What percentile corresponds to a salary of \$45,500?
(b) What salary corresponds to the 50th percentile?
(c) What salary do you require to be in the top third?
(d) To what percentile does a salary of \$49,000 correspond? Use linear interpolation.

Exercise 3-18 For a given sample size $n = 3$, answer the following:

(a) What is the range of percentiles that can be calculated? *Hint*: Find the percentile corresponding to the minimum data point, the median data point, and the maximum data point.
(b) *Conjecture*: How would you estimate the 10th percentile?

Quartiles

The ***first quartile*** corresponds to the 25th percentile and is denoted: Q_1.
The ***second quartile*** corresponds to the 50th percentile and is denoted: Q_2.
The ***third quartile*** corresponds to the 75th percentile and is denoted: Q_3.

Notes

(1) The median is the same as the 50th percentile, which is the same as the second quartile.

(2) In some contexts the term ***first quartile*** refers to the entire range from the lowest value up to Q_1, and similarly for the other quartiles. For example, "the household is in the 3rd quartile of income" is taken to mean the household income is between the 50th and 75th percentiles, not that it is exactly at the 75th percentile.

3.3.6 RANDOM VARIABLES AND PERCENTILES

The concepts of the previous several sections on percentile measures pertain mainly to data sets. But they can be extended to probability distributions of random variables, although there are some technical difficulties with the general case. However, in cases where the random variable X arises from a probability experiment with equally likely outcomes the extension is straightforward. One simply lists each value of X for each outcome (including repeated values of X), arranged in increasing order.

Example 3.3-8 Random Variable Percentiles

A random variable X has the following probability distribution:

x	2	4	6	7	12
$\Pr(X = x)$	$\frac{3}{7}$	$\frac{1}{7}$	$\frac{1}{7}$	$\frac{1}{7}$	$\frac{1}{7}$

(a) Associate X with a probability experiment with equally likely outcomes.

(b) Calculate the expected value (mean) of X.

(c) Calculate the median, mode, and midrange of X.

(d) Calculate the first quartile and third quartile of X.

(e) Calculate the 60th percentile of X.

Solution

(a) The random variable may be associated with the experiment of fraternity or sorority members eating pizza, (see Section 3.3.1). There are 7 members (the common denominator of the probabilities) whose slice consumptions are, respectively, and in increasing order: 2, 2, 2, 4, 6, 7, 12.

(b) From Section 3.3.1, $\mu_X = \dfrac{2+2+2+4+6+7+12}{7} = \dfrac{35}{7} = 5$ slices.

(c) The *median* = 50th percentile. The median corresponds to the 4th data point: $50 = 100 \cdot \frac{i}{7+1}$ implies $i = 4$. Since $x_4 = 4$, the median equals 4. The most common outcome is 2. The midrange is $\frac{2+12}{2} = 7$.

(d) The first quartile is $x_2 = 2$ and the third quartile is $x_6 = 7$.

(e) $60 = \frac{100i}{8}$ and so $i = 4.8$. Since $x_4 = 4$ and $x_5 = 6$, the 60th percentile is 80% of the way from 4 to 6, which turns out to be $4 + (6-4)(.8) = 5.6$.

Exercise 3-19 Using the graph of the cumulative distribution function below, find the (a) mean, (b) median, (c) mode, (d) midrange, and (e) third quartile for the random variable.

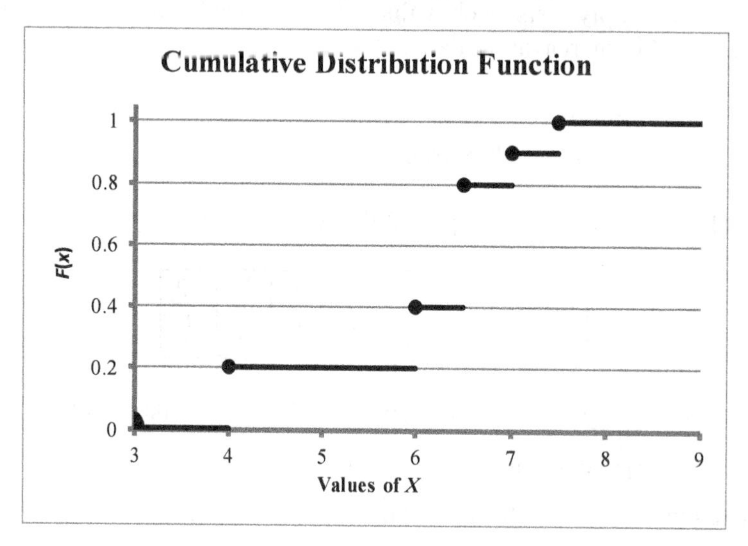

Exercise 3-20 Roll a pair of fair dice and let S denote the sum.

(a) Find the probability distribution for S.
(b) Calculate the expected value (mean) of S.
(c) Calculate the median, mode, and midrange of S.
(d) Calculate the first quartile and third quartile of S.

Exercise 3-21 Bunny Hop High staged a dance in the gymnasium. It was attended by m boys and n girls. Boys reported a mean number of 12.3 distinct dance partners and girls reported a mean number of 3.5 distinct dance partners. Assuming the reports are accurate, what can you say about the relative sized of m and n?

3.4 MEASURES OF DISPERSION

In section 3.3, we discussed measures of central tendency (mean, median, midrange, and mode) in an effort to pick a single number best characterizing the distribution. A little reflection will demonstrate that you cannot really characterize a distribution with a single number. An old joke has a statistician with his head in a pizza oven and his feet in a bucket of ice. When asked how he felt, the statistician replied that on average he felt fine. A stoic fellow, to be sure, but if prodded he might have volunteered that while the average was acceptable, the ***spread*** in temperature was a bit extreme.

In this section, we will discuss various statistics that describe how the data is spread out. Measures of spread, in combination with measures of central tendency, go a long way toward characterizing a given probability distribution. These combined statistics are important in setting admission standards, defining IQ ranges, setting qualifying time for major events like the Boston marathon, and ranking people by height and/or weight.

3.4.1 RANGE, INTER-QUARTILE RANGE

The simplest measure of data spread is the range, which is just the greatest data point (maximum) less the least data point (minimum).

Maximum, Minimum, Range, and Inter-quartile Range

If $x_1, x_2, ..., x_n$ are a collection of n data points listed in ascending order, then

(a) The ***minimum*** value is $\min(x_1, x_2, ..., x_n) = x_1$.

(b) The ***maximum*** value is $\max(x_1, x_2, ..., x_n) = x_n$.

(c) The ***range*** of the data is
$range(x_1, x_2, ..., x_n) = \max(x_1, x_2, ..., x_n) - \min(x_1, x_2, ..., _n) = x_n - x_1$.

(d) The ***inter-quartile range*** **(IQR)** is $IQR = Q_3 - Q_1$, where Q_3 is the third quartile and Q_1 is the first quartile.

Example 3.4-1 More Descriptive Statistics

Consider the listed heights of fifteen basketball players: 5'8", 5'9", 5'10", 5'10", 6'0", 6'0", 6'0", 6'2", 6'3", 6'3", 6'4", 6'5", 6'6", 6'9", and 7'1".

(a) Find the mean, median, mode, and midrange.
(b) Find the minimum and maximum.
(c) Find first, second, and third quartiles.
(d) Find the 90th percentile.
(e) Find the range and inter-quartile range.

Solution
We begin by converting all height to inches and listing them in ascending order: 68, 69, 70, 70, 72, 72, 72, 74, 75, 75, 76, 77, 78, 81, and 85. If the random variable X represents the height of a basketball player, then

(a) $\mu_X = 74.2\overline{6}$ inches. **Note**: It is common to round to one more decimal place than the data, $\overline{X} = 74.3$. The median is $x_8 = 74$ inches. The mode is 72 inches, which occurs three times. The midrange equals $\frac{x_1 + x_{15}}{2} = \frac{68 + 85}{2} = 76.5$ inches.

(b) The minimum equals $\min(x_1, x_2, \cdots, x_n) = x_1 = 68$ inches, and the maximum is $\max(x_1, x_2, \cdots, x_n) = x_{15} = 85$ inches.

(c) $Q_1 = x_4 = 70$, $Q_2 = x_8 = median = 74$, and $Q_3 = x_{12} = 77$.

(d) $90 = 100 \cdot \frac{i}{15+1}$ implies $i = 14.4$. The 90th percentile is 40% of the way between $x_{14} = 81$ and $x_{15} = 85$, which is $81 + .4(85-81) = 82.6$ inches.

(e) The range is 85 − 68 = 17 inches and $IQR = Q_3 - Q_1 = 77 - 70 = 7$ inches. ❐

Exercise 3-22 Consider the heights of seven female lacrosse players: 1 meter 53 centimeters, 1m 54cm, 1m 62cm, 1m 62cm, 1m 68cm, 1m 81cm, and 1m 83cm.

(a) Find the mean, median, mode, and midrange.

(b) Find the minimum and maximum.

(c) Find first, second, and third quartiles.

(d) Find the 40th percentile.

(e) Find the range and inter-quartile range.

3.4.2 VARIANCE

The ***range*** is quick to calculate and is an understandable measure of how spread out the data is, but there are problems with this simple statistic. First, it is volatile. We can have one million people all with height between 5'9" and 5'11" (hence a range of 2") but if we add one person 6'5", the range for this collection of a million and one people increases by a factor of 4 to 8". Is this a useful statistic? Most of the people are within one inch of 5'10".

Now consider a second group of 1 million people, half of whom are 5'9" and the other half are 6'5". This group also has a range of 8" but the make-up of this group is quite different than that of the group with only one tall person. One way to attempt to decrease this volatility is to talk about inter-quartile range. This statistic has similar deficiencies.

The ***variance*** of a distribution (or its square root, which is called the ***standard deviation***) is the most commonly used measure of the spread. It is calculated as the expected square deviation from the mean. In formal terms a transformed random variable $Y = (X - \mu_X)^2$ is created, and the variance of X is the expected value of Y. In practical terms, the variance is calculated according to the rule for transformations given in Section 3.3.1. Variance is often denoted using the symbol σ_X^2 ("sigma-squared") and the standard deviation is denoted by σ_X.

Variance of a Discrete Random Variable

If X is a discrete random variable with mean μ_X and probability function $\Pr(X{=}x_i) = p(x_i)$, then the ***variance*** of the random variable X is

$$Var[X] = \sigma_X^2 = \sum_{x_i} (x_i - \mu_X)^2 \cdot p(x_i).$$

If X arises from a data set of n equally likely outcomes, then the formula reduces to,

$$\sigma_X^2 = \frac{1}{n}\sum_{i=1}^{n}(x_i - \mu_X)^2.$$

In still more practical terms, the variance is usually easiest to calculate using the following:

The Variance Formula

$$Var[X] = E[X^2] - E[X]^2 = \sum_{x_i} x_i^2 p(x_i) - \left(\sum_{x_i} x_i p(x_i)\right)^2 = \sum_{x_i} x_i^2 p(x_i) - (\mu_X)^2$$

The derivation of this formula is accomplished by expanding the binomial $(X-\mu_X)^2 = X^2 - 2X\mu_X + {\mu_X}^2$ and applying the expectation term by term. Remember that μ_X is constant.

$$Var[X] = E[(X-\mu_X)^2] = E[X^2 - 2\mu_X X + \mu_X^2]$$

$$= E[X^2] - 2\mu_X E[X] + E[\mu_X^2] \text{ since expectation is a linear operator.}$$

$$= E[X^2] - 2\mu_X^2 + \mu_X^2 = E[X^2] - \mu_X^2 \text{ since } \mu_X = E[X].$$

Note

The variance formula, $E[X^2] - (E[X])^2$ is very elegant appearing (and useful) but the nuanced placement of the exponent 2 on the right-hand side can lead to confusion. $E[X^2]$ means square the values of X first and then take the expectation, whereas $(E[X])^2$ means take the expectation of X first and then square. As we have seen, these two quantities are generally different. Indeed, if they were always equal the variance would always be zero and thus not a very interesting statistic.

Example 3.4-2 The Variance Formula

Calculate the mean and the variance of the "pizza" random variable X from Section 3.3 with the following distribution:

x	2	4	6	7	12
$\Pr(X = x)$	3/7	1/7	1/7	1/7	1/7

Recall that $E[X]$ and $E[X^2]$ have been calculated previously, so we could just apply the variance formula directly. We shall, however, incorporate the entire calculation in the following tabular approach:

x	$p(x)$	$x \cdot p(x)$	$x^2 \cdot p(x)$
2	3/7	6/7	12/7
4	1/7	4/7	16/7
6	1/7	6/7	36/7
7	1/7	7/7	49/7
12	1/7	12/7	144/7
Totals:	1	35/7	257/7

The totals are $\sum p(x) = 1$, $E[X] = \sum x \cdot p(x) = \frac{35}{7} = 5$, and $E[X^2] = \sum x^2 \cdot p(x) = \frac{257}{7}$. $Var[X] = \frac{257}{7} - \left(\frac{35}{7}\right)^2 = \frac{574}{49} \approx 11.7$. □

Exercise 3-23 Consider the experiment of having three children with genders equally likely from Example 3.1-1. Let G be the random variable representing the number of girls. Calculate (a) the mean and (b) the variance of G.

It will prove useful in subsequent chapters to have a formula for calculating the variance of a linear transformation of a random variable. Suppose $Y = aX + b$ where a and b are real numbers. Then $E[Y] = a \cdot \mu_X + b$, and,

$$\begin{aligned} Var[Y] &= \sum_{x_i} \left((ax_i + b) - (a\mu_X + b)\right)^2 p(x_i) = \sum_{x_i} \left(a(x_i - \mu_X)\right)^2 p(x_i) \\ &= a^2 \sum_{x_i} (x_i - \mu_X)^2 p(x_i) = a^2 Var[X]. \end{aligned}$$

Mean and Variance of a Linear Transformation

Let X be a discrete random variable and let $Y = a \cdot X + b$, where a and b are real numbers. Then,

1. $E[Y] = E[a \cdot X + b] = a \cdot E[X] + b$.
2. $Var[Y] = Var[a \cdot X + b] = a^2 \cdot Var[X]$.

It makes sense that the variance of Y should not depend on the translation constant b, since variance is a measure of spread between values. Translating all of the values simultaneously does not affect this spread. Also, since variance is a squaring operation, it seems reasonable that the coefficient a is squared.

Exercise 3-24 Roll a pair of fair dice. Let S denote the sum of the two dice. Calculate the variance of S.

Exercise 3-25 The number of classes taken by individuals in a group of students is summarized by: 2, 8, 3, 1, 1, 0, 2, 24, 8, 2, and 5. Calculate the mean, mode, midrange, median, range, inter-quartile range, and variance for the number of classes.

Exercise 3-26 The ages of mathematics faculty are as follows. Calculate the variance for these ages.

x = age	32	39	45	57	62
$\Pr[X = x]$	.16	.04	♥	.10	.28

Exercise 3-27 The ages of students enrolled in *Probability and Statistics* has a cumulative probability distribution given by:

x	17	18	19	20	21	22	23
$F(x)$	0	.18	.47	.85	.94	.99	1.00

(a) Calculate the expected value for X.

(b) Calculate the variance for the discrete random variable X.

3.4.3 STANDARD DEVIATION

The standard deviation of a probability distribution is merely the square root of the variance. The standard deviation is more intuitive to use than variance since it is denominated in the same units as the underlying random variable. Suppose that there were 500 data points with mean 10 and standard deviation 3. As a general rule, the majority[3] of those 500 data points will be within one standard deviation of the mean; that is, between 7 and 13. And *almost all* of those data points are between 4 and 16 (two standard deviations). It is rare that something is three or more standard deviations from the mean, as you will observe by experience and future results.

We want you to develop your intuitive common sense when it comes to these calculations. You can often tell by looking at a set of data the approximate values for the mean and standard deviation. When confronted with a collection of data, take a few seconds to peruse the values and make a mental estimate of the mean and the standard deviation. Then calculate the mean and standard deviation and compare them to your estimates. Indeed, in any mathematical context it is wise to inspect your results for face validity. Does my calculated answer seem plausible? If, for example, you deposited $5000 in a bank account for two years earning 6% interest compounded continuously, then you should predict that the amount in the

[3] It depends on the distribution. For the ***normal distribution*** (a symmetric bell-shaped curve to be studied in detail in Chapter 6), 68.26% of the data falls within one standard deviation of the mean, 95.44% within two standard deviations, and 99.74% within three standard deviations. These percentages are often used as a rule-of-thumb in general, although they apply exactly only to a normal distribution.

account after 2 years is slightly more than \$5600. If you get more than \$800 million[4], then you should know that something is amiss. The correct answer is \$5,637.48.

Standard Deviation

Let X be a discrete random variable. The *standard deviation* of X is given by,

$$\sigma_X = \sqrt{Var[X]} = \sqrt{\sum_{x_i}(x_i - \mu_X)^2 \cdot p(x_i)} = \sqrt{E[X^2] - E[X]^2}.$$

Exercise 3-28 A high school math teacher gave a probability examination to her 12 students. Scores were 78, 48, 69, 102, 78, 93, 69, 84, 96, 59, 87, and 93.

(a) Find the mean, median, mode, and midrange.
(b) Find the minimum and maximum.
(c) Find the 80th percentile.
(d) Find the variance and standard deviation.

Exercise 3-29 SOA EXAM P Sample Exam Questions #64
A probability distribution of the claim sizes for an auto insurance policy is given in the table below. What percentage of claims is within one standard deviation of the mean claim size?

Claim Size	20	30	40	50	60	70	80
Probability	0.15	0.10	0.05	0.20	0.10	0.10	0.30

(A) 45% (B) 55% (C) 68% (D) 85% (E) 100%

Exercise 3-30 Referring to the Exercise 3-29, what percentage of the claims are within one standard deviation of the claim size mode?

3.4.4 STANDARDIZED RANDOM VARIABLES AND Z-SCORES

Have you ever heard the cliché "You cannot compare apples to oranges?" This is not quite accurate. If you are a statistician, you can compare apples to oranges, at least figuratively speaking. If "apples" and "oranges" represent two different random variables, such as ACT scores and SAT scores, then we can make comparisons by ***standardizing*** the different test scores.

Let X be a discrete random variable and define $Z = \frac{X-\mu_X}{\sigma_X} = \left(\frac{1}{\sigma_X}\right)X - \frac{\mu_X}{\sigma_X}$ as a linear transformation of X. Using the formulas for the mean and the variance of a linear transformation, we see that,

---

[4] $A(2) = 5000e^{6(2)} = \$813,773,957.10$. This mistake happens if you use a calculator or formula and an interest rate of 6 instead of $6\% = .06$.

$$\mu_Z = E[Z] = \left(\frac{\mu_X}{\sigma_X}\right) - \left(\frac{\mu_X}{\sigma_X}\right) = 0$$

and

$$\sigma_Z^2 = Var[Z] = \left(\frac{1}{\sigma_X}\right)^2 \sigma_X^2 = 1.$$

Standardized Random Variable

Let X be a discrete random variable and let $Z = \frac{X - \mu}{\sigma}$. Then Z is called the ***standardization of X.*** The random variable Z always has mean equal to 0 and standard deviation equal to 1.

In statistics, the standardized values of a random variable X are often called z-scores. The notation follows from the use of the letter Z to represent the standard normal distribution random variable – the symmetric bell-shaped *continuous* distribution with mean of zero and standard deviation of one. The benefit of z-scores is that they convert x-values to the number of standard deviations above (or below) the mean of the random variable X. In that way we can compare the results of ACT and SAT scores, even though they are given on totally different scales.

Definition: z-score

Let x be a value from a probability distribution with mean μ and standard deviation σ. Then the ***z-score*** for x is defined to be:

$$z = \frac{x - \mu}{\sigma}.$$

Example 3.4-3 Standardized Test Scores

A university has two sections of calculus. Professor Tuffin Fair teaches one of these sections and her historical mean score on an examination is 80 with a standard deviation of 10. The other section is taught by Dr. E.Z. Grader. His class average is 92 with a standard deviation of 4. You are enrolled in Dr. Fair's class (excellent choice). You study and do your homework (an even better choice). You take an exam and earn an 87. Your roommate is enrolled in the other class, has never opened her book, and believes that the Library is the name of a bar. Your roommate scored an 88 and is irritating you to no end because she claims to have performed better than you. Did she?

Solution

Your z-score is $z = \frac{87-80}{10} = .7$, which is quite a bit above average (mean). Your roommate's z-score is $z = \frac{88-92}{4} = -1$, which is a full standard deviation below average. You in fact did far better than your less studious roommate. ❐

While it is possible to glean useful information comparing the z-scores of disparate random variables, this procedure is most meaningful when the random variables in question are approximately symmetric and bell-shaped, that is to say, close to being so-called *normal* distributions.

Example 3.4-4 The z-score for Gas Prices

In Pittsburgh in 2007 the mean price of a gallon of 87 octane unleaded gasoline was \$2.949 with a standard deviation of 10 cents. The local Sheetz station sold the same grade of gasoline for \$3.099.

(a) Find the $z-$score of gas purchased at the Sheetz.

(b) Is it unreasonable to pay \$3.549 for gasoline in Chicago?

Solution

(a) $z = \frac{3.099 - 2.949}{.10} = 1.5$. One-and-a-half standard deviations above average is significantly above average for a typical distribution of prices.

(b) As for the price of gasoline in Chicago, we cannot say. It is inappropriate to compare gasoline prices in Pittsburgh to those in Chicago. We need to know the mean and standard deviation of gasoline prices in Chicago. It was possible that \$3.549 for gasoline in Chicago was the lowest price for gasoline in the entire Windy City. ❐

Exercise 3-31 Assume the I.Q. test scores are standardized to have mean 100 and standard deviation 16.

(a) What is the z-score of a woman with a 125 I.Q.?

(b) If someone has a z-score of $z = -1.75$, then find his I.Q.

Exercise 3-32 Jen and Ryan are having a friendly competition to find the "biggest" fruit. Jen visits her aunt in Phoenix and finds a 16.4 ounce navel orange. Ryan scours Janowski's apple orchard and discovers a 19 ounce red delicious apple. Who found the most impressive (relatively largest) piece of fruit? Use $\sigma_{\text{Orange}} = 2.85$ and $\sigma_{\text{Apple}} = 2.848$. Following are some (sample) random weights of selected fruit:

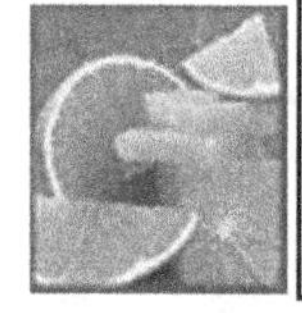

Navel Orange
14.0 oz, 12.0 oz, 10.5 oz, 9.3 oz, 8.4 oz, 7.6 oz, 6.7 oz, 5.9 oz, 5.4 oz, and 4.8 oz.

Red Delicious Apple
16.0 oz, 15.0 oz, 13.5 oz, 11.3 oz, 10.4 oz, 9.6 oz, 9.6 oz, 7.9 oz, and 7.4 oz.

3.4.5 CHEBYCHEV'S THEOREM

Chebychev's Theorem (a.k.a. Tchebychev or Tchebycheff) is a general result that applies to *most* discrete random variables (and as we shall see in future chapters, most continuous probability distributions as well). The theorem basically asserts that for a given positive number k there is a universal upper bound for the probability that a random variable X will be more than k standard deviations away from its mean. This upper bound depends only on k, not on the distribution of the random variable X. For example, the probability that a random variable is more than two standard deviations away from its mean is less than or equal to ¼. This holds for *any* random variable X, no matter how bizarre its probability distribution may be.

We will approach the Chebychev Theorem using a simpler inequality that holds for non-negative valued random variables. Markov's inequality relates probabilities to expectation and provides loose, but useful, bounds for the cumulative distribution function of a random variable.

Markov Inequality

Let Y be a discrete random variable taking only non-negative values, with finite mean μ_Y. Then for any $a > 0$,

$$\Pr[Y > a] \le \frac{\mu_Y}{a}.$$

Proof

$$\Pr[Y > a] = \sum_{y_i > a} p(y_i) \le \sum_{y_i > a} \frac{y_i}{a} p(y_i) \qquad \left(\text{since } \frac{y_i}{a} \ge 1\right)$$

$$= \frac{1}{a} \sum_{y_i > a} y_i p(y_i) \le \frac{1}{a} \underbrace{\sum_{y_i}}_{\text{sum over all } y_i} y_i p(y_i) = \frac{\mu_Y}{a}.$$

Chebychev's Theorem

Let X be a discrete random variable with finite mean μ_X and standard deviation σ_X. Let k be greater than 1. Then the probability that X is more than k standard deviations from the mean, μ_X, is less than or equal to $\frac{1}{k^2}$. That is,

$$\Pr\left(X < \mu_X - k \cdot \sigma_X \text{ or } X > \mu_X + k \cdot \sigma_X\right) = \Pr\left(|X - \mu| > k \cdot \sigma_X\right) \le \frac{1}{k^2}.$$

Equivalently,

$$\Pr\left(\mu_X - k \cdot \sigma_X \le X \le \mu_X + k \cdot \sigma_X\right) = \Pr\left(|X - \mu| \le k \cdot \sigma_X\right) \ge 1 - \frac{1}{k^2}.$$

Proof

Let

$$Y = \left(\frac{X - \mu_X}{\sigma_X}\right)^2.$$

Then

$$\mu_Y = E[Y] = E\left[\left(\frac{X-\mu_X}{\sigma_X}\right)^2\right]$$
$$= \frac{1}{\sigma_X^2}E\left[(X-\mu_X)^2\right] = \frac{\sigma_X^2}{\sigma_X^2} = 1.$$

We now apply the Markov inequality to Y with $a = k^2$. This results in,

$$\Pr\left[\left(\frac{X-\mu_X}{\sigma_X}\right)^2 > k^2\right] \le \frac{1}{k^2},$$

or equivalently,

$$\Pr\left[(X-\mu_X)^2 > k^2\sigma_X^2\right] \le \frac{1}{k^2}.$$

The last expression is equivalent to

$$\Pr\left[|X-\mu_X| > k\sigma_X\right] \le \frac{1}{k^2}.$$

Example 3.4-5 Chebyshev's Theorem

Suppose there is a large set of data with mean 50 and standard deviation 10. Find the smallest symmetric interval about the mean that is certain to contain at least 75% of the data points.

Solution

As in Section 3.3, let X be the random variable whose values are equally likely data points. We apply the Chebyshev Theorem in the form,

$$\Pr\left(\mu_X - k\cdot\sigma_X \le X \le \mu_X + k\cdot\sigma_X\right) = \Pr\left(|X-\mu| \le k\cdot\sigma_X\right) \ge 1-\frac{1}{k^2}.$$

We want the value of k such that $1-\frac{1}{k^2} = .75$. Solving, we find that $k = 2$. Now, substituting into the statement of the Chebyshev Theorem above, we have,

$$\Pr\left(50-2\cdot10 < X < 50+2\cdot10\right) = \Pr\left(30 < X < 70\right) > 1-\frac{1}{2^2} = 0.75.$$

In words, this says that there is at least a 75% chance that a data point chosen at random is between 30 and 70. Since all data points are equally likely, this means that at least 75% of the data points are between 30 and 70. ❐

Exercise 3-33 Complete the following table:

Number of standard deviations from the mean	2	3	4	5	10	25	100
Probability within k standard deviations of the mean	>.75				>.99		

In Chapter 4 and Chapter 7, we will see that this bound is quite conservative for most distributions.

Sometimes a data set has an observation that is several standard deviations below or above the mean. Such extreme values are called ***outliers***. Errors could have been made in observing or recording the data. Extra attention should be paid to outliers. Their accuracy should be verified and one needs to make a decision whether or not to include them when calculating statistics such as the sample mean or the sample standard deviation. For example, suppose that you were interested in the average net worth of a person living in Washington state. You choose to sample 100 people and calculate the *mean*. In the event that Bill Gates is included in your sample, you will have a huge average value that will greatly distort the information you seek. In this case, it may be best to eliminate Bill Gates from your sample. Be forewarned that you should not ignore data merely because you dislike it (unless you are working for a cigarette manufacturer).

Outliers

We define an outlier to be any data point with a z-score less than $z = -3$ or greater than $z = 3$.

3.4.6 COEFFICIENT OF VARIATION

The ***coefficient of variation*** is a statistic that measures the relative variability of a random variable – the ratio of the standard deviation to the mean. The coefficient of variation is a dimensionless number that is used in reliability theory and renewal theory. When comparing between data sets with different units or means, one could use the coefficient of variation for comparison instead of the standard deviation.

Coefficient of Variation

A random variable X with mean μ and standard deviation σ has ***coefficient of variation***

$$\frac{100 \cdot \sigma}{\mu}\%.$$

Example 3.4-6 Coefficient of Variation

Seven enrollment councilors ate the following number of chocolate candies:

x	2	4	6	7	12
$\Pr(X = x)$	3/7	1/7	1/7	1/7	1/7

Calculate the coefficient of variation for a random variable with this distribution.

Solution

Previously we calculated $\mu_X = 5$ and $Var[X] = \frac{574}{49}$ for the pizza example with the same distribution. So the coefficient of variation is $\frac{3.42}{5} = .6845 = 68.45\%$. Since the coefficient of variation is less than one, this distribution is considered to have relatively low variance. ❐

Exercise 3-34 Calculate the coefficient of variation for navel oranges and red delicious apples for Exercise 3-32 in section 3.4.4.

Exercise 3-35 Roll a pair of fair dice. Let X denote the sum. Calculate the coefficient of variation for X.

3.5 CONDITIONAL EXPECTATION AND VARIANCE

For a discrete probability distribution we now have the tools to calculate the mean and variance in straightforward fashion. An automobile insurance company, for example, would calculate the average insurance claim and the standard deviation of claims for a block of business, and use that information to set the premium for a new customer. After a year or two the insurance company has more information about the policyholder; primarily, the accident ***experience*** of the driver as well as information about payment history and modifications to the policy. In Section 2.6 we saw that how the probability of an accident in year 3 might be conditioned on the number of accidents in years 1 and 2. Given that a driver was involved in three accidents during the two year stretch, then the company would assign a higher probability that this driver will be involved in an accident in the third year. Using this new, so called conditional distribution, the insurance company could recalculate the expected number and variance for the number of accidents of the policyholder. Then they would be able to raise this reckless (or simply, unlucky) driver's rate or drop coverage all together.

Example 3.5-1 Conditional Distributions

A family has three children. Assume that the gender of the child is equally likely to be male or female. Let G denote the number of female children.

(a) Write down the probability distribution for G.

(b) Calculate the mean and variance of G.

(c) Given that the family has at least one female, find the conditional distribution for the number of girls.

(d) Calculate $E[G \mid G > 0]$ and $Var[G \mid G > 0]$.

(e) Calculate the coefficient of variation for G and for the conditional random variable $G \mid G > 0$.

Solution

Note that Parts (a) and (b) constitute Exercise 3-23.

(a)

g	0	1	2	3
$\Pr[G=g]$	$\frac{1}{8}$	$\frac{3}{8}$	$\frac{3}{8}$	$\frac{1}{8}$

(b) $E[G] = 0\left(\frac{1}{8}\right)+1\left(\frac{3}{8}\right)+2\left(\frac{3}{8}\right)+3\left(\frac{1}{8}\right) = \frac{12}{8} = \frac{3}{2} = 1.5$

$$E[G^2] = 0^2\left(\tfrac{1}{8}\right)+1^2\left(\tfrac{3}{8}\right)+2^2\left(\tfrac{3}{8}\right)+3^2\left(\tfrac{1}{8}\right) = 3.0$$

$$Var[G] = E[G^2]-\left(E[G]\right)^2 = 3-1.5^2 = .75$$

(c) Given at least one girl, the values of G are limited to 1, 2, or 3 and the conditional distribution of G is given by:

$G \mid G>0$	1	2	3
$\Pr[G \mid G>0]$	3/7	3/7	1/7

For example, $\Pr[G=1 \mid G>0] = \frac{\Pr\left[\{G=1\}\cap\{G>0\}\right]}{\Pr[G>0]} = \frac{\Pr[G=1]}{\Pr[G>0]} = \frac{3/8}{7/8} = \frac{3}{7}$.

The other calculations are similar.

(d) $E[G \mid G>0] = 1\left(\frac{3}{7}\right)+2\left(\frac{3}{7}\right)+3\left(\frac{1}{7}\right) = \frac{12}{7} = 1.7143$

$$E[G^2 \mid G>0] = 1^2\left(\tfrac{3}{7}\right)+2^2\left(\tfrac{3}{7}\right)+3^2\left(\tfrac{1}{7}\right) = \tfrac{24}{7} = 3.4286$$

$$Var[G \mid G>0] = E[G^2 \mid G>0]-\left(E[G \mid G>0]\right)^2 = \tfrac{24}{49} = .4898.$$

(e) For the random variable G, the coefficient of variation is $\frac{\sqrt{.75}}{1.5} = 57.735\%$.

For $G \mid G>0$, the coefficient of variation is $\frac{\sqrt{.4898}}{1.7143} = 40.825\%$. ❐

Notice that there are no really new ideas here. We simply calculate the mean and variance in the usual manner, but using the conditional probability distribution. We learned how to calculate conditional probabilities in Chapter 2. There are many applications using this idea. A person age 50 is promised a full pension at age 65 but leaves the company. What percentage of that pension is *fair* for the person to take with her given her current age of 50? Two people are playing tennis against one another. How does the likelihood that one player will win the match change if we know that he has won the first two sets? You bet your roommate \$20 on the outcome of the Steelers versus Titans football game. At halftime, the Titans are ahead by ten points. How much should your roommate accept from you to cancel the bet? That is, what will the roommate take with certainty, given the new information concerning the halftime score?

Exercise 3-36 The Chicago Cubs are playing a ***best-of five game series*** (the first team to win 3 games win the series and no other games are played) against the St. Louis Cardinals. Let X denote the total number of games played in the series. Assume that the Cubs wins 59% of their games versus their arch rival Cardinals and that the probability of winning a game is independent of other games. *Hint*: Draw a tree diagram. It may take awhile.

(a) Calculate the mean and standard deviation for X.

(b) Calculate the conditional mean and standard deviation for X given that the Cardinals win the first game.

Exercise 3-37 SOA EXAM P Sample Exam Questions #113
Two life insurance policies, each with a death benefit of 10,000 and a one-time premium of 500, are sold to a couple, one for each person. The policies will expire at the end of the tenth year. The probability that only the wife will survive at least ten years is 0.025, the probability that only the husband will survive at least ten years is 0.01, and the probability that both of them will survive at least ten years is 0.96.

What is the expected excess of premiums over claims, given that the husband survives at least ten years?

(A) 350 (B) 385 (C) 397 (D) 870 (E) 897

Exercise 3-38 Roll a pair of fair dice. Find the expected sum, given that at least one of the dice is a four.

Exercise 3-39

The distribution of loss due to fire damage to a warehouse is:

Amount of Loss	Probability
0	0.900
500	0.060
1,000	0.030
10,000	0.008
50,000	0.001
100,000	0.001

Given that a loss is greater than zero, calculate the expected amount of the loss.

Exercise 3-40 A junior actuary created a simple model for the amount of damage that will be caused by an insured in the coming year. Let D denote the amount of damage. She found the expected damage to be $E[D]=600$.

D	0	1,000	5,000	10,000
$\Pr(D)$	.82	.10		

(a) Find the missing probabilities.

(b) Find the standard deviation σ_D.

(c) Find the expected damage given that there is some damage.
That is, find $E[D \mid D>0]$.

(d) Find the conditional variance of the damage given that there is some damage.
That is, find $Var[D \mid D>0]$.

3.6 JOINTLY DISTRIBUTED RANDOM VARIABLES (ROUND 1)

The subject of ***jointly distributed*** random variables will be one of our major topics for study, covered in detail in Chapter 7. We will take a glimpse into the future by developing a few useful concepts concerning pairs of discrete random variables arising from the same probability experiment. The main goal is to derive formulas for calculating the means and variances of random variables that are expressed as sums of other, simpler, random variables.

To illustrate the general idea, let's return to the example of the three children with equally likely genders. Then the underlying sample space consists of eight equally likely outcomes $\{BBB, BBG, BGB, BGG, GBB, GBG, GGB, GGG\}$.

Example 3.6-1 Jointly Distributed Random Variables

Let X be the random variable that counts the number of girls on the first birth. In other words, X equals 0 or 1 (our model assumes no multiple births) depending on whether the first letter of an outcome word is B or G, respectively. Similarly, let Y equal the number of girls on the second birth. Again, Y takes the values 0 or 1, this time depending on whether the *second* letter of an outcome word is B or G, respectively. Since X and Y have the same sample space we can look at the new random variables defined and denoted by: $X+Y$ and XY.

(a) Write the probability functions for $X, Y, X+Y$ and XY.

(b) Calculate the means and the variances for these four random variables.

Solution

(a) The probabilities are straightforward to find. We illustrate with some sample calculations:

$$\Pr[X=1] = \Pr[\{GBB, GBG, GGB, GGG\}] = \frac{4}{8} = \frac{1}{2},$$

$$\Pr[X+Y=1] = \Pr[\{GBB, GBG, BGB, BGG\}] = \tfrac{4}{8} = \tfrac{1}{2},$$

$$\Pr[XY=1] = \Pr[\{GGB, GGG\}] = \tfrac{2}{8} = \tfrac{1}{4}.$$

Students are encouraged to check the complete probability tables given below.

(b) We observe that random variables X and Y are ***identically distributed.*** That is, they have the same probability function. Here are the complete tables, including the columns for mean and second moment:

x (or y) Values	**Prob**	$xp(x)$	$x^2 p(x)$
0	$\frac{1}{2}$	0	0
1	$\frac{1}{2}$	$\frac{1}{2}$	$\frac{1}{2}$
Total	1	$\frac{1}{2}$	$\frac{1}{2}$

$$E[X] = E[Y] = \tfrac{1}{2} \quad \text{and} \quad Var[X] = Var[Y] = \tfrac{1}{2} - \left(\tfrac{1}{2}\right)^2 = \tfrac{1}{4}.$$

$x+y$	**Prob**	$(x+y)p(x+y)$	$(x+y)^2 p(x+y)$
0	$\frac{1}{4}$	0	0
1	$\frac{1}{2}$	$\frac{1}{2}$	$\frac{1}{2}$
2	$\frac{1}{4}$	$\frac{1}{2}$	1
Total	1	1	$\frac{3}{2}$

$E[X+Y]=1$

$Var[X+Y] = \tfrac{3}{2} - 1^2 = \tfrac{1}{2}$

$x \cdot y$	**Prob**	$(x \cdot y)p(x \cdot y)$	$(x \cdot y)^2 p(x \cdot y)$
0	$\frac{3}{4}$	0	0
1	$\frac{1}{4}$	$\frac{1}{4}$	$\frac{1}{4}$
Total	1	$\frac{1}{4}$	$\frac{1}{4}$

$E[XY] = \tfrac{1}{4}$

$Var[XY] = \tfrac{1}{4} - \left(\tfrac{1}{4}\right)^2 = \tfrac{3}{16}$

❐

Note

$E[X+Y] = E[X]+E[Y]$, $Var[X+Y] = Var[X]+Var[Y]$, and $E[XY] = E[X] \cdot E[Y]$.

It is no coincidence that these things are all true. The first will always be true regardless of the random variables. The second two of these equalities follow from the *independence* of X and Y.

Intuitively, since X depends only on the first birth and Y depends only on the second birth, it seems plausible that in some sense X and Y are *independent.* Technically, we have only defined independence for events, but it is possible to extend the idea to pairs of random variables.

Joint Distribution and Independence for X and Y Discrete Random Variables

Let X and Y be random variables arising from the same discrete probability experiment. The **joint distribution** of X and Y is given by,

$$p(x,y)=\Pr[\{X=x\}\cap\{Y=y\}]$$

We say X and Y are **independent** if <u>for all</u> x and y the events $\{X=x\}$ and $\{Y=y\}$ are independent. That is,

$$p(x,y) = \Pr[\{X=x\}\cap\{Y=y\}] = \Pr[X=x]\cdot\Pr[Y=y] = p_X(x)\cdot p_Y(y)$$

In the above example it is easy to check that X and Y are independent according to this definition, although one would need to check the four separate cases $(x,y)=(0,0)$, $(0,1)$, $(1,0)$, and $(1,1)$. As an example, we calculate the case $(x,y)=(0,1)$:

$$\Pr[\{X=0\}\cap\{Y=1\}] = \Pr[\{BGB,BGG\}] = \frac{2}{8} = \frac{1}{4} = \left(\frac{1}{2}\right)\cdot\left(\frac{1}{2}\right) = \Pr[X=0]\cdot\Pr[Y=1].$$

The easiest way to convey all this information at once is with the following two-way table, which displays the joint distribution of X and Y:

$X\downarrow$	$Y\rightarrow$	0	1	$p_X(x)$
0		$\frac{1}{4}$	$\frac{1}{4}$	$\frac{1}{2}$
1		$\frac{1}{4}$	$\frac{1}{4}$	$\frac{1}{2}$
$p_Y(y)$		$\frac{1}{2}$	$\frac{1}{2}$	1

Here, the first and last columns represent the probability function of X, while the first and last rows represent the probability function of Y. The four numbers on the inside represent the so-called *joint* probability function $\Pr[\{X=x\}\cap\{Y=y\}]$ for the four cases (0,0), (0,1), (1,0), and (1,1). Then X and Y are independent if and only if the numbers on the *inside* are the products of the corresponding numbers on the *outside* margins.

Of course, it is quite possible to have a joint distribution in which the individual random variables are *not* independent. For example, let X be the number of girls on the first birth, and let Y be the number of girls on the first two births. Then the joint table is,

$X\downarrow$	$Y\rightarrow$	0	1	2	$p_X(x)$
0		$\frac{1}{4}$	$\frac{1}{4}$	0	$\frac{1}{2}$
1		0	$\frac{1}{4}$	$\frac{1}{4}$	$\frac{1}{2}$
$p_Y(y)$		$\frac{1}{4}$	$\frac{1}{2}$	$\frac{1}{4}$	1

Now, X and Y clearly fail to be independent since, for example,

$$0 = p(0,2) \neq p_X(0)p_Y(2) = \left(\frac{1}{2}\right)\left(\frac{1}{4}\right) = \frac{1}{8}.$$

Exercise 3-41 Suppose that a family has three children with equally likely outcomes. Let the random variable X denote the number of girls on the first birth, and Y denote the number of girls on the first two births. Compute the following:

(a) $E[X]$ and $Var[X]$

(b) $E[Y]$ and $Var[Y]$

(c) Write the probability function for $X+Y$ and use it to calculate $E[X+Y]$ and $Var[X+Y]$.

(d) Show that $Var[X+Y] \neq Var[X]+Var[Y]$.

The following two-way tabular display works in general for X and Y jointly distributed random variables from a common discrete probability experiment:

$X \setminus Y$	y_1	$\cdots$	y_n	$p_X(x)$
x_1	$p(x_1, y_1)$		$p(x_1, y_n)$	$p_X(x_1)$
$\vdots$				
x_m	$p(x_m, y_1)$		$p(x_m, y_n)$	$p_X(x_m)$
$p_Y(y)$	$p_Y(y_1)$		$p_Y(y_n)$	1

The first and last columns represent the probability distribution of X. As we shall formalize later, it is referred to as the ***marginal distribution*** for X. This seems a sensible appellation since it appears along the vertical margins of the table. Similarly, the first and last rows represent the marginal probability distribution of Y.

The probabilities on the margins (last column and last row) can easily be seen to be the sums of the probabilities in the respective row and column. For example,

$$p_X(x_i) = \Pr[X = x_i] = \Pr\left[\bigcup_{\left(\substack{\text{disjoint union} \\ \text{over all } j}\right)} \{X = x_i\} \cap \{Y = y_j\}\right] = \sum_j p(x_i, y_j),$$

with a similar calculation showing $p_Y(y_j) = \sum_i p(x_i, y_j)$.

The following theorem summarizes the main formulas for joint distributions that we will be using in the next several chapters.

Theorem

Let X and Y be random variables arising from the same probability experiment. Then,

(a) $E[X+Y]=E[X]+E[Y]$. This formula extends to sums of any length.

Further, if X and Y are **independent**, then

(b) $E[X \cdot Y]=E[X] \cdot E[Y]$, and

(c) $Var[X+Y]=Var[X]+Var[Y]$.

This formula extends to finite sums of any length provided the summands are pair-wise independent.

Note

We cannot over-emphasize the fact that the relationships (b) and (c) require the *independence* of *X* and *Y*.

Proof

(a)
$$
\begin{aligned}
E[X+Y] &= \sum_{i,j}(x_i+y_j)p(x_i,y_j) = \sum_i\sum_j x_i p(x_i,y_j)+\sum_i\sum_j y_j p(x_i,y_j) \\
&= \sum_i x_i \sum_j p(x_i,y_j)+\sum_j y_j \sum_i p(x_i,y_j) \\
&= \sum_i x_i p_X(x_i)+\sum_j y_j p_Y(y_j) \\
&= E[X]+E[Y].
\end{aligned}
$$

(b)
$$
\begin{aligned}
E[XY] &= \sum_{i,j} x_i y_j p(x_i,y_j) = \sum_{i,j} x_i y_j \overbrace{p_X(x_i)p_Y(y_j)}^{\text{using independence}} \\
&= \left(\sum_i x_i p_X(x_i)\right)\cdot\left(\sum_j y_j p_Y(y_j)\right) \\
&= E[X]\cdot E[Y].
\end{aligned}
$$

(c) We first calculate

$$
\begin{aligned}
E\left[(X+Y)^2\right] &= E[X^2+2XY+Y^2] \\
&\overset{\text{using (a)}}{=} E[X^2]+2E[XY]+E[Y^2] \\
&\overset{\text{using (b)}}{=} E[X^2]+\overbrace{2E[X]\cdot E[Y]}^{\text{independence used here}}+E[Y^2].
\end{aligned}
$$

Next, we expand

$$(E[X+Y])^2 = (E[X]+E[Y])^2 = E[X]^2 + 2E[X] \cdot E[Y] + E[Y]^2.$$

Then,

$$\begin{aligned} Var[X+Y] &= E[(X+Y)^2] - (E[X+Y])^2 \\ &= \left\{E[X^2] + 2E[X] \cdot E[Y] + E[Y^2]\right\} \\ &\qquad - \left\{E[X]^2 + 2E[X] \cdot E[Y] + E[Y]^2\right\} \\ &= \left\{E[X^2] - E[X]^2\right\} + \left\{E[Y^2] - E[Y]^2\right\} \\ &= Var[X] + Var[Y]. \end{aligned}$$

The extension of parts (a) and (c) to finite sums of length greater than two is a straightforward matter of mathematical induction. These formulas will be very useful in the study of special discrete random variables in the next chapter.

This constitutes a first, and somewhat limited, exposure to the subject of jointly distributed random variables. We have explicitly required that X and Y arise from the same underlying probability experiment. This enabled us to derive the joint distribution from the distributions of X and Y. In a later chapter, we will reverse the process and define the joint distribution first, and then derive the (marginal) distributions of X and Y from the joint distribution.

Example 3.6-2 Jointly Distributed Random Variables

Suppose that random variables X and Y are independent. Partial information for the joint distribution is given below. Find the standard deviation of $X+Y$.

$X \downarrow$ $Y \rightarrow$	5	15	25	$p_X(x)$
10				.40
20				
$p_Y(y)$	.20	.30		

Solution

We begin by filling in the missing information. (Marginal) probabilities must sum to one. By independence, the joint probabilities are products.

$X \downarrow$ $Y \rightarrow$	5	15	25	$p_X(x)$
10	.08	.12	.20	.40
20	.12	.18	.30	**.60**
$p_Y(y)$	.20	.30	**.50**	

With the joint distribution, we can directly compute

$$E[X+Y] = 15\times.08+25\times.12+35\times.2+25\times.12+35\times.18+45\times.3 = 34,$$

and

$$E\left[(X+Y)^2\right] = 15^2\times.08+25^2\times.12+35^2\times.2+25^2\times.12+35^2\times.18+45^2\times.3 = 1241.$$

Then,

$$Var[X+Y] = E\left[(X+Y)^2\right]-E[X+Y]^2 = 1241-34^2 = 85.$$

Thus, $\sigma_{X+Y} = \sqrt{85} = 9.22$.

Since X and Y are independent, we could have used the marginal probabilities for X and Y. Then, $E[X]=(10)(.4)+(20)(.6)=16$ and $E[X]=(10^2)(.4)+(20^2)(.6)=280$, so that $Var[X]=280-16^2=24$. Similarly, $Var[Y]=61$. Then, $Var[X+Y]=24+61=85$. ❐

Exercise 3-42 A pair of fair dice is tossed. Let X be the outcome on the first die and let Y be the outcome on the second die.

(a) Are X and Y independent? Explain.

(b) Calculate $Var[X+Y]$.

Exercise 3-43 A pair of fair dice is tossed. Let X represent the number of dice showing an even number, $X=0$, 1, or 2, and let Y represent the number of dice showing either a 1 or a 6, $Y=0$, 1, or 2.

(a) Calculate $E[X]$ and $Var[X]$.

(b) Calculate $E[Y]$ and $Var[Y]$.

(c) Write the probability function for $X+Y$ and calculate $E[X+Y]$ and $Var[X+Y]$.

(d) Does $Var[X+Y] = Var[X]+Var[Y]$? Explain why or why not.

Exercise 3-44 Consider the 3-child, equally likely genders, experiment. Let X_i be the number of girls on the ith birth ($X_i=0$ or 1 for $i=1,2,$ or 3).

(a) Show that $X=X_1+X_2+X_3$ is the random variable indicating the total number of girls in the three child family. **Note:** X is called a ***binomial random variable***.

(b) Calculate the mean and the variance of X_i. (Since the X_i's are identically distributed you need do the calculations only once.)

(c) Use the pair-wise independence of the X_i's and formulas (a) and (c) of the theorem to calculate the mean and the variance of X. Compare the results to the related exercise (Exercise 3-22) in Subsection 3.4.2.

Exercise 3-45 Suppose that we have the following information about basketball player shoe size, S, and player height, H.

		H = Height		
		68	**70**	**73**
S = Shoe Size	**8.5**	.25	.20	.15
	12	.05	.12	☺

(a) Find the expected height and expected shoe size of a basketball player.

(b) Find the expected shoe size, given that a player is 73 inches tall, $E[S|H=73]$.

(c) Find the conditional coefficient of variation for shoe size, given the player is 5'8" tall.

3.7 THE PROBABILITY GENERATING FUNCTION

We wish to touch on another technique, this one calculus based, for computing the means and variances of random variables. Let X be a random variable taking values from the whole numbers. As an example, consider once again the pizza consumption random variable. All of the information about the values of X and the calculation of its mean and variance is contained in the table from Section 3.4.2:

x	$\Pr(X=x)$	$xp(x)$	$x^2p(x)$
2	3/7	6/7	12/7
4	1/7	4/7	16/7
6	1/7	6/7	36/7
7	1/7	7/7	49/7
12	1/7	12/7	144/7
Total:	1	35/7	257/7

A compact method for conveying much of the same information is to construct a polynomial function using the values of X as powers and the probabilities as coefficients:

$$h_X(s) = \frac{3}{7}s^2 + \frac{1}{7}s^4 + \frac{1}{7}s^6 + \frac{1}{7}s^7 + \frac{1}{7}s^{12}.$$

Observe that $h_X(s) = E[s^X]$.

By taking the derivative with respect to s we get,

$$h'_X(s) = \frac{d}{ds}h_X(s) = (2)\left(\frac{3}{7}\right)s + (4)\left(\frac{1}{7}\right)s^3 + (6)\left(\frac{1}{7}\right)s^5 + (7)\left(\frac{1}{7}\right)s^6 + (12)\left(\frac{1}{7}\right)s^{11}.$$

It follows that $h'_X(1) = E[X]$. That is, we can determine the mean of a random variable X by evaluating the derivative of the probability generating function $h_X(s)$ (with respect to s) evaluated at $s = 1$.

Calculating the second derivative with respect to s gives,

$$h''_X(s) = (2\cdot 1)\left(\frac{3}{7}\right) + (4\cdot 3)\left(\frac{1}{7}\right)s^2 + (6\cdot 5)\left(\frac{1}{7}\right)s^4 + (7\cdot 6)\left(\frac{1}{7}\right)s^5 + (12\cdot 11)\left(\frac{1}{7}\right)s^{10}.$$

We see that $h''_X(1) = E\left[X\cdot(X-1)\right] = E[X^2] - E[X]$, so that

$$Var[X] = E[X^2] - E[X]^2 = h''_X(1) + h'_X(1) - \left(h'_X(1)\right)^2.$$

The Probability Generating Function for X

Let X be a random variable taking non-negative integer values.
Let $p_n = \Pr[X{=}n]$. Then the probability generating function for X is defined by,

$$h(s) = \sum_{i=0}^{\infty} p_i s^i = E[s^X].$$

(a) $E[X] = h'(1)$, and (b) $Var[X] = h''_X(1) + h'_X(1) - \left(h'_X(1)\right)^2$.

If X takes only a finite number of values with positive probability, then $h(s)$ is a polynomial (as in the example) and there will be no difficulties with computing the derivatives in (a) and (b). If $p_n > 0$ for *infinitely* many values n, then h'_X and h''_X involve infinite series, whose convergence is tied to the existence of $E[X]$ and $Var[X]$.

Exercise 3-46 Let X be the number of girls in the 3 child, equally likely gender, experiment, with probability distribution table,

i	$p_i = \Pr(X{=}i)$
0	1/8
1	3/8
2	3/8
3	1/8

(a) Show $h_X(s) = \left(\frac{1}{2} + \frac{1}{2}s\right)^3$ (expand using the binomial theorem).

(b) Calculate the mean and the variance of X by differentiating the probability generating function. Compare your results to the previous exercises in Subsection 3.4.2 and Section 3.6.

Exercise 3-47 Consider a random variable B with probability generating function $h_B(s) = (.3+.7s)^5$.

(a) Find $E[B]$.

(b) Find σ_B.

(c) Find the probability distribution for B.

Exercise 3-48 Consider a random variable G with probability generating function $h_G(t) = \frac{.8t}{1-.2t}$.

(a) Find $E[G]$.

(b) Find the coefficient of variation for G.

3.8 CHAPTER 3 SAMPLE EXAMINATION

1. Imagine rolling a pair of fair 5-sided dice with sides numbered 1-5. Let S be the random variable that represents the sum of the two dice.

   (a) Write down the probability distribution $p(s)$ and cumulative distribution function $F(S) = \Pr(S \le s)$.

   (b) Calculate the expected sum of the two dice, $E[S]$.

   (c) Calculate the variance $Var[S]$ and standard deviation σ_S for the sum.

   (d) Calculate the standard deviation for the sum given that the sum is less than 8.

2. Flip a fair coin until you get the first "head". Let X represent the number of flips before the first head appears. Calculate $E[X]$. **Note**: This is a ***geometric distribution*** that we will study in Section 4.3.

3. The number of cheeseburgers that a group of friends can eat is as follows: 7, 8, 4, 6, 9, 10, 2, 7, 8, 9, 3, 6, 9, 8, 4, 9, 6, 5, 9, 8, 7, 9, 0, and 3.

   (a) Find the mean, median, mode, and midrange.
   (b) Find the minimum and maximum.
   (c) Find the 80th percentile.
   (d) Find the variance and standard deviation.
   (e) Find the coefficient of variation.
   (f) Find the z-score of someone who could eat 11 cheeseburgers.

4. Suppose that the random variable X represents the value of a share of stock after one year. Consider the following probability distribution:

x	$p(x)$
0	.74
50	.12
100	.09
200	☺

(a) Calculate the expected value $E[X]$.

(b) Calculate the variance $Var[X]$ and standard deviation σ_X.

(c) If the stock cost \$30.00, should you buy a share?

(d) Calculate the expected gain of ten shares of stock that cost \$30 each, that is, $E[10X-300]$. *Hint*: Create a probability distribution for the value of your profit, $10X-300$, after one year.

(e) Calculate the variance of the gain for ten shares, that is, $Var[10X-300]$.

(f) Calculate the coefficient of variation for X.

5. **SOA/CAS Course 1 November 2000 #18**
Due to decreasing business, the amount an insurer expects to pay for claims will decrease at a constant rate of 5% per month indefinitely. This month the insurer paid 1000 in claims. What is the insurer's total expected amount of claims to be paid over the 30-month period that began this month?

(A) 13,922 (B) 14,707 (C) 14,922 (D) 15,707 (E) 15,922

6. Roll a pair of fair six-sided dice. *Hint*: You may wish to list the sample space.

(a) Calculate the conditional expected sum given that the sum is prime.

(b) Calculate the standard deviation of the sum given that the sum is prime.

7. Some people believe that baseball players get *hot*. One potential mathematical model is if a player had a hit in his last at-bat, then he will hit .300 in his next at-bat. If he failed to hit, then he will hit .250 in his next at-bat. Suppose that a player has 3 at-bats in a game and is coming off of a hit in his last at-bat in the previous game.

(a) Find the expected number of <u>hits</u> the player gets in the game.

(b) Find the standard deviation for the number of hits this player gets in the game.

8. The ages, X, of students enrolled in *Probability and Statistics* has cumulative probability distribution given by:

x	17	18	19	20	21	22	23
$F(x)$	0	.23	.48	.89	.94	.99	1.00

(a) Compute the mean, median, mode, and midrange for X.

(b) Compute the range, variance, and standard deviation for X.

(c) Given that a student is in his 20's, compute his expected age.

9. Suppose that a random variable X has probability generating function:

$$h_X(t) = .10 + .18t + .12t^2 + .37t^5 + .23t^8$$

(a) Find the mean of the random variable X.

(b) Find the standard deviation of the random variable X.

(c) What is the value of the third quartile?

10. The expected amount of damage done to a motorcycle is \$700 and has the following probability distribution.

D	**Pr(*D*)**
\$0	.73
\$2500	
\$5000	

(a) Find the conditional expected damage, given that there is positive damage.

(b) Compute the standard deviation of the damage, given that there is positive damage.

11. A survey has found that the number of televisions per household has cumulative probability distribution given by:

x	0	1	2	3	4	5
$F(x)$	0	.09	.41	.83	.96	1.0

(a) Compute the mean, median, and standard deviation.

(b) A local television station is running a promotion for families with more than two televisions in their home. The station pays \$75 if a family has 3 televisions, \$150 if a family has 4 televisions and \$225 if a family has 5 television sets. Given that a family has at least three televisions, compute their expected payment.

12. The lollipops from a sample of twelve bags of Charms Blow Pops™ were emptied onto a table and counted. The number of lollipops contained in each bag were: 15, 12, 13, 18, 14, 15, 15, 16, 14, 13, 15, and 13.

    (a) Compute the mean, median, mode, and midrange for the number of lollipops in a bag of Charms Blow Pops™.

    (b) Compute the inter-quartile range.

    (c) Little Susie opens her bag of Blow Pops™ and is upset to find only 10 lollipops. She believes that she has been cheated and wants to see if this bag is an outlier (three standard deviations or more below the mean). Was her bag of lollipops an outlier?

13. The amount of scholarship merit money awarded to incoming freshmen graphic-design majors has probability distribution given by:

Amount	0	500	1000	2500	3750	5000
Pr(*Amount*)	.09	☺	.15	.09	.05	.01

Calculate the mean and variance for the amount of merit money awarded to freshmen.

14. A local sorority is sponsoring a lottery. Each ticket has a value of W, a random variable taking the values 0, 10, or 100 dollars. The probability of winning \$100 is 10% and the variance of W is 73 times as large as the mean.

    (a) Find the probability distribution, the mean, and the standard deviation for W.

    (b) Find the expected value of a ticket given that it is a winner (either \$10 or \$100).

    (c) Find the coefficient of variation for W.

15. The random variable that models the return on a block of business (in millions) is denoted by R. Your boss wants to know the coefficient of variation for R. You lost the probability distribution, but remember the probability generating function for R to be $h_R(s) = (.8+.2s)^3$. Save your job.

16. **SOA EXAM P Sample Exam Questions #126**
Under an insurance policy, a maximum of five claims may be filed per year by a policyholder. Let p_n be the probability that a policyholder files n claims during a given year, where $n = 0,1,2,3,4,5$. An actuary makes the following observations:

(i) $p_n \geq p_{n+1}$ for $n = 0,1,2,3,4$.
(ii) The difference between p_n and p_{n+1} is the same for $n = 0,1,2,3,4$.
(iii) Exactly 40% of the policyholders file fewer than two claims during a given year.

Calculate the probability that a random policyholder will file more than three claims during a given year.

(A) 0.14 (B) 0.16 (C) 0.27 (D) 0.29 (E) 0.33

CHAPTER 4

SOME DISCRETE DISTRIBUTIONS

There are several named discrete probability distributions that are common enough to merit special attention. We have used many of these distributions as examples in previous chapters. Our goal is to classify these common probability distribution functions. We will then derive general formulas for the mean and the variance of random variables with these distributions. While it is important to learn many of the definitions and properties that follow, we encourage you to also concentrate on the techniques that we use to justify these properties. In particular, we will make extensive use of the methods discussed at the end of Chapter 3, particularly sums of random variables and probability generating functions.

4.1 THE DISCRETE UNIFORM DISTRIBUTION

The discrete uniform random variable is designed to model experiments such as those in Chapter 1, in which there are a finite number, n, of equally likely outcomes $\{x_1, x_2, ..., x_n\}$. We assign the probability $\frac{1}{n}$ to each of the elements in the sample space. Namely, $Pr(X{=}x_i) = \frac{1}{n}$ for $i = 1, 2, ..., n$. Here is an example.

Example 4.1-1 Discrete Uniform on 1 Through 10

Select a number at random (meaning that the outcomes are equally likely) from 1, 2, 3, 4, 5, 6, 7, 8, 9, and 10. Let X be the discrete random variable that denotes the number selected. Specifically, $Pr(X{=}x) = p(x) = \frac{1}{10}$ for $x = 1, 2, ..., 10$. Find the following:

(a) $Pr[X > 3]$,

(b) $E(X)$, and

(c) $Var(X)$.

Solution

(a) $Pr[X > 3] = Pr[X{=}4, \text{ or } X{=}5, \text{ or}..., X{=}10] = \frac{7}{10} = 1 - Pr[X \leq 3]$.

(b) $$E(X) = 1 \cdot \frac{1}{10} + 2 \cdot \frac{1}{10} + 3 \cdot \frac{1}{10} + \cdots + 10 \cdot \frac{1}{10}$$
$$= \frac{1}{10}(1 + 2 + 3 + \cdots + 10) = \frac{55}{10} = 5.5.$$

(c) $E[X^2] = 1^2 \cdot \frac{1}{10} + 2^2 \cdot \frac{1}{10} + 3^2 \cdot \frac{1}{10} + \cdots + 10^2 \cdot \frac{1}{10} = \frac{385}{10}$.

$$Var(X) = E[X^2] - (E[X])^2 = \frac{1}{10}(385) - 5.5^2 = 8.25.$$

□

To facilitate these sorts of calculations in general we next list a number of useful summation formulas for both finite sums and infinite series.

Finite Summation Formulas

(a) ***multiplication:*** $\sum_{i=1}^{n} a = \underbrace{a + a + \ldots + a}_{n \text{ times}} = n \cdot a,$

(b) ***arithmetic series:*** $\sum_{i=1}^{n} i = 1 + 2 + 3 + \cdots + n = \frac{n(n+1)}{2},$

(c) ***sums of squares:*** $\sum_{i=1}^{n} i^2 = \frac{n(n+1)(2n+1)}{6},$

(d) ***finite geometric series:***

$$\sum_{n=0}^{N} ax^n = a + ax + ax^2 + ax^3 + \ldots + ax^N = \frac{a(1-x^{N+1})}{1-x}, \text{ for any } x \neq 1.$$

Infinite Series

(e) ***infinite geometric series:***

$$\sum_{n=0}^{\infty} ax^n = a + ax + ax^2 + ax^3 + \cdots = \frac{a}{1-x} \text{ for } |x| < 1,$$

(f) ***exponential power series:***

$$\sum_{n=0}^{\infty} \frac{x^n}{n!} = 1 + \frac{x}{1!} + \frac{x^2}{2!} + \frac{x^3}{3!} + \cdots = e^x \text{ for all } x.$$

These formulas should be familiar from prior algebra and calculus classes.

The geometric series, in both the finite (d) and infinite (e) versions, is a particularly important tool. Recall that the terms in a geometric series consist of the increasing powers of a common factor, which is denoted by x in the above formulas. The x in the formula is often called the ***ratio*** of the series, because the ratio of consecutive terms is always equal to x.

A useful mnemonic device for remembering the summation formula for a geometric series is $\frac{first - next}{1 - ratio}$. *First* refers to the starting term of the series. *Next*, in the case of a finite geometric series, refers to the next term, ax^{N+1}, that would follow the last term in the series. In the case of an infinite series the *next* term would be the term at infinity, that is to say "x^∞," where $x^\infty = \lim_{N\to\infty} x^{N+1} = 0$. This is zero since $|x|$ must be less than one for convergence in the infinite case.

We provide the basic results for a general discrete uniform random variable next.

Discrete Uniform Distribution

A random variable X is said to have a ***discrete uniform distribution*** if its probability function is $\Pr(X = x) = p(x) = \frac{1}{n}$ for $x = 1, 2, \ldots, n$.

Mean and Variance of the Discrete Uniform Random Variable

Suppose that the random variable X has a ***discrete uniform distribution***.

(a) $E[X] = \frac{n+1}{2}$, and

(b) $Var[X] = \frac{n^2-1}{12}$.

Proof

(a) You are asked to verify $E[X] = \frac{n+1}{2}$ in Exercise 4-2.

(b)
$$Var[X] \overset{\text{variance formula}}{=} E[X^2] - (E[X])^2$$
$$= 1^2 \cdot \frac{1}{n} + 2^2 \cdot \frac{1}{n} + 3^2 \cdot \frac{1}{n} + \cdots + n^2 \cdot \frac{1}{n} - \left(\frac{n+1}{2}\right)^2 \quad \text{Use part a) for } E[X].$$
$$= \frac{1}{n}(1^2 + 2^2 + \cdots + n^2) - \left(\frac{n+1}{2}\right)^2$$
$$= \frac{1}{n}\left(\frac{n \cdot (n+1) \cdot (2n+1)}{6}\right) - \left(\frac{n+1}{2}\right)^2 \quad \text{Use sum of squares formula}$$
$$= (n+1)\left(\frac{2n+1}{6} - \frac{n+1}{4}\right) = (n+1)\left(\frac{(4n+2)-(3n+3)}{12}\right) = \frac{n^2-1}{12}.$$

Exercise 4-1 Use the principal of mathematical induction to prove the relationships for parts (a) through (d) in the finite summation formulas.

Exercise 4-2 Prove $E[X] = \frac{n+1}{2}$ for the discrete uniform distribution.

Exercise 4-3 Which of the following options would you prefer, under the assumption that more is better:

(a) One billion grains of rice, or

(b) One grain of rice on the first square of a checker board (64 squares total), 2 grains on the second square, 4 grains on the third square, 8 grain on the fourth square, and so forth, until all squares of the board have their requisite number of grains of rice?

Exercise 4-4 Suppose that the random variable X has the discrete uniform distribution for $x = 1, 2, ..., n$. Derive a formula for $E[X^3]$. This is called the ***third moment*** of X.

Exercise 4-5 Suppose that the random variable X has the discrete uniform distribution for $x = 1, 2, ..., n$. Derive a formula for ***skewness*** for discrete uniform distributions, where,

$$Skew[X] \overbrace{=}^{\text{definition}} E\left[(X-\mu)^3\right].$$

Exercise 4-6 Roll a single fair six-sided die. Let the random variable X denote the number of dots that are on the top side. Find

(a) $\Pr[X \geq 5]$.

(b) The mean.

(c) The median.

(d) The standard deviation for X.

Exercise 4-7 Let the random variable X denote the natural number selected at random from one to one hundred. Find

(a) $\Pr[68 < X \leq 85]$.

(b) The expected value of X.

(c) The mode of X.

(d) $Var(X)$.

Exercise 4-8 Select a number Y at random from $0, 1, 2, ..., 8$. Find

(a) $\Pr[Y \geq 5]$.

(b) $E[Y]$.

(c) The coefficient of variation for Y.

Note: $Y = X - 1$, where X has the discrete uniform distribution on $1, 2, 3, ..., 9$. Use the transformation relationship in conjunction with properties learned about the discrete uniform distribution.

Exercise 4-9 Select a number Z at random from $2, 4, 6, ..., 18$. Find

(a) $\Pr[Z \geq 5]$.

(b) The mean of the random variable Z.

(c) The standard deviation of Z.

Note: This time, $Z = 2X$ where X has the discrete uniform distribution on $1, 2, 3, ..., 9$.

Exercise 4-10 Select a number W at random from $50, 55, 60, 65, 70$. Find

(a) $\Pr[W \geq 55]$.

(b) $E[W]$.

(c) σ_W.

Hint: $W = 5X + 45$, where X has the discrete uniform distribution on $1, 2, 3, 4, 5$.

Exercise 4-11 If X has discrete uniform distribution for $x = 1, 2, ..., n$, then what general statement(s) can you make about the linear transformation $Y = aX + b$? Here a and b are real-valued constants.

Exercise 4-12 There are 21 students in class; each assigned an integer at random from 1 to 21.

(a) What is the probability that a student's number is prime?

(b) What is the probability that the student's number is larger than his expected number? That is, find $\Pr(X > E[X])$.

4.2 BERNOULLI TRIALS AND THE BINOMIAL DISTRIBUTION

A ***Bernoulli trial*** is an experiment that has two outcomes (for example, true-false; girl-boy; make-miss; on-off; 1-0; success-failure; yes-no). In order to classify all of these dichotomous situations as a single equivalent probability experiment, we use a generic discrete random variable, X, to represent it. It is traditional to adopt the convention that the value $X = 1$ is associated with ***success*** and the value $X = 0$ is associated with ***failure***. The assignment of the words *success* and *failure* to the two possible outcomes of the experiment is completely arbitrary.

Bernoulli Random Variable

Suppose that the random variable X has probability function given by $\Pr[X{=}1] = p$ and $\Pr[X{=}0] = q = 1 - p$. Then X is called a ***Bernoulli random variable*** with probability of success p.

Mean and Variance of a Bernoulli Random Variable

Suppose that X is a ***Bernoulli random variable*** with probability of success p. Then the mean and variance of the random variable are:

(a) $E[X] = p$, and

(b) $Var[X] = p \cdot q = p \cdot (1 - p)$.

Proof

The probability distribution and calculations are presented below in tabular form:

x	$p(x) = \Pr[X = x]$	$x \cdot p(x)$	$x^2 \cdot p(x)$
0	q	0	0
1	p	p	p
Totals	1	p	p

(a) $E[X] = 1 \cdot p + 0 \cdot (1-p) = p$, and

(b) $Var[X] = E[X^2] - E[X]^2 = p - p^2 = p \cdot (1-p) = p \cdot q$.

One of the most common discrete distributions is the ***binomial distribution***. It is the distribution that counts the number of successes in a series of n independent repetitions of the same underlying success/failure experiment. Each independent repetition, called a ***trial***, is described by a single Bernoulli random variable, with 1 representing success and 0 representing failure.

Consider then an underlying experiment that consists of n independent Bernoulli trials, each with the same probability of success p. By *independent* we mean, in the sense of Section 3.6, that if X_i and X_j $(i \neq j)$ are the Bernoulli random variables for the ith and j^{th} trials respectively, then X_i and X_j are independent as random variables. We let

$$Y = X_1 + \cdots + X_n.$$

Since each summand equals 1 if and only if the corresponding trial is a success, and 0 otherwise, the random variable Y counts the **total number of successes** in the n trials. Therefore Y takes the values $0, 1, 2, \ldots, n$.

Since expected value is a linear operation, we can express the expectation of Y in terms of the individual X_i's as follows:

$$\begin{aligned} E[Y] &= E[X_1 + X_2 + \cdots + X_n] \\ &= E[X_1] + E[X_2] + \cdots + E[X_n] \\ &= \underbrace{p + \cdots + p}_{n \text{ terms}} = n \cdot p. \end{aligned}$$

Since the summands are <u>independent</u> and $Var[X_i] = pq$, we see that

$$\begin{aligned} Var[Y] = Var\left[\sum_{i=1}^{n} X_i\right] &= Var[X_1 + X_2 + \cdots + X_n] \\ &= Var[X_1] + Var[X_2] + \cdots + Var[X_n] \quad \text{independence is used here.} \\ &= pq + pq + \cdots + pq = n \cdot p \cdot q. \end{aligned}$$

We have found the mean and variance for Y, but it remains to work out the probability function of Y. For this purpose, identify the outcomes of the experiment (n repeated Bernoulli trials) as all possible n-letter words using the letters S (success) and F (failure). Let y be a value of Y. That is, y is an integer between 0 and n. The event $\{Y = y\}$ consists of all of the *words* with exactly y successes (S's) and $(n - y)$ failures (F's). As we saw in Section 1.4.1, there are exactly ${}_nC_y = \binom{n}{y}$ such words (combinations), since we are choosing y positions out of the n possible positions for the successes.

What is the probability of such a word? Since the trials are independent, the probability of a word is the product of the probabilities of the individual letters that form this word. If there are y successes and $(n - y)$ failures, the probability of each specific word is $p^y q^{n-y}$. Since there are exactly $\binom{n}{y}$ such outcomes, each with probability $p^y q^{n-y}$, we have,

$$p(y) = \Pr(Y = y) = {}_nC_y \cdot p^y q^{n-y} = \binom{n}{y} \cdot p^y q^{n-y}.$$

We summarize our results as follows:

Binomial Random Variable

Let Y be the number of successes in n independent repetitions of a Bernoulli trial with probability of success p.

The random variable Y has probability function given by $\Pr(Y{=}y) = p(y) = \binom{n}{y} \cdot p^y \cdot q^{n-y}$ for $y = 0,1,2,\ldots,n$ and $0 \le p \le 1$.

The random variable Y is called a ***binomial random variable*** with ***parameters*** n and p.

Mean and Variance of a Binomial Random Variable

Suppose that Y is a ***binomial random variable*** with probability of success p and n trials. Then

(a) $\mu_Y = E[Y] = n \cdot p$, and

(b) $\sigma_Y^2 = Var[Y] = n \cdot p \cdot q = n \cdot p \cdot (1-p)$.

Example 4.2-1 Binomial Random Variable

Suppose that three people are selected at random for a focus group on sports and society. Let M denote the number of men in this three person sample. Assume that men constitute exactly 50.9% of the population. Calculate the probability distribution for M, the expected number of men surveyed, and the variance for the number of men surveyed.

Solution

For this problem we have $n=3$ and $p=.509$. Therefore the probability distribution is:

$$\Pr(M=0) = p(0) = {}_3C_0 \cdot .509^0 \cdot .491^3 = .118370771$$

$$\Pr(M=1) = p(1) = {}_3C_1 \cdot .509^1 \cdot .491^2 = .368130687$$

$$\Pr(M=2) = p(2) = {}_3C_2 \cdot .509^2 \cdot .491^1 = .381626313$$

$$\Pr(M=3) = p(3) = {}_3C_3 \cdot .509^3 \cdot .491^0 = .131872229$$

The mean is $\mu_M = E[M] = n \cdot p = 3 \cdot (.509) = 1.527$. The variance of the random variable M is $\sigma_M^2 = Var[M] = 3 \cdot (.509) \cdot (.491) = .749757$.

Exercise 4-13 Use the *probability distribution* alone (not the formulas) from Example 4.2-1 to verify that the mean is $\mu = 1.527$ and the variance is $\sigma^2 = .749757$ for a binomial random variable with $n=3$ and $p=.509$.

Exercise 4-14 Kobe Bryant is an eighty percent free-throw shooter. In a game against the Portland Trailblazers, Kobe went to the foul line twelve times. Let M denote the number of free-throws that Kobe makes. Assume his shots are independent. Calculate the probability distribution, $E[M]$, and $Var[M]$. What is the probability that Kobe misses more than two shots?

Exercise 4-15 Pi Dumbo is not very bright, but he makes up for it by playing video games all day. He sits for a 20-question multiple choice examination, each question having five possible answers. He fills in the answers on the bubble sheet randomly without bothering to read the questions. Let Y denote the number of questions that Pi Dumbo answers correctly. Find $\Pr[Y=10]$, $\Pr[Y \geq 2]$, $E[Y]$, and $Var[Y]$.

Exercise 4-16 The New York Yankees win 60% of their games. They are playing a three-game series against the Seattle Mariners. Find the probability that they win the series.

Note: We are aware that these games are not truly independent (pitching match-ups, injuries, quality of the opponent, home-field advantage), but we will use the binomial distribution as a reasonable ***model*** for what will actually happen. If our model does not accurately reflect what happens over time, then we will adjust our model and/or our assumptions. If you become adept at fitting models to data and making accurate predictions, then you will have a lucrative career in the gaming industry, or as an actuary in the insurance industry[1] two disparate businesses that have a lot in common mathematically.

---

[1] Or, as a sports statistician. Billy Beane, a baseball general manager and main subject for the book *Moneyball*, said, "At some point what's really going to happen is we are all going to employ *actuaries*, like insurance companies."

Exercise 4-17 Let the binomial random variable X have $n=3$ and $p=0.9$. Calculate $\Pr[X=1]$, $\Pr[2\leq X]$, $E[X]$, and $Var[X]$.

Exercise 4-18 Let the binomial random variable X have $n=9$ and $p=.4$. Calculate $\Pr[X=2]$, $\Pr[X>2]$, $E[X]$, and $Var[X]$.

Exercise 4-19 Let the binomial random variable X have $n=20$ and $p=.25$. Give a description of a real-life situation that the random variable would model. Calculate $\Pr[X=2]$, $\mu=E[X]$, $\Pr[X<\mu]$, and σ_X.

Exercise 4-20 SOA EXAM P Sample Exam Questions #28

A hospital receives 1/5 of its flu vaccine shipments from Company *X* and the remainder of its shipments from other companies. Each shipment contains a very large number of vaccine vials. For Company *X*'s shipments, 10% of the vials are ineffective. For every other company, 2% of the vials are ineffective. The hospital tests 30 randomly selected vials from a shipment and finds that one vial is ineffective. What is the probability that this shipment came from Company *X*?

(A) 0.10 (B) 0.14 (C) 0.37 (D) 0.63 (E) 0.86

Exercise 4-21 SOA/CAS Course 1 2000 Sample Examination #40

A small commuter plane has 30 seats. The probability that any particular passenger will not show up for a flight is 0.10, independent of other passengers. The airline sells 32 tickets for the flight. Calculate the probability that more passengers show up for the flight than there are seats available.

(A) 0.0042 (B) 0.0343 (C) 0.0382 (D) 0.1221 (E) 0.1564

Exercise 4-22 SOA EXAM P Sample Exam Questions #39

A company prices its hurricane insurance using the following assumptions.

(i) In any calendar year, there can be at most one hurricane.

(ii) In any calendar year, the probability of a hurricane is 0.05.

(iii) The number of hurricanes in any calendar year is independent of the number of hurricanes in any other calendar year.

Using the company's assumptions, calculate the probability that there are fewer than 3 hurricanes in a 20-year period.

(A) 0.06 (B) 0.19 (C) 0.38 (D) 0.62 (E) 0.92

Exercise 4-23 Consider a binomial random variable X with $n=10$ and $Var[X]=(.25)E[X]$.

(a) Find the probability of success p.
(b) Find the probability that $X=7$.

Exercise 4-24 Suppose that Alice is a 65% free-throw shooter and takes $n=5$ shot attempts. Let X denote the number of free-throws made. Find $\Pr(X > E[X])$.

Exercise 4-25 Provide an alternative proof of the formulas for the mean and the variance of a binomial random variable using the probability generating function method of Section 3.7.

Methodology: If Y is binomial with parameters n and p then,

$$h_Y(s) = \sum_{y=0}^{n} \Pr[Y=y]\cdot s^y = \sum_{y=0}^{n} ({}_nC_y p^y q^{n-y})s^y = \sum_{y=0}^{n} {}_nC_y (ps)^y q^{n-y}.$$

Use the binomial theorem to simplify the last sum. Use calculus and properties of probability generating functions to determine the mean and variance of the binomial distribution.

Exercise 4-26 CAS/SOA P Sample Exam Questions #31
A company establishes a fund of 120 from which to pay an amount, C, to any of its 20 employees who achieve a high performance level during the coming year. Each employee has a 2% chance of achieving a high performance level during the coming year, independent of any other employee.

Determine the maximum value of C for which the probability is less than 1% that the fund will be inadequate to cover all payments for high performance.

(A) 24 (B) 30 (C) 40 (D) 60 (E) 120

Exercise 4-27 SOA EXAM P Sample Exam Questions #96
A tour operator has a bus that can accommodate 20 tourists. The operator knows that tourists may not show up, so he sells 21 tickets. The probability that an individual will not show up is 0.02, independent of all other tourists. Each ticket costs 50, and is non-refundable if a tourist fails to show up. If a tourist shows up and a seat is not available, the tour operator has to pay 100 (ticket cost $+50$ penalty) to the tourist. What is the expected revenue of the tour operator?

(A) 935 (B) 950 (C) 967 (D) 976 (E) 985

4.3 THE GEOMETRIC DISTRIBUTION

In the previous section we focused on a particular number, n, of repeated independent Bernoulli trials. In this section we allow the individual trials to go on indefinitely, and turn our attention to a different random variable, namely, the *number of failures prior to the first success.*

To illustrate, imagine you are in a gymnasium shooting hoops. Being a persevering sort you are determined to make a three-point shot before leaving the gymnasium floor.

Example 4.3-1 Geometric Random Variable

Suppose that you make 60% of your three-point shots in practice and that your shots constitute independent Bernoulli trials. You shoot until you make a three-point shot. Let X denote the number of misses prior to your first success (made shot). Construct the probability distribution function for X.

Solution

Consider the useful probability tree diagram.

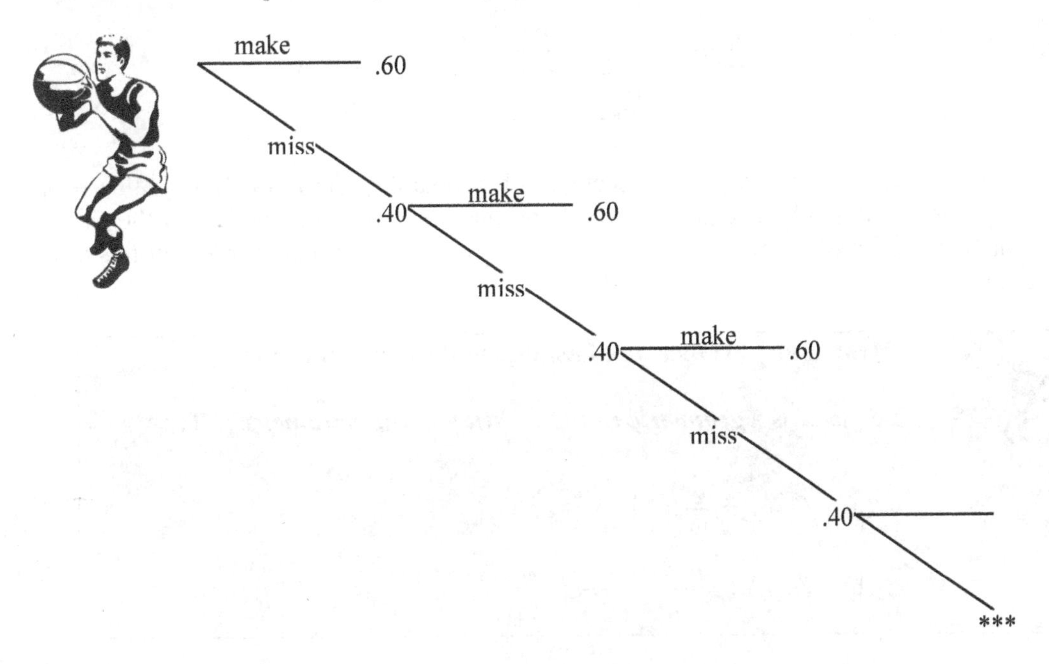

The tree diagram continues indefinitely. The resulting probabilities are:

X	0	1	2	3	4	⋯	k	⋯
Pr(X)	.6	$.6\times.4=.24$	$.6\times.4^2=.096$	$.6\times.4^3$	$.6\times.4^4$	⋯	$(.6)\times(.4)^k$	⋯

In general, we assume a series of repeated, independent, Bernoulli trials each with probability of success of p. We let X equal the *number of failures before the first success.* Then X can assume **any** integer value, starting with zero. As with the binomial random variable, the outcomes of the underlying experiment can be considered as words using the letters F and S. For the geometric random variable, the event $\{X = k\}$ consists of the single word $\{\underbrace{F \cdots F}_{k\ F\text{'s}} \cdot S\}$. The first several words are $\{S\}, \{FS\}, \{FFS\}$, and so forth. It follows from the independence of the trials that,

$$\Pr(X = k) = p \cdot (1-p)^k = p \cdot q^k \text{ for } k = 0,1,2,\ldots \text{ (refer to example 4.3-1)}$$

Geometric Random Variable

Consider a series of independent Bernoulli trials with probability of success p and let the random variable X be the number of *failures* before the first *success*. The random variable X has probability function given by

$$\Pr(X = k) = p \cdot (1-p)^k = p \cdot q^k \text{ for } k = 0,1,2,\ldots,\ q = 1-p \text{ and } 0 < p < 1,$$

and X is called a ***geometric random variable with parameter*** p.

Note:

$$\sum_{k=0}^{\infty} pq^k \underbrace{=}_{\substack{\text{geometric}\\\text{series}}} \frac{p}{1-q} = \frac{p}{p} = 1,$$

so that the *first success* will eventually occur (with probability one) at a finite value of k. Depending upon p, this might require a very large number of repetitions, but as the saying goes: if at first you don't succeed, try, try again. Perseverance will be rewarded, in this case, with certainty.

Mean and Variance of a Geometric Random Variable

Suppose that X is a ***geometric random variable with parameter*** p. Then

(a) $E[X] = \dfrac{q}{p} = \dfrac{1-p}{p}$, and

(b) $Var[X] = \dfrac{q}{p^2} = \dfrac{1-p}{p^2}$

We will compute the mean and variance for a geometric random variable using the ***probability generating function*** technique of section 3.7. Why? Because it is relatively simple to determine the probability generating function for a geometrically distributed random variable (and therefore the mean and variance) and it is relatively difficult to evaluate directly sums like,

$$E[X] = 0 \cdot p + 1 \cdot q \cdot p + 2 \cdot q^2 \cdot p + \cdots .$$

Since $\Pr(X=k) = p \cdot q^k$, we see that the probability generating function is

$$h(s) = \underbrace{\sum_{k=0}^{\infty} p_k s^k}_{\text{def'n of prob generating ftn}} = \sum_{k=0}^{\infty}\left(p \cdot q^k\right)s^k = \sum_{k=0}^{\infty} p(q^k s^k) = \sum_{k=0}^{\infty} p(q \cdot s)^k \underbrace{=}_{\text{geometric series formula}} \frac{p}{(1-qs)} = p(1-qs)^{-1},$$

$$h'(s) = (-1)p(1-qs)^{-2}(-q) = pq(1-qs)^{-2},$$

so

$$E[X] = h'(1) = pq(1-q)^{-2} = pq(p)^{-2} = \frac{q}{p}.$$

$$h''(s) = (-2)pq(1-qs)^{-3}(-q) = 2pq^2(1-qs)^{-3},$$

therefore

$$h''(1) = 2pq^2(p)^{-3} = 2\frac{q^2}{p^2}.$$

$$Var[X] = h''(1) + h'(1) - h'(1)^2 = 2\frac{q^2}{p^2} + \frac{q}{p} - \frac{q^2}{p^2} = \frac{q^2}{p^2} + \frac{q}{p}$$
$$= \frac{q^2 + pq}{p^2} = \frac{q(q+p)}{p^2} = \frac{q}{p^2}.$$

Example 4.3-2 Geometric Mean and Variance

Suppose that you are a 60% three-point shooter in practice. Shoot until you make a three-point shot. Let X denote the number of misses prior to your first make. Calculate the expected value and standard deviation for X.

Solution

$p = .6$, therefore, $E[X] = \frac{q}{p} = \frac{.4}{.6} = \frac{2}{3}$, $Var[X] = \frac{.4}{(.6)^2} = \frac{10}{9}$ and $\sigma_X = \frac{\sqrt{10}}{3}$. ❐

Example 4.3-3 The St. Petersburg Paradox

A seemingly foolproof gambling strategy is to double your bet after each loss until you ultimately win. To illustrate, suppose you can wager any amount A as often as you like on the flip of a fair coin. If the toss is heads you win A, if tails you lose A. Your modest goal is to experience a net gain of A with certainty. Is this possible?

Solution

For simplicity we let $A = \$1$. You begin by betting \$1 on the first toss. If you win, you go home happy and up a dollar. If you lose, you double your bet so that \$2 is wagered on the outcome of a second toss of the coin. If the second toss is heads then your net gain (taking into account your loss on the first toss) is $(\$2-\$1)=\$1,$ so you quit and go home up a dollar. If the second toss is tails, again then you double your prior wager of \$2 to \$4. This time, if you win, your net gain is $\$4-(\$1+\$2)-\$1,$ so that again you can quit ahead by \$1.

Using random variables, we let X be the number of tails (failures) before the first head appears. Then X has the geometric distribution with probability of success $p=\frac{1}{2}$. Let a second random variable W represent the amount wagered on the winning toss. Then W is a transformation of the random variable representing the first success, namely $W=2^X$. The following tables summarize the situation:

$X=k$	$k+1$ = # of Coin Toss	$\Pr(X=k)$	Wager, W	Net Proceeds
0	1	$\frac{1}{2^1}=\frac{1}{2}$	$2^0=1$	1
1	2	$\frac{1}{2^2}=\frac{1}{4}$	$2^1=2$	$1=(2-1)$
2	3	$\frac{1}{2^3}=\frac{1}{8}$	4	$1=(4-2-1)$
...	...	...	...	...
k	$k+1$	$\frac{1}{2}\cdot\left(\frac{1}{2}\right)^k=\left(\frac{1}{2}\right)^{k+1}$	2^k	1
...	...	...	...	...

Since it is certain that heads will eventually occur, you are guaranteed to win \$1. That is, $E[\text{net proceeds}] = (1)\left(\frac{1}{2}\right)+(1)\left(\frac{1}{2}\right)^2+(1)\left(\frac{1}{2}\right)^3+\cdots = 1.$ Next, we attempt to calculate the expected value of the amount of the winning wager.

$$E[W] = E[2^X] = \sum_{k=0}^{\infty}\left(2^k\right)\left(\frac{1}{2}\right)^{k+1} = \sum_{k=0}^{\infty}\left(\frac{1}{2}\right) = \infty.$$

The moral of the story is that you are assured of winning \$1 eventually, but you had better bring along a credit card with no limit to cover the expected amount wagered. ❒

Exercise 4-28 Eighty percent of an eagle's offspring are able to fly when they are pushed out of the nest for the first time. Let D denote the number of chicks that fail to fly prior to a chick having a successful flight. Calculate $\Pr[D<2]$, $E[D]$, and $E[D^2]$. Note: Such failure to fly would be considered by an Eaglet Insurer to be "bad experience."

Exercise 4-29 Ninety percent of your advances in a pub are rebuffed. Let R denote the number of times you are rejected before someone shows interest in you. Calculate the probability that you experience less than two rejections, your expected number of rejections, and σ_R. What assumptions are you making?

Exercise 4-30 The Society of Actuaries (SOA) passes 40% of the students who sit for their SOA Exam P. Assume sophomore Macon Love has a 40% chance of passing Exam P each time he sits for the exam. What is the expected number of attempts Macon must make prior to passing the examination?

Exercise 4-31 **(Geometric Distribution is *Memory-less*[2]):** Suppose that X is a geometric random variable with parameter (probability of success) p and let a and b be non-negative integers.

(a) Show that $\Pr(X \geq a+b \mid X \geq a) = \Pr(X \geq b)$.

(b) Show $E[X \mid X \geq a] = a + E[X]$.

Exercise 4-32 Given a geometric random variable X with standard deviation $\sigma_X = \frac{\sqrt{15}}{2}$, find $\Pr[(X < E[X])]$.

Exercise 4-33 Can there exist two distinct geometric random variables with the same variance (but different means)? Why or why not?

Exercise 4-34 You are given a geometric random variable X with the property that $\Pr(X=3) = 5 \cdot \Pr(X=6)$. Find $E[X]$.

Differences in notation and inconsistent definitions for a single idea are two of the more frustrating aspects of mathematics. Do you recall all of the ways to write the derivative,

$$f'(x) = \frac{dy}{dx} = D_x f(x) = y'$$

or the ways to write the complement of a set?

Some authors and textbooks define the geometric random variable slightly differently than the number of failures, X, *prior* to the first success. Their geometric random variable Y denotes the number of the trial *upon which* the first success occurs.

Note that always, $Y = X + 1$.

---

[2] In simple terms, the ***memory-less*** property implies what has happened in the past (misses/makes) has no effect on what will occur in the future. You will have the same mean, variance, and probabilities going forward. That is, if you are a 20% free-throw shooter and will shoot until you make a shot, then you expect to shoot 5 more times no matter how many misses have occurred previously. For example, even if you have already missed 8 shots, you still expect to shoot 5 more times.

Exercise 4-35 Define the probability distribution for the random variable Y that represents the trial upon which the first success occurs. Be sure to state the range of values Y may take, and write a general expression for $\Pr(Y = y)$.

Exercise 4-36 Let Y denote the number of the trial *upon which* the first success occurs, so that $Y = X + 1$, where X is a geometric random variable.

(a) Derive formulas for the expected value and variance of the random variable Y.

(b) Let a be a positive integer. Show that $E[Y \mid Y \geq a] = (a-1) + E[Y]$.

Exercise 4-37 Suppose that your bank roll is \$75.00 and your goal is to win \$5. Your strategy is to spin the roulette wheel and wager \$5 on black (18 red numbers, 18 black numbers, and 2 green *house* numbers). If you lose, you will double your wager until you eventually win \$5 or are out of money. Find the probability that you lose your \$75 bankroll.

Exercise 4-38 (*Not Geometric – Not Independent – Not Trivial*) Select a card from a standard deck *without* replacement until you get an ace. Let X denote the number of cards drawn prior to drawing that first ace.

(a) Find the probability distribution for X.

(b) Use a spreadsheet program to display the probability distribution.

(c) Calculate the mean and the standard deviation for X.

(d) Compare these results to a geometric random variable with $p = \frac{4}{52}$.

Exercise 4-39 SOA/CAS Course 1 2000 Sample Examination #7

As part of the underwriting process for insurance, each prospective policyholder is tested for high blood pressure. Let X represent the number of tests completed when the first person with high blood pressure is found. The expected value of X is 12.5. Calculate the probability that the sixth person tested is the first one with high blood pressure.

(A) 0.000 (B) 0.053 (C) 0.080 (D) 0.316 (E) 0.394

Exercise 4-40 SOA EXAM P Sample Exam Questions #48

An insurance policy on an electrical device pays a benefit of 4000 if the device fails during the first year. The amount of the benefit decreases by 1000 each successive year until it reaches 0. If the device has not failed by the beginning of any given year, the probability of failure during that year is 0.4. What is the expected benefit under this policy?

(A) 2234 (B) 2400 (C) 2500 (D) 2667 (E) 2694

Exercise 4-41 SOA EXAM P Sample Exam Questions #136

A fair die is rolled repeatedly. Let X be the number of rolls needed to obtain a 5 and Y the number of rolls needed to obtain a 6. Calculate $E(X \mid Y = 2)$. **Note:** See also Exercise 8.6 #15.

(A) 5.0 (B) 5.2 (C) 6.0 (D) 6.6 (E) 6.8

4.4 THE NEGATIVE BINOMIAL DISTRIBUTION

The negative binomial distribution is a generalization of the geometric distribution. You should verify that the previous results for geometric random variables are consistent with those we derive subsequently. The requirements for a negative binomial process are:

(a) The trials are identical.

(b) Each trial is independent of the other trials.

(c) The random variable M denotes the number of failures prior to the rth success.

(d) The probability of success is p and the probability of failure is $q = 1 - p$.

Example 4.4-1 Negative Binomial Distribution

Let us try to build our intuition on the basketball court. Coach Cramer insists that a player makes 10 free-throws before the player leaves practice. Assume the player is a seventy-five percent foul shooter and her free-throw shots are independent of one another. We let M denote the number of free-throws missed prior to the 10th success. Calculate the probabilities of the events $M = 0$, $M = 1$, and $M = 2$.

Solution

The minimum number of free-throws that must be attempted is 10, in which case all attempts are successes and our random variable M equals 0. Since this means the player makes her first ten shots, $\Pr(M=0) = .75^{10}$. Suppose now $M = 1$, meaning that the player makes her 10th shot with only one miss. This means she takes a total of 11 shots, missing exactly one of the first 10, and making the 11th shot attempt (her 10th success). Then, $\Pr(M = 1) = {}_{10}C_1 \cdot (.25) \cdot (.75)^{10}$, where the binomial coefficient ${}_{10}C_1$ counts the number of ways to select one of the first 10 shots for the miss. For the event $M = 2$, the basketball player has exactly two misses and she has her 10th success (make) on her 12th shot attempt. There are ${}_{11}C_2$ ways to choose the 2 misses out of the first 11 attempts, so that $\Pr(M = 2) = {}_{11}C_2 \cdot (.25)^2 \cdot (.75)^{10}$. ❐

In the general case, we let M be the number of failures prior to the rth success. The event $\{M = k\}$ means exactly k failures occurred prior to the rth success, and each outcome in this event is represented by a word of length $r + k$, consisting of exactly k failures in the first $(r + k - 1)$ slots, and ending with a success in the final, $(r + k)^{th}$ slot. There are $\binom{r+k-1}{k}$ such words. Since each word consists of r successes and k failures, the independence of the trials assures that each such word has probability $p^r \cdot q^k$. In general,

$$\Pr(M = k) = \binom{r+k-1}{k} \cdot p^r \cdot q^k \quad \text{for} \quad k = 0, 1, 2, \ldots .$$

Negative Binomial Random Variable

Consider a series of independent Bernoulli trials with probability of success p and let the random variable M be the number of *failures* before the rth *success*, where r is a positive integer.

Then the random variable M has probability distribution given by

$$p_k = \Pr(M=k) = \binom{r+k-1}{k} \cdot p^r \cdot q^k = \binom{r+k-1}{r-1} \cdot p^r \cdot (1-p)^k$$

for $k = 0,1,2,...,$ and $0 < p < 1$.

The random variable M is called a ***negative binomial random variable with parameters*** r and p.

The seemingly odd name of this random variable derives from the fact that the probability generating function for M is $h_M(s) = p^r(1-qs)^{-r}$. The power series for this function is actually the (infinite) binomial expansion of $(1-qs)$ to the power **negative** r.

Exercise 4-42 Find the power series expansion for $h_M(s) = p^r(1-qs)^{-r}$ and show that the coefficient of s^k is $\Pr(M=k) = {}_{r+k-1}C_k \cdot p^r \cdot q^k$.

Hint: The formula for the coefficient, a_k, of s^k is $a_k = \dfrac{h^{(k)}(0)}{k!}$; $k = 0,1,2,\ldots$

Exercise 4-43 Consider a random variable M with probability generating function $h_M(s) = p^r(1-qs)^{-r}$. Compute the mean and variance of M.

The easiest way to develop the formulas for the mean and the variance of a negative binomial is to decompose the random variable for the negative binomial distribution into a sum of r independent, identically distributed geometric random variables, $X_1, X_2, ..., X_r$. Here X_1 represents the number of failures before the first success. After the first success we start counting over again (recall the ***memory-less*** property from Exercise 4-31) and count the number of new failures prior to the second success. We continue in this fashion until we observe the rth success. It follows that

$$M = X_1 + X_2 + \cdots + X_r.$$

The mean and the variance for M follow directly from the formulas for a geometric random variable, the independence of the X_i's, and the summation formulas from Section 3.6.

Mean and Variance of a Negative Binomial Random Variable

Suppose that M is a ***negative binomial random variable with parameters*** r and p. Then

$$\text{(a)}\ E[M] = \frac{r \cdot q}{p}, \quad \text{and} \quad \text{(b)}\ Var[M] = \frac{r \cdot q}{p^2}.$$

Example 4.4-2 Make 10 Baskets continued:

Coach Cramer wants a particular 75% free-throw shooting basketball player to make 10 shots before leaving practice. We let M denote the number of misses she has prior to the 10th success. Find the mean and the variance of M.

Solution

Using the negative binomial distribution with $r = 10$ and $p = .75$, we have

$$E[M] = \frac{r \times q}{p} = \frac{10 \times .25}{.75} = \frac{10}{3} = 3.\overline{3} \quad \text{and} \quad Var[M] = \frac{r \times q}{p^2} = \frac{10 \times .25}{(.75)^2} = \frac{40}{9}.$$

❐

Exercise 4-44 Suppose that the probability of making a three pointer is $p = 35\%$.

(a) Calculate the probability that you make your third shot *on* your fifth attempt.
(b) Calculate the probability that you make your third shot *by* your fifth attempt.
(c) Calculate the expected value and variance for the *number of misses* that you have prior to making your third shot.
(d) Calculate the expected value for the *total number of shot attempts* that you take in order to make three shots.

Exercise 4-45 Klinger is flipping cards from his cot into a trashcan across the room. He makes 20% of his shots. Let Y denote the number of playing cards that miss the trashcan prior to Klinger making (i.e., landing in the trashcan) seven cards.

(a) Calculate $\Pr(Y = 8)$.
(b) Calculate $E[Y]$.
(c) Calculate the standard deviation for Y.
(d) What is the probability that Klinger can make 7 cards using a single deck?
Hint: You may wish to use a spreadsheet program to help with the calculations.

Exercise 4-46 Given a negative binomial random variable X with mean of six and a variance of 12, calculate $\Pr(X < 3)$.

Exercise 4-47 The formula for the probability distribution of a negative binomial random variable uses the factor ${}_{r+k-1}C_{r-1}$ that counts the number of ways to make $r-1$ shots in the first $r+k-1$ attempts. It is equivalent to count the number of ways to miss k shots in the first $r+k-1$ attempts. How would you revise the definition of the negative binomial distribution using misses instead of makes?

Exercise 4-48 SOA EXAM P Sample Exam Questions #43

A company takes out an insurance policy to cover accidents that occur at its manufacturing plant. The probability that one or more accidents will occur during any given month is 3/5. The number of accidents that occur in any given month is independent of the number of accidents that occur in all other months. Calculate the probability that there will be at least four months in which no accidents occur before the fourth month in which at least one accident occurs.

(A) 0.01 (B) 0.12 (C) 0.23 (D) 0.29 (E) 0.41

Exercise 4-49 SOA EXAM P Sample Exam Questions #140

Each time a hurricane arrives, a new home has a 0.4 probability of experiencing damage. The occurrences of damage in different hurricanes are independent. Calculate the mode of the number of hurricanes it takes for the home to experience damage from two hurricanes.

(A) 2 (B) 3 (C) 4 (D) 5 (E) 6

4.5 THE HYPER-GEOMETRIC DISTRIBUTION

In this section we will formalize a discussion of the hyper-geometric distribution. You already know about the distribution (perhaps not by this name) and can calculate the probabilities using combinations. We will motivate the general results by solving an example using our earlier methods.

Example 4.5-1 Hyper-geometric Distribution

Suppose that a class of chemistry majors has 6 men and 9 women. We will select three students at random to compete in a chemistry competition. Let the random variable W denote the number of women that are chosen for this competition. Calculate the probability distribution for W, the expected value of W, and the variance of W.

Solution

We must choose 3 from a population of 15 students. For the random variable W to equal zero, we must choose zero women from the nine available and three men from the six available.

Then, $\Pr[W=0]=\dfrac{{}_9C_0\cdot{}_6C_3}{{}_{15}C_3}=\dfrac{20}{455}$. The other cases are similar and are tabulated below.

W = #Women	0	1	2	3
Pr(W)	$\dfrac{{}_9C_0\cdot{}_6C_3}{{}_{15}C_3}=\dfrac{20}{455}$	$\dfrac{{}_9C_1\cdot{}_6C_2}{{}_{15}C_3}=\dfrac{135}{455}$	$\dfrac{{}_9C_2\cdot{}_6C_1}{{}_{15}C_3}=\dfrac{216}{455}$	$\dfrac{{}_9C_3\cdot{}_6C_0}{{}_{15}C_3}=\dfrac{84}{455}$

$$E[W] = 0\cdot\frac{20}{455}+1\cdot\frac{135}{455}+2\cdot\frac{216}{455}+3\cdot\frac{84}{455} = \frac{819}{455} = 1.8$$

$$Var[W] = \underbrace{0^2\cdot\tfrac{20}{455}+1^2\cdot\tfrac{135}{455}+2^2\cdot\tfrac{216}{455}+3^2\cdot\tfrac{84}{455}}_{E[W^2]}-\left(\tfrac{819}{455}\right)^2$$

$$=3.86-(1.8)^2 = .617142857.$$

❐

In general, the ***hyper-geometric*** distribution appears when a population of size $B+G$ has exactly two types of elements (e.g., boys and girls, republicans and democrats, cats and dogs, blue balls and green balls). There are B elements of the first type (boys) and G elements of the second type (girls). We refer in short to type B elements and type G elements, letting B and G stand for both the number and the type.

We will select a sample of size n without replacement. Then the random variable X represents the number of type B elements in the sample. Analogous to the binomial distribution, one may think of X as the number of successes. In this case, of course, the trials (individual selections) are not independent since sampling without replacement changes the composition of the population after each individual selection.

Hyper-geometric Random Variable

A finite population consists of B objects of type I and G objects of type II. Let X be the number of type I objects in a sample of size n selected without replacement. Then,

$$\Pr(X=k)=p_k=\frac{{}_BC_k\cdot {}_GC_{n-k}}{{}_{B+G}C_n},$$

with $0\le k\le B$ and $0\le n-k\le G$. Then X (the number of successes) is called a ***hyper-geometric random variable***.

Mean and Variance of the Hyper-geometric Random Variable

Suppose that X is a ***hyper-geometric random variable***. Then

(a) $\mu_X = E[X] = n\cdot\left(\frac{B}{B+G}\right)$, and

(b) $\sigma_X^2 = Var[X] = n\cdot\left(\frac{B}{B+G}\right)\cdot\left(\frac{G}{B+G}\right)\cdot\left(\frac{B+G-n}{B+G-1}\right)$.

Please don't let these formulas intimidate you. We hope that you are already comfortable computing the probability of events using combinations. You choose some boys from the population of boys, and some girls from the population of girls. As for the expected value and variance, they are what you would have guessed (almost). The ratio $\left(\frac{B}{B+G}\right)$ is just the probability of selecting a boy from the population. The ratio $\left(\frac{G}{B+G}\right)$ is the probability of selecting a girl from the whole population. And the annoying factor $\left(\frac{B+G-n}{B+G-1}\right)$ has to do with the fact that we are sampling from a finite population without replacement. Notice that this factor is close to 1 if the total population, $(B+G)$ is large relative to the sample size n.

To verify the formula for the expected value of X we will let X_k equal 1 if the kth object sampled is of type B, and 0 otherwise. Clearly, X_1 is a Bernoulli trial with $p = \frac{B}{B+G}$. Less clear, but still true, is the fact that **all** X_k are Bernoulli trials with the same $p = \frac{B}{B+G}$. This is not immediately obvious because the X_k are not independent. With each selection, the composition of the remaining population changes, depending on whether a B or a G was removed. However, the final composition of the sample doesn't depend on the order of the sampling. Imagine you do not observe the composition of the sample until it is complete. (This is like not peeking at your poker hand until all the cards have been dealt.) Then conditional probability doesn't come into play, and by symmetry, the kth object sampled is no more or less likely to be of type B than any other object selected.

Still not convinced? Then, draw a tree diagram for the second selection and use conditional probabilities to calculate:

$$\begin{aligned}
\Pr(X_2 = 1) &= \Pr(X_2 = 1 \mid X_1 = 0) \cdot \Pr(X_1 = 0) + \Pr(X_2 = 1 \mid X_1 = 1) \cdot \Pr(X_1 = 1) \\
&= \frac{B}{B+G-1} \cdot \frac{G}{B+G} + \frac{B-1}{B+G-1} \cdot \frac{B}{B+G} \\
&= \frac{BG + (B-1)B}{(B+G-1)(B+G)} \\
&= \frac{B(B+G-1)}{(B+G-1)(B+G)} \\
&= \frac{B}{B+G}.
\end{aligned}$$

Since $X = X_1 + X_2 + \cdots + X_n$, it follows that $E[X] = n \cdot \frac{B}{B+G}$. Unfortunately, we cannot play the same game to calculate $Var[X]$ since the X_k are dependent. That calculation will have to wait until we deal with more general sums in Chapter 8.

Example 4.5-2 (Example 4.5-1 continued)

Suppose that a class of chemistry majors has 6 boys and 9 girls. A group of three students is chosen and the random variable W denotes the number of women in the group of three. Calculate the expected value and the variance of W using our theorem.

Solution

$$\mu = E[W] = n \cdot \left(\frac{G}{G+B} \right) = 3 \cdot \left(\frac{9}{15} \right) = 1.8$$

and

$$\sigma^2 = Var[W] = 3 \cdot \left(\frac{9}{15} \right) \cdot \left(\frac{6}{15} \right) \cdot \left(\frac{12}{14} \right) = .617142857.$$

❐

Exercise 4-50 The Humane Society has 27 dogs and 14 cats. They are going to select five animals at random to sell for scientific experiments. Let D denote the number of dogs selected for scientific experiments. Calculate $\Pr(D<3)$, $E[D]$, and σ_D.

Exercise 4-51 Suppose that women in the U.S. Senate constitute a minority of 16, versus a majority of 84 men. Suppose that a seven-member committee is selected at random. Calculate the probability that the committee has a majority of women.

Exercise 4-52 At Maryjane College in Clarksville, there are 75 administrators and 25 faculty. A committee of eight hapless souls was assembled at random from the 100 admin and faculty in order to verify compliance with still another mindless bureaucratic mandate from the State Department of Education. College President Wound Uptight was outraged about the *unfairness* of the committee, as 6 of its members were from the administration (and thus could not attend the weekend golf junket for administrators). He demanded that the committee be reconstituted *more fairly*, with at least four members of the faculty selected. Comment to President Uptight.

Exercise 4-53 Convince yourself that the hypergeometric distribution is approximately equal to the binomial distribution, with $p=\frac{G}{B+G}$, (G is being treated as the success) when $B+G$ is large and n small.

Exercise 4-54 Use a spreadsheet program to plot the probability distributions of the following:

(a) X is a binomial random variable with $n=5$ and $p=.4$

(b) Y is a hypergeometric random variable with $n=5$, $B=20$, $G=30$.

(c) Z is a hypergeometric random variable with $n=5$, $B=400$, $G=600$.

(d) Compute and compare the expected values and variances of X, Y, and Z.

Exercise 4-55 SOA EXAM P Sample Exam Questions #132
A store has 80 modems in its inventory, 30 coming from Source A and the remainder from Source B. Of the modems from Source A, 20% are defective. Of the modems from Source B 8% are defective. Calculate the probability that exactly two out of a random sample of five modems from the store's inventory are defective.

(A) 0.010 (B) 0.078 (C) 0.102 (D) 0.105 (E) 0.125

4.6 THE POISSON DISTRIBUTION

We now come to one of the most fascinating of the standard discrete random variables. The Poisson distribution, named for Siméon D. Poisson (1781-1840), is used to count the number of times a random and sporadically occurring phenomenon actually occurs over a period of observation. Here are a few of the many and varied instances in which the Poisson random variable comes into play:

(a) misprints in a manuscript,

(b) phone calls coming into a call-service center,

(c) customers arriving at the take-out window,

(d) α-particles emitted by a radioactive substance,

(e) Supreme Court vacancies,

(f) break-outs of war,

(g) nineteenth century Prussian cavalrymen kicked to death by their horse. (No, really! There is data, as we shall soon see).

These are events that certainly can and do occur, but would seem at first glance to be inherently so unpredictable that nothing useful could be said. The events may be rare, in some cases even a bit bizzare. The source of the fascination surrounding the Poisson distribution is that, despite the random and sporadic nature of these events, we can still bring a semblance of predictability to their occurrence patterns.

Naturally, we are unable to predict exactly when the next several Supreme Court vacancies will occur. Even actuaries armed with mortality tables cannot foresee individual fortunes any better than carnival tarot-card readers. But with the help of past observations, we can use the Poisson random variable to make remarkably good predictions about the *frequency* of occurrences in the future.

The key to making these predictions is to have a number, λ (a Greek letter pronounced "lamb-da"), for the overall ***rate*** of occurrence. For example, in (a), we might know that there are 500 misprints in a 250 page manuscript, which translates into an average rate of $\lambda = 2$ misprints per page. How many pages will have no misprints at all? How many pages will have more than 3 misprints? Or, suppose we know that up until the year 1932, Supreme Court vacancies occurred at the overall rate of about one every two years. This is, $\lambda = \frac{1 \text{ vacancy}}{2 \text{ years}} = \frac{1}{2}$. Can we then predict the distribution of vacancies over the next 70 years? These are the sorts of questions we can answer using the Poisson distribution.

4.6.1 THE POISSON PROBABILITY FUNCTION

Before giving the Poisson probability formula, we review some properties of the exponential function, $f(x) = e^x$.

1. The power series expansion of the exponential function is (formula (f) in Section 4.1)

$$e^x = 1+\frac{x}{1!}+\frac{x^2}{2!}+\frac{x^3}{3!}+\cdots = \sum_{n=0}^{\infty}\frac{x^n}{n!}.$$

2. $\lim_{n\to\infty}\left(1+\frac{x}{n}\right)^n = e^x.$

In the following definition, the random variable Z is the number of occurrences of a sporadically occurring event over a given period, where the rate of occurrence per period is λ.

Poisson Random Variable

Suppose that the random variable Z has probability function given by

$$\Pr(Z=k)=e^{-\lambda}\frac{\lambda^k}{k!} \quad \text{for } k=0,1,2,\dots \text{ and } \lambda>0.$$

Then Z is called a ***Poisson random variable with parameter*** λ.

Each probability is positive since $e^{-\lambda}$ is positive, $k!$ is positive, and $\lambda>0$. In order to verify that this is a valid probability distribution, we need to show that the probabilities sum to one:

$$\sum_{k=0}^{\infty}p_k = \sum_{k=0}^{\infty}e^{-\lambda}\frac{\lambda^k}{k!} = e^{-\lambda}\sum_{k=0}^{\infty}\frac{\lambda^k}{k!} \underbrace{=}_{\text{using (1)}} e^{-\lambda}\cdot e^{\lambda} = e^0 = 1.$$

Mean and Variance of a Poisson Random Variable

Suppose that Z is a ***Poisson random variable with parameter*** λ. Then

(a) $E[Z]=\lambda$, and
(b) $Var[Z]=\lambda$.

Proof

We will calculate the probability generating function, $h(s)$, for a Poisson random variable with parameter λ. Recall from Section 3.7 that we can determine the mean and second moment of the random variable by evaluating the first and second derivatives of $h(s)$ and evaluating at $s=1$.

$$h(s) = \sum_{k=0}^{\infty}p_k s^k = \sum_{k=0}^{\infty}\left(e^{-\lambda}\frac{\lambda^k}{k!}\right)s^k = e^{-\lambda}\sum_{k=0}^{\infty}\frac{(\lambda s)^k}{k!} = e^{-\lambda}e^{\lambda s}.$$

Then,

$$h'(s)=\lambda e^{-\lambda}e^{\lambda s} \quad \text{and} \quad h''(s)=\lambda^2 e^{-\lambda}e^{\lambda s}.$$

We see that $h'(1)=\lambda$ and $h''(1)=\lambda^2$. Thus,

$$E[Z] = h'(1) = \lambda \quad \text{and} \quad Var[Z] = h''(1) + h'(1) - h'(1)^2 = \lambda.$$

The Poisson distribution has the interesting property that its mean and its variance are both the same (and equal to the parameter λ).

Example 4.6-1 Poisson Distribution

Suppose that, on average, the number of misprints occur at the rate of 3 per page. Select a random page and let the random variable Z denote the number of misprints on that page. We assume Z is a Poisson random variable with parameter $\lambda = 3$. Find $\Pr(Z=k)$ for $k = 0,1,2,3,4,$ and 8. Find $E[Z]$ and σ_Z.

Solution

$E[Z] = \lambda = 3$ and $\sigma_Z = \sqrt{Var[Z]} = \sqrt{3}$.

Z	0	1	2	3	4	8
$\Pr(Z=k)$	$\frac{e^{-3} \cdot 3^0}{0!} = .0498$	$\frac{e^{-3} \cdot 3^1}{1!} = .1494$	$\frac{e^{-3} \cdot 3^2}{2!} = .2240$	.2240	.1680	.0081

❒

Exercise 4-56 Suppose that X has Poisson distribution with $\Pr(X=0) = .001007785$. Find the expected value of X.

Exercise 4-57 Suppose that, on average, a bushel of 100 apples contains three apples with a worm. Select a bushel of 100 apples at random. Let *Worm* be the random variable that represents the number of apples with a worm. Assume that the number of apples that contain a worm can be modeled by a binomial random variable with parameters $n = 100$ and $p = 0.03$. Find $\Pr(Worm = k)$ for $k = 0,1,2,3,4,$ and 8. Compare these probabilities with the Poisson model.

Exercise 4-58 During the lunch rush, vehicles arrive at the local Wendy's at an average rate of $\lambda = 2.4$ cars/minute. Find the most likely number of cars (mode) that arrive during the next minute.

Exercise 4-59 Suppose $Z \sim Pois(\lambda)$ and $E[Z^2] = 12$. Find $\Pr(Z \le 1)$.

Exercise 4-60 SOA EXAM P Sample Exam Questions #67

A baseball team has scheduled its opening game for April 1. If it rains on April 1, the game is postponed and will be played the next day that it does not rain. The team purchases insurance against rain. The policy will pay 1000 for each day, up to 2 days, that the opening game is postponed. The insurance company determines that the number of consecutive days of rain beginning on April 1 is a Poisson random variable with mean 0.6. What is the standard deviation of the amount the insurance company will have to pay?

(A) 668 (B) 699 (C) 775 (D) 817 (E) 904

4.6.2 POISSON PROCESSES

In order to understand why the Poisson distribution is used to model random and sporadically occurring events, we will take an informal glance at the concept of a Poisson process. Imagine that certain phenomena, for example Supreme Court vacancies, occur at random points in time. Suppose that the overall rate of occurrence is λ per unit period.

In a Poisson process the rate λ for a **unit** period can be scaled to a rate $\lambda \cdot h$ for an interval of arbitrary length h (this is called the ***scalability property***). For example, a mean rate of $\lambda = .5$ Supreme Court vacancies per year is equivalent to $\lambda_2 = 1$ vacancy per 2-year period, or $\lambda_{1/4} = \frac{1}{8}$ vacancies per quarter-year, or $\lambda_{10} = 5$ vacancies per 10-year period.

To qualify as a Poisson process with mean rate of occurrence λ per unit period, three properties must be satisfied. We can loosely express these properties in terms of small subintervals I each of length $h > 0$. Then a ***Poisson process*** must satisfy,

(1) For h sufficiently small, the probability of more than one occurrence in an interval of length h is negligible. For example, for Supreme Court vacancies ($\lambda = \frac{1}{2}$ per year), we would say the probability of one vacancy *today* is $\lambda_{1/365} = \frac{1}{730}$, and the probability of two is on the order of $\left(\frac{1}{730}\right)^2$, a much smaller number. If you believe that two justices could resign today, then your interval h is not sufficiently small. Reduce the time interval to minutes or seconds.

(2) The *expected* number of occurrences in an interval of length h (sufficiently small) is approximately equal to λh (the ***scalability*** property)

(3) If I_1 and I_2 are non-overlapping intervals of length h (sufficiently small) then the number of occurrences in I_1 and I_2 are independent.

These properties allow us to partition a unit period into n subintervals each of length $h = \frac{1}{n}$ (h can be made sufficiently small if n is sufficiently large). This is the same idea used in calculus to define the definite integral. We then derive the Poisson probabilities as the limiting case of binomial distributions as the subintervals in the partition become smaller and smaller. The argument goes as follows.

Let X_i be the number of occurrences in the ith subinterval. By property (1), X_i can be treated as a Bernoulli trial with probability of success (an occurrence) equal to p. From property (2), we see that $p = E[X_i] = \Pr(X_i = 1) = \lambda \cdot h = \frac{\lambda}{n}$. Property (3) allows us to treat the n individual Bernoulli trials in the n subintervals as independent.

Now, we let the random variable Z denote the number of occurrences in a given unit period of a Poisson process. The preceding paragraph implies $Z \approx X_1 + X_2 + \cdots + X_n$, a sum of independent, identically distributed Bernoulli trials with $p = \lambda \cdot h = \frac{\lambda}{n}$. Thus Z approximately satisfies the conditions for a binomial random variable. For such a random variable,

$$\Pr(Z = k) \approx {}_nC_k \cdot p^k \cdot (1-p)^{n-k} = \frac{\overbrace{n(n-1)\cdots(n-k+1)}^{k \text{ factors}}}{k!} \cdot \left(\frac{\lambda}{n}\right)^k \cdot \left(1 - \frac{\lambda}{n}\right)^{n-k}.$$

We factor out $\frac{\lambda^k}{k!}$ and pair up the n^k in the denominator with the k factors $n(n-1)\cdots(n-k+1)$ in the numerator, to obtain,

$$P[Z = k] \approx \left(\frac{\lambda^k}{k!}\right)\left(\frac{n}{n}\right)\left(\frac{n-1}{n}\right)\cdots\left(\frac{n-k+1}{n}\right)\left(1 - \frac{\lambda}{n}\right)^{-k}\left(1 - \frac{\lambda}{n}\right)^n.$$

For n sufficiently large (with respect to k) all factors except for the first and the last are approximately one. The factor $\left(\frac{\lambda^k}{k!}\right)$ does not depend on n. Using property (2) of facts about e^x (listed at the beginning of this subsection) the limit of the last factor is,

$$\lim_{n \to \infty}\left(1 - \frac{\lambda}{n}\right)^n = e^{-\lambda}.$$

Therefore,

$$\Pr(Z{=}k) \approx e^{-\lambda} \cdot \frac{\lambda^k}{k!}.$$

A perfectly rigorous treatment would show that as n goes to infinity, the above approximation converges to exactly $e^{-\lambda}\frac{\lambda^k}{k!}$. In other words, we have that the number of occurrences, Z, in a unit period of a Poisson process is the Poisson random variable with parameter λ. Here, λ is the mean rate of occurrence per unit period. This demonstration shows that the Poisson distribution is the limiting distribution of binomial random variables where $n \to \infty$ while $n \cdot p \to \lambda$. In fact, the Poisson distribution is often used to approximate a binomial random variable when n is large and p is small (see Subsection 4.6.5). This is because calculating probabilities for a Poisson random variable is easier than calculating probabilities using the binomial formula when n is large.

4.6.3 POISSON PROCESS DATA SETS

Do the various phenomena (a) – (g) mentioned at the beginning of this section occur as a result of a Poisson process? If yes, then we are justified in modeling the number of occurrences using a Poisson random variable with the appropriate parameter λ. The first step, when we have a random and sporadically occurring event, is to collect some data in order to estimate the rate of occurrence[3], λ. In the case of misprints in a manuscript, we would count the total number of misprints and estimate our parameter as

$$\lambda = \frac{\text{total number of misprints}}{\text{total number of pages}}.$$

Next, we compare the actual past occurrence patterns, as recorded, to the predicted frequencies using the Poisson distribution with parameter λ. Close agreement between them serves to validate the theoretical model of the data as resulting from a Poisson process.

Comparison of Automobile Accident Models

We will illustrate the approach using an example based upon automobile accidents at a certain junction. Suppose records show that 100 accidents have occurred over 100 months, so that the mean rate of occurrence is $\lambda = 1$ accident per month. There is no guarantee that accidents occur according to a Poisson process. Here we present four different accident-pattern models, all with the overall rate of $\lambda = 1$ accident per month and a maximum of four accidents in a single month. For each, we let Z be the associated random variable taking values 0, 1, 2, 3, and 4 with probabilities calculated from the frequencies in the second column:

Pattern (1): Gradually Decreasing Accidents per Month

k = number of accidents	months with k accidents	Total Accidents
0	45	0
1	25	25
2	15	30
3	15	45
4	0	0
Total:	100	100

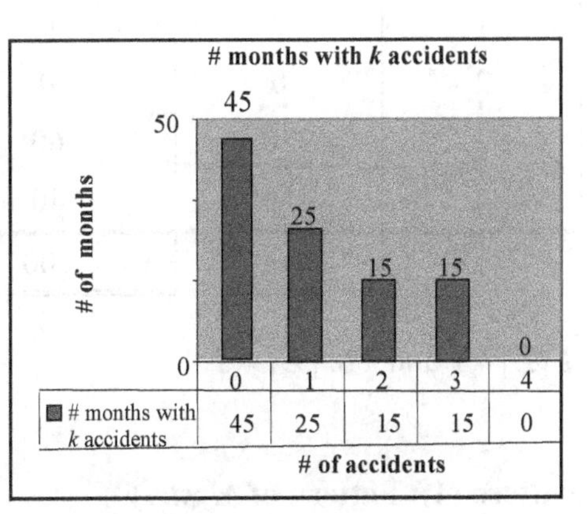

$$E[Z] = \frac{0\cdot 45 + 1\cdot 25 + 2\cdot 15 + 3\cdot 15 + 4\cdot 0}{100} = 1.$$

$$E[Z^2] = \frac{0^2\cdot 45 + 1^2\cdot 25 + 2^2\cdot 15 + 3^2\cdot 15 + 4^2\cdot 0}{100} = 2.2.$$

$$Var[Z] = 2.2 - 1.0^2 = 1.2.$$

[3] This is a simple instance of estimating model parameters, an important topic in statistics.

Pattern (2): Exactly One Accident Each Month

k =number of accidents	months with k accidents	Total Accidents
0	0	0
1	100	100
2	0	0
3	0	0
4	0	0
Total:	100	100

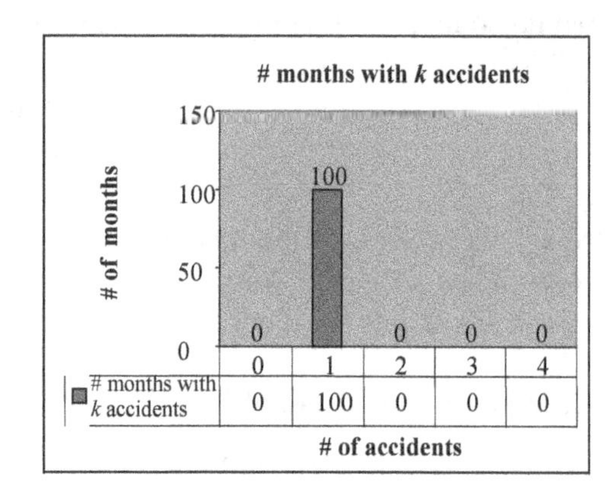

$E[Z]=1$ and $Var[Z]=0$.

Pattern (3): Zero or Multiple Accidents Each Month

k = number of accidents	months with k accidents	Total Accidents
0	70	0
1	0	0
2	0	0
3	20	60
4	10	40
Total:	100	100

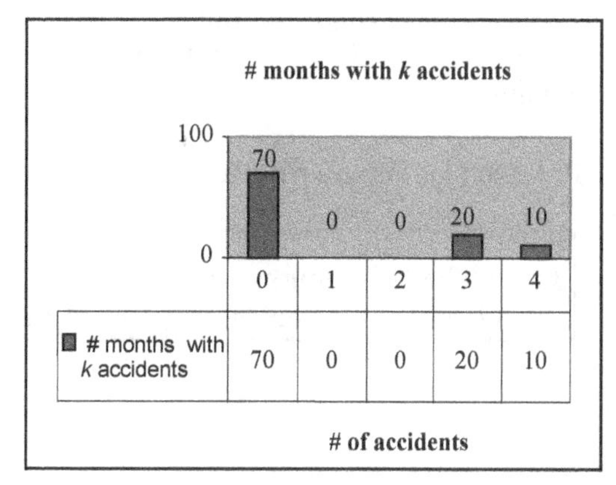

$E[Z]=1$ and $Var[Z]=2.4$.

Pattern (4): Pattern of *Mystery*

k = number of accidents	months with k accidents	Total accidents
0	36	0
1	38	38
2	18	36
3	6	18
4	2	8
Total:	100	100

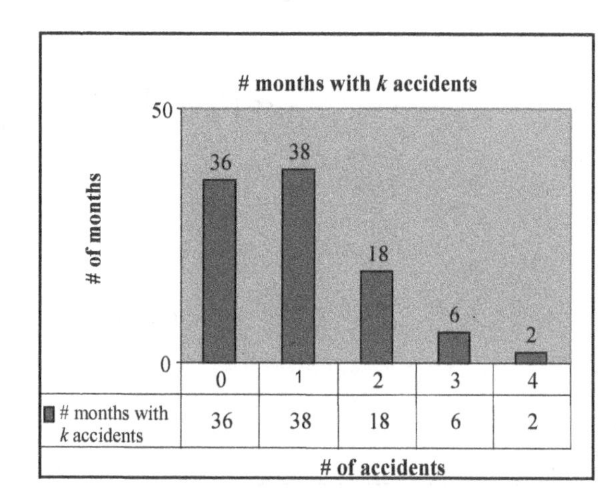

$E[Z]=1$ and $Var[Z]=0.96$.

Which of these patterns is most likely to have resulted from observing a Poisson process with $\lambda = 1$? A quick decision can be reached based on the fact that the mean and the variance of a Poisson random variable are equal. The mean and variance of Pattern (4) are the closest to being equal. Pattern (1) is the next closest but has too many accident-free months to be a Poisson process with $\lambda = 1$. Pattern (2) has accidents happening at a uniform rate of 1 per month, which contradicts being *random and sporadic*, and leads to a variance of zero. Pattern (3) represents a process in which accidents happen in clusters, with mostly accident-free months accompanied by a few really bad months.

The theoretical probability distribution of accidents based upon the assumption that accidents happen according to a Poisson process with $\lambda = 1$ is:

k = number of accidents	$p_k = e^{-1}\frac{1}{k!}$	Predicted number of months with k accidents
0	0.367879	36.79
1	0.367879	36.79
2	0.183940	18.39
3	0.061313	6.13
4	0.015328	1.53
5 or more	0.003660	0.37
Total:	1.00	100

The predicted number of months with k accidents (in the last column) is calculated on the premise that the number of months with k accidents is binomial with $n = 100$ and $p = p_k = \frac{e^{-1} 1^k}{k!}$. Therefore, the expected number of months with k accidents is E[months with k accidents] $= n \cdot p_k$. We see that the predicted frequencies fit the actual frequencies in Pattern (4) quite closely, so that we would be justified in believing that the data in Pattern (4) arose from a Poisson process.

The data for Prussian cavalry deaths caused by horse kicking was made famous by Ladislaus Josephovich Bortkiewicz, who published a book in 1898 about the Poisson distribution called (English translation) *The Law of Small Numbers*[4]. The data was collected over a 20 year timeframe (1861-1880) for ten cavalry corps, a total of 200 periods, where a period is taken as a cavalry corps-year. The data shows a total of 122 such deaths over the 20 years, leading to a rate $\lambda = \frac{122}{200} = .61$ deaths per corps-year.

Example 4.6-2 The Prussian Cavalry

Generate the probability distribution of the Poisson random variable with parameter $\lambda = \frac{122}{200} = .61$. Use the model to predict the number of corps-years (out of 200) that will result in k deaths by horse-kicking, for $k = 0, 1, 2, 3,$ and 4. Compare this model's predictions with the actual historical data tabulated in the last column below.

[4] Das Gesetz der Kleinen Zahlen, *Monatshefte für Mathematik*, vol. 9, 1898.

Solution

The calculations are contained in the following table.

k = number of deaths by horse-kicking	$p_k = e^{-0.61}\frac{(0.61)^k}{k!}$	Predicted number of corps-years $(200 \cdot p_k)$	***Actual (historical) number of corps-years***
0	0.5434	108.7	***109***
1	0.3314	66.3	***65***
2	0.1011	20.2	***22***
3	0.0206	4.1	***3***
4	0.0031	0.6	***1***
5 or more	0.0004	0.1	***0***
Total:	1.0000	200.0	***200***

The following chart compares the model predictions for the number of years with k deaths $(k = 0,12,...)$ with the actual historical data.

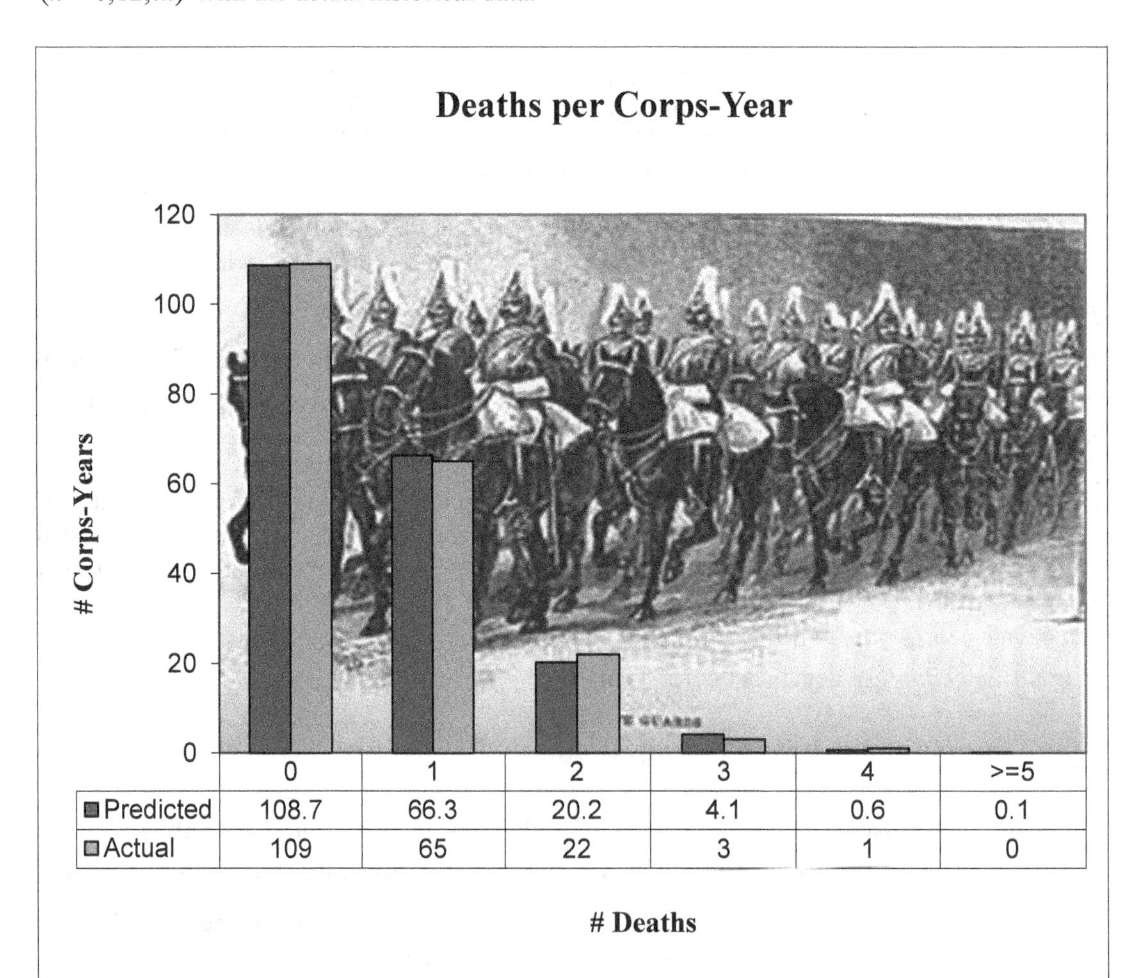

Our Poisson model, based on $\lambda = .61$, predicts corps-year frequencies that are amazingly close to the actual frequencies. How close? The Chi-Squared test[5] is a widely accepted statistical measure of goodness-of-fit, that is, how well the model predicts what actually happens. It shows that there is a better than 95% chance that the data arose from a Poisson process with parameter $\lambda = .61$. ❐

Exercise 4-61 The following data set summarizes Supreme Court Vacancies from (1790-1932)[6].

(a) Calculate the mean and variance of the data.
(b) Construct a prediction table based upon the Poisson random variable with an appropriate rate λ.
(c) Compare the predicted frequencies to the actual frequencies. Do you think the data arose from a Poisson process?

Number of Supreme Court Vacancies, *k*	0	1	2	3	4	5 or more
Actual number of years from (1790-1932) with *k* vacancies	87	47	6	3	0	0
Poisson Prediction of number of years with k vacancies						

Exercise 4-62 During World War II, England experienced numerous hits by flying-bombs attached to primitive rockets. Were the hits random and sporadic, or were the German rockets capable of being guided toward strategic targets? To analyze this, a portion of London was partitioned into 576 square sectors (of area equal to 0.25 square kilometers) and the number of hits in each sector was recorded.

(a) Calculate the mean and variance of the data.
(b) Construct a prediction table based upon the Poisson random variable with an appropriate rate λ.
(c) Compare the predicted frequencies to the actual frequencies. Do you think the data arose from a Poisson process?

Bomb hits on London[7], *k*	0	1	2	3	4	5 or more
Number of sectors with *k* bomb hits	229	211	93	35	7	1
Poisson Prediction of number of sectors with *k* bomb hits						

---

[5] $\chi^2 = \sum \frac{(\text{Observed} - \text{Expected})^2}{\text{Expected}}$.

[6] This example is based on W.A. Wallis, "The Poisson Distribution and the Supreme Court," *Journal of the American Statistical Association*, Vol. 31 (1936) 376-380. Wallis covered the period 1837-1932. The data for the example here is extracted from: www.supremecourtus.gov/about/members.pdf.

[7] The data for this example is reported in R.D. Clarke, "An Application of the Poisson Distribution," *Journal of the Institute of Actuaries*, 72 (1946), pp. 48.

Exercise 4-63 Suppose that our editor makes an average of two and one-half typographical errors per page. You select a page of this text at random. Let T denote the number of typographical errors on that randomly chosen page. If typos occur as a result of a Poisson process, find

(a) The probability distribution for T.
(b) The mode for T.
(c) The $Var[T]$.

Exercise 4-64 Suppose that our editor makes an average of two and one-half typographical errors per page. You select a 30-page chapter of this text at random and let C denote the total number of typos in the chapter. If typos occur as a result of a Poisson process, find

(a) Find μ_C.
(b) Find $\Pr(C = 70)$.

Exercise 4-65 The number of automobile accidents at the corner of Wall and Street is assumed to have Poisson distribution with a mean of five per week. Let A denote the number of automobile accidents that will occur next week. Find

(a) $\Pr[A < 3]$.
(b) The median of A.
(c) σ_A.

Exercise 4-66 If Z is a Poisson random variable with mean n (a positive integer), then show that $\Pr[Z = n] = \Pr[Z = (n-1)]$.

Note: The probability distribution is at a maximum here.

Exercise 4-67 If Z is a Poisson random variable with mean λ, where λ is not an integer, then when is the probability distribution at its maximum? That is, find the mode.

Exercise 4-68 Davis, age 10, loves to attend baseball games. He is thrilled to catch a foul ball, get a souvenir T-shirt from the rally monkey, receive an autograph from a major league player, or anything similar. Suppose that there are 100 such souvenirs distributed randomly to 4936 fans who attended a Milwaukee Brewers' spring training game at Maryvale park in Phoenix, Arizona. Calculate the probability that Davis got three souvenirs (a Richie Sexson baseball hat, an inning-ending game ball, and a Kevin Young cracked bat[8]). Explain to Davis how fortunate he was.

---

[8] Davis Hull was sporting all three souvenirs.

Exercise 4-69 SOA EXAM P Sample Exam Questions #30

An actuary has discovered that policyholders are three times as likely to file two claims as to file four claims. If the number of claims filed has a Poisson distribution, what is the variance of the number of claims filed?

(A) $1/\sqrt{3}$ (B) 1 (C) $\sqrt{2}$ (D) 2 (E) 4

4.6.4 SUMS OF POISSON RANDOM VARIABLES

Consider a new random variable S which is the **sum** of two independent Poisson random variables. An important property of Poisson distributions is that this sum is itself a Poisson random variable. The mean rate of occurrence for the sum, $E[S]=\lambda_S$, is the sum of the contributing mean rates.

Suppose, for example, you are observing arrivals at the drive-up window of a bank. Let M denote the number of males arriving in a fifteen minute period, and let F be the number of females arriving in the same fifteen minute period. Assume males and females arrive independently according to Poisson processes. If males arrive at the rate of three (3) per fifteen minutes and females arrive at the rate of four (4) per fifteen minutes, then the total number of arrivals at the bank drive-up window is $S=M+F$, a Poisson random variable with a mean rate of arrivals equal to seven (7) per fifteen minute period.

While this result appears obvious and intuitive for Poisson random variables, it is not always the case that the sum of two random variables of one type will result in a random variable of that same type.

Sum of Independent Poisson Random Variables

Suppose that Z_i are independent ***Poisson random variables with mean*** λ_i for $i=1,2$. Then $Z=Z_1+Z_2$ is a Poisson random variable with mean (parameter)

$$E[Z] = E[Z_1+Z_2] = \lambda_1+\lambda_2.$$

We will discuss sums of random variables in Chapter 8, and will defer the formal treatment until then. We include the discussion here in order to complete the picture of the Poisson random variable.

Exercise 4-70 Suppose that the number of accidents at a certain intersection has Poisson distribution with a mean of two per week. Calculate the probability that the number of accidents in a four-week period is less than five. Calculate the expected value and variance of the number of accidents in a four-week period.

Exercise 4-71 An excellent typist makes an average of one error for every 20 pages of text. One of her products is 300 pages in length.

(a) How many mistakes would you predict?

(b) What is the likelihood that there are fewer than 5 errors in her 300 page product?

Exercise 4-72 Jack E. disdains profanity. He swears only once per month (assume 30 days per month). Calculate the probability that Jack E. swears twice today.

We offered a warning that what appears reasonable and intuitive may not always be the case. It should be clear that if you pick a random number from $\{1,2,3\}$ today (that is, the discrete uniform distribution) and pick a second number at random from $\{1,2,3\}$ tomorrow, then the random variable representing the sum of those two numbers is **not** a discrete uniform random variable. The next exercise explores this situation.

Exercise 4-73 Pick two random numbers (with replacement) from $\{1,2,3\}$. Let S denote the sum of the two numbers.

(a) Use a tree diagram to calculate the probability distribution for S.

(b) Is this a discrete uniform distribution?

(c) Calculate the $\Pr\left[(S < E[S])\right]$.

Exercise 4-74 SOA EXAM P Sample Exam Questions #50

A company buys a policy to insure its revenue in the event of major snowstorms that shut down business. The policy pays nothing for the first such snowstorm of the year and \$10,000 for each one thereafter, until the end of the year. The number of major snowstorms per year that shut down business is assumed to have a Poisson distribution with mean 1.5. What is the expected amount paid to the company under this policy during a one-year period?

(A) 2,769 (B) 5,000 (C) 7,231 (D) 8,347 (E) 10,578

4.6.5 POISSON APPROXIMATION TO THE BINOMIAL DISTRIBUTION

We mentioned earlier that the *derivation* of the Poisson probability function showed that binomial probabilities with large n and small p can be fairly accurately approximated using the Poisson distribution. Here is how:

Suppose that Y is a *binomial random variable* with probability of success p (small) and n trials (n large). Define Z as a *Poisson random variable with parameter* $\lambda = np$. Notice that $E[Z]=\lambda=np=E[Y]$, and $Var[Z]=\lambda=np\approx np(1-p)=E[Y]$. Since p is small, $q=(1-p)\approx 1$. Therefore, Z and Y will have similar distributions. Calculating the exact binomial probabilities can be time consuming or difficult, especially if the number of trials is large. It is reasonable to approximate the calculation of

$$\Pr(Y=k) = {}_nC_k \cdot p^k \cdot q^{n-k} \quad \text{with} \quad \Pr(Z=k) = \frac{e^{-np}\cdot(np)^k}{k!}.$$

Exercise 4-75 The probability of hitting the bull's-eye is 5%. Eighty (80) shots are taken, each shot constituting an independent Bernoulli trial. Calculate the probability that the number of bull's-eyes is 0, 1, 2, 3, 4, or 5 using both the binomial model, and the associated Poisson model.

4.7 SUMMARY COMPARISON OF DISCRETE DISTRIBUTIONS

In this chapter we studied several common discrete probability distributions – discrete uniform, binomial, geometric, negative binomial, hypergeometric, and Poisson. They are all relatively simple discrete distributions. Can we use these distributions to model real-life situations? Or rather, do real-life phenomena satisfy the assumptions of the model?

For example, automobile accidents are not always independent (imagine hazardous driving conditions in a widespread winter snowstorm, or the parking lot after a Steelers game). Likewise, we do not believe hits in baseball or made shots in basketball are independent repetitions of the same Bernoulli trial (player ability to perform varies from moment to moment). On the other hand, we have seen, for example, that the Poisson process provides a close fitting model for some of the most unlikely real-world phenomena.

Understanding the requirements (definitions) and characteristics (means and variances) of these distributions forms a nice starting point in trying to *fit* the appropriate distributions to real life data.

Exercise 4-76 Use a spreadsheet program to make bar graphs of the following probability distributions:

(a) Discrete uniform random variable with $n = 5$.

(b) Binomial random variable with $n = 10$ and $p = .3$.

(c) Hypergeometric random variable with $n = 5, B = 10,$ and $G = 30$.

(d) Poisson random variable with $\lambda = 3$.

(e) Geometric random variable with $p = .25$.

(f) A negative binomial random variable with $r = 3$ and $p = .5$.

Exercise 4-77 When choosing the distribution to fit to data, one often focuses on the fundamental statistics of the mean and variance. Complete the following matrix and learn the results:

Random Variable	Mean	Variance	Mean (<, =, >) Variance
Poisson			
Binomial			
Negative Binomial			

Exercise 4-78 Suppose that you collected some data about the number of coins found under the driver's seat of cars. You computed the expected number of coins to be 5 with a standard deviation of 4.

(a) Which distribution (Poisson, binomial, or negative binomial) is the likely candidate to attempt to fit to the data?
(b) What parameter(s) would you choose?

Exercise 4-79 Michael, the produce manager at the local Albertson's supermarket store, would like you to estimate the probability that a customer will purchase 5 or more bunches of bananas. He has collected the following data:

Number of bunches of bananas	0	1	2	3	4	5 or more
Frequency	49	35	12	3	1	0

(a) Calculate the mean and variance for the aforementioned data.
(b) Which distribution (Poisson, binomial, or negative binomial) would you use to model the data? Which parameter(s) would you choose?
(c) If 1000 customers visit the store, estimate the probability that someone will purchase 5 or more bunches of bananas.

Exercise 4-80 Consider the following discrete distributions:

(1) Discrete uniform random variable with $n = 10$.
(2) Binomial random variable with $n = 10$ and $p = .6$.
(3) Hypergeometric random variable with $n = 5, B = 10,$ and $G = 30$.
(4) Poisson random variable with $\lambda = 4$.
(5) Geometric random variable with $p = .25$.
(6) A negative binomial random variable with $r = 3$ and $p = .5$.

For each of these distributions, calculate the probability that:

(a) The random variable is within one standard deviation of the mean.
(b) The random variable is within two standard deviations of the mean.
(c) The random variable is within three standard deviations of the mean.

Note: For the normal distribution, these probabilities are 68.26%, 95.44%, and 99.74% respectively.

4.8 CHAPTER 4 SAMPLE EXAMINATION

1. The Coca-Cola® *Live the Magic*™ *game* had game pieces on the lids of .5 liter plastic bottles of Coca-Cola® products. They advertised that 1 in 12 wins a free 20 ounce bottle of Coke. Your statistics teacher purchased 18 Cokes, independently and with equal probabilities of winning a free Coke.

   (a) What is the probability that he won exactly 4 free Cokes?
   (b) If X denotes the number of free Cokes won by your statistics teacher, find σ_X.

2. Roll a pair of fair dice.

   (a) Find the probability that on an individual roll, the sum is 5 or 11.
   (b) Find the probability that the first time that you roll a sum of 5 or 11 is on your 7th roll of the pair of dice.
   (c) Find the probability that the third time that you roll a sum of 5 or 11 is on your 7th roll of the dice.

3. I am not excited about grading exams. I would rather jam a dull stick into my leg. I will therefore randomly assign your grade by picking an integer uniformly from 77 to 100 (inclusively).

   (a) What is the probability that you get a "B-/B/B+" on this exam (a score between and including 80 to 89)?
   (b) What is your expected score on this exam?
   (c) What is the variance of scores on this exam?

4. At the fall picnic for mathematics education students, Big Blind Al brought 12 Bud Lights, 6 Cokes, 16 Diet Cokes, and 1 Sprite. Four (4) students show up at the picnic and Blind Al randomly gathers five total beverages from the cooler. Please answer the following:

   (a) What is the probability that 2 of the 5 beverages are Bud Lights?
   (b) What is the expected number of Bud Lights that Big Blind Al grabs?
   (c) If X denotes the number of Bud Lights that Big Blind Al selects, then calculate $E[5X^2 - 6X + 3]$.

5. Your Aunt Cathy Daring suffers from allergies and sneezes randomly, sporadically, and explosively, startling small children and setting dogs to howling in a quarter-mile radius. Her nephew, a graduate student at Cal Tech, used seismographic equipment to track the frequency of the violent expulsions, and determined she sneezed on average 48 times per day (24 hour period). Suppose that Aunt Cathy is driving your family to school. The trip takes 2 hours. Further suppose that the number of sneezes follows the Poisson distribution. Find the following:

   (a) The average (mean) number of expected sneezes in the 2-hour period.
   (b) The standard deviation of the number of sneezes in the 2-hour period.
   (c) The probability that Aunt Cathy can make it through her entire 2-hour road trip without sneezing.
   (d) The probability that she sneezes at least three times during her road trip.

6. Find $E[(3X+4)(7-2X)]$ if X has a binomial distribution with $n = 50$ and $p = .96$.

7. An engineering student, Ms. Klutz, injures herself at a rate of 3 times per month. Suppose that the Poisson distribution models the number of times that she injures herself.

(a) Calculate the probability that she injures herself exactly 2 times next month.
(b) Calculate the probability that she injures herself exactly 5 times next year.

8. **SOA EXAM P Sample Exam Questions #14**
In modeling the number of claims filed by an individual under an automobile policy during a three-year period, an actuary makes the simplifying assumption that for all integers $n \geq 0$, $p_{n+1} = \frac{1}{5} p_n$, where p_n represents the probability that the policyholder files n claims during the period. Under this assumption, what is the probability that a policyholder files more than one claim during the period?

(A) 0.04 (B) 0.16 (C) 0.20 (D) 0.80 (E) 0.96

9. **SOA EXAM P Sample Exam Questions #41**
A study is being conducted in which the health of two independent groups of ten policyholders is being monitored over a one-year period of time. Individual participants in the study drop out before the end of the study with probability 0.2 (independently of the other participants). What is the probability that at least 9 participants complete the study in one of the two groups, but not in both groups?

(A) 0.096 (B) 0.192 (C) 0.235 (D) 0.375 (E) 0.469

10. **SOA EXAM P Sample Exam Questions #32**
A large pool of adults earning their first driver's license includes 50% low-risk drivers, 30% moderate-risk drivers, and 20% high-risk drivers. Because these drivers have no prior driving record, an insurance company considers each driver to be randomly selected from the pool. This month, the insurance company writes 4 new policies for adults earning their first driver's license. What is the probability that these 4 will contain at least two more high-risk drivers than low-risk drivers?

(A) 0.006 (B) 0.012 (C) 0.018 (D) 0.049 (E) 0.073

11. Suppose that X has a binomial distribution with $E[X] = 8$ and standard deviation $\sigma_X = \sqrt{4.8}$. Find $\Pr(X = 10)$.

12. **SOA EXAM P Sample Exam Questions #142**

An auto insurance company is implementing a new bonus system. In each month, if a policyholder does not have an accident, he or she will receive a 5.00 cash-back bonus from the insurer. Among the 1,000 policyholders of the auto insurance company, 400 are classified as low risk drivers and 600 are classified as high-risk drivers. In each month, the probability of zero accidents for high-risk drivers is 0.80 and the probability of zero accidents for low-risk drivers is 0.90.

Calculate the expected bonus payment from the insurer to the 1000 policyholders in one year.

(A) 48,000 (B) 50,400 (C) 51,000 (D) 54,000 (E) 60,000

13. **SOA EXAM P Sample Exam Questions #151**

From 27 pieces of luggage, an airline luggage handler damages a random sample of four. The probability that exactly one of the damaged pieces of luggage is insured is twice the probability that none of the damaged pieces are insured.

Calculate the probability that exactly two of the four damaged pieces are insured.

(A) 0.06 (B) 0.13 (C) 0.27 (D) 0.30 (E) 0.31

14. **SOA EXAM P Sample Exam Questions #152**

Automobile policies are separated into two groups: low-risk and high-risk. Actuary Rahul examines low-risk policies, continuing until a policy with a claim is found and then stopping. Actuary Toby follows the same procedure with high-risk policies. Each low-risk policy has a 10% probability of having a claim. Each high-risk policy has a 20% probability of having a claim. The claim statuses of polices are mutually independent.

Calculate the probability that Actuary Rahul examines fewer policies than Actuary Toby.

(A) 0.2857 (B) 0.3214 (C) 0.3333 (D) 0.3571 (E) 0.4000

15. **SOA EXAM P Sample Exam Questions #153**

Let X represent the number of customers arriving during the morning hours and let Y represent the number of customers arriving during the afternoon hours at a diner.

You are given:

1) X and Y are Poisson distributed.
2) The first moment of X is less than the first moment of Y by 8.
3) The second moment of X is 60% of the second moment of Y.

Calculate the variance of Y.

(A) 4 (B) 12 (C) 16 (D) 27 (E) 35

CHAPTER 5

CALCULUS, PROBABILITY, AND CONTINUOUS DISTRIBUTIONS

At the beginning of this text we alluded to the distinction between discrete and continuous probability models. For the former, the number of outcomes is either ***finite*** or ***countably infinite***, meaning that the outcomes can be listed as $1^{st}, 2^{nd}, 3^{rd}$, and so on. Through the first four chapters we have dealt almost exclusively with discrete models, and we developed the language of random variables and distributions for that case. This required relatively little recourse to calculus.

It is now time to take up the subject of continuous random variables, which, by contrast, relies heavily on the tools of calculus. Continuous models have outcomes along a continuum, or interval, on the real number line. The resulting ***events*** are subsets of the real numbers. Although more complicated events are possible, for our purposes events are almost always simple intervals of the form (a,b), or $(a,b]$, and so forth, with endpoints either included or excluded. Since there is no way to create a list of all of the possible outcomes (because the set of points along an interval of real numbers is ***uncountable***), we cannot display the values and probabilities of the random variable in tabular form - a technique that formed the mainstay of Chapters 3 and 4. Instead, we define probabilities and display the distribution graphically, using ***functions***, generally of the type that will be familiar from calculus courses.

5.1 CUMULATIVE DISTRIBUTION FUNCTIONS

There are two closely related functions that are used to represent a continuous random variable *X*. The first to be discussed is called the ***Cumulative Distribution Function*** (CDF) and is denoted by $F_X(x)$. We always use capital letters to denote distribution functions. The subscript *X* is used to identify the particular random variable and may be omitted if there is no likelihood of confusion.

Definition of the Cumulative Distribution Function (CDF)

The cumulative distribution function of a random variable *X* is defined by,

$$F_X(x) = \Pr(-\infty, x] = \Pr[-\infty < X \le x].$$

That is, the value of the CDF of X at the point x is the probability of the event that the random variable X takes a value less than or equal to x. We will address the question of where these probabilities originate momentarily.

We demonstrate first how to express the probability of an interval, $(a,b]$, [1] in terms of the cumulative distribution function $F_X(x)$. Suppose that a and b are real numbers with $a < b$. Observe that the interval $(-\infty,b]$ can be decomposed into two disjoint intervals $(-\infty,b] = (-\infty,a] \underbrace{\cup}_{\text{disjoint}} (a,b]$. By the additive property of probabilities, we have

$$\begin{aligned} \Pr(-\infty,b] &= \Pr\{(-\infty,a] \cup (a,b]\} \\ &= \Pr(-\infty,a] + \Pr(a,b]. \end{aligned}$$

Rearranging terms, we have

$$\begin{aligned} \Pr(a,b] &= \Pr[a < X \le b] \\ &= \Pr(-\infty,b] - \Pr(-\infty,a] \\ &= F_X(b) - F_X(a). \end{aligned}$$

Similarly, basic properties of probability assure that the CDF is a non-decreasing function starting at 0 from $x = -\infty$ and increasing monotonically to 1 at $x = \infty$. To summarize,

Properties of All Cumulative Distribution Functions

(a) $F_X(x)$ is a non-decreasing function.

(b) $\lim_{x \to -\infty} F_X(x) = 0$.

(c) $\lim_{x \to \infty} F_X(x) = 1$.

(d) $\Pr(a < X \le b) = F_X(b) - F_X(a)$.

There is nothing in the definition or properties of the CDF that relates exclusively to ***continuous*** random variables. Indeed, the CDF exists for all random variables, whether discrete, continuous or a ***mixture*** (discussed later in Section 5.4) of both discrete and continuous random variables. Surprisingly, there is not an exact consensus among authors about the precise meaning of a *continuous* random variable. We can, however, make a useful, if somewhat imprecise, distinction based on the CDF of the random variable. If $F_X(x)$ is a continuous function for all real numbers x (loosely speaking, no jumps in the graph), then we will say that X is a continuous random variable. By contrast, if $F_X(x)$ is a step function

[1] We follow the standard convention of using an open parenthesis to signify an open interval boundary point, and a closed bracket to signify a closed interval boundary point. Thus, for example, $(a,b] = \{x \mid a < x \le b\}$.

(horizontal lines punctuated by discrete vertical jumps) then our random variable X is discrete. Any other function satisfying properties (a) – (d) constitutes the CDF for a ***mixed*** distribution.

We illustrate this with some examples.

Example 5.1-1 CDF for a Discrete Random Variable

The simplest discrete random variable is a Bernoulli trial. Consider such a trial with probability of success $p = 0.6$. Our random variable X takes the two values 0 (failure, with probability 0.4) and 1 (success, with probability 0.6). The resulting CDF is the following step function:

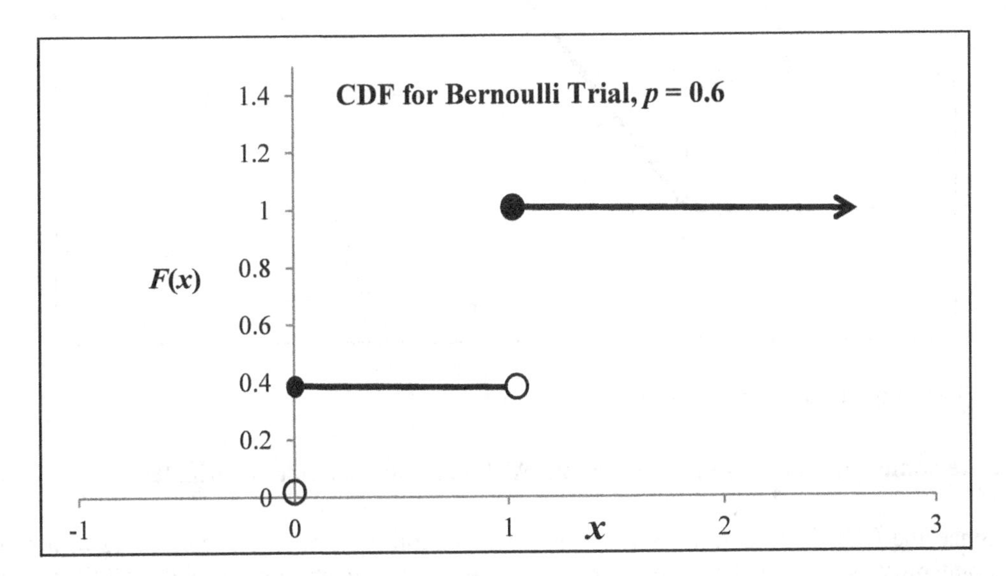

The simplest example of a continuous probability model (described briefly in Chapter 1), involves selecting a number at random along the interval of real numbers from zero to one (hereafter referred to as the ***unit interval***). The random variable X takes the value of the number selected. An underlying physical probability experiment might consist of spinning a fair and balanced wheel whose circumference is calibrated with the unit interval. A fixed pointer will indicate the resulting value of X once the wheel stops.

To avoid any possibility of ambiguity between the outcomes 0 and 1 (same point on the circumference) we will always assign 0 to that point[2]. That is, the domain of X is taken to be $[0,1)$. This crude experiment is more commonly simulated on computers with sophisticated pseudo-random number generators.

Example 5.1-2 Continuous Random Number Generator X on the Unit Interval $[0,1)$.

The events for this experiment are intervals $(a,b]$ of real numbers. Since we are selecting numbers between 0 and 1, we should have $\Pr(0 \le X < 1) = 1$. Since we are choosing randomly,

[2] As noted in the following, the probability of a spin actually landing on that exact point is zero.

the probability of an interval $(a,b]$ contained within the unit interval should depend only on its length, not its position. In other words, we should have $\Pr(a,b] = \Pr[a < X \le b] = b-a$. Therefore, the CDF for our random variable X should be defined piece-wise by $F_X(x) = x$ for $0 \le x < 1$ with $F_X(x) = 0$ to the left of the origin, and $F_X(x) = 1$ to the right of the unit interval (i.e., for all $x \ge 1$).

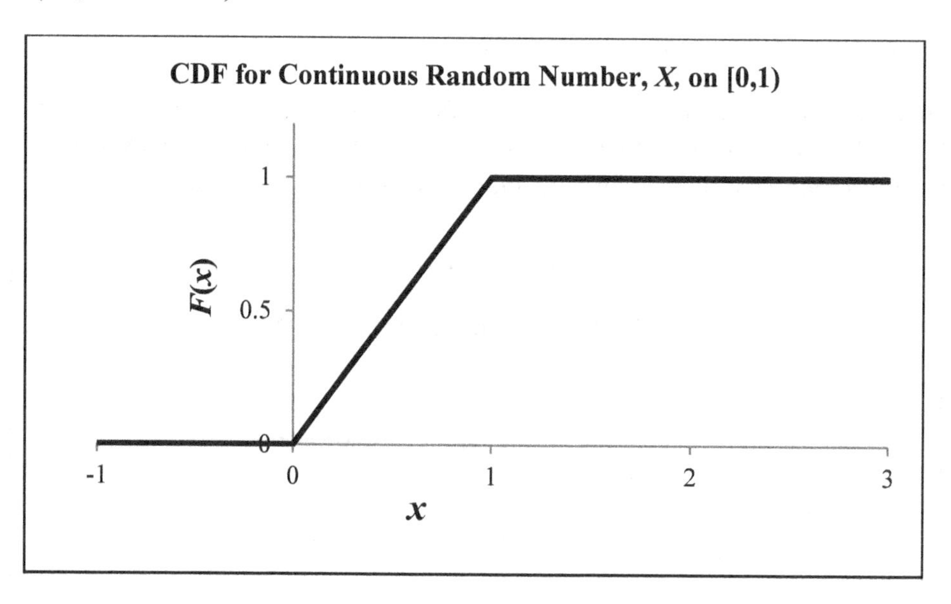

Note: The graph of $F_X(x)$ is continuous for all x.

Here are some important observations about ALL continuous random variables:

- Since the CDF is a continuous function, the probability of an event consisting of a single point must be zero. Otherwise, there would be a jump in the graph of the CDF. In other words, $\Pr[\{a\}] = \Pr(a,a] = F(a) - F(a) = 0$ for all values a. A more precise statement would involve left and right-hand limits at $x = a$. The continuity of $F_X(x)$ assures us that these limits are equal, so that $\Pr[\{a\}] = 0$.

- Therefore, for a continuous random variable X, the probability of an event $(a,b]$ is the same whether the endpoints are included or not included. Consequently, we will no longer feign concern as to whether we write an interval event for a continuous random variable as $(a,b]$, or $[a,b]$, or (a,b), etc.

If we have a mixed random variable (partly continuous and partly discrete), then the endpoints may be of substantial concern.

- When defining $F_X(x) = \Pr(-\infty, x] = \Pr[X \le x]$, we deferred the question of from whence the probability function arose. It turns out that we can reverse the definition and assign probabilities using (d) directly from any function $F(x)$ that satisfies the first three properties (a-c) of a CDF. We then ***define*** the random variable X and the corresponding probability function as that emerging from $F(x)$. In particular, $\Pr[E]$ is defined, at least for E an interval, by property (d). That, it turns out, will suffice, for then $\Pr[E]$ can be ***extended*** to a wide collection of events, E. This is best accomplished by means of some advanced mathematics called measure theory. Since the events we are most often concerned with at this level are intervals, or simple unions of intervals, we need not be overly concerned about the subtleties of measure theory.

- One interesting oddity that results from the previous point is that the probability of our theoretical random number generator picking a rational number is zero! This is because the set of rational numbers on the unit interval is "small," in set theoretic terms. This is indeed peculiar, since all of the random number generators on computers are constrained by the finite number of decimal places available to always choose rational numbers. But these are, of course, just real-world simulations of our idealized random number generator.

5.2 DENSITY FUNCTIONS

Let X be a continuous random variable, with CDF $F_X(x)$. We make the further assumption that the CDF is ***differentiable*** on the domain for which $0 < F_X(x) < 1$. The derivative of $F_X(x)$, denoted by the lower case $f_X(x)$, is called the ***density function*** of the random variable X.

Definition of the Density Function of a Random Variable X

If X is a random variable with a differentiable cumulative distribution function $F_X(x)$, then the density function for X is given by:

$$f_X(x) = F'_X(x).$$

For example, if X is the random number generator on the unit interval in example 5.1-2, then we have $F_X(x) = x$ for $0 \le x \le 1$ and therefore, $f_X(x) = \frac{d}{dx}x = 1$, for $0 < x < 1$. The density function for X is identically 1 on the unit interval and identically zero outside the unit interval.

The alert student will note that $f_X(x)$ is not defined at $x = 0$ and $x = 1$, since $F_X(x)$ is not differentiable at those points (there are *corners* in the graph there).

The relationship between derivatives and indefinite integrals is studied in calculus courses. The further connection to probability distributions is embedded in the ***Fundamental Theorem of Calculus***, which relates indefinite integrals (or, as they are sometimes called, antiderivatives) to definite integrals via the calculation of areas under graphs. The relationship $f_X(x) = F'_X(x)$ is rendered, using indefinite integral notation, into $F_X(x) = \int f_X(t)\,dt + C$. We choose the value of the arbitrary constant of integration, C, that will assure that $\lim_{x \to \infty} F_X(x) = 1$. This is equivalent to writing $F_X(x)$ as the definite integral $F_X(x) = \int_{-\infty}^{x} f_X(t)\,dt$.

Relationships Between the CDF and the Density Function for *X*

1. The density function is calculated from the CDF by differentiating:

$$f_X(x) = F'_X(x),$$

2. The CDF is calculated from the density function by integrating:

$$F_X(x) = \int_{-\infty}^{x} f_X(t)\,dt.$$

Note
In formula (2) the upper endpoint of the definite integral is the independent variable x, and the variable of integration is denoted by a different letter, in this case t. This is important, as it is incorrect, indeed nonsensical, to have the variable of integration the same as one of the endpoints of integration. This is not just a notational nicety, for confusing the two will lead to incorrect relationships (and wrong answers) for the CDF.

Example 5.2-1 Calculate the Density Function from the CDF

Find the probability density function for the random variable Y with cumulative distribution function

$$F(y) = \begin{cases} 0 & \text{if } y < 0 \\ 1 - e^{-y^2} & \text{if } y \geq 0 \end{cases}.$$

Solution
Observe that $F(y)$ satisfies the properties of a cumulative distribution function. That is, $F(y)$ is increasing from 0 to 1. The corresponding density function is found by differentiation. Remember to use the chain rule.

$$f(y) = F'(y) = \begin{cases} 0 & \text{if } y < 0 \\ 2ye^{-y^2} & \text{if } y \geq 0 \end{cases}.$$

The density function is identically zero outside the interval $[0, \infty)$ of non-negative real numbers. In such a situation, we say that Y is a ***non-negative*** random variable. ❐

Convention: It is common to omit defining the parts of the CDF that are zero or one. In the aforementioned example, one would typically write $F(y) = 1 - e^{-y^2}$ for

$y \geq 0$. Similarly, one would write $f(y) = 2ye^{-y^2}$ for $y > 0$, implying that $f(y) = 0$ for $y \leq 0$. We will say the random variable Y "lives" on the interval $[0, \infty)$, as a shorthand description of the situation. In advanced mathematics the interval where Y lives is referred to as the ***support***.

Example 5.2-2 Calculate the CDF from a Density Function

The non-negative random variable X has density function given by $f(x) = \frac{1}{3}e^{-x/3}$ for $x \geq 0$. Find the CDF of the random variable X.

Solution

$$F(x) = \int_{-\infty}^{x} f(t)dt = \int_{-\infty}^{0} 0dt + \frac{1}{3}\int_{0}^{x} e^{-\frac{1}{3}t}\, dt = -e^{-(1/3)t}\Big|_{0}^{x} = 1 - e^{-(1/3)x}; \quad 0 \leq x < \infty.$$

Thus,

$$F(x) = \begin{cases} 1 - e^{-x/3} & \text{for } x \geq 0 \\ 0 & \text{for } x < 0 \end{cases}.$$

□

Note: This is an example of what is known as the ***exponential*** family of distributions, which we will study in detail in chapter 6.

The properties listed above for a cumulative distribution function translate into corresponding properties for a probability density function.

Properties of the Probability Density Function

(1) $f_X(x) \geq 0$. That is, all density functions are non-negative.

(2) $\int_{-\infty}^{\infty} f_X(x)dx = 1$;

(3) $\Pr[a \leq X \leq b] = \int_{a}^{b} f_X(x)dx$.

The first property follows from the definition of derivative and the fact that the CDF is non-decreasing. The second property allows us to think of probability as area under a graph, with total probability (area) equal to one. The third property completes that thought by relating the probability of an event (the interval from a to b) to the area under the density function graph between $x = a$ and $x = b$.

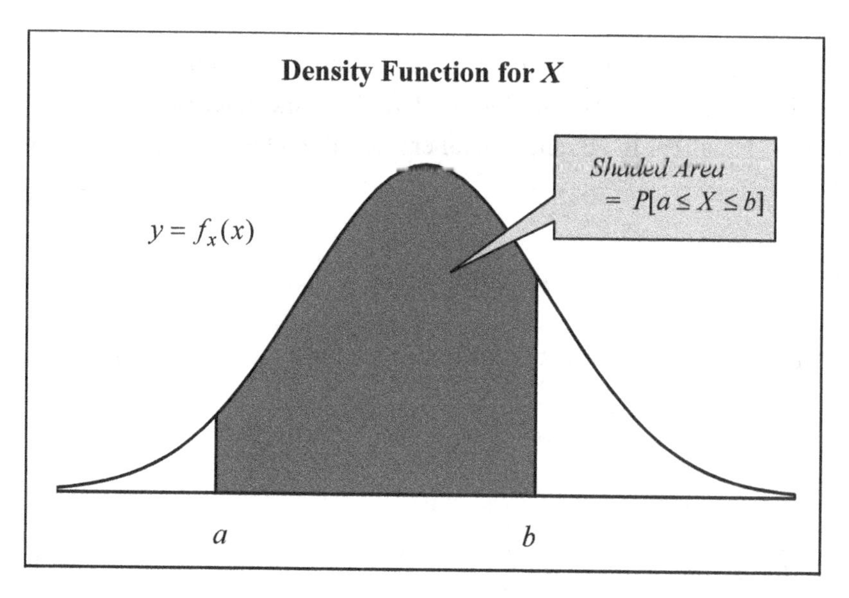

As previously noted, for a continuous random variable the probability of a single point is zero. This can now be thought of as the area under the curve from $x = a$ to $x = a$, or the zero "area" of a one dimensional line under the graph. However the value $f(x)$ of the density function at x (in other words, the height of the graph at x) can be used as a surrogate for the "probability of x." More precisely, $f(x)$ is the instantaneous ***rate*** at which probability is being accumulated at x.

A short interval from x to $x + dx$ may have positive probability that can be calculated as

$$\Pr\left[x \le X \le x + dx\right] = \int_x^{x+dx} f(t)dt \approx \underbrace{f(x)}_{\substack{\text{height of} \\ \text{density function}}} \cdot \underbrace{dx}_{\substack{\text{length of} \\ \text{interval}}} .$$

The last approximation is obtained by thinking of the area under the density function from x to $x + dx$ as the area of a skinny rectangle of height $f(x)$ and width dx. Just as in the derivation of the definite integral in calculus, we can *discretize* the problem of calculating area (probability) by partitioning the interval $[a,b]$ into small subintervals and summing the areas of the resulting skinny rectangles. The limit of the resulting sum is the definite integral used to calculate probability.

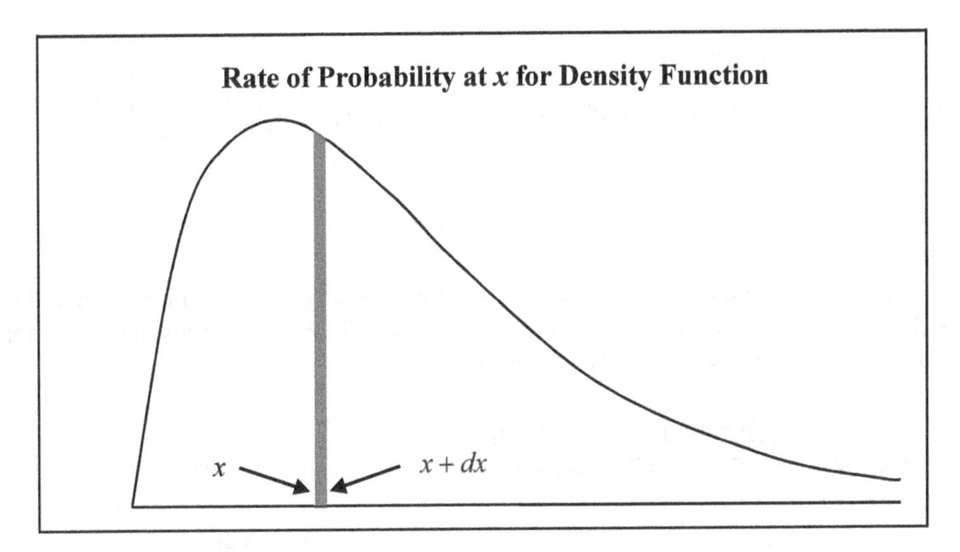

We now have two methods for calculating the probability of an interval event, one using the CDF, $\Pr(a \le X \le b) = F(b) - F(a)$, and one using the density function, $\Pr(a \le X \le b) = \int_a^b f(x)dx$. The fact that these two are equivalent is a manifestation of the Fundamental Theorem of Calculus. Which is easiest to use in solving problems? The answer generally depends on what information is given and how many events require probability calculations. Often, it is a matter of personal choice. We will illustrate the ideas above with a number of examples.

For purposes of problem solving, we note that if $g(x)$ is a non-negative function and has a finite definite integral (greater than zero), then one can choose k so that $f(x) = k \cdot g(x)$ is a valid density function. The constant k is chosen to assure that the total probability equals one, that is, $\int_{-\infty}^{\infty} f(x)dx = 1$. This means that,

$$k = \frac{1}{\int_{-\infty}^{\infty} g(x)\,dx}.$$

Most density functions for continuous random variables are identically zero outside of a single interval (A, B), which is called the ***domain***[3] of the random variable X. As previously, we will say that the underlying random variable x ***lives*** on that interval (A, B). It follows that the CDF satisfies $F(A) = 0$ and $F(B) = 1$. If A or B is infinite, then this must be understood in the usual calculus sense of a limit (for example, $\lim_{x \to \infty} F(x) = 1$ in the case $B = \infty$).

Typical exercises might involve finding the proportionality constant k; determining if $f(x)$ is a valid probability density function; or computing properties such as the mean and variance of the underlying random variable X.

[3] Possibly, $A = -\infty$ and/or $B = \infty$.

Example 5.2-3 Calculate the CDF from the Density Function

Consider a random variable X with density function $f(x)=\begin{cases} c\cdot x^3\cdot(1-x) & \text{for } 0\le x\le 1 \\ 0 \text{ otherwise} \end{cases}$.

Find $\Pr(X>.6)$.

> **Note**
>
> Like the random number generator, this random variable is a member of the ***Beta*** family of distributions, which live on the unit interval. We will study these in more detail in Section 6.6. We will solve the problem two ways:

Solution Method 1: Density Function Method:
We first find the constant c from the fact that total probability must equal 1.

$$1 = c\int_0^1 x^3(1-x)\,dx = c\int_0^1 (x^3-x^4)\,dx = c\cdot\left[\frac{x^4}{4}-\frac{x^5}{5}\right]\Bigg|_0^1 = \frac{c}{20}.$$

Therefore $c=20$.

$$\Pr(X>.6) = \Pr[.6<X<1] = 20\int_{.6}^1 x^3(1-x)\,dx = 20\left[\frac{x^4}{4}-\frac{x^5}{5}\right]\Bigg|_{.6}^1 = .6630.$$

Solution Method 2: CDF Method:
We find the cumulative distribution function, $F(X)$. For x, $0\le x\le 1$,

$$F(x) = c\int_0^x t^3(1-t)\,dt = c\left[\frac{x^4}{4}-\frac{x^5}{5}\right].$$

Next, we evaluate c using the fact that $F(1)=1$.

$$1 = F(1) = c\left[\frac{1}{4}-\frac{1}{5}\right] = \frac{c}{20}.$$

It follows that $c=20$ and $F(x)=5x^4-4x^5$. Finally,

$$\Pr(X>.6) = 1-\Pr[X\le .6] = 1-F(.6) = 1-\left[5(.6^4)-4(.6^5)\right] = .6630.$$ ❐

These procedures work equally well when one or both of the endpoints are infinite, although these calculations involve limits.

Example 5.2-4 Evaluate *k* and Calculate a Probability

The density function for a random variable X is given by $f(x)=\frac{k}{(x+10)^2};\ 0\le x<\infty$. Calculate $\Pr[X\le 10]$.

Solution

Using the CDF method, we find

$$F(x) = k\int_0^x (t+10)^{-2}\,dt = \left.\frac{k(t+10)^{-1}}{-1}\right|_0^x = k\left[\frac{1}{10}-\frac{1}{x+10}\right] \text{ for } 0 \le x < \infty.$$

Then $1 = F(\infty) = \lim_{x\to\infty} k\left[\frac{1}{10}-\frac{1}{x+10}\right] = k\frac{1}{10}$. So the unknown constant is $k = 10$ and the CDF is $F(x) = 1-\frac{10}{x+10}$. Then, $\Pr[X \le 10] = F(10) = \frac{1}{2}$. □

Example 5.2-5 Calculate a Conditional Probability

SOA Exam P Sample Exam Questions #36

A group insurance policy covers the medical claims of the employees of a small company. The value, V, of the claims made in one year is described by $V = 100{,}000Y$ where Y is a random variable with density function

$$f(y) = \begin{cases} k(1-y)^4 & \text{for } 0<y<1 \\ 0 & \text{otherwise} \end{cases}.$$

and k is a constant. What is the conditional probability that V exceeds 40,000, given that V exceeds 10,000?

(A) 0.08 (B) 0.13 (C) 0.17 (D) 0.20 (E) 0.51

Solution

In terms of the random variable Y, the problem can be formulated as: calculate $\Pr[Y > .4 \mid Y > .1] = \frac{\Pr[Y>.4]}{\Pr[Y>.1]}$. We will find the CDF of Y in order to evaluate these probabilities.

$$F(y) = \int_0^y k(1-t)^4\,dt = \left.\frac{k(1-t)^5}{-5}\right|_0^y = \frac{k}{5}\left[1-(1-y)^5\right];\ 0 \le y \le 1.$$

Then $1 = F(1) = \frac{k}{5}$ implies $k = 5$ and $F(y) = 1-(1-y)^5$. Thus,

$$\Pr[Y > y] = 1-F(y) = (1-y)^5.$$

Finally,

$$\Pr[Y > .4 \mid Y > .1] = \frac{\Pr[Y > .4]}{\Pr[Y > .1]} = \frac{.6^5}{.9^5} = \left(\frac{2}{3}\right)^5 = .1317.$$ □

Notes

(1) Because the question involves a conditional probability, the constant k will cancel out top and bottom, and needn't have been calculated.

(2) Using the given density function, the calculation is,

$$\Pr[Y > .4 \mid Y > .1] = \frac{\Pr[Y > .4]}{\Pr[Y > .1]} = \frac{k\int_{0.4}^{1}(1-y)^4\, dy}{k\int_{0.1}^{1}(1-y)^4\, dy} = \frac{.6^5}{.9^5} = .1317.$$

Exercise 5-1 Consider the random variable X with cumulative distribution function

$$F(x) = \begin{cases} 0 & \text{if } x < 0 \\ 1 - e^{-4x} & \text{if } x \geq 0 \end{cases}.$$

(a) Show that $F(x)$ has the properties of a cumulative distribution function.

(b) Find the probability density function $f(x)$.

(c) Find $\Pr(1 < x \leq 2)$ using both the distribution function and the density function.

(d) Evaluate the conditional probability $\Pr\left[X > 2 \mid X > 1\right]$.

Exercise 5-2 Consider the random variable X with cumulative distribution function,

$$F_X(x) = \begin{cases} 0 & \text{if } x < 0 \\ x^3 & \text{if } 0 \leq x \leq 1. \\ 1 & \text{if } x > 1 \end{cases}$$

(a) Show that $F(x)$ has the properties of a cumulative distribution function.

(b) Find the probability density function $f(x)$.

(c) Find $\Pr(.1 < X \leq .5)$ using both the distribution function and the density function.

(d) Evaluate the conditional probability $\Pr(.1 < X \mid X \leq .5)$.

Exercise 5-3 Consider the random variable Y with probability density function

$$f(y) = ke^{-3y} \text{ for } 0 \le y < \infty.$$

(a) Find the value of k using the density function method.

(b) Evaluate k by finding the CDF and using $F(\infty) = 1$.

(c) Find $\Pr(2 < y \le 3)$ using both the cumulative distribution function and the density function.

(d) Find $\Pr(Y > -3)$.

Exercise 5-4 Suppose the afternoon temperature on the beaches of Honolulu in June is a random variable denoted by T. Further suppose the temperature is *uniformly* distributed between 82° F and 90° F. In other words, the density function for T is constant on the interval [82,90]. (Random variables that have ***continuous uniform distribution*** will be discussed in detail in section 6.1.)

(a) What is the average (mean) temperature? That is, find $E[T]$.

(b) What is the probability that the temperature is exactly 87°F?

(c) What is the probability that a digital thermometer that rounds to the nearest degree displays a temperature reading of 87°F? That is, find $\Pr(86.5 < T < 87.5)$.

Exercise 5-5 SOA Exam P Sample Exam Questions #38
An insurance company insures a large number of homes. The insured value, X, of a randomly selected home is assumed to follow a distribution with density function

$$f(x) = 3x^{-4}; \; 1 < x < \infty.$$

Given that a randomly selected home is insured for at least 1.5, what is the probability that it is insured for less than 2?

(A) 0.578 (B) 0.64 (C) 0.704 (D) 0.829 (E) 0.875

Exercise 5-6 SOA Exam P Sample Exam Questions #33
The loss due to a fire in a commercial building is modeled by a random variable X with density function

$$f(x) = \begin{cases} 0.005(20-x) & \text{for } 0 < x < 20 \\ 0 & \text{otherwise} \end{cases}.$$

Given that a fire loss exceeds 8, what is the probability that it exceeds 16?

(A) $\frac{1}{25}$ (B) $\frac{1}{9}$ (C) $\frac{1}{8}$ (D) $\frac{1}{3}$ (E) $\frac{3}{7}$

Exercise 5-7 SOA Exam P Sample Exam Questions #35

The lifetime of a machine part has a continuous distribution on the interval $(0,40)$ with probability density function *f*, where $f(x)$ is proportional to $(10+x)^{-2}$.

What is the probability that the lifetime of the machine part is less than 5?

(A) 0.03 (B) 0.13 (C) 0.42 (D) 0.58 (E) 0.97

5.3 GREAT EXPECTATIONS

Procedures for finding expected values and variances for continuous random variables are similar to those for discrete distributions discussed in Chapter 3. The main difference is that instead of sums we use definite integrals.

The most general statement of the underlying principle involves calculating the expected value of a function, or transformation, of a random variable. This is essentially a restatement of the rule discussed in Subsection 3.3.1.

The Expectation of a Transformed Random Variable

Let X be a continuous random variable with density function $f(x)$. The expected value of a transformation $g(X)$ of X, is calculated by,

$$E[g(X)] = \int_{-\infty}^{\infty} g(x)\cdot f(x)\,dx.$$

If the random variable X lives on the interval (A,B), then we use the limits of integration from A to B since the density function $f(x)$ is identically zero outside this interval.

Example 5.3-1 Expected Value of X

Let X be the random variable whose density is given by $f(x)=20x^3(1-x);\ 0\le x\le 1$. Calculate the expected value $E[X]$.

Solution

In this case, $g(x)=x$ and X lives on $0\le x\le 1$, so the expected value is

$$E[X] = \int_0^1 x\cdot f(x)\,dx = 20\int_0^1 x^4(1-x)\,dx = 20\int_0^1 (x^4-x^5)\,dx = 20\left[\frac{1}{5}-\frac{1}{6}\right] = \frac{20}{30} = \frac{2}{3}. \quad ❒$$

Example 5.3-2 Expected Value of a Transformed Random Variable

Let X have density function $f(x) = 2e^{-2x}$; $0 \le x < \infty$. Calculate the expected value of e^X.

Solution

$$E[e^X] = \int_0^\infty e^x \cdot (2e^{-2x})\,dx = 2\int_0^\infty e^{-x}\,dx = 2\frac{e^{-x}}{-1}\bigg|_0^\infty = 2(1-0) = 2. \quad \square$$

Example 5.3-3 Expected Value of X

Let X have density function $f(x) = 2e^{-2x}$; $0 \le x < \infty$. Calculate the expected value of X.

Solution

$E[X] = \int_0^\infty x \cdot (2e^{-2x})\,dx = \int_0^\infty 2x \cdot e^{-2x}\,dx$. This integral requires integration-by-parts. This is a common integral type and we shall give some shortcuts in the next chapter. Let $u = 2x$ and $dv = e^{-2x}dx$. Then we get $du = 2dx$ and $v = \frac{e^{-2x}}{-2}$.

$$E[X] = u \cdot v\Big|_0^\infty - \int_0^\infty v\,du = -xe^{-2x}\Big|_0^\infty + \int_0^\infty e^{-2x}dx = \left\{-xe^{-2x} - \frac{e^{-2x}}{2}\right\}\bigg|_0^\infty = \frac{1}{2}. \quad \square$$

The integrals we encounter in examples like the last two are termed *improper* in calculus, meaning that a limit of integration is infinite. The mathematically precise way to evaluate these is to write $\int_0^\infty f(x)dx = \lim_{T\to\infty} \int_0^T f(x)dx$ and evaluate any indeterminate expressions using L'Hôpital's rule. We will omit all these niceties when the limit is clear. In particular, you should know that exponential functions dominate polynomials in the limit, so that, for example, $\lim_{T\to\infty} Te^{-2T} = 0$.

Example 5.3-4 Expected Value of a Transformation

Let X be uniformly distributed on the interval $[-2,6]$. That is, suppose the density function is

$$f(x) = \begin{cases} \frac{1}{8} & \text{for } -2 \le x \le 6 \\ 0 & \text{otherwise} \end{cases}.$$

Calculate the expected value of $g(X) = X^3 + \sqrt{X+2}$.

Solution

$$E\left[X^3+\sqrt{X+2}\right] = \int_{-\infty}^{\infty}(x^3+\sqrt{x+2})\cdot f(x)\,dx = \int_{-2}^{6}(x^3+\sqrt{x+2})\cdot\frac{1}{8}\,dx$$

$$= \frac{1}{8}\left(\frac{x^4}{4}+\frac{2(x+2)^{3/2}}{3}\right)\Bigg|_{x=-2}^{x=6}$$

$$= \frac{1}{8}\left(\frac{6^4}{4}+\frac{2(6+2)^{3/2}}{3}-\frac{(-2)^4}{4}-\frac{2\cdot 0^{3/2}}{3}\right)$$

$$= 41.89.$$

❐

Exercise 5-8 Consider a random variable X with density function

$$f(x) = \begin{cases} \frac{x^3}{64} & \text{for } 0 \le x \le 4 \\ 0 & \text{otherwise} \end{cases}.$$

(a) Calculate the mean of X.
(b) Calculate the second moment of the random variable X.
(c) Calculate the cumulative probability distribution function $F_X(x)$.
(d) Find $\Pr(2 < x \le 3)$.

Exercise 5-9 SOA Exam P Sample Exam Questions #55

An insurance company's monthly claims are modeled by a continuous, positive random variable X, whose probability density function is proportional to $(1+x)^{-4}$, where $0 < x < \infty$. Determine the company's expected monthly claims.

(A) $\frac{1}{6}$ (B) $\frac{1}{3}$ (C) $\frac{1}{2}$ (D) 1 (E) 3

Exercise 5-10 Consider a continuous random variable X with probability density function given by $f(x) = 4\cdot x\cdot(1-x^2)$, for $0 \le x \le 1$. Calculate $E[\sqrt{X}]$.

5.3.1 THE VARIANCE FORMULA

As with discrete random variables, the two most important characteristics of a continuous distribution are the mean and the variance. The mean is synonymous with $E[X]$, often denoted by the Greek letter μ, pronounced "mew," or μ_X (to reinforce that we are talking about the mean of the random variable X) and calculated as described above.

Variance is defined, as previously, to be the expected square deviation from the mean, and is most often calculated by the use of the variance formula.

Variance of a Continuous Random Variable *X*

The ***variance*** of a random variable X is denoted by $Var[X]$ (or σ_X^2) and is defined as

$$Var[X] \overbrace{=}^{\text{definition}} E\left[(X-\mu_X)^2\right] = \int_{-\infty}^{\infty}(x-\mu_X)^2 f_X(x)\,dx.$$

Variance Formula: $Var[X] = E[X^2]-E[X]^2 = \int_{-\infty}^{\infty} x^2 f(x)\,dx - \mu_X^2.$

The derivation of the variance formula is identical with the discrete version presented in Subsection 3.4.2 and we will not repeat it here. It is usually easier to use the variance formula instead of the definition when computing the variance.

Example 5.3-5 Mean and Variance of the Random Number Generator *X*

Find the mean and the variance of the random variable X whose density function is $f(x)=1$ for $0<x<1$.

Solution

In Example 5.1-2, we showed that the CDF of the random number generator in the unit interval is given by $F(x)=x;\ 0\le x\le 1$, so that the density function is $f(x)=1;\ 0<x<1$. Therefore, X is the random number generator and the mean of this random variable is,

$$E[X] = \int_0^1 x\,dx = \left.\frac{x^2}{2}\right|_0^1 = \frac{1}{2}.$$

Next, the second moment of X is $E[X^2] = \int_0^1 x^2\,dx = \left.\frac{x^3}{3}\right|_0^1 = \frac{1}{3}.$

And finally,

$$Var[X] = E[X^2]-E[X]^2 = \frac{1}{3}-\left(\frac{1}{2}\right)^2 = \frac{1}{3}-\frac{1}{4} = \frac{1}{12}.$$ ❐

Example 5.3-6 Variance of X

Let X be the random variable whose density is given by $f(x) = 20x^3(1-x);\ 0 \le x \le 1$. Calculate $Var[X]$.

Solution

From Example 5.3-1, we already determined that the mean of X is $\mu_X = \frac{2}{3}$. We calculate

$$\begin{aligned} E[X^2] = \int_0^1 x^2 f(x)\,dx &= 20\int_0^1 x^5(1-x)\,dx \\ &= 20\int_0^1 (x^5 - x^6)\,dx \\ &= 20\left[\frac{1}{6} - \frac{1}{7}\right] = \frac{20}{42} = \frac{10}{21}. \end{aligned}$$

Then

$$\sigma_X^2 = E[X^2] - (\mu_X)^2 = \frac{10}{21} - \left(\frac{2}{3}\right)^2 = \frac{2}{63}.$$ ❒

As previously, the ***standard deviation***, σ_X, of X is the square root of the variance.

Other useful formulas from Chapter 3 that carry over without change concern the mean and variance of a linear transformation of a continuous random variable X.

The Mean and Variance of a Linear Transformation

Let X denote any random variable and suppose $Y = aX + b$ where a and b are real-valued constants. Then,

(1) $E[Y] = a \cdot E[X] + b$, and

(2) $Var[Y] = a^2 \cdot Var[X]$.

Exercise 5-11 Consider the random variable X with density function $f(x) = \frac{x^3}{64}$, for $0 \le x \le 4$ (see Exercise 5-8). Calculate the variance of X.

Exercise 5-12 Consider a continuous random variable X with probability density function given by $f(x) = \begin{cases} c \cdot x & \text{for } 0 \le x \le 1 \\ 0 & \text{otherwise} \end{cases}$.

Calculate the coefficient of variation of X.

Exercise 5-13 Consider a continuous random variable X with probability density function given by $f(x)=\begin{cases}4\cdot x\cdot(1-x^2) & \text{for } 0\le x\le 1\\ 0 & \text{otherwise}\end{cases}$.

Calculate the standard deviation of X.

Exercise 5-14 Consider a continuous random variable X with probability density function given by $f(x)=\begin{cases}\frac{1}{7} & \text{for } 1\le x\le 8\\ 0 & \text{otherwise}\end{cases}$.

(a) Calculate the standard deviation of X.

(b) Calculate $E\left[\sqrt{X+1}\right]$.

Exercise 5-15 Consider a random variable Y with cumulative distribution function

$$F_Y(y)=\begin{cases}0 & \text{for } y\le 2\\ \frac{1}{4}\left\{\frac{-y^2}{4}+3y-5\right\} & \text{for } 2<y<6\\ 1 & \text{otherwise}\end{cases}.$$

(a) Find the probability density function of Y.

(b) Find the mean of Y.

(c) Graph the probability density function for Y.

Exercise 5-16 SOA Exam P Sample Exam Questions #60

A recent study indicates that the annual cost of maintaining and repairing a car in a town in Ontario averages 200 with a variance of 260. If a tax of 20% is introduced on all items associated with the maintenance and repair of cars (*i.e.*, everything is made 20% more expensive), what will be the variance of the annual cost of maintaining and repairing a car?

(A) 208 (B) 260 (C) 270 (D) 312 (E) 374

Hint: If C denotes the cost random variable, then $C_{\text{new}}=1.2\cdot C_{\text{old}}$ and the linear transformation formula for variance can be used.

5.3.2 THE MODE OF A CONTINUOUS DISTRIBUTION

The concept of mode is carried over from discrete distributions to stand for the value x of greatest likelihood. Of course, for a continuous random variable all (individual) points are equally likely, with probability zero. The translation that is generally adopted is to find the point where the *rate* of probability accumulation is greatest. This amounts to finding the point(s) where the value of the probability density function is greatest.

The Mode of a Continuous Random Variable X

Let X have probability density function $f(x)$. The ***mode*** of X is defined as the value(s), x_{mode}, that maximizes $f(x)$. If there are more than two values of x that maximize $f(x)$ we will say a mode does not exist[4].

The mode can often be found using calculus tools, that is, by differentiating $f(x)$ and setting the derivative to zero to locate the critical point that represents the global maximum, x_{mode}. Always, in solving max-min problems in calculus, it is important to check the endpoints to make sure you are actually finding the global maximum.

Example 5.3-7 Calculate a Mode

Let X be the random variable whose density is given by $f(x) = 20x^3(1-x)$ for $0 \le x \le 1$. Find x_{mode}.

Solution

Differentiating the density function gives $f'(x) = 20[3x^2 - 4x^3]$. Setting $f'(x) = 0$, we find the two critical points, $x = 0$ and $x = \frac{3}{4}$. Since $f(x)$ is zero at the endpoints, we must have $x_{\text{mode}} = \frac{3}{4}$. □

Example 5.3-8 Calculate a Mode

Let Z have density function $f(z) = 2e^{-2z}$ for $0 \le z < \infty$. Calculate the mode of Z.

Solution

In this case $f'(z) = -4e^{-2z}$ is never zero (it is always negative, implying that the density function is always decreasing for $z > 0$). It follows that the mode occurs at the left end-point of the domain. That is, $z_{\text{mode}} = 0$. □

Exercise 5-17 Find the mode(s) of the random variable X with density function

$$f(x) = 6x(1-x) \text{ on } 0 \le x \le 1.$$

Exercise 5-18 Find the mode(s) of the random variable T with density function

$$f(t) = 7.5t - 18.75t^2 + 15t^3 - 3.75t^4 \text{ on } 0 \le t \le 2.$$

---

[4] How many modes are too many? There is not universal agreement as to when the mode is undefined. We will adopt the convention that more than two is too many. Most would agree that the density function $f(x) = 1;\ 0 \le x \le 1$, which attains its maximum over the entire interval, has no identifiable mode. Fortunately this conundrum rarely arises in practice.

5.3.3 MEDIANS AND PERCENTILES

Here is an instance where the concept for continuous random variables is actually easier than for discrete random variables.

Percentiles of a Continuous Random Variable X

Suppose p is a number between 0 and 1. The $100p^{th}$ percentile of X is the point x_p such that $p = \Pr\left[X \le x_p\right]$.

The 50th percentile is called the ***median***, and is denoted by $x_{.5}$.

Solving for percentiles of any random variable is conceptually straightforward when the CDF of X is available. Since $p = \Pr[X \le x_p] = F(x_p)$, finding percentiles is a matter of solving the equation $p = F(x_p)$ for x_p in terms of p. The resulting equation may be hard to solve in practice, requiring graphing utilities, sophisticated software, Newton's method, or more complicated numerical approximation methods. But the principle is simple, illustrated by the following schematic of a CDF graph.

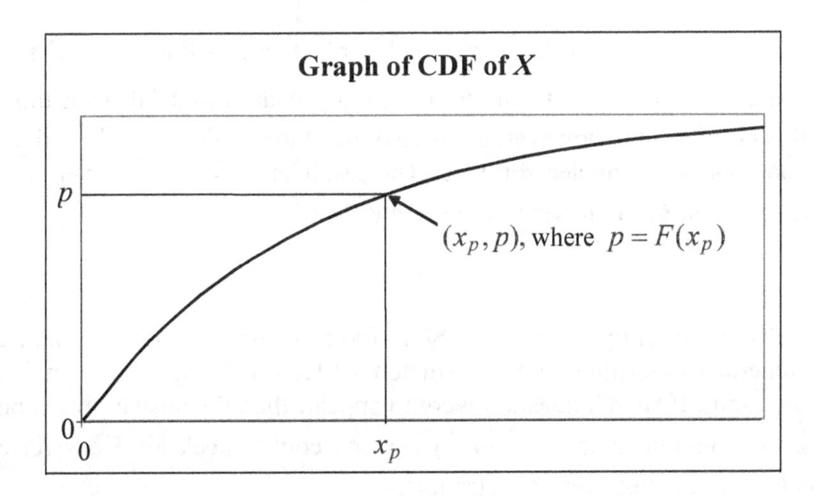

Example 5.3-9 Percentiles for the Random Number Generator X

Find the $100p^{th}$ percentile of X.

Solution

Since $F(x) = x$, $x_p = p$ for all p. For example, the median $x_{.5} = .5 = \frac{1}{2}$. The 75th percentile is $x_{.75} = .75$, and so forth. ❐

Example 5.3-10 Percentiles

The non-negative random variable X has density $f(x)=\frac{1}{3}e^{-(1/3)x}$; $0\le x<\infty$. Find a closed formula expression for the $100p^{th}$ percentile of X.

Solution
In Example 5.2-2 we calculated the CDF for this random variable:

$$F(x) = 1-e^{-(1/3)x};\ 0\le x<\infty.$$

Therefore, for a given percentile p, we must solve the equation $p=1-e^{-(1/3)x}$ for x. Using the natural logarithm we find $x_p=-3\ln(1-p)$. For example, the median is $x_{.5}=-3\ln(0.5)=2.079$ and the 90th percentile is $x_{.9}=-3\ln(.1)=6.908$. ❐

Example 5.3-11 Median

Let X be the random variable whose density is given by $f(x)=20x^3(1-x)$; $0\le x\le 1$. Find the median of X.

Solution
In Example 5.2-3 we determined that $F(x)=5x^4-4x^5$. Thus, to find the median we need to solve the equation $.5=5x^4-4x^5$. Even an algebra superstar would falter on this 5th degree polynomial. It is easy enough, however, to approximate the solution with the aid of a graphing calculator or appropriate computer software. The result is 0.686, which can be checked by evaluating $5(.686)^4-4(.686)^5$ to see how close this is to 0.5. ❐

Notes

With the calculators currently allowed on SOA Exam P, this would be a hard calculation, requiring a numerical algorithm such as Newton's Method. Such a problem is unlikely to come up on the exam. If such a question were to appear, then the answer might be in symbol form (that is, the solution to $.5=5x^4-4x^5$), or one could check all 5 answer choices. Of course, if you have to do this, you are in trouble.

Example 5.3-12 The mean, median and mode for X

Let X have probability density function $f(x)=2e^{-2x}$; $0\le x<\infty$. Calculate and compare the mean, the median and the mode of X.

Solution
From Example 5.3-3, the mean of X is $\mu_X=\frac{1}{2}$. From Example 5.3-8, the mode is zero. To calculate the median, we need to find the CDF and solve $F(x_{.5})=50\%$.

$F(x)=\int_0^x 2e^{-2t}\,dt=1-e^{-2x}$. Solving $1-e^{-2x}=0.5$ gives $x_{.5}=-\frac{\ln .5}{2}=.3466$.

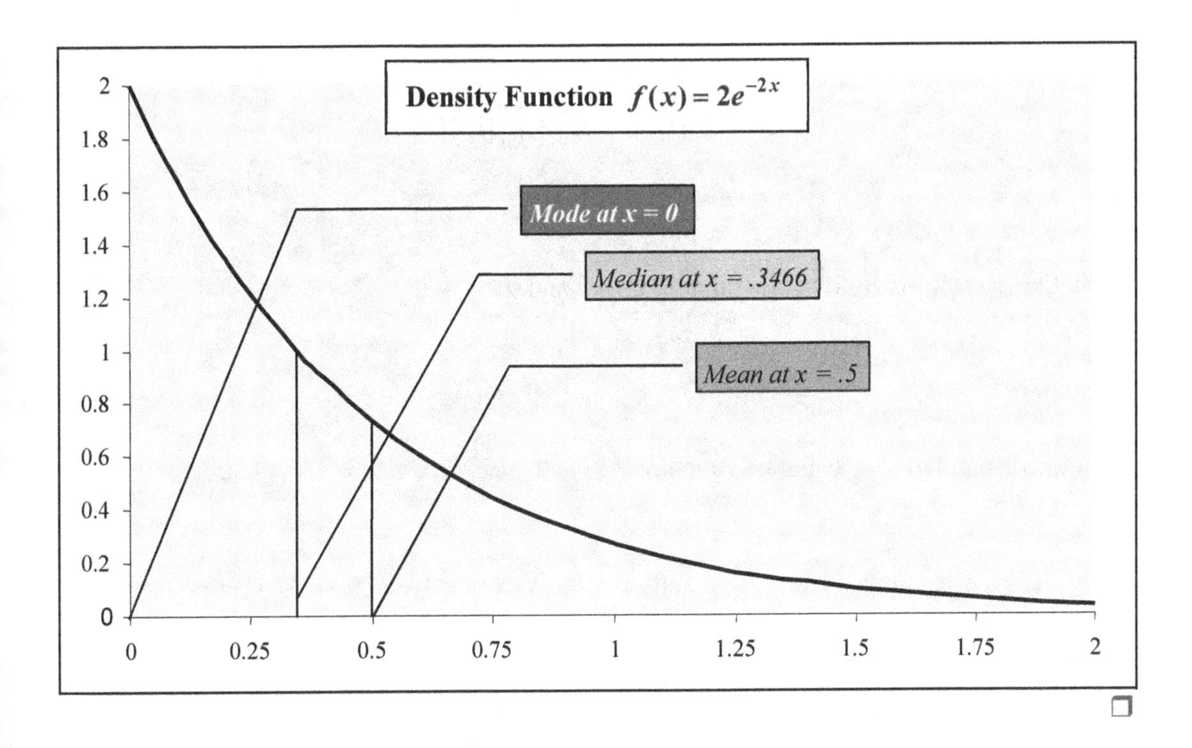

5.3.4 CALCULATING THE EXPECTED VALUE OF *X* WITH THE CDF

Just as probabilities can be calculated from either the density function or the CDF, it is also possible to calculate expected values with $f(x)$ or $F(x)$. In this subsection we present an alternative formula for calculating the mean $E[X]$ that uses the CDF rather than the density. The advantage is that the resulting integral *might* be easier to evaluate.

Calculating $E[X]$ Using the CDF

Let X be a non-negative random variable living on the interval (A, B). Then

$$E[X] = A + \int_A^B [1 - F(x)]\, dx.$$

The proof of this relationship uses integration-by-parts. Once derived, the formula can often be used to *avoid* an integration-by-parts that may arise when using the density function $f(x)$.

Proof

$$E[X] \overbrace{=}^{\text{definition}} \int_A^B x \cdot f(x)\, dx.$$

We let $u = x$ and $dv = f(x)\, dx$. Then $du = dx$ and $v = F(x) + C$.

We set $C = -1$, so that $v = -[1 - F(x)]$. Completing the integration-by-parts, we get

$$\begin{aligned} E[X] &= u\cdot v\Big|_A^B - \int_A^B v\cdot du \\ &= -x\cdot[1-F(x)]\Big|_A^B + \int_A^B [1-F(x)]dx \\ &= A + \int_A^B [1-F(x)]dx. \end{aligned}$$

In the last step we used the fact that $F(B)=1$ and $F(A)=0$.

Example 5.3-13 Expected Value of *X* Using the CDF

Let X have cumulative distribution function $F(x)=1-e^{-2x}$ for $x\geq 0$. Calculate μ_X.

Solution

Using the CDF formula, with $A=0$ and $B=\infty$, we have $1-F(x) = 1-(1-e^{-2x}) = e^{-2x}$,

and we get $E[X] = 0+\int_0^\infty e^{-2x}\,dx = \left.\frac{e^{-2x}}{-2}\right|_0^\infty = \frac{1}{2}.$

Here we note that $f(x)=2e^{-x/2}$, for $0\leq x<\infty$. In Example 5.3-3, we calculated $E[X]$ using this density function and integration-by-parts. ❐

Example 5.3-14 Expected Value of *X* Using the CDF

Let X be the random variable whose density is given by $f(x)=20x^3(1-x);\ 0\leq x\leq 1$. Calculate the expected value $E[X]$ using the CDF method.

Solution

By way of comparison, in Example 5.3-1 we determined that $E[X]=\frac{2}{3}$ using the density function. In Example 5.2-3 we showed the CDF for this random variable is $F(x)=5x^4-4x^5$ on the unit interval. So

$$E[X]=\int_0^1[1-F(x)]dx=\int_0^1[1-5x^4+4x^5]dx = \left. x-x^5+\frac{4x^6}{6}\right|_0^1 = \frac{2}{3}.$$ ❐

Example 5.3-15 **(See Exercise 5-9)**

An insurance company's monthly claims are modeled by a continuous, positive random variable X, whose probability density function is proportional to $(1+x)^{-4}$, where $0<x<\infty$. Determine the company's expected monthly claims.

Solution

We will now solve this using the CDF method.

$$F(x)=\int_0^x k(1+t)^{-4}\,dt = \left.\frac{k(1+t)^{-3}}{-3}\right|_0^x = \frac{k}{3}-\frac{k(1+x)^{-3}}{3}.$$

Then $1 = F(\infty) = \frac{k}{3}$ implies $k = 3$ and $F(x) = 1-(1+x)^{-3}$.

So, $E[X] = \int_0^\infty (1+x)^{-3}\,dx = \left.\frac{(1+x)^{-2}}{-2}\right|_0^\infty = \frac{1}{2}.$ ❐

Note: Solving this using the density function would require integration-by-parts.

Exercise 5-19 Consider a continuous random variable Z with probability density function given by $f(z)=2z$ for $0\le z\le 1$. Calculate:

(a) The cumulative probability distribution function $F_Z(z)$.

(b) The mean, μ_Z.

(c) The mode for Z.

(d) The median value of Z.

(e) The 78th percentile of the random variable, $z_{.78}$.

(f) The inter-quartile range (the difference between the 75th and 25th percentiles).

Exercise 5-20 Consider a continuous random variable X with probability density function given by $f(x)=\begin{cases} c\cdot x\cdot(1-x^2) & \text{for } 0\le x\le 1 \\ 0 & \text{otherwise}\end{cases}$.

Calculate:

(a) The value of the constant c.

(b) The cumulative probability distribution function $F(x)$.

(c) The mean value of X.

(d) The mode for X.

(e) The median.

(f) $x_{.21}$, the 21st percentile of the random variable.

Exercise 5-21 Consider a random variable Z with probability density function given by $f(z) = \frac{|\sin z|}{4}$ on the interval $0 \le z \le 2\pi$. Find the mean, median, and mode for Z.

Exercise 5-22 Create your own random variables that have zero, one, and two modes respectively.

Exercise 5-23 Consider a continuous random variable W with probability density function that is constant on the interval $[0,20]$. Calculate,

(a) The density function $f(w)$.

(b) The cumulative probability distribution function $F(w)$.

(c) The mean, $E[W]$. Solve this in three different ways!

(d) The mode for W.

(e) The median value of W.

(f) $w_{.69}$, the 69th percentile.

Exercise 5-24 Consider a continuous random variable X with probability density function given by $f(x) = c \cdot x$ for $1 \le x \le 5$, zero otherwise.

(a) Calculate the mean, median, and mode of X.

(b) Graph the density function and denote the mean, median, and mode on the graph.

Exercise 5-25 **SOA/CAS Course 1 2000 Sample Examination #39**

The loss amount, X, for a medical insurance policy has cumulative distribution function

$$F(x) = \begin{cases} 0 & x < 0 \\ \frac{1}{9}\left(2x^2 - \frac{x^3}{3}\right) & 0 \le x \le 3. \\ 1 & x > 3 \end{cases}$$

Calculate the mode of the distribution.

(A) $\frac{2}{3}$ (B) 1 (C) $\frac{3}{2}$ (D) 2 (E) 3

Exercise 5-26 SOA Exam P Sample Exam Questions #59

An insurer's annual weather-related loss, X, is a random variable with density function

$$f(x) = \begin{cases} \dfrac{2.5(200)^{2.5}}{x^{3.5}} & \text{for } x > 200 \\ 0 & \text{otherwise} \end{cases}.$$

Calculate the difference between the 30th and 70th percentiles of X.

(A) 35 (B) 93 (C) 124 (D) 231 (E) 298

Exercise 5-27 SOA Exam P Sample Exam Questions #45

Let X be a continuous random variable with density function

$$f(x) = \begin{cases} \frac{|x|}{10} & \text{for } -2 \leq x \leq 4 \\ 0 & \text{otherwise} \end{cases}.$$

Calculate the expected value of X.

(A) $\frac{1}{5}$ (B) $\frac{3}{5}$ (C) 1 (D) $\frac{28}{15}$ (E) $\frac{12}{5}$

5.4 MIXED DISTRIBUTIONS

It is easy enough to conjure up probability experiments leading to random variables that are partly continuous and partly discrete. These are called ***mixed distributions***, and they occasionally come up in insurance applications, as we will see in the next section. For example, it is reasonable to assume that models for life expectancy would be continuous distributions. But since it is extremely unlikely for one to live past (for example) 105 years of age, models can be defined (or forced) to have an ***ultimate*** age of 105. In the model we have a discrete point at age 105 with $\Pr\{105\} = p$, where in the real-world, $p = \Pr(\text{Age at Death} > 105)$ is a positive number, however small.

For a simple example, imagine a spinning wheel in which a sector constituting 75% of the area is painted gray and the remaining 25% is painted green. Let the circumference of the gray sector be calibrated using the unit interval. Then let the random variable $X = x$ if the pointer is at x in the gray sector when the wheel stops. If the pointer is anywhere in the green sector, then X is equal to one.

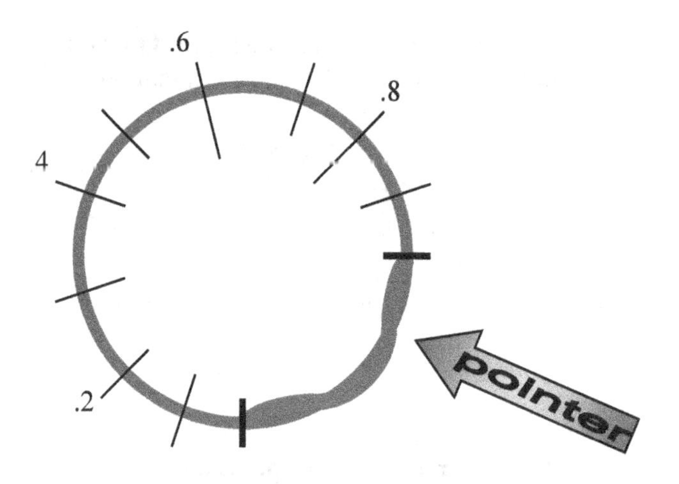

The random variable X is described conditionally in terms of the underlying experimental outcome. Let W be a Bernoulli trial random variable, with *success* meaning the pointer is in the gray area, and *failure* meaning the pointer is in the green area. The probability table for W is given by,

Event	W	Probability
Gray	1	$\frac{3}{4}$
Green	0	$\frac{1}{4}$

Given $W=1$ (gray sector), $X=Y$, where Y is the continuous random-number-generating random variable described in Examples 5.1-2 and 5.3-5. That is, $F_Y(x)=x;\ \ 0\le x\le 1$. Given $W=0$ (green sector), $X=Z$, where Z is the simple discrete random variable taking only the value one (i.e., $\Pr[Z=1]=1$). In other words, 75% of the time X is the continuous random variable Y, and 25% of the time X is the discrete random variable Z.

In Example 5.4-1 below we will derive the the mean and the variance of the (mixed) spinning wheel random variable X. The easiest way to approach this is to consider X in the context of a broader class of distributions called ***mixtures***.[5] This is a useful topic for advanced study, but for now we consider only the special case of so-called ***two-point mixtures***.

Two-Point Mixture Distributions

Let Y and Z be any two given random variables with CDFs F_Y and F_Z, respectively, and let p be a given number such that $0<p<1$. For any real number x define,

$$F_X(x)=pF_Y(x)+(1-p)F_Z(x).$$

The resulting random variable X is called a ***two-point mixture*** of Y and Z with mixing weights p and $1-p$.

[5] Here the term ***mixture*** refers to the 75-25 mixture of Y and Z, not the fact that one is continuous and the other discrete.

Note: This does NOT imply that $X = pY + (1-p)Z$, which is an entirely different concept.

Since F_Y and F_Z satisfy properties (a), (b) and (c) for CDFs, it is easy to verify that F_X does as well, with property (d) giving the underlying probabilities for the random variable *X*. If *Y* and *Z* are both continuous random variables, then differentiating both sides of the CDF formula shows that,

$$f_X(x) = pf_Y(x) + (1-p)f_Z(x).$$

It follows that,

$$E\left[X^n\right] = \int_{-\infty}^{\infty} x^n f_X(x)dx = p\int_{-\infty}^{\infty} x^n f_Y(x)dx + (1-p)\int_{-\infty}^{\infty} x^n f_Z(x)dx = pE\left[Y^n\right] + (1-p)E\left[Z^n\right].$$

Appropriate modifications to this argument for the discrete or mixed case show that this holds in general.

Calculating Moments of Two-Point Mixture Distributions

Let *X* be a ***two-point mixture*** of *Y* and *Z* with mixing weights *p* and $1-p$. Then,

$$E\left[X^n\right] = pE\left[Y^n\right] + (1-p)E\left[Z^n\right]$$

Example 5.4-1 Expectation and Variance of the Spinning Wheel Random Variable

Find the mean and the variance of the random variable in the above spinning wheel example.

Solution

The spinning wheel random variable *X* can be written as the two point mixture of *Y* and *Z*, with mixing weights .75 and .25, respectively. Here *Y* is the random-number generator on [0,1) and *Z* is the constant random variable with $\Pr\left[Z=1\right]=1$. This means the CDF for *X* is given by,

$$F_X(x) = .75F_Y(x) + .25F_Z(x), \text{ where,}$$

$$F_Y(x) = \begin{cases} 0 & \text{for } x \le 0 \\ x & \text{for } 0 < x < 1 \\ 1 & \text{for } 1 \le x \end{cases} \quad \text{and} \quad F_Z(x) = \begin{cases} 0 & \text{for } x < 1 \\ 1 & \text{for } 1 \le x \end{cases}$$

We first check that the above mixture X is indeed the spinning wheel random variable defined conditionally at the beginning of this section. First, let the Bernoulli trial random variable $W=1$ (gray sector), so that $0 \le x < 1$. Then $F_Z(x)=0$ and $F_Y(x)=x$, so that $F_X(x)=.75 \cdot x+.25 \cdot 0=.75x$. If $W=0$ (green sector) then $X=1$ and $F_X(x)$ $=.75F_Y(1)+.25F_Z(1)=.75 \cdot 1+.25 \cdot 1=1$. The graph of F_X is given below:

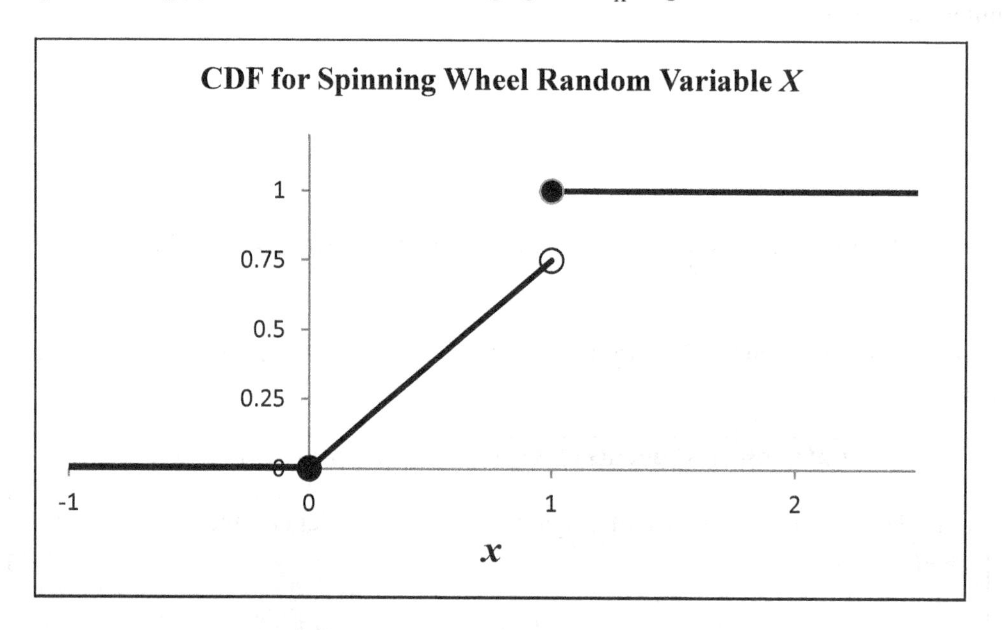

The CDF is continuous everywhere except at $x=1$, where it has a jump of height 0.25. The jump in the graph at $x=1$ corresponds to the discrete part Z (green sector) and the continuous part Y (gray sector) is represented by the smooth straight line portion of the graph for x on the interval $[0,1)$.

The moments of X can be easily calculated using the moments formula for mixtures, above. In Example 5.3.5 showed that $E[Y]=1/2$ and $E\left[Y^2\right]=1/3$, and trivially, $E[Z]=1$ and $E\left[Z^2\right]=1$. Thus,

$$E[X]=\frac{3}{4}E[Y]+\frac{1}{4}E[Z]=\frac{3}{4}\cdot\frac{1}{2}+\frac{1}{4}\cdot 1=\frac{5}{8}$$

And

$$E\left[X^2\right]=\frac{3}{4}E\left[Y^2\right]+\frac{1}{4}E\left[Z^2\right]=\frac{3}{4}\cdot\frac{1}{3}+\frac{1}{4}\cdot 1=\frac{1}{2}$$

So, $Var[X]=\frac{1}{2}-\left(\frac{5}{8}\right)^2=\frac{7}{64}$.

Note: Beware, the mixture formulas work for moments, but not variances. That is to say,

$$\frac{7}{64} = Var[X] \neq \frac{3}{4}Var[Y] + \frac{1}{4}Var[Z] = \frac{3}{4}\cdot\left[\frac{1}{3} - \left(\frac{1}{2}\right)^2\right] + \frac{1}{4}\cdot 0 = \frac{1}{16}.$$

In this example the random variable lives on the unit interval, and has a discrete jump only at the right-hand endpoint. Clearly, more complicated examples are possible, arising, for example, from more elaborate paintings and calibrations of the spinning wheel.

We can also calculate the expected value and variance of a mixed random variable directly from its CDF. We use the probability density function $f(x)$ and an integral on the continuous part of the distribution, and the discrete point probability at a jump point(s). We illustrate using the spinning wheel random variable X. From the above we see that the direct formula for the CDF of X is given by,

$$F(x) = \begin{cases} 0 & \text{if } x \leq 0 \\ \frac{3}{4}x & \text{if } 0 \leq x < 1 \quad \text{(the pointer is in the Gray sector)} \\ 1 & \text{if } 1 \leq x \quad \text{(the pointer is in the Green sector)} \end{cases}$$

Then on the continuous part (gray) the density function $f(x) = F'(x) = \frac{3}{4}$, and on the discrete part (green), $\Pr[X = 1] = 1/4$, so that,

$$E[X] = \underbrace{\int_0^1 x\underbrace{\left(\frac{3}{4}\right)}_{f(x)}dx}_{\textit{continuous part}} + \underbrace{(1)\cdot\Pr[X=1]}_{\textit{discrete part}} = \frac{3}{8} + \frac{1}{4} = \frac{5}{8}.$$

Similarly,

$$E[X^2] = \underbrace{\int_0^1 x^2\left(\frac{3}{4}\right)dx}_{\text{continuous part}} + \underbrace{(1^2)\left(\frac{1}{4}\right)}_{\text{discrete part}} = \frac{1}{4} + \frac{1}{4} = \frac{1}{2}.$$

Again, it follows that $Var[X] = \frac{1}{2} - \left(\frac{5}{8}\right)^2 = \frac{7}{64}$.

Exercise 5-28 SOA Exam P Sample Exam Questions #62

A random variable X has the cumulative distribution function

$$F(x) = \begin{cases} 0 & \text{for } x < 1 \\ \frac{x^2-2x+2}{2} & \text{for } 1 \leq x < 2. \\ 1 & \text{for } x \geq 2 \end{cases}$$

Calculate the variance of X.

(A) $\frac{7}{12}$ (B) $\frac{1}{8}$ (C) $\frac{5}{36}$ (D) $\frac{4}{3}$ (E) $\frac{23}{12}$

Exercise 5-29 A driver has a 90% chance of going through the next year without incurring a loss. Given she does incur a loss, the amount, X, of the loss has density function $f(x) = 2e^{-2x}$; $0 < x < \infty$. Calculate the expected loss for the year.

Exercise 5-30 You design an insurance policy that pays a random amount $\text{Payment} = 1000 \cdot A$, where A denotes the age at death. A is assumed to have a continuous uniform distribution on $[50,110]$ (that is, a constant density function). Modify this policy to pay an amount $\text{Modified Payment} = \begin{cases} 1000 \cdot A & \text{if } A < 90 \\ 100,000 & \text{if } A \geq 90 \end{cases}$.

(a) Find the expected value and standard deviation of the original payment.
(b) Find the expected value and standard deviation of the modified payment.

5.5 APPLICATIONS TO INSURANCE: DEDUCTIBLES AND CAPS

There are several insurance concepts that can be modeled using some of the preceding ideas. For this discussion we will focus on two related random variables. The first is the *actual loss amount random variable*, to be denoted by X. This is the monetary amount of a covered loss suffered by an insurance customer. As a rule, this will be a continuous random variable living on an interval (A, B); $0 \leq A < B < \infty$. The second is the *payment random variable*, Y, reflecting the actual *benefit*, that is, the amount the insurance company pays out to compensate for the loss.

Insurance companies have several contractual arrangements to assure that the payment, Y, is different from (less than) the actual loss, X. The first of these is the ***deductible***, which will be familiar to anyone who has purchased auto or health insurance (among many other types).

5.5.1 DEDUCTIBLE INSURANCE

Consider a given loss amount random variable X living on the interval (A, B); $0 \leq A < B < \infty$. We denote the deductible amount by d, where $A \leq d < B$. Let Y be the payment random variable resulting from the deductible d. Then the relationship between X and Y is given by,

$$Y = \begin{cases} 0 & \text{if } A \leq X < d \\ X - d & \text{if } d \leq X < B \end{cases}.$$

Frequently, a problem will call for the expected value of Y, the insurance payment. Basically, this is an application of the transformation formula, and we can use the density $f_X(x)$ of the actual loss X to calculate,

$$E[Y] = \int_A^d 0 \cdot f_X(x)\,dx + \int_d^B (x-d) \cdot f_X(x)\,dx = \int_d^B (x-d) \cdot f_X(x)\,dx.$$

Note: $E[Y] \neq E[X] - d$, which is a common mistake.

Example 5.5-1 Deductible Insurance

The loss amount X has density $f_X(x) = \frac{1}{1000}$; $0 \le x \le 1000$. (In Chapter 6 you will learn to refer to this as the uniform distribution on [0,1000].) The insurance policy has a deductible amount of 100 per loss.

(a) Calculate the expected value of the loss.
(b) Calculate the expected amount of the benefit payment. That is, find the mean amount of the payment made by the insurance company.
(c) Calculate the expected amount of the portion of loss not covered by insurance.

Solution

In this example we have $A = 0$ and $B = 1000$.

(a) $$E[X] = \int_0^{1000} x\left(\frac{1}{1000}\right) dx = \left(\frac{1}{1000}\right)\left(\frac{x^2}{2}\right)\Bigg|_{x=0}^{1000} = 500.$$

(b) Let Y be the payout amount (benefit). Then, from the preceding discussion,

$$E[Y] = \int_{100}^{1000} (x-100)\frac{1}{1000}\, dx = \frac{1}{1000}\frac{(x-100)^2}{2}\Bigg|_{100}^{1000} = \frac{900^2}{2000} = 405.$$

(c) Let Z be the amount of a loss not covered by insurance. Then

$$Z = \begin{cases} X & \text{if } 0 \le x < 100 \\ 100 & \text{if } 100 \le x < 1000 \end{cases}.$$

Thus,

$$\begin{aligned} E[Z] &= \int_0^{100} x\frac{1}{1000}\, dx + \int_{100}^{1000} 100\frac{1}{1000}\, dx \\ &= \frac{1}{1000}\frac{x^2}{2}\Bigg|_0^{100} + 100 \cdot \Pr[100 \le X \le 1000] \\ &= 5 + 100 \cdot \frac{9}{10} = 95. \end{aligned}$$

❐

Note

A moment's thought shows that the sum of the last two answers must be the expected loss, which is 500. So part (c) could have been calculated as the difference between the answers to (a) and (b).

5.5.2 CAPPED INSURANCE

A second common practice for reducing the payment random variable Y is to ***cap*** the covered loss at a given level, C, which is less than the maximum actual loss B. In this instance the relationship between Y and X is given by,

$$Y = \begin{cases} X & \text{if } A \leq X < C \\ C & \text{if } C \leq X < B \end{cases}.$$

An example of this type of modification to the loss random variable X may be seen in part (c) of Example 5.5-1. In that example, the amount Z of the total loss X that is *not* covered by insurance under a deductible of 100 is exactly the same as the portion of the loss X capped at 100.

To calculate the expected payment under a cap of C we proceed as before, treating Y as a transformation of X. Then,

$$E[Y] = \int_A^C x f_X(x)\,dx + \int_C^B C f_X(x)\,dx = \int_A^C x f_X(x)\,dx + C\int_C^B f_X(x)\,dx,$$

so that,

$$E[Y] = \int_A^C x f_X(x)\,dx + C\Pr[X > C].$$

Example 5.5-2 Capped Insurance Payment

You are given a loss amount X (the same as in Example 5.5-1) with density function $f_X(x) = \frac{1}{1000}$; $0 \leq x \leq 1000$. There is a policy limit of 500. Calculate the expected claim payment if there is no deductible.

Solution

If Y is the payment then,

$$Y = \begin{cases} X & \text{if } 0 \leq X < 500 \\ 500 & \text{if } 500 \leq X < 1000 \end{cases}.$$

Therefore,

$$\begin{aligned} E[Y] &= \int_0^{500} x f_X(x)\,dx + \int_{500}^{1000} 500 f_X(x)\,dx \\ &= \int_0^{500} x\left(\frac{1}{1000}\,dx\right) + \int_{500}^{1000} 500\left(\frac{1}{1000}\,dx\right) \\ &= \left(\frac{500^2}{2}\right)\left(\frac{1}{1000}\right) + (500)\left(\frac{1}{1000}\right)(1000-500) \\ &= 375. \end{aligned}$$

❐

Note: The second integral equals $500 \cdot \Pr[X \geq 500]$.

Example 5.5-3 Capped Insurance Payment

You are given a loss amount X with density function $f_X(x) = \frac{3}{x^4};\ 1 \le x < \infty$.

(a) Calculate the total expected loss.

(b) Calculate the expected benefit paid if there is a cap of 10 on the insurance.

(c) Calculate the expected benefit paid if there is a deductible of 10.

Solution

In this instance, $A = 1$ and $B = \infty$.

(a) $$E[X] = \int_1^\infty x f_X(x)\,dx = \int_1^\infty x(3x^{-4})\,dx$$

$$= 3\int_1^\infty x^{-3}\,dx = -\frac{3}{2}x^{-2}\Big|_{x=1}^{\infty} = -\frac{3}{2}(0-1) = \frac{3}{2}$$

(b) If Y is the payment, then

$$Y = \begin{cases} X & \text{if } 1 \le X < 10 \\ 10 & \text{if } 10 \le X < \infty \end{cases}.$$

Therefore,

$$E[Y] = \int_1^{10} x f_X(x)\,dx + \int_{10}^\infty 10 f_X(x)\,dx$$

$$= 3\int_1^{10} x^{-3}\,dx + 10\int_{10}^\infty 3x^{-4}\,dx$$

$$= \left(\frac{3}{2}\right)\left(-x^{-2}\right)\Big|_{x=1}^{10} + (10)\left(-x^{-3}\right)\Big|_{x=10}^{\infty}$$

$$= \left(\frac{3}{2}\right)\left(1-\frac{1}{100}\right) + (10)\left(\frac{1}{1000}\right) = 1.495.$$

(c) The payment benefit in the case of a deductible is given by,

$$Z = \begin{cases} 0 & \text{if } 1 \le X < 10 \\ X-10 & \text{if } 10 \le X < \infty \end{cases}.$$

Therefore, we can evaluate $E[Z] = \int_{10}^\infty (x-10) f_X(x)\,dx = \int_{10}^\infty (x-10)(3x^{-4})\,dx$ directly. However, here is a shortcut that avoids another tedious integration. Using the reasoning employed in Example 5.5-1 (c), we can recognize that the portion of the loss X ***not*** paid under the deductible (and hence, absorbed by the insured) is X capped at 10. Thus, we can calculate the answer to (c) as the difference between (a) and (b). That is,

$$E[Z] = E[X] - E[Y] = 1.5 - 1.495 = .005.$$

❐

Notes

(1) In this rather contrived example, the insured is much better off with a cap of 10 rather than a deductible of 10. With the cap the expected insurance payment is $\frac{1.495}{1.5} = 99.\bar{6}\%$ of the expected loss amount. The situation is just the reverse in the case of a deductible of 10.

(2) This duality between caps and deductibles will be explored more closely in the next subsection.

Example 5.5-4 Capped Insurance Payment

You are given a loss amount X with density function $f_X(x) = 2e^{-2x};\ 0 \le x < \infty$.

(a) Calculate the total expected loss.

(b) Calculate the expected benefit paid if there is a cap of $C = \ln 2$ on the insurance.

Solution

In this instance, $A = 0$ and $B = \infty$.

(a) We showed in Example 5.3-3, using integration-by-parts, that for this density function we have

$$E[X] = \int_0^{\infty} (x)(2e^{-2x})\, dx = \frac{1}{2}.$$

(b) Let Y be the payment under a cap of $\ln 2$. Then

$$Y = \begin{cases} X & \text{if } 0 \le X < \ln 2 \\ \ln 2 & \text{if } \ln 2 \le X < \infty \end{cases}.$$

and

$$E[Y] = \int_0^{\ln 2} (x)(2e^{-2x})\, dx + \int_{\ln 2}^{\infty} (\ln 2)(2e^{-2x})\, dx.$$

The first integral here also requires integration-by-parts to evaluate. Let $u = x$ and $dv = 2e^{-2x}\, dx$. Then $du = dx$ and $v = -e^{-2x}$, so that,

$$\begin{aligned}
E[Y] &= \underbrace{-xe^{-2x}\Big|_{x=0}^{\ln 2} + \int_0^{\ln 2} e^{-2x}\, dx}_{\text{integration by parts}} + \underbrace{\ln 2}_{\text{cap}} \cdot \underbrace{\int_{\ln 2}^{\infty} (2e^{-2x})\, dx}_{\text{probability that loss exceeds the cap of } \ln 2} \\
&= -xe^{-2x}\Big|_{x=0}^{\ln 2} + \left(\frac{e^{-2x}}{-2}\right)\Bigg|_{x=0}^{\ln 2} + (\ln 2)\left(\frac{2e^{-2x}}{-2}\right)\Bigg|_{x=\ln 2}^{\infty} \\
&= (-\ln 2)(e^{-2\ln 2}) - \left(\frac{1}{2}\right)(e^{-2\ln 2} - 1) + (\ln 2)(e^{-2\ln 2}) \\
&= \left(\frac{1}{2}\right)(1 - e^{-2\ln 2}) = \left(\frac{1}{2}\right)(1 - e^{\ln 2^{-2}}) = \left(\frac{1}{2}\right)(1 - 2^{-2}) = \frac{3}{8}.
\end{aligned}$$

□

Note

We will show in the following subsection how to avoid using integration-by-parts by expressing the integrals in terms of the loss CDF, $F_X(x)$.

Before moving on, we point out that the payment random variable Y in both the deductible and the capped case constitutes a distribution of mixed type. In the former case we have $\Pr[Y=0] = \Pr[X \le d]$, which is generally non-zero, and in the latter case $\Pr[Y=C] = \Pr[X \ge C] = p$, also generally non-zero. A comparison between the CDFs of X and Y in the case of a cap C can be seen in the following graphs:

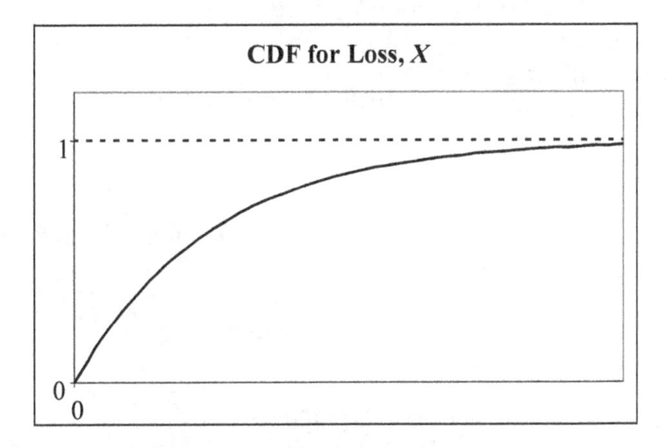

We see that the CDF for Y has a discrete jump at C. The fact that we are able to treat Y as a transformation of the ***continuous*** loss random variable X means that we are able to calculate expected values directly in terms of the density function $f_X(x)$ for X without having to worry about the jump points.

5.5.3 THE CDF METHOD FOR DEDUCTIBLES AND CAPS

The above examples may suggest that there is a hidden relationship between deductibles and caps. In this section we will cast some additional light on this affair, revealing in fact that deductibles and caps are in reality two sides of the same coin. This will be facilitated by employing the alternative calculation for expectation in terms of the cumulative distribution function $F_X(x)$ for the loss random variable X.

We begin, as above, with a continuous loss random variable X living on the interval (A, B); $0 \le A < B$. Recall from Subsection 5.3.4 that in terms of $F_X(x)$ we have,

$$E[X] = A + \int_A^B [1 - F_X(x)]\, dx.$$

In the following we introduce some streamlined notation[6] for the payment random variables and derive compact formulas in terms of the CDF of X for capped benefits, benefits with a deductible, and a combined policy with both a deductible and a cap.

Caps and Deductibles in Terms of $F_X(x)$

Let X be a continuous loss random variable with domain (A,B); $0 \le A < B$, and let C be any number such that $A < C < B$.

Let $Y^C = \begin{cases} X & \text{if } A \le X < C \\ C & \text{if } C \le X < B \end{cases}$ (benefit capped at C), and

Let $Y_C = \begin{cases} 0 & \text{if } A \le X < C \\ X-C & \text{if } C \le X < B \end{cases}$ (benefit with deductible of C)

Let $Y_d^u = \begin{cases} 0 & \text{if } A \le X < d \\ X-d & \text{if } d \le X < u \\ u-d & \text{if } u \le X < B \end{cases}$ (benefit with deductible of d and cap of u)

(1) $X = Y^C + Y_C$

(2) $E[Y^C] = A + \int_A^C [1-F_X(x)]\,dx$

(3) $E[Y_C] = \int_C^B [1-F_X(x)]\,dx$

(4) $Y_d^u = Y^u - Y^d$

(5) $E[Y_d^u] = \int_d^u [1-F_X(x)]\,dx$

Proof

(1) $Y^C + Y_C = \begin{cases} X+0 & \text{if } A \le X < C \\ C+(X-C) & \text{if } C \le X < B \end{cases} = \begin{cases} X & \text{if } A \le X < C \\ X & \text{if } C \le X < B \end{cases} = X.$

(2) Using the density function method from Subsection 5.5.2, we have

$$E[Y^C] = \int_A^C x\,f_X(x)\,dx + C\Pr[X > C].$$

Now, use integration-by-parts with $u = x$ and $dv = f_X(x)\,dx$. Then $du = dx$ and $v = F_X(x) + k$, where k is a constant of integration to which we assign the value -1, so that $v = -[1-F_X(x)]$. Thus,

[6] Advanced manuals and texts for actuarial Exam C often use the notation $X \wedge C$ for capped payments, and $(X-C)_+$ for deductible payments. We will content ourselves with a more intuitive system sufficient for our introductory treatment here.

$$E[Y^C] = -x[1-F_X(x)]\Big|_{x=A}^{C} + \int_A^C [1-F_X(x)]\,dx + C\Pr[X>C].$$

Since $F_X(A)=0$ and $[1-F_X(C)]=\Pr[X>C]$, we have,

$$\begin{aligned} E[Y^C] &= -C[1-F_X(C)] + A[1-F_X(A)] + \int_A^C [1-F_X(x)]\,dx + C\cdot\Pr[X>C] \\ &= -C\Pr[X>C] + A + \int_A^C [1-F_X(x)]\,dx + C\cdot\Pr[X>C] \\ &= A + \int_A^C [1-F_X(x)]\,dx. \end{aligned}$$

(3) From (1), $X = Y^C + Y_C$, so that $Y_C = X - Y^C$. It follows that,

$$\begin{aligned} E[Y_C] = E[X]-E[Y^C] &= \left(A+\int_A^B [1-F_X(x)]\,dx\right) - \left(A+\int_A^C [1-F_X(x)]\,dx\right) \\ &= \int_C^B [1-F_X(x)]\,dx \end{aligned}$$

(4) This follows because,

$$\begin{aligned} Y_d^u &= \begin{cases} 0 & \text{if } A \le X < d \\ X-d & \text{if } d \le X < u \\ u-d & \text{if } u \le X < B \end{cases} = \begin{cases} X-X & \text{if } A \le X \le d \\ X-d & \text{if } d < X < u \\ u-d & \text{if } u \le X < B \end{cases} \\ &= \begin{cases} X & \text{if } A \le X \le d \\ X & \text{if } d < X < u \\ u & \text{if } u \le X < B \end{cases} - \begin{cases} X & \text{if } A \le X \le d \\ d & \text{if } d < X < u \\ d & \text{if } u \le X < B \end{cases} \\ &= \begin{cases} X & \text{if } A \le X < u \\ u & \text{if } u \le X < B \end{cases} - \begin{cases} X & \text{if } A \le X \le d \\ d & \text{if } d < X < B \end{cases} \\ &= Y^u - Y^d \end{aligned}$$

(5) Thus, from (4) and (2) we have,

$$\begin{aligned} E[\,Y_d^u\,] &= E[Y^u] - E[Y^d] \\ &= \left(A + \int_A^u [1 - F_X(x)]\,dx\right) - \left(A + \int_A^d [1 - F_X(x)]\,dx\right) \\ &= \int_d^u [1 - F_X(x)]\,dx. \end{aligned}$$

Notes

(1) We can think of the capped insurance payment Y^C as the portion of the total loss X that is *not* covered by insurance under the policy Y_C governed by a deductible amount of C. Then (1) simply says the total loss can be decomposed into the portion absorbed by the insured and the portion covered by insurance.

(2) Because of (1) we can calculate any one of the quantities X, Y^C, Y_C directly from the other two.

(3) This approach, using $F_X(x)$, is preferable to using the density function $f_X(x)$ in those situations where the latter method will require integration-by-parts.

Example 5.5-5 Using the Loss CDF (Example 5.5-4 Revisited)

You are given a loss amount X with density function $f_X(x) = 2e^{-2x};\ 0 \le x < \infty$.

(a) Calculate the total expected loss.
(b) Use the CDF of X to calculate the expected benefit paid if there is a cap of $C = \ln 2$ on the insurance.

Solution

(a) In Example 5.3-12 we showed $F(x) = 1 - e^{-2x}$ and in Example 5.3-13 we calculated

$$E[X] = \int_0^\infty [1 - F_X(x)]\,dx = \frac{1}{2}.$$

(b) Using formula (2) with $A = 0$ and $C = \ln 2$ we have,

$$\begin{aligned} E[Y^C] &= \int_0^{\ln 2} \left[1 - (1 - e^{-2x})\right] dx \\ &= \int_0^{\ln 2} e^{-2x}\,dx \\ &= \left(\frac{1}{2}\right)(1 - e^{-2\ln 2}) = \left(\frac{1}{2}\right)(1 - e^{\ln 2^{-2}}) = \left(\frac{1}{2}\right)(1 - 2^{-2}) = \frac{3}{8}. \end{aligned}$$

❐

Example 5.5-6 Using the Loss CDF for Caps and Deductible Combined

You are given a loss amount X with density function $f_X(x) = \dfrac{(3)(100)^3}{(x+100)^4}$; $0 \le x < \infty$.

(a) Assume there is a cap of C on the benefit paid. Calculate a general expression in terms of C for the expected benefit paid under the cap.

(b) Calculate the total expected loss.

(c) Calculate the expected benefit paid if there is a deductible of 10.

(d) Calculate the expected benefit paid if there is a deductible of 10 and a cap of 100.

Solution

Note that performing these calculations using the density function would lead to integrals of the type $\int x(x+100)^{-4}\,dx$, which require integration-by-parts to evaluate. Therefore, it is expedient to derive the CDF of X and use the CDF method of evaluation. We begin by finding the CDF for X:

$$\begin{aligned}
F_X(x) &= \int_0^x (3)(100)^3(t+100)^{-4}\,dt \\
&= -(100)^3(t+100)^{-3}\Big|_{t=0}^{x} \\
&= -(100)^3\left[(x+100)^{-3} - (100)^{-3}\right] \\
&= 1-(100)^3(x+100)^{-3} \\
&= 1-\left(\frac{100}{x+100}\right)^3.
\end{aligned}$$

(a) Using formula (2) with $A = 0$, we have,

$$\begin{aligned}
E[Y^C] &= \int_0^C [1-F_X(x)]\,dx \\
&= \int_0^C (100)^3(x+100)^{-3}\,dx \\
&= \frac{(100)^3}{-2}(x+100)^{-2}\Big|_{x=0}^{C} = \left(\frac{(100)^3}{-2}\right)\left[(C+100)^{-2} - (100)^{-2}\right] \\
&= \left(\frac{100}{-2}\right)(100)^2\left[(C+100)^{-2} - (100)^{-2}\right] \\
&= \left(\frac{100}{2}\right)\left[1-\left(\frac{100}{C+100}\right)^2\right] \\
&= (50)\left[1-\left(\frac{100}{C+100}\right)^2\right]
\end{aligned}$$

(b) The total expected loss can be calculated as the capped expected benefit with $C = \infty$. This leads to $E[X] = E[Y^{\infty}] = 50$.

(c) From (a), we have the expected value of the corresponding capped insurance is given by

$$E[Y^{10}] = (50)\left[1-\left(\frac{100}{10+100}\right)^2\right] = 8.68.$$

Hence,

$$E[Y_{10}] = E[X]-E[Y^{10}] = 50-(50)\left[1-\left(\frac{100}{10+100}\right)^2\right] = (50)\left(\frac{100}{10+100}\right)^2 = 41.32.$$

(d) First, $E[Y^{100}] = (50)\left[1-\left(\frac{100}{100+100}\right)^2\right] = 37.50$. Now, $Y_d^u = Y^u - Y^d$, so that,

$$E[Y_{10}^{100}] = E[Y^{100}]-E[Y^{10}] = 37.50-8.68 = 28.82.$$

Alternatively, we could calculate directly (from formula (5)),

$$E[Y_{10}^{100}] = \int_{10}^{100}[1-F_X(x)]\,dx = \int_{10}^{100}\left(\frac{100}{x+100}\right)^3 dx.$$

❐

Note
The loss X in this example has a ***Pareto*** distribution, which will be studied in Subsection 6.7.2.

Exercise 5-31 SOA Exam P Sample Exam Questions #53

An insurance policy reimburses a loss up to a benefit limit of 10. The policyholder's loss, Y, follows a distribution with density function:

$$f(y) = \begin{cases} \frac{2}{y^3} & \text{for } y>1 \\ 0 & \text{otherwise} \end{cases}.$$

What is the expected value of the benefit paid under the insurance policy?

(A) 1.0 (B) 1.3 (C) 1.8 (D) 1.9 (E) 2.0

Exercise 5-32 SOA/CAS Course 1 2000 Sample Examination #34

Under a group insurance policy, an insurer agrees to pay 100% of the medical bills incurred during the year by employees of a small company, up to a maximum total of one million dollars. The total amount of bills incurred, X, has probability density function

$$f(x) = \begin{cases} \frac{x(4-x)}{9} & \text{for } 0 < x < 3 \\ 0 & \text{otherwise} \end{cases}.$$

where x is measured in millions. Calculate the total amount, in millions of dollars, the insurer would expect to pay under this policy.

(A) 0.120 (B) 0.301 (C) 0.935 (D) 2.338 (E) 3.495

Exercise 5-33 SOA Exam P Sample Exam Questions #40

An insurance policy pays for a random loss X subject to a deductible of C, where $0 < C < 1$. The loss amount is modeled as a continuous random variable with density function

$$f(x) = \begin{cases} 2x & \text{for } 0 < x < 1 \\ 0 & \text{otherwise} \end{cases}.$$

Given a random loss X, the probability that the insurance payment is less than 0.5 is equal to 0.64. Calculate C.

(A) 0.1 (B) 0.3 (C) 0.4 (D) 0.6 (E) 0.8

Exercise 5-34 SOA Exam P Sample Exam Questions #51

A manufacturer's annual losses follow a distribution with density function

$$f(x) = \begin{cases} \frac{2.5(0.6)^{2.5}}{x^{3.5}} & \text{for } x > 0.6 \\ 0 & \text{otherwise} \end{cases}.$$

To cover its losses, the manufacturer purchases an insurance policy with an annual deductible of 2. What is the mean of the manufacturer's annual losses not paid by the insurance policy?

(A) 0.84 (B) 0.88 (C) 0.93 (D) 0.95 (E) 1.00

Exercise 5-35 SOA Exam P Sample Exam Questions #56

An insurance policy is written to cover a loss, X, where X has a uniform distribution on $[0,1000]$. At what level must a deductible be set in order for the expected payment to be 25% of what it would be with no deductible?

(A) 250 (B) 375 (C) 500 (D) 625 (E) 750

Exercise 5-36 SOA Exam P Sample Exam Questions #63

The warranty on a machine specifies that it will be replaced at failure or age 4, whichever occurs first. The machine's age at failure, X, has density function

$$f(x) = \begin{cases} \frac{1}{5} & \text{for } 0 < x < 5 \\ 0 & \text{otherwise} \end{cases}.$$

Let Y be the age of the machine at the time of replacement. Determine the variance of Y.
Hint: The age at replacement is the age at failure capped at 4.

(A) 1.3 (B) 1.4 (C) 1.7 (D) 2.1 (E) 7.5

Exercise 5-37 A manufacturer's annual losses follow a distribution with density function

$$f(x) = \begin{cases} \frac{x}{5000} & \text{for } 0 \le x \le 100 \\ 0 & \text{otherwise} \end{cases}.$$

The manufacturer takes out insurance to cover the losses with an annual deductible of 10. Calculate the expected value of the manufacturer's losses NOT paid by the insurer.

Exercise 5-38 The future lifetime T in years for a 25 year-old is modeled by a constant density function on $[0, 75]$. A policy pays T if T is less than or equal to 40 and pays 40 if T is greater than 40 (policy benefit capped at 40). Calculate the expected benefit. ***Hint***: It is less than 40.

Exercise 5-39 Let X be the loss random variable with probability density function $f(x) = \frac{2}{(100)^2}(100 - x);\ 0 \le x \le 100$. The deductible amount is 10. **Given that a loss exceeds the deductible,** calculate the expected insurance payment.

Hint: Account for the conditional probability.

Exercise 5-40 SOA Exam P Sample Exam Questions #52

An insurance company sells a one-year automobile policy with a deductible of 2. The probability that the insured will incur a loss is 0.05. If there is a loss, the probability of a loss amount N is $\frac{K}{N}$, for $N = 1, \ldots, 5$ and K a constant. These are the only possible loss amounts and no more than one loss can occur. Determine the net premium for the policy.

(A) 0.031 (B) 0.066 (C) 0.072 (D) 0.110 (E) 0.150

Exercise 5-41 SOA Exam P Sample Exam Questions #129

The cumulative distribution function for health care costs experienced by a policyholder is modeled by the function

$$F(x) = \begin{cases} 1 - e^{-x/100} & \text{for } x > 0 \\ 0 \text{ otherwise} \end{cases}.$$

The policy has a deductible of 20. An insurer reimburses the policyholder for 100% of the health care costs between 20 and 120 less the deductible. Health care costs above 120 are reimbursed at 50%.

Let G be the cumulative distribution function of reimbursements given that the reimbursement is positive. Calculate $G(115)$.

(A) 0.683 (B) 0.727 (C) 0.741 (D) 0.757 (E) 0.777

5.6 THE MOMENT GENERATING FUNCTION

In section 3.7, we introduced a technique for transforming a discrete distribution on the non-negative integers into a power series. The power series was named the ***probability generating function***, and it was used for calculating the ***moments*** of the underlying distribution (random variable). By moments, we mean moments about zero, so that the kth moment is $E[X^k]$ for $k = 0,1,2,\ldots$. The main purpose of introducing the probability generating function was to simplify the calculations of mean and variances of standard discrete probability distributions.

In this section we will present a related transformation, called the ***Moment Generating Function* (MGF)**, that serves the same purpose but works equally well for discrete or continuous random variables. Once again, the idea is to transform the distribution into a calculus function, and use derivatives to calculate the moments of the underlying random variable.

Definition of the Moment Generating Function

Let X be any random variable. The ***moment generating function*** (MGF) of X is denoted by $M_X(t)$ and is defined by,

$$M_X(t) = E_X[e^{tX}].$$

Observe that $Y = e^{tX}$ is a transformation of the random variable X and we can calculate its expected value by using the transformation formula. The subscript on the expected value is written to remind the reader to take the expected value with respect to the random variable X.

The resulting MGF will be a function of the ordinary calculus variable t.

If X is a ***discrete*** random variable taking values on the non-negative integers then,

$M_X(t) = E[e^{tX}] = \sum_{k=0}^{\infty} e^{tk} p_k$ where $p_k = \Pr[X=k]$. This is just the result of expressing the calculation in the by now familiar tabular form, as in Chapter 3:

k	$e^{t \cdot k}$	$\Pr[X=k]$	$E[e^{tX}]$
0	1	p_0	p_0
1	e^t	p_1	$p_1 e^t$
2	e^{2t}	p_2	$p_2 e^{2t}$
$\vdots$	$\vdots$	$\vdots$	$\vdots$
Totals		1	$\sum_k p_k e^{kt}$

Example 5.6-1 MGF of a Discrete Random Variable

Find the moment generating function for the discrete random variable X with probability distribution:

X	2	5
$\Pr(X=x)$	.4	.6

Solution

k	$e^{t \cdot k}$	$\Pr[X=k]$	$E[e^{tX}]$
2	e^{2t}	.4	$.4e^{2t}$
5	e^{5t}	.6	$.6e^{5t}$
Totals		1	$.4e^{2t} + .6e^{5t}$

$M_X(t) = E[e^{tX}] = .4e^{2t} + .6e^{5t}$. ❐

Example 5.6-2 MGF for a Bernoulli Trial Random Variable

Find the MGF, $M_X(t)$, of the random variable X, which takes the value 1 with probability p, and takes the value 0 with probability $q = 1 - p$.

Solution

k	$e^{t \cdot k}$	$\Pr[X=k]$	$E[e^{tX}]$
0	1	q	q
1	e^t	p	pe^t
Totals		1	$q + pe^t$

Thus, $M_X(t) = q + pe^t$.

For a continuous random variable X living on the interval (A, B), the MGF is calculated as an integral using the density function: $M_X(t) = \int_A^B e^{tx} \cdot f_X(x)dx$.

Example 5.6-3 MGF for a Continuous Random Variable

Let X have the density function $f(x)=e^{-x}$ where $0 \le x < \infty$. Calculate the MGF of X.

Solution

$$M_X(t) = E[e^{tx}] = \int_0^{\infty} e^{tx} \cdot f(x)\,dx = \int_0^{\infty} e^{tx}(e^{-x}\,dx) = \int_0^{\infty} e^{-(1-t)x}\,dx = \overbrace{\left.\frac{e^{-(1-t)x}}{-(1-t)}\right|_{x=0}^{\infty}}^{\text{This step requires that } t<1}$$

Therefore $M_X(t)=\frac{1}{1-t}$ for all $t<1$. ❒

Note

The improper integral in this example can only be evaluated when t is less than 1. This illustrates that the MGF may be defined only for a restricted range of t-values. There are instances where the MGF may not be defined for any value t.

It should be emphasized that the moment generating function of a random variable X is just a tool and is rarely, if ever, an end in itself. As a tool, it is useful for two main reasons. First, it establishes a one-to-one correspondence between distributions and associated calculus functions. This is helpful because it is sometimes necessary to identify a distribution by recognizing its MGF. Second, as the name implies, it is a genuine aid in calculating moments, especially the first two, which establish the mean and the variance of the underlying random variable (using the variance formula).

Although this tool has other important theoretical applications in the rigorous development of mathematical probability, for now it will be used mainly for the two purposes described above.

We next list the main properties of moment generating functions (MGFs):

Properties of the Moment Generating Function:

(1) **Moments:** $M_X^{(k)}(0)=E[X^k]$, (The superscript (k) means the kth derivative.)

(2) **Linear Transformation:** If $Y=aX+b$, then $M_Y(t)=e^{bt}M_X(at)$.

(3) **Sums of Independent Random Variables:** If $X_1,\ldots,X_n$ are independent random variables and $S=X_1+\cdots+X_n$, then

$$M_S(t) = M_{X_1}(t)\cdot M_{X_2}(t)\cdots M_{X_n}(t).$$

(4) **Corollary to (3):** If $X_1,\ldots,X_n$ are independent random variables, all with common distribution X, then $M_S(t)=\left[M_X(t)\right]^n$.

We will provide some comments and explanations, followed by a number of examples:

(1) This is the property that allows us to calculate moments of the random variable X. For a *plausibility* argument, consider the following. By definition, $M_X(t) = E[e^{tX}]$ is a function of t. When we calculate the derivative of this function with respect to t (details below), we will use the fact that, "the derivative of an expected value is the expected value of the derivative." This is straightforward when X is discrete and the expected value is just a sum. In this case the statement amounts to, "the derivative of a sum is the sum of the derivatives." The argument works as well for integrals, and so we have,

$$M'_X(t) = \frac{d}{dt}[M_X(t)] = \frac{d}{dt}E[e^{tX}] = \overbrace{E\left[\frac{d}{dt}e^{t\cdot X}\right]}^{\substack{\text{Taking the derivative} \\ \text{inside the expectation} \\ \textit{can} \text{ be formally justified}}} = E[Xe^{t\cdot X}].$$

Note that in differentiating with respect to t, we treated the random variable X in the exponent as the constant coefficient of the variable t.

Taking successive derivatives leads to $M_X^{(k)}(t) = E[X^k e^{tX}]$. Hence, evaluating at $t = 0$, $M_X^{(k)}(0) = E[X^k \cdot e^{0\cdot X}] = E[X^k \cdot 1] = E[X^k]$, which leads to property (1).

(2) If $Y = aX + b$, then $M_Y(t) \overset{\text{def'n}}{=} E[e^{(aX+b)\cdot t}] = E_X[e^{aXt+bt}]$

$$= E_X[e^{bt} \cdot e^{(at)\cdot X}] = e^{bt} \cdot M_X(at).$$

(3) We have given informal discussions previously about "independent" random variables. For independent events, "the probability of an intersection is the product of the probabilities." For random variables this translates into "the expected value of a product is the product of the expected values." Employing this idea,

$$M_S(t) \overset{\text{def'n}}{=} E\left[e^{t\cdot(X_1+\cdots+X_n)}\right] = E\left[\overbrace{e^{t\cdot(X_1)} \cdot e^{t\cdot(X_2)} \cdots e^{t\cdot(X_n)}}^{\text{properties of exponents}}\right]$$

$$= \overbrace{E\left[e^{t\cdot(X_1)}\right] \cdot E\left[e^{t\cdot(X_2)}\right] \cdots E\left[e^{t\cdot(X_n)}\right]}^{\text{by independence of random variables } X_1, X_2, \cdots, X_n}$$

$$= M_{X_1}(t) \cdot M_{X_2}(t) \cdots M_{X_n}(t).$$

Example 5.6-4 Calculating Mean and Variance Using the MGF

(a) For the Bernoulli trial random variable in Example 5.6-2 use the MGF to calculate the mean and variance of X.

(b) For the continuous random variable in Example 5.6-3 use the MGF to calculate mean and variance.

Solution

For (a), the random variable X takes the value 1 with probability p, and takes the value 0 with probability $q=1-p$. We determined that $M_X(t)=q+pe^t$. Taking derivatives of the MGF with respect to t, $M'_X(t)=pe^t$ and $M''_X(t)=pe^t$. Then

$$E[X] = M'_X(0) = pe^t\Big|_{t=0} = p.$$

Next,

$$E[X^2] = M''_X(0) = p.$$

Using the variance formula, $Var[X] = E[X^2]-(\mu_X)^2 = p-p^2 = p(1-p) = pq$.

For (b), the random variable X was defined to have the density function $f(x)=e^{-x}$, where $0\le x<\infty$. We determined that $M_X(t)=\frac{1}{1-t}$. So, $M'_X(t)=(1-t)^{-2}$ and $M''_X(t)=2(1-t)^{-3}$. Evaluating these derivatives at $t=0$, yields $E[X]=M'_X(0)=1$, $E[X^2]=M''_X(0)=2$, and $Var[X] = 2-1^2 = 1$. ❐

A useful shortcut in calculating mean and variance is to employ logarithmic differentiation. That is, instead of differentiating $M_X(t)$, differentiate $\ln M_X(t)$.

Shortcuts Formulas for Expectation and Variance

Let X be a random variable with MGF $M_X(t)$. Define $h(t)=\ln(M_X(t))$.

(1) $E[X]=h'(0)$, and

(2) $Var[X]=h''(0)$.

Proof

We begin by observing, $M_X(0)=E[e^{0\cdot X}]=E[1]=1$. We next find $h'(t)$ using the rule for differentiating the natural logarithm, and remembering to use the chain rule. We get, $h'(t)=\frac{M'_X(t)}{M_X(t)}$. Evaluating at $t=0$,

$$h'(0) = \frac{M'_X(0)}{M_X(0)} = \frac{E[X]}{1} = E[X].$$

For the second derivative, we use the quotient rule to obtain

$$h''(t) = \frac{M_X(t)\cdot M''_X(t)-[M'_X(t)]^2}{[M_X(t)]^2}.$$

Finally we see

$$h''(0) = \frac{1\cdot E[X^2]-E[X]^2}{1^2} = Var[X].$$

Example 5.6-5 Shortcut Formulas

Let X have density function $f(x) = 2e^{-2x}$; $0 \le x < \infty$. Calculate the MGF for X and use the shortcut formulas to find μ_X and σ_X^2.

Solution

$$M_X(t) = E[e^{t \cdot X}] = \int_0^\infty e^{t \cdot x} \cdot (2e^{-2x}\, dx) = \int_0^\infty 2e^{-(2-t) \cdot x}\, dx = \left. \frac{2e^{-(2-t) \cdot x}}{-(2-t)} \right|_{x=0}^{\infty} = \frac{2}{2-t}.$$

The MGF exists for all $t < 2$.

$$h(t) = \ln(M_X(t)) = \ln\left(\frac{2}{2-t}\right) = \overbrace{\ln 2 - \ln(2-t)}^{\text{property of } \ln(x)}.$$

Then, $h'(t) = (2-t)^{-1}$ and $h''(t) = (2-t)^{-2}$. Then

$$E[X] = h'(0) = (2-0)^{-1} = \frac{1}{2}$$

and

$$Var[X] = h''(0) = (2-0)^{-2} = 2^{-2} = \frac{1}{4}.$$ ❐

This was far easier (see Exercise 5-42 below) than computing moments using the MGF directly.

Exercise 5-42 Consider the random variable X with density function $f(x) = 2e^{-2x}$, where $0 \le x < \infty$. This is the random variable in Example 5.6-5. Calculate μ_X and σ_X directly using the density function.

Exercise 5-43 For the following discrete random variable X, find the mean and the variance using the MGF method.

X	2	5
$\Pr(X = x)$	.4	.6

Exercise 5-44 Suppose that Y is a Bernoulli random variable with probability of success $p = 70\%$.

(a) Find the moment generating function for Y.
(b) Find $E[Y]$ and $Var[Y]$ using the MGF method. Did you use the regular or shortcut method?

Exercise 5-45 Suppose the discrete random variable X has probability distribution:

X	-1	2	5
$\Pr(X=x)$	.5	.2	☺

(a) Find the moment generating function for X.
(b) Find $E[X]$ using the MGF for X, $M_X(t)$.
(c) Define a new random variable $Y=3\cdot X-4$. Find $M_Y(t)$ using property (2) (Linear Transformations) of moment generating functions.

Exercise 5-46 Suppose that a random variable X has moment generating function $M_X(t) = .6+.3e^t+.1e^{2t}$.

(a) Calculate $E[X]$ and $Var[X]$.
(b) Write the probability distribution for X.
(c) Find the mode, median, and midrange for X.

Exercise 5-47 Consider a random variable Z with MGF given by $M_Z(t)=(.2+.8e^t)^5$.

(a) Find the mean and standard deviation for the random variable Z.
(b) Define a new random variable as $W=3-7Z$. Find the MGF of the random variable W. Compute μ_W and σ_W.

Exercise 5-48 SOA Exam P Sample Exam Questions #98

Let X_1, X_2, X_3 be a random sample from a discrete distribution with probability function

$$p(x) = \begin{cases} \frac{1}{3} & \text{for} \quad x=0 \\ \frac{2}{3} & \text{for} \quad x=1. \\ 0 & \text{otherwise} \end{cases}$$

Determine the moment generating function, $M_Y(t)$, of the product $Y=X_1\cdot X_2\cdot X_3$.

(A) $\frac{19}{27}+\frac{8}{27}e^t$ (B) $1+2e^t$ (C) $\left(\frac{1}{3}+\frac{2}{3}e^t\right)^3$ (D) $\frac{1}{27}+\frac{8}{27}e^{3t}$ (E) $\frac{1}{3}+\frac{2}{3}e^{3t}$

Hint: Work out the probability distribution for Y first.

Exercise 5-49 Let X_1, X_2, X_3 be defined as in Exercise 5-48. Determine the MGF for $W = X_1-2X_2+3X_3-4$.

Exercise 5-50 Let X have density function $f(x)=\frac{1}{3}e^{-(1/3)x}$; $0 \le x < \infty$. Calculate the MGF for X and use the shortcut formulas to find the mean and variance of X.

Exercise 5-51 SOA Exam P Sample Exam Questions #57

An actuary determines that the claim size for a certain class of accidents is a random variable, X, with moment generating function $M_X(t)=\frac{1}{(1-2500t)^4}$. Determine the standard deviation of the claim size for this class of accidents.

(A) 1,340 (B) 5,000 (C) 8,660 (D) 10,000 (E) 11,180

Exercise 5-52 SOA Exam P Sample Exam Questions #137

Let X and Y be identically distributed independent random variables such that the moment generating function of $X+Y$ is

$$M(t) = 0.09e^{-2t} + 0.24e^{-t} + 0.34 + 0.24e^{t} + 0.09e^{2t},$$

for $-\infty < t < \infty$.

Calculate $\Pr[X \le 0]$.

(A) 0.33 (B) 0.34 (C) 0.50 (D) 0.67 (E) 0.70

5.6.1 THE MGF FOR THE BINOMIAL DISTRIBUTION

Let X be a Bernoulli trial random variable, with the probability of success equal to p. In Example 5.6-2, we showed the MGF for X is given by $M_X(t)=q+pe^t$. The binomial distribution with n trials, each with probability of success p, can be represented with the random variable $S=\sum_{i=1}^{n} X_i$, a sum of independent, identically distributed Bernoulli trials. Then S represents the total number of successes in n trials. Therefore, the MGF for S can be written using property (4) for moment generating functions.

Moment Generating Function of a Binomial

Let S be a binomial random variable with n trials and probability of success p. Then

$$M_S(t) = (q+pe^t)^n.$$

This provides us with still another method for calculating the formulas for the mean and the variance of the binomial distribution. The easiest way to accomplish this is by means of the shortcut formulas. Let

$$h(t) \overset{\text{def'n}}{=} \ln\left(M_S(t)\right) = \ln(q+pe^t)^n = n \cdot \ln(q+pe^t).$$

$$h'(t) = \frac{npe^t}{q+pe^t},$$

so

$$E[S] = h'(0) = \frac{n \cdot p \cdot 1}{q+p \cdot 1} = n \cdot p.$$

$$h''(t) = \frac{(q+pe^t)(npe^t)-(npe^t)(pe^t)}{(q+pe^t)^2},$$

so

$$Var[S] = h''(0) = \frac{np-np^2}{1^2} = np(1-p) = npq.$$

We made use of the fact that $q+p=1$.

5.6.2 THE MGF FOR THE GEOMETRIC AND NEGATIVE BINOMIAL

In a series of independent Bernoulli trials with probability of success p, we let X denote the number of failures prior to the first success. Then, from Chapter 4 we have that X is a geometric random variable. Recall the probabilities for the random variable X are given by,

$$p_k = \Pr[X=k] = p \cdot (1-p)^k = p \cdot q^k \quad \text{for } k = 0,1,2,\ldots,\infty.$$

Moment Generating Function of a Geometric Random Variable

Let X be the number of failures before the first success in a sequence of independent identical Bernoulli trials with probability of success p. Then X is a geometric distribution with MGF given by,

$$M_X(t) = \frac{p}{1-q \cdot e^t}.$$

Proof

$$M_X(t) = \sum_{k=0}^{\infty} e^{kt} p_k = \sum_{k=0}^{\infty} e^{kt}(pq^k) = \sum_{k=0}^{\infty} p(qe^t)^k = \frac{p}{1-qe^t},$$

where the last step makes use of the formula for the sum of a geometric series, with ratio qe^t.

Exercise 5-53 Use the MGF method (regular or shortcut) to find formulas for the mean and the variance of the geometric random variable.

Recall from Chapter 4 that the negative binomial is the generalization of the geometric random variable. In a series of independent Bernoulli trials, we let S denote the number of failures prior to the rth success. Then S is a negative binomial random variable and can be expressed as, $S = X_1 + \cdots + X_r$, where the X_i's are independent and have identical geometric distributions.

We can use property (4) of MGFs to state the following result:

Moment Generating Function for the Negative Binomial Variable

Let S be the number of failures before the rth success in a sequence of independent identical Bernoulli trials with probability of success p. Then S is a negative binomial distribution with MGF given by,

$$M_S(t) = \left(\frac{p}{1-qe^t}\right)^r.$$

Exercise 5-54 Use the MGF method (regular or shortcut) to find formulas for the mean and the variance of the negative binomial random variable. Note: The shortcut method is much easier.

Exercise 5-55 Suppose a random variable V has MGF $M_V(t) = \left(\frac{0.3}{1-.7e^t}\right)^4$.

(a) Find the expected value and standard deviation for V.
(b) Find $\Pr[V = 2]$.

5.6.3 THE MGF FOR THE POISSON DISTRIBUTION

Let X be a Poisson random variable with mean rate of occurrence λ. From Chapter 4, we have that the probability function of X is $p_k = \Pr[X{=}k] = e^{-\lambda}\frac{\lambda^k}{k!}$, for $k = 0,1,2,\ldots$.

Moment Generating Function for the Poisson Distribution

If X is Poisson with parameter λ, then the MGF for X is given by,

$$M_X(t) = e^{\lambda\cdot(e^t-1)}.$$

Proof

$$M_X(t) = \sum_{k=0}^{\infty} e^{k\cdot t} p_k = \sum_{k=0}^{\infty} e^{k\cdot t}\left(e^{-\lambda}\frac{\lambda^k}{k!}\right) = e^{-\lambda}\sum_{k=0}^{\infty}\frac{(\lambda e^t)^k}{k!} = e^{-\lambda}e^{\lambda e^t} = e^{\lambda\cdot(e^t-1)}.$$

Here, we made use of the infinite series expansion for e^x, $e^x = \sum_{k=0}^{\infty}\frac{x^k}{k!}$, with $x = \lambda e^t$.

Exercise 5-56 Suppose $X \sim \text{Poisson}(\lambda = 6)$.

(a) Find $M_X(t)$.

(b) Let $Y = 2X+3$. Find $M_Y(t)$.

Exercise 5-57 Use the MGF method (regular or shortcut) to find formulas for the mean and the variance of the Poisson random variable with mean λ.

Note: The shortcut method is so much easier it isn't even close.

Exercise 5-58 A random variable Z has MGF $M_Z(t) = e^{3.7(e^t - 1)}$. Find the mode of Z.

5.7 CHAPTER 5 SAMPLE EXAMINATION

1. The continuous random variable X has expected value $E[X] = .6$ and density function

$$f(x) = \begin{cases} ax + bx^2 & \text{for } 0 < x < 1 \\ 0 \text{ otherwise} \end{cases}.$$

(a) Find the constants a and b.
(b) Find the $\Pr(X \leq .4)$.
(c) Calculate σ_X.
(d) Calculate the mode of the random variable X.

2. Consider a continuous random variable X with cumulative distribution function

$$F(x) = \begin{cases} 0 & \text{if } x < 0 \\ 1 - e^{-3x} & \text{if } x \geq 0 \end{cases}.$$

(a) Calculate $\Pr(X \leq .5)$.
(b) Determine measures of central tendency (mean, median, and mode).
(c) Calculate the 65th percentile for the random variable X.
(d) Calculate the second moment of the random variable.
(e) Calculate the standard deviation of X.

3. Consider a random variable Y with density function given by

$$f(y) = \begin{cases} \frac{3(1-y^2)}{4} & \text{if } -1 \le y \le 1 \\ 0 & \text{otherwise} \end{cases}.$$

(a) Calculate $\Pr(Y \le .5)$.
(b) Determine measures of central tendency (mean, median, and mode).
(c) Calculate the 65th percentile for the random variable Y.
(d) Calculate the variance for the random variable Y.

4. Consider a random variable T with density function given by

$$f(t) = \begin{cases} .2 & \text{if } 6 \le t \le 11 \\ 0 & \text{otherwise} \end{cases}.$$

Find the coefficient of variation for T.

5. Suppose $E[X]=2$, $E[X^3]=9$, and $E\left[(X-2)^3\right]=0$. Calculate the $Var(X)$.

6. **SOA Exam P Sample Exam Questions #68**
An insurance policy reimburses dental expense, X, up to a maximum benefit of 250. The probability density function for X is:

$$f(x) = \begin{cases} ce^{-0.004x} & \text{for } x \ge 0 \\ 0 & \text{otherwise} \end{cases}$$

where c is a constant.

Calculate the median benefit for this policy.

(A) 161 (B) 165 (C) 173 (D) 182 (E) 250

7. Compute the mean, mode, and standard deviation for the density function in Question 6.

8. **SOA/CAS Course 1 May 2003 Examination #34**
The lifetime of a machine part has continuous distribution on the interval (0, 40) with probability density function $f(x)$, where $f(x)$ is proportional to $(10+x)^{-2}$. Calculate the probability that the lifetime of the machine part is less than 6.

(A) 0.04 (B) 0.15 (C) 0.47 (D) 0.53 (E) 0.94

9. Calculate the mean, median, and mode for the density function in Question 8.

10. The continuous random variable X has density function

$$f(x) = \begin{cases} ax(x-1)^2 & 0 < x < 1 \\ 0 & \text{otherwise} \end{cases}.$$

(a) Find the constant a.
(b) Find the $\Pr(X > .80)$.
(c) Calculate the variance of X.
(d) Calculate the mode of the random variable X.

11. Consider a continuous random variable X with cumulative distribution function

$$F(x) = \Pr(X \le x) = \begin{cases} 0 & \text{if } x < 0 \\ 1 - e^{-x} & \text{if } x \ge 0 \end{cases}.$$

(a) Calculate $\Pr(X \le .8)$.
(b) Calculate the mode for the random variable X.
(c) Calculate the mean for the random variable X.
(d) Calculate the 80th percentile for the random variable X.
(e) Calculate the standard deviation of X.

12. Suppose the random variable X represents the amount of damage done to a rusty Pontiac Sunfire during the 2006-2007 calendar year. Ninety-two percent of the time, the vehicle will sustain no damage. Otherwise, we model the damage to be uniformly distributed (that is, a constant density function) on the interval $[500, 3000]$.

(a) Calculate the expected damage to the car.
(b) If insurance were purchased with a \$1000 deductible, calculate the average insurance payment.

13. The continuous random variable X has density function

$$f(x) = \begin{cases} 5e^{-5x} & \text{for } x \ge 0 \\ 0 & \text{otherwise} \end{cases}.$$

(a) Compute the moment generating function for X, $M_X(t)$.
(b) Use $M_X(t)$ to compute the expected value for X.

14. Consider a random variable Y with cumulative probability function

$$F(y) = \Pr(Y \le y) = \begin{cases} 0 & \text{for } y < 2 \\ \frac{(y-2)^2}{2} + .3 & \text{for } 2 \le y < 3. \\ 1 & \text{for } y \ge 3 \end{cases}$$

(a) Graph $F(y)$.

(b) Calculate the expected value of Y.

15. Suppose a random variable W has MGF given by $M_W(t) = \frac{.6}{1-.4e^t}$.

(a) Find σ_W.

(b) Find $\Pr[W = 3]$.

16. **SOA Exam P Sample Exam Questions #133**
A man purchases a life insurance policy on his 40th birthday. The policy will pay 5000 only if he dies before his 50th birthday and will pay 0 otherwise. The length of lifetime, in years, of a male born the same year as the insured has cumulative distribution function

$$F(t) = \begin{cases} 0 & \text{for } t \le 0 \\ 1 - e^{\frac{1-1.1^t}{1000}} & \text{for } t > 0 \end{cases}.$$

Calculate the expected payment to the man under this policy.

(A) 333 (B) 348 (C) 421 (D) 549 (E) 574

CHAPTER 6

SOME CONTINUOUS DISTRIBUTIONS

In this chapter we will delve more deeply into some of the standard continuous random variables that come up repeatedly in applications. From Chapter 5 we know that a continuous random variable X can be characterized by either its density function $f(x)$ or its cumulative distribution function $F(x)$. In the following we will define the common named random variables by either their density function or CDF, and provide formulas for means, variances, and moment generating functions.

6.1 UNIFORM RANDOM VARIABLES

A ***uniform*** random variable is the continuous version of discrete distributions with equally likely outcomes. As we have noted, for a continuous random variable the probability of a single point is always zero. So we need to capture the "equally likely outcomes" aspect differently. We accomplish this by requiring that the density function be constant on its domain.

In Chapter 5 (see Example 5.1-2) we discussed the random number generator on the unit interval $[0,1]$. We will refer to this random variable U as the ***standard uniform*** distribution. Its properties follow.

Standard Uniform

We say U is the ***standard uniform*** random variable if it has density function

$$f(u)=1;\ 0\leq u\leq 1\ \ (\text{and } f(u)=0 \text{ elsewhere}).$$

Properties

(1) The cumulative distribution function (CDF) for U is

$$F(u) = \Pr(U\leq u) = \begin{cases} 0 & -\infty<u<0 \\ u & 0\leq u\leq 1 \\ 1 & 1<u<\infty \end{cases}$$

Note: In the future we will use the short-hand notation $F(u)=u;\ 0\leq u\leq 1$, the rest being understood.

(2) The mean of the standard uniform random variable is $\mu=E[U]=\frac{1}{2}$.

(3) The variance is $\sigma_U^2=Var[U]=\frac{1}{12}$.

The CDF follows from the usual integration of the density function. For the mean and the variance, we compute the first and second moments of the standard uniform random variable.

$$E[U] = \int_0^1 u \cdot 1\, du = \frac{1}{2}$$

and

$$E[U^2] = \int_0^1 u^2 \cdot 1\, du = \frac{1}{3}.$$

Using the variance formula, $Var[U] = E[U^2] - (E[U])^2 = \frac{1}{3} - \left(\frac{1}{2}\right)^2 = \frac{1}{12}.$

We will define the ***general uniform random variable*** as a linear transformation of the standard uniform random variable. To be more precise, assume X is to be uniform on the interval $[A, B]$, and let U represent the standard uniform on $[0,1]$. Then we can define X by the relationship,

$$X = A + (B - A) \cdot U.$$

Note that as U varies from 0 to 1, X varies from A to B. This formula is equivalent to the formula for U in terms of X,

$$U = \frac{X - A}{B - A},$$

which can be thought of as the ***standardizing*** of the random variable X.

The properties of a general uniform random variable can now be derived directly from those of the standard uniform. We begin with the CDF for X. For any value x such that $A \leq x \leq B$ we have,

$$F_X(x) = \Pr[X \leq x] = \Pr\left[A + (B - A) \cdot U \leq x\right] = \Pr\left[U \leq \frac{x - A}{B - A}\right] = F_U(u),$$

where $u = \frac{x-A}{B-A}$. Since x is between A and B it follows that u is between 0 and 1, so that

$$F_X(x) = F_U(u) = F_U\left(\frac{x - A}{B - A}\right) = \frac{x - A}{B - A}; \quad A \leq x \leq B.$$

The density function is found by differentiating the distribution function with respect to x:

$$f_X(x) = F_X'(x) = \frac{1}{B - A}; \quad A \leq x \leq B.$$

Thus, we see that the density is indeed constant on the domain of X and that constant is the reciprocal of the length of the domain. This is precisely what is required to assure that the total probability equals one. In this instance the total probability is the area of the rectangular region from $x = A$ to $x = B$ (an interval of length $B - A$) and height $\frac{1}{B-A}$.

The probability of an event of the form $[a,b]$ with $A \le a \le b \le B$ can be found from either the density function $f(x)$ or the cumulative distribution function $F(x)$, and it takes the value $\Pr[a \le X \le b] = \int_a^b \frac{1}{B-A}\,dx = F(b)-F(a) = \frac{b-a}{B-A}.$

This is equivalent to saying the probability of an event is proportional to its length and that it can be calculated as the ratio of the length of the event to the length of the domain.

Formulas for the mean and the variance of a general uniform random variable follow directly from the corresponding formulas for the standard uniform:

$$E[X] = E\left[A+(B-A)\cdot U\right] = A+(B-A)\cdot E[U] = A+(B-A)\cdot\frac{1}{2} = \frac{A+B}{2}.$$

Using the variance formula for a linear transformation, $Var[a\cdot Y+b] = a^2 \cdot Var[Y]$, we have

$$Var[X] = Var\left[A+(B-A)\cdot U\right] = (B-A)^2 \cdot Var[U] = \frac{(B-A)^2}{12}.$$

In words, the mean of X is the midpoint of the domain and the variance is the length of the domain squared over 12. We summarize these results:

The Uniform Random Variable *X* on the Interval $[A,B]$

We say *X* has the ***uniform distribution*** on $[A,B]$, denoted $X \sim Unif[A,B]$, if the density function is given by,

$$f_X(x) = \frac{1}{B-A};\quad A \le x \le B.$$

Properties

(1) The CDF is given by $F_X(x) = \frac{x-A}{B-A};\ A \le x \le B.$ Equivalently, X could be defined by this CDF.

(2) $\mu_X = E[X] = \frac{A+B}{2}.$

(3) $\sigma_X^2 = Var[X] = \frac{(B-A)^2}{12}.$

(4) If $A \le a \le b \le B$, then $\Pr[a \le X \le b] = \frac{b-a}{B-A} = \frac{\text{Length of Event}}{\text{Length of Domain}}.$

Example 6.1-1 Uniform Distribution

Pick a real number at random from the interval $[0,10]$. Let X denote the random number.

(a) Find the density function and the CDF of the resulting random variable.
(b) Find the mean and the standard deviation.
(c) Find the probability that the chosen number is between e and π.

Solution

The underlying distribution is the continuous uniform distribution on $[0,10]$.

(a) The density function is $f_X(x) = \begin{cases} \frac{1}{10} & \text{for } 0 \le x \le 10 \\ 0 & \text{otherwise} \end{cases}$.

The CDF is $F_X(x) = \Pr(X < x) = \begin{cases} 0 & \text{if } x < 0 \\ \frac{x}{10} & \text{if } 0 \le x \le 10. \\ 1 & \text{if } x > 10 \end{cases}$

Using our short-hand notation, we would just write $F(x) = \frac{x}{10}$ for $0 \le x \le 10$.

(b) The mean equals $E[X] = \frac{0+10}{2} = 5$ and the variance is $Var[X] = \frac{(10-0)^2}{12} = \frac{25}{3}$.

Therefore the standard deviation is $\sigma_X = \frac{5\sqrt{3}}{3} = 2.887$.

(c) $\Pr[e < X < \pi] = \frac{\pi - e}{10} = .0423$. ❒

Example 6.1-2 Waiting for a Bus

Assume the time of arrival of a bus is uniformly distributed on the interval from 12:00 noon to 12:30 pm.

(a) Ellen arrives at the bus stop at precisely noon. Calculate the probability she waits at least 15 minutes for the bus to arrive.
(b) Calculate the probability she waits until at least 12:25 pm.
(c) Alan arrives at 12:10 and learns from Ellen that the bus has not yet arrived. Calculate the probability Alan waits at least 15 minutes for the bus.

Solution

Let T denote the arrival time of the bus in minutes with $T = 0$ corresponding to 12:00 noon. Since Ellen arrives at noon, her waiting time equals T. The random variable T has a continuous uniform distribution on the interval $[0,30]$.

(a) $\Pr[T > 15] = \frac{30-15}{30-0} = 50\%$. Does it make sense that half of the time Ellen will wait for at least 15 minutes?

(b) $\Pr[T > 25] = \frac{30-25}{30-0} = \frac{1}{6} = 0.1667.$

(c) This is a conditional probability since we are given that $T > 10$. Therefore, if X denotes Alan's waiting time, then $X = T-10$, ***given*** that $10 < T \leq 30$. To determine this probability we recall how we compute conditional probabilities:

$$\begin{aligned}\Pr[X > 15] &= \Pr[T-10 > 15 | 10 < T \leq 30] \\ &= \Pr\left[T > 25 | 10 < T \leq 30\right] \\ &= \frac{\Pr\left[(25 < T \leq 30) \cap (10 < T \leq 30)\right]}{P[10 < T \leq 30]} \\ &= \frac{\Pr[25 < T \leq 30]}{\Pr[10 < T \leq 30]} = \frac{\frac{30-25}{30}}{\frac{30-10}{30}} = \frac{5}{20} = 0.25.\end{aligned}$$

It is easy to show, using a general form of this calculation, that Alan's wait time $X = T-10$ is uniformly distributed on the interval $[0, 20]$. ❐

Exercise 6-1 Suppose the random variable D has uniform distribution on the interval $[60, 90]$. This could be a simple *model* for your age at death. Pleasant dreams!

(a) Find the density and distribution functions for D.
(b) Find the mean and variance of D.
(c) Find the 69th percentile and the first quartile of D.
(d) Find the probability that D is within one standard deviation of its mean.

Exercise 6-2 Suppose the random variable Y is uniformly distributed on the interval $[-2, 5]$.

(a) Find the density function $f_Y(y)$ and cumulative distribution function $F_Y(y)$.
(b) Find the mean and variance of Y.
(c) Find the median and mode of Y.
(d) Find the probability that Y is negative, $\Pr(Y < 0)$.
(e) Find the probability that Y is equal to one, $\Pr(Y = 1)$.

Exercise 6-3 The cable company claims that their installer will arrive at sometime between 12:00 noon and 5:00 pm. Assume that he arrives according to the uniform distribution. Let the random variable T denote the installer's arrival time.

(a) Find the density and distribution functions for T.
(b) Find the mean and the standard deviation of T.
(c) Find the probability that the cable guy arrives between 3:00 pm and 4:00 pm.
(d) Find the conditional probability that the cable guy arrives between 3:00 pm and 4:00 pm, given that he has not arrived by 2:00 pm.

Exercise 6-4 Suppose the random variable X has a uniform distribution on the interval $[a,b]$.

(a) Find the probability that X is within one standard deviation of its mean.
(b) Find the probability that X is within two standard deviations of its mean.

Exercise 6-5 Johnny-Come-Lately is always late to class. He arrives uniformly between 5 and 15 minutes late each day. Define the random variable A as the number of minutes by which Johnny is late. Compute the mean and standard deviation of A.

Exercise 6-6 SOA EXAM P Sample Exam Questions #65
The owner of an automobile insures it against damage by purchasing an insurance policy with a deductible of 250. In the event that the automobile is damaged, repair costs can be modeled by a uniform random variable on the interval $(0,1500)$. Determine the standard deviation of the insurance payment in the event that the automobile is damaged.

(A) 361 (B) 403 (C) 433 (D) 464 (E) 521

6.2 THE EXPONENTIAL DISTRIBUTION

The exponential distribution is closely related to the Poisson process and Poisson random variable described in Section 4.6. Recall that a Poisson process is a way of modeling certain random and sporadically occurring phenomena in which the overall average, or mean, rate of occurrence is λ per unit time. In Chapter 4 we discussed the Poisson random variable Y, which models the *number* of occurrences in a given unit time period.

Note that Y takes integer values $(0,1,2,\cdots)$ and hence is a ***discrete*** random variable. The object of interest now is a new random variable X that measures the ***time*** to the next occurrence. This new random variable X takes values along the continuum $[0,\infty)$, and hence is a ***continuous*** random variable. As we will see shortly the density function for X involves exponential functions, and hence the name exponential distribution.

Before investigating the properties of the exponential distribution and its relationship to the Poisson we present a review on integrating exponential functions.

6.2.1 INTEGRATION REVIEW

In the following we will be making use of integration techniques for ***improper*** integrals with exponential-type integrands. What makes them improper is the fact that the range of integration is from 0 to *infinity*. We touched on this in Chapter 5 in some of the examples, and we wish to have some applicable formulas available for later use. We will present a basic result now, and will give more general results later in the chapter when we discuss variations on the exponential distribution.

Integration Formula

Let $b > 0$ be a positive constant. Then $\int_0^\infty e^{-bx}\,dx = \frac{1}{b}$.

Proof

$$\int_0^\infty e^{-bx}dx = \overbrace{\lim_{M\to\infty}\int_0^M e^{-bx}dx}^{\text{definition of improper integral}} = \lim_{M\to\infty}\left[\frac{e^{-bx}}{-b}\Bigg|_{x=0}^{x=M}\right] = \lim_{M\to\infty}\left[\frac{e^{-bM}-1}{-b}\right] = \frac{0-1}{-b} = \frac{1}{b}.$$

6.2.2 EXPONENTIAL AND POISSON RELATIONSHIP

Consider now a Poisson process with mean rate of occurrence per unit time of λ, pronounced "lambda." In Chapter 4 we showed that if Y represents the *number* of occurrences in a given unit time period, then Y is a Poisson random variable with probabilities $\Pr[Y{=}k] = e^{-\lambda}\frac{\lambda^k}{k!}$ for $k = 0,1,2,\ldots$.

The ***exponential*** random variable X is used to model the *time* until the next occurrence, which we will often refer to as the ***waiting time***.

We will proceed by deriving the CDF of X, the time until the next occurrence in a Poisson process. The key to this derivation is the *scalability* property (property (2) for a Poisson process, in Subsection 4.6.2), which says that the number of occurrences in a time interval of length h is also a Poisson random variable, denoted by Y_h, with mean rate of occurrence equal to $\lambda \cdot h$.

To calculate the CDF for X, let x represent a non-negative value of time.

$$\begin{aligned} F_X(x) = \Pr[X \le x] &= 1 - \Pr[X > x] \\ &= 1 - \Pr[\text{Time to next occurrence} > x] \\ &= 1 - \Pr\big[\text{No occurrences in the interval } [0,x)\big] \\ &= 1 - \Pr[Y_x{=}0] = 1 - e^{-\lambda x}\cdot\frac{\lambda^0}{0!} = 1 - e^{-\lambda x}. \end{aligned}$$

Here Y_x is the <u>number</u> of occurrences in the time interval $[0,x)$, a <u>discrete</u> Poisson random variable with parameter $\lambda \cdot x$. Since X lives on the interval $[0,\infty)$ we have $F_X(x) = 0$ for negative values of x. Thus, for $x \ge 0$ we have,

$$F_X(x) = 1 - e^{-\lambda x}.$$

Note that $F_X(0) = 0$, that $F_X(x)$ is increasing and $\lim_{x\to\infty} F_X(x) = 1$, so that $F_X(x)$ indeed has the requisite properties of a CDF. Note also that the starting time in the above derivation is completely arbitrary. The waiting time to the next occurrence has the same distribution regardless of when the process comes under observation. This leads to the so-called ***memoryless*** property of the exponential distribution, which we illustrate below in Example 6.2-3 and the subsequent discussion.

We denote the reciprocal of λ with the symbol $\beta = \frac{1}{\lambda}$. If λ is the mean rate of occurrence, then the reciprocal, β, is the ***average wait time until the next occurrence*** in the Poisson process. As it turns out, this is the same as the average time *between* occurrences.

For example, if λ equals 10 occurrences per hour then β equals 1/10 of an hour (6 minutes), the average time between occurrences. As we shall see, β is in fact the mean of X.

6.2.3 PROPERTIES OF THE EXPONENTIAL RANDOM VARIABLE

The previous discussion leads us to define the exponential distribution, whose properties are summarized below.

Properties of the Exponential Random Variable *X*

Let $\beta = \frac{1}{\lambda}$ (we represent the following formulas in terms of both λ and β). We say the non-negative random variable X has an ***exponential distribution*** with mean β, (abbreviated $X \sim Exp(\beta)$) if the CDF of X takes the form,

$$F_X(x) = 1 - e^{-\lambda x} = 1 - e^{-(1/\beta)x};\ 0 \le x < \infty.$$

(a) The density function of an exponential random variable is:

$$f_X(x) = \lambda e^{-\lambda \cdot x} = \frac{1}{\beta} e^{-(1/\beta)x};\ 0 \le x < \infty.$$

(b) The mean of the random variable is: $E[X] = \frac{1}{\lambda} = \beta$.

(c) The variance is: $Var[X] = \sigma_X^2 = \frac{1}{\lambda^2} = \beta^2$.

(d) The Moment Generating Function (MGF) is: $M_X(t) = \frac{\lambda}{\lambda - t} = \frac{1}{1 - \beta t};\ t < \lambda$.

X can be thought of as the ***time until the next occurrence*** in a Poisson process. In this case λ represents the mean rate of occurrence and β represents the mean time between occurrences.

Proof

The formula for the CDF was derived above. The formula for the density function follows immediately by differentiating the distribution function. The derivation of the MGF is as follows (note that x is the variable of integration and that t is treated as a constant in the integration. The resulting MGF is a function of t only):

$$M_X(t) \overset{\text{def'n}}{=} E[e^{t\cdot X}] = \int_0^\infty e^{tx}\cdot f_X(x)\,dx$$

$$= \int_0^\infty e^{tx}\cdot\lambda\cdot e^{-\lambda x}\,dx$$

$$= \lambda\int_0^\infty e^{-(\lambda-t)x}\,dx = \overbrace{\frac{\lambda}{-(\lambda-t)}e^{-(\lambda-t)x}\Big|_{x=0}^{\infty}}^{\text{valid since } t<\lambda} = \frac{\lambda}{\lambda-t} = \frac{1}{1-\beta t}.$$

This function is valid for all $t < \lambda$.
We can now determine the mean and variance from the MGF, which we do using the shortcut formulas from Section 5.6. Recall if we define $h(t) = \ln(M_X(t))$, then the mean is $h'(0) = E[X]$ and variance is $h''(0) = V[X]$.[1]

$$h(t) = \ln\left(\frac{\lambda}{\lambda-t}\right) = \ln(\lambda) - \ln(\lambda - t).$$

$$h'(0) = E[X] = \frac{1}{\lambda-t}\Big|_{t=0} = \frac{1}{\lambda} = \beta$$

and

$$h''(0) = \frac{1}{(\lambda-t)^2}\Big|_{t=0} = \frac{1}{\lambda^2} = \beta^2.$$

Example 6.2-1 The Median of an Exponential Random Variable

Find the median, $x_{.5}$, for an exponential random variable with a mean of β.

Solution

We solve the required equation by using the natural logarithm. $0.5 = F(x_{.5}) = 1 - e^{-1/\beta(x_{.5})}$ implies that $-\frac{1}{\beta}(x_{.5}) = \ln .5 = \ln\frac{1}{2} = -\ln 2$, which implies that $x_{.5} = \beta\cdot\ln 2$.

The median of an *exponential* random variable with mean β is $\boxed{x_{.5} = \beta\cdot\ln 2.}$ ❐

Example 6.2-2 Graphing an Exponential Random Variable

Graph the density function for the exponential distribution with mean 3. Label the mean, median, and mode on your graph.

Solution

Since the mean is 3, the parameter $\beta = 3$. The median is $x_{.5} = \beta\cdot\ln 2 \approx 2.08$. Since the density function $f(x) = \frac{1}{3}e^{-(1/3)x}$ is a decreasing function, the mode occurs at $x = 0$. ❐

---

[1] We occasionally will use $V[X]$ as a convenient shorthand for $Var[X]$ in the following.

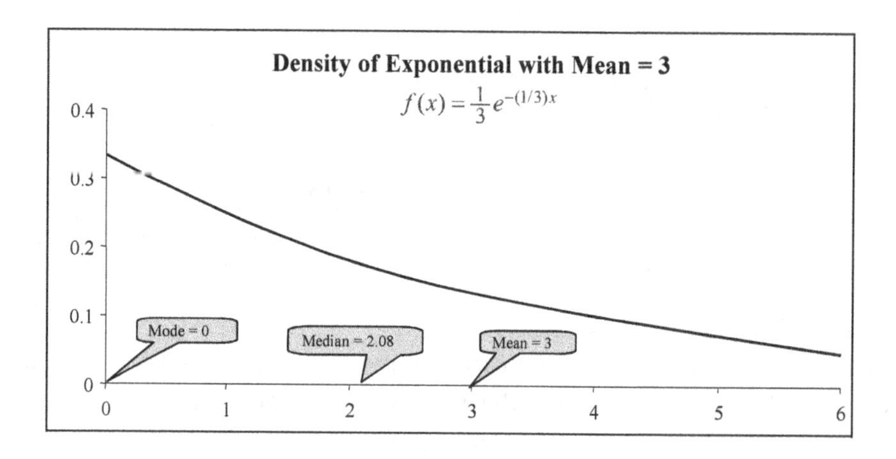

Example 6.2-3 Waiting Time in a Poisson Process

Suppose buses arrive at a certain stop according to a Poisson process with a mean rate of 4 per hour.

(a) Ellen arrives at the bus stop at 12:00 noon with no bus in sight. Calculate the probability she waits until at least 12:15 pm for the bus.

(b) Calculate the probability she waits until at least 12:25 for the bus.

(c) Alan arrives at 12:10 and sees Ellen still waiting at the bus stop. Alan does not know how long Ellen has been waiting. Calculate the probability that Alan waits until at least 12:25 pm for the bus.

(d) Assume that when Alan arrives at 12:10 he politely asks Ellen how long she has been waiting and she answers 10 minutes. Calculate the probability (from Alan's perspective with this new information) that he waits until at least 12:25 pm for the bus.

Solution

We are given that bus arrivals constitute a Poisson process with a mean rate $\lambda = 4$ per hour. Therefore the mean waiting time is $\beta = \frac{1}{4}$ hour $=$ 15 minutes. Now let X be the random variable representing the time in minutes until the next occurrence (in this case the arrival of a bus). We know that X is an ***exponentially distributed*** random variable with cumulative distribution function $F(x) = 1 - e^{-(1/15)x}$; $0 \le x < \infty$.

(a) The probability Ellen waits until at least 12:15 is
$\Pr[X > 15] = 1 - F(15) = 1 - (1 - e^{-(1/15)\cdot(15)}) = e^{-(1/15)\cdot(15)} = e^{-1} = 0.3679 \approx 37\%$.

(b) The probability that Ellen waits until at least 12:25 (i.e., the bus arrives after 12:25) is
$\Pr[X > 25] = 1 - F(25) = e^{-(1/15)(25)} = e^{-5/3} = 0.1889 \approx 19\%$.

(c) Since Alan's waiting time clock starts at 12:10 the probability he waits until at least 12:25 is $\Pr[X > 15] = 1 - F(15) = e^{-(1/15)(15)} = e^{-1} = 0.3679$. Do you observe that the probability that Alan must wait for at least 15 minutes is equal to the probability that Ellen must wait for 15 minutes (see part (a))?

(d) This time, with the additional information, Alan's waiting time clock starts at 12:00 and the probability he waits until at least 12:25 is the probability $X > 25$ minutes conditioned on the known event $X > 10$ minutes. So we calculate

$$\begin{aligned}\Pr[X > 25 | X > 10] &= \frac{\Pr[(X > 25) \cap (X > 10)]}{\Pr[X > 10]} \\ &= \frac{\Pr[X > 25]}{\Pr[X > 10]} \\ &= \frac{e^{-(1/15)(25)}}{e^{-(1/15)(10)}} = e^{-1} = 0.3679.\end{aligned}$$

□

The fact that (a), (c) and (d) all yield the same answer is somewhat baffling. It says that the probability of waiting at least 15 minutes is independent of when the clock starts. This is an inherent feature of the Poisson process and the resultant exponential waiting time random variable. It is referred to as the ***memory-less*** property of exponential random variables. In general, the current waiting time until the next occurrence is independent of the time elapsed since the last occurrence. Contrast this memory-less property of the exponential distribution with the behavior of the uniform distribution, also with a mean of 15 minutes, in Example 6.1-2.

This property is formulated in general terms in the following box, using conditional probability.

Memory-less Property of the Exponential Distribution

Let X be an exponential random variable with mean β. Then

$$\Pr[X > b] = \Pr[X > a + b \mid X > a].$$

In words, the probability of waiting at least time b is the same as the probability of waiting an additional time b given that time a has already elapsed.

Proof

$$\begin{aligned}\Pr[X > b] &= 1 - F(b) = 1 - [1 - e^{-(1/\beta)(b)}] \\ &= e^{-(1/\beta)(b)} \\ &= \frac{e^{-(1/\beta)(a+b)}}{e^{-(1/\beta)(a)}} = \Pr[X > a + b | X > a].\end{aligned}$$

Exercise 6-7 Show that the mode for any exponential distribution occurs at $x = 0$.

Note: This implies that regardless of the mean of an exponentially distributed random variable, the "most likely" outcome occurs at time zero.

Exercise 6-8 Find the standard deviation of the random variable Y with cumulative probability distribution function $F_Y(y) = \begin{cases} 0 & \text{for } y < 0 \\ 1 - e^{-2y} & \text{for } y \geq 0 \end{cases}$.

Exercise 6-9 Consider a random variable Z that is distributed according to the exponential distribution with mean 5.

(a) Find the standard deviation of Z.
(b) Find the median of Z.
(c) Find the mode of Z.
(d) Find $\Pr(1 < Z < 3)$.
(e) Find $\Pr(-1 < Z < 5)$.

Exercise 6-10 A package of light bulbs is advertised to last for an average of 750 hours. Assume the lifetime of a light bulb is modeled by the exponential distribution. Further assume that average represents the mean. Let T denote the number of hours that the light bulb lasts.

(a) Find the density function $f_T(t)$ and distribution function $F_T(t)$ for T.
(b) Find the probability that the light bulb burns out before the average life expectancy of the bulb. That is, find $\Pr(T < 750)$.
(c) Find the time T when there is a fifty percent chance that the bulb has burned out. That is, find the median (the value m such that $\Pr(T < m) = .50$.)

Exercise 6-11 Show $E[X | X > \phi] = E[X] + \phi$.

Note: This is the memory-less property in terms of expected values. It implies that you are wasting your time changing light bulbs prior to them burning out. That is, an old light bulb has the same remaining life expectancy as a new bulb if light bulbs last according to an exponential distribution.

Exercise 6-12 Consider an exponential distribution with *median* equal to seven. Calculate

(a) The mean.
(b) The mode.
(c) The parameter lambda.
(d) The probability that the random variable is less than 5.

Exercise 6-13 Find the inter-quartile range of an exponential random variable with standard deviation of 10.

Exercise 6-14 Suppose that $X \sim Exp(7)$. Find $E[X^3]$.

Exercise 6-15 Suppose that a random variable X is distributed according to the exponential distribution with variance equal to 9. Find $E[X^4]$.

Exercise 6-16 Suppose that $Y \sim Exp(3)$. Compute $E[7Y^2 - 11Y^6]$.

Exercise 6-17 SOA EXAM P Sample Exam Questions #69

The time to failure of a component in an electronic device has an exponential distribution with a median of four hours. Calculate the probability that the component will work without failing for at least five hours.

(A) 0.07 (B) 0.29 (C) 0.38 (D) 0.42 (E) 0.57

Exercise 6-18 SOA EXAM P Sample Exam Questions #29

The number of days that elapse between the beginning of a calendar year and the moment a high-risk driver is involved in an accident is exponentially distributed.

An insurance company expects that 30% of high-risk drivers will be involved in an accident during the first 50 days of a calendar year. What portion of high-risk drivers is expected to be involved in an accident during the first 80 days of a calendar year?

(A) 0.15 (B) 0.34 (C) 0.43 (D) 0.57 (E) 0.66

Exercise 6-19 SOA EXAM P Sample Exam Questions #37

The lifetime of a printer costing 200 is exponentially distributed with mean 2 years. The manufacturer agrees to pay a full refund to the buyer if the printer fails during the first year following the purchase, and a one-half refund if it fails during the second year. If the manufacturer sells 100 printers, how much should it expect to pay in refunds?

(A) 6,321 (B) 7,358 (C) 7,869 (D) 10,256 (E) 12,642

Exercise 6-20 SOA EXAM P Sample Exam Questions #70

An insurance company sells an auto insurance policy that covers losses incurred by a policy holder, subject to a deductible of 100. Losses incurred follow an exponential distribution with mean 300. What is the 95th percentile of actual losses that exceed the deductible?

(A) 600 (B) 700 (C) 800 (D) 900 (E) 1000

Exercise 6-21 SOA EXAM P Sample Exam Questions #47

A piece of equipment is being insured against early failure. The time from purchase until failure of the equipment is exponentially distributed with mean 10 years. The insurance will pay an amount x if the equipment fails during the first year, and it will pay $0.5x$ if failure occurs during the second or third year. If failure occurs after three years, no payment will be made. At what level must x be set if the expected payment made under this insurance is to be 1000?

(A) 3858 (B) 4449 (C) 5382 (D) 5644 (E) 7325

Exercise 6-22 SOA EXAM P Sample Exam Questions #130

The value of a piece of factory equipment after three years of use is $100(0.5)^X$ where X is a random variable having moment generating function

$$M_X(t) = \frac{1}{1-2t} \text{ for } t < \frac{1}{2}.$$

Calculate the expected value of this piece of equipment after three years of use.

(A) 12.5 (B) 25.0 (C) 41.9 (D) 70.7 (E) 83.8

Exercise 6-23 CAS Exam 3 Fall 2005 #20

Losses follow an exponential distribution with parameter θ. For a deductible of 100, the expected payment per loss is 2000. For a deductible of 100, the expected payment per loss is 2,000. Which of the following represents the expected payment per loss for a deductible of 500?

(A) θ (B) $\theta(1-e^{-500/\theta})$ (C) $2{,}000e^{-400/\theta}$ (D) $2{,}000(e^{-5/\theta})$ (E) $2{,}000\frac{1-e^{-500/\theta}}{1-e^{-100/\theta}}$

Exercise 6-24 Losses follow an exponential distribution with parameter θ. For an insurance policy with a payment cap of 100, the expected payment per loss is 20. Which of the following represents the expected payment per loss for a cap of 500?

(A) θ (B) $\theta(1-e^{-500/\theta})$ (C) $20e^{-400/\theta}$ (D) $20(e^{-5/\theta})$ (E) $20\frac{(1-e^{-500/\theta})}{(1-e^{-100/\theta})}$

6.3 THE NORMAL DISTRIBUTION

Many random phenomena in nature have distributions whose density function possesses a symmetric bell-shaped form (see Figure 6.1). Population characteristics of all kinds, from height or weight to standardized test scores, are frequently modeled using a random variable whose density function conforms to this familiar bell-shaped curve, known as the normal distribution. Each normal random variable is completely determined by its mean μ and variance σ^2 (or standard deviation). Put another way, normal distributions constitute a two parameter family of random variables.

Notation: $X \sim N(\mu, \sigma^2)$ will be our notation for a normally distributed random variable with mean equal to μ and variance equal to σ^2.

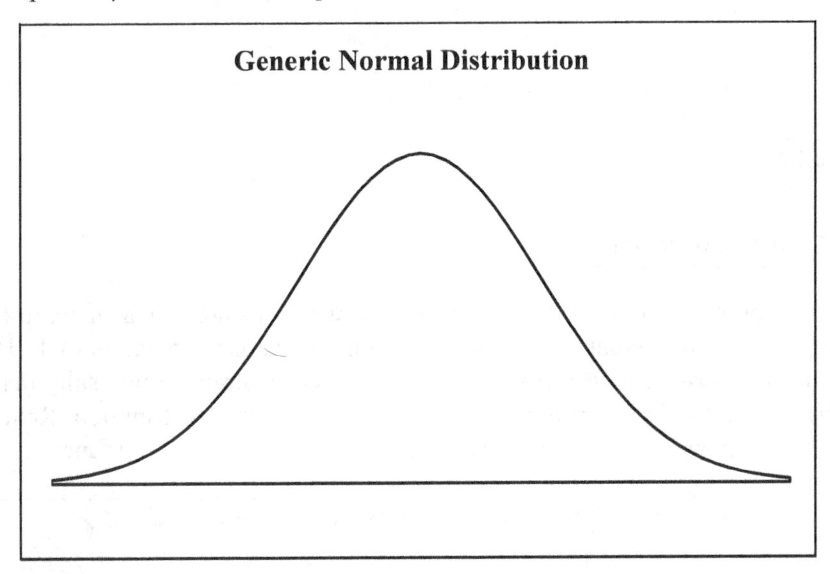

Figure 6.1

Later in this chapter we will see that normal distributions arise as approximations to sums or averages of independent observations from almost any random variable (These are called **Central Limit Theorems**). This is one reason normal distributions are so common in nature. Before we begin with the properties of normal distributions we provide another integration review that allows us to calculate the constant in the formula for the density function of a normal distribution.

6.3.1 INTEGRATION REVIEW

Integration Formula

$$\int_{-\infty}^{\infty} e^{-(z^2/2)}\,dz = \sqrt{2\pi}.$$

Note

The graph of the integrand is the bell-shaped curve centered at 0 and asymptotic to the x-axis toward both $-\infty$ and ∞. It is not straight-forward to evaluate this improper integral because there does not exist a closed formula expression for the anti-derivative. Consequently, it is necessary to resort to some skullduggery involving double integrals and polar coordinates to arrive at the result. We outline the calculation below:

We begin by assuming the improper integral $\int_{-\infty}^{\infty} e^{-(z^2/2)}\,dz$ converges to some number that we call A. Then $A^2 = \left[\int_0^{\infty} e^{-(z^2/2)}\,dz\right] \cdot \left[\int_0^{\infty} e^{-(z^2/2)}\,dz\right]$. The z is just a dummy variable for integration. We can just as easily use x and y. Thus,

$$A^2 = \left[\int_{-\infty}^{\infty} e^{-x^2/2} dx\right] \cdot \left[\int_{-\infty}^{\infty} e^{-y^2/2}\, dy\right] = \overbrace{\int_{-\infty}^{\infty}\int_{-\infty}^{\infty} e^{-(1/2)(x^2+y^2)}\, dx\, dy}^{\text{iterated double integral}}$$

$$= \overbrace{\int_0^{2\pi}\int_0^{\infty} e^{-(1/2)r^2}\, r dr\, d\theta}^{\text{change to polar coordinates}} = \overbrace{\int_0^{2\pi}\left[\int_0^{\infty} e^{-u}\, du\right] d\theta}^{\text{substitute } u = r^2/2} = 2\pi.$$

Thus, $A = \sqrt{2\pi}$.

6.3.2 THE STANDARD NORMAL

Paralleling our approach to the uniform distribution, we will start with a standard form of the normal, centered (mean, median, and mode) at 0 with a standard deviation of 1. This will be defined as the ***standard normal random variable***, which is almost universally denoted as Z. We begin by stating the basic properties of the standard normal distribution. Remember that any continuous random variable is defined by its density (or distribution) function.

Properties of the Standard Normal Random Variable, *Z*

(a) The density function is: $f_Z(z) = \frac{1}{\sqrt{2\pi}} e^{-(z^2/2)}$ for $-\infty < z < \infty$.

(b) The CDF is: $F_Z(z) = \int_{-\infty}^{z} f_Z(u)\, du$.

Note: A common alternative notation for the CDF uses Φ (the capital Greek letter ***phi***). Thus,

$$F_Z(z) = \Phi(z) = \Pr[Z \leq z] = \int_{-\infty}^{z} f_Z(u)\, du = \int_{-\infty}^{z} \frac{1}{\sqrt{2\pi}} e^{-(u^2/2)}\, du.$$

(c) The mean equals: $\mu_Z = 0$.

(d) The variance equals: $\sigma_Z^2 = 1$.

(e) The moment generating function is: $M_Z(t) = e^{t^2/2}$.

Proof

From the integration formula above, we see that $\int_{-\infty}^{\infty} \frac{1}{\sqrt{2\pi}} e^{-(z^2/2)}\, dz = 1$. Since $f(z) > 0$ for all real values z, $f(z)$ is a valid probability density function.

We proceed next to the derivation of the formula for the moment generating function of the standard normal random variable Z. We will use the MGF to quickly compute the mean and variance, in parts (c) and (d) of the property box.

$$M_Z(t) \overset{\text{definition}}{=} E\left[e^{t \cdot Z}\right] = \int_{-\infty}^{\infty} e^{tz} \cdot \frac{1}{\sqrt{2\pi}} e^{-(z^2/2)} dz$$

$$= \int_{-\infty}^{\infty} \frac{1}{\sqrt{2\pi}} e^{-(1/2)\left(z^2 - 2tz\right)} dz$$

$$= \int_{-\infty}^{\infty} \frac{1}{\sqrt{2\pi}} e^{-(1/2)\overbrace{(z^2 - 2tz + t^2 - t^2)}^{\text{complete the square}}} dz$$

$$= \int_{-\infty}^{\infty} \frac{1}{\sqrt{2\pi}} e^{-(1/2)(z^2 - 2tz + t^2)} \cdot e^{t^2/2} \cdot dz$$

$$= \frac{e^{t^2/2}}{\sqrt{2\pi}} \cdot \int_{-\infty}^{\infty} e^{-(1/2)\left(z^2 - 2tz + t^2\right)} dz$$

$$= \left(\frac{e^{t^2/2}}{\sqrt{2\pi}} \cdot \int_{-\infty}^{\infty} e^{-(1/2)(z-t)^2} dz \right)$$

$$\overset{\left\{\substack{\text{using the} \\ \text{substitution} \\ u = z - t}\right\}}{=} e^{t^2/2} \cdot \overbrace{\left(\int_{-\infty}^{\infty} \frac{1}{\sqrt{2\pi}} e^{-(1/2)u^2} du \right)}^{\{\text{equals 1 from integration formula above}\}} = e^{t^2/2}.$$

We can now calculate the mean and variance using the shortcut formulas based on the natural logarithm of the MGF, $h(t) = \ln\left(M_Z(t)\right)$ (see section 5.6). Let $h(t) = \ln\left(e^{t^2/2}\right) = \frac{t^2}{2}$. Thus, $h'(t) = t$ and $h''(t) = 1$.

(a) $E[Z] = h'(0) = 0$.

(b) $Var[Z] = h''(0) = 1$.

The omission of a closed formula for the CDF in (b) of the property box for Z is unavoidable. There is no simple formula for the anti-derivative of $f(z)$. An important consequence of this fact is that there is no direct way to calculate probabilities involving the standard normal distribution. Instead, we must rely on standard normal tables or software tools on calculators or computers.

In pictorial form $\Phi(z)$ is the area under the standard normal curve to the left of z, as illustrated below for $z = 1$.

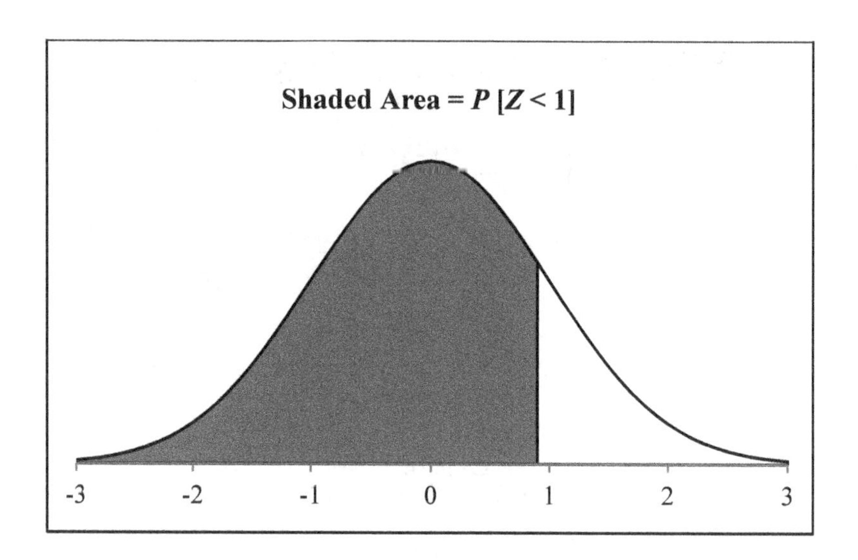

Using Tables to Find Values of $\Phi(z)$:

1. Most tables tabulate the CDF, $\Phi(z)$, for non-negative values of z only, relying on symmetry to find $\Phi(z)$ for negative z. The result is $\Phi(-z) = 1 - \Phi(z)$ for $z \geq 0$. We will illustrate this shortly.
2. Many versions of the standard normal table tabulate only the $\Pr[0 \leq Z \leq z]$, which, by symmetry, equals $\Phi(z) - \frac{1}{2}$.
3. A typical table will list the non-negative values of z to one decimal place down the left-hand column, with 10 adjacent columns showing the second decimal place for z, with headings .00, .01,…,.09.

This table extract shows the look-up for $z = 1.96$, with $\Phi(1.96) = \Pr[Z \leq 1.96] = 0.9750$.

z	…	.05	.06	.07	…
⋮					
1.8		0.9678	0.9686	0.9693	
1.9		0.9744	0.9750	0.9756	
2.0		0.9798	0.9803	0.9808	
⋮					

Here are some examples of using tables for finding the probabilities of events involving Z.

Example 6.3-1 Calculating Probabilities for Z using the Table

(a) Find $\Phi(-2) = \Pr[Z \leq -2]$.
(b) Find $\Pr[-2 \leq Z \leq 1]$.

Solution

It is always a good idea to sketch a quick schematic of the area needed, at least until you become adept at the look-ups, in order to confirm that you are doing the calculation correctly. For example, in (a),

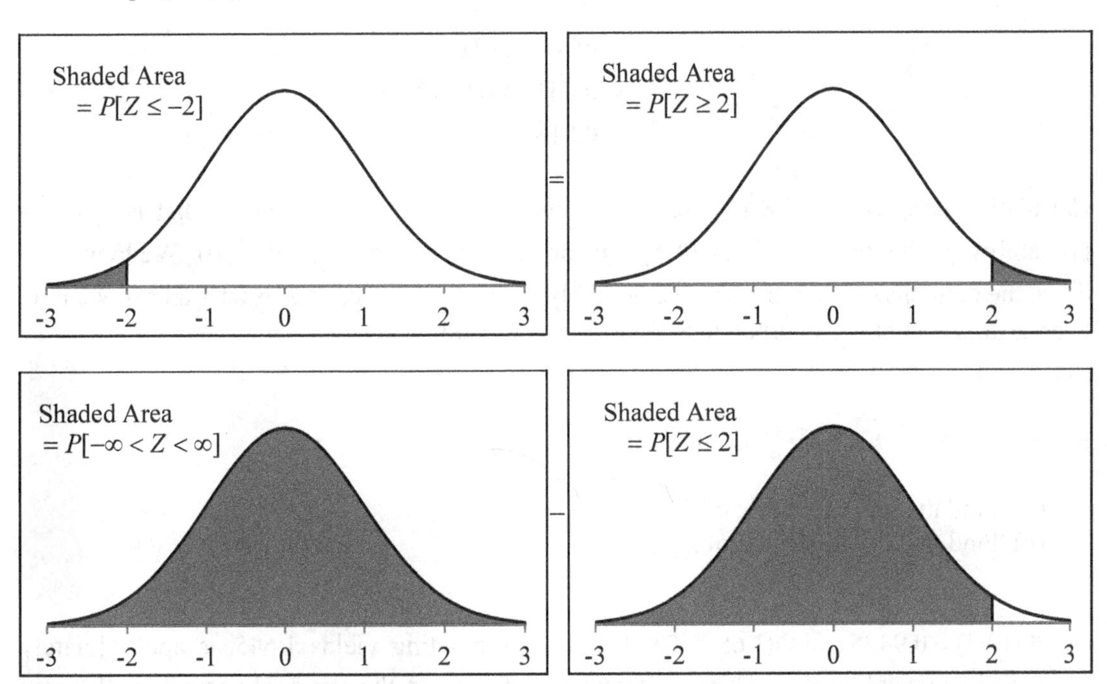

$$\Phi(-2) = \Pr[Z \le -2] = 1 - \Phi(2) = 1 - 0.9772 = 0.0228.$$

The same argument applied in general shows why $\Phi(-z) = 1 - \Phi(z)$ for $z \ge 0$.

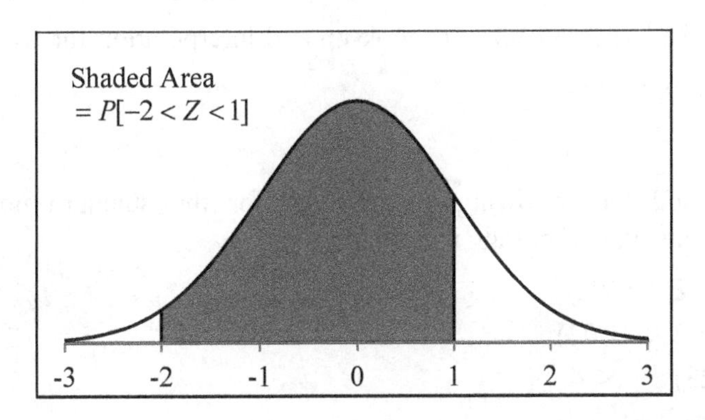

For (b),

$$\begin{aligned}\Pr[-2 \le Z \le 1] &= \Phi(1) - \Phi(-2) \\ &= \underbrace{\Phi(1)}_{\Pr(Z \le 1)} - \underbrace{[1 - \Phi(2)]}_{\Pr(Z \le -2) = \Pr(Z \ge 2)} \\ &= \Phi(1) + \Phi(2) - 1 \\ &= 0.8413 + 0.9772 - 1 \\ &= 0.8185.\end{aligned}$$

Calculating ***percentiles*** of Z requires a "backwards" look-up in the table. That is, given a probability, p, find the z such that $\Phi(z) = p$, or equivalently, $z = z_p = \Phi^{-1}(p)$. We look for p down the columns of probabilities; we probably won't find percentiles exactly and so we may need to approximate or interpolate to find the corresponding z_p.

Example 6.3-2 Finding Percentiles of *Z*

(a) Find the 95th percentile of Z.
(b) Find the 20th percentile of Z.

Solutions

(a) $\Phi(1.64) = 0.9495$ and $\Phi(1.65) = 0.9505$, so interpolating yields 1.645 as approximately the 95th percentile of Z. This is a famous landmark on the standard normal, well worth remembering.

(b) The median (50th percentile) is $z_{.50} = 0$ and $z_{.20} < z_{.50}$, so the 20th percentile, $z_{.20}$, is negative. Thus, $0.20 = \Phi(z_{.20}) = 1 - \Phi(-z_{.20})$. Then, $\Phi(-z_{.20}) = 0.80$. This means that $-z_{.20} = z_{.80} = 0.842$ (using backward look-up and interpolation for $z_{.80}$). Therefore, $z_{.20}$ is equal to $-z_{.80} = -0.842$. ❐

Exercise 6-25 Find the following probabilities for the standard normal distribution, $Z \sim N(0,1)$. Draw an appropriate picture.

(a) $\Pr(-2.98 < Z \le .99)$
(b) $\Pr(-1.75 < Z < -1.04)$
(c) $\Pr(Z \ge -.23)$
(d) $\Pr(|Z| < .73)$
(e) $\Pr(|Z| > 2.05)$

Exercise 6-26 Find the following z_α values from the standard normal distribution, $Z \sim N(0,1)$, that yield the given probabilities. Draw an appropriate picture.

(a) $\Pr(Z \geq z_\alpha) = .10$
(b) $\Pr(Z \geq z_\alpha) = .05$
(c) $\Pr(Z \geq z_\alpha) = .01$
(d) $\Pr(Z \geq z_\alpha) = .95$
(e) $\Pr(Z < z_\alpha) = .75$
(f) $\Pr(Z \geq z_\alpha) = 2.64$

Exercise 6-27

(a) Find the standard normal z-value so that the area that lies to the right of z is .2119.
(b) Find the standard normal z-value (or z-score) so that the area between $-z$ and z is .9030.

6.3.3 THE GENERAL NORMAL DISTRIBUTION

A random variable X is said to be ***normally distributed*** if it can be written as a linear transformation of the standard normal random variable Z. In other words, the family of general normal distributions consists of all X satisfying $X = a \cdot Z + b$. Since the density function of Z is symmetric about zero, we have the distribution of Z equal to the distribution of $-Z$. Consequently, we can assume that a is a non-negative number. Also, if $a = 0$ then our random variable $X = a \cdot Z + b$ reduces to the constant b, so we will restrict a to the positive numbers in the following.

Properties of the Normal Distribution

Let $X = a \cdot Z + b$ where $a > 0$ and $-\infty < b < \infty$.

(1) The mean is: $E[X] = \mu_X = b$.

(2) The variance is: $Var[X] = \sigma_X^2 = a^2$. This implies that the standard deviation is $\sigma_X = a$.

(3) $X = \sigma_X \cdot Z + \mu_X$ if and only if $Z = \frac{X - \mu_X}{\sigma_X}$. This is denoted as, $X \sim N(\mu_X, \sigma_X^2)$.

(4) The CDF is: $F_X(x) = \Phi\left(\frac{x-\mu_X}{\sigma_X}\right)$ for all $-\infty < x < \infty$. We need tables or software to evaluate this.

(5) The density function is $f_X(x) = \frac{1}{\sigma_X \sqrt{2\pi}} e^{-\frac{1}{2}\left(\frac{x-\mu_X}{\sigma_X}\right)^2}$ for all $-\infty < x < \infty$.

(6) The MGF is $M_X(t) = e^{\mu_X t + (1/2)\sigma_X^2 t^2}$.

(7) $\Pr[c \le X \le d] = \Phi\left(\frac{d-\mu_X}{\sigma_X}\right) - \Phi\left(\frac{c-\mu_X}{\sigma_X}\right)$.

(8) If $X \sim N(\mu_X, \sigma_X^2)$ and $Y \sim N(\mu_Y, \sigma_Y^2)$ are **independent** random variables, then $(X+Y) \sim N(\mu_X + \mu_Y, \sigma_X^2 + \sigma_Y^2)$.

Proof

Properties (1) and (2) follow easily from the basic rules for linear transformations. Property (3) is simple algebra and shows that the standard normal Z is the standardized form of X.

(4) $$F_X(x) = \Pr[X \le x] = \Pr[\sigma_X Z + \mu_X \le x] = \overbrace{\Pr\left[Z \le \frac{x-\mu_X}{\sigma_X}\right]}^{\text{since } \sigma_X > 0} = \Phi\left(\frac{x-\mu_X}{\sigma_X}\right).$$

(5) With $z = \frac{x-\mu_X}{\sigma_X}$, we calculate the density function $f_X(x) = F_X'(x)$ using the chain rule from calculus:

If $F_X(x) = \Phi(z)$ then

$$\frac{d}{dx} F_X(x) = \frac{d\Phi}{dz} \cdot \frac{dz}{dx} = f_Z(z) \cdot \frac{1}{\sigma_X} = \frac{1}{\sigma_X \sqrt{2\pi}} e^{-\frac{1}{2}\left(\frac{x-\mu_X}{\sigma_X}\right)^2}.$$

Property (6) is a direct consequence of the formula for the MGF of a linear transformation (Section 5.6). Property (7) follows immediately from (4) and the relationship $z = \frac{x-\mu_X}{\sigma_X}$. This is how questions involving the probabilities of events for the general normal are reduced to the standard normal and the associated tables. Finally, property (8) follows because the MGF of the random variable $X + Y$ is

$$M_{X+Y}(t) = \underbrace{M_X(t) \cdot M_Y(t)}_{X \text{ and } Y \text{ are independent}} = \left(e^{\mu_X t + (1/2)\sigma_X^2 t^2}\right) \cdot \left(e^{\mu_Y t + (1/2)\sigma_Y^2 t^2}\right)$$
$$= e^{(\mu_X + \mu_Y)t + (1/2)\left(\sigma_X^2 + \sigma_Y^2\right)t^2}.$$

We recognize this as a MGF for a normally distributed random variable with mean $\mu_X + \mu_Y$ and variance $\sigma_X^2 + \sigma_Y^2$.

We see from (4) or (5) that within the normal family of distributions, the density and distribution functions are completely determined by μ_X and σ_X (or σ_X^2). We say that the normal distributions constitute a two-parameter family, using the parameters μ_X and σ_X^2. This is often abbreviated (as in (3) above) by writing $X \sim N(\mu_X, \sigma_X^2)$.

Example 6.3-3 General Normal Distribution

Suppose that the random variable X is normally distributed with mean equal to 5 and standard deviation equal to 2, that is, $X \sim N(5, 2^2)$. Find $\Pr(X < 9.3)$ and $\Pr(-.2 < X < 6.9)$.

Solution

We begin by computing the related z-scores. Let Z denote the standard normal distribution. If $x = 9.3$, then

$$z = \frac{x-\mu}{\sigma} = \frac{9.3-5}{2} = 2.15.$$

$$\Pr(X < 9.3) = \Pr(Z < 2.15) = .9842.$$

$$\Pr(-.2 < X < 6.9) = \Pr(-2.6 < Z < .95)$$

$$= \Phi(.95) - [1 - \Phi(2.6)] = .8289 + .9953 - 1 = .8243.$$ ❐

Example 6.3-4 General Normal Reverse Lookup

Suppose that $X \sim N(50, 8)$. Find the value x such that $\Pr(X > x) = .025$.

Solution

First, let z be the value of Z such that $\Pr(Z > z) = .025$.

This is equivalent to $\Pr(Z \le z) = 1 - .025 = .975$. Thus, $z = z_{.975} = 1.96$.

Solving $1.96 = \frac{x-50}{8}$ implies that $x = 65.68$. We see that 65.68 is 1.96 standard deviations above the mean of X. ❐

Exercise 6-28 Suppose that the height for college centers is normally distributed with mean of 6 feet 7 inches and standard deviation of 2.5 inches.

(a) If a college center were selected at random, find the probability that he is less than 7 feet tall.

(b) Roy Williams only recruits the tallest 20% of college centers. How tall must a center be in order to be recruited by the hall-of-fame coach?

Exercise 6-29 A person must score in the upper 2% of the population on an IQ test to qualify for membership in MENSA, the international high-IQ society. If IQ scores are normally distributed with mean equal to 100 and standard deviation of 15, what score must a person get to qualify for MENSA?

Exercise 6-30 The Boston marathon sets a qualifying standard to only allow the best 17% of marathon runners in their race. Suppose that competitive marathon race times are normally distributed with mean of 3 hours and 20 minutes with a standard deviation of 24 minutes.

(a) What is the slowest time that will qualify for the Boston marathon?
(b) What time would the slowest 1.5% of competitive marathon racers finish?

Exercise 6-31 The average size of a rainbow trout in Red Fish Lake is normally distributed with a mean of 11.7 inches and a standard deviation of 3.9 inches.

(a) What is the probability that the next fish you catch is at least 16 inches long?
(b) How large must the fish be to be in the top 1% of all trout in Red Fish Lake?

Exercise 6-32 The average age of a mathematics faculty member is 56.2 years. Assume that age is normally distributed with variance of 25.

(a) Calculate the probability that a mathematics faculty member is less than 60.
(b) Calculate the probability that a mathematics faculty member is between 48 and 59.
(c) What is the age range of the youngest 10% of the mathematics faculty?

Exercise 6-33 On a probability examination, grades were distributed according to the normal distribution with mean of 73 and standard deviation of 12.

(a) Determine the corresponding z-scores of students who earned 69, 79, and 96.
(b) Determine the scores of students who earned z-scores of $z = -1.62$ and $z = .47$.

Exercise 6-34 The official size of the circumference for men's NCAA basketballs is specified to be 30 inches. Variability in manufacturing produces balls whose circumferences are normally distributed with mean 30 inches and standard deviation .07 inches.

(a) Find the probability that the circumference of a ball is between 29.95 and 30.05 inches.
(b) What circumference do the largest 5% of the basketballs possess?

Exercise 6-35 A chemistry teacher grades on the normal curve. The top 8% earn A's, the next 17% earn B's, the middle 50% earn C's, the bottom 8% earn F's, and the remaining students earn D's. Her grades had mean 58 and standard deviation 17.

(a) What grade would Bob receive if he averaged 48?
(b) What are the range of scores to earn A's, B's, etc.?

Exercise 6-36 Assume that IQ scores are normally distributed with mean 100 and standard deviation 16. Find the probability that a person chosen at random has an IQ score between 50 and 70.

6.3.4 THE LOGNORMAL DISTRIBUTION

We say the random variable Y has a ***lognormal distribution*** if the transformed random variable $X = \ln(Y)$ is normally distributed with mean μ_X and variance σ_X^2. This is equivalent to writing $Y = e^X$, where $X \sim N(\mu_X, \sigma_X^2)$. Note that the lognormal distributions also constitute a two-parameter family, but the parameters are specified in terms of the mean and variance of the normal random variable X. We also see from the properties of the exponential function that Y is a non-negative random variable regardless of the parameters μ_X and σ_X^2.

A typical application of the lognormal occurs in modeling exponential growth or decay when the rate of growth, rather than being fixed, is treated as a normal random variable. The lognormal distribution is used as a model for insurance claim distributions given that a claim has been filed. In finance or actuarial science Y could represent the value of one dollar after investment for a unit period of time at a continuous rate of interest X, which is random, but normally distributed. Then $Y = e^X$ has lognormal distribution. The graph of the density function for Y in the case that $X = Z \sim N(0,1)$ appears as follows:

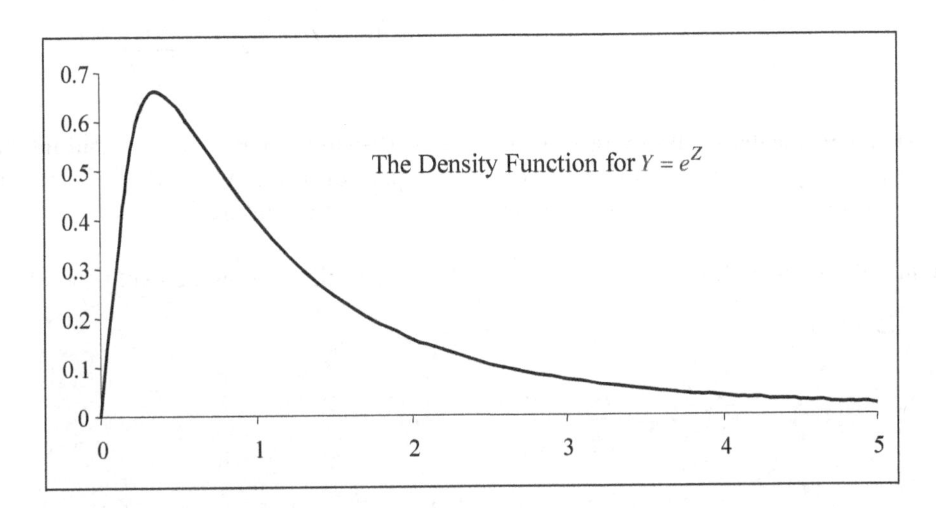

Properties of the Lognormal Distribution

Let $Y = e^X$ where $X \sim N(\mu_X, \sigma_X^2)$. Then Y is said to have a ***lognormal*** distribution with parameters μ_X and σ_X^2.

(1) The mean is $E[Y] = e^{\mu_X + (1/2)\sigma_X^2}$.

(2) The variance is $Var[Y] = e^{2\mu_X + \sigma_X^2} \cdot (e^{\sigma_X^2} - 1)$.

(3) The CDF for Y is $F_Y(y) = \Phi\left(\frac{\ln y - \mu_X}{\sigma_X}\right)$; $0 < y < \infty$.

(4) The density function for Y is $f_Y(y) = \frac{1}{y\sigma_X\sqrt{2\pi}} e^{-\frac{1}{2}\left(\frac{\ln y - \mu_X}{\sigma_X}\right)^2}$; $0 < y < \infty$.

(5) For $0 < a < b$, $\Pr[a < Y \le b] = \Phi\left(\frac{\ln b - \mu_X}{\sigma_X}\right) - \Phi\left(\frac{\ln a - \mu_X}{\sigma_X}\right)$.

Proof

Properties (1) and (2) make use of the MGF for X. Recall that

$$M_X(t) = E[e^{t\cdot X}] = e^{\mu_X \cdot t + \frac{\sigma_X^2 \cdot t^2}{2}}.$$

Now,

$$E[Y] = E[e^X] = M_X(1) = e^{\mu_X + (1/2)\sigma_X^2}.$$

Also,

$$E[Y^2] = E[e^{2X}] = M_X(2) = e^{2\mu_X + 2\sigma_X^2}.$$

Using the variance formula and a little algebra yields (2).

$$(3)\quad F_Y(y) = \Pr[Y \le y] = \Pr[e^X \le y] \overbrace{=}^{\text{key step}} \Pr[X \le \ln y]$$
$$= \Pr\left[Z \le \frac{\ln y - \mu_X}{\sigma_X}\right] = \Phi\left(\frac{\ln y - \mu_X}{\sigma_X}\right).$$

Typical questions involving the lognormal distribution involve computing the probability of various events. This technique for finding the CDF is widely applicable and learning it now will serve you well in the future.

(4) For the density formula, let $u = \frac{\ln y - \mu_X}{\sigma_X}$. Then, differentiating using the chain rule gives

$$f_Y(y) = F_Y'(y) = \frac{d\Phi}{du} \cdot \frac{du}{dy}$$
$$= \frac{1}{\sqrt{2\pi}} e^{-\frac{1}{2}(u)^2} \cdot \frac{1}{y\sigma_X} = \frac{1}{y\sigma_X\sqrt{2\pi}} e^{-\frac{1}{2}\left(\frac{\ln y - \mu_X}{\sigma_X}\right)^2}.$$

(5) This is just a restatement of property (7) of the normal distribution:

$$\Pr[a < Y \le b] = \Pr[\ln a < X \le \ln b] = \Phi\left(\frac{\ln b - \mu_X}{\sigma_X}\right) - \Phi\left(\frac{\ln a - \mu_X}{\sigma_X}\right).$$

Example 6.3-5 Lognormal

Suppose that $X \sim N(\mu = -3, \sigma^2 = 4)$. Let the random variable $Y = e^X$.

(a) Find $E[Y]$.

(b) Find the standard deviation of the random variable Y.

(c) Determine the $\Pr(Y > .5)$.

(d) Find the median of the lognormal random variable Y.

Solution

(a) $E[Y] = e^{\mu + (\sigma^2/2)} = e^{-3+(4/2)} = e^{-1} = .3679.$

(b) $Var[Y] = e^{2\mu+\sigma^2}(e^{\sigma^2} - 1) = e^{-6+4} \cdot (e^4 - 1) = 7.2537,$ so $\sigma_Y = \sqrt{7.2537} = 2.6933.$

(c) $\Pr(Y > .5) = \Pr(e^X > .5) = \Pr(X > \ln .5) = \overbrace{\Pr\left(Z > \frac{\ln .5 - (-3)}{2}\right)}^{\text{standardizing } X \sim N(-3,2^2)} = \Pr(Z > 1.15) = .1244.$

(d) The median $y_{.5}$ is the solution to $\frac{1}{2} = \Pr(Y \le y_{.5}) = \Pr(e^X \le y_{.5}) = \Pr(X \le \ln y_{.5}).$

This happens when $\ln y_{.5} = \mu_X = -3,$ which occurs at $y_{.5} = e^{-3}.$ ❐

Exercise 6-37 Suppose that Y is distributed according to a lognormal distribution with parameters $\mu = 5.1$ and $\sigma = 1.2$.

(a) Find $E[Y]$ and $Var[Y]$.

(b) Find $\Pr(150 < Y < 250)$.

(c) Find the value Y that corresponds to the top 10%.

Exercise 6-38 Suppose that Y is distributed according to a lognormal distribution with $E[Y] = 2$ and $Var[Y] = 12$.

(a) Find the parameters μ and σ.

(b) Find the probability $\Pr(1 < Y < 3)$.

Exercise 6-39 Suppose that L is distributed according to a lognormal distribution with parameters $\mu = .1$ and $\sigma = .4$.

(a) Find the expected value and variance of the random variable L.

(b) Find $\Pr(.9 < L < 1.5)$.

(c) Find the value L that corresponds to the bottom 5%.

6.4 THE LAW OF AVERAGES AND THE CENTRAL LIMIT THEOREM

We come now to one of the most important results in modern probability and statistics, the Central Limit Theorem. We will express this result in the language of statistics, using the idea of a random sample.

Early in the text we relied on the notion of a probability experiment and discussed the probabilities of *events* that may result from performing this experiment.

Later, we introduced *random variables* as numerical outcomes determined by performing a probability experiment. Subsequently, we saw that a random variable could be characterized by its underlying *distribution* (a probability function or density, or a CDF), wholly divorced from the underlying probability experiment.

Let us return for a moment to the idea of the underlying experiment that gives rise to the random variable, X. Say for example, the experiment is to throw a dart randomly at the unit interval (or equivalently, spin a wheel calibrated from 0 to 1) and that the value of X is the resulting point. In this case X happens to be the standard uniform random variable and the experiment amounts to choosing a number at random from $[0,1]$. We will employ this framework in describing random samples.

6.4.1 RANDOM SAMPLES

In statistics the random variable X of interest is often called the ***population*** random variable. The population has mean μ_X and standard deviation σ_X. Each time we perform the experiment (throwing the dart, for example) we obtain a particular value for X, which we call an ***observation***. If we repeat the experiment n times, the resulting n observations of X, denoted $X_1, X_2, \ldots, X_n$, or $(X_i)_{i=1}^n$, are called a ***random sample*** of size n chosen from X.

Of course, each repetition of the experiment is considered to be independent of the others, and each repetition is conducted under identical circumstances. This means that the observations $X_1, X_2, \ldots, X_n$ in the random sample are ***independent, identically distributed*** random variables, each with the same distribution as the population X. This is abbreviated by writing $(X_i)_{i=1}^n$ I.I.D. $\sim X$. In particular each of the observations, X_i, has the same mean μ_X and variance σ_X^2 as the population random variable X.

Example 6.4-1 Random Sample

Let X be the result of tossing a dart at the unit interval $[0,1]$. Let $(X_i)_{i=1}^n$ be the random sample resulting from n independent tosses of the dart.

(a) Write the mean and the variance of the population X.

(b) Let $S = \sum_{i=1}^{n} X_i$. Write the mean and the variance of S.

(c) Let $T = \frac{1}{n}\sum_{i=1}^{n} X_i$. Write the mean and the variance of T.

Solution

(a) From Section 6.1 we have that $\mu_X = \frac{1}{2}$ and $\sigma_X^2 = \frac{1}{12}$.

(b) Using standard formulas from Chapter 3 for the sum of random variables, we have $\mu_S = E\left[\sum_{i=1}^{n} X_i\right] = E[X_1] + E[X_2] + \cdots + E[X_n] = n \cdot \frac{1}{2}$. Since the $(X_i)_{i=1}^n$ are independent, it follows similarly that $\sigma_S^2 = \frac{n}{12}$.

(c) We use the fact that T is the linear transformation $T = \frac{1}{n} \cdot S$ of S, to determine

$$E[T] = \left(\frac{1}{n}\right) \cdot E[S] = \left(\frac{1}{n}\right) \cdot \left(\frac{n}{2}\right) = \frac{1}{2}$$

and

$$\sigma_T^2 = \left(\frac{1}{n}\right)^2 \cdot Var[S] = \left(\frac{1}{n}\right)^2 \left(\frac{n}{12}\right) = \frac{1}{12 \cdot n}.$$

❐

Notes:

(1) The random variable T in (c) is called the ***sample mean*** and is often denoted by $\overline{X}$.

(2) It is important to keep in mind that the sample mean is random, whereas the population mean is fixed.

(3) Observe that the sample mean T and the population X random variables have the same expected value of $\frac{1}{2}$, regardless of the sample size n, but that the variance of T decreases as the sample size increases.

Before stating the main results of this section we will summarize the above discussion about random samples of an arbitrary random variable X in general terms.

Terminology and Properties for Random Samples

Let X be a random variable with finite mean μ_X and variance σ_X^2. We refer to X as the ***population*** random variable.

Let $X_1, X_2, \ldots, X_n$ be independent, identically distributed observations of X abbreviated as $(X_i)_{i=1}^n$ I.I.D. $\sim X$. Then $(X_i)_{i=1}^n$ is called a ***random sample*** of X.

Let $S=\sum_{i=1}^{n} X_i$ and $\overline{X} = \frac{1}{n}\cdot\sum_{i=1}^{n} X_i$. **Note:** $\overline{X}$ is called the ***sample mean***. Then

(1) $E[S]=\mu_S=n\cdot\mu_X$ and $Var[S]=n\cdot\sigma_X^2$. Equivalently, $\sigma_S=\sqrt{n}\cdot\sigma_X$.

(2) $E[\overline{X}]=\mu_X$ and $Var[\overline{X}]=\sigma_{\overline{X}}^2=\frac{\sigma_X^2}{n}$. Equivalently, $\sigma_{\overline{X}}=\frac{\sigma_X}{\sqrt{n}}$.

The formulas in (1) and (2) are easily derived from general principles as in Example 6.4.1. The concept embodied in the formulas of (2) is generally referred to as The Law of Averages, or The Law of Large Numbers, which we elaborate upon next.

6.4.2 CHEBYSHEV AND THE LAW OF AVERAGES[2]

We see from property (2) above for random samples that the sample mean $\overline{X}$, as a random variable, has the same mean as the population random variable X, but that as the sample size increases, the variance (or equivalently, the standard deviation) goes to zero. Take the example of throwing a dart at the unit interval. A single throw can be all over the map, so to speak. But the *average* of a large number of throws will be very close to 1/2, the expected value of X. In popular parlance this idea is often referred to as "the law of averages." The same thing occurs with coin-tossing. In a single toss of a coin the proportion of heads can only be 0 or 1 and the outcome either way is as far as possible from the mean of 1/2. Likewise, in a small number of tosses of a fair coin, the proportion of heads can vary widely, but as the number of tosses (the sample size) increases, the proportion of heads converges to 1/2, in a sense to be made more precise below. This, again, is an instance of the law of averages, or as it is also called, the ***Law of Large Numbers***.

We will demonstrate a version of this using a general form of Chebyshev's Theorem, first discussed for discrete random variables in Subsection 3.4.5.

In Chapter 3 we proved Chebyshev's Theorem by first using a simpler inequality for non-negative random variables, called the ***Markov Inequality***. This is very easily adapted to the continuous case by replacing the discrete sums with integrals.

Markov Inequality

Let Y be a random variable taking only non-negative values, with density function $f_Y(y)$ and mean μ_Y. Then for any $a>0$,

$$\Pr[Y>a]\leq\frac{\mu_Y}{a}.$$

[2] "The Law of Averages" is a popular term that vaguely, and often incorrectly, invokes the concept studied in this section. Our intention is to give it a more precise meaning and to put it on a sounder footing. The term, "Law of Large Numbers" is an alternative phrase, which is more accurate, but less familiar.

Proof

$$\Pr[Y > a] = \int_a^\infty f_Y(y)\,dy$$

$$\leq \overbrace{\int_a^\infty \left(\frac{y}{a}\right) f_Y(y)\,dy}^{\text{since } \frac{y}{a}\geq 1 \text{ on } [a,\infty)}$$

$$= \frac{1}{a}\int_a^\infty y\cdot f_Y(y)\,dy$$

$$\leq \frac{1}{a}\int_0^\infty y\cdot f_Y(y)dy = \frac{\mu_Y}{a}.$$

Chebyshev's Theorem

Let X be a random variable with finite mean μ_X and standard deviation σ_X. Let $k > 1$. Then the probability that X is more than k standard deviations from the mean μ_X is less than or equal to $\frac{1}{k^2}$.

That is, $\Pr(X < \mu_X - k\cdot\sigma_X \text{ or } X > \mu_X + k\cdot\sigma_X) = \Pr(|X-\mu| > k\cdot\sigma_X) \leq \frac{1}{k^2}$.

Equivalently,

$$\Pr(\mu_X - k\cdot\sigma_X \leq X \leq \mu_X + k\cdot\sigma_X) = \Pr(|X-\mu| \leq k\cdot\sigma_X) \geq 1-\frac{1}{k^2}.$$

The proof of Chebyshev's Theorem using the Markov Inequality is identical to the one given in Chapter 3.

The version of the law of averages given below is called the ***Weak Law of Large Numbers***. The "weak" refers to the type of convergence. Basically, it says that there is a very high probability that the sample mean will be very close to the population mean, provided that the sample size n is sufficiently large. There is also a "strong" version of the Law of Large Numbers, so strong in fact that we are lacking the muscle to wrestle with it at the level of this book.

The Weak Law of Large Numbers

Let X be a population random variable with finite mean μ_X and variance σ_X^2.

Let $(X_i)_{i=1}^n$ be a random sample from X with sample mean $\overline{X} = \frac{1}{n}\cdot\sum_{i=1}^n X_i$. Let ε and δ be two given positive numbers. Then for a sample size n sufficiently large,

$$\Pr\left[\left|\overline{X} - \mu_X\right| \leq \varepsilon\right] \geq 1-\delta.$$

Before giving the proof we make a few observations:

(1) The quantities ε and δ should be thought of as *arbitrarily* small. $\left|\overline{X}-\mu_X\right| \le \varepsilon$ says that random sample means $\overline{X}$ are close (within ε) to the fixed population mean μ_X. If you wanted, for example, to make sure that these sample means are within ε of the population mean at least 99.9% of the time, then $\delta = .001$.

(2) In statistics the quantity ε, which measures the discrepancy between the sample mean and the true population mean, is called the ***margin of error***. The probability $1-\delta$ is called the ***confidence level***.

(3) If, for example, $\varepsilon = .01$ and $\delta = .05 = 5\%$, then the law of large numbers says that for a sufficiently large sample size we can be 95% confident that the margin of error is less than 0.01.

(4) Finally, note that the required sample size n for the given ε and δ depends only on the variance of X, not on a complete knowledge of the distribution of X.

Proof

First, we choose k sufficiently large so that $\frac{1}{k^2} \le \delta$. Second, choose n sufficiently large so that $k \cdot \sigma_{\overline{X}} = k \cdot \frac{\sigma_X}{\sqrt{n}} \le \varepsilon$. When Chebyshev's Theorem is applied to the sample mean $\overline{X}$, we have,

$$\Pr\left[\left|\overline{X}-\mu_X\right| \le \varepsilon\right] \ge \Pr\left[\left|\overline{X}-\mu_X\right| \le k \cdot \frac{\sigma_X}{\sqrt{n}} = k \cdot \sigma_{\overline{X}}\right] \ge 1 - \frac{1}{k^2} \ge 1 - \delta.$$

Example 6.4-2 The Law of Large Numbers

Let X be any random variable (the population) with a finite mean and a variance of 4. Find the smallest sample size (using Chebyshev's Theorem) for which we can be 96% confident that the sample mean differs from the population mean by less than 0.01.

Solution

First, we see $\delta = .04$. Paralleling the above proof, $\frac{1}{k^2} = \delta = .04$ implies $k = 5$. Then, with $\varepsilon = 0.01$ we choose n so that $k\sigma_{\overline{X}} = k\frac{\sigma_X}{\sqrt{n}} = 5 \cdot \frac{2}{\sqrt{n}} = 0.01$, which implies $n = 1{,}000{,}000$. ❐

A random sample of size one million is rather large as these things go. As we shall see below, using the Central Limit Theorem (sample mean version) we can achieve the same result with a far smaller sample size. As it stands now though, a sample size of one million will ensure that we can be 96% confident that the sample mean $\overline{X}$ is within 0.01 for *any* distribution with a variance of 4 (or less).

6.4.3 CENTRAL LIMIT THEOREM (SUM VERSION)

The Central Limit Theorem (CLT) says more; it is essentially a statement about the actual distributions of S and $\overline{X}$. The CLT asserts that as the sample size n increases, the distributions of the sum S and the mean $\overline{X}$ become increasingly bell-shaped, that is to say, normal, in appearance. Although the two cases are mathematically virtually identical, we will present them as two separate versions. This will help in identifying the two most common types of applications of the Central Limit Theorem.

Central Limit Theorem (Sum Version)

Let X be a population random variable with finite mean μ_X and finite variance σ_X^2 and let $(X_i)_{i=1}^n$ be a **random sample** of X. Let $S = \sum_{i=1}^{n} X_i$.

Then for large sample sizes n the distribution of S is approximately normally distributed with mean $\mu_S = n \cdot \mu_X$ and variance $\sigma_S^2 = n \cdot \sigma_X^2$.

Notes

(1) This formulation of the Central Limit Theorem is fairly easy to understand and to apply to problem solving, but leaves a lot to be desired in terms of mathematical precision. We will give a more rigorous statement below and then outline the steps for a proof.

(2) How large does the sample size n need to be in order to approximate S with the normal distribution? The answer depends on the distribution of the population random variable X and how good an approximation is required. A general rule-of-thumb found in statistics books is a sample size of at least 30. *Caveat: This is a rule-of-thumb only and may not always produce reliable results.*

(3) If the population X is itself a normally distributed random variable, then property (8) of the Normal Distribution (Subsection 6.3.3) shows that S is also normal and the word "approximately" can be deleted.

To illustrate the power of the central limit theorem as an approximation tool, we develop an example using a population random variable X that is as far as practicable from normal. A useful candidate is the exponential distribution, which is highly skewed and asymmetric.

In the following graphs we use a population random variable X that is exponentially distributed with a mean of one. Then for $S = \sum_{i=1}^{n} X_i$ we have $\mu_S = n$ and $\sigma_S^2 = n$. We compare the graph of the density function for S with the graph of W, the density function for the approximating ***normal*** distribution. The normal approximation W has mean and variance both equal to n, the same as S; that is, $W \sim N(n,n)$.

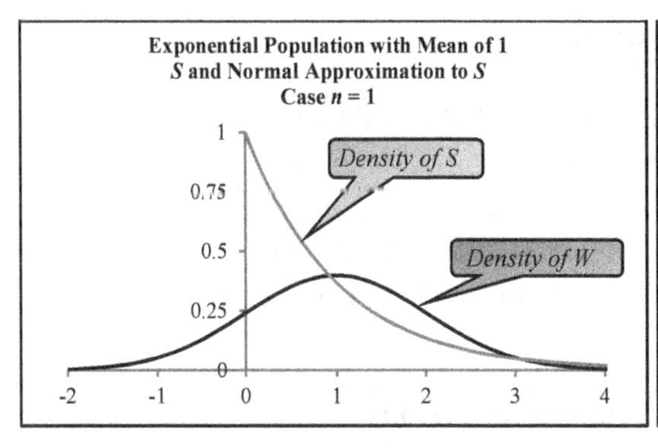

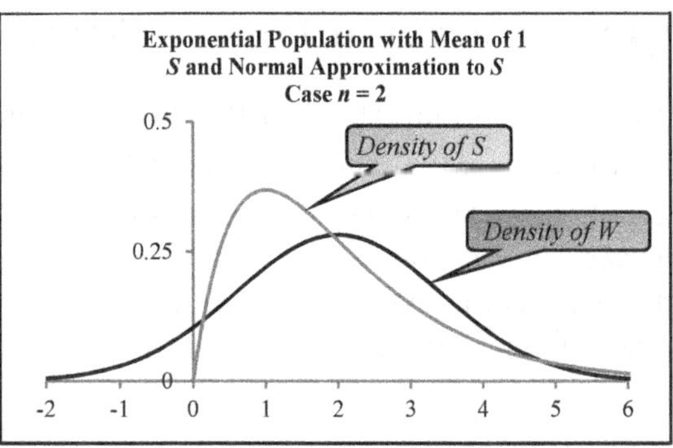

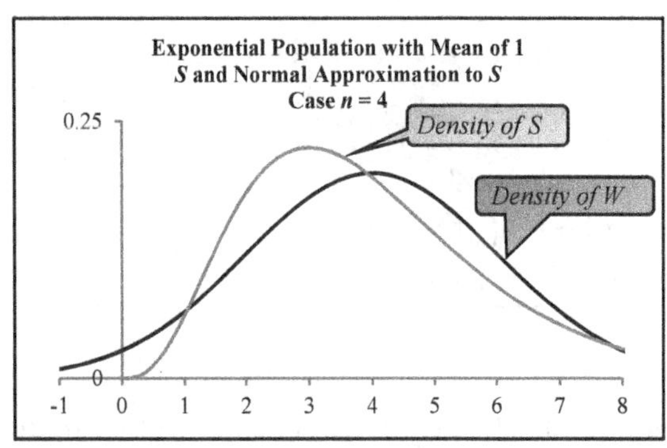

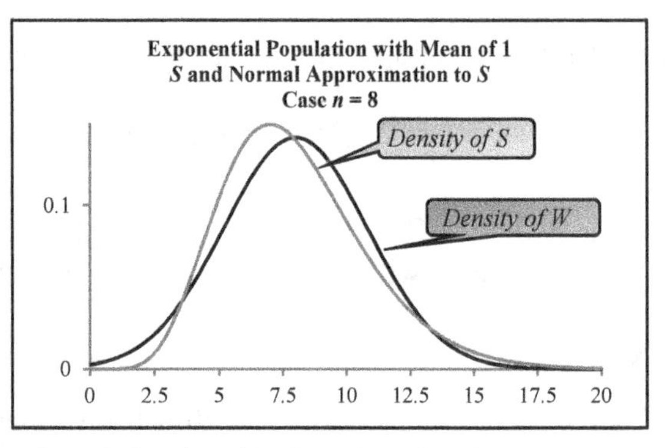

We see that even for relatively small n the graphs of the density function for the sum S are quickly becoming more normal in appearance. The exact distribution of the sum S is of the Gamma type, which we will formally meet in Section 6.5.

Example 6.4-3 Central Limit Theorem with Exponential Population

People arrive at the check-out counter according to a Poisson process with a mean waiting time of one minute between arrivals. Estimate the probability that the 8th arrival occurs between time 7 and time 9.

Solution

Here $S = X_1 + X_2 + \cdots + X_8$ is the total time elapsed until the 8th arrival, where X_i, the waiting time for the ith arrival is exponentially distributed with a mean of one. Then $\mu_S = 8$ and $\sigma_S^2 = 8$ so that $\sigma_S = \sqrt{8} = 2.8284$. We let $W \sim N(8,8)$. We determine the probability by standardizing and using the normal table:

$$\Pr[7 \leq S \leq 9] \underbrace{\approx}_{\text{CLT}} \Pr[7 \leq W \leq 9] = \Pr\left[\frac{7-8}{\sqrt{8}} \leq Z \leq \frac{9-8}{\sqrt{8}}\right] = \Pr[-.35 \leq Z \leq .35] = .2736.$$

❐

Notes

(1) For comparison, the graphs for S and W in this example are precisely those illustrated in the graph for the case $n = 8$ just above.

(2) The exact distribution for S is of the gamma type, which we will study in the next section. The probabilities for the gamma and the normal can be calculated precisely using the

built-in Microsoft Excel® functions GAMMADIST and NORMDIST. The exact values are $\Pr[7 \le S \le 9] = .2748$ and $\Pr[7 \le W \le 9] = .2763$. This results in a relative error[3] of about half-of-one percent by using the CLT approximation to S. And this is with a small sample size $(n = 8)$ and a highly skewed population.

6.4.4 THE CONTINUITY CORRECTION

There is no reason that the population random variable need be continuous. Indeed many of the most common applications of the Central Limit Theorem occur with discrete population random variables like the Poisson or the binomial. If X is a discrete random variable taking integer values only, then we make the transition from discrete sum S to continuous normal random variable W with mean μ_S and variance σ_S^2 as follows:

$$\Pr[S = k] = \Pr\left[k - \frac{1}{2} \le S \le k + \frac{1}{2}\right] \approx \Pr\left[k - \frac{1}{2} \le W \le k + \frac{1}{2}\right].$$

This has the effect of expanding the event under consideration by a half-unit in each direction. This procedure is called the ***continuity correction***, although a more appropriate name would be the "discrete correction."[4]

Example 6.4-4 Normal Approximation to the Binomial Distribution

Jamie is a 70% free-throw shooter. Suppose that Jamie attempts 100 independent free-throws.

(a) Using the Central Limit Theorem, calculate the approximate probability that she makes between 60 and 75 of her 100 free-throws.
(b) Calculate the approximate probability she makes more than 70 of her 100 free-throws.

Solution

In this example the population random variable is a Bernoulli trial – that is, the result of a single attempt. So X is a random variable taking the values 0 or 1. The mean is $\mu_X = p = 0.7$ and the variance is $\sigma_X^2 = pq = (.7 \times .3) = .21$, so the standard deviation is $\sigma_X = \sqrt{pq} = \sqrt{0.7 \cdot 0.3} = 0.4583$. The number of successes (made free-throws) in 100 tries is the binomial random variable given by,

$$S = \sum_{i=1}^{100} X_i \text{ with } \mu_S = (100)(.7) = 70,\ \sigma_S^2 = (100)(.21) = 21.$$

---

[3] By *relative error* we mean $\frac{|\Pr[7 \le S \le 9] - \Pr[7 \le W \le 9]|}{\Pr[7 \le S \le 9]} = \frac{|.2748 - .2763|}{.2748} = .0055 = 0.55\%$

[4] One could devise a more general continuity correction, applicable to any discrete distribution X, by using the midpoints between values of X. We will confine our discussion of the correction to the most common situation, where X takes integer values and the gaps are one unit.

Then S is approximated by the normal random variable $W \sim N(70,21)$ and, using the "continuity correction," we have for (a),

$$\begin{aligned}\Pr[60 \le S \le 75] &= \Pr[59.5 \le S \le 75.5] \\ &\approx \Pr[59.5 \le W \le 75.5] \\ &= \Pr\left[\frac{59.5-70}{\sqrt{21}} \le Z \le \frac{75.5-70}{\sqrt{21}}\right] = \Pr\left[-2.29 \le Z \le 1.20\right] = 87.4\%.\end{aligned}$$

Part (b) is similar, but take note that the event is making *more than* 70, so that the continuity correction changes the left-hand endpoint to 70.5:

$$\begin{aligned}\Pr[70 < S] &= \Pr[71 \le S < \infty] \\ &\underbrace{=}_{\substack{\text{continuity}\\\text{correction}}} \Pr[70.5 \le S < \infty] \\ &\underbrace{\approx}_{\text{CLT}} \Pr[70.5 \le W < \infty] \\ &= \Pr\left[\frac{70.5-70}{\sqrt{21}} \le Z\right] = \Pr[.11 \le Z] = 1-.5438 = 45.62\%.\end{aligned}$$

The exact answers, computed by summing the appropriate binomial probabilities, are 87.39% and 46.23%, respectively. ❐

Example 6.4-5 Normal Approximation to the Binomial

Let S be a binomially distributed random variable with $n = 20$ and $p = 0.3$. Approximate $\Pr(S \ge 8)$ using the normal distribution and sketch the histogram for S with the appropriate normal density function overlaid.

Solution

$$E[S] = n \cdot p = 20 \cdot 0.3 = 6 \quad \text{and} \quad \sigma_S^2 = n \cdot p \cdot q = 20 \cdot 0.3 \cdot 0.7 = 4.2$$

$$\Pr[8 \le S] = \Pr(7.5 \le S) \approx \Pr\left(Z \ge \frac{7.5-6}{\sqrt{4.2}}\right) = \Pr(Z \ge .73) = .2327$$

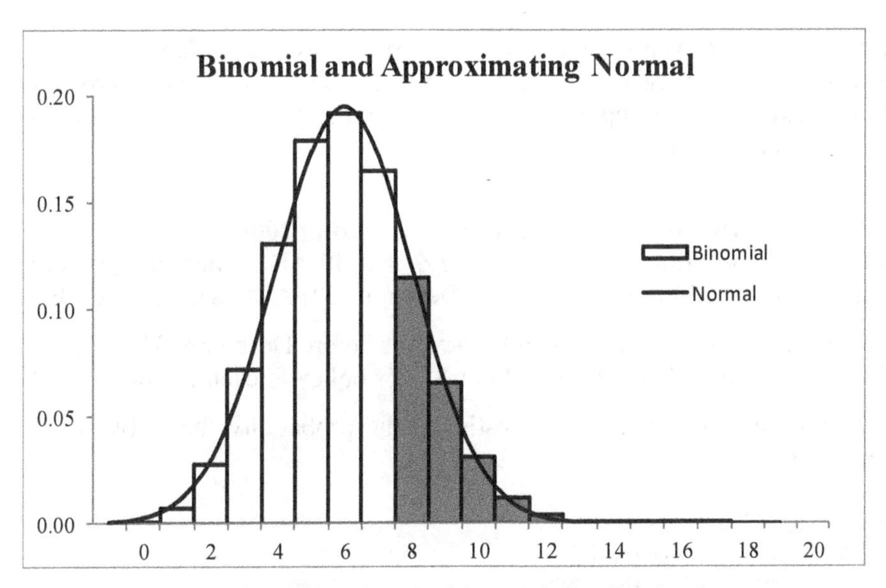

The exact answer for this probability using the binomial distribution is

$$\Pr(X \geq 8) = \Pr(X = 8) + \cdots + \Pr(X = 20) = 0.227728.$$ ❐

The reasoning of the last two examples can be extended to approximating binomial probabilities with the normal distribution in general. Again, the accuracy of the approximation depends on the parameters n and p. There is a rule-of-thumb for this particular case that says that the approximation is good if the variance, $n \cdot p \cdot q$, of S is at least 10.

The general statement for the continuity correction, as illustrated in the above examples, is given next.

Approximating the Binomial Using the Central Limit Theorem

Let S be a binomial random variable counting the number of successes in n trials with probability of success p. Then

$$\Pr[a \leq S \leq b] \approx \Pr\left[\frac{a-0.5-np}{\sqrt{npq}} \leq Z \leq \frac{b+0.5-np}{\sqrt{npq}}\right].$$

Exercise 6-40 Given that $X \sim Binomial(20, 0.3)$, approximate $\Pr(X = 8)$ using the normal distribution. Compare the approximation to the exact probability using the binomial distribution.

Exercise 6-41 Assume we flip a fair coin 100 times.
(a) Use the normal approximation to the binomial distribution to approximate the probability of getting more than 60 heads.
(b) Estimate the probability of getting exactly 60 heads. Compare these approximations to the exact probabilities.

Exercise 6-42 US Airways found that 6.2% of its passengers fail to show up for their scheduled flight. If they have 150 reservations, use the normal approximation to the binomial distribution to approximate the probability that 5 or more passengers fail to show up for their flight.

Exercise 6-43 **CAS Exam 3, Segment 3L** **Spring 2008** **#1**
Agent ABC is informed by insurer XYZ that if ABC succeeds in selling XYZ's insurance to at least 56 customers before December 31, ABC will receive a bonus.

- ABC will make a sales pitch to 500 customers before December 31.
- ABC's probability of success in selling XYZ's policy to each customer is 10%.

Using the Central Limit Theorem, estimate the probability that ABC will succeed in earning the bonus.

(A) Less than 18%
(B) At least 18% but less than 22%
(C) At least 22% but less than 26%
(D) At least 26% but less than 30%
(E) At least 30%

6.4.5 CENTRAL LIMIT THEOREM (SAMPLE MEAN VERSION)

We continue our discussion of the Central Limit Theorem with a reformulation in terms of the sample mean, which we denote in conventional fashion as $\overline{X}$.

Central Limit Theorem (Sample Mean Version)

Let X be a population random variable with finite mean μ_X and finite variance σ_X^2 and let $(X_i)_{i=1}^n$ be a **random sample** of X. Let $\overline{X} = \frac{1}{n}\sum_{i=1}^{n} X_i$ denote the ***sample mean***. Then for large sample sizes n, the distribution of $\overline{X}$ is approximately normal[5] with mean $\mu_{\overline{X}} = \mu_X$ and variance $\sigma_{\overline{X}}^2 = \frac{\sigma_X^2}{n}$. Notation: $\overline{X} \sim N\left(\mu_X, \frac{\sigma_X^2}{n}\right)$.

The expressions for the mean and the variance of $\overline{X}$ are from the general properties of a random sample given in Subsection 6.4.1.

Example 6.4-6 **Central Limit Theorem for Sample Means**

Jamie is a 70% free-throw shooter. Calculate an estimate for the probability she makes between 68% and 72% of her next 400 free-throws.

---

[5] If X is normal then so is $\overline{X}$, and the word "approximately" can be dispensed with in this case.

Solution

The population random variable X is the same as Example 6.4-4, the Bernoulli random variable with $\mu_X = p = 0.7$ and standard deviation $\sigma_X = \sqrt{pq} = 0.4583$. Now $\bar{X} = \frac{1}{400}\sum_{i=1}^{400} X_i$ is the *proportion* of made free-throws and the problem is to calculate $\Pr[.68 \le \bar{X} \le .72]$. By the Central Limit Theorem, $\bar{X}$ is approximately normal, with mean of 0.7 and standard deviation $\sigma_{\bar{X}} = \frac{\sigma_X}{\sqrt{n}} = \sqrt{\frac{pq}{n}} = \sqrt{\frac{(.7)(.3)}{400}} = .0229$. Thus,

$$\Pr[.68 \le \bar{X} \le .72] \approx \Pr\left[\frac{.68-.70}{.0229} \le Z \le \frac{.72-.70}{.0229}\right] = \Pr[|Z| \le .87] = 61.73\%.$$

In general, when the population is a Bernoulli trial, as in the above example, we know that $S = \sum_{i=1}^{n} X_i$ has binomial distribution. Hence, the sample average random variable $\bar{X} = \frac{1}{n} \cdot S$ is called a ***binomial proportion***. Note that we do not use the continuity correction here. Although $\bar{X}$ is indeed discrete, since it does not take integer values we will not apply the continuity correction.[6]

Example 6.4-7 CLT and Sample Size (Example 6.4-2 Revisited)

Let X be any random variable (the population) with a finite mean and a variance of 4. Estimate the smallest sample size (using the Central Limit Theorem) for which we can be 96% confident that the sample mean differs from the population mean by less than 0.01.

Solution

We wish to find the n that assures $\Pr\left[\left|\bar{X}-\mu_X\right| \le 0.01\right] \ge .96$. We standardize by using $\sigma_{\bar{X}} = \frac{\sigma_X}{\sqrt{n}}$ and replacing $\frac{\bar{X}-\mu_X}{\sigma_{\bar{X}}}$ with its standard normal approximation:

$$.96 = \Pr\left[\left|\bar{X}-\mu_X\right| \le 0.01\right] = \Pr\left[\frac{\left|\bar{X}-\mu_X\right|}{\sigma_{\bar{X}}} \le \frac{0.01}{\sigma_{\bar{X}}} = \frac{0.01}{\frac{\sigma_X}{\sqrt{n}}} = \frac{0.01}{2} \cdot \sqrt{n}\right]$$

$$\approx \Pr\left[|Z| \le \frac{0.01}{2} \cdot \sqrt{n}\right].$$

This requires the backward lookup for $\Pr[|Z| \le z_\alpha] = .96$, which results in $z_\alpha = 2.05$. Thus, $2.05 = \frac{0.01}{2} \cdot \sqrt{n}$ which yields $n = 168{,}100$. This is a large sample size[7] but considerably less than the one million required using Chebyshev's Theorem in Example 6.4-2. ❐

[6] This could result in a small discrepancy between the Central Limit Theorem (sum version, using the continuity correction) and the Central Limit Theorem applied to proportions, as in this example.

[7] Keep in mind it is an *estimate* for the necessary sample size, and is based on the applicability of the Central Limit Theorem for this unknown population random variable. Since the sample sizes in these examples are quite large, the resulting estimates are likely quite reliable.

Example 6.4-8 Estimating the Sample Mean

Suppose patrons at the local pub have ages that are distributed according to an unknown "tavern" distribution with mean 32.8 and standard deviation of 7.4 years. We randomly sample $n = 25$ people leaving the local pub and determine their ages. Find $\Pr(\overline{X} < 31)$.

Solution

$$\mu_{\overline{X}} = 32.8$$

and

$$\sigma_{\overline{X}} = \frac{\sigma_X}{\sqrt{n}} = \frac{7.4}{\sqrt{25}} = 1.48.$$

$$\begin{aligned}
\Pr(\overline{X} < 31) &= \Pr\left[\frac{\overline{X} - \mu_X}{\sigma_X / \sqrt{n}} < \frac{31 - 32.8}{1.48}\right] \\
&\approx \Pr\left(Z < \frac{31 - 32.8}{1.48}\right) \\
&= \Pr(Z < -1.22) \\
&= .1112.
\end{aligned}$$

❒

Example 6.4-9 Reliability of the Sample Mean

In a large philosophy class, grades are distributed according to a normal distribution with mean 82 and standard deviation of 9.4. We plan to sample 64 students and compute the sample mean of their scores. Find an interval, symmetric about the population mean of 82, such that the sample mean will land in that interval 90% of the time.

Solution

For the standard normal distribution, the solution z_α to $\Pr(-z_\alpha < Z < z_\alpha) = .9$ is $z_\alpha = 1.645$. Assuming by the Central Limit Theorem that $\frac{\overline{X} - \mu}{\sigma/\sqrt{n}}$ has approximately the standard normal distribution Z, we can solve

$$z_\alpha = 1.645 = \frac{\overline{X} - \mu}{\sigma / \sqrt{n}} = \frac{\overline{X} - 82}{9.4 / \sqrt{64}}$$

to give $\overline{X} = 83.93$.

Similarly, solving $-z_\alpha = -1.645 = \frac{\overline{X} - \mu}{\sigma/\sqrt{n}} = \frac{\overline{X} - 82}{9.4/\sqrt{64}}$ we have $\overline{X} = 80.07$.

Therefore $\Pr(80.07 < \overline{X} < 83.93) = \Pr(-z_\alpha < Z < z_\alpha) = 90\%$. That is, ninety percent of the time the sample average of 64 students' philosophy class grades will be between approximately 80 and 84. ❒

Exercise 6-44 Suppose you and 74 of your closest friends ($n = 75$ total) each randomly select a real number from 50 to 100. Let X_i denote the uniformly distributed random number on the interval [50,100] that the ith person selected. Let $\overline{X} = \frac{X_1 + X_2 + \cdots + X_{75}}{75}$ denote the sample mean.

(a) Find $E[X_i]$ and $Var[X_i]$.

(b) Find $E[\overline{X}]$ and $Var[\overline{X}]$.

(c) Compute $\Pr[72 < X_i < 77]$ and $\Pr[72 < \overline{X} < 77]$.

Exercise 6-45 A large population has mean 100 and standard deviation 16.

(a) What is the probability that the sample mean will be within ±2 of the population mean if the sample size is $n = 100$?

(b) What is the probability that the sample mean will be within ±2 of the population mean if the sample size is $n = 400$?

(c) What is the advantage of a larger sample size?

Exercise 6-46 SOA EXAM P Sample Exam Questions #87

In an analysis of healthcare data, ages have been rounded to the nearest multiple of 5 years. The difference between the true age and the rounded age is assumed to be uniformly distributed on the interval from −2.5 years to 2.5 years. The healthcare data are based on a random sample of 48 people. What is the approximate probability that the mean of the rounded ages is within 0.25 years of the mean of the true ages?

(A) 0.14 (B) 0.38 (C) 0.57 (D) 0.77 (E) 0.88

Exercise 6-47 SOA EXAM P Sample Exam Questions #81

Claims filed under auto insurance policies follow a normal distribution with mean 19,400 and standard deviation 5,000. What is the probability that the average of 25 randomly selected claims exceeds 20,000?

(A) 0.01 (B) 0.15 (C) 0.27 (D) 0.33 (E) 0.45

Exercise 6-48 SOA EXAM P Sample Exam Questions #83

A company manufacturers a brand of light bulb with a lifetime in months that is normally distributed with mean 3 and variance 1. A consumer buys a number of these bulbs with the intention of replacing them successively as they burn out. The light bulbs have independent lifetimes. What is the smallest number of bulbs to be purchased so that the succession of light bulbs produces light for at least 40 months with probability at least 0.9772?

(A) 14 (B) 16 (C) 20 (D) 40 (E) 55

Exercise 6-49 SOA EXAM P Sample Exam Questions #80
A charity receives 2025 contributions. Contributions are assumed to be independent and identically distributed with mean 3125 and standard deviation 250. Calculate the approximate 90th percentile for the distribution of the total contributions received.

(A) 6,328,000 (B) 6,338,000 (C) 6,343,000 (D) 6,784,000 (E) 6,977,000

Exercise 6-50 A random variable W has probability density function $f(w)=12w^2-12w^3$ for $0\le w\le 1$. Let $S=W_1+W_2+\cdots+W_{225}$ be the sum of $n=225$ independent observations.

(a) Compute the expected value and variance of S.
(b) Estimate $\Pr(S<141)$.

Exercise 6-51 SOA EXAM P Sample Exam Questions #85
The total claim amount for a health policy follows a distribution with density function

$$f(x)=\frac{1}{1000}e^{-x/1000} \text{ for } 0<x.$$

The premium for the policy is set at 100 over the expected total claim amount. If 100 policies are sold, what is the approximate probability that the insurance company will have claims exceeding the premiums collected?

(A) 0.001 (B) 0.159 (C) 0.333 (D) 0.407 (E) 0.460

Exercise 6-52 SOA EXAM P Sample Exam Questions #82
An insurance company issues 1250 vision care insurance policies. The number of claims filed by a policyholder under a vision care insurance policy during one year is a Poisson random variable with mean 2. Assume the numbers of claims filed by distinct policyholders are independent of one another. What is the approximate probability that there is a total of between 2450 and 2600 claims during a one-year period?

(A) 0.68 (B) 0.82 (C) 0.87 (D) 0.95 (E) 1.00

6.4.6 OUTLINE FOR A PROOF OF THE CENTRAL LIMIT THEOREM

A rigorous proof is beyond the level of this book, but we will describe the main ideas. We continue with the same terminology and notation. Thus, let X be a population random variable with finite mean μ_X and finite variance σ_X^2 and let $(X_i)_{i=1}^n$ be a **random sample** of X, with $S=\sum_{i=1}^{n} X_i$. As in the above, we have $\mu_S=n\mu_X$ and $\sigma_S^2=n\sigma_X^2$. We denote the standardized version of S as $W_n=\frac{S-\mu_S}{\sigma_S}=\frac{S-n\mu_X}{n^{1/2}\sigma_X}$, using the subscript n on W_n to emphasize the dependence on the sample size n. In the statement of the Central Limit Theorem we say "for large n, S is approximately normal." This is equivalent to saying that W_n is approximately the standard normal. A more precise statement is that the CDF of W_n converges to the CDF of Z as n goes to infinity. That is, if $F_n(z)$ is the CDF of W_n and $\Phi(z)$ is the CDF of Z, then

$$\lim_{n\to\infty} F_n(z) = \Phi(z) \text{ for all } z.$$

This limit is hard to establish directly, but turns out to be easier (although still not easy) to show for the corresponding moment-generating functions. Thus, one of the keys to proving the Central Limit Theorem is to "upload" the limit statement to the world of MGFs, prove it in this alternative universe, and then "download" the result back to the world of CDFs. This relies heavily on the fact, mentioned previously in Section 5.6, that there is a one-to-one correspondence between distributions and MGFs. This is the case for a large class of probability distributions, including virtually all of the ones that we use in practice.

Once the limit statement is transformed into the corresponding statement for MGFs, the proof is completed by showing that the MGFs of W_n converge to $e^{t^2/2}$ (which we recognize as the MGF for the standard normal Z) as n goes to infinity.

This step of the proof requires the evaluation of some complicated limits using L'Hospital's Rule from calculus. For completeness, we show the details next, although the argument is rather technical and can be skipped on a first reading.

We begin by standardizing S in two steps: First, we write

$$S - \mu_S = S - n\mu_X = \sum_{i=1}^{n}(X_i - \mu_X) = \sum_{i=1}^{n} U_i,$$

where $(U_i)_{i=1}^{n}$ are independent and identically distributed as $U = X - \mu_X$. Then $E[U] = 0$ and $Var[U] = \sigma_X^2$. Second, we take

$$W_n = \frac{S - \mu_S}{\sigma_S} = \frac{1}{n^{1/2}\sigma_X} \cdot \sum_{i=1}^{n} U_i.$$

Next, we use properties (2) and (4) of MGFs from section 5.6 to express the MGF of W_n in terms of the MGF of U, to obtain

$$M_{W_n}(t) = \left[M_U\left(\frac{t}{n^{1/2}\sigma_X} \right) \right]^n.$$

We next want to take the limit of this expression as n goes to infinity, a procedure that is facilitated by taking natural logarithms and using L'Hospital's Rule. Thus, taking logs, we have,

$$\ln M_{W_n}(t) = n \ln\left[M_U\left(\frac{t}{n^{1/2}\sigma_X}\right)\right] = n \ln M_U(s),$$

where

$$s = \frac{t}{n^{1/2}\sigma_X} = \frac{t}{\sigma_X} n^{-1/2}.$$

Note that $\lim_{n\to\infty} s = 0$. For brevity, let $h(s) = \ln M_U(s)$. From section 5.6 we know that

$$h(0) = \ln 1 = 0,$$

$$h'(0) = \frac{dh}{ds}(0) = E[U] = 0,$$

and

$$h''(0) = \frac{d^2h}{ds^2}(0) = Var[U] = \sigma_X^2.$$

We now calculate $\lim_{n\to\infty}\left(\ln M_{W_n}(t)\right) = \lim_{n\to\infty} n h(s) = \lim_{n\to\infty} \frac{h(s)}{n^{-1}}$. This is an indeterminate limit of the form $\frac{0}{0}$ so we apply L'Hospital's Rule (and the chain rule) to obtain,

$$\begin{aligned}
\lim_{n\to\infty}\left(\ln M_{W_n}(t)\right) &= \lim_{n\to\infty} \frac{h(s)}{n^{-1}} = \lim_{n\to\infty} \frac{\frac{dh}{dn}}{-n^{-2}} \\
&= \lim_{n\to\infty} \frac{\frac{dh}{ds}\cdot\frac{ds}{dn}}{-n^{-2}} \\
&= \lim_{n\to\infty} \frac{h'(s)\left(-\frac{1}{2}\frac{t}{\sigma_X}n^{-3/2}\right)}{-n^{-2}} = \frac{1}{2}\frac{t}{\sigma_X}\lim_{n\to\infty}\frac{h'(s)}{n^{-1/2}}.
\end{aligned}$$

This is again an indeterminate limit of the form $\frac{0}{0}$, so once again we apply L'Hospital's Rule (and the chain rule) to obtain,

$$\begin{aligned}
\lim_{n\to\infty}\left(\ln M_{W_n}(t)\right) &= \frac{1}{2}\frac{t}{\sigma_X}\lim_{n\to\infty}\frac{h''(s)\left(-\frac{1}{2}\frac{t}{\sigma_X}n^{-3/2}\right)}{-\frac{1}{2}n^{-3/2}} \\
&= \frac{1}{2}\frac{t^2}{\sigma_X^2}\lim_{n\to\infty} h''(s) = \frac{1}{2}\frac{t^2}{\sigma_X^2}h''(0) \\
&= \frac{1}{2}\frac{t^2}{\sigma_X^2}\cdot\sigma_X^2 = \frac{t^2}{2}.
\end{aligned}$$

The conclusion is that $\lim_{n\to\infty}\left(M_{W_n}(t)\right) = e^{t^2/2}$, or equivalently W_n converges to Z as n goes to infinity.

6.5 THE GAMMA DISTRIBUTION

The family of gamma distributions generalizes the exponential distributions studied in Section 6.2. The density function for an exponential random variable has the form $f(x) = \frac{1}{\beta}e^{-(1/\beta)x}$ for $x \geq 0$ and employs the single parameter β. The extension to the gamma family is achieved by adding a second parameter α and constructing a new density function taking the form $f(x) = c \cdot x^{\alpha-1}e^{-(1/\beta)x}$; for $0 \leq x < \infty$. This reduces to the exponential density function if $\alpha = 1$. Determining the value of the constant c that assures this is a density function (total area under the curve equal to one) requires some further terminology from calculus.

6.5.1 THE GAMMA FUNCTION

The gamma function, a construct from integral calculus, is used as a generalization of the familiar operation $n!$ to "factorials" for non-integer valued positive numbers. The formula used to define this function makes use of an improper integral and is denoted by an upper-case gamma, $\Gamma(\alpha)$. The definition and properties follow.

The Definition of the Gamma Function, $\Gamma(\alpha)$

$$\Gamma(\alpha) = \int_0^\infty x^{\alpha-1} e^{-x}\, dx; \quad 0 < \alpha < \infty.$$

Some Properties of the Gamma Function

(1) ***Iteration Formula***: $\Gamma(\alpha) = (\alpha-1)\cdot\Gamma(\alpha-1)$ for all $\alpha > 1$.
(2) If $\alpha = n$ is a positive integer, then $\Gamma(n) = (n-1)!$
(3) $\Gamma\left(\frac{1}{2}\right) = \sqrt{\pi}$.

The definition of $\Gamma(\alpha)$ for $0 < \alpha < 1$ is a doubly improper integral since $\alpha < 1$ means that the integrand spikes to ∞ at $x = 0$. But the integral remains convergent, so that the definition is meaningful; property (3) is a special instance of this case.

To derive property (1) we use integration-by-parts with $u = x^{\alpha-1}$ and $dv = e^{-x}dx$. Then we find $du = (\alpha-1)x^{\alpha-2}\,dx$ and $v = -e^{-x}$.

$$\Gamma(\alpha) = \int_0^\infty x^{\alpha-1}e^{-x}\,dx = -x^{\alpha-1}e^{-x}\Big|_{x=0}^{\infty} + (\alpha-1)\int_0^\infty x^{\alpha-2}e^{-x}\,dx$$

$$= -(0-0) + (\alpha-1)\int_0^\infty x^{(\alpha-1)-1}e^{-x}\,dx = (\alpha-1)\cdot\Gamma(\alpha-1).$$

Property (2) uses mathematical induction on the iteration formula of property (1). Here is an informal description of the procedure. First, note that $\Gamma(1) = \int_0^\infty e^{-x}\,dx = 1 = 0!$ (Recall that $0! = 1$). Using property (1), we have,

$$\Gamma(2) = (2-1)\Gamma(2-1) = (1)\Gamma(1) = 1 \cdot 0! = 1!.$$

Next,

$$\Gamma(3) = (3-1)\Gamma(3-1) = 2\Gamma(2) = 2 \cdot 1! = 2!,$$

then

$$\Gamma(4) = (4-1)\Gamma(4-1) = 3\Gamma(3) = 3 \cdot 2! = 3!.$$

and so forth, step-by-step through the non-negative integers.

Property (3) requires the integral substitution $x = \frac{u^2}{2}$ and recognition that the resulting integral is one-half of the area under the standard normal density function.

$$\Gamma(1/2) \overbrace{=}^{\text{def'n}} \int_0^\infty x^{-1/2} e^{-x}\,dx = \overbrace{\int_0^\infty \left(\frac{u^2}{2}\right)^{-1/2} e^{-(u^2/2)}\,(u\,du)}^{\text{substitute } x=(u^2/2),\ dx=u\,du}$$

$$= \sqrt{2}\int_0^\infty e^{-(u^2/2)}\,du = \sqrt{2}\left(\frac{1}{2}\sqrt{2\pi}\right) = \sqrt{\pi}.$$

Exercise 6-53 Determine the exact numerical values of each of the following.

(a) $\Gamma(3.5)$.

(b) $\Gamma(5)$.

(c) $\Gamma(2)$.

Exercise 6-54 Use integration by parts to compute $\Gamma(2)$ from the definition.

Exercise 6-55 Evaluate the indefinite integral $\int_0^\infty 5t^3 e^{-t}\,dt$. This is easy. Think about the definition of the Gamma function.

6.5.2 DEFINITION AND PROPERTIES FOR THE GAMMA FAMILY

We are now in a position to define the gamma distribution, a two-parameter family of random variables whose density functions have a "polynomial" factor ($x^{\alpha-1}$) and an "exponential" factor ($e^{-(1/\beta)x}$). As with the exponential random variable, we write the formulas in both of the two most commonly encountered formats, that is, using either λ or β for the second parameter, where $\lambda = \frac{1}{\beta}$. The parameter λ is referred to as the "rate" parameter, since, as we shall see, it plays the same role as the rate of occurrence in a Poisson process. The reciprocal form β is often called the "scale" parameter.

Definition of the Gamma Distribution

A non-negative random variable X is said to have a ***Gamma distribution***, denoted $X \sim \Gamma(\alpha,\beta)$, if the density function for X has the form

$$f_X(x) = \frac{1}{\beta^\alpha \cdot \Gamma(\alpha)} \cdot x^{\alpha-1} e^{-(1/\beta)x} \text{ for } 0 \le x < \infty, \ \alpha > 0, \text{ and } \beta = \frac{1}{\lambda} > 0.$$

Equivalently, $f_X(x) = \dfrac{\lambda^\alpha}{\Gamma(\alpha)} \cdot x^{\alpha-1} e^{-\lambda x}$ for $0 \le x < \infty$, $\alpha > 0$, and $\lambda > 0$.

Properties of the Gamma Distribution

Let $X \sim \Gamma(\alpha,\beta)$ be a Gamma random variable with parameters $\alpha > 0$, and $\beta = \frac{1}{\lambda} > 0$.

(1) The mean of the random variable: $E[X] = \frac{\alpha}{\lambda} = \alpha \cdot \beta$.

(2) The variance: $Var[X] = \frac{\alpha}{\lambda^2} = \alpha \cdot \beta^2$.

(3) The MGF of X: $M_X(t) = \left(\frac{\lambda}{\lambda - t}\right)^\alpha = \left(\frac{1}{1-\beta t}\right)^\alpha$. This MGF is defined for all $t < \lambda$.

(4) Let X and Y be independent Gamma random variables with $X \sim \Gamma(\alpha_X,\beta)$ and $Y \sim \Gamma(\alpha_Y,\beta)$. Then $X+Y \sim \Gamma(\alpha_X+\alpha_Y,\beta)$.

(5) Suppose $X \sim \Gamma(\alpha,\beta)$ and $Y = k \cdot X$, where $k > 0$. Then $Y \sim \Gamma(\alpha, k \cdot \beta)$.

(6) For any $\alpha, \beta > 0$, $\int_0^\infty x^{\alpha-1} e^{-(1/\beta)x}\, dx = \beta^\alpha \cdot \Gamma(\alpha)$

Proof

We begin by verifying $f_X(x) = \dfrac{1}{\beta^\alpha \Gamma(\alpha)} \cdot x^{\alpha-1} e^{-(1/\beta)x}$ is in fact a valid density function. We need to verify that $f_X(x) \ge 0$ and that the total area under the graph equals one. The former is obvious from the formula for $f_X(x)$ and the latter follows from the definition of the gamma function:

$$\int_0^\infty x^{\alpha-1} e^{-(1/\beta)x}\, dx = \overbrace{\int_0^\infty (\beta u)^{\alpha-1} e^{-u} (\beta\, du)}^{\left\{\substack{\text{substitute } u=(1/\beta)x, \text{ or equivalently,} \\ x=\beta u,\ dx=\beta du}\right\}} = \beta^\alpha \int_0^\infty u^{\alpha-1} e^{-u}\, du = \beta^\alpha \cdot \Gamma(\alpha).$$

We next derive the MGF of property (3):

$$M_X(t) \overbrace{=}^{\text{def'n}} E[e^{t\cdot X}] = \int_0^{\infty} e^{t\cdot x} f_X(x)dx$$

$$= \frac{1}{\beta^{\alpha}\cdot\Gamma(\alpha)}\cdot\int_0^{\infty} e^{t\cdot x}x^{\alpha-1}e^{-(1/\beta)x}\,dx$$

$$= \frac{1}{\beta^{\alpha}\cdot\Gamma(\alpha)}\cdot\int_0^{\infty} x^{\alpha-1}e^{t\cdot x-(1/\beta)x}\,dx$$

$$= \frac{1}{\beta^{\alpha}\cdot\Gamma(\alpha)}\cdot\int_0^{\infty} x^{\alpha-1}e^{-((1/\beta)-t)\cdot x}\,dx$$

Now define γ by $\frac{1}{\gamma} = \frac{1}{\beta}-t = \frac{1-\beta t}{\beta}$. Then,

$$M_X(t) = \frac{1}{\beta^{\alpha}\Gamma(\alpha)}\cdot\overbrace{\int_0^{\infty} x^{\alpha-1}e^{-(1/\gamma)x}\,dx}^{\left\{\frac{1}{\gamma}=\frac{1}{\beta}-t=\frac{1-\beta t}{\beta}\right\}} \overbrace{=}^{\{\text{definition of }\Gamma(\alpha,\gamma)\}} \frac{\gamma^{\alpha}\Gamma(\alpha)}{\beta^{\alpha}\Gamma(\alpha)} \overbrace{=}^{\left\{\gamma=\frac{\beta}{1-\beta t}\right\}} \left(\frac{1}{1-\beta t}\right)^{\alpha} = \left(\frac{\lambda}{\lambda-t}\right)^{\alpha}.$$

Note that for the improper integral to be convergent we must have $\gamma > 0$, which means $t < \frac{1}{\beta} = \lambda$. We can now deduce properties (1) and (2) by differentiating the natural logarithm of the MGF. $h(t) = \ln[M_X(t)] = \ln(1-\beta t)^{-\alpha} = -\alpha\cdot\ln(1-\beta t)$.

$$h'(t) = \alpha\cdot\beta\cdot(1-\beta t)^{-1}$$

so

$$E[X] = h'(0) = \alpha\cdot\beta.$$

$$h''(t) = \alpha\cdot\beta^2\cdot(1-\beta t)^{-2}$$

so

$$Var[X] = h''(0) = \alpha\cdot\beta^2.$$

Properties (4) and (5) can be easily demonstrated using properties of moment generating functions. For (4), we let $S = X+Y$. Since X and Y are independent,

$$M_S(t) = M_X(t)\cdot M_Y(t) = \left(\frac{\lambda}{\lambda-t}\right)^{\alpha_X}\cdot\left(\frac{\lambda}{\lambda-t}\right)^{\alpha_Y} = \left(\frac{\lambda}{\lambda-t}\right)^{\alpha_X+\alpha_Y}.$$

The punch line is that we recognize $\left(\frac{\lambda}{\lambda-t}\right)^{\alpha_X+\alpha_Y}$ as the MGF of a Gamma distribution with parameters $\alpha_X+\alpha_Y$ and $\beta = \frac{1}{\lambda}$, and so $S \sim \Gamma(\alpha_X+\alpha_Y, \beta)$.

For (5), if $Y = k \cdot X$ then $M_Y(t) = M_X(k \cdot t) = \left(\frac{1}{1-\beta \cdot (k \cdot t)}\right)^{\alpha} = \left(\frac{1}{1-(k \cdot \beta) \cdot t}\right)^{\alpha}$. Again we recognize this as the MGF for $Y \sim \Gamma(\alpha, k \cdot \beta)$.

Property (6) follows from the fact that $f_X(x) = \frac{1}{\beta^{\alpha} \cdot \Gamma(\alpha)} \cdot x^{\alpha-1} e^{-(1/\beta)x}$ is a density function, so that $1 = \int_0^{\infty} \frac{1}{\beta^{\alpha} \cdot \Gamma(\alpha)} \cdot x^{\alpha-1} e^{-(1/\beta)x}\, dx$. Property (6) follows immediately.

The following graph shows several gamma density functions with $\beta = 2$.

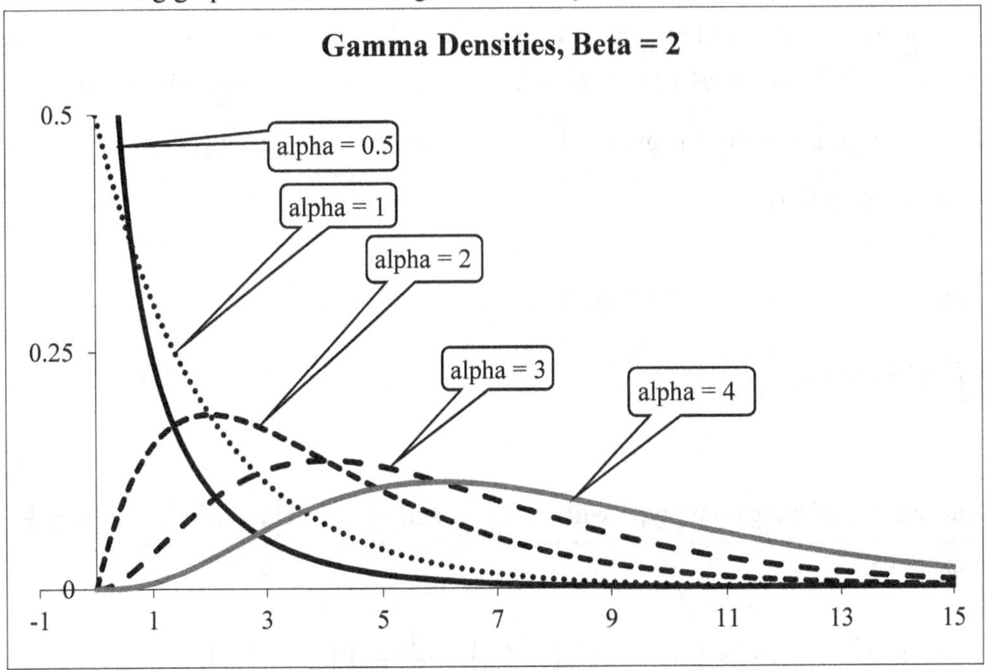

Notes

(1) The case $\alpha = 1$, $\beta = 2$ is an exponential random variable with mean $\beta = 2$.

(2) The case $\alpha = 0.5, \beta = 2$ is an important one in statistics. The resulting distribution is called the ***chi-squared***, with one degree of freedom. We will elaborate on this later.

(3) The densities for higher values of α are less skewed in appearance and in fact are becoming closer to normal distributions.

(4) Property (5) is the reason β is called the "scale" parameter. The parameter α is sometimes referred to as the "shape" parameter. The graph shows how different values of α affect the shape of the gamma density for the case $\beta = 2$.

(5) Property (6) is useful for effortlessly evaluating definite integrals with integrands of the general Gamma type.

Example 6.5-1 Recognize a Gamma

Consider a random variable X with probability density function

$$f_X(x) = \begin{cases} \frac{xe^{-x/3}}{9} & \text{for } 0 \le x < \infty \\ 0 & \text{otherwise} \end{cases}.$$

Compute the mean and standard deviation, $E[X]$ and σ_X.

Solution

We recognize that $f_X(x)$ is the density function for $X \sim \Gamma(2,3)$. Hence $E[X] = \alpha \cdot \beta = 2 \cdot 3 = 6$, $Var[X] = \alpha \cdot \beta^2 = 2 \cdot 3^2 = 18$, so the standard deviation for X is $\sigma_X = 3\sqrt{2}$. If you actually integrated $\int_0^\infty x^2 \frac{xe^{-x/3}}{9}\, dx$ to compute $E[X^2]$, then you must have a lot of spare time. ❐

Example 6.5-2 Evaluate a Definite Integral of the Gamma Type

Find $\int_0^\infty x^{3/2} e^{-2x}\, dx$.

Solution

The integrand is of the gamma type with $\alpha - 1 = \frac{3}{2}$ and $\frac{1}{\beta} = 2$. Thus, $\alpha = \frac{5}{2}$ and $\beta = \frac{1}{2}$, and by property (6),

$$\int_0^\infty x^{3/2} e^{-2x}\, dx = \left(\frac{1}{2}\right)^{5/2} \cdot \Gamma\left(\frac{5}{2}\right) = \left(\frac{1}{2}\right)^{5/2} \cdot \frac{3}{2} \cdot \frac{1}{2} \cdot \Gamma\left(\frac{1}{2}\right) = \left(\frac{1}{2}\right)^{5/2} \cdot \frac{3}{4} \cdot \sqrt{\pi}.$$ ❐

6.5.3 COMPARING THE GAMMA AND LOGNORMAL

Both the gamma and lognormal distributions apply to non-negative random variables, and their density functions have similar graphs, at least in the case $\alpha > 1$ for the gamma. Both are widely used as models for insurance loss distributions.

There are, however, certain important, if somewhat subtle, differences. A statistician would say the lognormal has a "heavier tail" than the gamma, meaning that $\Pr[X > a]$ goes to zero (as a goes to infinity) more slowly for a lognormal distribution than it does for a gamma distribution. Thus, in insurance applications, the lognormal would be better for modeling losses where a greater weight is desired for events of the form $[X > a]$, especially where a represents a catastrophic level of loss.

The tail weight of a distribution is a qualitative statement that can be made more precise using advanced notions. However, we will content ourselves here with a simple example to compare the gamma and the lognormal distributions.

Let Y have a lognormal distribution, $Y = e^X$, with $X \sim N\left(\mu = 0, \sigma^2 = \frac{1}{4}\right)$. Using the formulas for the mean and variance of the lognormal we have $\mu_Y = 1.1331$ and $\sigma_Y^2 = .3647$. We take a gamma random variable T with the same mean and variance. This requires our parameters to be $\alpha = 3.5208$ and $\beta = .3218$. The resulting densities are superimposed on the graph below:

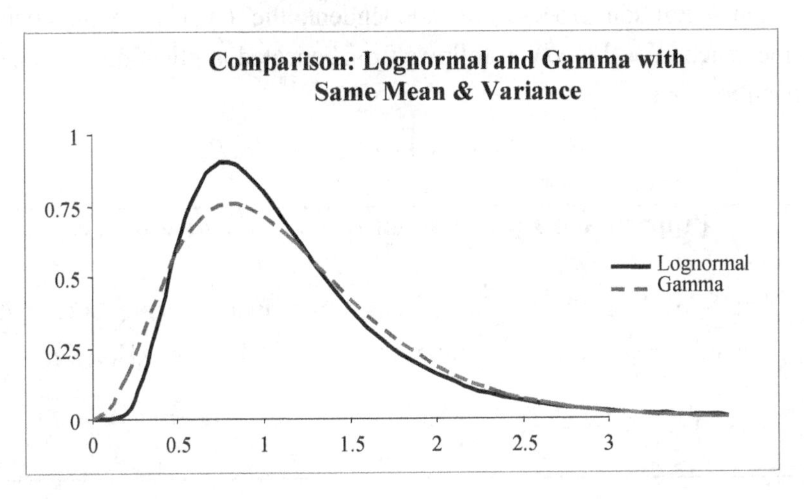

The next graph shows a blow-up of the tails for $x > 3$. This illustrates the heavier tail of the lognormal distribution.

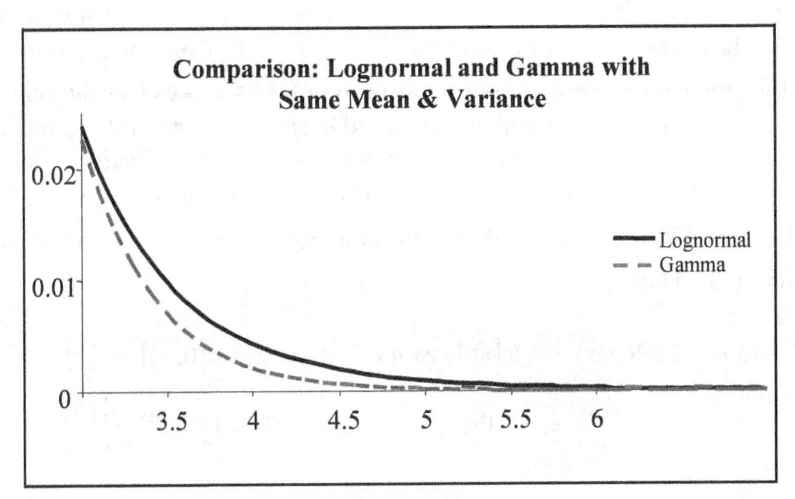

6.5.4 GAMMA AND POISSON CONNECTIONS

Consider a Poisson process with λ as the mean rate of occurrence per unit time. Using the terminology of section 6.2, we let Y denote the discrete random variable for the number of occurrences in a unit period, and we let X be the continuous random variable for the time until the next occurrence. Then Y is a Poisson random variable with parameter λ and X is an exponential random variable with mean $\beta = \frac{1}{\lambda}$.

What about the time, S, until the nth occurrence? We can think of S as a sum, $S = X_1 + X_2 + \cdots + X_n$, where X_1 represents the time to the first occurrence, X_2 represents the *additional* time from X_1 to the second occurrence, and so forth. Because of the "memory-less" property of the exponential random variable, the $(X_i)_{i=1}^n$ are identically distributed as exponentials, each with mean $\beta = \frac{1}{\lambda}$. Also, since events in non-overlapping time-intervals in a Poisson process are independent, the $(X_i)_{i=1}^n$ are independent random variables. Since each $X_i \sim \Gamma(1, \beta)$, it follows from repeated application of property (4) of the gamma distribution that,

$$S = X_1 + X_2 + \cdots + X_n \sim \Gamma(n, \beta).$$

Properties of $\Gamma(\alpha, \beta)$ when $\alpha = n$, a Positive Integer

1. Let S be the time until the nth occurrence in a Poisson process with mean rate of $\lambda = \frac{1}{\beta}$ occurrence(s) per unit period. Then $S \sim \Gamma(n, \beta)$.

2. The CDF $F_S(x) = \frac{\lambda^n}{(n-1)!}\int_0^x t^{n-1} e^{-\lambda t}\, dt = 1 - e^{-\lambda x}\sum_{k=0}^{n-1}\frac{(\lambda x)^k}{k!}$.

Proof

The first property follows from the discussion in the preceding paragraph. For the second, we have the integral formula for $F_S(x)$ directly from integrating the density function for the gamma. We use the λ form in order to make clear the relation to the associated Poisson random variable. We could establish the equality to the sum using repeated or tabular integration by parts, but we prefer a probabilistic explanation. For a given $x > 0$ let Y_x denote the number of occurrences in the time period $(0, x]$. From the scalability property of the Poisson we know that Y_x is Poisson with parameter $\lambda \cdot x$. Thus,

$$\begin{aligned} F_S(x) &= \Pr[S \le x] = \Pr\big[\text{at least } n \text{ occurrences in } (0, x]\big] = \Pr[Y_x \ge n] \\ &= 1 - \Pr[Y_x = 0, 1, \ldots, n-1] = 1 - e^{-\lambda x}\sum_{k=0}^{n-1}\frac{(\lambda x)^k}{k!}. \end{aligned}$$

Property (2) establishes a closed formula, albeit a complicated one in general, for the CDF of a gamma distribution with parameter α equal to a positive integer. This integral formula is sometimes referred to as an ***incomplete Gamma function.***

Example 6.5-3 The CDF of a Gamma (Example 6.4-3 Revisited)

People arrive at the check-out counter according to a Poisson process with a mean waiting time of one minute between arrivals. Calculate the exact probability that the 8th arrival occurs between time 7 and time 9.

Solution

Let $S = X_1 + X_2 + \cdots + X_8$ be the total time elapsed until the 8th arrival, where X_i, the waiting time for the ith arrival, is exponentially distributed with a mean of one. Recall that in Example 6.4-3 we approximated S using the CLT in order to calculate the required probability. We now recognize that $S \sim \Gamma(8,1)$ and use the CDF formula above. Then,

$$F_S(x) = \frac{1}{(8-1)!}\int_0^x t^{8-1}e^{-t}\,dt = 1-e^{-x}\sum_{k=0}^{8-1}\frac{(x)^k}{k!} = 1-e^{-x}\sum_{k=0}^{7}\frac{x^k}{k!}.$$

Now,

$$\begin{aligned}\Pr[7 \le S \le 9] &= \left[1-e^{-9}\sum_{k=0}^{7}\frac{9^k}{k!}\right]-\left[1-e^{-7}\sum_{k=0}^{7}\frac{7^k}{k!}\right]\\ &= e^{-7}\sum_{k=0}^{7}\frac{7^k}{k!}-e^{-9}\sum_{k=0}^{7}\frac{9^k}{k!}\\ &= .5987-.3239\\ &= .2748\end{aligned}$$

❐

Exercise 6-56 Your obnoxious calculus teacher throws insults in your general direction according to a Poisson distribution with a mean of four insults per month. (Assume for simplicity sake that one month is exactly 4 weeks.)

(a) Find the probability that your teacher insults you exactly three times this month.
(b) Find the probability that your teacher does not insult you this week.
(c) Find the probability that your teacher insults you at least once this week.
(d) Find the probability that the first time your teacher insults you is next week.
(e) Find the probability that more than 5 weeks elapse before the hurling of the 4th insult.

Exercise 6-57 The number of times that your campus computer network fails has a Poisson distribution with a mean of once per week.

(a) Find the probability that the network fails twice this week.
(b) How many days must pass before you are at least 50% confident that the campus computer network has failed?

Exercise 6-58 Suppose that a random variable X is distributed according to the gamma distribution with mean 10 and variance 20.

(a) Find the parameters α and β.
(b) Find the probability density function for X.
(c) Find $\Pr(X < 6)$.
(d) Compute $E[4X-5X^2]$.

Exercise 6-59 Suppose that a random variable X is distributed according to the gamma distribution with parameters $\alpha = 6$ and $\beta = 2$. That is, $X \sim \Gamma(6,2)$.
(a) Compute the mean and variance for X.
(b) Compute $E[X^3]$.
(c) Compute $E\left[\sqrt{X}\right]$.

Exercise 6-60 Your dog does his business according to a Poisson process with an average of once per day. You have only 3 plastic bags.
(a) Determine the probability that the three bag supply will last through the two-day weekend.
(b) Let T be the time (in days) that the 4th business trip occurs. Write the density function for T.
(c) Calculate $\Pr[T < 2]$.

6.5.5 THE DISTRIBUTION OF Z^2

The measurement of a quantity, say the volume of water in a cup, or the population of a country, or the length of a squirming laboratory rat, is subject to error. These errors are usually equally likely to underestimate as to overestimate, and in fact tend to follow a normal distribution. In statistics we often work with the squares of the sums of errors. We do that to assure that positive and negative errors don't cancel each other out in the sum. Consequently, it becomes important to understand the distribution of the square of a normally distributed random variable.

This understanding begins with the square, Z^2, of the standard normal random variable, $Z \sim N(0,1)$. This is a non-negative random variable that, as we shall see below, follows a gamma distribution with parameters $\alpha = \frac{1}{2}$ and $\beta = 2$. It goes by the name chi-square.

Properties of the Square of the Standard Normal

Let $Z \sim N(0,1)$ be the standard normal random variable.

(1) Then, $Z^2 \sim \Gamma\left(\frac{1}{2},2\right)$. This distribution is called ***chi-square***[8] ***with 1 degree of freedom***, and denoted $\chi^2(1)$.

(2) Sum of Z^2: Let $S = Z_1^2 + Z_2^2 + \cdots + Z_n^2$, where $(Z_i)_{i=1}^n$ are independent standard normal random variables. Then $S \sim \Gamma\left(\frac{n}{2},2\right)$. This distribution is called ***chi-square with n degrees of freedom***, and denoted $\chi^2(n)$.

(3) $E[S] = n$ and $Var[S] = 2n$.

[8] It is a matter of some dispute whether the correct pronunciation of the symbol χ^2 should be *chi-square* or *chi-squared*. Both are widely used. We shall use them interchangeably with the understanding that (a) we are multiple authors, and (b) we plead shameless recourse to the Ralph Waldo Emerson dictum that "*A foolish consistency is the hobgoblin of little minds...*"

Proof

We begin by deriving the CDF of Z^2. For $x>0$,

$$F(x)=\Pr[Z^2 \le x] = \Pr\left[-\sqrt{x} \le Z \le \sqrt{x}\right] = 1-2\left\{1-\Pr\left[Z \le x^{1/2}\right]\right\} = -1+2\Phi\left(\sqrt{x}\right).$$

We obtain the density function $f(x)$ by differentiation and using the chain rule with $u=x^{1/2}$.

$$f(x) = F'(x) = \frac{dF}{du}\cdot\frac{du}{dx} = 2\left(f_Z(u)\right)\left(\frac{1}{2}x^{-1/2}\right)$$

$$= 2\left(\frac{1}{\sqrt{2\pi}}e^{-\frac{(x^{1/2})^2}{2}}\right)\left(\frac{1}{2}x^{-1/2}\right) = \frac{1}{\sqrt{2\pi}}x^{-1/2}e^{-(1/2)x}.$$

This is the density function for $\Gamma\left(\frac{1}{2},2\right)$. The fact that $S \sim \Gamma\left(\frac{n}{2},2\right)$ follows immediately from the properties of the gamma distribution (see Subsection 6.5.2).

Example 6.5-4 Probability for a Chi-Squared Random Variable

Suppose that $X \sim \chi^2(8)$. Find $\Pr(X>10)$.

Solution

We know that the density function for $X \sim \chi^2(8) = \Gamma(4,2)$ is

$$f_X(x) = \begin{cases} \frac{x^3e^{-x/2}}{96} & \text{for } x>0 \\ 0 & \text{otherwise} \end{cases}.$$

We can always use the density function and multiple or tabular integration-by-parts to find $\Pr(X>10) = \int_{10}^{\infty} f_X(x)\,dx = \int_{10}^{\infty}\frac{x^3e^{-x/2}}{96}\,dx = .2650$. We could also evaluate this definite integral with a math package like Maple or a graphing calculator like the TI-83. Tabular integration is tedious and math programs are like black boxes. There is another solution that uses the connection to a Poisson process.

Alternative Solution

We can think of X as the waiting time for the $\alpha = 4^{th}$ occurrence in a Poisson process with mean rate of $\lambda = \frac{1}{2}$. Let Y_{10} be the number of occurrences in the interval $[0,10]$.

Then by scalability, Y_{10} is Poisson with rate $\lambda_{10} = (10)\left(\frac{1}{2}\right) = 5$. So,

$$\begin{aligned}\Pr[X > 10\} &= \Pr[4^{th} \text{ occurrence later than time } 10] \\ &= \Pr\left[\text{at most 3 occurrences in } [0,10]\right] \\ &= \Pr[Y_{10} \leq 3] = \sum_{k=0}^{3} \frac{e^{-5}5^k}{k!} = e^{-5}\left(1 + 5 + \tfrac{25}{2} + \tfrac{125}{6}\right) = .2650.\end{aligned}$$

This is essentially the formula in (2) (the incomplete Gamma function) of the Properties of $\Gamma(\alpha, \beta)$ with α an integer in Section 6.5.4. ❒

Exercise 6-61 Suppose that $(Z_i)_{i=1}^{100}$ is an independent and identically distributed sample from the standard normal distribution, $N(0,1)$. Let $S = \sum_{i=1}^{100} Z_i^2$. Find $\Pr\left(S < 5 \cdot \sqrt{2} \cdot \sigma_S\right)$.

(a) Use the Central Limit Theorem to approximate the probability,

(b) Use the Microsoft Excel® function CHIDIST,

(c) Use the Microsoft Excel® function GAMMADIST.

The applications of the chi-squared distribution in statistics will be explored more fully in Chapters 9-11.

6.6 THE BETA FAMILY OF DISTRIBUTIONS

We saw in the previous section that the gamma family constituted a two-parameter family of non-negative random variables, generating a wide variety of shapes according to the values of the parameters α and β. In this section we introduce another two-parameter family of distributions living on the unit interval [0,1].

The form of the density function is $c \cdot x^{\alpha-1}(1-x)^{\beta-1}$; $0 \leq x \leq 1$. The case $\alpha = 1$ and $\beta = 1$ produces the standard uniform random variable with $c = 1$. Once again, the problem of determining a formula for the standardizing constant c requires some integration theory.

6.6.1 THE BETA FUNCTION

Definition of the Beta Function

Given $\alpha > 0$ and $\beta > 0$, the ***beta function*** is defined as the following definite integral: $B(\alpha, \beta) = \int_0^1 x^{\alpha-1}(1-x)^{\beta-1} dx.$

Properties of the Beta Function

(1) $B(\alpha, \beta) = B(\beta, \alpha)$.

(2) $B(\alpha, \beta) = 2\int_0^{\pi/2} (\sin^{2\alpha-1}\theta)(\cos^{2\beta-1}\theta) d\theta.$

(3) $B(\alpha,\beta)=\dfrac{\Gamma(\alpha)\cdot\Gamma(\beta)}{\Gamma(\alpha+\beta)}$.

Note: This property will often be used to determine the scale constant c for the beta density function.

Proof

(1) Follows immediately from the substitution $u=1-x$.

(2) Makes use of the substitution $x=\sin^2\theta$, which implies that $1-x=\cos^2\theta$, and $dx=2\sin\theta\cos\theta d\theta$, with the interval of integration $[0,1]$ with respect to x transformed to $[0,\pi/2]$ with respect to θ.

(3) Makes use of double integrals and the switch from rectangular to polar coordinates. From the definition of the gamma function we have,

$$\begin{aligned}
\Gamma(\alpha)\cdot\Gamma(\beta) &= \left(\int_0^\infty x^{\alpha-1}e^{-x}\,dx\right)\cdot\left(\int_0^\infty y^{\beta-1}e^{-y}\,dy\right)\\
&= \int_0^\infty\int_0^\infty x^{\alpha-1}y^{\beta-1}e^{-(x+y)}\,dxdy\\
&= \overbrace{\int_0^\infty\int_0^\infty u^{2\alpha-2}v^{2\beta-2}\,e^{-(u^2+v^2)}\,(2u\,du)(2v\,dv)}^{\text{substitute } x=u^2,\ y=v^2}\\
&= 4\int_0^\infty\int_0^\infty u^{2\alpha-1}v^{2\beta-1}\,e^{-(u^2+v^2)}\,du\,dv\\
&= \overbrace{4\int_0^\infty\int_0^{\pi/2}(r^{2\alpha-1}\cos^{2\alpha-1}\theta)(r^{2\beta-1}\sin^{2\beta-1}\theta)e^{-(r^2)}r\,dr\,d\theta}^{\text{polar coordinates } u=r\cos\theta,\ v=r\sin\theta}\\
&= \left(2\int_0^\infty r^{2\alpha+2\beta-1}\,e^{-r^2}\,dr\right)\left(2\int_0^{\pi/2}\cos^{2\alpha-1}\theta\sin^{2\beta-1}\theta\,d\theta\right)\\
&= \left(\int_0^\infty r^{2\alpha+2\beta-2}\,e^{-r^2}\,(2rdr)\right)\left(2\int_0^{\pi/2}\cos^{2\alpha-1}\theta\sin^{2\beta-1}\theta\,d\theta\right)\\
&= \left(\overbrace{\int_0^\infty x^{\alpha+\beta-1}\,e^{-x}\,dx}^{\text{substitute } x=r^2}\right)\left(2\int_0^{\pi/2}\cos^{2\alpha-1}\theta\sin^{2\beta-1}\theta\,d\theta\right)\\
&= \Gamma(\alpha+\beta)B(\beta,\alpha) = \Gamma(\alpha+\beta)B(\alpha,\beta).
\end{aligned}$$

Exercise 6-62 Show that $B(4,2)=B(2,4)$ from the definition of the beta function (integrate). Graph the two definite integrals and see if the areas are equal.

Exercise 6-63 Find the values of $B(5,8)$ and $B(3,1.5)$.

6.6.2 THE BETA FAMILY OF DISTRIBUTIONS

Definition of the Beta Distribution

A random variable X is said to belong to the ***Beta family*** with parameters α and β if $\alpha > 0$ and $\beta > 0$ the density function is given by,

$$f(x) = \begin{cases} \dfrac{x^{\alpha-1}\cdot(1-x)^{\beta-1}}{B(\alpha,\beta)} = \dfrac{\Gamma(\alpha+\beta)}{\Gamma(\alpha)\cdot\Gamma(\beta)}\cdot x^{\alpha-1}\cdot(1-x)^{\beta-1} & \text{for } 0 < x < 1 \\ 0 & \text{otherwise.} \end{cases}$$

Properties of Beta Distributions

Suppose X has beta distribution with parameters α and β.

(1) The mean equals: $E[X] = \dfrac{\alpha}{\alpha+\beta}$.

(2) The variance equals: $Var[X] = \dfrac{\alpha\beta}{(\alpha+\beta)^2(\alpha+\beta+1)}$.

Proof

$$\begin{aligned} E[X] &= \int_0^1 x f(x)\,dx \\ &= \frac{\Gamma(\alpha+\beta)}{\Gamma(\alpha)\Gamma(\beta)}\int_0^1 x^{\alpha}(1-x)^{\beta-1}\,dx \\ &= \frac{\Gamma(\alpha+\beta)}{\Gamma(\alpha)\Gamma(\beta)}\frac{\Gamma(\alpha+1)\Gamma(\beta)}{\Gamma(\alpha+\beta+1)} = \frac{\alpha}{\alpha+\beta}. \end{aligned}$$

Similarly,

$$\begin{aligned} E[X^2] &= \int_0^1 x^2 f(x)\,dx \\ &= \frac{\Gamma(\alpha+\beta)}{\Gamma(\alpha)\Gamma(\beta)}\frac{\Gamma(\alpha+2)\Gamma(\beta)}{\Gamma(\alpha+\beta+2)} = \frac{\alpha(\alpha+1)}{(\alpha+\beta)(\alpha+\beta+1)}. \end{aligned}$$

The formula for the variance can now be calculated from $Var[X] = E[X^2] - (E[X])^2$ with the application of a little algebraic elbow-grease.

We sketch the graphs of several beta distributions in Figure 6.2 on the following page. Observe what happens to the graphs near $x = 0$ for large values α and near $x = 1$ for large values of β.

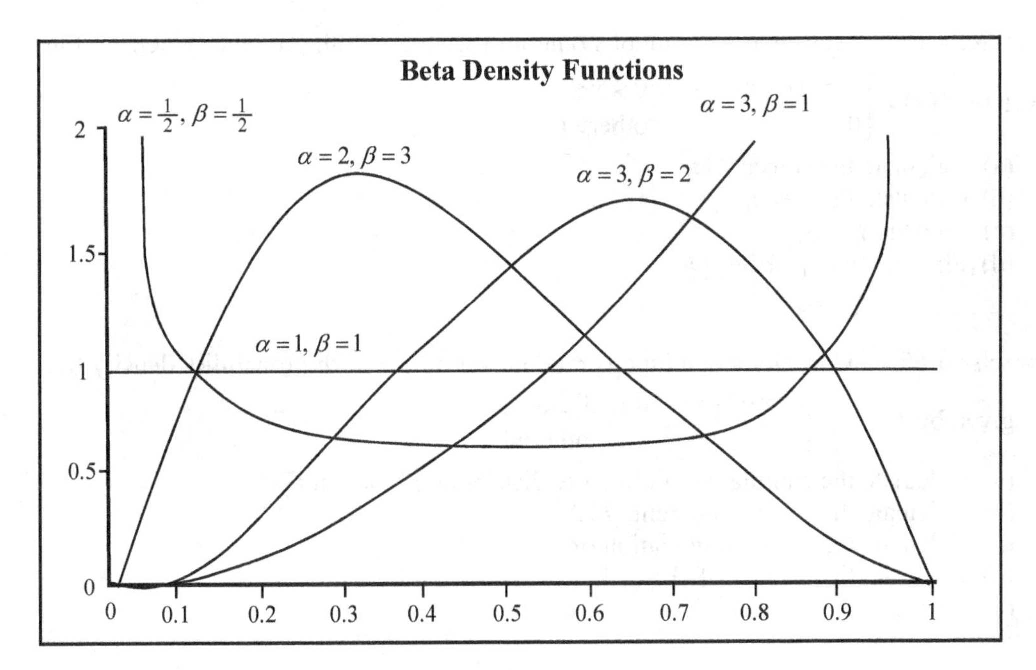

Figure 6.2

Example 6.6-1 Beta Distribution

In the previous chapter, we had a random variable X with density function

$$f(x) = \begin{cases} c \cdot x^3 \cdot (1-x), & \text{for } 0 \le x \le 1 \\ 0 & \text{otherwise} \end{cases}.$$

(a) What is the scaling constant c?
(b) What is $E[X]$?
(c) Find the $\Pr(.4 < X \le .7)$.

Solution

(a) We now recognize this random variable to be a beta distribution with parameters $\alpha = 4$ and $\beta = 2$. Since $B(4,2) = \frac{\Gamma(4)\cdot\Gamma(2)}{\Gamma(4+2)} = \frac{3!\cdot 1!}{5!} = \frac{1}{20}$, we have $c = 20$.

(b) Method 1: By direct integration: $E[X] = \int_0^1 x \cdot 20 \cdot x^3 \cdot (1-x)dx = \frac{2}{3}$.

Method2: By recognizing the integral as another beta density function:

$$E[X] = \int_0^1 x \cdot 20 \cdot x^3 \cdot (1-x)\,dx = 20 \cdot \int_0^1 x^4 \cdot (1-x)\,dx$$

$$= 20 \cdot \mathrm{B}(5,2) = 20 \cdot \frac{\Gamma(5)\cdot\Gamma(2)}{\Gamma(7)} = 20 \cdot \frac{4!\cdot 1!}{6!} = \frac{20}{6\cdot 5} = \frac{2}{3}.$$

(c) $\Pr(.4 < X \le .7) = \int_{.4}^{.7} 20 \cdot x^3 \cdot (1-x)\,dx = 20\left[\frac{x^4}{4} - \frac{x^5}{5}\right]_{x=.4}^{.7} = .4412.$ ❐

Exercise 6-64 Consider a continuous random variable X with probability density function given by $f(x)=\begin{cases} c\cdot x\cdot(1-x)^5 & \text{for } 0\le x\le 1 \\ 0 & \text{otherwise.}\end{cases}$

(a) Calculate the constant c.
(b) Calculate the mean.
(c) Find $\Pr(X>.5)$.
(d) Find the first quartile, Q_1.

Exercise 6-65 Consider a continuous random variable X with probability density function given by $f_X(x)=\begin{cases} c\cdot x\cdot(1-x), & \text{for } 0\le x\le 1 \\ 0 & \text{otherwise.}\end{cases}$

(a) Calculate the cumulative probability distribution function $F_X(x)$.
(b) Calculate the second moment, $E[X^2]$.
(c) Calculate the standard deviation of X, σ_X.
(d) Calculate the median of X.
(e) Calculate the mode of X.

Exercise 6-66 Consider a continuous random variable X with probability density function given by $f(x)=\begin{cases} c\cdot x^{3/2}\cdot(1-x)^2 & \text{for } 0\le x\le 1 \\ 0 & \text{otherwise.}\end{cases}$

(a) Calculate the constant c.
(b) Calculate the mean.
(c) Calculate the standard deviation of X.
(d) Find $\Pr(.4\le X\le .9)$.

Exercise 6-67 Suppose that $X\sim Beta(3,4)$. Compute $E[7X-5X^6]$.

Exercise 6-68 The amount a college senior spends on beer per month (in thousands of \$US) is modeled by the random variable Y with density function

$$f_Y(y) = \begin{cases} c\cdot(1-y)^4 & \text{for } 0\le y\le 1 \\ 0 & \text{otherwise} \end{cases}.$$

(a) Is this a beta distribution? If so, what are the parameters?
(b) Calculate the scaling constant c.
(c) Find the cumulative distribution function $F_Y(y)$.
(d) How much money should this student ask his parents for so that he is 75% confident he will have enough funds for beer?
(e) What are the mean and standard deviation for Y?
(f) Find $\Pr(.8<Y<1.4)$.

Exercise 6-69 Suppose that $Y\sim Beta(2,5)$. Find the median and the 95th percentile for Y.

6.7 MORE CONTINUOUS DISTRIBUTIONS

There are numerous other continuous distributions that come up in practice, besides the main ones we have already covered. In this section we investigate a few of these, and introduce a new way of characterizing non-negative continuous random variables, called the hazard-rate, or failure rate, function.

In the applications that follow we consider the non-negative random variable X to be the total lifetime of some object or person, measured from birth to death at some random future time. This applies equally to the "age-at-death" of a person or creature, or the elapsed time from start of service to failure for a mechanical or electrical component. Since X takes values in time we will use t to denote those values and write the CDF of X as $F(t) = \Pr[X \le t]$ (the probability of death or failure within time t), and the associated density function as $f(t) = F'(t)$. In this context it is customary to introduce the ***survival function***, $S(t)$, which is the complement to the CDF $F(t)$. That is to say, $S(t) = 1 - F(t) = \Pr[X > t]$, the probability of surviving beyond age t.

This terminology is common in actuarial science (***life contingencies***) and elsewhere, although mathematically $S(t)$ and $F(t)$ are interchangeable ideas. Psychologically speaking, however, it is perhaps more upbeat when discussing mortality to couch things in terms of survival rather than its opposite.

6.7.1 THE HAZARD RATE

In discussing these life-and-death issues it is commonplace to speak of ***mortality*** rates. A mortality rate in usual parlance is tied to a particular age, t, and refers to the probability of death in the next time period for someone (or something) that has survived to age t.

The hazard rate is a mathematically precise way of capturing the idea of a mortality rate for continuously varying age-at-death random variables. Before giving the formal definition we will illustrate the concept by conducting a simple thought experiment. Imagine a population cohort of 100 individuals, all born at the same time, and all with the same general health and life prospects going forward.

Although similar or identical at birth, we don't expect these 100 souls to all die at precisely the same moment in time. Rather, they will die off randomly, following what amounts to a probability distribution. Some will succumb early to illness; others may suffer accidental deaths; and still others will embrace healthy life styles and sedate professions like, for example, actuary, thereby increasing the likelihood of prolonged life.

If we observe these lives over time we can describe the results with a function $L(t)$ which equals the number of survivors at age t. Then, by standardizing, we can define the associated survival function $S(t) = \frac{1}{100} L(t)$. Clearly, $S(0) = 1$, and absent any resurrections, $S(t)$ is a non-increasing function. Also, assuming no one is immortal, $\lim_{t \to \infty} S(t) = 0$. Therefore,

$F(t) = 1 - S(t)$ is a CDF for an age-at-death random variable, X. In this case X has a discrete distribution since there are at most 100 distinct values in time that X can take.

We can describe $E[X]$ as the life expectancy of a newborn, and we can define a mortality rate at a certain age. For example, the mortality rate at age 65 equals $\frac{L(65)-L(66)}{L(65)}$, the fraction of (living) 65 year-olds who will die in the next year. In probability-speak, this is the conditional probability of dying within the next year, given survival to age 65. That is,

$$\begin{aligned}\Pr\left[65 < X \le 66 \middle| X > 65\right] &= \frac{\Pr[65 < X \le 66]}{\Pr[X > 65]} \\ &= \frac{F(66) - F(65)}{1 - F(65)} \\ &= \frac{S(65) - S(66)}{S(65)} \\ &= \frac{L(65) - L(66)}{L(65)}.\end{aligned}$$

We now extend this notion of mortality rate as a conditional probability at a given age to the case of a continuous age-at-death random variable. Because of the continuity we will use the conditional probability **rate** at a given age.

Recall, in Chapter 5 we spoke of the density function $f(x)$ of a random variable X as representing the "instantaneous" **rate** of probability at $X = x$. In our present context we would say that for small time intervals dt,

$$\Pr[t < X \le t + dt] = S(t) - S(t + dt) \approx -S'(t)dt = f(t)dt,$$

which can be interpreted as saying the probability of a small interval is approximately $rate \times length = f(t)\,dt$.

The related concept we are after here is the "**relative** instantaneous" **rate** of death (or failure) at time t, where the "relative" part refers to the rate being conditioned upon survival up to time t. Thus,

$$\Pr\left[t < X \le t + dt \middle| X > t\right] = \frac{S(t) - S(t + dt)}{S(t)} \approx -\frac{S'(t)dt}{S(t)} = \frac{f(t)dt}{S(t)},$$

and

$$\lim_{dt \to 0} \frac{\Pr\left[t < X \le t + dt \middle| X > t\right]}{dt} = \frac{f(t)}{S(t)}.$$

In actuarial science, for continuous age-at-death random variables, this is called the ***force of mortality*** (rather than ***mortality rate***) at age t.

In more general contexts this conditional probability rate is called either the hazard rate, or the failure rate. We adopt the former in the formal definition that follows.

Definition of the Hazard Rate

Let X represent a (non-negative) age-at-death random variable, with CDF $F(t)$ and density $f(t)$. The ***hazard rate***, or ***failure rate*** at time t is given by

$$\mu(t) = \frac{f(t)}{S(t)} = \frac{f(t)}{1-F(t)}.$$

The key idea is embodied in the conditional probability statement,

$$\Pr\left[t < X \le t+dt \middle| X > t\right] = \frac{f(t)dt}{S(t)} = \frac{f(t)}{1-F(t)} \cdot dt = (\text{hazard rate}) \times (\text{length}).$$

In actuarial mathematics the alternative notation μ_t is used for the force of mortality.

Example 6.7-1 When Age-at-Death is Exponential

Let the age-at-death random variable X have an exponential distribution with mean β. Calculate the hazard rate (force of mortality) at any given time t.

Solution

$F(t) = 1-e^{-(1/\beta)t} = 1-e^{-\lambda t}$ $\left(\lambda = \frac{1}{\beta}\right)$ and the density function for X is $f(t) = \lambda \cdot e^{-\lambda t}$. Thus the hazard rate is $\mu(t) = \frac{f(t)}{1-F(t)} = \frac{\lambda e^{-\lambda t}}{e^{-\lambda t}} = \lambda$. □

If death or failure is the occurrence in a Poisson process then the hazard rate reduces to the constant Poisson mean rate of occurrence. Under a fixed rate this would say, for example, that a 90-year-old sky-diver who smokes would have the same force of mortality (mortality rate) as a healthy 25 year-old actuary with no life-threatening proclivities. This is another manifestation of the memory-less property of the exponential distribution. While this isn't a practical model for most biological lifetimes it often works well enough for electrical systems.

We conclude this section by showing how to reconstruct a probability distribution from the hazard rate. This shows that the hazard rate can be used instead of the CDF or density function to characterize the distribution of a continuous non-negative random variable.

Finding the CDF from the Hazard Rate

Let X be a non-negative random variable with hazard rate $\mu(t)$. Then the CDF for X is given by

$$F(x) = 1-e^{-\int_0^x \mu(t)dt}.$$

Proof

$$\mu(t) = \frac{F'(t)}{1-F(t)} = \frac{-S'(t)}{S(t)} = -\frac{d}{dt}[\ln S(t)].$$

So

$$d[\ln S(t)] = -\mu(t)dt.$$

By integrating both sides we get,

$$\int_0^x d[\ln S(t)] = [\ln S(t)]\Big|_0^x$$

$$= \ln S(x) - \ln S(0) = \ln S(x) - \ln 1 = \ln S(x) = -\int_0^x \mu(t)dt.$$

Taking exponentials, $S(x) = e^{-\int_0^x \mu(t)\,dt}$ and the conclusion follows since $F(x) = 1 - S(x)$.

Note
Any integrable non-negative function $\mu(t)$ can serve as a hazard rate for a random variable provided $\lim_{x\to\infty}\left(\int_0^x \mu(t)dt\right) = \infty$.

6.7.2 THE PARETO DISTRIBUTION

To try out our new tool for defining probability distributions we introduce the hazard rate function for the ***Pareto*** family of distributions. We give two versions of the Pareto. In the first, or Type I, a random variable belongs to the Pareto family if it lives on the interval $[\beta, \infty)$ for a given $\beta > 0$ and has a hazard rate of the form $\mu(t) = \frac{\alpha}{t}$ for a given $\alpha > 0$. Since the hazard rate diminishes as t increases the Pareto is not realistic as a model for mortality. Under this model, with the parameters fixed, the 90-year-old sky-diver would have a better chance of surviving the next year than the healthy 25-year-old actuary. The Pareto is used by economists for modeling things like wealth distributions. In economics and marketing there is a rule-of-thumb called the 80-20 rule. This would say, for example, the wealthiest 20% of the population possesses 80% of the wealth, or that the top 20% of beer consumers consume 80% of the beer.

Properties of the Pareto Distribution, Type I[9]

Let X be a random variable on the interval $[\beta, \infty)$ with hazard rate given by $\mu(t) = \frac{\alpha}{t}$; $\alpha > 0$, $\beta > 0$ and $\beta \le t < \infty$.

(1) The CDF of X is: $F(x) = 1 - \left(\frac{\beta}{x}\right)^\alpha$; for $\beta \le x < \infty$.

(2) The density of X is: $f(x) = \frac{\alpha\beta^\alpha}{x^{\alpha+1}}$: for $\beta \le x < \infty$.

(3) If $\alpha > 1$, then the mean equals: $E[X] = \frac{\alpha\beta}{\alpha-1}$.

(4)If $\alpha > 2$, then the variance equals:

$$Var[X] = \frac{\alpha\beta^2}{\alpha-2} - \left(\frac{\alpha\beta}{\alpha-1}\right)^2 = \frac{\alpha\beta^2}{(\alpha-2)(\alpha-1)^2}.$$

[9] This is called the single-parameter Pareto in some texts.

Proof

$$\int_{\beta}^{x} \mu(t)\,dt = \int_{\beta}^{x} \frac{\alpha}{t}\,dt = \alpha(\ln x - \ln\beta) = \ln\left(\frac{x}{\beta}\right)^{\alpha},$$

so

$$F(x) = 1 - e^{-\ln(x/\beta)^{\alpha}} = 1 - \left(\frac{\beta}{x}\right)^{\alpha}.$$

Now (2) follows by differentiating $F(x)$, and (3) and (4) are straight-forward integrations of power functions.

Example 6.7-2 Pareto Distribution

Suppose that the random variable X is distributed according to the Pareto distribution with parameters $\alpha = 7$ and $\beta = 3$.

(a) Graph the density function.
(b) Find the expected value and standard deviation of X.
(c) Find $\Pr(X < 4)$.

Solution

(a) The density function is $f(x) = \frac{7 \cdot 3^7}{x^8};\quad x \geq 3$. Realize that $f(x) = 0$ for $x < \beta = 3$.

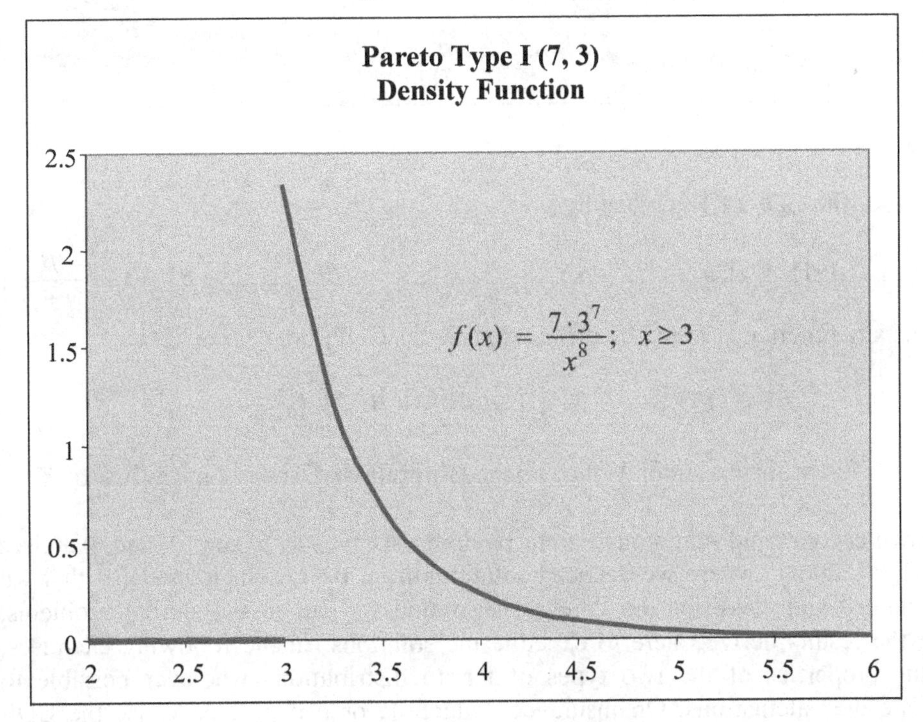

(b) The mean is $E[X] = \frac{\alpha \cdot \beta}{\alpha - 1} = \frac{7 \cdot 3}{6} = 3.5$.

The variance is $Var[X] = \frac{\alpha \cdot \beta^2}{\alpha - 2} - \left(\frac{\alpha \cdot \beta}{\alpha - 1}\right)^2 = \frac{7 \cdot 9}{5} - 3.5^2 = .35$, which implies that the standard deviation equals $\sigma_X = .5916$.

(c) $\Pr(X < 4) = 1 - \left(\frac{3}{4}\right)^7 = .8665$.

There is a second type of Pareto distribution, which is a simple modification of the first type. The advantage of the second type is that the distribution lives on the interval $[0, \infty)$ rather than the subset $[\beta, \infty)$ of the non-negative real numbers.

Properties of the Pareto Distribution, Type II[10]

Let X be a random variable on the interval $[\beta, \infty)$ with a Pareto Distribution of Type I and let $Y = X - \beta$. Then $0 \le Y < \infty$ and Y is called a Pareto Distribution of Type II.

(1) The CDF of Y is: $F_Y(y) = 1 - \left(\frac{\beta}{y+\beta}\right)^\alpha$ for $0 \le y < \infty$.

(2) The density of Y is: $f_Y(y) = \frac{\alpha\beta^\alpha}{(y+\beta)^{\alpha+1}}$ for $0 \le y < \infty$.

(3) If $\alpha > 1$, then $E[Y] = \frac{\beta}{\alpha - 1}$.

(4) If $\alpha > 2$, then $Var[Y] = \frac{\alpha\beta^2}{\alpha - 2} - \left(\frac{\alpha\beta}{\alpha - 1}\right)^2$.

Proof

For $y \ge 0$, the CDF of Y is given by,

$$F_Y(y) = \Pr[Y \le y] = \Pr[X - \beta \le y] = \Pr[X \le y + \beta] = F_X(y+\beta) = 1 - \left(\frac{\beta}{y+\beta}\right)^\alpha.$$

The density function is found by differentiating the CDF. For (3) we have

$$E[Y] = E[X - \beta] = \frac{\alpha\beta}{\alpha - 1} - \beta = \frac{\beta}{\alpha - 1}.$$

Since Y is just a translation of X, the variance formula of Y is the same as that of X.

We have encountered numerous Pareto random variables as examples and exercises in earlier sections, where we deduced solutions from basic principles. Now that we have named and classified the Pareto distribution we can revisit similar problems, using the results derived here to expedite the solutions. In the following exercises, use the properties of the two types of Pareto distributions wherever possible to facilitate the calculations. On insurance deductible or cap problems, use the CDF method for the most efficient approach.

[10] This is also referred to as the two-parameter Pareto.

Exercise 6-70 Suppose that the random variable X is distributed according to the Pareto type I distribution with parameters $\alpha = 5$ and $\beta = 2$.

(a) Find $\Pr(3 < X < 4)$.

(b) Find $E[X]$.

(c) Find $E[7X - 3X^2]$.

Exercise 6-71 For the Pareto type I random variable with parameters $\alpha = 3.5$ and $\beta = 4$, find the following:

(a) $E[X]$

(b) $Var[X]$

(c) The median of X.

(d) $\Pr(3 \le X \le 9)$

Exercise 6-72 Let X be a loss random variable of Pareto type II with $\alpha = 3$ and $\beta = 5$.

(a) Find the 90th percentile of X.

(b) The payment random variable is based on a deductible of 2. Calculate the expected payment.

Exercise 6-73 Let the density function for the loss random variable X be given by $f(x) = \frac{(3)(2000)^3}{(x+2000)^4}$; $x > 0$.

(a) Find the expected value of X.

(b) Calculate the probability that X is greater than 1000.

(c) Find the median.

(d) The payment random variable is based on a cap of 3000. Calculate the expected payment.

Exercise 6-74 The lifetime in years is a random variable X whose density is $f(x) = \frac{c}{x^4}$; $x > 3$.

(a) Find the expected lifetime.

(b) Calculate $\Pr[X < 6 \mid X > 4]$.

(c) Find the median.

Exercise 6-75 **SOA EXAM P** **Sample Exam Questions** **#61**

An insurer's annual weather-related loss, X, is a random variable with density function

$$f(x) = \begin{cases} \frac{2.5(200)^{2.5}}{x^{3.5}} & \text{for } x > 200 \\ 0 & \text{otherwise} \end{cases}.$$

Calculate the difference between the 25th and 75th percentiles of X.

(A) 124 (B) 148 (C) 167 (D) 224 (E) 298

Exercise 6-76 SOA EXAM P Sample Exam Questions #55

An insurance company's monthly claims are modeled by a continuous, positive random variable X, whose probability density function is proportional to $(1+x)^{-4}$ where $0<x<\infty$. Determine the company's expected monthly claims.

(A) $\frac{1}{6}$ (B) $\frac{1}{3}$ (C) $\frac{1}{2}$ (D) 1 (E) 3

Exercise 6-77 SOA EXAM P Sample Exam Questions #51

A manufacturer's annual losses follow a distribution with density function,

$$f(x) = \frac{(2.5)(0.6)^{2.5}}{x^{3.5}}; \quad 0.6<x<\infty$$

To cover its losses, the manufacturer purchases an insurance policy with an annual deductible of 2. What is the mean of the manufacturer's annual losses not paid by the insurance policy? (Hint: Losses NOT paid is equivalent to a cap of 2.)

(A) .84 (B) .88 (C) .93 (D) .95 (E) 1.00

6.7.3 THE WEIBULL DISTRIBUTION

The hazard rate function for the ***Weibull distribution*** is given by $\mu(t)=\alpha\beta t^{\alpha-1}$ for $\alpha>1$, $\beta>0$, and $t>0$. Since the hazard increases as t increases, it leads to more realistic age-at-death random variables and is often used in life contingency examples.

Properties of the Weibull Distribution

Let X be a random variable on the interval $[0,\infty)$ with hazard rate given by $\mu(t)=\alpha\beta t^{\alpha-1}$ $\alpha>1$, $\beta>0$ and $0\le t<\infty$.

(1) The CDF of X is given by $F(x) = 1-e^{-\beta x^{\alpha}}$; for $0\le x<\infty$.

(2) The density of X is given by $f(x)=\alpha\beta x^{\alpha-1}e^{-\beta x^{\alpha}}$; for $0\le x<\infty$.

(3) $E[X] = \beta^{-(1/\alpha)}\Gamma\left(1+\frac{1}{\alpha}\right)$.

(4) $Var[X] = \beta^{-(2/\alpha)}\left[\Gamma\left(1+\frac{2}{\alpha}\right)-\Gamma\left(1+\frac{1}{\alpha}\right)^2\right]$

Proof

For (1), we have $\int_0^x \alpha\beta t^{\alpha-1}\,dt=\beta x^{\alpha}$, and so $F(x)=1-e^{-\beta x^{\alpha}}$. For (2), the density function for X follows from differentiating the CDF found in (1). For (3) and (4) we first derive a general expression for the kth moment.

$$E[X^k] = \int_0^{\infty} x^k(\alpha\beta x^{\alpha-1}e^{-\beta x^{\alpha}})\,dx = \overbrace{\beta\int_0^{\infty} x^k e^{-\beta x^{\alpha}}(\alpha x^{\alpha-1}dx) = \beta\int_0^{\infty} u^{(k/\alpha)}e^{-\beta u}\,du}^{\text{substitution } u=x^{\alpha},\ du=\alpha x^{\alpha-1}dx}.$$

The last integral is of the Gamma type with parameters α' and β', where,

$$\alpha' - 1 = \frac{k}{\alpha} \text{ and } \beta' = \frac{1}{\beta}.$$

Therefore,

$$E[X^k] = \beta \cdot (\beta')^{\alpha'} \cdot \Gamma(\alpha') = \beta \cdot \left(\frac{1}{\beta}\right)^{1+(k/\alpha)} \cdot \Gamma\left(1 + \frac{k}{\alpha}\right) = \beta^{-(k/\alpha)} \cdot \Gamma\left(1 + \frac{k}{\alpha}\right).$$

The formulas for the mean and the variance of X now follow easily.

Exercise 6-78 Let X be a Weibull random variable with parameters $\alpha = 3$ and $\beta = 2$.
(a) Find $\Pr(X > 1)$.
(b) Write an expression for the expected value of X.

6.8 CHAPTER 6 SAMPLE EXAMINATION

1. The number of ounces of coffee that a patron of Viola's restaurant consumes is modeled by a random variable B that has a uniform distribution on the interval $[0, 200]$.

   (a) Determine the measures of central tendency (mean, median, mode, and midrange) for the random variable B.
   (b) Determine measures of spread (range, inter-quartile range, variance, and standard deviation).

2. Suppose that Dr. Goodgrader fails $p = 35\%$ of his students. She now has a class of 200 students. Use the normal approximation to the binomial distribution to estimate the probability that she fails 66 or fewer students.

3. A random variable A is assumed to have an exponential distribution with first quartile equal to 100. Find the third quartile.

4. Suppose that the random variable C has a Gamma distribution with parameters $\alpha = 4$ and $\beta = 6$. Find $E[C^5]$.

5. A random variable D has density function proportional to $x^3 \cdot (1 - x)$ on the interval $0 \le x \le 1$. Find the mode of D.

6. Suppose that L has a lognormal distribution with parameters $\mu = 5.1$ and $\sigma = 1.2$. Determine the median of L.

7. David went on 64 sales trips last year. For each trip the sales revenue is a normal random variable with a mean of 7 and a standard deviation of 3. Assuming the sales revenue random variables are independent, calculate the 90th percentile of annual sales revenue resulting from the 64 trips.

8. The IQ of actuarial science majors is assumed to be normally distributed with mean 112 and standard deviation of 14. In a class of 19 students, find the probability that the mean IQ of all 19 students is greater than 120.

9. **SOA EXAM P Sample Exam Questions #127**
Automobile losses reported to an insurance company are independent and uniformly distributed between 0 and 20,000. The company covers each loss subject to a deductible of 5,000.

   Calculate the probability that the total payout on 200 reported losses is between 1,000,000 and 1,200,000.

   (A) 0.0803 (B) 0.1051 (C) 0.1799 (D) 0.8201 (E) 0.8575

10. The density function for a continuous loss random variable is proportional to $(x+2100)^{-5}$ for $0 \le x < \infty$. The insurance payment is based on a deductible of 300. Find the expected payment.

11. **SOA Exam P Sample Exam Questions #18**
Two instruments are used to measure the height, h, of a tower. The error made by the less accurate instrument is normally distributed with mean 0 and standard deviation $0.0056h$. The error made by the more accurate instrument is normally distributed with mean 0 and standard deviation $0.0044h$. Assuming the two measurements are independent random variables, what is the probability that their average value is within $0.005h$ of the height of the tower?

    (A) 0.38 (B) 0.47 (C) 0.68 (D) 0.84 (E) 0.90

12. **SOA Exam P Sample Exam Questions #150**
An automobile insurance company issues a one-year policy with a deductible of 500. The probability is 0.8 that the insured automobile has no accident and 0.0 that the automobile has more than one accident. If there is an accident, the loss before application of the deductible is exponentially distributed with mean 3000. Calculate the 95th percentile of the insurance company payout on this policy.

(A) 3466 (B) 3659 (C) 4159 (D) 8487 (E) 8987

13. **SOA Exam P Sample Exam Questions #147**
The amount of a claim that a car insurance company pays out follows an exponential distribution. By imposing a deductible of d, the insurance company reduces the expected claim payment by 10%.

Calculate the percentage reduction on the variance of the claim payment.

(A) 1% (B) 5% (C) 10% (D) 20% (E) 25%

CHAPTER 7

MULTIVARIATE DISTRIBUTIONS

Often we wish to study several random variables simultaneously. For example, throughout the book we have been making use of sums of two or more random variables. We first encountered this in Section 3.6, where we worked with several different random variables arising from the same probability experiment. In this chapter we present a more complete framework for studying these, so-called, ***multivariate distributions***, alternatively called ***jointly distributed*** random variables.

Here, we will take a more general approach than in Chapter 3, allowing us to transcend our earlier reliance on a common underlying probability experiment. An important consideration in our previous encounters with multiple random variables involved independence. We might, for example, be looking at summing random variables arising from observing independent repetitions of the same experiment. An important subtext for our studies here involves the interaction between the random variables, which can range from independence to total dependence. We will be presenting the machinery for quantifying and calculating this.

We will begin first with joint distribution tables for discrete distributions - first introduced in Section 3.6 - defining the concepts of marginal and conditional distributions. We then proceed to adapt these concepts to the case of jointly distributed continuous random variables. While the concepts we encounter can be applied to any number of jointly distributed random variables, we will concentrate mainly on pairs of random variables. Similar results for higher dimensions can be derived by extension of the two-dimensional case.

7.1 JOINT DISTRIBUTIONS FOR DISCRETE RANDOM VARIABLES

In Chapter 3 we defined discrete random variables and generally portrayed their distributions in a tabular format. In Section 3.6 we saw that two discrete random variables arising from the same experiment could be displayed in a 2-dimensional table. For our purposes here we will take the 2-dimensional table, with suitable restrictions, as the definition for a discrete joint distribution of two random variables.

Joint Probability Distribution

Let R be a set of ordered pairs in the plane of the form (x_i, y_j); $i = 1, \ldots, m$, $j = 1, \ldots, n$. Let p be a function on R satisfying:

(1) $p(x_i, y_j) \geq 0$ for all pairs (x_i, y_j).

(2) $\sum_{i=1}^{m}\sum_{j=1}^{n} p(x_i, y_j) = \sum_{j=1}^{n}\sum_{i=1}^{m} p(x_i, y_j) = 1.$

Let X be the random variable taking on values (x_i); $i = 1, \ldots, m$, with probability distribution given by $p_X(x_i) = \sum_{j=1}^{n} p(x_i, y_j)$ and let Y be the random variable taking values (y_j); $j = 1, \ldots, n$, with probability distribution given by $p_Y(y_j) = \sum_{i=1}^{m} p(x_i, y_j)$.

Then X and Y are said to be ***jointly distributed*** on R with probability distribution given by

$$\Pr[X = x_i, Y = y_j] = p(x_i, y_j).$$

The individual random variables X and Y are referred to as the ***marginal distributions***

Notes

(1) Properties (1) and (2) assure that the joint distribution is in fact a probability distribution.

(2) Property (2) also assures that the individual probability distributions for X and Y each sum to one. This is because, in the case of X,

$$\sum_{i=1}^{m} p_X(x_i) = \sum_{i=1}^{m}\left[\sum_{j=1}^{n} p(x_i, y_j)\right] = \sum_{i=1}^{m}\sum_{j=1}^{n} p(x_i, y_j) = 1.$$

A similar argument holds for Y.

(3) It is possible for m and/or n to be infinite, as might be the case in dealing with, for example, Poisson or geometric random variables.

The information for a discrete joint distribution can be neatly summarized in tabular form as follows:

		Y			
		y_1	$\cdots$	y_n	$p_X(x)$
X	x_1	$p(x_1, y_1)$		$p(x_1, y_n)$	$p_X(x_1)$
	$\vdots$				
	x_m	$p(x_m, y_1)$		$p(x_m, y_n)$	$p_X(x_m)$
	$p_Y(y)$	$p_Y(y_1)$		$p_Y(y_n)$	1

The pairs (x_i, y_j) in R constitute the sample space, or outcomes, for a probability experiment. The dark green cells in the table contain the individual probabilities of the outcomes and constitute the joint probability distribution. The marginal distribution $p_X(x_i) = \sum_{j=1}^{n} p(x_i, y_j)$ for X is tabulated along the right margin, as the sums of the probabilities in the corresponding rows. Similarly, the marginal distribution for Y is displayed along the bottom margin, with $p_Y(y_j) = \sum_{i=1}^{m}(x_i, y_j)$, the sum along the corresponding column.

Example 7.1-1 Discrete Joint Distributions

At Simple Choices Insurance Company, customers choose from 3 levels of automobile liability coverage: 1, 3, or 5 million (state law requires at least 1 million). They may also choose from 3 levels of personal liability coverage: 0, 1, or 4 million. Let X be the level of auto liability selected (in millions) and let Y be the level of personal liability selected (again, in millions). Then (X, Y) is a joint probability distribution with $p(x, y)$ representing the probability that a customer chooses the levels (x, y) of coverage. The joint distribution is shown in the following table:

		Y		
		0	**1**	**4**
X	**1**	.10	.05	.15
	3	.05	.20	.25
	5	.15	.00	.05

(a) Verify that this is in fact a joint distribution table.

(b) Calculate $\Pr(X = 3, Y = 4)$.

(c) Write the marginal distributions for X and Y.

Solution

		Y			
		0	**1**	**4**	$p_Y(y)$
X	1	.10	.05	.15	.30
	3	.05	.20	.25	.50
	5	.15	.00	.05	.20
	$p_Y(y)$	.30	.25	.45	1.00

X	$p_Y(y)$
1	.30
3	.50
5	.20
	1.00

Y	$p_Y(y)$
0	.30
1	.25
4	.45
	1.00

(a) We note that all the $p(x_i, y_j) \geq 0$ and sum to 1.00 (lower right hand corner).

(b) $p(3,4) = \Pr(X = 3, Y = 4) = .25$. The marginal distribution for X is obtained by summing the rows, while the marginal for Y is obtained by summing the columns. Marginal distributions can be displayed individually as well, as shown alongside the joint table. ❐

The calculations for expected values, moments, and more general functions of X and Y are straightforward, stemming from the following procedure:

Calculating Expected Values

Let X and Y be jointly distributed with probability distribution

$$p(x_i, y_j);\ i = 1, \ldots\ldots, m,\ j = 1, \ldots, n.$$

If $f(x, y)$ is a real-valued function, then

$$E[f(X,Y)] = \sum_{i=1}^{m}\sum_{j=1}^{n} f(x_i, y_j)p(x_i, y_j) = \sum_{j=1}^{n}\sum_{i=1}^{m} f(x_i, y_j)p(x_i, y_j).$$

It is easy to show that if $f(x, y)$ is a function of either X alone or Y alone, then the expected value can be calculated using the appropriate marginal with the same result.

Example 7.1-2 Expectations with Discrete Joint Distributions

A fair coin is tossed twice. Let X be the number of heads on the first toss. Let Y be the number of heads on the first two tosses.

(a) Calculate the joint distribution table for X and Y.
(b) Using the joint table, calculate the quantities, $E[X]$, $Var[X]$, $E[Y]$, and $Var[Y]$.
(c) Calculate the marginal distributions for X and Y. Use them to recalculate the quantities in (b).
(d) Calculate $E[X+Y]$ and $Var[X+Y]$.
(e) Show that $E[X+Y] = E[X] + E[Y]$, but that $Var[X+Y] \neq Var[X] + Var[Y]$.

Solution

(a) The joint distribution table is easily calculated based on the four equally likely outcomes to the underlying experiment of tossing a coin twice:

		Y		
		0	**1**	**2**
X	**0**	.25	.25	0
	1	0	.25	.25

(b) From the above definition, we take $f(x, y) = x$ and evaluate,

$$E[X] = \sum_{i=1}^{2}\sum_{j=1}^{3} x_i p(x_i, y_j) = \left[0\cdot.25 + 0\cdot.25 + 0\cdot 0\right] + \left[1\cdot 0 + 1\cdot.25 + 1\cdot.25\right] = .50.$$

Similarly, with $f(x, y) = x^2$, we obtain

$$E[X^2] = \left[(0)^2(.25) + (0)^2(.25) + (0)^2(0)\right] + \left[(1^2)(0) + (1^2)(.25) + (1^2)(.25)\right] = 0.50.$$

It follows that $Var[X] = 0.50-(0.50)^2 = 0.25$.

Again, using the definition above with $f(x,y)=y$ and $f(x,y)=y^2$ we find

$$E[Y] = \sum_{i=1}^{2}\sum_{j=1}^{3} y_j p(x_i,y_j)$$
$$= [(0)(.25)+(1)(.25)+(2)(0)]+[(0)(0)+(1)(.25)+(2)(.25)] = 1.0.$$

$$E[Y^2] = \sum_{i=1}^{2}\sum_{j=1}^{3} y_j^2 p(x_i,y_j)$$
$$= \left[(0^2)(.25)+(1^2)(.25)+(2^2)(0)\right]+\left[(0^2)(0)+(1^2)(.25)+(2^2)(.25)\right]=1.5,$$

so that $Var[Y]=1.5-1^2=0.5$.

We chose to sum horizontally first (the inner sum, with respect to *j*), then vertically (the outer sum, with respect to *i*). Clearly, reversing the order of summation between *i* and *j* would produce the same results.

(c) We calculate the marginal distributions of X and Y and display the individual tables for the first and second moments:

		Y			
		0	**1**	**2**	$p_X(x)$
X	**0**	.25	.25	0	0.50
	1	0	.25	.25	0.50
$p_Y(y)$		0.25	0.50	0.25	1.00

X	$p_X(x)$	$E[X]$	$E[X^2]$
0	.5	0	0
1	.5	.5	.5
Total	1.0	.5	.5

Y	$p_Y(y)$	$E[Y]$	$E[Y^2]$
0	.25	0	0
1	.50	.50	.50
2	.25	.50	1.00
Total	1.00	1.00	1.50

Thus, we see that using the marginal distributions produces the same moments for X and Y as the joint distribution, and therefore, of course, the same expectations and variances.

(d) We use $f(x,y)=x+y$ and $f(x,y)=(x+y)^2$ to calculate $E[X+Y]$ and $E\left[(X+Y)^2\right]$.

$$E[X+Y] = \sum_{i=1}^{2}\sum_{j=1}^{3}(x_i+y_j)p(x_i,y_j)$$
$$= [(0+0)(.25)+(0+1)(.25)+(0+2)(0)]$$
$$+[(1+0)(0)+(1+1)(.25)+(1+2)(.25)] = 1.5.$$
$$E\left[(X+Y)^2\right] = \sum_{i=1}^{2}\sum_{j=1}^{3}(x_i+y_j)^2 p(x_i,y_j)$$
$$= \left[(0+0)^2(.25)+(0+1)^2(.25)+(0+2)^2(0)\right]$$
$$+\left[(1+0)^2(0)+(1+1)^2(.25)+(1+2)^2(.25)\right] = 3.5.$$

It follows from the variance formula that $Var[X+Y] = 3.5-1.5^2 = 1.25$.

For (e) we see that $E[X+Y] = 1.5 = .5+1.0 = E[X]+E[Y]$. On the other hand, $Var[X+Y] = 1.25 \neq .75 = .25+.50 = Var[X]+Var[Y]$.

Note

We see that the means and variances of the marginal distributions could be calculated either from the joint distribution (as in (b)), or from the individual marginal distributions (as in (c)). We also make the observation, from the original statement of the problem, that $X \sim Binomial(p=.5, n=1)$ and $Y \sim Binomial(p=.5, n=2)$. Thus, we can calculate the means and variances directly from the standard formulas. Of course, each of these methods is valid; that is to say, all roads may lead to Rome, but some routes are shorter than others. ❐

Example 7.1-3 Expectations

Compute $E\left[2X-\sqrt{Y}\right]$ if the random variables X and Y are discrete with joint probability function as follows:

		Y		
		0	**1**	**4**
	1	.10	.05	.15
X	**3**	.05	.20	.25
	5	.15	.00	.05

Solution

$$E\left[2X-\sqrt{Y}\right] = \left(2\cdot1-\sqrt{0}\right)\cdot.1+\left(2\cdot1-\sqrt{1}\right)\cdot.05+\left(2\cdot1-\sqrt{4}\right)\cdot.15$$
$$+\cdots+\left(2\cdot5-\sqrt{4}\right)\cdot.05 = 4.45.$$ ❐

Example 7.1-4 Infinite Discrete Joint Distribution

Consider random variables X and Y with joint probability distribution $p(x,y) = \frac{e^{-4}3^x}{x! \cdot y!}$ for $x = 0,1,2,\ldots$ and $y = 0,1,2,\ldots$.

(a) Show that $p(x,y)$ is a joint probability distribution.

(b) Find the marginal probabilities $p_Y(y) = \sum_{i=1}^{\infty} p(x_i, y)$.

Solution

As a warm-up computation, $\Pr(X=2, Y=3) = p(2,3) = \frac{e^{-4}3^2}{2! \cdot 3!} = 0.013737$.

The probabilities need to be positive (they are) and sum to one. Recall that $e^{\alpha} = \sum_{n=0}^{\infty} \frac{\alpha^n}{n!} = 1 + \alpha + \frac{\alpha^2}{2!} + \frac{\alpha^3}{3!} + \cdots$.

$$
\begin{aligned}
\sum_{x,y} p(x,y) &= \sum_{x=0}^{\infty}\sum_{y=0}^{\infty} \frac{e^{-4}3^x}{x! \cdot y!} = \sum_{x=0}^{\infty}\left(\frac{e^{-4}3^x}{x!}\sum_{y=0}^{\infty}\frac{1}{y!}\right) \\
&= \sum_{x=0}^{\infty}\left(\frac{e^{-4}3^x}{x!} \cdot e^1\right) \\
&= \sum_{x=0}^{\infty}\left(\frac{e^{-3}3^x}{x!}\right) = e^{-3} \cdot \sum_{x=0}^{\infty}\frac{3^x}{x!} = e^{-3} \cdot e^3 = 1.
\end{aligned}
$$

Similarly,

$$
p_Y(y) = \sum_{x=0}^{\infty} p(x,y) = \sum_{x=0}^{\infty} \frac{e^{-4}3^x}{x! \cdot y!} = \frac{e^{-4}}{y!}\sum_{x=0}^{\infty}\frac{3^x}{x!} = \frac{e^{-4}}{y!} \cdot e^3 = \frac{e^{-1}}{y!}
$$

for all $y = 0,1,2,\ldots$. ❐

Exercise 7-1 Consider random variables X and Y with joint probability distribution, $p(x,y) = \frac{e^{-4}3^x}{x! \cdot y!}$ for $x = 0,1,2,\ldots$ and $y = 0,1,2,\ldots$.

(a) Find the marginal probabilities for X, $p_X(x) = \sum_{j=1}^{\infty} p(x, y_j)$.

(b) Compute the mean of the random variable X.

Example 7.1-5 Joint Distributions

Given that X and Y are discrete random variables with joint probability distribution:

		Y		
		0	**1**	**4**
	1	.10	.05	.15
X	**3**	.05	.20	.25
	5	.15	.00	.05

(a) Calculate the marginal probability functions for X and Y.
(b) Calculate the expected value of X from both the joint distribution and the marginal distribution for X.

Solution
(a) The joint table with marginal distributions displayed is:

		Y			
		0	**1**	**4**	$p_X(x)$
	1	.10	.05	.15	**.30**
X	**3**	.05	.20	.25	**.50**
	5	.15	.00	.05	**.20**
$p_Y(y)$		.30	.25	.45	1.00

Therefore the marginal distributions are:

Y	$p_Y(y)$
0	.30
1	.25
4	.45
	1.00

X	$p_X(x)$
1	.30
3	.50
5	.20
	1.00

(b) Using the joint probability function, we have

$$\begin{aligned} E[X] &= \sum\sum x \cdot p(x,y) \\ &= 1\cdot p(1,0)+1\cdot p(1,1)+1\cdot p(1,4) \\ &\qquad +3\cdot p(3,0)+3\cdot p(3,1)+\cdots+5\cdot p(5,4) \\ &= 1\cdot(.10)+1\cdot(.05)+1\cdot(.15)+3\cdot(.05) \\ &\qquad +3\cdot(.20)+3\cdot(.25)+5\cdot(.15)+5\cdot(.00)+5\cdot(.05) \\ &= \mathbf{1\cdot(.30)+3\cdot(.50)+5\cdot(.20)} \\ &= 2.80. \end{aligned}$$

Using our marginal distribution for X,

$$\begin{aligned} E[X] &= \sum x \cdot p_X(x) \\ &= 1\cdot p_X(1)+3\cdot p_X(3)+5\cdot p_X(5) = \mathbf{1\cdot(.30)+3\cdot(.50)+5\cdot(.20)} = 2.80. \end{aligned}$$ □

Exercise 7-2 Regarding the random variables in Example 7.1-5, compute $E[Y^2]$ using both the joint probability distribution and the marginal probability distribution for Y.

Exercise 7-3 Suppose X and Y are discrete random variables with $p(x,y) = k \cdot (x+2y)$ for $x = 0,1,2,3$ and $y = 1,2$.

(a) Determine the constant k.

(b) Find $E[\sqrt{Y}]$.

Exercise 7-4 **SOA EXAM P** **Sample Exam Questions** **#100**

A car dealership sells 0, 1, or 2 luxury cars on any day. When selling a car, the dealer also tries to persuade the customer to buy an extended warranty for the car. Let X denote the number of luxury cars sold in a given day, and let Y denote the number of extended warranties sold.

$$P(X=0, Y=0) = \tfrac{1}{6} \qquad P(X=1, Y=0) = \tfrac{1}{12}$$

$$P(X=1, Y=1) = \tfrac{1}{6} \qquad P(X=2, Y=0) = \tfrac{1}{12}$$

$$P(X=2, Y=1) = \tfrac{1}{3} \qquad P(X=2, Y=2) = \tfrac{1}{6}$$

What is the variance of X? Author's note: Solve this problem two different ways; (1) using marginal probabilities and (2) directly.

(A) 0.47 (B) 0.58 (C) 0.83 (D) 1.42 (E) 2.58

Exercise 7-5 A cooler has 6 Gatorades®, 2 colas, and 4 waters. You select three beverages from the cooler at random. Let B denote the number of Gatorades® selected and let C denote the number of colas selected. For example, if you grabbed a cola and two waters, then $C = 1$ and $B = 0$.

(a) Construct a joint probability distribution for B and C.

(b) Find the marginal distribution $p_B(b)$.

(c) Compute $E[C]$.

(d) Compute $E[3B - C^2]$.

7.2 CONDITIONAL DISTRIBUTIONS – THE DISCRETE CASE

When working with a joint distribution it may be the case that we know the value of one of the marginal random variables, and seek the distribution of the other conditioned on that knowledge. In Example 7.1-2, the experiment consists of tossing a fair coin twice. Suppose we observe the outcome of the first toss to be heads, and therefore know that $X=1$. How does this affect our perception of *Y*, the number of heads on the first two tosses? Clearly, this changes the distribution of *Y* since, among other things, $Y=0$ is now impossible. The resulting new distribution of *Y* is called the ***conditional distribution***, given $X=1$.

Conditional Distributions - The Discrete Case

If X and Y are discrete random variables with joint probability distribution $p(x_i, y_j);\ i=1,\ldots,m,\ j=1,\ldots,n$, then the ***conditional probability distribution for X given*** $Y=y_j$ is

$$p_X(x_i \mid Y=y_j) = \Pr\left[X=x_i \mid Y=y_j\right] = \frac{\Pr\left[\{X=x_i\}\cap\{Y=y_j\}\right]}{\Pr\left[\{Y=y_j\}\right]} = \frac{p(x_i, y_j)}{p_Y(y_j)}$$

for $i=1,\ldots,m$.

Similarly, the ***conditional probability distribution for Y given*** $\boldsymbol{X=x_i}$ is

$$p_Y(y_j \mid X=x_i) = \frac{p(x_i, y_j)}{p_X(x_i)}$$

for $j=1,\ldots,n$.

This is most easily understood in terms of the joint distribution table, where the conditional distribution of *X* given $Y=y_j$ is represented by the jth column (see below). Since $p_Y(y_j)$ is the sum of the jth column, it follows that

$$\sum_{i=1}^{m} p(x_i \mid Y=y_j) = \sum_{i=1}^{m} \frac{p(x_i, y_j)}{p_Y(y_j)} = \frac{1}{p_Y(y_j)} \sum_{i=1}^{m} p(x_i, y_j) = 1,$$

so that this does in fact constitute a probability distribution. Analogously, the conditional distribution of *Y* given $X=x_i$ is represented by the ith row of the table.

				Y			
		y_1	$\cdots$	y_j	$\cdots$	y_n	$p_X(x)$
	x_1	$p(x_1, y_1)$	$\cdots$	$p(x_1, y_j)$	$\cdots$	$p(x_1, y_n)$	$p_X(x_1)$
X	$\vdots$		$\cdots$		$\cdots$		
	x_m	$p(x_m, y_1)$	$\cdots$	$p(x_m, y_j)$	$\cdots$	$p(x_m, y_n)$	$p_X(x_m)$
	$p_Y(y)$	$p_Y(y_1)$		$p_Y(y_j)$		$p_Y(y_n)$	1

Example 7.2-1 Conditional Distribution

A fair coin is tossed twice. Let X be the number of heads on the first toss. Let Y be the number of heads on the first two tosses. Calculate,

(a) The conditional distribution of Y given that $X=1$.
(b) The expected value of Y given that $X=1$.

Solution
(a) From Example 7.1-2, the joint distribution table is:

		Y			
		0	**1**	**2**	$p_X(x)$
X	**0**	.25	.25	0	0.50
	1	0	.25	.25	0.50
$p_Y(y)$		.25	.50	.25	1.00

The conditional distribution of Y given that $X=1$ is represented by the high-lighted row. We display this as:

Y	$p(1,y)$	$p(Y\mid X=1)$
0	0	0
1	.25	.50
2	.25	.50
Totals	.50	1.00

The $p(Y\mid X=1)$ column is the conditional distribution of Y given that $X=1$. It is just the second column (copied from the $X=1$ row of the joint table) scaled by dividing by $p_X(1)=.50$.

(b) We can calculate $E[Y\mid X=1]$ in tabular form:

Y	$p(Y\mid X=1)$	$Y\cdot p(Y\mid X=1)$
0	0	0
1	.50	.50
2	.50	1.00
Totals	1.00	1.50

Thus, $E[Y\mid X=1] \;=\; 0\cdot 0+1\cdot .50+2\cdot(.50) \;=\; 1.50.$ ❐

Of course, in this simple example we know at once that given $X=1$, Y is equally likely to be one or two, depending on whether the second toss is tails or heads. So the expected value of 1.5 is hardly unexpected, so to speak.

Example 7.2.2 Conditional Distribution (Example 7.1-5 continued)

Given that X and Y are discrete random variables with joint probability distribution:

		Y		
		0	**1**	**4**
	1	.10	.05	.15
X	**3**	.05	.20	.25
	5	.15	.00	.05

(a) Calculate the conditional distribution of X given $Y=4$.
(b) Calculate the expected value of X given $Y=4$.

Solution

(a) We display the calculations in tabular form beginning with the $p(x,4)$ column taken directly from the joint distribution table:

X	$p(x,4)$	$p(X\mid Y=4)$	$X \cdot p(X\mid Y=4)$
1	.15	$\frac{.15}{.45}=\frac{3}{9}$	$\frac{3}{9}$
3	.25	$\frac{.25}{.45}=\frac{5}{9}$	$\frac{15}{9}$
5	.05	$\frac{.05}{.45}=\frac{1}{9}$	$\frac{5}{9}$
Totals:	0.45	$\frac{9}{9}=1$	$\frac{23}{9}$

(b) So $E[X\mid Y=4] = \frac{23}{9}$.

A shortcut form of the calculation makes use of the fact that the $p(x,4)$ can be scaled by any number k to eliminate decimals or fractions, provided one remembers to divide by the column total at the conclusion. For example, scaling by 100 to eliminate decimals yields:

X	$p(x,4)\cdot k$	$X \cdot p(x,4)\cdot k$
1	15	15
3	25	75
5	5	25
Totals:	45	115

Thus, $E[X\mid Y=4] = \frac{115}{45} = \frac{23}{9}$. ❐

Exercise 7-6 Given discrete random variables X and Y with joint probability function:

		Y			
		0	1	2	3
X	1	.17	.05	.11	.04
	2	.00	.13	.23	.06
	3	.10	.01	.02	.08

(a) Find $p(x|Y=3)$.
(b) Find $E[X|Y=3]$.
(c) Find $Var[X|Y=3]$.
(d) Find $Var[X]$.
(e) Find $E[Y|X=2]$.
(f) Find $E[\sqrt{Y}|X=3]$.

Exercise 7-7 SOA/CAS Course 1 2000 Sample Examination #37

An insurance contract reimburses a family's automobile accident losses up to a maximum of two accidents per year. The joint probability distribution for the number of accidents of a three person family (X,Y,Z) is

$$p(x,y,z) = k(x+2y+z)$$

Where $x=0,1,\ y=0,1,2,\ z=0,1,2$ and x, y, and z are the number of accidents incurred by X, Y, and Z respectively. Determine the expected number of un-reimbursed accident losses given that X is not involved in any accidents.

(A) $\frac{5}{21}$ (B) $\frac{1}{3}$ (C) $\frac{5}{9}$ (D) $\frac{46}{63}$ (E) $\frac{7}{9}$

Exercise 7-8 SOA EXAM P Sample Exam Questions #114

A diagnostic test for the presence of a disease has two possible outcomes: 1 for disease present and 0 for disease not present. Let X denote the disease state of a patient, and let Y denote the outcome of the diagnostic test. The joint probability function of X and Y is given by:

$$P(X=0,Y=0) = 0.800 \qquad P(X=1,Y=0) = 0.050$$
$$P(X=0,Y=1) = 0.025 \qquad P(X=1,Y=1) = 0.125$$

Calculate $Var(Y|X=1)$.

(A) 0.13 (B) 0.15 (C) 0.20 (D) 0.51 (E) 0.71

Exercise 7-9 SOA EXAM P Sample Exam Questions #131

Let N_1 and N_2 represent the number of claims submitted to a life insurance company in April and May, respectively. The joint probability functions of N_1 and N_2 is

$$p(n_1,n_2) = \begin{cases} \frac{1}{3}\left(\frac{1}{4}\right)^{n_1-1} e^{-n_1}(1-e^{-n_1})^{n_2-1} & \text{for } n_1 = 1,2,3,\ldots \text{ and } n_2 = 1,2,3,\ldots \\ 0 & \text{otherwise} \end{cases}$$

Calculate the expected number of claims that will be submitted to the company in May if exactly 2 claims were submitted in April.

(A) $\frac{3}{16}(e^2-1)$ (B) $\frac{3}{16}e^2$ (C) $\frac{3e}{4-e}$ (D) e^2-1 (E) e^2

7.3 INDEPENDENCE – DISCRETE CASE

Recall that two events E and F are ***independent*** if $\Pr[E \mid F] = \Pr[E]$, or equivalently, $\Pr[E \cap F] = \Pr[E] \cdot \Pr[F]$. We have previously extended the concept of independence to random variables, at least in an informal sense, by requiring all events related to one random variable be independent of events related to the other. We now provide a more rigorous definition for the independence of two random variables by invoking the joint and marginal probability distributions.

Independence for Joint Discrete Distributions

Let X and Y be discrete random variables with joint probability distribution $p(x_i, y_j);\ i = 1,\ldots,m,\ j = 1,\ldots,n$. We say X and Y are ***independent*** if for all (x_i, y_j) we have

$$p(x_i, y_j) = p_X(x_i) \cdot p_Y(y_j).$$

Notes

(1) In words, the joint probability distribution can be ***decomposed*** as the product of the marginal distributions.

(2) In terms of the joint table, this means that all the numbers on the inside are the product of the associated numbers on the margin.

(3) This definition of independence is equivalent to saying that any conditional distribution of X for a given value of Y will be the same as the marginal distribution for X (in other words, independent of the given value of Y).

To see this last statement, we assume X and Y are independent. Then

$$p(x_i \mid Y = y_j) = \frac{p(x_i, y_j)}{p_Y(y_j)} = \frac{\overbrace{p_X(x_i) \cdot p_Y(y_j)}^{\text{independence used here}}}{p_Y(y_j)} = p_X(x_i) \quad \text{for} \quad i = 1,\ldots,m.$$

A very useful consequence of the decomposition property above is the following:

Independence and Expectation

Let X and Y be discrete independent random variables with joint probability distribution $p(x_i, y_j);\ i = 1, \ldots, m,\ j = 1, \ldots, n$. Then,

$$E[X \cdot Y] = E[X] \cdot E[Y].$$

Proof

Using the above decomposition property we have

$$E[X \cdot Y] = \sum_{i=1}^{m} \sum_{j=1}^{n} x_i y_j p(x_i, y_j)$$

$$= \sum_{i=1}^{m} \sum_{j=1}^{n} x_i y_j p_X(x_i) \cdot p_Y(y_j) = \sum_{i=1}^{m} x_i p_X(x_i) \cdot \sum_{j=1}^{n} y_j p_Y(y_j) = E[X] \cdot E[Y].$$

Example 7.3-1 Independence (Example 7.1-5 continued)

Given that X and Y are discrete random variables with joint probability distribution:

		Y		
		0	**1**	**4**
	1	.10	.05	.15
X	**3**	.05	.20	.25
	5	.15	.00	.05

Determine whether or not X and Y are independent.

Solution

From Example 7.1-5 we have the joint and marginal distribution table:

		Y			
		0	**1**	**4**	$p_X(x)$
	1	.10	.05	.15	.30
X	3	.05	.20	.25	.50
	5	.15	.00	.05	.20
	$p_Y(y)$	.30	.25	.45	1.00

If any inside cell is not the product of the associated numbers on the margins then X and Y fail to be independent. In this case we need only check the upper left-hand cell where $X = 1$ and $Y = 0$.

$$p(1,0) = .10 \neq .09 = .30 \cdot .30 = p_X(1) \cdot p_Y(0).$$

Thus, X and Y are **not** independent (they are ***dependent*** random variables).

Alternate Solution

Computing expected values we find $E[X]=2.8$, $E[Y]=2.05$, and $E[XY]=5.25$. We know that the random variables are dependent since $E[X\cdot Y] \neq E[X]\cdot E[Y]$. ❒

Example 7.3-2 Independence

Determine whether or not X and Y are independent, with the joint distribution given by:

		Y: 0	Y: 1	Y: 4
X	2	$\frac{1}{4}$	$\frac{1}{3}$	$\frac{1}{12}$
	3	$\frac{1}{8}$	$\frac{1}{6}$	$\frac{1}{24}$

Solution

The joint and marginal distribution table is:

		Y: 0	Y: 1	Y: 4	$p_X(x)$
X	2	$\frac{1}{4}$	$\frac{1}{3}$	$\frac{1}{12}$	$\frac{2}{3}$
	3	$\frac{1}{8}$	$\frac{1}{6}$	$\frac{1}{24}$	$\frac{1}{3}$
	$p_Y(y)$	$\frac{3}{8}$	$\frac{1}{2}$	$\frac{1}{8}$	1

To verify independence we need to establish that ***each*** of the six (white) cells of the joint distribution is the product of the corresponding numbers on the margins. A little arithmetic with fractions will demonstrate this to be the case. ❒

Exercise 7-10 Suppose that X and Y are independent discrete random variables with marginal distributions provided.

		Y: 1	Y: 2	Y: 3	$p_X(x)$
X	−1				.4
	0				.5
	5				.1
	$p_Y(y)$	.6	.3	.1	1.00

Find $E[X\cdot Y]$.

7.4 COVARIANCE AND CORRELATION

We have seen that joint discrete distributions may or may not lead to independent marginal distributions. We now describe a tool that allows us to quantify the degree to which the marginal distributions may be dependent upon one another. The concept that describes the extent to which two random variables are interrelated is called the ***covariance***.

Suppose for example that X represents the height of a person and Y represents the weight of the person. One would conjecture that there is a *positive* relationship between the random variables. That is, if the person is taller than average, then he is more likely to weigh more than average. Using the definition below, we shall see that this leads to a positive covariance. Alternatively, suppose that X represents the number of children in a family and Y represents the disposable income. One would conjecture that there is a *negative* relationship between these random variables. That is, the larger the family the smaller the disposable income. In this instance we would then find a negative value for the covariance.

Covariance

If X and Y are jointly distributed random variables then the ***covariance between X and Y*** is defined to be

$$Cov[X,Y] = E[(X-\mu_X)(Y-\mu_Y)].$$

Note

If the values of X are larger than its mean μ_X at the same time that the values of Y are larger than its mean μ_Y, then the expression $(X-\mu_X)(Y-\mu_Y)$ will be positive. Similarly if the values of X and Y are simultaneously less than their respective means, then the expression $(X-\mu_X)(Y-\mu_Y)$ will still be positive. Such a situation would produce a positive covariance.

Conversely, if X and Y tend to run in opposite directions, then when one is above average the other will tend to be below average, leading to a negative number for $(X-\mu_X)(Y-\mu_Y)$. This situation would produce a negative covariance.

Just as with variance calculations, there is a shortcut formula for computing covariance. We will refer to this as the ***covariance formula***.

Covariance Formula

$$Cov[X,Y] = E[X \cdot Y] - E[X] \cdot E[Y].$$

Proof

$$\begin{aligned} Cov[X,Y] &= E\big[(X-\mu_X)(Y-\mu_Y)\big] \\ &= E[XY - X\mu_Y - Y\mu_X + \mu_X\mu_Y] \\ &= E[XY] - \mu_Y \cdot E[X] - \mu_X \cdot E[Y] + \mu_X \cdot \mu_Y \\ &= E[XY] - E[Y]\cdot E[X] - E[X]\cdot E[Y] + E[X]\cdot E[Y] \\ &= E[XY] - E[X]\cdot E[Y]. \end{aligned}$$

Notes

(1) From both the definition and the covariance formula we see that

$$Cov\big[X,X\big] = Var[X].$$

(2) In this computation, we made no reference as to whether the random variables are discrete or continuous or of mixed type. The calculation is completely general.

(3) If $Y = a \cdot X + b$ then it follows immediately from the definition of covariance that

$$\boxed{Cov[X,Y] = Cov[X, a\cdot X + b] = a\cdot Var[X].}$$

(4) If X and Y are independent, then $Cov[X,Y] = 0$. This is an immediate consequence of the covariance formula and the fact that $E[X\cdot Y] = E[X]\cdot E[Y]$ when X and Y are independent. We will see in Example 7.4-2 below that a covariance of zero does **not** necessarily imply independence.

Example 7.4-1 Covariance Calculation

Find the covariance between X and Y with joint probability function:

		Y		
		0	**1**	**4**
	1	.10	.05	.15
X	**3**	.05	.20	.25
	5	.15	.00	.05

Solution

From Example 7.1-5 we know that $E\big[X\big] = 2.80$ and can easily calculate $E\big[Y\big] = 2.05$. Then,

$$\begin{aligned} Cov(X,Y) &= E[X\cdot Y] - E[X]\cdot E[Y] \\ &= \{1\cdot 0\cdot(.10) + 1\cdot 1\cdot(.05) + 1\cdot 4\cdot(.15) + \cdots + 5\cdot 4\cdot(.05)\} - [2.80]\cdot[2.05] \\ &= 5.25 - 5.74 \\ &= -.49. \end{aligned}$$

❐

Example 7.4-2 Covariance and Independence

A game consists of tossing a fair coin twice. If the outcome is either two heads or two tails then the player wins one dollar. Otherwise, the player wins zero. Let X denote the number of heads resulting from tossing the coin twice and let Y represent the payoff.

(a) Work out the joint distribution table for X and Y.
(b) Calculate the covariance of X and Y.
(c) Determine whether or not X and Y are independent.

Solution

From the statement we have that X has the binomial distribution with parameters $n=2$, $p=\frac{1}{2}$ and we have the conditional distributions of Y given $X=0,1$, or 2. From this we can work out the joint probability distribution for X and Y. For example,

$$p(0,0) = p_Y(0|X=0)\cdot p_X(0) = \Pr[Y=0|X=0]\cdot\Pr[X=0] = (0)\cdot\left(\tfrac{1}{4}\right) = 0.$$

Similarly,

$$p(0,1) = p_Y(1|X=0)\cdot p_X(0) = \Pr[Y=1|X=0]\cdot\Pr[X=0] = (1)\cdot\left(\tfrac{1}{4}\right) = \frac{1}{4}.$$

Reasoning in this fashion we see the complete joint and marginal distribution table is given by,

		Y		
		0	**1**	$p_X(x)$
X	**0**	0	$\frac{1}{4}$	$\frac{1}{4}$
	1	$\frac{1}{2}$	0	$\frac{1}{2}$
	2	0	$\frac{1}{4}$	$\frac{1}{4}$
	$p_Y(y)$	$\frac{1}{2}$	$\frac{1}{2}$	1

We now have that $E[X]=1$ and $E[Y]=\frac{1}{2}$. Also, the only non-zero contribution to $E[XY]$ is when $X=2, Y=1$, yielding $E[XY]=(2)(1)\left(\frac{1}{4}\right)=\frac{1}{2}$. Thus,

$$Cov[X,Y] = E[XY]-E[X]\cdot E[Y] = 0.$$

Finally, since $p(0,0)=0\neq\left(\frac{1}{2}\right)\left(\frac{1}{4}\right)=\frac{1}{8}=p_X(0)p_Y(0)$, we see that while X and Y have zero covariance, they are not independent. ❐

Exercise 7-11 Use the definition of covariance to prove the result in Note 3 above: If $Y = a\cdot X+b$, then $Cov[X,Y] = Cov[X,a\cdot X+b] = a\cdot Var[X]$.

Exercise 7-12 Roll a pair of fair 3-sided dice. Let *f* denote the number of dots on the first die and let *T* denote the total number of dots. Determine the $Cov(F,T)$.

Exercise 7-13 Suppose $X \sim Binomial(n=10, p=.2)$. Calculate the covariance of X and $2X$.

It is often necessary to calculate the variance of linear combinations of random variables. We have done this previously, calculating the variance of sums of *independent* random variables, in which case the variance of a sum is the sum of the variances. We will now treat the general case for the variance of a linear combination of two jointly distributed random variables. A more general treatment of variances and covariances for sums of random variables will be given in Chapter 8.

The Variance of a Linear Combination

Let X and Y be jointly distributed (discrete, continuous, or mixed) random variables. Then,

$$Var[aX+bY] = a^2 \cdot Var[X] + b^2 \cdot Var[Y] + 2a \cdot b \cdot Cov[X,Y].$$

If X and Y are independent, then

$$Var[aX+bY] = a^2 \cdot Var[X] + b^2 \cdot Var[Y].$$

Proof

We have $E[aX+bY] = a\mu_X + b\mu_Y$, and so, using the definition of variance,

$$\begin{aligned} Var[aX+bY] &= E\left[\{(aX+bY)-(a\mu_X+b\mu_Y)\}^2\right] \\ &= E\left[\{a(X-\mu_X)+b(Y-\mu_Y)\}^2\right] \\ &= E\left[a^2(X-\mu_X)^2 + 2ab(X-\mu_X)(Y-\mu_Y) + b^2(Y-\mu_Y)^2\right] \\ &= a^2 Var[X] + b^2 Var[Y] + 2ab Cov[X,Y]. \end{aligned}$$

If X and Y are independent, then $Cov[X,Y]=0$, so the second statement follows immediately.

Exercise 7-14 Suppose $E[X]=-3$, $E[X^2]=13$, $Var[Y]=20$, $E[Y]=4$, and $E[XY]=7$. Find $Var[5X-9Y]$.

Exercise 7-15 SOA EXAM P Sample Exam Questions #99

An insurance policy pays a total medical benefit consisting of two parts for each claim. Let X represent the part of the benefit that is paid to the surgeon, and let Y represent the part that is paid to the hospital. The variance of X is 5000, the variance of Y is 10,000, and the variance of the total benefit, $X+Y$, is 17,000.

Due to increasing medical costs, the company that issues the policy decides to increase X by a flat amount of 100 per claim and to increase Y by 10% per claim.

Calculate the variance of the total benefit after these revisions have been made.

(A) 18,200 (B) 18,800 (C) 19,300 (D) 19,520 (E) 20,670

Hint: What is $Var[(X+100)+1.1Y]$?

Exercise 7-16 SOA EXAM P Sample Exam Questions #107

Let X denote the size of a surgical claim and let Y denote the size of the associated hospital claim. An actuary is using a model in which $E(X)=5$, $E(X^2)=27.4$, $E(Y)=7$, $E(Y^2)=51.4$, and $Var(X+Y)=8$.

Let $C_1=X+Y$ denote the size of the combined claims before the application of a 20% surcharge on the hospital portion of the claim, and let C_2 denote the size of the combined claims after the application of that surcharge. Calculate $Cov(C_1, C_2)$.

(A) 8.80 (B) 9.60 (C) 9.76 (D) 11.52 (E) 12.32

Exercise 7-17 Find the covariance between X and Y with joint probability function:

		Y			
		0	**1**	**2**	**3**
	1	.17	.05	.11	.04
X	**2**	.00	.13	.23	.06
	3	.10	.01	.02	.08

Exercise 7-18 What is $Cov[3X,-7Y]$ in terms of $Cov[X,Y]$?

Exercise 7-19 A box contains 3 slips of paper labeled 1, 2, and 3, respectively. Two slips are drawn at random. Let X be the number on the first slip drawn and let Y be the number on the second slip drawn.

(a) Find $Cov(X,Y)$ under the assumption that the sample is drawn with replacement.

(b) Find $Cov(X,Y)$ under the assumption that the sample is drawn without replacement.

Exercise 7-20 A box contains 3 red balls and 2 black balls. A sample of size 2 is drawn without replacement. Let X be the number of red and let Y be the number of black.

(a) Find $Var[X]$.

(b) Find $Var[Y]$.

(c) Find $Cov(X,Y)$.

(d) Find $Var[X+Y]$.

Exercise 7-21 A box contains 4 balls numbered 1 through 4. One ball is selected at random.

Let $X=1$ if ball 1 or ball 2 is selected, 0 otherwise;
Let $Y=1$ if ball 1 or ball 3 is selected, 0 otherwise;
Let $Z=1$ if ball 1 or ball 4 is selected, 0 otherwise.

(a) Find the joint distribution of X and Y. Are X and Y independent?

(b) Calculate $Var[X+Y+Z]$.

(c) Calculate $E[X \cdot Y \cdot Z]$.

(d) Calculate $E[X] \cdot E[Y] \cdot E[Z]$.

(e) Are X, Y, and Z independent?

The magnitude of the covariance is affected by the units used to scale the random variables X and Y. In order to come up with a more absolute, unit-free, measure we can standardize the calculation by dividing by the product of the standard deviations. The resulting quantity is called the correlation coefficient.

Correlation

If X and Y are jointly distributed random variables then the ***correlation coefficient between X and Y*** is defined to be

$$\rho = \frac{Cov[X,Y]}{\sigma_X \cdot \sigma_Y}.$$

We assume, of course, that the standard deviations of X and Y are non-zero so that the correlation coefficient is well-defined. In the following we show that the correlation is always a number between negative one and positive one, and that if one random variable is a linear transformation of the other random variable, then the correlation coefficient is exactly equal to -1 or 1.

Properties of Correlation

(1) For any random variables X and Y, the correlation coefficient ρ satisfies,

$$-1 \leq \rho \leq 1.$$

(2) Let $Y = aX + b;\ a \neq 0$. Then $\rho = \begin{cases} 1 & \text{if } a > 0 \\ -1 & \text{if } a < 0 \end{cases}.$

Proof

Let $W = X - \lambda \cdot Y$, where λ is a real number whose value will be chosen presently. The variance of every random variable is non-negative. Using the formula for the variance of a linear expression, we have

$$0 \leq Var[W] = Var[X - \lambda Y] = Var[X] + \lambda^2 \cdot Var[Y] - 2 \cdot \lambda \cdot Cov[X,Y].$$

Now we substitute $\lambda = \frac{Cov[X,Y]}{Var[Y]}$, which shows that

$$0 \leq Var[X] + \left(\frac{Cov[X,Y]}{Var[Y]}\right)^2 \cdot Var[Y] - 2\left(\frac{Cov[X,Y]}{Var[Y]}\right) Cov[X,Y]$$

$$= Var[X] - \frac{(Cov[X,Y])^2}{Var[Y]}.$$

Rearranging this and dividing both sides of the inequality by $Var[X]$ shows $\frac{(Cov[X,Y])^2}{Var[X] \cdot Var[Y]} \leq 1$, and the first result follows upon taking square roots of both sides.

For property (2) we note that $Var[Y] = a^2 \cdot Var[X]$ which implies $\sigma_Y = |a| \cdot \sigma_X$. Next, $Cov[X,Y] = Cov[X, aX + b] = a \cdot \sigma_X^2$. Therefore,

$$\rho = \frac{Cov[X,Y]}{\sigma_X \sigma_Y} \overset{\text{def'n}}{=} \frac{a \cdot \sigma_X^2}{|a| \cdot \sigma_X^2} = \frac{a}{|a|} = \begin{cases} 1 & \text{if } a > 0 \\ -1 & \text{if } a < 0 \end{cases}.$$

Exercise 7-22 Find the correlation coefficient between X and Y with joint probability function:

		Y		
		–1	**1**	**3**
	1	.10	.05	.15
X	**2**	.25	.10	☺
	3	.15	.05	.05

Exercise 7-23 What is the correlation between a random variable X and a second random variable $Y = -3X$? That is, find $Cov[X,Y] = Cov[X,-3X]$.

Exercise 7-24 SOA Course 1 November 2000 #17

A stock market analyst has recorded the daily sales revenue for two companies over the last year and displayed them in the histograms below.

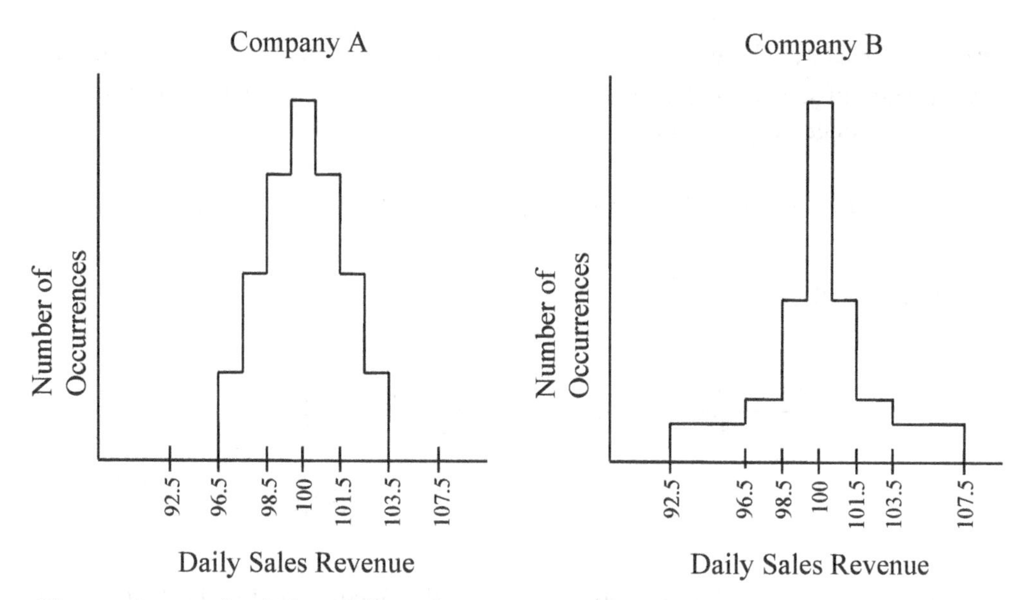

The analyst noticed that daily sales revenue above 100 for Company A was always accompanied by daily sales revenue below 100 for Company B, and vice versa. Let X denote the daily sales revenue for Company A and let Y denote the daily sales revenue for Company B, on some future day. Assuming that for each company the daily sales revenues are independent and identically distributed, which of the following is true?

(A) $Var(X) > Var(Y)$ and $Var(X+Y) > Var(X)+Var(Y)$.

(B) $Var(X) > Var(Y)$ and $Var(X+Y) < Var(X)+Var(Y)$.

(C) $Var(X) > Var(Y)$ and $Var(X+Y) = Var(X)+Var(Y)$.

(D) $Var(X) < Var(Y)$ and $Var(X+Y) > Var(X)+Var(Y)$.

(E) $Var(X) < Var(Y)$ and $Var(X+Y) < Var(X)+Var(Y)$.

Hints: The statement that *The analyst noticed that daily sales revenue above 100 for Company A was always accompanied by daily sales revenue below 100 for Company B, and vice versa* means that $Cov(X,Y) < 0$. Also, the more spread out the data is from the mean, the greater the variance (and standard deviation).

7.5 JOINT DISTRIBUTIONS FOR CONTINUOUS RANDOM VARIABLES

In this section we extend all of the ideas discussed in Section 7.1 for discrete joint distributions to the continuous case. As we saw, a discrete joint distribution lives on a domain consisting of a two-by-two grid of cells. In the continuous analog to the discrete case the joint distribution has a two-variable density function that lives on a domain consisting of a region R in the $x-y$ plane. When the region R is rectangular, of the form $R=[a,b]\times[c,d]$, the analogy to the discrete case is fairly straightforward, with integration replacing summation in most of the formulas of Section 7.1.

Things become trickier when the region R is not a simple rectangle. While in principle the shape of R can be completely arbitrary, for practical purposes we restrict ourselves to regions that are bounded by curves given by ordinary calculus-type functions of the form $y = h(x)$, or $x = g(y)$. As we shall see shortly, we can then employ the double or multiple integral techniques encountered in multivariate calculus.

Joint Density Function on *R*

Let R be a region in the $x-y$ plane and let $f(x,y)$ be a function on the plane satisfying,

(1) $f(x,y) \geq 0$ for all x and y,

(2) $f(x,y) \equiv 0$ outside of R, and

(3) $\iint_R f(x,y)\,dA = \int_{-\infty}^{\infty}\int_{-\infty}^{\infty} f(x,y)\,dx\,dy = \int_{-\infty}^{\infty}\int_{-\infty}^{\infty} f(x,y)\,dy\,dx = 1.$

Let X be the random variable with density function defined by,

$$f_X(x) = \int_{-\infty}^{\infty} f(x,y)\,dy$$

for all values of x.

Let Y be the random variable with density function defined by,

$$f_Y(y) = \int_{-\infty}^{\infty} f(x,y)\,dx$$

for all values y.

Then X and Y are said to be ***jointly distributed*** on R with joint density function given by $f(x,y)$, and the individual random variables X and Y are called the ***marginal distributions***, with ***marginal density functions*** $f_X(x)$ and $f_Y(y)$, respectively.

Notes

(1) In (3), the probability increment $f(x,y)dA$, is analogous to the incremental probability $f(x)dx$ in the single variable case. We use dA to denote an increment of area in the plane. The double integral is evaluated as an iterated integral, and $dA = dx \cdot dy$ or $dA = dy \cdot dx$ depending on the order chosen for the calculation.

(2) The marginal density function for X, $f_X(x)$, is a function of x alone. Similarly, the marginal density function for Y, $f_Y(y)$, is a function of y alone.

(3) There are several ways to conceptualize the probability density function, $f(x,y)$. For one, it can be viewed as a 3-dimensional graph over the region R in the plane. Since $f(x,y) \geq 0$, the graph constitutes a "roof" over the floor R, and $\iint_R f(x,y)\,dA$ is the volume of the space between roof and floor. Then increments of probability, $f(x,y)dA$,

can be visualized as narrow columns of incremental volume, and, because of (3), the total volume is one.

(4) Alternatively, $f(x,y)$ can be thought of as the continuously varying two-dimensional density of a thin laminate plate (called a lamina in physics and calculus). Then, $f(x,y)dA$ is an increment of mass (probability), with the region R having a total mass (probability) of one, as represented by the double integral in property (3) above.

(5) The marginal densities can be thought of as the continuous analogs of the tabular discrete case. They are continuous sums (that is, integrals) along vertical lines for X and horizontal lines for Y. They can be thought of as cross-sectional areas of the volume described in note 2.

(6) While we have used limits of integration from $-\infty$ to ∞ to capture the general case, the actual limits of integration used in calculations depend on the boundary curves for R (since by property (2) $f(x,y) \equiv 0$ outside R). Thus, an alternative notation for the marginal density of X would be $f_X(x) = \int_{R(x)} f(x,y)dy$, indicating that the actual limits of integration are along the vertical line $R(x)$ representing the intersection of R with the associated full vertical line (see Figure 7.5.1 below). Likewise, we could write $f_Y(y) = \int_{R(y)} f(x,y)\,dx$ for the corresponding horizontal integration yielding the marginal density for Y.

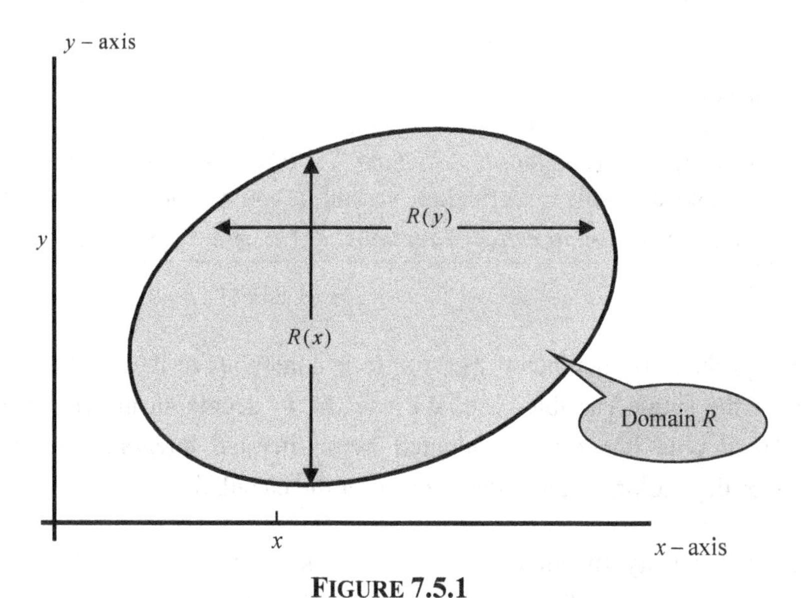

FIGURE 7.5.1

(7) The integrals $\int_{R(x)} f(x,y)\,dy$ and $\int_{R(y)} f(x,y)\,dx$ defining $f_X(x)$ and $f_Y(y)$ respectively are exactly the same as the *inside* integrals used in the iterated double integral formulas for $\iint_R f(x,y)\,dA$. For example, in evaluating a double integral over the rectangle $[a,b]\times[c,d]$ we can imagine fixing an *x-value* and integrating vertically (dy) from $y=c$ to $y=d$ for the *inside* integral, and then integrating horizontally (dx) from $x=a$ to $x=b$ for the *outside* integral (see Figure 7.5.2(A)). This traces out the region

R as a series of tall, thin vertical rectangles of thickness dx (like books on a shelf). In this case the inside integral produces $f_X(x)$. The opposite order of integration traces out R as a stack of narrow horizontally-oriented rectangles of thickness dy (like a deck of cards viewed edge-on), with $f_Y(y)$ as the inside integral (see Figure 7.5.2(B)).

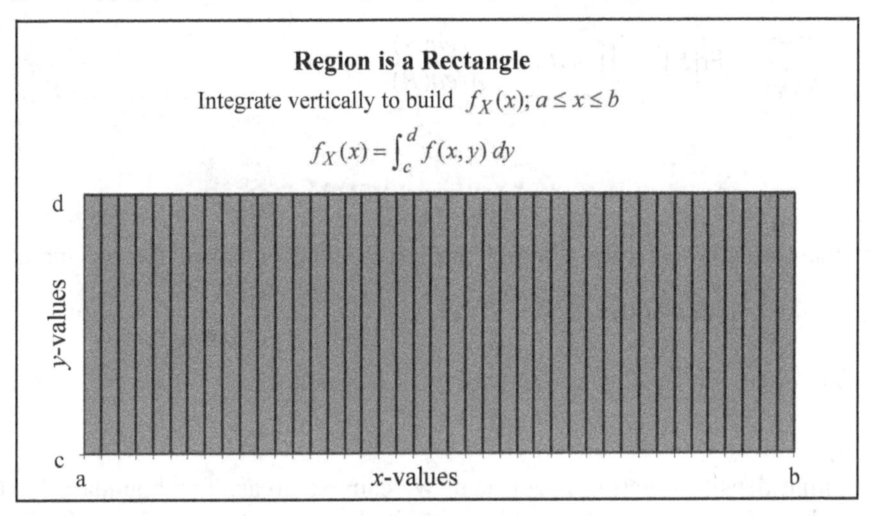

FIGURE 7.5.2(A)

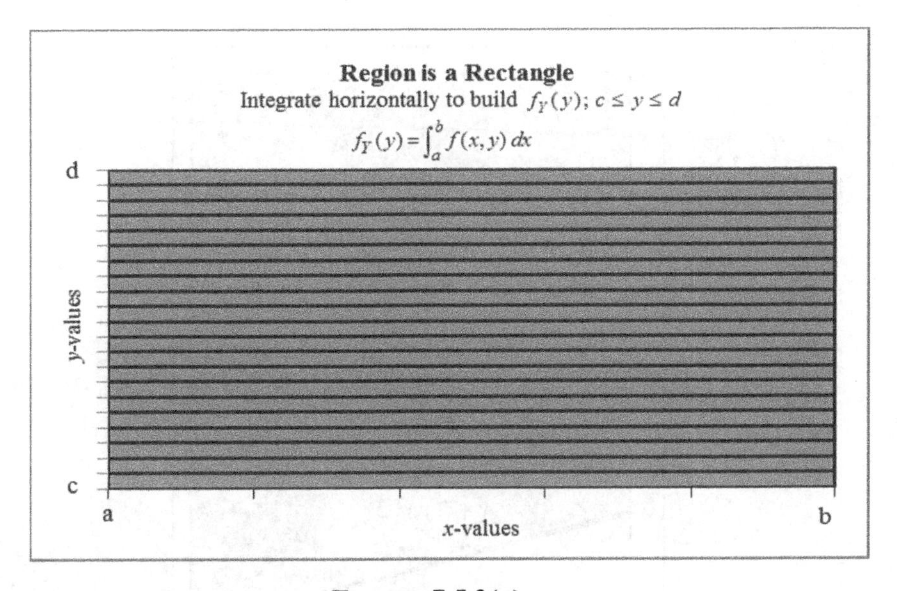

FIGURE 7.5.2(B)

(8) An event associated with the joint distribution may be visualized as a subset S of the region R. To calculate the probability of the event S we integrate the joint density function over $S: \Pr(S) = \iint_S f(x,y)dA$. The actual limits of integration will again depend on the shape and boundary curves describing S.

(9) The case where $f(x, y) = k$ (a constant) over R is analogous to a uniform distribution for a single random variable. In this instance we have $k = \frac{1}{Area(R)}$ since $1 = \iint_R f(x, y)\,dA = \iint_R k\,dA = k\iint_R dA = k \cdot Area(R)$. Also in this case,

$$\Pr[S] = \iint_S k\,dA = \frac{Area(S)}{Area(R)}.$$

Example 7.5-1 Probability of an Event – Uniform Case

Suppose that random variables X and Y are jointly distributed on the rectangle R given by $[0,2]\times[0,5]$, with joint density function $f(x, y) = \begin{cases} .1 & \text{for } 0 \le x \le 2, 0 \le y \le 5 \\ 0 & \text{otherwise.} \end{cases}$

Find $\Pr(S)$, where S is the event described by $X + 2Y > 2$.

Solution

Since the joint density function is constant, we can use areas. The boundary for the event S consists of boundary lines for R together with the boundary line $x + 2y = 2$. The event S is the gray region (see below) in R above the line $x + 2y = 2$.

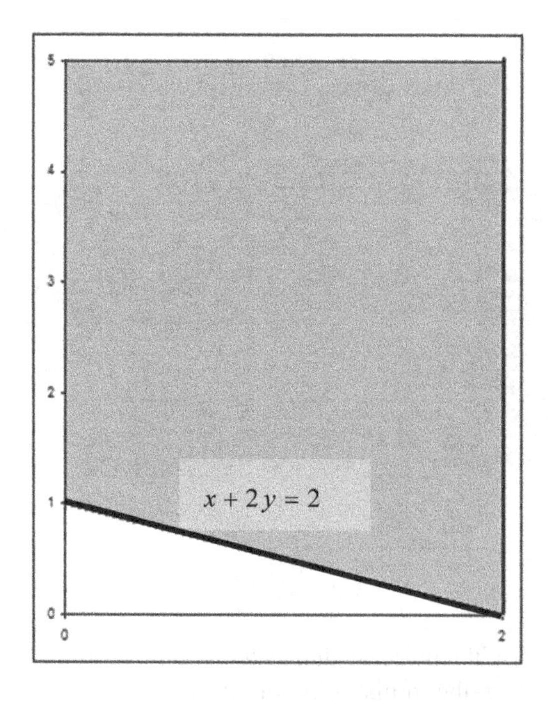

FIGURE 7.5.3

Then $\Pr(S) = \frac{Area(S)}{Area(R)} = 1 - \frac{Area(\text{unshaded triangle})}{Area(R)} = 1 - \frac{1}{10} = 0.9.$

Alternate Solution
Calculating probability using the joint density function, we have,

$$\Pr(S) = \Pr(X+2Y>2) = \iint_S f(x,y)dx\,dy$$
$$= \int_0^2 \int_{1-x/2}^5 .1\,dy\,dx = \int_0^2 (.4+.05x)\,dx = .9.$$ ❐

Example 7.5-2 Marginal Densities – Uniform, Rectangular Case

For the joint distribution in Example 7.5-1, calculate the marginal density functions for X and Y.

Solution

$$f_X(x) = \int_{R(x)} f(x,y)\,dy = \int_0^5 0.1dy = 0.5 \text{ for } 0 \le x \le 2,$$

and

$$f_Y(y) = \int_{R(y)} f(x,y)dx = \int_0^2 0.1dx = 0.2 \text{ for } 0 \le y \le 5.$$ ❐

Notes

(1) As one might have expected, X and Y are uniform (constant density functions) on the intervals $[0,2]$ and $[0,5]$, respectively.

(2) One should not assume that the joint distribution being uniform on R guarantees that the marginal distributions are also uniformly distributed. This is only the case if R is a rectangular region. We shall see counter-examples shortly.

Example 7.5-3 Joint Distribution, (Rectangular Case continued)

Suppose that random variables X and Y are jointly distributed on the rectangle R given by $[0,2]\times[0,5]$, with joint density function $f(x,y)=\frac{x}{10}$ on R, zero otherwise.

(a) Calculate the marginal density functions for X and Y.
(b) Calculate $\Pr[X+2Y>2]$.

Solution
For $0 \le x \le 2$, we have

$$f_X(x) = \int_{R(x)} f(x,y)dy = \int_0^5 \frac{x}{10}\,dy = \frac{x}{2},$$

and for $0 \le y \le 5$,

$$f_Y(y) = \int_{R(y)} f(x,y)\,dx = \int_0^2 \frac{x}{10}\,dx = \left.\frac{x^2}{20}\right|_0^2 = \frac{1}{5}.$$

Since the joint density function is no longer constant we cannot use areas for the calculation in (b). Instead we must integrate over the gray shaded region in Figure 7.5.3. We will visualize the region as a "shelf of books," meaning the inside integral uses dy and the outside uses dx. For the inside integral the line $R(x)$ has a variable lower boundary given by $x+2y=2$ which implies $y=1-\frac{x}{2}$. Thus,

$$\Pr[X+2Y>2] = \int_0^2\int_{1-\frac{x}{2}}^5 \frac{x}{10}\,dy\,dx = \frac{1}{10}\int_0^2 xy\Big|_{y=1-\frac{x}{2}}^5 dx = \frac{1}{10}\int_0^2 x\left[5-\left(1-\frac{x}{2}\right)\right]dx = \frac{14}{15}.$$

❒

We can calculate moments and more general expectations in the usual fashion.

Expected Values – Continuous Case

Let X and Y be jointly distributed on the region R with joint density function $f(x,y)$. Let $G(x,y)$ be a given function on R. Then,

$$E[G(X,Y)] = \iint_R G(x,y)\cdot f(x,y)\,dA$$

Note

If G is a function of X (or Y) alone then $E[G(X)]$ can be calculated as a single integral in the usual way using the marginal density function for X. That is, $E[G(X)]=\int_{-\infty}^{\infty} G(x)\,f_X(x)\,dx$ (or similarly, $E[G(Y)]=\int_0^{\infty} G(y)\,f_Y(y)\,dy$). This is because

$$E[G(X)] = \iint_R G(x)\cdot f(x,y)\,dA = \int_{-\infty}^{\infty}\int_{-\infty}^{\infty} G(x)f(x,y)\,dy\,dx$$

$$= \int_{-\infty}^{\infty} G(x)\left[\int_{-\infty}^{\infty} f(x,y)\,dy\right]dx = \int_{-\infty}^{\infty} G(x)f_X(x)\,dx.$$

Example 7.5-4 Expected Values (Example 7.5-3 continued)

Suppose that random variables X and Y are jointly distributed on the rectangle R given by $[0,2]\times[0,5]$, with joint density function $f(x,y)=\frac{x}{10}$ on R. Calculate the expected values and variances for X and Y, and the covariance of X and Y.

Solution

We begin with $E[X]$. From the definition above and the related iterated double integrals we have, with $G(x,y)=x$,

$$E[X] = \iint_R x\cdot f(x,y)\,dA = \iint_R \frac{x^2}{10}\,dA = \frac{1}{10}\int_0^5\int_0^2 x^2\,dx\,dy = \frac{1}{10}\int_0^2\int_0^5 x^2\,dy\,dx.$$

However, since we already know from Example 7.1-3 (a) the marginal density function for X, $f_X(x) = \frac{x}{2}$ for $0 \le x \le 2$, we can make use of the note following the definition to reduce the calculation to a single integration:

$$E[X] = \int_{-\infty}^{\infty} x\, f_X(x)\, dx = \int_0^2 x \cdot \frac{x}{2}\, dx = \frac{1}{2}\int_0^2 x^2\, dx = \frac{4}{3}.$$

Similarly,

$$E[X^2] = \int_0^2 x^2 \cdot \frac{x}{2}\, dx = \frac{1}{2}\int_0^2 x^3\, dx = 2,$$

so that

$$Var[X] = 2 - \left(\frac{4}{3}\right)^2 = \frac{2}{9}.$$

Also,

$$E[Y] = \int_0^5 y\, f_Y(y)\, dy = \int_0^5 y \cdot \frac{1}{5}\, dy = \frac{1}{5}\int_0^5 y\, dy = \frac{5}{2},$$

$$E[Y^2] = \frac{1}{5}\int_0^5 y^2\, dy = \frac{25}{3},$$

and thus,

$$Var[Y] = \frac{25}{3} - \left(\frac{5}{2}\right)^2 = \frac{25}{12}.$$

Of course, we could have also recognized that the marginal distribution of the random variable Y is uniformly distributed on $[0,5]$, and made use of the standard formulas for uniform distributions in Chapter 6.

Finally, to calculate the covariance we use the covariance formula, $Cov[X,Y] = E[X \cdot Y] - E[X] \cdot E[Y]$, from Section 7.4, which is equally valid for continuous distributions. To calculate $E[X \cdot Y]$ we use $G(x,y) = xy$, and since this is a function of both x and y we must use a double integral formula:

$$E[XY] = \iint_R xy \cdot f(x,y)\, dA = \int_0^2 \int_0^5 xy \cdot \frac{x}{10}\, dy\, dx$$

$$= \frac{1}{10}\int_0^2 x^2 \left(\frac{y^2}{2}\Big|_{y=0}^{5} \right) dx = \frac{1}{10}\int_0^2 \frac{25x^2}{2}\, dx = \frac{10}{3},$$

and so

$$Cov[X,Y] = \frac{10}{3} - \left(\frac{4}{3}\right)\left(\frac{5}{2}\right) = 0.$$ ❐

We shall see later that the zero covariance is a consequence of X and Y being independent continuous random variables.

As noted previously, a joint density function $f(x,y)$ living on a region R of the $x-y$ plane may be represented as a 3-dimensional graph, with probability then construed as the volume under the graph $z = f(x,y)$. This construct is entirely analogous to the one variable situation where we associate probability with area under the graph of the density function. Because 3-dimensional graphs are notoriously difficult to draw (and higher dimensional situations impossible), this perspective is less often used in multivariate situations. We provide a tractable example below.

Example 7.5-5 Joint Density on a Rectangle

Let $f(x,y) = \begin{cases} 1.75 - x - .5y & \text{for } 0 \le x \le 1,\ 0 \le y \le 1 \\ 0 & \text{otherwise.} \end{cases}$

(a) Sketch the graph and verify that $f(x,y)$ is a joint density function.
(b) Calculate $\Pr(X < Y)$.
(c) Find the marginal density function $f_X(x)$.
(d) Calculate $E[X]$ two ways.
  (i) Using the joint density function $f(x,y)$.
  (ii) Using the marginal density function $f_X(x)$.

Solution
The joint density function looks like

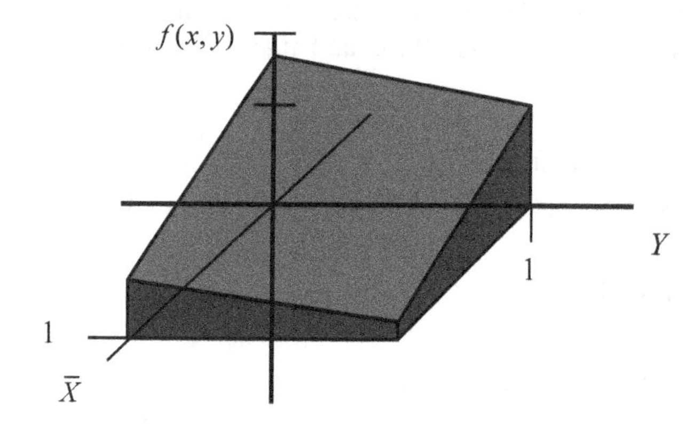

(a) We need to verify that $f(x,y)$ is always positive and that the total volume (probability) equals one. On the square $0 \le x \le 1,\ 0 \le y \le 1$, the smallest value of the function $f(x,y)$ is $.25 > 0$ (it occurs at $x = 1, y = 1$). Furthermore,

$$\int_{-\infty}^{\infty}\int_{-\infty}^{\infty} f(x,y)\,dx\,dy = \int_0^1\int_0^1 (1.75-x-.5y)\,dx\,dy$$

$$= \int_0^1 \left(1.75x - \frac{x^2}{2} - .5xy\right)\Bigg|_{x=0}^{x=1} dy = \int_0^1 (1.25-.5y)\,dy$$

$$= \left(1.25y - \frac{y^2}{4}\right)\Bigg|_{y=0}^{y=1} = (1.25-.25)-(0-0)$$

$$= 1.$$

(b) We integrate over the region where the random variable X is less than the random variable Y. That is, $0 \le x < y$ and $0 \le y \le 1$.

$$\Pr(X<Y) = \int_0^1\int_0^y (1.75-x-.5y)\,dx\,dy$$

$$= \int_0^1 \left(1.75x - \frac{x^2}{2} - .5xy\right)\Bigg|_{x=0}^{x=y} dy = \int_0^1 (1.75y-y^2)\,dy$$

$$= \frac{7y^2}{8} - \frac{y^3}{3} = \left(\frac{7}{8}-\frac{1}{3}\right)-(0-0) = \frac{13}{24}.$$

(c) The marginal density is computed as follows:

$$f_X(x) = \int_{-\infty}^{\infty} f(x,y)\,dy = \int_0^1 (1.75-x-.5y)\,dy$$

$$= \left(1.75y - xy - \frac{y^2}{4}\right)\Bigg|_{y=0}^{y=1} = 1.50-x; \quad 0 \le x \le 1.$$

(d) Computing the mean of X.

(i) From the joint density function $f(x,y)$:

$$E[X] = \int_{-\infty}^{\infty}\int_{-\infty}^{\infty} x\cdot f(x,y) = \int_0^1\int_0^1 x\cdot(1.75-x-.5y)\,dx\,dy$$

$$= \int_0^1 \left(\frac{7x^2}{8} - \frac{x^3}{3} - \frac{x^2y}{4}\right)\Bigg|_{x=0}^{x=1} dy$$

$$= \int_0^1 \left(\frac{13}{24} - \frac{y}{4}\right) dy$$

$$= \left(\frac{13y}{24} - \frac{y^2}{8}\right)\Bigg|_{y=0}^{y=1} = \left(\frac{13}{24}-\frac{1}{8}\right)-(0-0) = \frac{10}{24} = \frac{5}{12}.$$

(ii) From the marginal density function in (c):

$$E[X] = \int_{-\infty}^{\infty} x \cdot f_X(x)\,dx = \int_0^1 x \cdot (1.5 - x)\,dx$$

$$= \int_0^1 \left(\frac{3}{2}x - x^2\right) dx$$

$$= \left(\frac{3x^2}{4} - \frac{x^3}{3}\right)\Bigg|_{x=0}^{x=1} = \left(\frac{3}{4} - \frac{1}{3}\right) - (0-0) = \frac{5}{12}.$$ ❐

Example 7.5-6 Joint Distribution, Non-Rectangular Region

Let $f(x, y) = k \cdot x$ be a joint density function on the region R in the plane described by $0 \le x \le y \le 1$.

(a) Calculate the value of the constant k.
(b) Find the marginal density functions for X and Y.
(c) Calculate the expected values and variances for the marginal random variables X and Y.

Solution

The region R consists of the shaded triangle in the diagram below:

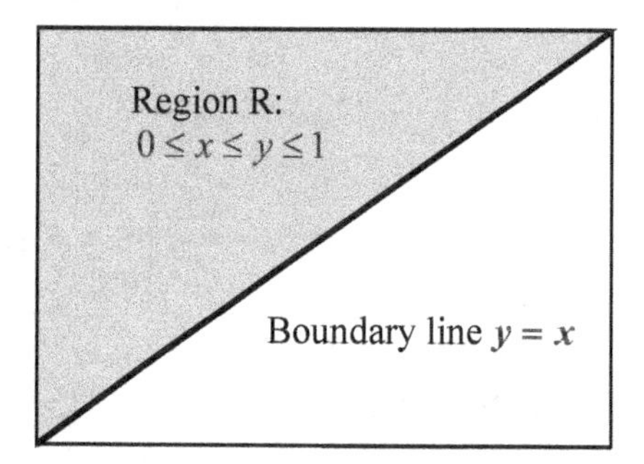

(a) To calculate the constant k we must evaluate $1 = \iint_R f(x, y)\,dA = k\iint_R x\,dA.$ To evaluate this as an iterated double integral we can either integrate in the order $dy\,dx$ (vertical, then horizontal) or the reverse, $dx\,dy$ (horizontal, then vertical). We will illustrate both, beginning with $dy\,dx$:

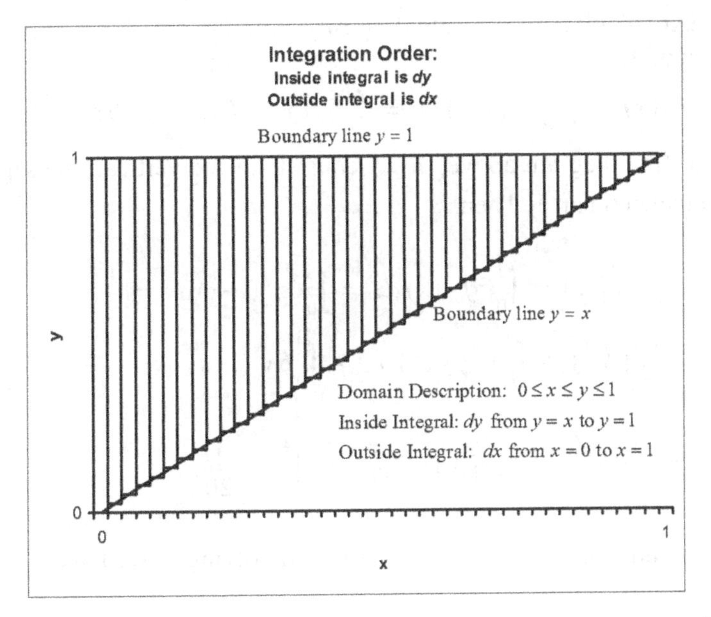

$$1 = k\int_0^1\int_x^1 x\,dy\,dx = k\int_0^1\left(x\,y\Big|_{y=x}^1\right)dx = k\int_0^1 x(1-x)\,dx = k\left[\frac{x^2}{2}-\frac{x^3}{3}\right]_0^1 = \frac{k}{6}.$$

Therefore $k = 6$.

The same double integral but with the order reversed is illustrated by:

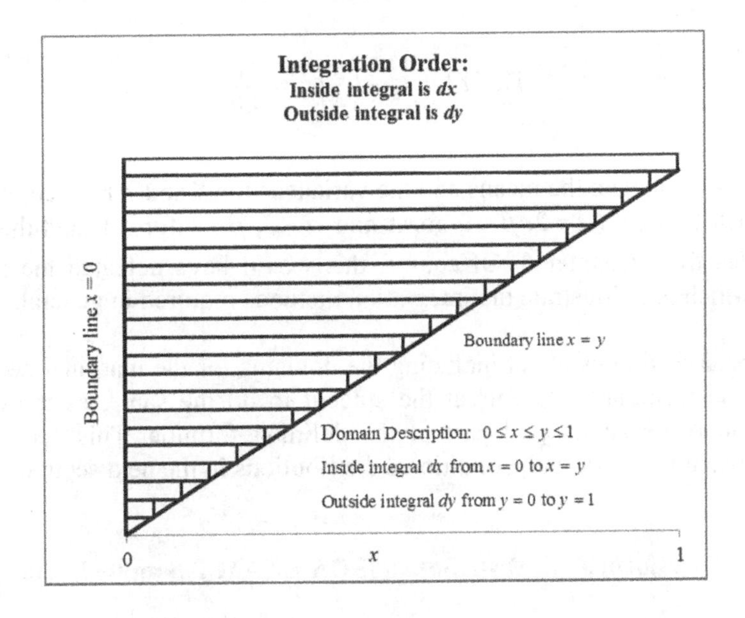

The corresponding calculation is,

$$1 = k\int_0^1\int_0^y x\,dxdy = k\int_0^1\left(\frac{x^2}{2}\Big|_{x=0}^y\right)dy = k\int_0^1\frac{y^2}{2}dy = \frac{k}{6},$$

which implies $k = 6$.

(b) and (c) We next calculate the marginal density function for X and use it to compute the mean and variance of X.

$$f_X(x) = \int_{R(x)} f(x,y)\,dy = \int_x^1 6x\,dy = 6x(1-x); \;\; 0 \le x \le 1.$$

Note that from the first diagram above, $R(x)$ consists of the vertical line segment $x \le y \le 1$, for all values x between 0 and 1. Then,

$$E[X] = \int_0^1 x\,f_X(x)\,dx = \int_0^1 6x^2(1-x)\,dx = \frac{1}{2},$$

$$E[X^2] = \int_0^1 x^2\,f_X(x)\,dx = \int_0^1 6x^3(1-x)\,dx = \frac{3}{10},$$

and

$$Var[X] = \frac{3}{10} - \left(\frac{1}{2}\right)^2 = \frac{1}{20}.$$

Referring to the second diagram for the calculations involving Y, we have,

$$f_Y(y) = \int_{R(y)} f(x,y)dx = \int_0^y 6x\,dx = 3y^2; \;\; 0 \le y \le 1,$$

$$E[Y] = \int_0^1 y\,f_Y(y)\,dy = \int_0^1 y\cdot(3y^2)\,dy = 3\int_0^1 y^3\,dy = \frac{3}{4},$$

$$E[Y^2] = 3\int_0^1 y^4\,dy = \frac{3}{5}.$$

Then,

$$Var[Y] = \frac{3}{5} - \left(\frac{3}{4}\right)^2 = \frac{3}{80}.$$

❒

Notes

(1) We could also calculate the means and the variances for X and Y by recognizing that they are Beta distributions ($\alpha = 2$, $\beta = 2$ for X and $\alpha = 3$, $\beta = 1$ for Y), and then make use of the formulas from Chapter 6. Of course, this would have defeated the purpose of the example, which is to illustrate the integration methods required in general.

(2) We have been fastidious about including the domains for the marginal density functions alongside the formulas. We entreat the student to do the same, as the domain of the density function is an integral part of the defining formula. This becomes especially important when we work with conditional distributions in the next section.

Example 7.5-7 Uniform Joint Distribution (SOA EXAM P Sample Exam Question #92)

Two insurers provide bids on an insurance policy to a large company. The bids must be between 2000 and 2200. The company decides to accept the lower bid if the two bids differ by 20 or more. Otherwise, the company will consider the two bids further.

Assume that the two bids have a joint density function given by $f(x,y)=\frac{1}{40,000}$ on the square domain $R=[2000,2200]\times[2000,2200]$. Determine the probability that the company considers the two bids further.

Solution

To solve the problem directly requires evaluating (for $2000 \le x, y \le 2200$) $\Pr(|X-Y|<20)$ $=\Pr(-20<X-Y<20)$, which is pictured below as consisting of the shaded region with the 3 sub-pieces labeled *A*, *B*, and *C*.

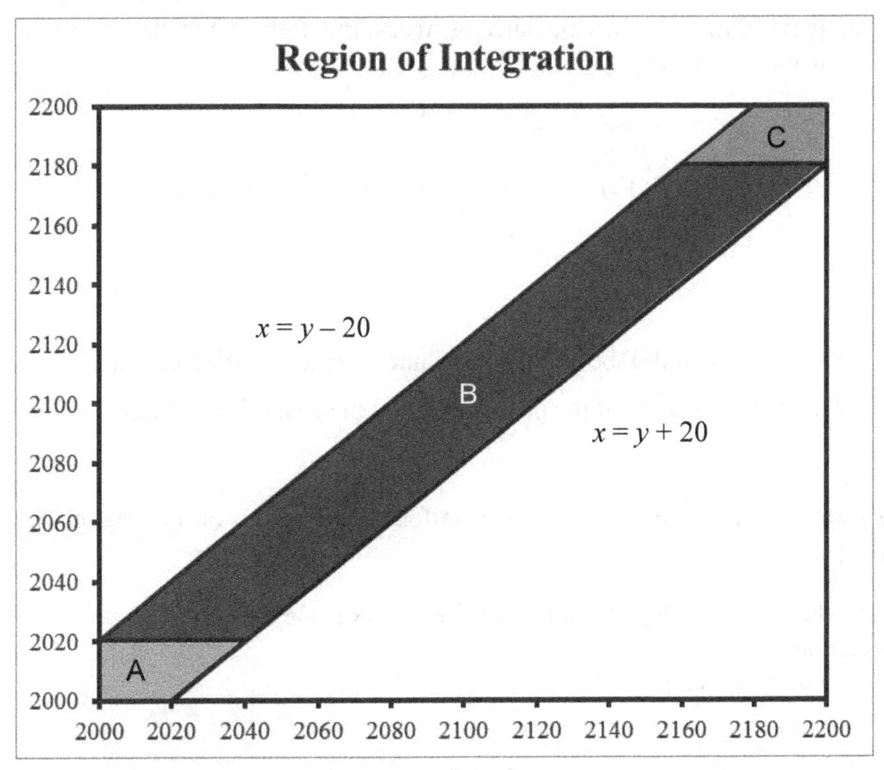

To calculate this directly requires breaking the integral into 3 separate parts, each with different limits of integration:

$$\iint_{A+B+C} f(x,y)\,dx\,dy = \iint_{\text{Region }A} f(x,y)\,dx\,dy + \iint_{\text{Region }B} f(x,y)\,dx\,dy + \iint_{\text{Region }C} f(x,y)\,dx\,dy$$

$$= \int_{2000}^{2020}\int_{2000}^{y+20} f(x,y)\,dx\,dy + \int_{2020}^{2180}\int_{y-20}^{y+20} f(x,y)\,dx\,dy + \int_{2180}^{2200}\int_{y-20}^{2200} f(x,y)\,dx\,dy. \quad \square$$

Notes

(1) From the diagram, this integral is different from $\int_{2000}^{2200}\int_{y-20}^{y+20} f(x,y)\,dx\,dy$, which has pieces at the lower and upper ends protruding outside the domain of the joint distribution.

(2) If we change the order of integration to $dy\,dx$, then the sub-regions are a bit different.

(3) The joint density function is $f(x,y)=\frac{1}{40{,}000}$, so the answer could be found by integrating this constant joint density over each of the 3 sub-regions as indicated.

(4) Of course, it would be easier to find the probability of the complementary region, which consists of the two un-shaded isosceles right triangles.

(5) Easier still, is to recognize that since the joint density is a constant, the joint distribution is uniformly distributed on the domain $R=[2000,2200]\times[2000,2200]$. This means that probabilities can be calculated as **areas**, the preferred method of solution. Thus, the answer is given by

$$1-\overbrace{(2)}^{\text{two triangles}}\frac{\overbrace{\left(\tfrac{1}{2}\right)(180)^2}^{\text{area of each triangle}}}{\underbrace{(200)^2}_{\text{total area of } R}}=1-\left(\frac{9}{10}\right)^2=0.19.$$

Exercise 7-25 Let X and Y be jointly distributed random variables with $f(x,y)=\frac{1}{6}$ for $0\le x\le 2$ and $0\le y\le 3$. Find the probability that their sum is less than 1.

Exercise 7-26 Let X and Y have a joint uniform distribution on the region described by $0\le y\le 1-x^2;\ -1\le x\le 1$.

(a) Find the marginal density functions for X and Y. Be sure to give the domains of the functions.
(b) Compute $E[X]$ and $E[Y]$.

Exercise 7-27 SOA EXAM P Sample Exam Questions #79

A device contains two components. The device fails if either component fails. The joint density function of the lifetimes of the components, measured in hours, is $f(s,t)$, where $0<s<1$ and $0<t<1$. What is the probability that the device fails during the first half hour of operation?

(A) $\int_0^{0.5}\int_0^{0.5} f(s,t)\,ds\,dt$

(B) $\int_0^{1}\int_0^{0.5} f(s,t)\,ds\,dt$

(C) $\int_{0.5}^{1}\int_{0.5}^{1} f(s,t)\,ds\,dt$

(D) $\int_0^{0.5}\int_0^{1} f(s,t)\,ds\,dt+\int_0^{1}\int_0^{0.5} f(s,t)\,ds\,dt$

(E) $\int_0^{0.5}\int_{0.5}^{1} f(s,t)\,ds\,dt+\int_0^{1}\int_0^{0.5} f(s,t)\,ds\,dt$

Exercise 7-28 SOA/CAS Course 1 2000 Sample Examination #4

Let X and Y be random losses with joint density function

$$f(x,y) = e^{-(x+y)} \text{ for } x>0 \text{ and } y>0.$$

An insurance policy is written to reimburse $X+Y$. Calculate the probability that the reimbursement is less than 1.

(A) e^{-2} (B) e^{-1} (C) $1-e^{-1}$ (D) $1-2e^{-1}$ (E) $1-2e^{-2}$

Exercise 7-29 SOA EXAM P Sample Exam Questions #97

Let T_1 and T_2 represent the lifetimes in hours of two linked components in an electronic device. The joint density function for T_1 and T_2 is uniform over the region defined by $0 \le t_1 \le t_2 \le L$ where L is a positive constant. Determine the expected value of the sum of the squares of T_1 and T_2.

(A) $\frac{L^2}{3}$ (B) $\frac{L^2}{2}$ (C) $\frac{2L^2}{3}$ (D) $\frac{3L^2}{4}$ (E) L^2

Exercise 7-30 SOA EXAM P Sample Exam Questions #89

The future lifetimes (in months) of two components of a machine have the following joint density function:

$$f(x,y) = \begin{cases} \frac{6}{125{,}000}(50-x-y) & \text{for } 0<x<50-y<50 \\ 0 & \text{otherwise} \end{cases}.$$

What is the probability that both components are still functioning 20 months from now?

(A) $\frac{6}{125{,}000}\int_0^{20}\int_0^{20}(50-x-y)\,dy\,dx$

(B) $\frac{6}{125{,}000}\int_{20}^{30}\int_{20}^{50-x}(50-x-y)\,dy\,dx$

(C) $\frac{6}{125{,}000}\int_{20}^{30}\int_{20}^{50-x-y}(50-x-y)\,dy\,dx$

(D) $\frac{6}{125{,}000}\int_{20}^{50}\int_{20}^{50-x}(50-x-y)\,dy\,dx$

(E) $\frac{6}{125{,}000}\int_{20}^{50}\int_{20}^{50-x-y}(50-x-y)\,dy\,dx$

Exercise 7-31 SOA EXAM P Sample Exam Questions #91

An insurance company insures a large number of drivers. Let X be the random variable representing the company's losses under collision insurance, and let Y represent the company's losses under liability insurance. X and Y have joint density function

$$f(x,y) = \begin{cases} \frac{2x+2-y}{4} & \text{for } 0<x<1 \text{ and } 0<y<2 \\ 0 & \text{otherwise} \end{cases}.$$

What is the probability that the total loss is at least 1?

(A) 0.33 (B) 0.38 (C) 0.41 (D) 0.71 (E) 0.75

Exercise 7-32 SOA EXAM P Sample Exam Questions #118

Let X and Y be continuous random variables with joint density function

$$f(x,y) = \begin{cases} 15y & \text{for } x^2 \le y \le x \\ 0 & \text{otherwise} \end{cases}.$$

Let $g(y)$ be the marginal density function of Y. Which of the following represents $g(y)$?

(A) $g(y) = \begin{cases} 15y & \text{for } x^2 \le y \le x \\ 0 & \text{otherwise} \end{cases}$

(B) $g(y) = \begin{cases} \frac{15y^2}{2} & \text{for } x^2 < y < x \\ 0 & \text{otherwise} \end{cases}$

(C) $g(y) = \begin{cases} \frac{15y^2}{2} & \text{for } 0 < y < 1 \\ 0 & \text{otherwise} \end{cases}$

(D) $g(y) = \begin{cases} 15y^{3/2}(1-y^{1/2}) & \text{for } x^2 < y < x \\ 0 & \text{otherwise} \end{cases}$

(E) $g(y) = \begin{cases} 15y^{3/2}(1-y^{1/2}) & \text{for } 0 < y < 1 \\ 0 & \text{otherwise} \end{cases}$

Exercise 7-33 Is the following a density function? If so, calculate $E[X]$ and $\Pr(X < .6)$.

$$f(x,y) = \begin{cases} 1.3-.8x-.4y & \text{for } 0 \le x \le 1,\ 0 \le y \le 2 \\ 0 & \text{otherwise} \end{cases}.$$

Exercise 7-34 Jack the rabbit is hopping around an enclosed fenced field that is 40 feet wide and 50 feet long. His location is uniformly distributed throughout the field. What is the probability that at a given time, Jack is more than 10 feet from a fence?

Exercise 7-35 Jack the rabbit is hopping around an enclosed fenced field that is 40 feet wide and 80 feet long. His location is uniformly distributed throughout the field. Stockton the dog is attached to a dog-run centered at the northwest corner of the fence that limits his movements to the shape of an ellipse, that extends south 20 feet and east 30 feet. If Stockton can catch Jack, he will eat the rabbit. Meanwhile, in the neighboring yard the local bully, Sling-Shot Sammy, is in his tree house looking for prey. He can hit and kill any animal that is in the southeast corner of the field, midway along the south edge of the field on a straight line up to the northeast corner. What is the probability that at a given time, Jack the rabbit is in a safe place in the field?

Exercise 7-36 SOA EXAM P Sample Exam Questions #94

Let T_1 be the time between a car accident and the claim report to the insurance company. Let T_2 be the time between the report of the claim and payment of the claim. The joint density function of T_1 and T_2, $f(t_1,t_2)$, is constant over the region $0 < t_1 < 6$, $0 < t_2 < 6$, $t_1 + t_2 < 10$, and zero otherwise. Determine $E[T_1 + T_2]$, the expected time between a car accident and payment of the claim.

(A) 4.9 (B) 5.0 (C) 5.7 (D) 6.0 (E) 6.7

Exercise 7-37 SOA/CAS Course 1 2000 Sample Examination #31

Let X and Y be random losses with joint density function.

$$f(x,y) = 2x \text{ for } 0 < x < 1 \text{ and } 0 < y < 1.$$

An insurance policy is written to cover the loss $X + Y$. The policy has a deductible of 1. Calculate the expected payment under the policy.

(A) $\frac{1}{4}$ (B) $\frac{1}{3}$ (C) $\frac{1}{2}$ (D) $\frac{7}{12}$ (E) $\frac{5}{6}$

Exercise 7-38 SOA EXAM P Sample Exam Questions #122

A device contains two circuits. The second circuit is a backup for the first, so the second is used only when the first has failed. The device fails when and only when the second circuit fails. Let X and Y be the times at which the first and second circuits fail, respectively. X and Y have joint probability density function

$$f(x,y) = \begin{cases} 6e^{-x}e^{-2y} & \text{for } 0<x<y<\infty \\ 0 & \text{otherwise} \end{cases}$$

What is the expected time at which the device fails?

(A) 0.33 (B) 0.50 (C) 0.67 (D) 0.83 (E) 1.50

Exercise 7-39 SOA EXAM P Sample Exam Questions #121

Let X represent the age of an insured automobile involved in an accident. Let Y represent the length of time the owner has insured the automobile at the time of the accident. Let X and Y have joint probability density function

$$f(x,y) = \begin{cases} \frac{1}{64}(10-xy^2) & \text{for } 2\le x\le 10 \text{ and } 0\le y\le 1 \\ 0 & \text{otherwise} \end{cases}$$

Calculate the expected age of an insured automobile involved in an accident.

(A) 4.9 (B) 5.2 (C) 5.8 (D) 6.0 (E) 6.4

Exercise 7-40 SOA EXAM P Sample Exam Questions #138

A machine consists of two components, whose lifetimes have joint density function

$$f(x,y) = \begin{cases} \frac{1}{50} & \text{for } x>0, y>0, x+y<10 \\ 0 & \text{otherwise} \end{cases}$$

The machine operates until both components fail. Calculate the expected operational time of the machine.

(A) 1.7 (B) 2.5 (C) 3.3 (D) 5.0 (E) 6.7

7.6 CONDITIONAL DISTRIBUTIONS – THE CONTINUOUS CASE

In this section we extend the concept of conditional distributions to the continuous case. In the discrete case we obtained the conditional distribution for Y given $X = x_i$ by restricting our attention to the column of the joint probabilities corresponding to $X = x_i$ and then scaling those probabilities to assure a sum of one. We can use the same strategy in the continuous case in order to define the conditional distribution of Y given $X = x$.

In the generic diagram for a joint distribution over a general region R (reproduced in Figure 7.6.1) we use the vertical line segment $R(x)$ in place of the vertical column of the discrete table, and restrict our attention to the joint density along $R(x)$.

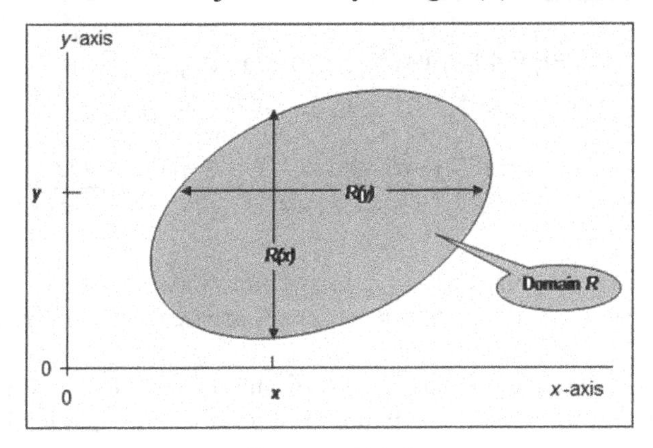

FIGURE 7.6.1

Our aim is to define the conditional density function of Y given by $f_Y(y|X=x) = k \cdot f(x,y)$, where:

(i) The given value x is held fixed,
(ii) the y-values vary along the vertical line segment $R(x)$,
(iii) the constant k is chosen to assure a total probability of one.

Using (iii), we have,

$$1 = \int_{R(x)} f_Y(y|X=x)\,dy = \int_{R(x)} k\, f(x,y)\,dy = k\int_{R(x)} f(x,y)\,dy = k\, f_X(x)$$

where the last integral is by definition $f_X(x)$, the marginal density function of X evaluated at the given x. The same reasoning applies to the conditional density of X given $Y = y$ and leads to the following definitions.

Conditional Density Functions

Let R be a region in the plane with joint density function $f(x,y)$. Then the ***conditional density function for Y given $X = x$*** is

$$f_Y(y|X=x) = \frac{f(x,y)}{f_X(x)}$$

for y-values such that $(x,y) \in R(x)$.

The ***conditional density function for X given $Y = y$*** is

$$f_X(x|Y=y) = \frac{f(x,y)}{f_Y(y)}$$

for all x-values such that $(x,y) \in R(y)$.

Notes

(1) Another way of looking at this definition, using incremental events and the definition of conditional probability, is that

$$f_Y(y \mid X=x)\,dy \underbrace{\approx}_{\text{for small } dx} \Pr[y \le Y \le y+dy \mid x \le X \le x+dx]$$

$$= \frac{\Pr[y \le Y \le y+dy \text{ and } x \le X \le x+dx]}{\Pr[x \le X \le x+dx]} \approx \frac{f(x,y)\,dx\,dy}{f_X(x)\,dx} = \frac{f(x,y)}{f_X(x)}\,dy.$$

(2) The conditional density $f_Y(y \mid X=x)$ is a function of y alone. The value $X=x$ is fixed, and is treated as a constant whenever we are dealing with $f_Y(y \mid X=x)$.

(3) It is of vital importance to specify the domains, $R(x)$ or $R(y)$, upon which the conditional density function lives. Inattention to this detail can lead to fatal errors.

(4) Conditional moments are calculated using conditional densities in the usual manner. For example, the conditional expectation of Y given $X=x$ is evaluated as,

$$E[Y \mid X=x] = \int_{R(x)} y \cdot f_Y(y \mid X=x)\,dy.$$

Example 7.6-1 Conditional Densities (Example 7.5-6 continued)

Let $f(x,y)=6x$ be a joint density function on the triangular region R in the plane described by $0 \le x \le y \le 1$.

(a) Calculate the conditional density of Y given $X=\frac{1}{4}$.

(b) Calculate the conditional expectation and variance of Y given $X=\frac{1}{4}$.

(c) Calculate $\Pr\left[Y \le \frac{1}{2} \middle| X=\frac{1}{4}\right]$.

(d) Calculate $\Pr\left[Y \le \frac{1}{2} \middle| X \ge \frac{1}{4}\right]$.

(e) Calculate the conditional density of X given $Y=\frac{3}{4}$.

(f) Calculate the conditional expectation and variance of X given $Y=\frac{3}{4}$.

Solution

In Example 7.5-6 we showed that the marginal density function for X is $f_X(x)=6x(1-x)$; $0 \le x \le 1$. The conditional distribution of Y given $X=\frac{1}{4}$ lives on $R\left(\frac{1}{4}\right)$ (see Figure 7.6.2), which is specified by the vertical line $\frac{1}{4} \le y \le 1$.

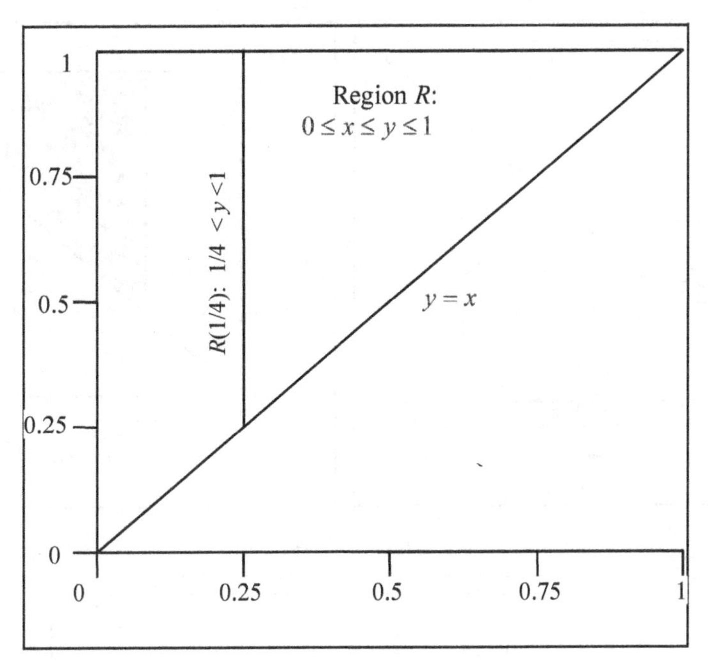

FIGURE 7.6.2

Therefore,

(a) $f_Y\left(y \mid X=\frac{1}{4}\right) = \dfrac{f\left(\frac{1}{4},y\right)}{f_X\left(\frac{1}{4}\right)} = \dfrac{(6)\left(\frac{1}{4}\right)}{(6)\left(\frac{1}{4}\right)\left(1-\frac{1}{4}\right)} = \frac{4}{3}; \quad \frac{1}{4} \le y \le 1.$

Note that $f_Y\left(y \mid X=\frac{1}{4}\right)$ is constant, and therefore the underlying conditional random variable is uniformly distributed on the interval $\frac{1}{4} \le y \le 1$. This is a consequence of the fact that the joint density function on R depends on x alone. Thus, using the formulas for a uniform distribution we have,

(b) $E\left[Y \mid X=\frac{1}{4}\right] = \dfrac{\frac{1}{4}+1}{2} = \frac{5}{8}$ and $Var\left[Y \mid X=\frac{1}{4}\right] = \dfrac{\left(1-\frac{1}{4}\right)^2}{12} = \frac{3}{64}.$

(c) Since $f_Y\left(y \mid X=\frac{1}{4}\right)$ is uniform on $\frac{1}{4} \le y \le 1$,

$$\Pr\left[Y \le 1/2 \mid X=\frac{1}{4}\right] = \frac{\frac{1}{2}-\frac{1}{4}}{1-\frac{1}{4}} = \frac{1}{3}.$$

(d) Although (d) looks similar to (c), the approach is quite different. Since we are not given a specific value for X we cannot use a conditional distribution for the calculation. Instead, we must use conditional probability with the given event $\left[X \ge \frac{1}{4}\right]$.

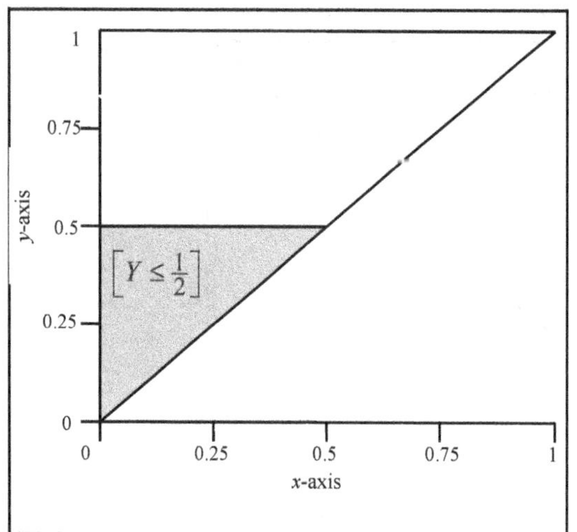

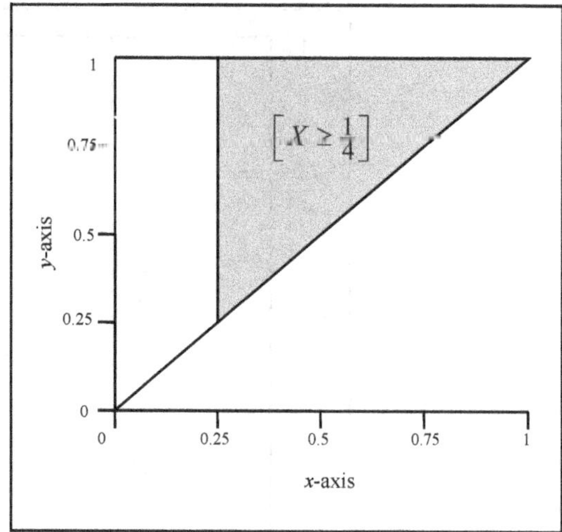

FIGURE 7.6.3

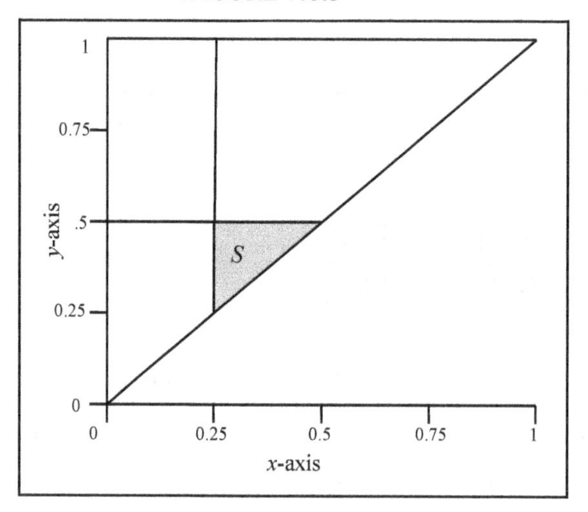

FIGURE 7.6.4

If $S = \left[Y \le \frac{1}{2}\right] \cap \left[X \ge \frac{1}{4}\right]$ (see Figures 7.6.3 and 7.6.4) then S is the subset of R consisting of the triangle with vertices $\left(\frac{1}{4}, \frac{1}{4}\right), \left(\frac{1}{4}, \frac{1}{2}\right)$ and $\left(\frac{1}{2}, \frac{1}{2}\right)$ and described by $\frac{1}{4} \le x \le y \le \frac{1}{2}$. Hence,

$$\Pr[S] = \iint_S f(x,y)\, dA = \int_{1/4}^{1/2} \int_x^{1/2} 6x\, dy\, dx$$

$$= 6\int_{1/4}^{1/2} x\left(\tfrac{1}{2} - x\right) dx = \int_{1/4}^{1/2} (3x - 6x^2)\, dx = \frac{1}{16},$$

and

$$\Pr\left[X \ge \tfrac{1}{4}\right] = \int_{1/4}^{1} f_X(x)\, dx = \int_{1/4}^{1} 6x(1-x)\, dx = 3x^2 - 2x^3\Big|_{1/4}^{1} = \frac{27}{32}.$$

Thus,

$$\Pr\left[Y \le \tfrac{1}{2} \mid X \ge \tfrac{1}{4}\right] = \frac{\Pr[S]}{\Pr\left[X \ge \frac{1}{4}\right]} = \frac{\frac{1}{16}}{\frac{27}{32}} = \frac{2}{27}.$$

(e) Given $Y = \frac{3}{4}$, we have (see Figure 7.6.5) that the conditional density for X lives on the horizontal line $R\left(\frac{3}{4}\right)$: $0 \le x \le \frac{3}{4}$.

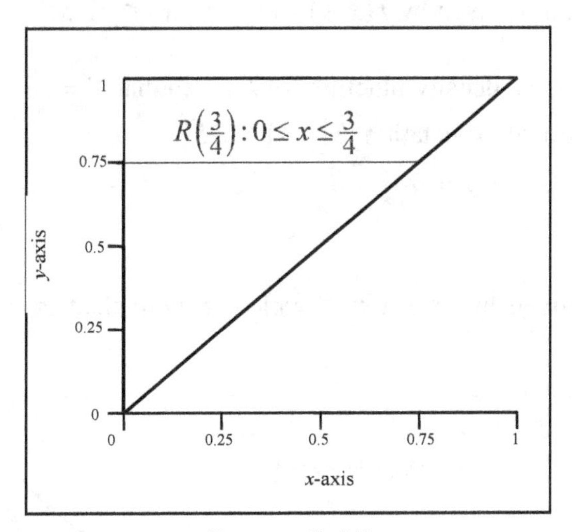

FIGURE 7.6.5

We have from Example 7.5-6 that $f_Y(y) = 3y^2;\ 0 \le y \le 1$, so

$$f_X\left(x \mid Y = \frac{3}{4}\right) = \frac{f\left(x, \frac{3}{4}\right)}{f_Y\left(\frac{3}{4}\right)} = \frac{6x}{(3)\left(\frac{3}{4}\right)^2} = \frac{32x}{9};\ 0 \le x \le \frac{3}{4}.$$

(f) Using the density function $f_X\left(x \mid Y = \frac{3}{4}\right) = \frac{32x}{9};\ 0 \le x \le \frac{3}{4}$, we have,

$$E\left[X \mid Y = \frac{3}{4}\right] = \int_{R(3/4)} x\, f_X\left(x \mid Y = \frac{3}{4}\right) dx$$

$$= \int_0^{3/4} (x)\left(\frac{32x}{9}\right) dx = \frac{32}{9}\int_0^{3/4} x^2\, dx = \frac{32}{9}\cdot\frac{x^3}{3}\Bigg|_0^{3/4} = \frac{1}{2},$$

and,

$$E\left[X^2 \mid Y = \frac{3}{4}\right] = \int_{R(3/4)} x^2\, f_X\left(x \mid Y = \frac{3}{4}\right) dx$$

$$= \int_0^{3/4} x^3\, dx \frac{32}{9} = \frac{32}{9}\cdot\frac{x^4}{4}\Bigg|_0^{3/4} = \frac{9}{32},$$

so that,

$$Var\left[X \mid Y = \frac{3}{4}\right] = \frac{9}{32} - \left(\frac{1}{2}\right)^2 = \frac{1}{32}.$$

❒

Example 7.6-2 Conditional Density Function (Exercise 7-32 continued)

The joint density function is given by $f(x,y)=15y$ for $x^2 \le y \le x$.

(a) Find the conditional density function for X given that $Y = y$.
(b) Find the conditional expectation $E[X \mid Y=.25]$.
(c) Calculate $\Pr[X \le 0.4 \mid Y=.25]$.

Solution

(a) The domain R is given by $x^2 \le y \le x$, which is equivalent to $y \le x \le y^{1/2}$ (see Figure 7.6.6).

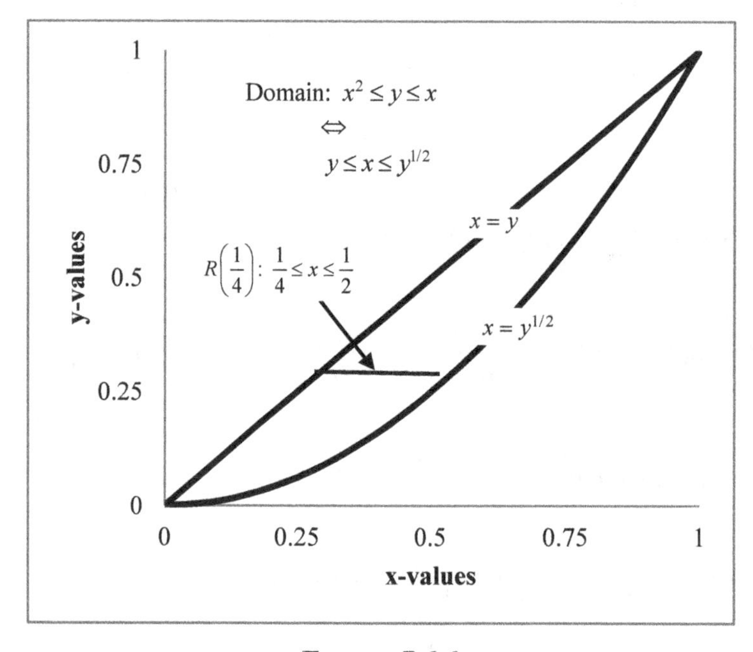

FIGURE 7.6.6

By definition, $f(x \mid Y=y) = \frac{f(x,y)}{g(y)} = \frac{15y}{g(y)};\ y < x < y^{1/2}$, where $g(y)$ is the marginal density function for Y evaluated at the given value $Y = y$. Referring to Figure 7.6.6. we see that,

$$g(y) = \int_y^{y^{1/2}} f(x,y)\,dx = \int_y^{y^{1/2}} 15y\,dx = 15y \cdot x\Big|_{x=y}^{x=y^{1/2}} = 15y(y^{1/2}-y);\ 0 \le y \le 1,$$

(which is the same as answer (E) for Exercise 7-32).

It follows that,

$$f(x \mid Y=y) = \frac{f(x,y)}{g(y)} = \frac{15y}{15y(y^{1/2}-y)} = \frac{1}{y^{1/2}-y};\ y < x < y^{1/2}.$$

Before continuing with parts (b) and (c) we pause to make a few observations.

Notes

(1) Giving the domain of $f(x|Y=y)$, $(y<x<y^{1/2})$ is just as important as stating the formula for the conditional density.

(2) It is equally important to take the time to sketch the graphs in these problems. Otherwise, you run a very substantial risk of getting the limits of integration wrong, and getting the domains wrong as well.

(3) Observe that in this problem the conditional density function $f(x|Y=y)$ is a constant (remember, $Y=y$ is given and fixed), and so, for each given value of y, $X|Y=y$ is a uniform random variable on the interval $y<x<y^{1/2}$. We will exploit this fact in parts (b) and (c).

For parts (b) and (c) you are given that $Y=.25=\frac{1}{4}$. Then,

$$f\left(x|Y=\tfrac{1}{4}\right) = \frac{1}{y^{1/2}-y};\ y<x<y^{1/2}\Bigg|_{y=1/4} = \frac{1}{\frac{1}{2}-\frac{1}{4}} = 4;\ \tfrac{1}{4}<x<\tfrac{1}{2}.$$

Since the conditional distribution is uniform on $\frac{1}{4}<x<\frac{1}{2}$, we have,

(b) $E[X|Y=.25]$ is the midpoint of the interval $= \dfrac{\frac{1}{4}+\frac{1}{2}}{2} = \dfrac{3}{8}$.

(c) Again, since $X|Y=\frac{1}{4}$ is uniform on the interval $\frac{1}{4}<x<\frac{1}{2}$ of length 0.25, we can calculate,

$$\Pr[X\le 0.4|Y=.25] = \frac{.4-.25}{.5-.25} = \frac{.15}{.25} = \frac{3}{5} = .6.$$ ❒

Exercise 7-41 Let X and Y have a joint uniform distribution inside the circle of radius 2, centered at the origin.

(a) Find the marginal densities for X and Y. Be sure to give the domains of the density functions.
(b) Find the conditional density of Y given that $X=1$.
(c) Find the probability that $Y>.8$ given that $X=1$.

Exercise 7-42 Alice purchases a policy from Wonderland Insurance to help defray the consequences of a loss random variable X. You are given that X is uniformly distributed over the unit interval $[0,1]$. Wonderland has a delightfully capricious claims settlement process, supervised by the Mad Hatter. The insurance payment Y is chosen randomly (that is, uniformly) from the interval $[0,x]$ for each given loss value, $X=x$. Customer complaints are handled by the Queen of Hearts.

(a) Find the joint density $f(x,y)$ of X and Y.

(b) Given that her loss X is one-half, find the probability that Alice's claim settlement Y is more than one-quarter.

(c) Given that Mr. Hatter settles for one-quarter, what is the probability that Alice's loss was more than one-half?

(d) Find the expected net loss (*Net Loss* $= X-Y$) experienced by Alice.

(e) What is the expected value of the loss Alice experienced given that Mr. Hatter settled for $Y=\frac{3}{4}$?

(f) What is the expected loss Alice experiences if she files a complaint?

Exercise 7-43 The joint density of X and Y is given by $f(x,y)=6xy(2-x-y)$; $0\le x\le 1,\ 0\le y\le 1$. Calculate the expected value of X given that $Y=y$.

Exercise 7-44 The joint density of X and Y is given by $f(x,y)=\frac{e^{-y}}{y}$; $0<x<y$, $0<y<\infty$. Calculate $E[X^2\,|\,Y=y]$.

Exercise 7-45 X and Y have joint uniform distribution on the region in the first quadrant bounded by the curve $y=4-x^2$. Find the standard deviation of Y given $X=1$.

Exercise 7-46 SOA EXAM P Sample Exam Questions #77
A device runs until either of two components fails, at which point the device stops running. The joint density function of the lifetimes of the two components, both measured in hours, is

$$f(x,y) = \frac{x+y}{8} \quad \text{for } 0<x<2 \text{ and } 0<y<2.$$

What is the probability that the device fails during its first hour of operation?

(A) 0.125 (B) 0.141 (C) 0.391 (D) 0.625 (E) 0.875

Exercise 7-47 SOA EXAM P Sample Exam Questions #125

The distribution of Y, given X, is uniform on the interval $[0, X]$. The marginal density of X is

$$f(x) = \begin{cases} 2x & \text{for } 0 < x < 1 \\ 0 & \text{otherwise} \end{cases}.$$

Determine the conditional density of X, given $Y = y$, where positive.

(A) 1 (B) 2 (C) $2x$ (D) $\frac{1}{y}$ (E) $\frac{1}{1-y}$

7.7 INDEPENDENCE AND COVARIANCE IN THE CONTINUOUS CASE

As in the discrete case, we will identify independence by the factorization of the joint density function into the product of the marginal density functions. This, as before, is equivalent to having a conditional density function be the same as the marginal density function for any given value of the second random variable.

Independent Continuous Random Variables

If X and Y are the marginal random variables for the continuous joint density function $f(x, y)$ on the region R, then the random variables are ***independent*** if

$$f(x,y) = f_X(x) \cdot f_Y(y) \text{ for all } x \text{ and } y; \ -\infty < x, y < \infty.$$

Otherwise the random variables are said to be ***dependent***.

Note

In order for X and Y to be independent it is necessary that the region R be a simple rectangle with sides parallel to the coordinate axes. For, if R is not such a rectangle then there will exist pairs (x, y) *outside* of R but with $f_X(x) > 0$ and $f_Y(y) > 0$. In such a case, $0 = f(x, y) \neq f_X(x) f_Y(y) > 0$, so that X and Y cannot be independent.

Most of the definitions and properties involving covariance and correlation carry over from the discrete case without change. For convenience we reiterate them here.

Covariance and Correlation Properties

Let $f(x,y)$ be a joint density function on the region R in the plane and let X and Y be the resulting marginal random variables.

(1) $Cov[X,Y] \overbrace{=}^{\text{def'n}} E[(X-\mu_X)(Y-\mu_Y)] = E[X\cdot Y]-E[X]\cdot E[Y]$.

(2) If X and Y are independent, then $E[X\cdot Y] = E[X]\cdot E[Y]$.

(3) If X and Y are independent, then $Cov[X,Y]=0$.

(4) $Var[aX+bY] = a^2Var[X]+b^2Var[Y]+2abCov[X,Y]$.

(5) If $Cov[X,Y]=0$ (in particular, if X and Y are independent), then $Var[X+Y] = Var[X]+Var[Y]$.

(6) More generally, if $(X_i)_{i=1}^m$ are pair-wise independent[1] and $S=\sum_{i=1}^m X_i$, then

$$Var[S] = \sum_{i=1}^m Var[X_i] = Var[X_1]+Var[X_2]+\cdots+Var[X_m].$$

(7) The ***correlation between X and Y*** is defined to be $\rho_{X,Y} = \dfrac{Cov[X,Y]}{\sigma_X\cdot\sigma_Y}$.

For any pair X and Y, $-1\le\rho_{XY}\le 1$.

The only item requiring proof is (2), which proceeds as in the discrete case, but with integrals instead of sums. If X and Y are independent then R is a rectangle of the form $(a,b)\times(c,d)$ (where these endpoints could be infinite). Then

$$\begin{aligned}E[X\cdot Y] = \iint_R xy\, f(x,y)\,dA &= \iint_R xy\, f_X(x)f_Y(y)\,dA \\ &= \int_a^b\int_c^d \big(x f_X(x)\big)\big(y f_Y(y)\big)\,dy\,dx \\ &= \left(\int_a^b x f_X(x)\,dx\right)\cdot\left(\int_c^d y f_Y(y)\,dy\right) = E[X]\cdot E[Y].\end{aligned}$$

Example 7.7-1 Correlation (Example 7.5-6 continued)

Let $f(x,y)=6x$ be a joint density function on the region R in the plane described by $0\le x\le y\le 1$. Calculate the correlation coefficient between X and Y.

---

[1] As in property (5), the conclusion holds assuming only that $Cov\left[X_i,X_j\right]=0$ for all i,j with $i\neq j$. However the result is mainly applied under the assumption of independence, or pairwise independence, and is simpler to state in that case.

Solution

This is a good opportunity for the reader to take an inventory of what needs to be found in order to answer this example:

(1) The definition of correlation, $\rho = \dfrac{Cov(X,Y)}{\sigma_X \cdot \sigma_Y}$.

(2) The covariance formula, $Cov(X,Y) = E[X \cdot Y] - E[X] \cdot E[Y]$, which will require a double integral to calculate $E[X \cdot Y]$.

(3) The marginal density functions $f_X(x)$ and $f_Y(y)$ in order to determine the means and the variances of X and Y, respectively. Fortunately, this groundwork has been laid in Example 7.5-6, from which we have:

$$E[X] = \frac{1}{2},\ Var[X] = \frac{1}{20},\ E[Y] = \frac{3}{4},$$

and

$$Var[Y] = \frac{3}{80}.$$

Since we can choose whichever order of integration we want we will use $dx\,dy$:

$$\begin{aligned} E[XY] &= \int_0^1 \int_0^y (xy)(6x)\,dx\,dy = \int_0^1 \int_0^y 6x^2 y\,dx\,dy \\ &= \int_0^1 (6y)\left(\frac{x^3}{3}\bigg|_{x=0}^{y} \right) dy \\ &= \int_0^1 2y^4\,dy = \frac{2}{5}. \end{aligned}$$

Then,

$$Cov[X,Y] = E[XY] - E[X] \cdot E[Y] = \frac{2}{5} - \frac{1}{2} \cdot \frac{3}{4} = \frac{1}{40}.$$

Finally,

$$\rho = \frac{Cov[X,Y]}{\sigma_X \cdot \sigma_Y} = \frac{\frac{1}{40}}{\sqrt{\frac{1}{20} \cdot \frac{3}{80}}} = \frac{1}{\sqrt{3}}.$$ ❐

Example 7.7-2 Continuous Joint Distribution

Suppose that X and Y have a joint uniform distribution on the triangle with endpoints $(-2,0)$, $(0,2)$, and $(2,0)$.

(a) Find the joint density function, $f(x,y)$.

(b) Find the marginal density functions $f_X(x)$ and $f_Y(y)$.

(c) Find the conditional density function $f_X(x|Y=1)$.

(d) Are X and Y independent?

(e) Calculate the correlation coefficient, ρ, for X and Y.

Solution

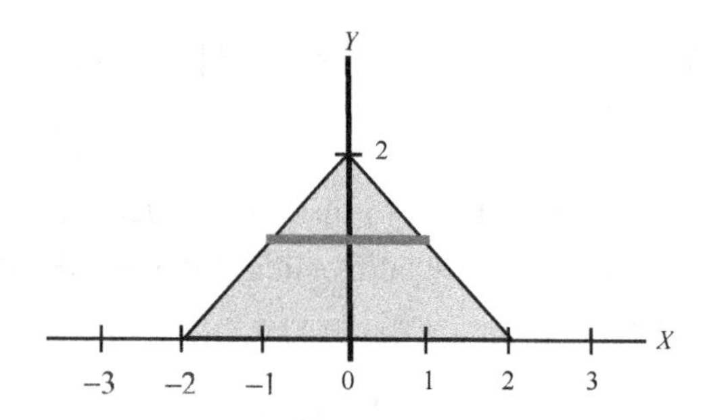

(a) Since X and Y have a joint uniform distribution, probabilities are just ratios of areas. In particular, the joint density function is constant and must equal 1 divided by the area of the triangle, which is 4. The two diagonal sides of the triangle are described by $x = y-2; 0 \le y \le 2$, and $x = 2-y; 0 \le y \le 2$, respectively. Thus,

$$f(x,y) = \frac{1}{4}, \quad 0 \le y \le 2 \text{ and } y-2 \le x \le 2-y$$

(b) $$f_X(x) = \begin{cases} \int_0^{x+2} \frac{1}{4}dy = \frac{x+2}{4} & \text{if } -2 \le x \le 0 \\ \int_0^{2-x} \frac{1}{4}dy = \frac{2-x}{4} & \text{if } 0 < x \le 2 \end{cases}$$

$$f_Y(y) = \begin{cases} \int_{y-2}^{2-y} \frac{1}{4}\,dx = \frac{4-2y}{4} & \text{if } 0 \le y \le 2 \\ 0 & \text{otherwise} \end{cases}$$

(c) $$f_X(x|Y=1) = \frac{f(x,1)}{f_Y(1)} = \begin{cases} \frac{1/4}{1/2} = .5 & \text{for } -1 \le x \le 1 \\ 0 & \text{otherwise} \end{cases}$$

(d) No. For example, $\frac{1}{4} = f\left(0,\frac{1}{2}\right) \ne f_X(0)\cdot f_Y\left(\frac{1}{2}\right) = \frac{1}{2}\cdot\frac{3}{4} = \frac{3}{8}$.

Remember that $f(x,y)$ must be defined on a rectangle for random variables X and Y to *possibly* be independent.

(e) By symmetry, $Cov[X,Y] = 0$. Or, going through the calculation,

$$\int_0^2\int_{y-2}^{2-y}(xy)\left(\frac{1}{4}\right)dx\,dy = \frac{1}{8}\int_0^2\left[(2-y)^2-(y-2)^2\right]y\,dy = 0,$$

since $(2-y)^2 = (y-2)^2$.

Since the covariance of X and Y is zero, the correlation is as well. ❐

Note: This provides another example, this time with continuous distributions, of a pair of dependent random variables with zero covariance.

Exercise 7-48 John and Mary arrive under the clock tower independently. Let X be John's arrival time and let Y be Mary's arrival time. If John arrives first and Mary is not there then he will leave. If Mary arrives first then she will wait up to one hour before leaving. John's arrival time, X, is exponentially distributed with mean of 1. Mary's arrival time has density $f(y)=\frac{2}{9}y;\ 0\leq y\leq 3$. Calculate the probability that they will meet.

Exercise 7-49 Let X be uniformly distributed on the interval $[0,2]$ and Y be uniformly distributed on the interval [0,1], and suppose that the random variables are independent.

(a) Find the joint density function $f(x,y)$.

(b) Derive the marginal density function for Y from the joint density function.

(c) Compute $E\left[\sqrt{X}-Y^3\right]$.

Exercise 7-50 SOA EXAM P Sample Exam Questions #105

Let X and Y be continuous random variables with joint density function

$$f(x,y) = \begin{cases} \frac{8}{3}xy & \text{for } 0\leq x\leq 1,\ x\leq y\leq 2x \\ 0 & \text{otherwise} \end{cases}.$$

Calculate the covariance of X and Y.

(A) 0.04 (B) 0.25 (C) 0.67 (D) 0.80 (E) 1.24

Exercise 7-51 SOA EXAM P Sample Exam Questions #111

Once a fire is reported to a fire insurance company, the company makes an initial estimate, X, of the amount it will pay to the claimant for the fire loss. When the claim is finally settled, the company pays an amount, Y, to the claimant. The company has determined that X and Y have the joint density function

$$f(x,y) = \frac{2}{x^2(x-1)}y^{-(2x-1)/(x-1)} \quad x>1,\ y>1.$$

Given that the initial claim estimated by the company is 2, determine the probability that the final settlement amount is between 1 and 3.

(A) $\frac{1}{9}$ (B) $\frac{2}{9}$ (C) $\frac{1}{3}$ (D) $\frac{2}{3}$ (E) $\frac{8}{9}$

Exercise 7-52 Suppose that X and Y are distributed uniformly on the triangle with endpoints $(-2,0)$, $(0,2)$, and $(2,0)$. Find $E[Y^2|X=1]$.

Exercise 7-53 Suppose that X and Y are uniformly distributed on the unit circle.

(a) Find the joint density function.

(b) Find $\Pr\left[(X^2+Y^2)>.5\right]$.

(c) Find $\Pr[X<Y]$.

(d) Find general formulas for the conditional density functions $f_X(x|Y=y)$ and $f_Y(y|X=x)$.

(e) Are X and Y independent?

Exercise 7-54 SOA EXAM P Sample Exam Questions #101

The profit for a new product is given by $Z=3X-Y-5$. X and Y are independent random variables with $Var[X]=1$ and $Var[Y]=2$. What is the variance of Z?

(A) 1 (B) 5 (C) 7 (D) 11 (E) 16

Exercise 7-55 SOA EXAM P Sample Exam Questions #90

An insurance company sells two types of auto insurance policies: Basic and Deluxe. The time until the next Basic Policy claim is an exponential random variable with mean two days. The time until the next Deluxe Policy claim is an independent exponential random variable with mean three days. What is the probability that the next claim will be a Deluxe Policy claim?

(A) 0.172 (B) 0.223 (C) 0.400 (D) 0.487 (E) 0.500

Exercise 7-56 SOA EXAM P Sample Exam Questions #104

A joint density function is given by

$$f(x,y) = \begin{cases} kx & \text{for } 0<x<1,\ 0<y<1 \\ 0 & \text{otherwise} \end{cases},$$

where k is a constant. What is the $Cov(X,Y)$?

(A) $-\frac{1}{6}$ (B) 0 (C) $\frac{1}{9}$ (D) $\frac{1}{6}$ (E) $\frac{2}{3}$

Exercise 7-57 SOA EXAM P Sample Exam Questions #115

The stock prices of two companies at the end of any given year are modeled with random variables X and Y that follow a distribution with joint density function,

$$f(x,y) = \begin{cases} 2x & \text{for } 0<x<1,\ x<y<x+1 \\ 0 & \text{otherwise} \end{cases}.$$

What is the conditional variance of Y given that $X = x$?

(A) $\frac{1}{12}$ (B) $\frac{7}{6}$ (C) $x+\frac{1}{2}$ (D) $x^2-\frac{1}{6}$ (E) $x^2+x+\frac{1}{3}$

Exercise 7-58 SOA EXAM P Sample Exam Questions #110

Let X and Y be continuous random variables with joint density function

$$f(x,y) = \begin{cases} 24xy & \text{for } 0<x<1 \text{ and } 0<y<1-x \\ 0 & \text{otherwise} \end{cases}.$$

Calculate $\Pr\left[Y<X \middle| X=\frac{1}{3}\right]$.

(A) $\frac{1}{27}$ (B) $\frac{2}{27}$ (C) $\frac{1}{4}$ (D) $\frac{1}{3}$ (E) $\frac{4}{9}$

Exercise 7-59 SOA EXAM P Sample Exam Questions #119

An auto insurance policy will pay for damage to both the policyholder's car and the other driver's car in the event that the policyholder is responsible for an accident. The size of the payment for damage to the policyholder's car, X, has a marginal density function of 1 for $0<x<1$. Given $X = x$, the size of the payment for damage to the other driver's car, Y, has conditional density of 1 for $x<y<x+1$. If the policyholder is responsible for an accident, what is the probability that the payment for damage to the other driver's car will be greater than 0.500?

(A) $\frac{3}{8}$ (B) $\frac{1}{2}$ (C) $\frac{3}{4}$ (D) $\frac{7}{8}$ (E) $\frac{15}{16}$

Exercise 7-60 SOA/CAS 2000 Sample Examination #18

Let $F(x)$ represent the fraction of payroll earned by the highest paid fraction x of employees in a company (for example, $F(0.2)=0.5$ means that the highest paid 20% of workers earn 50% of the payroll). Gini's index of inequality, G, is one way to measure how evenly payroll is distributed among all employees and is defined as follows:

$$G = 2\int_0^1 \left|x-F(x)\right| dx$$

In a certain company, the distribution of payroll is described by the density function:

$$f(x) = 3(1-x)^2\,;\ 0 \le x \le 1.$$

Calculate G for this company.

(A) 0.0 (B) 0.4 (C) 0.5 (D) 1.0 (E) 1.5

7.8 THE MULTINOMIAL DISTRIBUTION

In this section we present a particular example of a discrete multivariate distribution. Recall that a binomial random variable describes the number of successes in independent repetitions of an experiment with just two outcomes, success and failure. As we saw in Chapter 2, multinomial coefficients are generalizations of binomial coefficients, and may be applied to the situation where there are more than two outcomes to the experiment.

Multinomial Random Variables

Consider a probability experiment with r distinct outcomes, and that $p_1, p_2, \ldots, p_r$ represent the probabilities of the respective outcomes, so that the p_i's are positive and sum to one. Consider n independent repetitions of the experiment and let X_i represent the number of times outcome i occurs. Then,

(1) $X_1 + X_2 + \cdots + X_r = n.$

(2) Each X_i has binomial distribution with parameters n and p_i.

(3) Given non-negative integers k_i; $i = 1, \ldots, r$, with $k_1 + \cdots + k_r = n$, then

$$\Pr[X_1 = k_1, X_2 = k_2, \ldots, X_r = k_r] = \binom{n}{k_1 \; k_2 \cdots k_r} p_1^{k_1} p_2^{k_2} \cdots p_r^{k_r}.$$

(4) For $i \neq j$, $Cov\left[X_i, X_j\right] = -np_i p_j$

Proof

Property (1) is self-evident. Property (2) follows immediately from regarding the ith outcome as a success and any other outcome as a failure. Property (3) is a straightforward extension of the reasoning for a binomial distribution to the multinomial case. One can consider the outcome of n repetitions of the experiment as an n-letter word with the letters $A_1, A_2, \ldots, A_r$ representing the particular outcomes. Then basic combinatorial principles show that the number of n-letter words with A_1 appearing k_1 times, A_2 appearing k_2 times, and so forth, is the multinomial coefficient $\binom{n}{k_1 \; k_2 \cdots k_r}$. By independence, the probability of each such word is $p_1^{k_1} p_2^{k_2} \cdots p_r^{k_r}$. It then follows that

$$\Pr[X_1 = k_1, X_2 = k_2, \ldots, X_r = k_r] = \binom{n}{k_1 \; k_2 \cdots k_r} p_1^{k_1} p_2^{k_2} \cdots p_r^{k_r}$$

The demonstration of property (4) will be deferred to Chapter 8.

7.9 BIVARIATE NORMAL DISTRIBUTIONS

A noteworthy special case of jointly distributed random variables involves the multivariate normal distribution. This is a topic best dealt with in general using the tools of linear algebra, but because of its importance we will treat the two-dimensional case. The integrals and formulas are complex algebraically, but the results are very useful in statistics, where normal distributions abound.

The formula for the joint density function is given by,

$$f(x,y) = \frac{1}{2\pi\sigma_X\sigma_Y\sqrt{1-\rho^2}} \cdot e^{-\frac{1}{2\cdot(1-\rho^2)}\left[\left(\frac{x-\mu_X}{\sigma_X}\right)^2 - 2\rho\left(\frac{x-\mu_X}{\sigma_X}\right)\left(\frac{y-\mu_Y}{\sigma_Y}\right)+\left(\frac{y-\mu_Y}{\sigma_Y}\right)^2\right]}; \quad -\infty < x, y < \infty.$$

While the notation used for all of the parameters is suggestive, and as we shall see, consistent with their usual meanings, there is no reason, a priori, to believe this. What we establish in the proofs below is that this is in fact a proper joint density function, yielding normal marginal distributions for X and Y with parameters μ_X, σ_X and μ_Y, σ_Y, respectively, and with correlation coefficient ρ. We summarize these properties next. Proofs are included for completeness but may be skipped by those readers prepared to accept the results on faith.

Properties of the Bivariate Normal Distribution

$$f(x,y) = \frac{1}{2\pi\sigma_X\sigma_Y\sqrt{1-\rho^2}} e^{-\frac{1}{2(1-\rho^2)}\left[\left(\frac{x-\mu_X}{\sigma_X}\right)^2 - 2\rho\left(\frac{x-\mu_X}{\sigma_X}\right)\left(\frac{y-\mu_Y}{\sigma_Y}\right)+\left(\frac{y-\mu_Y}{\sigma_Y}\right)^2\right]}$$

constitutes a joint probability distribution in the $x-y$ plane, for which,

(1) The marginal density distribution for X is normal with mean μ_X and variance σ_X^2.

(2) The marginal density distribution for Y is normal with mean μ_Y and variance σ_Y^2.

(3) The correlation coefficient for X and Y is ρ.

(4) The conditional density function $f_X(x \mid Y=y)$ is normal with,

   (a) $E[X \mid Y=y] = \mu_X + \frac{\sigma_X}{\sigma_Y}\rho(y-\mu_Y)$, and

   (b) $Var[X \mid Y=y] = \sigma_X^2(1-\rho^2)$.

(5) The conditional density function $f_Y(y \mid X=x)$ is normal with,

   (a) $E[Y \mid X=x] = \mu_Y + \frac{\sigma_Y}{\sigma_X}\rho(x-\mu_X)$, and

   (b) $Var[Y \mid X=x] = \sigma_Y^2(1-\rho^2)$.

(6) X and Y are independent if and only if $\rho = 0$.

Proof

The formula for the joint density uses the standardized versions of x and y, so we will abbreviate with $u=\frac{x-\mu_X}{\sigma_X}$ and $v=\frac{y-\mu_Y}{\sigma_Y}$. In the integrals that follow, this change of variables implies that $dx=\sigma_X\,du$ and $dy=\sigma_Y\,dv$. Recall the functional form of the density function for the standard normal is $f_Z(z)=\frac{1}{\sqrt{2\pi}}e^{-(z^2/2)}$. In the calculations below we will be making use of the following results concerning the standard normal:

$$\text{(i)}\ \int_{-\infty}^{\infty} f_Z(z)dz \;=\; \frac{1}{\sqrt{2\pi}}\int_{-\infty}^{\infty} e^{-(z^2/2)}\,dz \;=\; 1$$

$$\text{(ii)}\ \mu_Z=\int_{-\infty}^{\infty} z\,f_Z(z)dz \;=\; \frac{1}{\sqrt{2\pi}}\int_{-\infty}^{\infty} z\,e^{-(z^2/2)}\,dz \;=\; 0$$

$$\text{(iii)}\ \sigma_Z^2 \;=\; \int_{-\infty}^{\infty}(z-\mu_Z)^2 f_Z(z)dz \;=\; \frac{1}{\sqrt{2\pi}}\int_{-\infty}^{\infty} z^2\,e^{-(z^2/2)}\,dz \;=\; 1.$$

First, we must demonstrate that $\int_{-\infty}^{\infty}\int_{-\infty}^{\infty} f(x,y)\,dy\,dx \;=\; 1$, and we begin with the "inside" integral, $g(x)=\int_{-\infty}^{\infty} f(x,y)dy$, which will provide the marginal density function for X. Thus,

$$g(x) \;=\; \int_{-\infty}^{\infty}\frac{1}{2\pi\sigma_X\sigma_Y\sqrt{1-\rho^2}}\,e^{-\frac{1}{2(1-\rho^2)}\left[\left(\frac{x-\mu_X}{\sigma_X}\right)^2-2\rho\left(\frac{x-\mu_X}{\sigma_X}\right)\left(\frac{y-\mu_Y}{\sigma_Y}\right)+\left(\frac{y-\mu_Y}{\sigma_Y}\right)^2\right]}\,dy$$

$$=\; \frac{1}{2\pi\sigma_X\sigma_Y\sqrt{1-\rho^2}}\int_{-\infty}^{\infty} e^{-\frac{1}{2(1-\rho^2)}[(u)^2-2\rho(u)(v)+(v)^2]}\,\sigma_Y\,dv.$$

We now complete the square on v in the integrand by writing

$$(u)^2-2\rho(u)(v)+(v)^2 \;=\; v^2-2\rho uv+\rho^2u^2+(1-\rho^2)u^2 \;=\; (v-\rho u)^2+(1-\rho^2)u^2.$$

Thus,

$$g(x) \;=\; \frac{1}{2\pi\sigma_X\sqrt{1-\rho^2}}\int_{-\infty}^{\infty} e^{-\frac{1}{2(1-\rho^2)}\left[(v-\rho u)^2+(1-\rho^2)u^2\right]}\,dv$$

$$=\; \frac{e^{-(u^2/2)}}{2\pi\sigma_X\sqrt{1-\rho^2}}\int_{-\infty}^{\infty} e^{-\frac{1}{2(1-\rho^2)}\left[(v-\rho u)^2\right]}\,dv.$$

If we now make the substitution

$$z^2 = \frac{1}{1-\rho^2}\left[(v-\rho u)^2\right],$$

then

$$z = \frac{1}{\sqrt{1-\rho^2}}(v-\rho u) \text{ and } dv = \sqrt{1-\rho^2}\,dz.$$

Then,

$$g(x) = \frac{e^{-(u^2/2)}}{2\pi\sigma_X\sqrt{1-\rho^2}}\int_{-\infty}^{\infty} e^{-(z^2/2)}\sqrt{1-\rho^2}\,dz$$

$$= \frac{e^{-(u^2/2)}}{\sqrt{2\pi}\,\sigma_X}\int_{-\infty}^{\infty}\frac{1}{\sqrt{2\pi}}e^{-(z^2/2)}dz = \frac{e^{-(u^2/2)}}{\sqrt{2\pi}\,\sigma_X} = \frac{1}{\sqrt{2\pi}\,\sigma_X}e^{-\frac{1}{2}\left(\frac{x-\mu_X}{\sigma_X}\right)^2}.$$

This shows that $g(x) = f_X(x)$ is indeed normal with parameters μ_X and σ_X^2. Finally,

$$\int_{-\infty}^{\infty}\int_{-\infty}^{\infty} f(x,y)\,dy\,dx = \int_{-\infty}^{\infty} g(x)\,dx = 1$$

since, by the above, $g(x) = f_X(x)$ is a probability density function. By symmetry it also follows that $f_Y(y) = \frac{1}{\sqrt{2\pi}\,\sigma_Y}e^{-\frac{1}{2}\left(\frac{y-\mu_Y}{\sigma_Y}\right)^2}$.

Next, we calculate

$$Cov[X,Y] = \int_{-\infty}^{\infty}\int_{-\infty}^{\infty}(x-\mu_X)(y-\mu_Y)f(x,y)\,dy\,dx$$

$$= \frac{1}{2\pi\sigma_X\sigma_Y\sqrt{1-\rho^2}}\int_{-\infty}^{\infty}\int_{-\infty}^{\infty}(x-\mu_X)(y-\mu_Y)$$

$$\times e^{-\frac{1}{2(1-\rho^2)}\left[\left(\frac{x-\mu_X}{\sigma_X}\right)^2 - 2\rho\left(\frac{x-\mu_X}{\sigma_X}\right)\left(\frac{y-\mu_Y}{\sigma_Y}\right) + \left(\frac{y-\mu_Y}{\sigma_Y}\right)^2\right]}\,dx\,dy$$

$$= \frac{1}{2\pi\sigma_X\sigma_Y\sqrt{1-\rho^2}}\int_{-\infty}^{\infty}\int_{-\infty}^{\infty}(u\sigma_X)(v\sigma_Y)e^{-\frac{1}{2(1-\rho^2)}\left[u^2-2\rho uv+v^2\right]}(\sigma_X du)(\sigma_Y dv)$$

$$= \frac{\sigma_X\sigma_Y}{2\pi\sqrt{1-\rho^2}}\int_{-\infty}^{\infty}\int_{-\infty}^{\infty} uv\,e^{-\frac{1}{2(1-\rho^2)}\left[u^2-2\rho uv+\rho^2v^2+(1-\rho^2)v^2\right]}\,du\,dv$$

$$= \frac{\sigma_X\sigma_Y}{2\pi\sqrt{1-\rho^2}}\int_{-\infty}^{\infty}\int_{-\infty}^{\infty} uve^{-(v^2/2)}\,e^{-\frac{1}{2(1-\rho^2)}[u^2-2\rho uv+\rho^2v^2]}\,du\,dv$$

$$= \frac{\sigma_X\sigma_Y}{2\pi\sqrt{1-\rho^2}}\int_{-\infty}^{\infty} ve^{-(v^2/2)}\int_{-\infty}^{\infty} u\,e^{-\frac{1}{2(1-\rho^2)}[u-\rho v]^2}\,du\,dv.$$

Now we substitute $z^2 = \frac{1}{(1-\rho^2)}[u-\rho v]^2$. This implies $u = z\sqrt{1-\rho^2}+\rho v$ and $du = \sqrt{1-\rho^2}\,dz$.

$$\begin{aligned}
Cov[X,Y] &= \frac{\sigma_X\sigma_Y}{2\pi\sqrt{1-\rho^2}}\int_{-\infty}^{\infty} ve^{-(v^2/2)}\left[\int_{-\infty}^{\infty}\left(z\sqrt{1-\rho^2}+\rho v\right)e^{-(z^2/2)}\sqrt{1-\rho^2}\,dz\right]dv \\
&= \frac{\sigma_X\sigma_Y}{2\pi\sqrt{1-\rho^2}}\int_{-\infty}^{\infty} ve^{-(v^2/2)}\left[(1-\rho^2)\sqrt{2\pi}\int_{-\infty}^{\infty}\frac{1}{\sqrt{2\pi}}ze^{-(z^2/2)}\,dz\right. \\
&\qquad \left.+\rho v\sqrt{1-\rho^2}\sqrt{2\pi}\int_{-\infty}^{\infty}\frac{1}{\sqrt{2\pi}}e^{-(z^2/2)}\,dz\right]dv \\
&= \frac{\sigma_X\sigma_Y}{2\pi\sqrt{1-\rho^2}}\int_{-\infty}^{\infty} ve^{-(v^2/2)}\left[0+\rho v\sqrt{1-\rho^2}\sqrt{2\pi}\,(1)\right]dv \\
&= \sigma_X\sigma_Y\rho\int_{-\infty}^{\infty}\frac{1}{\sqrt{2\pi}}v^2e^{-(v^2/2)}\,dv = \sigma_X\sigma_Y\rho.
\end{aligned}$$

Thus $\rho = \frac{Cov[X,Y]}{\sigma_X\sigma_Y}$ is indeed the correlation coefficient.

By definition, the conditional density function of, say, Y given $X = x$ is calculated as,

$$\begin{aligned}
f_Y(y\,|\,X=x) = \frac{f(x,y)}{f_X(x)} &= \frac{\dfrac{1}{2\pi\sigma_X\sigma_Y\sqrt{1-\rho^2}}e^{-\frac{1}{2(1-\rho^2)}\left[(u)^2-2\rho(u)(v)+(v)^2\right]}}{\dfrac{1}{\sqrt{2\pi}\,\sigma_X}e^{-(1/2)(u)^2}} \\
&= \frac{e^{-\frac{1}{2(1-\rho^2)}\left[u^2-2\rho uv+v^2-(1-\rho^2)u^2\right]}}{\sqrt{2\pi}\,\sigma_Y\sqrt{1-\rho^2}} \\
&= \frac{e^{-\frac{1}{2(1-\rho^2)}\left[v^2-2\rho uv+\rho^2u^2\right]}}{\sqrt{2\pi}\,\sigma_Y\sqrt{1-\rho^2}} \\
&= \frac{e^{-\frac{1}{2(1-\rho^2)}[v-\rho u]^2}}{\sqrt{2\pi}\,\sigma_Y\sqrt{1-\rho^2}} \\
&= \frac{e^{-\frac{1}{2(1-\rho^2)}\left[\left(\frac{y-\mu_Y}{\sigma_Y}\right)-\rho\left(\frac{x-\mu_X}{\sigma_X}\right)\right]^2}}{\sqrt{2\pi}\,\sigma_Y\sqrt{1-\rho^2}} = \frac{e^{-\frac{1}{2}\left[\frac{y-\left(\mu_Y+\frac{\sigma_Y}{\sigma_X}\rho(x-\mu_X)\right)}{\sigma_Y\sqrt{1-\rho^2}}\right]^2}}{\sqrt{2\pi}\,\sigma_Y\sqrt{1-\rho^2}}.
\end{aligned}$$

This shows properties 5(a) and 5(b), with properties 4(a) and 4(b) following similarly.

Finally, for property (6), the independence of X and Y always implies $\rho = 0$. Conversely, if $\rho = 0$ then the density formula for the bivariate normal factors into the product of the density functions for X and Y, showing that they are independent.

Note: Property (6) is noteworthy for the fact that with X and Y normal, the covariance (or equivalently, the correlation) being zero is a sufficient condition for independence. We have seen examples (please refer to Examples 7.4-2 and 7.7-2) where the covariance is zero but the random variables are not independent.

Example 7.9-1 Bivariate Normal

Let $f(x,y) = \frac{5}{3\pi} e^{-\frac{25}{18}\left[(x-3)^2 - \frac{16}{5}(x-3)(y+1) + 4(y+1)^2\right]}$; $-\infty < x, y < \infty$. Show that this is a bivariate normal density function.

(a) Find the parameters of the marginal random variables, X and Y.
(b) Find the correlation coefficient between X and Y.
(c) Draw a 3-dimensional graph of $f(x,y)$.
(d) Draw a 2-dimensional graph showing the level sets of $f(x,y)$.

Solution

The general form of the bivariate normal density function is

$$f(x,y) = \frac{1}{2\pi\sigma_X\sigma_Y\sqrt{1-\rho^2}} e^{-\frac{1}{2(1-\rho^2)}\left[\left(\frac{x-\mu_X}{\sigma_X}\right)^2 - 2\rho\left(\frac{x-\mu_X}{\sigma_X}\right)\left(\frac{x-\mu_Y}{\sigma_Y}\right) + \left(\frac{x-\mu_Y}{\sigma_Y}\right)^2\right]},$$

so by inspection we have, $\mu_X = 3$, $\mu_Y = -1$, $\sigma_X = 1$, and $\sigma_Y = \frac{1}{2}$. Then $\frac{16}{5} = \frac{2\rho}{\sigma_X\sigma_Y} = 4\rho$ implies $\rho = \frac{4}{5}$. Next, we verify that $\frac{25}{18} = \frac{1}{2(1-\rho^2)}$ and that $\frac{5}{3\pi} = \frac{1}{2\pi\sigma_X\sigma_Y\sqrt{1-\rho^2}}$, so that $f(x,y)$ is indeed the appropriate bivariate density function.

Figure 7.9.1 contains the graphs of $f(x,y)$. Note that the 3-dimensional graph is a bell centered along the pole $(\mu_X, \mu_Y) = (3,-1)$. The cross-sectional curves shown in the figure represent the un-scaled versions of the conditional density functions of X for a given value of Y. The level sets in the second graph are where the quadratic form $\left[(x-3)^2 - \frac{16}{5}(x-3)(y+1) + 4(y+1)^2\right]$ is constant. These are ellipses in the plane. Since the correlation coefficient $\rho = \frac{4}{5}$, X and Y are highly positively correlated and the resulting ellipses are fairly elongated. If $\rho = 0$ and $\sigma_X = \sigma_Y$, then the cross-product term drops out of the quadratic form and the level sets will be circular.

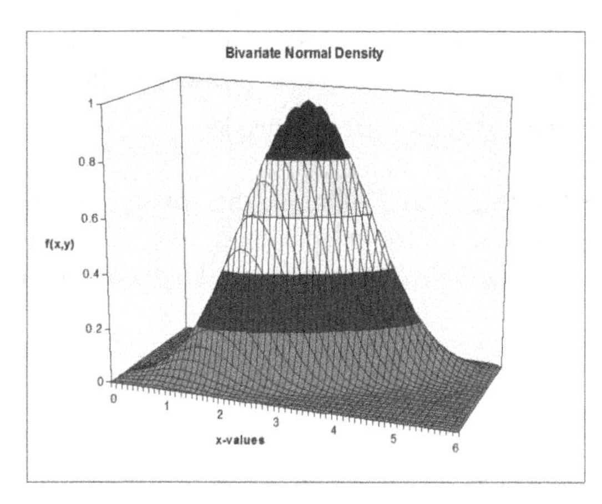

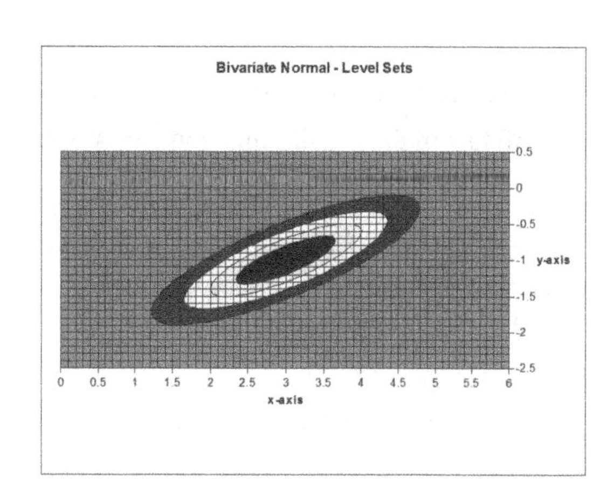

FIGURE 7.9.1

Exercise 7-61 Suppose random variables X and Y have bivariate normal distribution with parameters $\mu_X = -2, \sigma_X = 3, \mu_Y = 4, \sigma_Y = 5,$ and with correlation coefficient $\rho = .4$. Find the standard deviation for X given $Y = 6$.

Exercise 7-62 Suppose random variables X and Y have bivariate normal distribution with parameters $\mu_X = 8, \sigma_X = 3, \mu_Y = -3, \sigma_Y = 5,$ and with correlation coefficient $\rho = -.3$. Find $\Pr(X < 9 | Y = -2)$.

Exercise 7-63 Suppose $f(x,y) = \dfrac{1}{2.4\pi} e^{-\frac{1}{2(.36)}\left[x^2 - 1.6x\left(\frac{y+1}{2}\right) + \left(\frac{y+1}{2}\right)^2\right]}$ for all x and y.

(a) Find ρ.

(b) Determine $\Pr(X > .5)$.

Exercise 7-64 SOA EXAM P Sample Exam Questions #42

For Company A there is a 60% chance that no claim is made during the coming year. If one or more claims are made, the total claim amount is normally distributed with mean 10,000 and standard deviation 2,000. For Company B there is a 70% chance that no claim is made during the coming year. If one or more claims are made, the total claim amount is normally distributed with mean 9,000 and standard deviation 2,000. Assume the total claim amounts of the two companies are independent. What is the probability that, in the coming year, Company B's total claim amount will exceed Company A's total claim amount?

(A) 0.180 (B) 0.185 (C) 0.217 (D) 0.223 (E) 0.240

7.10 MOMENT GENERATING FUNCTION FOR A JOINT DISTRIBUTION

The computations of expectations involving powers of both X and Y in a joint distribution can sometimes be facilitated by employing the joint moment generating function. As in the single variable case, we can relate these joint moments to differentiation of the moment generating function. In the case of joint distributions the moment generating function has two independent variables and we use partial derivatives to generate the appropriate moments.

Joint Moment Generating Function

Let X and Y be jointly distributed random variables. We define the joint moment generating function $M_{X,Y}(s,t)$ by

$$M_{X,Y}(s,t) = E[e^{sX+tY}].$$

Then

(1) $$E[X^m Y^n] = \frac{\partial^{(m+n)}\left[M_{X,Y}(0,0)\right]}{\partial s^{(m)}\,\partial t^{(n)}}.$$

(2) X and Y are independent if and only if $M_{X,Y}(s,t) = M_X(s)\cdot M_Y(t)$.

Proof: The demonstration of (1) is similar to the single random variable case and depends on the fact that, "the derivative of a sum is the sum of the derivatives." This holds for the continuous case (definite integrals) as well as the discrete case. Using this observation, we see that,

$$\frac{\partial\left[M_{X,Y}(s,t)\right]}{\partial s} = E\left[X\cdot e^{sX+tY}\right] \text{ and } \frac{\partial\left[M_{X,Y}(s,t)\right]}{\partial t} = E\left[Y\cdot e^{sX+tY}\right].$$

Thus, successive partial derivatives with respect to s and t bring down successive products of X or Y, respectively. The result is,

$$\frac{\partial^{(m+n)}\left[M_{X,Y}(s,t)\right]}{\partial s^{(m)}\,\partial t^{(n)}} = E\left[X^m\cdot Y^n\cdot e^{sX+tY}\right].$$

Then,

$$\frac{\partial^{(m+n)}\left[M_{X,Y}(0,0)\right]}{\partial s^{(m)}\,\partial t^{(n)}} = E\left[X^m\cdot Y^n\right].$$

For property (2) we note first that if X and Y are independent then,

$$M_{X,Y}(s,t) = E\left[e^{sX+tY}\right] = E\left[e^{sX}\cdot e^{tY}\right] = \overbrace{E\left[e^{sX}\right]\cdot E\left[e^{tY}\right]}^{\text{using the independence of } X \text{ and } Y} = \overbrace{M_X(s)\cdot M_Y(t)}^{\text{defn of the MGF}}.$$

The fact that the factorization of the moment-generating function implies independence relies on the one-to-one correspondence between joint distributions and their moment-generating functions[2].

[2] This uniqueness property, analogous to the one-variable case, can be proven in more advanced courses.

Assume now that $M_{X,Y}(s,t) = M_X(s) \cdot M_Y(t)$. The right-hand side is the joint MGF that would result if X and Y were independent. It follows by the uniqueness of the joint MGF that $M_{X,Y}(s,t)$ is the joint MGF for a pair X and Y which must, indeed, be independent.

Example 7.10-1 Bivariate Moment Generating Function

Let U and V be independent standard normal random variables. Let $X = U+V$ and $Y = U-V$. Calculate $E[X^2 Y^2]$.

Solution

Recall that the MGF for the standard normal Z is given by $M_Z(t) = e^{(t^2/2)}$. Now, by definition, we have the joint moment generating function,

$$\begin{aligned} M_{X,Y}(s,t) &= E[e^{sX+tY}] \\ &= E[e^{s(U+V)+t(U-V)}] \\ &= E[e^{(s+t)U+(s-t)V}] \\ &= E[e^{(s+t)U} e^{(s-t)V}] \\ &= \overbrace{E\left[e^{(s+t)U}\right]E\left[e^{(s-t)V}\right]}^{\text{Since } U \text{ and } V \text{ independent}} \\ &= M_Z(s+t)M_Z(s-t) = e^{(s+t)^2/2}e^{(s-t)^2/2} = e^{s^2+t^2}. \end{aligned}$$

Now, taking 4 partial derivatives, we find,

$$E[X^2 Y^2] = \frac{\partial^{(4)} M_{X,Y}(0,0)}{\partial s^{(2)}\, \partial t^{(2)}} = (16s^2t^2 + 8t^2 + 8s^2 + 4)e^{s^2+t^2}\Big|_{(0,0)} = 4. \quad \square$$

Exercise 7-65 SOA EXAM P Sample Exam Questions #58

A company insures homes in three cities, J, K, and L. Since sufficient distance separates the cities, it is reasonable to assume that the losses occurring in these cities are independent. The moment generating functions for the loss distributions of the cities are:

$$M_J(t) = (1-2t)^{-3} \qquad M_K(t) = (1-2t)^{-2.5} \qquad M_L(t) = (1-2t)^{-4.5}$$

Let X represent the combined losses from the three cities. Calculate $E(X^3)$.

(A) 1,320 (B) 2,082 (C) 5,760 (D) 8,000 (E) 10,560

Exercise 7-66 SOA EXAM P Sample Exam Questions #95

X and Y are independent random variables with common moment generating function $M(t) = e^{t^2/2}$. Let $W = X+Y$ and $Z = Y-X$. Determine the joint moment generating function, $M(t_1,t_2)$, of W and Z.

(A) $e^{2t_1^2+2t_2^2}$ (B) $e^{(t_1-t_2)^2}$ (C) $e^{(t_1+t_2)^2}$ (D) $e^{2t_1 t_2}$ (E) $e^{t_1^2+t_2^2}$

7.11 CHAPTER 7 SAMPLE EXAMINATION

1. Suppose $E[X]=2$, $E[X^2]=\frac{9}{2}$, $E[Y]=1$, $E\left[Y^2\right]=\frac{3}{2}$, and $Var\left[X-3Y\right]=4$.

   Calculate $Cov\left[X,Y\right]$.

2. Let X and Y have a joint probability density function that is uniformly distributed on the triangle connecting the points $(-4,0)$, $(0,2)$, and $(4,0)$.

   (a) Write the joint density function $f(x,y)$.

   (b) Find $E[X]$ and $E[Y]$.

   (c) Find $\Pr(X<.5\,|\,Y>1)$.

   (d) Find $E[X\,|\,Y=.3]$.

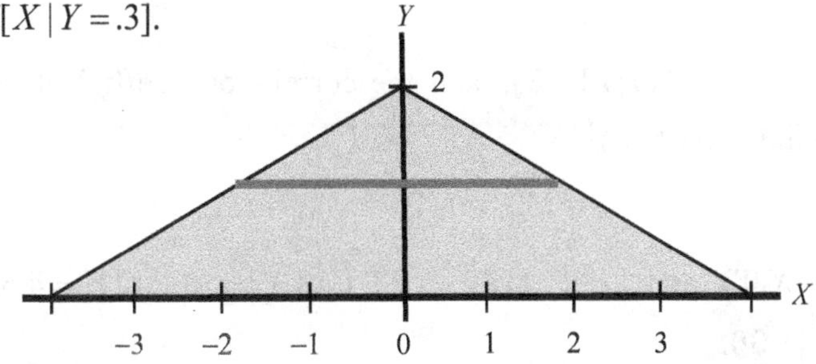

3. Let X and Y have joint probability density function given by

$$f(x,y) = \begin{cases} x+y & \text{for } 0 \le x \le 1, \quad 0 \le y \le 1 \\ 0 & \text{otherwise} \end{cases}.$$

   (a) Find the marginal density functions $f_X(x)$ and $f_Y(y)$.

   (b) Find $E[X]$ and $E[Y]$.

   (c) Are X and Y independent or dependent? Why?

   (d) Find the conditional density $f(x\,|\,Y=y)$.

   (e) Find $E[X\,|\,Y=.7]$.

   (f) Find $E[7X-3Y]$.

   (g) Find the correlation ρ.

   (h) Find $Var[7X-3Y]$.

4. Let $p(x,y)=k(x^2+xy)$ for $x=1,2$ and $y=1,2,3$.
   (a) Find the constant k.
   (b) Construct the joint probability distribution $p(x,y)$.
   (c) Find the marginal probabilities of X and Y, $p_X(x)$ and $p_Y(y)$.
   (d) Find $E[X]$ and $E[Y]$.
   (e) Are X and Y independent or dependent? Why?
   (f) Find the conditional probability $p(x\mid y)$.
   (g) Find $E[X\mid Y=2]$.
   (h) Find $E[XY]$.
   (i) Find the correlation ρ.
   (j) Find $V[X]$ and $V[Y]$.
   (k) Find the covariance of X and Y.
   (l) Find $Var[7X-3Y]$.

5. Suppose $E[X]=1$, $Var[X]=2$, $E[Y]=-3$, $Var[Y]=4$, $E[Z]=0$, $Var[Z]=5$, $Cov[X,Y]=.5$, $Cov[X,Z]=-1$, and $Cov[Y,Z]=2$.
   (a) Find $E[2X-Y+4Z]$.
   (b) Find $Var[2X-Y+4Z]$.

6. Suppose $\sigma_X=2$, $Var[Y]=25$, and the correlation coefficient between X and Y is $\rho=\frac{-3}{5}$. Find $Cov(X,Y)$.

7. Find $Cov(X,Y)$ and ρ if $E[X]=-2$, $Var[X]=10$, $E[Y]=4.5$, $E[3XY-5]=50$ and $E[Y^2]=30$.

8. **SOA EXAM P Sample Exam Questions #112**
   A company offers a basic life insurance policy to its employees, as well as a supplemental life insurance policy. To purchase the supplemental policy, an employee must first purchase the basic policy. Let X denote the proportion of employees who purchase the basic policy, and Y the proportion of employees who purchase the supplemental policy. Let X and Y have the joint density function $f(x,y)=2(x+y)$ on the region where the density is positive. Given that 10% of the employees buy the basic policy, what is the probability that fewer than 5% buy the supplemental policy?

   (A) 0.010 (B) 0.013 (C) 0.108 (D) 0.417 (E) 0.500

9. **SOA EXAM P Sample Exam Questions #106**
Let X and Y denote the values of two stocks at the end of a five-year period. X is uniformly distributed on the interval $(0,12)$. Given $X = x$, Y is uniformly distributed on the interval $(0,x)$. Determine $Cov(X,Y)$ according to this model.

(A) 0 (B) 4 (C) 6 (D) 12 (E) 24

10. **SOA EXAM P Sample Exam Questions #116**
An actuary determines that the annual numbers of tornadoes in counties P and Q are jointly distributed as follows:

		Annual Number of Tornadoes in County Q			
		0	1	2	3
Annual Number of Tornadoes in County P	0	0.12	0.06	0.05	0.02
	1	0.13	0.15	0.12	0.03
	2	0.05	0.15	0.10	0.02

Calculate the conditional variance of the annual number of tornadoes in county Q, given that there are no tornadoes in county P.

(A) 0.51 (B) 0.84 (C) 0.88 (D) 0.99 (E) 1.76

11. **SOA EXAM P Sample Exam Questions #117**
A company is reviewing tornado damage claims under a farm insurance policy. Let X be the portion of a claim representing damage to the house and let Y be the portion of the same claim representing damage to the rest of the property. The joint density function of X and Y is,

$$f(x,y) = \begin{cases} 6[1-(x+y)] & \text{for } x>0, y>0, x+y<1 \\ 0 & \text{otherwise} \end{cases}$$

Determine the probability that the portion of a claim representing damage to the house is less than 0.2.

(A) 0.360 (B) 0.480 (C) 0.488 (D) 0.512 (E) 0.520

12. **SOA EXAM P Sample Exam Questions #88**
The waiting time for the first claim from a good driver and the waiting time for the first claim from a bad driver are independent and follow exponential distributions with means 6 years and 3 years, respectively. What is the probability that the first claim from a good driver will be filed within 3 years and the first claim from a bad driver will be filed within 2 years?

(A) $\frac{1}{18}(1-e^{-2/3}-e^{-1/2}+e^{-7/6})$

(B) $\frac{1}{18}e^{-7/6}$

(C) $1-e^{-2/3}-e^{-1/2}+e^{-7/6}$

(D) $1-e^{-2/3}-e^{-1/2}+e^{-1/3}$

(E) $1-\frac{1}{3}e^{-2/3}-\frac{1}{6}e^{-1/2}+\frac{1}{18}e^{-7/6}$

13. **SOA EXAM P Sample Exam Questions #109**
A company offers earthquake insurance. Annual premiums are modeled by an exponential random variable with mean 2. Annual claims are modeled by an exponential random variable with mean 1. Premiums and claims are independent. Let X denote the ratio of claims to premiums. What is the density function of X?

(A) $\frac{1}{2x+1}$ (B) $\frac{2}{(2x+1)^2}$ (C) e^{-x} (D) $2e^{-2x}$ (E) xe^{-x}

14. **SOA EXAM P Sample Exam Questions #123**
You are given the following information about N, the annual number of claims for a randomly selected insured:

$$P(N=0)=\frac{1}{2} \qquad P(N=1)=\frac{1}{3} \qquad P(N>1)=\frac{1}{6}$$

Let S denote the total annual claim amount for an insured. When $N=1$, S is exponentially distributed with mean 5. When $N>1$, S is exponentially distributed with mean 8.

Determine $\Pr(4<S<8)$.

(A) 0.04 (B) 0.08 (C) 0.12 (D) 0.24 (E) 0.25

15. **SOA EXAM P** **Sample Exam Questions** **#84**

Let X and Y be the number of hours that a randomly selected person watches movies and sporting events, respectively, during a three-month period. The following information is known about X and Y:

$E(X)$	50
$E(Y)$	20
$Var(X)$	50
$Var(Y)$	30
$Cov(X,Y)$	10

One hundred people are randomly selected and observed for these three months. Let T be the total number of hours that these one hundred people watch movies or sporting events during this three-month period. Approximate the value of $\Pr(T < 7100)$.

(A) 0.62 (B) 0.84 (C) 0.87 (D) 0.92 (E) 0.97

16. **SOA EXAM P** **Sample Exam Questions** **#78**

A device runs until either of two components fails, at which point the device stops running. The joint density function of the lifetimes of the two components, both measured in hours, is

$$f(x,y) = \frac{x+y}{27} \text{ for } 0 < x < 3 \text{ and } 0 < y < 3.$$

Calculate the probability that the device fails during its first hour of operation.

(A) 0.04 (B) 0.41 (C) 0.44 (D) 0.59 (E) 0.96

17. **SOA EXAM P** **Sample Exam Questions** **#93**

A family buys two policies from the same insurance company. Losses under the two policies are independent and have continuous uniform distributions on the interval from 0 to 10. One policy has a deductible of 1 and the other has a deductible of 2. The family experiences exactly one loss under each policy. Calculate the probability that the total benefit paid to the family does not exceed 5.

(A) 0.13 (B) 0.25 (C) 0.30 (D) 0.32 (E) 0.42

18. **SOA EXAM P Sample Exam Questions #124**

The joint probability density for X and Y is

$$f(x,y) = \begin{cases} 2e^{-(x+2y)} & \text{for } x>0,\ y>0 \\ 0 & \text{otherwise} \end{cases}.$$

Calculate the variance of Y given that $X>3$ and $Y>3$.

(A) 0.25 (B) 0.50 (C) 1.00 (D) 3.25 (E) 3.50

19. **SOA EXAM P Sample Exam Questions #144**

A client spends X minutes in an insurance agent's waiting room and Y minutes meeting with the agent. The joint density function of X and Y can be modeled by

$$f(x,y) = \begin{cases} \frac{1}{800} e^{\frac{-x}{40}} e^{\frac{-y}{20}} & \text{for } x>0, y>0 \\ 0, & \text{otherwise} \end{cases}$$

[Modified] Find an expression that represents the probability that a client spends less than 60 minutes at the agent's office.

20. **SOA EXAM P Sample Exam Questions #145**

New dental and medical plan options will be offered to state employees next year. An actuary uses the following density function to model the joint distribution of the proportion X of state employees who will choose Dental Option 1 and the proportion Y who will choose Medical Option 1 under the new plan options:

$$f(x,y) = \begin{cases} 0.50 & \text{for } 0<x<.5 \text{ and } 0<y<.5 \\ 1.25 & \text{for } 0<x<.5 \text{ and } .5<y<1 \\ 1.50 & \text{for } .5<x<1 \text{ and } 0<y<.5 \\ 0.75 & \text{for } .5<x<1 \text{ and } .5<y<1 \\ 0 & \text{otherwise} \end{cases}$$

Calculate $Var[Y \mid X = 0.75]$

(A) 0.000 (B) 0.061 (C) 0.076 (D) 0.083 (E) 0.141

CHAPTER 8

A PROBABILITY POTPOURRI

8.1 THE DISTRIBUTION OF A TRANSFORMED RANDOM VARIABLE

On a number of occasions in earlier chapters we have looked into the properties of a random variable Y that is given by a transformation $Y = g(X)$ of a known random variable X. Deriving the distribution of Y from that of X can be challenging in general, but there are a number of techniques available that can be used to determine various properties of Y such as its moments or its density function.

One procedure, familiar from earlier work with continuous distributions, is the cumulative distribution function (CDF) method. The idea is to derive the CDF $F_Y(y)$ of Y using the CDF $F_X(x)$ of the random variable X and the functional relationship $Y = g(X)$. The basic setup for starting this procedure is,

$$F_Y(y) = \Pr[Y \le y] = \Pr[g(X) \le y],$$

with subsequent steps depending on the specific nature of $g(X)$. In the continuous case the density function for Y, $f_Y(y)$, can then be found by differentiating $F_Y(y)$. Specific examples of this technique can be found in Subsection 6.2.2 in the derivation of the exponential distribution as the waiting time random variable for a Poisson process, and in Subsection 6.5.5 in the derivation of χ^2 as the distribution for $g(Z) = Z^2$ (Z being the standard normal random variable). Here is another example.

Example 8.1-1 The CDF Technique

Let X be the random variable whose density function is given by $f_X(x) = 2x;\ 0 \le x \le 1$. Find the density function $f_Y(y)$ of $Y = e^X$.

Solution

Here $g(X) = e^X$ and the random variable Y lives on the interval $[g(0), g(1)] = [e^0, e^1] = [1, e]$. The CDF for X, determined from its given density function, is $F_X(x) = \int_0^x 2t\,dt = x^2$; $0 \le x \le 1$. Now, for any y, $1 \le y \le e$,

$$F_Y(y) = \Pr[Y \le y] = \Pr[g(X) \le y] = \Pr[e^X \le y] = \Pr[X \le \ln y] = F_X(\ln y) = (\ln y)^2.$$

Next, we find $f_Y(y) = F_Y'(y) = \frac{2\ln y}{y};\ 1 \le y \le e.$ □

Notes

(1) A crucial step in the derivation of $F_Y(y)$ was going from the event $[e^X \le y]$ to the equivalent event $[X \le \ln y]$. This was facilitated by the fact that the transformation $y = e^x$ is *invertible* with inverse function $x = g^{-1}(y) = \ln y$.

(2) Please observe that the formula $(\ln y)^2$ for the CDF of Y is **not** the same as $\ln y^2 = 2\ln y$.

Exercise 8-1 Verify that $f_Y(y) = \begin{cases} \frac{2\ln y}{y} & \text{for } 1 \le y \le e \\ 0 & \text{otherwise} \end{cases}$ is a valid density function.

Example 8.1-2 The CDF Technique

Suppose that X is a continuous uniform random variable on the interval [4,16] and that $Y = \sqrt{X}$. Find the variance of Y.

Solution

It should be clear that the density function for Y lives on the interval [2,4].
The CDF for the random variable X is $F_X(x) = \Pr(X \le x) = \frac{x-4}{12};\ 4 \le x \le 16$.

The distribution function for the random variable Y is found by our standard technique $F_Y(y) = \Pr(Y \le y) = \Pr(\sqrt{X} \le y) = \Pr(X \le y^2) = F_X(y^2) = \frac{y^2-4}{12}$ for $4 \le y^2 \le 16$. Since $Y = \sqrt{X}$ is always positive, this corresponds to $2 \le y \le 4$. Therefore, the probability density function for the random variable Y is given by

$$f_Y(y) = F_Y'(y) = \begin{cases} \frac{y}{6} & \text{if } 2 \le y \le 4 \\ 0 & \text{otherwise} \end{cases}.$$

$$Var[Y] = E[Y^2] - \left(E[Y]\right)^2 = \int_2^4 y^2 \cdot \frac{y}{6}dy - \left(\int_2^4 y \cdot \frac{y}{6}dy\right)^2 = 10 - \left(\frac{28}{9}\right)^2 = .321. \quad \square$$

Note

In this example we could also compute the variance of Y using only the density function for X:

$$E[Y] = E[\sqrt{X}] = \int_4^{16} x^{1/2} f_X(x)dx = \int_4^{16} x^{1/2}\frac{1}{12}dx = \frac{28}{9}$$

and

$$E[Y^2] = E[X] = \frac{4+16}{2} = 10.$$

Thus, $Var[Y] = E[Y^2] - \left(E[Y]\right)^2 = 10 - \left(\frac{28}{9}\right)^2 = .321$.

8.1.1 THE TRANSFORMATION FORMULA

The approach followed in Examples 8.1-1 and 8.1-2, where the transformation formulas $y = e^x$ and $y = \sqrt{x}$ had inverse functions, can be pursued in general to derive a useful formula yielding the density function $f_Y(y)$ directly in terms of the density function $f_X(x)$.

The key to this is to observe that the transformation $y = g(x)$ has an inverse $x = h(y) = g^{-1}(y)$ providing either,

(a) $y = g(x)$ is a monotonically *increasing* function, or
(b) $y = g(x)$ is a monotonically *decreasing* function.

In either case the graph of $y = g(x)$, or equivalently $x = h(y) = g^{-1}(y)$, creates a one-to-one pairing of points (x, y) on this graph. The two cases can be represented by the following schematic drawings in Figure 8.1.1.

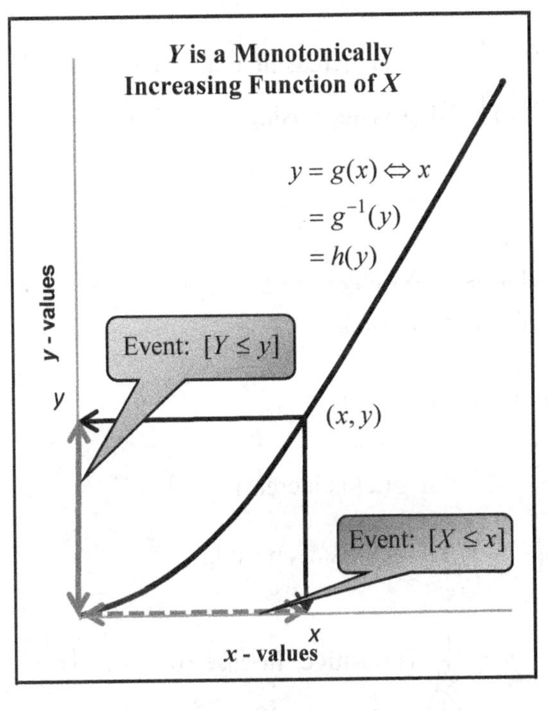

Case (a)

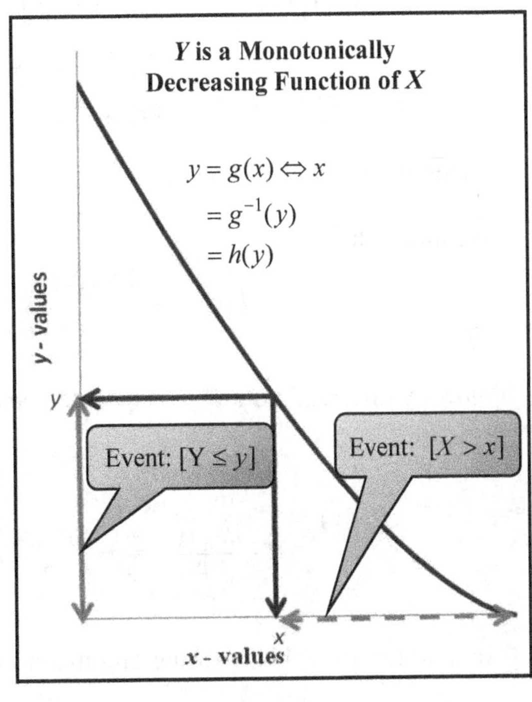

Case (b)

FIGURE 8.1.1

In case (a), where y is an increasing function of x, the event $[Y \le y]$, shown by the solid green arrow the y-axis, is equivalent to the event $[X \le x]$, shown in the dashed green arrow on the x-axis, providing x and y correspond to ordered pairs on the graph of $y = g(x)$. In case (b), we see that the decreasing function $y = g(x)$ makes the event $[X > x]$ equivalent to the event $[Y \le y]$.

The Transformation Formula

Let X be a continuous random variable with density function $f_X(x)$ and let $Y = g(X)$ be a monotonically increasing *or* decreasing differentiable transformation. Then the density function $f_Y(y)$ is given by

$$f_Y(y) = f_X(x) \cdot \left|\frac{dx}{dy}\right|,$$

where $y = g(x)$. Since $g(x)$ is monotonic, it has an inverse function $x = g^{-1}(y)$.

Proof

Suppose that the value y is given and define $x = g^{-1}(y)$ to be the corresponding value of x. Then,

$$F_Y(y) = \Pr[Y \le y] = \Pr[g(X) \le y]$$
$$= \begin{cases} \Pr\left[X \le x = g^{-1}(y)\right] & \text{if } g \text{ is increasing (case (a))} \\ \Pr\left[X > x = g^{-1}(y)\right] & \text{if } g \text{ is decreasing (case (b))} \end{cases}$$

(refer again to Figure 8.1.1).

It follows that,

$$F_Y(y) = \begin{cases} F_X(x) & \text{if } g \text{ is increasing} \\ 1 - F_X(x) & \text{if } g \text{ is decreasing} \end{cases}$$

Now by differentiating with respect to y and using the chain rule,

$$f_Y(y) = \begin{cases} \dfrac{dF_X(x)}{dx} \cdot \dfrac{dx}{dy} & \text{if } g(x) \text{ is increasing} \\ -\dfrac{dF_X(x)}{dx} \cdot \dfrac{dx}{dy} = \dfrac{dF_X(x)}{dx} \cdot \left(-\dfrac{dx}{dy}\right) & \text{if } g(x) \text{ is decreasing} \end{cases}$$

In case (a) $g(x)$ is increasing and therefore $\frac{dx}{dy} = \frac{1}{\frac{dy}{dx}}$ is positive. In case (b) $g(x)$ is decreasing and therefore $\frac{dx}{dy}$ is negative (meaning that $-\frac{dx}{dy}$ is positive). In either case $f_Y(y) = f_X(x) \cdot \left|\frac{dx}{dy}\right|$.

Example 8.1-3 The Transformation Formula (continuation of Example 8.1-1)

Let X be the random variable whose density function is given by $f_X(x) = 2x;\ 0 \le x \le 1$. Use the transformation formula to find the density function $f_Y(y)$ of $Y = e^X$.

Solution

The transformation is $y = g(x) = e^x$. This implies $x = \ln y$ and therefore $\frac{dx}{dy} = \frac{1}{y}$ for $1 \le y \le e$.

By the transformation formula,

$$f_Y(y) = f_X(x) \cdot \left|\frac{dx}{dy}\right| = 2x \cdot \left|\frac{1}{y}\right| = 2\ln y \cdot \left|\frac{1}{y}\right| = \begin{cases} \frac{2\ln y}{y} & \text{for } 1 \le y \le e \\ 0 & \text{otherwise} \end{cases}.$$ ❐

Notes

(1) The final answer should be expressed exclusively in terms of y, using the relationship $x = \ln y$.

(2) Finding the domain of the random variable Y is as important as finding the function $f_Y(y)$. Please be careful and thorough.

Example 8.1-4 The Transformation Formula

Suppose that the future lifetime X of a compact fluorescent bulb is exponentially distributed with a mean of 60 months. An improvement in the manufacturing leads to a light bulb with a 35% better future lifetime. Let $Y = 1.35 \cdot X$ denote the future lifetime of the improved bulb.

(a) Calculate $E[Y]$.
(b) Calculate $F_Y(y)$, the CDF of Y.
(c) Calculate the density function $f_Y(y)$ by differentiating the result of (b) and confirm by using the transformation formula.
(d) Calculate $\Pr(Y \le 50)$.

Solution

(a) The distribution of Y is not required. We can calculate expected values of linear transformations immediately.

$$E[Y] = E[1.35 \cdot X] = 1.35 \cdot E[X] = 1.35 \cdot 60 = 81 \text{ months.}$$

(b) Since the random variable X is exponentially distributed with mean 60, we have

$$F_X(x) = 1 - e^{(-x/60)} \text{ for } x > 0.$$

Then

$$F_Y(y) = \Pr[Y \le y] = \Pr[1.35 \cdot X \le y]$$
$$= \Pr\left[X \le \frac{y}{1.35}\right] = F_X\left(\frac{y}{1.35}\right) = 1 - e^{-\frac{(y/1.35)}{60}} = 1 - e^{-\frac{y}{81}}$$

for $y > 0$.

We recognize the distribution function $F(y) = 1 - e^{-y/81}$ for $y > 0$ as that of an exponentially distributed random variable with a mean of 81. That is, we know exactly the random variable Y.

(c) We have $f_Y(y) = F_Y'(y) = \frac{1}{81}e^{-y/81}$; $y > 0$. Of course this is the density function for an exponential random variable of mean 81. To verify this result using the transformation formula, we have the transformation $y = g(x) = 1.35x$, which implies,

$$x = \frac{y}{1.35} \text{ and } \frac{dx}{dy} = \frac{1}{1.35}.$$

Thus,

$$f_Y(y) = f_X(x) \cdot \left|\frac{dx}{dy}\right| = \frac{1}{60}e^{-\frac{x}{60}} \cdot \left|\frac{1}{1.35}\right| = \frac{1}{(60)(1.35)}e^{-\frac{y}{(60)(1.35)}} = \frac{1}{81}e^{-\frac{y}{81}};\ y > 0.$$

(d) We found the CDF for Y in part (b). $\Pr(Y \le 50) = F_Y(50) = 1 - e^{-50/81} = .4606$. ❒

Example 8.1-5 Scaling an Exponentially Distributed Random Variable

Let X be an exponential random variable with a mean of β and let $Y = cX$, where c is a positive constant. Show that Y is also an exponential random variable, with a mean of $c\beta$. (This gives a general version of the previous example.)

Solution

The density for X is $f_X(x) = \frac{1}{\beta}e^{-(1/\beta)x}$; $x \ge 0$ and the transformation is $y = c \cdot x$, or equivalently, $x = \frac{1}{c}y$, so that $\frac{dx}{dy} = \frac{1}{c}$. Using the transformation formula,

$$f_Y(y) = f_X(x) \cdot \left|\frac{dx}{dy}\right| = \frac{1}{\beta}e^{-\frac{1}{\beta}x} \cdot \frac{1}{c} = \frac{1}{\beta}e^{-\frac{1}{\beta}\left(\frac{1}{c}y\right)} \cdot \frac{1}{c} = \frac{1}{(c \cdot \beta)}e^{-\frac{1}{(c \cdot \beta)}y},$$

for $y \ge 0$.

We recognize $f_Y(y)$ as the density function of an exponentially distributed random variable with mean $c \cdot \beta$. ❒

Note

We obtained this result previously for the general Gamma family of random variables, using moment-generating functions. The exponential distribution is of course the special case of the Gamma with parameter $\alpha = 1$. Please refer to Chapter 6, property (5) in the Property Box for Gamma distributions.

Example 8.1-6 The Transformation Formula (continuation of Example 8.1-2)

Suppose that X is a continuous uniform random variable on the interval [4,16] and that $Y = \sqrt{X}$. Find the density function of Y.

Solution

$Y = g(X) = \sqrt{X}$ is monotonically increasing for $4 \le x \le 16$ and $y = x^{1/2}$ implies $x = y^2$ and $\frac{dx}{dy} = 2y$. Also, the density function for X is $f_X(x) = \frac{1}{16-4} = \frac{1}{12}$; $4 \le x \le 16$. Therefore,

$$f_Y(y) = f_X(y^2) \cdot |2y| = \frac{1}{12} \cdot 2|y| = \begin{cases} \frac{y}{6} & \text{for } 2 \le y \le 4 \\ 0 & \text{otherwise} \end{cases}.$$

❐

Exercise 8-2 Suppose that the random variable X is uniformly distributed on the interval $[0,1]$. Let the random variable $Y = e^X$.

(a) Find the probability density function for Y.
(b) Find the expected value of Y.

Exercise 8-3 Suppose that the random variable X has an exponential distribution with mean one. Let the random variable $Y = \sqrt{X}$.

(a) Find the probability density function for Y.
(b) Find the expected value of Y.
(c) Calculate the variance of Y.

Exercise 8-4 Suppose that the random variable X has an exponential distribution with mean four. Let the random variable $Y = \ln(X+1)$.

(a) Find the probability density function for Y.
(b) Find the expected value of Y.

Exercise 8-5 SOA EXAM P Sample Exam Questions #73

An actuary models the lifetime of a device using the random variable $Y = 10 \cdot X^{0.8}$, where X is an exponential random variable with mean 1 year.

Determine the probability density function $f(y)$, for $y > 0$, of the random variable Y.

(A) $10y^{0.8}e^{-8y^{-0.2}}$

(B) $8y^{-0.2}e^{-10y^{0.8}}$

(C) $8y^{-0.2}e^{-(0.1y)^{1.25}}$

(D) $(0.1y)^{1.25}e^{-0.125(0.1y)^{0.25}}$

(E) $0.125(0.1y)^{0.25}e^{-(0.1y)^{1.25}}$

Exercise 8-6 Suppose that the random variable X is uniformly distributed on the interval $[1,5]$. Let the random variable $Y = \frac{1}{X}$. Find the following:

(a) The cumulative distribution function for Y.
(b) The density function for Y.
(c) The expected value of Y.
(d) The variance and standard deviation of Y.
(e) Is $E[Y] = \frac{1}{E[X]}$ or $Var[Y] = \frac{1}{Var[X]}$?

Exercise 8-7 If X is uniformly distributed on the interval $[1,5]$ and $Y = e^X$, then find $f_Y(8)$.

Exercise 8-8 **SOA EXAM P** **Sample Exam Questions** **#71**

The time, T, that a manufacturing system is out of operation has cumulative distribution function

$$F(t) = \begin{cases} 1 - \left(\frac{2}{t}\right)^2 & \text{for } t > 2 \\ 0 & \text{otherwise} \end{cases}.$$

The resulting cost to the company is $Y = T^2$. Determine the density function of Y, for $y > 4$.

(A) $\frac{4}{y^2}$ (B) $\frac{8}{y^{3/2}}$ (C) $\frac{8}{y^3}$ (D) $\frac{16}{y}$ (E) $\frac{1024}{y^5}$

Exercise 8-9 **SOA EXAM P** **Sample Exam Questions** **#72**

An investment account earns an annual interest rate R that follows a uniform distribution on the interval $(0.04, 0.08)$. The value of a 10,000 initial investment in this account after one year is given by $V = 10{,}000e^R$.

Determine the cumulative distribution function, $F(v)$, of V for values of v that satisfy $0 < F(v) < 1$.

(A) $\frac{10{,}000^{v/10{,}000} - 10{,}408}{425}$

(B) $25e^{v/10{,}000} - 0.04$

(C) $\frac{v - 10{,}408}{10{,}833 - 10{,}408}$

(D) $\frac{25}{v}$

(E) $25\left[\ln\left(\frac{v}{10{,}000}\right) - 0.04\right]$

Exercise 8-10 SOA EXAM P Sample Exam Questions #74

Let T denote the time in minutes for a customer service representative to respond to 10 telephone inquiries. T is uniformly distributed on the interval with endpoints 8 minutes and 12 minutes. Let R denote the average rate, in customers per minute, at which the representative responds to inquiries.

Which of the following is the density function of the random variable R on the interval $\left(\frac{10}{12} \le r \le \frac{10}{8}\right)$?

(A) $\frac{12}{5}$ (B) $3-\frac{5}{2r}$ (C) $3r-\frac{5\ln(r)}{2}$ (D) $\frac{10}{r^2}$ (E) $\frac{5}{2r^2}$

Exercise 8-11 SOA EXAM P Sample Exam Questions #75

The monthly profit of Company I can be modeled by a continuous random variable with density function f. Company II has a monthly profit that is twice that of Company I.

Determine the probability density function of the monthly profit of Company II.

(A) $\frac{1}{2}f\left(\frac{x}{2}\right)$ (B) $f\left(\frac{x}{2}\right)$ (C) $2f\left(\frac{x}{2}\right)$ (D) $2f(x)$ (E) $2f(2x)$

8.1.2 THE DISTRIBUTION OF THE SUM OF RANDOM VARIABLES

We can often use the CDF method to identify the distribution of a sum $S+T$ of two given random variables S and T. In the next section we will summarize situations in which it is possible to immediately recognize the distribution type of a sum of independent random variables when the summands come from the same distribution family. Our focus here will be on situations where the distribution type of the sum is not readily apparent.

Example 8.1-7 The Distribution of an Independent Sum

Let S and T be independent standard uniform random variables and define $X = S+T$. Find the density function of the sum X.

Solution

Both S and T are uniformly distributed on the interval $[0,1]$ and they are independent so that the joint density function is given by $f(s,t)=1;\ 0 \le s,t \le 1$. This is to say the joint distribution of S and T is uniform on the unit square $[0,1]\times[0,1]$. The sum X must take values between 0 and 2. Using techniques for joint distributions in Chapter 7, we fix a value of x between 0 and 2, and calculate

$$F_X(x) = \Pr[X \le x] = \Pr[S+T \le x].$$

There are two cases to consider:

Case (1) $0 \le x \le 1$:

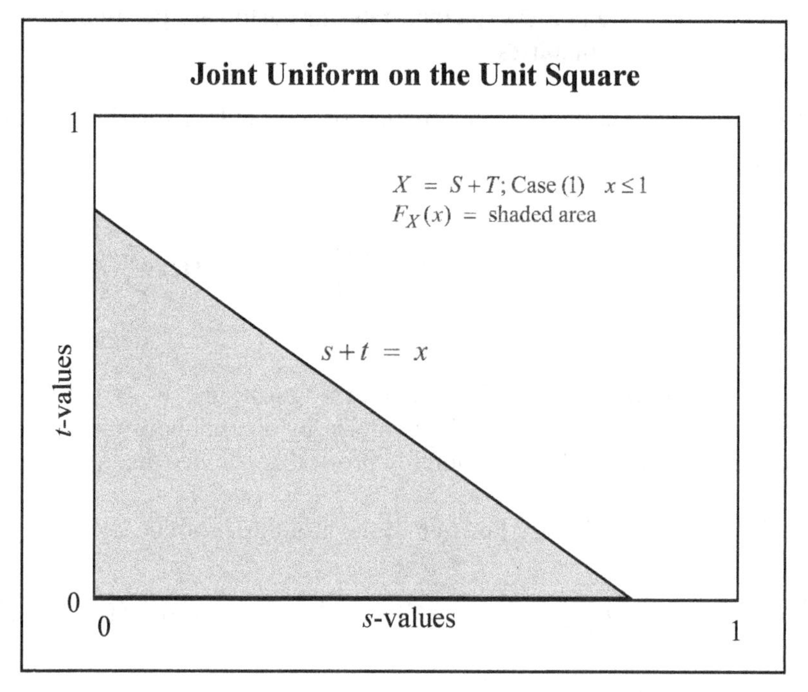

In this case $F_X(x) = \Pr[X \le x] = \Pr[S+T \le x]$ is the shaded area of the triangle. The hypotenuse line is $s + t = x$, which has s and t intercepts both equal to x. Hence, the area of the green triangle is $\frac{1}{2}x^2$.

Case (2) $1 < x \le 2$:

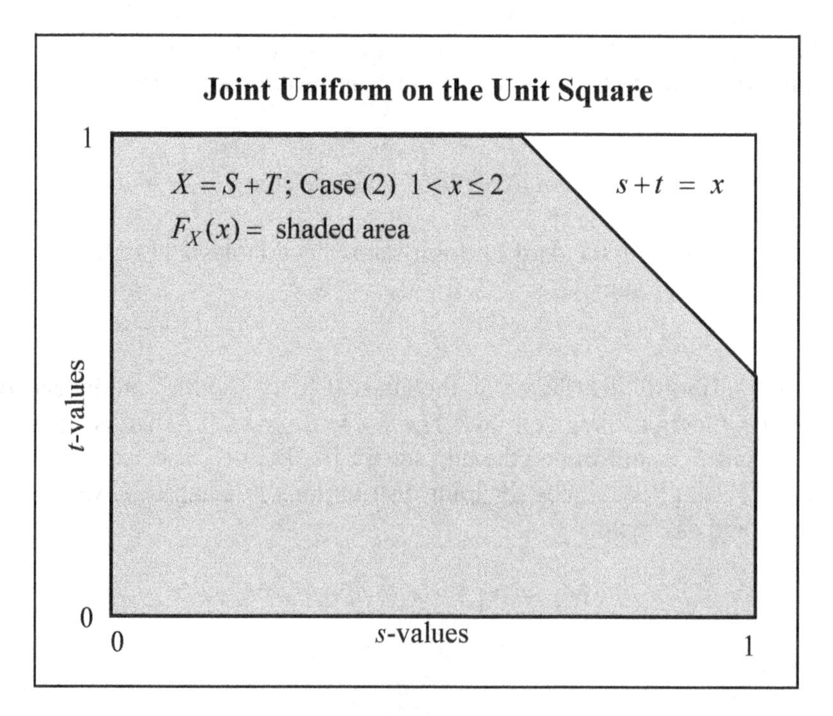

The complement to the shaded area is the non-shaded triangle at the upper right. The hypotenuse is still the line $s+t=x$, but which now intersects the vertical line $s=1$ at height $t=x-1$. Therefore, the length of the vertical side of the non-shaded triangle is $1-(x-1)=2-x$. By symmetry, the length of the horizontal side is also $2-x$. Thus,

$$F_X(x) = \Pr[X \le x] = \Pr[S+T \le x]$$

is the area of the shaded region $= 1-\frac{1}{2}(2-x)^2$.

So,

$$F_X(x) = \begin{cases} \frac{1}{2}x^2 & \text{for } 0 \le x \le 1 \\ 1-\frac{1}{2}(2-x)^2 & \text{for } 1 < x \le 2 \end{cases},$$

and

$$f_X(x) = \begin{cases} x & \text{for } 0 \le x \le 1 \\ 2-x & \text{for } 1 < x \le 2 \end{cases}.$$

Here is a graph of the density function $f_X(x)$ on the interval $[0,2]$:

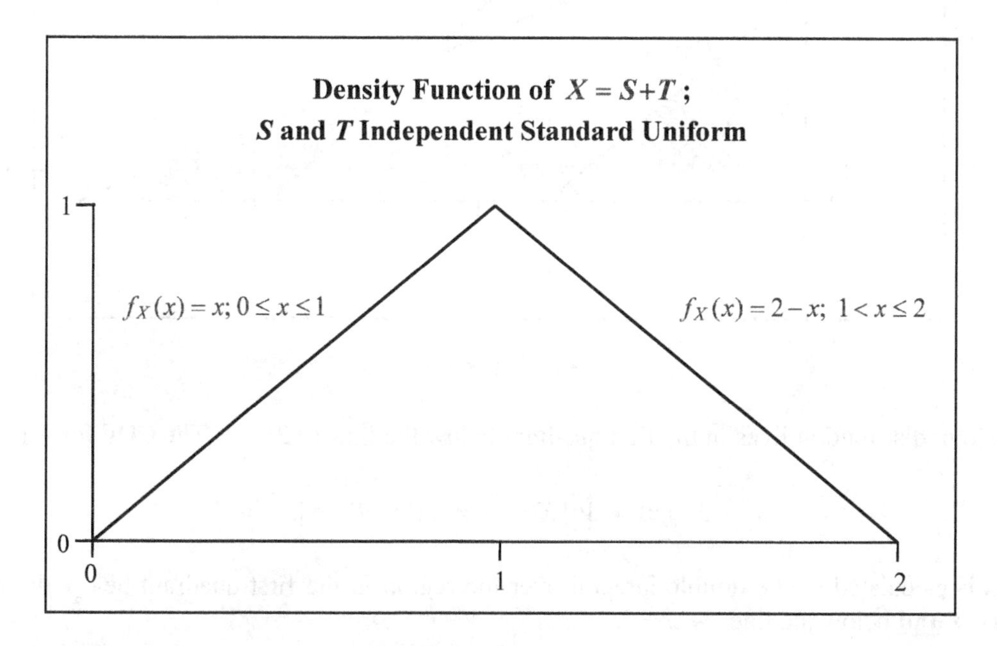

Note that X is **not** uniformly distributed on $[0,2]$. This is called a ***triangular*** distribution.

The procedure for sums of random variables that are not necessarily independent is quite similar, as the following example illustrates.

Example 8.1-8 A Non-Independent Sum

Let S and T be jointly distributed with joint density $f(s,t)=\frac{1}{3}(4s+t)$ living on the domain $s,t\geq 0,\ s+2t\leq 2$. Let $X=S+T$. Set up the double integrals used to find the CDF $F_X(x)$ for the random variable X.

Solution

From the Figure 8.1.2 we see that once again, there are two cases for $s+t=x$, one for $0\leq x\leq 1$ and a second for $1<x\leq 2$.

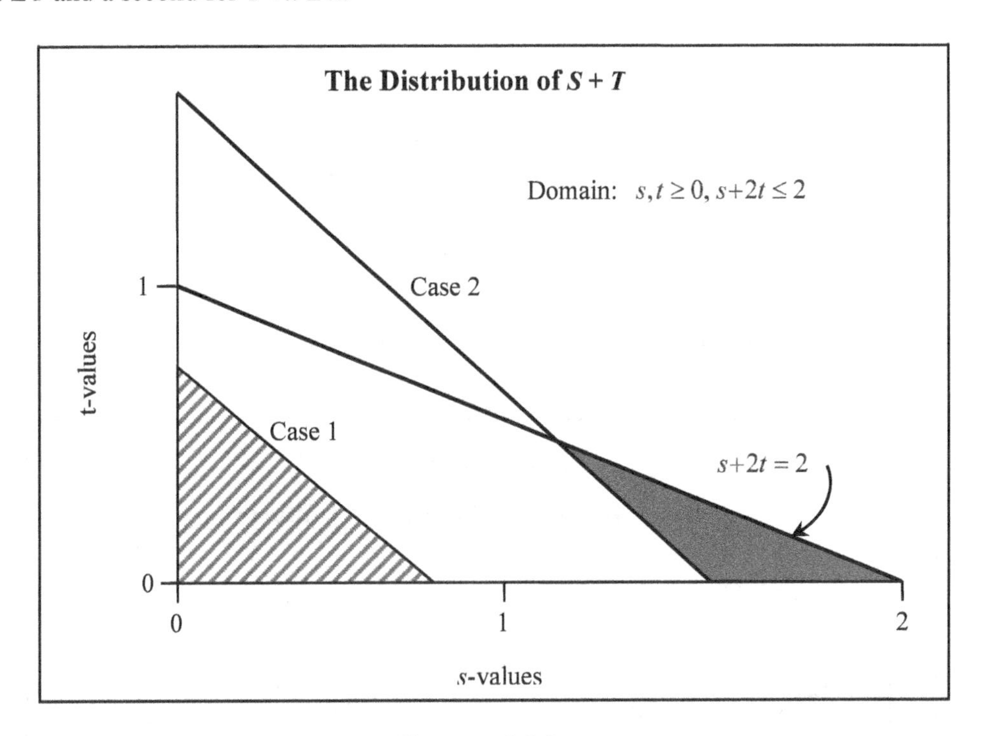

FIGURE 8.1.2

The joint distribution lives in the first quadrant below the line $s+2t=2$. The CDF for X is,

$$F_X(x) = \Pr[X\leq x] = \Pr[S+T\leq x].$$

This is calculated as the double integral over the region in the first quadrant below the line $s+t=x$ **and** below the line $s+2t=2$.

Case (1): $0\leq x\leq 1$. In this case the region below $s+t=x$ is entirely contained in the joint distribution domain below the line $s+2t=2$. This is the striped region in the sketch. Then,

$$F_X(x) = \Pr[X\leq x] = \Pr[S+T\leq x] = \int_0^x\int_0^{x-t} f(s,t)\,ds\,dt = \frac{1}{3}\int_0^x\int_0^{x-t}(4s+t)\,ds\,dt.$$

Case (2): $1<x\le 2$. In this case the line $s+t=x$ intersects the domain boundary line $s+2t=2$. Simple algebra shows this point of intersection to be at $s=2x-2$ and $t=2-x$. Thus, $F_X(x)=\Pr[X\le x]=\Pr[S+T\le x]$ is the double integral of the joint density function over the 4-sided polygon with vertices at $(0,0),(0,1),(2,x),$ and $(2x-2,2-x)$. This probability is best calculated by integrating over the complementary region shaded green in the diagram. Thus,

$$\begin{aligned} F_X(x)=\Pr[X\le x] &= \Pr[S+T\le x] \\ &= 1-\int_0^{2-x}\int_{x-t}^{2-2t} f(s,t)\,ds\,dt = 1-\frac{1}{3}\int_0^{2-x}\int_{x-t}^{2-2t}(4s+t)\,ds\,dt. \end{aligned}$$

For the record,

$$F_X(x) = \begin{cases} \frac{5}{18}x^3 & 0\le x\le 1 \\ \frac{1}{18}(10-36x+42x^2-11x^3) & 1<x\le 2 \end{cases}$$

The ambitious student is invited to check these calculations, an exercise not recommended for the faint of heart.

8.1.3 THE CONVOLUTION INTEGRAL

We return to the situation where S and T are independent. In this instance we can make use of a short-cut formula called the ***convolution integral*** to calculate the density function of the sum. We assume here that S and T are both non-negative so that the joint distribution lives on the first quadrant. For example, S and T could represent two independent loss random variables and we are interested in the distribution of the total loss, which is given by $X=S+T$.

The Convolution Integral

Let S and T be independent non-negative random variables with density functions $f_S(s)$ and $f_T(t)$. Let $X=S+T$. Then the density function for X is given by

$$f_X(x) = \int_0^x f_S(s)\cdot f_T(x-s)\,ds.$$

Proof

Since S and T are independent we have the joint density function given by $f(s,t)=f_S(s)\cdot f_T(t)$. We proceed as usual by first calculating the CDF for X, $F_X(x)$:

$$\begin{aligned} F_X(x) = \Pr[X\le x] &= \Pr[S+T\le x] \\ &= \int_0^x\int_0^{x-s} f(s,t)\,dt\,ds = \int_0^x\int_0^{x-s} f_S(s)f_T(t)\,dt\,ds \\ &= \int_0^x f_S(s)\cdot\left[\int_0^{x-s} f_T(t)\,dt\right]ds \\ &= \int_0^x f_S(s)\cdot\overbrace{\left[F_T(x-s)-F_T(0)\right]}^{\text{definition of CDF of } T}ds \\ &= \int_0^x f_S(s)\cdot\left[F_T(x-s)\right]ds. \end{aligned}$$

The last step follows because T is a non-negative random variable and so,

$$F_T(0) = \Pr(T \leq 0) = 0.$$

The formula can now be derived by differentiating both sides with respect to x. This operation is more subtle than may meet the eye since x appears in both the upper limit of integration and inside the integral. Thus, the differentiation requires the chain rule for partial derivatives. We omit the details.

Notes

(1) An easy way to remember the formula – and to think of it – is to regard the integrand as the continuous sum of probability rates for the event $S+T=x$, conditioned on $S=s$. Since S and T are independent, this is just the product of the probability rates for $S=s$ and $T=x-s$.

(2) Integrals of this type arise in a variety of contexts, and since the separate density formulas for S and T in the integrand are intertwined, they are called convolutions.

(3) The convolution integral works best when the density functions $f_S(s)$ and $f_T(t)$ are defined by one formula on the interval $[0,\infty)$. In Example 8.1-7, for example, the densities equal one on the interval $[0,1]$ and zero on $[1,\infty)$. This would require decomposing the convolution integral into multiple ranges, which can be a very complicated process.

Example 8.1-9 The Convolution Formula

Let S and T be independent exponentially distributed random variables with density functions $f_S(s)=\lambda e^{-\lambda s}$ and $f_T(t)=\mu e^{-\mu s}$, respectively, where $\lambda \neq \mu$. (The following exercise deals with the case $\lambda = \mu$.) Find the density function $f_X(x)$ for $X=S+T$.

Solution

$$\begin{aligned}
f_X(x) &= \int_0^x f_S(s)\cdot f_T(x-s)ds \\
&= (\lambda\cdot\mu)\cdot\int_0^x e^{-\lambda s}\cdot e^{-\mu(x-s)}ds \\
&= \lambda\cdot\mu\cdot e^{-\mu x}\cdot\int_0^x e^{-\lambda s}\cdot e^{\mu s}\,ds \\
&= \lambda\cdot\mu\cdot e^{-\mu x}\cdot\int_0^x e^{-(\lambda-\mu)s}ds \\
&= \frac{\lambda\cdot\mu\cdot e^{-\mu x}}{-(\lambda-\mu)}\cdot e^{-(\lambda-\mu)s}\Bigg|_{s=0}^{s=x} = \frac{\lambda\cdot\mu\cdot e^{-\mu x}}{(\lambda-\mu)}\cdot[1-e^{-(\lambda-\mu)x}] = \frac{\lambda\cdot\mu}{(\lambda-\mu)}\cdot[e^{-\mu x}-e^{-\lambda x}].
\end{aligned}$$

❐

Example 8.1-10 SOA EXAM P Sample Exam Questions #108

A device containing two key components fails when both components fail. The lifetimes, T_1 and T_2, of these components are independent with common density function $f(t) = e^{-t}$, $t > 0$. The cost, *X*, of operating the device until failure is $2T_1 + T_2$.

Which of the following is the density function of X for $x > 0$?

(A) $e^{-x/2} - e^{-x}$ (B) $2(e^{-x/2} - e^{-x})$ (C) $\frac{x^2 e^{-x}}{2}$ (D) $\frac{e^{-x/2}}{2}$ (E) $\frac{e^{-x/3}}{3}$

Solution

The lifetimes T_1 and T_2 are independent exponential random variables, both with $\lambda = 1$. From Example 8.1-5 we see that $2T_1$ is an exponentially distributed random variable with mean 2. Hence, it has density $f_{2T_1}(t) = \frac{1}{2}e^{-(1/2)t}$. Now Example 8.1-9 applies with $\lambda = 1$ and $\mu = \frac{1}{2}$, so that $f_X(x) = \frac{\lambda\mu}{(\lambda - \mu)}[e^{-\mu x} - e^{-\lambda x}] = e^{-(1/2)x} - e^{-x}$; $x \geq 0$. ❒

Note

This is not to suggest that the result of Example 8.1-9 should be memorized in order to work a problem of this type. This answer can be derived from scratch using the general CDF technique for a sum and then differentiating. The best course would be to use the convolution formula (which should be learned) along with the recognition that $2T_1$ is also an exponential random variable. Using the convolution formula is less labor-intensive since it requires only a single integration, whereas the general method requires a double integration followed by a derivative.

A variation on the convolution integral for sums concerns ***ratios*** of independent random variables. Applications of this arise occasionally in statistics, as we shall have an opportunity to see in Chapter 9.

Convolution Integral for the Ratio of Two Random Variables

Let *S* and *T* be independent non-negative random variables with density functions $f_S(s)$ and $f_T(t)$. Let $X = \frac{T}{S}$. Then the density function for *X* is given by

$$f_X(x) = \int_0^\infty f_S(s) \cdot f_T(x \cdot s) \cdot s \cdot ds \text{ for all } x > 0.$$

Proof

We proceed in the usual way, by first setting up a double integral for the CDF of *X*. For any $x > 0$ we have $F_X(x) = \Pr[X \leq x] = \Pr\left[\frac{T}{S} \leq x\right] = \Pr[T \leq x \cdot S]$. The event $T \leq x \cdot S$

can be pictured as the region R_x in the first quadrant of the (s,t) plane described by $R_x: 0 \le t \le x \cdot s,\ 0 \le s < \infty$.

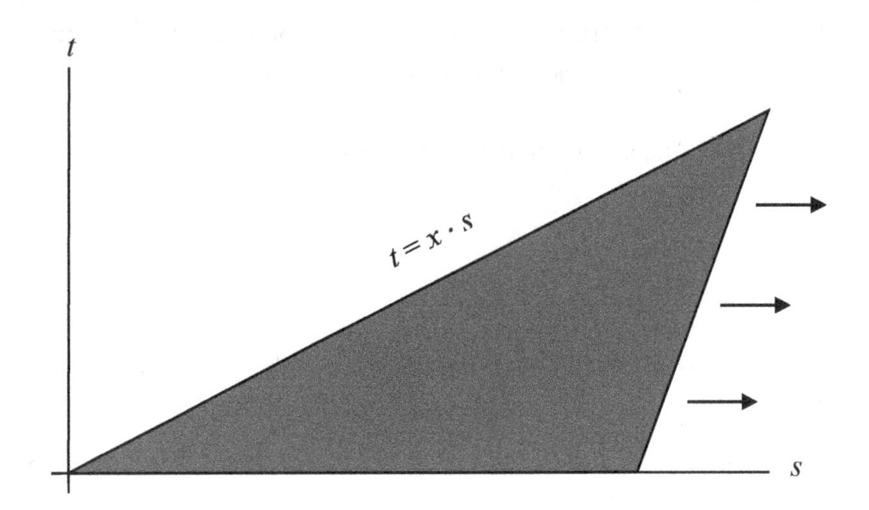

Thus,

$$F_X(x) = \Pr[T \le x \cdot S] = \int_0^\infty \int_0^{xs} f_S(s) \cdot f_T(t) \cdot dt \cdot ds$$

$$= \int_0^\infty f_S(s) \cdot \overbrace{\left[F_T(x \cdot s) - F_T(0)\right]}^{\text{defn of CDF of } T} \cdot ds = \int_0^\infty f_S(s) \cdot F_T(x \cdot s) \cdot ds.$$

We now differentiate both sides with respect to x. Note that by the chain rule, $\frac{d}{dx}\left[F_T(x \cdot s)\right] = f_T(x \cdot s) \cdot \frac{d}{dx}(x \cdot s) = f_T(x \cdot s) \cdot s$. The result is that,

$$f_X(x) = \int_0^\infty f_S(s) \cdot f_T(x \cdot s) \cdot s \cdot ds.$$

Example 8.1-11 A Convoluted Quotient

Let S and T be independent identically distributed exponential random variables, each with a mean of one. Let $X = \frac{T}{S}$. Find the density function of X.

Solution

We have $f_S(s) = e^{-s}$ and $f_T(t) = e^{-t}$ for $s, t \ge 0$. For $x > 0$ we have,

$$f_X(x) = \int_0^\infty f_S(s) \cdot f_T(x \cdot s) \cdot s \cdot ds = \int_0^\infty e^{-s} \cdot e^{-xs} \cdot s \cdot ds = \int_0^\infty s \cdot e^{-(1+x)s} ds.$$

The last integral can be solved using integration-by-parts, or more easily, by recognizing the integrand as that of a gamma distribution with $\alpha = 2$ and $\beta = \frac{1}{1+x}$. Thus,

$$f_X(x) = \int_0^\infty s \cdot e^{-(1+x)s} ds = \beta^\alpha \Gamma(\alpha) = \left(\frac{1}{1+x}\right)^2 \quad \text{for } x > 0.$$

Exercise 8-12 (The case $\lambda = \mu$) Suppose that S and T are independent identically distributed exponential random variables with rate parameter λ. Use the convolution integral to find the density function $f_X(x)$ for $X = S+T$.

Note

The result, $f_X(x) = \lambda^2 x e^{-\lambda x}$, is the density function for the Gamma random variable $X \sim \Gamma\left(2, \frac{1}{\lambda}\right)$, which can be construed as the sum of two independent exponentials, both with mean $\beta = \frac{1}{\lambda}$.

Exercise 8-13 Suppose that S is uniformly distributed on $[0,2]$ and T is an independent (from S) random variable with density $f_T(t) = 2t;\ 0 \le t \le 1$. Find the density function $f_X(x)$ for the sum $X = S+T$.

Exercise 8-14 Suppose that S is an exponentially distributed random variable with mean 1 and T is an independent exponential random variable with mean 2.

(a) Find the density function for $f_X(x)$ for the sum $X = S+T$.

(b) Use $f_X(x)$ to compute $E[X]$.

Exercise 8-15 **SOA EXAM P** **Sample Exam Questions** **#102**

A company has two electric generators. The time until failure for each generator follows an exponential distribution with mean 10. The company will begin using the second generator immediately after the first one fails. What is the variance of the total time that the generators produce electricity?

(A) 10 (B) 20 (C) 50 (D) 100 (E) 200

8.1.4 SIMULATING VALUES OF THE RANDOM VARIABLE X

In this subsection we demonstrate a useful technique for simulating observations from a random variable. Imagine, for example, you are tasked with designing the most efficient customer service system you can for a large retailer. You build an elaborate computer model for customer flow from the parking lot to the checkout lines. Naturally, everything depends on knowing when and how often customers arrive. But these critical inputs to your model are completely random and unknowable, at least in exact terms.

One way to address this dilemma is to assume your customers arrive according to some probability distribution (a Poisson process would be a good choice for this example). If you could simulate random arrivals from this distribution on the computer, you could then run your model multiple times with these random inputs in order to study the resulting patterns of outputs. This would allow you to experiment with different parameters and to study ways you might improve the design for the customer service system.

To simulate observations from X means to generate random numbers that replicate the distribution for X. The ***Inverse Transform Method*** can be used to generate this random sample of observations from X. In broad outline, the procedure is to first generate a random sample from the standard uniform distribution on the interval [0,1]. These numbers are interpreted as random probabilities p. In practice, this set of random numbers will be provided through a random number generating algorithm on a computer or calculator, or, moving down the technology ladder, from a table of random numbers.

The second step is to compute the corresponding percentile values x_p from the cumulative distribution function of X by solving $F_X(x_p) = p$. The set of values x_p will then represent the simulated observations of the random variable X.

This procedure is most easily illustrated using a continuous random variable X whose CDF F_X has an inverse function. In Chapter 5 (Subsection 5.3.3) we indicated the method for finding percentile values for a continuous random variable using the CDF. If p is a number between 0 and 1 representing the probability of the event $[X \leq x_p]$, then x_p is by definition the value of X at the $100p^{th}$ percentile. In other words, $p = \Pr[X \leq x_p] = F_X(x_p)$ (see Figure 8.1.3).

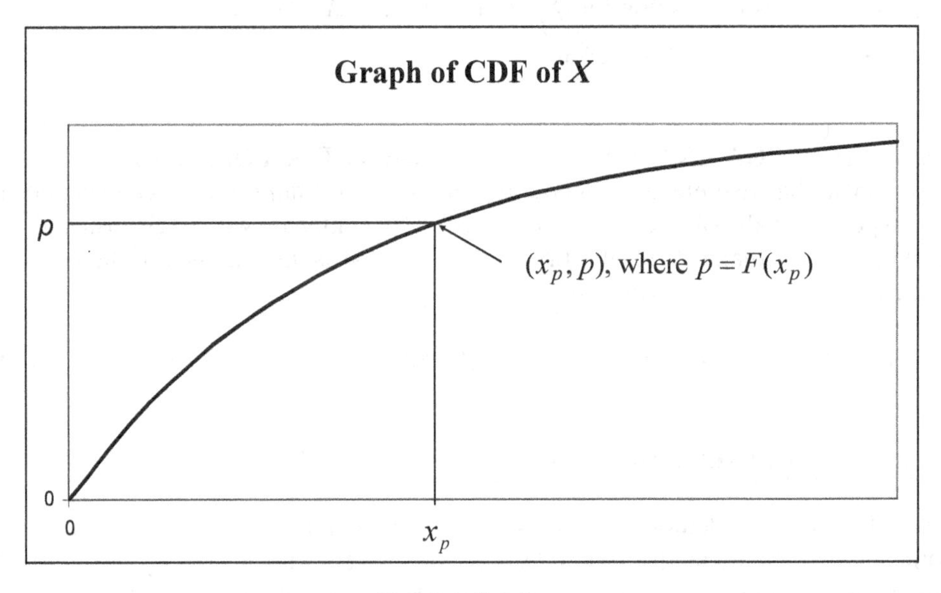

FIGURE 8.1.3

Assuming that the CDF $F_X(x)$ is continuous and strictly increasing, we can use the graph $y = F_X(x)$ to create the ordered pair p and x_p. Solving the equation $p = F_X(x_p)$ for x_p amounts to finding the inverse function $F_X^{-1}(x)$.

The procedure for a discrete random variable X is similar. In this case the corresponding CDF $F_X(x)$ is a step function, so that finding the inverse is not quite as straightforward.

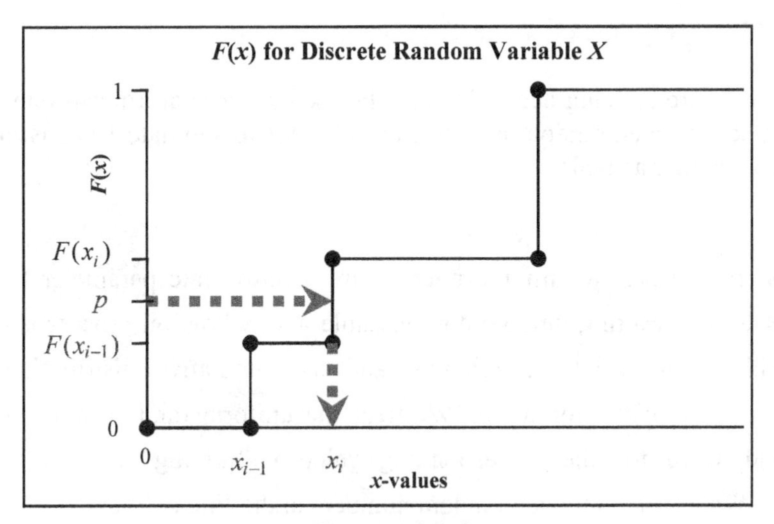

FIGURE 8.1.4

Let x_{i-1} and x_i be consecutive values of X with non-zero probability (see Figure 8.1.4). Then, $F_X(x_i) - F_X(x_{i-1}) = \Pr[X \le x_i] - \Pr[X \le x_{i-1}] = \Pr[X = x_i] = p_i$. We assign a random number p to the X value of x_i if and only if it falls in the interval $\left(F_X(x_{i-1}), F_X(x_i)\right]$. Then the probability of selecting the outcome x_i is $\Pr\left\{\left(F_X(x_{i-1}), F_X(x_i)\right]\right\} = p_i$, which is exactly what we are trying to accomplish.

Example 8.1-12 Discrete Inverse Transformation Method

You are provided with the following random sample from the standard uniform: .28918, .69578, .88231, .33276, .70997, and .79936. Simulate outcomes from a Poisson random variable X with mean $\lambda = 2$. Recall $\Pr(X{=}x) = \frac{e^{-\lambda}\lambda^x}{x!}$ for $x = 0, 1, 2, \ldots$.

Solution
Poisson random variables have a discrete distribution. We first find the cumulative probabilities $F_X(x) = \Pr(X \le x)$.

x = outcome	Pr($X = x$)	Cumulative Probabilities
0	.13533528	.13533528
1	.27067057	.40600585
2	.27067057	.67667642
3	.18044704	.85712346
4	.09022352	.94734698
5	.03608941	.98343639
6 or more	.01656361	1.0000000

Random numbers ~ *Uniform* [0,1]	Simulated Poisson Experiment
.28918	**1**
.69578	**3**
.88231	**4**
.33276	**1**
.70997	**3**
.79936	**3**

As an illustration, the number .69578 in the interval $\left(F_X(2), F_X(3)\right] = (.676676, .857123]$. Therefore, the corresponding simulated outcome for the Poisson random variable is 3. ❐

Example 8.1-13 Continuous Inverse Transform Method

Assume customers are arriving according to a Poisson process at a mean rate of one every two minutes. Use the given random numbers in Table 8.1 to simulate 10 consecutive waiting times between customer arrivals.

Solution

Since arrivals are one per two minute interval, the Poisson rate parameter for one minute periods is $\lambda = \frac{1}{2}$. The waiting time random variable X in a Poisson process is exponentially distributed with mean $\beta = \frac{1}{\lambda} = 2$ minutes, and its cumulative distribution function is $F_X(x) = \Pr(X \le x) = 1 - e^{-x/2}$ for $x \ge 0$. We treat the uniform random number $U = u$ as a probability and solve for the corresponding value x. Solving $u = 1 - e^{-x/2}$ results in $x = -2\ln(1-u)$. We are given the ten random numbers in the first column. ❐

Note: Since U is uniformly distributed in the interval $[0,1]$, so is $1-U$. Therefore, when simulating an exponential random variable with mean θ, we often use $x = -\theta \cdot \ln(u)$ as it is equivalent to $x = -\theta \cdot \ln(1-u)$.

TABLE 8.1

GIVEN Random numbers $\sim$ *Uniform*[0,1]	**Simulated Exponential Experiment $x = -2\ln(1-u)$**	**Alternate Simulated Exponential Experiment $x = -2\ln(u)$**
.28918	**0.68267209**	**2.48141189**
.69578	**2.38000831**	**0.72544352**
.88231	**4.27940246**	**0.25042362**
.33276	**0.80921096**	**2.20066754**
.70997	**2.47505918**	**0.68506513**
.79936	**3.21248604**	**0.44788774**
.56865	**1.68167091**	**1.12898030**
.05859	**0.12075306**	**5.67438249**
.90106	**4.62648335**	**0.20836686**
.31595	**0.75944853**	**2.30434261**

The numbers in either the second or the third column represent the times in minutes between customer arrivals. If we were to generate thousands of simulated arrivals using the same procedure we would find that a histogram of the outcomes would closely resemble (that is, have nearly the same distribution) as a superimposed graph of the density function $f_X(x) = \frac{1}{2}e^{-(1/2)x}$ for $x > 0$.

Returning to the general theory, in the case of a continuous random variable X it remains to show that random numbers from the standard uniform distribution are actually transformed by F_X^{-1} into a random sample from X. In other words, if we choose a large number of observations by repeatedly solving $p_i = F_X(x_i)$ for x_i, then the resulting frequency chart or histogram of the x_i's will closely replicate the density function of X. If U is defined to be the standard uniform random variable then this amounts to showing $F_X^{-1}(U) = X$, or equivalently, $F_X(X) = U$. This is easily accomplished using the Transformation Formula from Subsection 8.1.1.

The Transformation $U = F_X(X)$

Let X be a continuous random variable with strictly increasing CDF $F_X(x)$. If $U = F_X(X)$, then U is the standard uniform random variable on $[0,1]$.

Proof

We need to show that $f_U(u) = 1$ for $0 \le u \le 1$. We observe that by the properties of a CDF, U takes values between 0 and 1. The transformation is given by $u = F_X(x)$, which establishes a pairing of u and x using the graph of $u = F_X(x)$, We next differentiate $u = F_X(x)$ implicitly with respect to u using the chain-rule:

$$1 = \frac{du}{du} = \frac{d}{dx}\left(F_X(x)\right)\frac{dx}{du} = f_X(x)\frac{dx}{du}.$$

But the transformation formula gives $f_U(u) = f_X(x) \cdot \left|\frac{dx}{du}\right|$. Since $u = F_X(x)$ is increasing, we know that $\frac{dx}{du}$ is positive so that $\left|\frac{dx}{du}\right| = \frac{dx}{du}$. Thus,

$$f_U(u) = f_X(x) \cdot \left|\frac{dx}{du}\right| = f_X(x)\frac{dx}{du} = 1.$$

Exercise 8-16 Simulate the outcome of the roll of two fair six-sided dice. Our random variable is the sum. Use the random numbers .28918, .69578, .88231, .33276, and .70997 and the discrete inverse transformation method to simulate this experiment five times.

Exercise 8-17 Simulate the outcome (sum) of the roll of two loaded six-sided dice. Both dice are loaded such that the probability of a side is proportional to the number of dots on the side. Use the random numbers .28918, .69578, .88231, .33276, and .70997 and the discrete inverse transformation method to simulate this experiment five times.

Exercise 8-18 Use the random numbers .79936, .56865, .05859, .90106, and .31595 to simulate outcomes of the random variable $X \sim \textit{Uniform}[-5, 25]$.

Exercise 8-19 Let B denote a binomial random variable with parameters $n = 10$ and $p = .7$. You may think of B as representing the number of made free-throws in 10 attempts from a 70% free-throw shooter. Use the random numbers .79936, .56865, .05859, .90106, and .31595 to simulate values of B.

Exercise 8-20 Simulate an exponential distribution with mean equal to three.

(a) Using a spreadsheet program, perform one hundred simulations.
(b) Calculate the mean and variance of your 100 simulated outcomes.
(c) Compare your results to the mean and variance of the random variable being simulated.

Exercise 8-21 A random variable W has cumulative distribution function $F_W(w) = 1-e^{-.03w^2}$ for $0 \le w < \infty$. This is the Weibull distribution with parameters $\alpha = 2$ and $\beta = .03$. You wish to simulate the outcomes of this random variable using the inverse transformation method. The following are random numbers chosen from $Uniform[0,1]$: .345, .690, .122, .998, .222, .509, .294, .890, .573, and .007.

(a) Calculate the simulated values for the random variable W.
(b) Calculate the mean and the standard deviation of the simulated outcomes.
(c) Compare your results to the mean and the standard deviation of the random variable being simulated.

Exercise 8-22 Given a random variable X with density function

$$f(x) = \begin{cases} cx(1-x)^2 & \text{for } 0 \le x \le 1 \\ 0 & \text{otherwise} \end{cases}.$$

(a) Find the constant c.
(b) Find the cumulative distribution function $F_X(x)$.
(c) Use the random numbers: .345, .690, .122, .998, .222, .509, .294, .890, .573, and .007 to calculate the simulated values for the random variable X. **Note:** You will need a graphing calculator to solve $x = F_X{}^{-1}(u)$.
(d) Calculate the mean and the standard deviation of the simulated outcomes.
(e) Compare your results to the mean and the standard deviation of the random variable being simulated.

Exercise 8-23 Bert is a 60 year-old mathematician who would like to simulate his age at death. He will use the random numbers .28918, .69578, .88231, .33276, .70997, .79936, .56865, .05859, .90106, and .31595 in the simulation.

(a) Simulate Bert's age at death if his mortality is distributed $Uniform[60,100]$.
(b) Simulate Bert's age at death if his future lifetime is distributed $Exp[\mu = 20]$.
(c) Ernie is Bert's business partner, who wishes to purchase a $100,000 life insurance policy payable on the date of Bert's death. Ernie uses the continuous discount rate of $\delta = 6\%$. Use the simulated future lifetimes in (a) and (b) to estimate the present valuc of this life insurance policy (see hint below).
(d) Calculate the theoretical expected present value of the life insurance policy under both mortality assumptions.

Hint: $PV[\text{insurance policy}] = E[100{,}000e^{-\delta \cdot t}] = \int 100{,}000e^{-\delta \cdot t} \cdot f(t)\,dt$

8.1.5 ORDER STATISTICS

Another application of the CDF method involves ***order statistics***. The context concerns a random sample $(X_i)_{i=1}^n$ consisting of n independent observations from a random variable X. Here X is considered the ***population*** random variable. For example, as we have seen previously, X could be a test score on a standardized test, which has a normal distribution. Then each X_i represents the score of a particular individual who constitutes part of a randomly selected sample of size n. Recall that the X_i's are independent and each has the same distribution as the population random variable X.

Order statistics refer to the distributions that arise from looking at the test scores from a random sample arranged in ascending (lowest to highest) order. The most common questions will involve the distribution of the highest (or lowest) score from a random sample of given size. Highest or lowest may also be called the maximum, or minimum, of the random sample. Other questions could concern the distribution of the median, or "middle" score, or the distribution of the 2nd highest score, and so forth.

Clearly, the distribution of, for example, the highest score in a random sample of size n $(n > 1)$ will differ from the distribution of the population. For one thing, the mean of the highest score random variable must be greater than the mean of the population. But the distribution type may differ markedly as well.

Assume the population random variable X has CDF $F_X(x)$ and density function $f_X(x)$. The object is to find formulas for the CDF and/or density of an order statistic (the maximum observation, for example), in terms of $F_X(x)$ and $f_X(x)$.

To keep the notation simple, let $W = Max(X_i)_{i=1}^n$ and $V = Min(X_i)_{i=1}^n$. Typical problems might involve computing the expected values of W or V or the probability of various events depending on W or V. The following box lists formulas for the CDF and densities of W or V that are worth remembering. Even better though, is remembering the process by which the formulas are derived, since the process will work even in unconventional problems for which the specific formulas may not exactly fit. The "process" is nothing more than the CDF method we have been discussing in this section.

Distributions of the Maximum and Minimum of a Random Sample

Let $(X_i)_{i=1}^n$ be a random sample from a population X with CDF $F_X(x)$ and density function $f_X(x)$. Let $W = Max(X_i)_{i=1}^n$ and $V = Min(X_i)_{i=1}^n$.

(1) The CDF for W is given by $F_W(x) = (F_X(x))^n$.

(2) The CDF for V is given by $F_V(x) = 1 - (1 - F_X(x))^n$.

(3) The density function for W is given by $f_W(x) = n \cdot f_X(x) \cdot (F_X(x))^{n-1}$.

(4) The density function for V is given by $f_V(x) = n \cdot f_X(x) \cdot (1 - F_X(x))^{n-1}$.

Proof

Note that the maximum of a set of n observations is less than or equal to x if and only if all observations are less than or equal to x. Thus,

$$\begin{aligned} F_W(x) = P[W \le x] &= \Pr\left[Max(X_i)_{i=1}^n \le x\right] \\ &= \Pr[\text{all observations} \le x] \\ &\overset{\text{by independence}}{=} \prod_{i=1}^n \Pr[X_i \le x] = \prod_{i=1}^n F_{X_i}(x) = \left(F_X(x)\right)^n. \end{aligned}$$

A similar argument helps us find the distribution function of the minimum.

$$\begin{aligned} F_V(x) = \Pr[V \le x] &= \Pr\left[Min(X_i)_{i=1}^n \le x\right] \\ &= \Pr[\text{at least one observation} \le x] \\ &= 1 - \Pr[\text{all observations } > x] \\ &= 1 - \Pr\{[X_1 > x] \cap [X_2 > x] \cap \cdots \cap [X_n > x]\} \\ &\overset{\text{independence}}{=} 1 - \prod_{i=1}^n \Pr(X_1 > x) = 1 - \prod_{i=1}^n \left(1 - F_{X_i}(x)\right) = 1 - \left(1 - F_X(x)\right)^n. \end{aligned}$$

We obtain the density functions for W or V in the usual manner by differentiating:

$$f_W(x) = n \cdot F_X'(x) \cdot \left(F_X(x)\right)^{n-1} = n \cdot f_X(x) \cdot \left(F_X(x)\right)^{n-1},$$

and

$$f_V(x) = 0 - n \cdot (-1) \cdot F_X'(x) \cdot \left(1 - F_X(x)\right)^{n-1} = n \cdot f_X(x) \cdot \left(1 - F_X(x)\right)^{n-1}.$$

Note

These formulas work equally well for a discrete population random variable, with the density function replaced by the probability function. A useful – if somewhat imprecise – way to think of the density formulas in the continuous case is in terms of continuous probability rates. In this sense, formula (3) says roughly that the probability (rate) $f_W(x)$ that x is the maximum of n independent observations is the product of the probability (rate) that one of the observations equals x and the remaining $n-1$ observations are less than or equal to x. The n out front represents the n ways to choose one particular observation (from $X_1, X_2, \ldots, X_n$) to take the value x. This can be expressed with more rigor using small intervals $(x, x+dx)$ and taking limits, but hopefully, you get the idea behind the formula. In a similar vein formula (4) gives the probability rate for x to be the minimum as the product of the n ways to choose a particular observation with the probability rate that the one observation equals x and the remaining $n-1$ observations are greater than x.

Example 8.1-14 SOA EXAM P Sample Exam Questions #76 (Expanded)

Claim amounts for wind damage to insured homes are independent random variables with common density function

$$f(x) = \begin{cases} \frac{3}{x^4} & \text{for } x > 1 \\ 0 & \text{otherwise} \end{cases}$$

where x is the amount of a claim in thousands.

(a) Find the probability that a claim is below average (i.e., less than the mean).
(b) Suppose 3 claims will be made. What is the expected value of the largest of the three claims?
(c) Suppose 3 claims will be made. What is the expected value of the smallest of the three claims?

Solution

The CDF for a single claim is given by $F(x) = \Pr[X \le x] = \int_1^x \frac{3}{t^4}\,dt = 1 - x^{-3}$ for $1 < x < \infty$.
Since $f(x) = F'(x) = 3x^{-4}$, we have $E[X] = \int_1^\infty x(3x^{-4}dx) = 3\int_1^\infty x^{-3}\,dx = \frac{3}{2} = 1.5$ thousand $= 1500$.

(a) $\Pr[X \le 1.5] = F(1.5) = 1 - \left(\frac{3}{2}\right)^{-3} = 1 - \frac{8}{27} = \frac{19}{27}$.

(b) This is the actual sample exam problem. Let $W = Max(X_i)_{i=1}^3$. We calculate $E[W]$ using the density function $f_W(x)$. From formula (3) we have

$$\begin{aligned} f_W(x) &= n \cdot f_X(x) \cdot (F_X(x))^{n-1} \\ &= (3)\left(\frac{3}{x^4}\right)(1 - x^{-3})^2 = 9(x^{-4} - 2x^{-7} + x^{-10});\ x \ge 1. \end{aligned}$$

$$\begin{aligned} E[W] &= 9\int_1^\infty x(x^{-4} - 2x^{-7} + x^{-10})\,dx \\ &= 9\int_1^\infty (x^{-3} - 2x^{-6} + x^{-9})\,dx = 9\left(\frac{1}{2} - \frac{2}{5} + \frac{1}{8}\right) = 2.025 \text{ thousand } = 2025. \end{aligned}$$

(c) Let $V = Min(X_i)_{i=1}^3$. For variety, we calculate $E[V]$ using the CDF $F_V(x)$. We have

$$F_V(x) = 1 - (1 - F_X(x))^n = 1 - \left(1 - (1 - x^{-3})\right)^3 = 1 - x^{-9};\ 1 < x < \infty.$$

Then,

$$\begin{aligned} E[V] &= 1 + \int_1^\infty \left[1 - (1 - x^{-9})\right]dx = 1 + \int_1^\infty x^{-9}\,dx \\ &= 1 + \frac{1}{8} = \frac{9}{8} = 1.125 \text{ thousand } = 1125. \end{aligned}$$

Alternatively, we can use the density for V. From the formula we have

$$f_V(x) = 3f(x)[1-F(x)]^2 = 3(3x^{-4})[1-(1-x^{-3})]^2$$
$$= 9x^{-4}(x^{-3})^2 = 9x^{-10},\ 1 < x < \infty,$$

which could also be obtained by differentiating $F_V(x) = 1-x^{-9}$. Then

$$E[V] = \int_1^\infty x(9x^{-10}\,dx) = 9\int_1^\infty x^{-9}\,dx = \frac{9}{8} = 1.125 \text{ thousand} = 1125. \quad \square$$

Notes

(1) The claim random variable X has a Pareto, Type I distribution, with $\alpha = 3$, $\beta = 1$, and the relevant formulas could be used to write down the CDF and the expected value of X.

(2) The expected amount of a single claim is 1.5 (thousand), and this is bracketed by the expected amount of the smallest of three claims (1.125) and the expected amount of the largest of three claims (2.025).

It is possible, using the same reasoning behind the formulas for the minimum and maximum order statistics, to write expressions for the CDF and density of a general order statistic.

CDF and Density Function for the kth Order Statistic

Let $(X_i)_{i=1}^n$ be a random sample from a continuous population X with CDF $F(x)$ and density function $f(x)$. Let $(Y_i)_{i=1}^n$ be the same observations written in ascending order from Y_1 (smallest) to Y_n (largest). Let Y_k be the kth order statistic $(k = 1, 2, \ldots, n)$.

(1) The CDF of Y_k is given by,

$$F_{Y_k}(y) = \sum_{i=k}^{n} \binom{n}{i} F(y)^i \cdot [1-F(y)]^{n-i}.$$

(2) The density function of Y_k is given by,

$$f_{Y_k}(y) = \frac{n!}{(n-k)!(k-1)!} F(y)^{k-1} f(y)[1-F(y)]^{n-k}.$$

Note: Y_k can be thought of as the k-th smallest observation or, equivalently, the $(n-k)$-th largest observation.

For (1), let y be given, and let us say that an observation X that is less than or equal to y is a success. Then $Y_k \le y$ if there are at least k successes. Thus, using the binomial distribution with $p = \Pr[X \le y] = F(y)$ we have

$$F_{Y_k}(y) = \Pr[Y_k \le y] = \Pr[\text{at least } k \text{ successes}] = \sum_{i=k}^{n} \binom{n}{i} F(y)^i \cdot [1 - F(y)]^{n-i}.$$

The density function in (2) could be derived by differentiation, but we will simply present an intuitively appealing rationale: We consider the probability rate for the event $Y_k = y$. For each of the n observations there are three possible outcomes. Either the observation is less than y, equal to y, or greater than y. This amounts to a multinomial distribution with $Y_k = y$ being the event that $k-1$ observations are less than y, one observation equals y, and the remaining $n-k$ observations are greater than y. The formula for $f_{Y_k}(y)$ represents the multinomial probability of this event, with the multinomial coefficient $\binom{n}{n-k \;\; 1 \;\; k-1} = \frac{n!}{(n-k)!(k-1)!}$. Once again, to make this plausibility argument rigorous we would need to replace the event $Y_k = y$ (which has zero probability) with the event $[y \le Y_k \le y+dy]$, divide through by dy to obtain a rate, and then take the limit as dy goes to zero. This argument works equally well in the discrete case, with probabilities, rather than probability rates, and with the density function replaced by the probability function.

Example 8.1-15 The kth Largest Observation

A random variable X has density $f(x) = \frac{4}{x^5}$; $1 \le x < \infty$. A random sample of seven observations is taken.

(a) Find the density and CDF for the distribution of the 6th largest (equivalently, 2nd smallest) observation.
(b) Find the expected value of the 6th largest observation.
(c) Find the probability that the 6th largest observation is more than 1.1.

Solution

(a) The CDF for X is given by $F(x) = 1 - x^{-4}$; $1 \le x < \infty$. Let Y_2 be the second order statistic, that is, the 2nd smallest (equivalently, 6th largest).

In general, $f_{Y_k}(y) = \frac{n!}{(n-k)!(k-1)!} F(y)^{k-1} f(y)[1 - F(y)]^{n-k}$.

Then for $1 \le y < \infty$, with $k = 2$ and $n = 7$,

$$\begin{aligned} f_{Y_2}(y) &= \frac{7!}{5!} F(y)^1 f(y)[1 - F(y)]^5 \\ &= 42(1 - y^{-4})(4y^{-5})(y^{-4})^5 = 168(y^{-25} - y^{-29}). \end{aligned}$$

The CDF is $F_{Y_2}(y) = \int_1^y f_{Y_2}(t)dt = 1 - 7y^{-24} + 6y^{-28}; 1 \le y < \infty$.

(b) Calculating the expected value from the CDF gives

$$E[Y_2] = 1+\int_1^{\infty}\left[1-F_{Y_2}(y)\right]dy = 1+\int_1^{\infty}[7y^{-24}-6y^{-28}]dy$$

$$= 1+\frac{7}{23}-\frac{6}{27} = \frac{224}{207} = 1.0821.$$

This, of course, can also be calculated using $\int_1^{\infty} y f_{Y_2}(y)\,dy$.

(c) Finally, $\Pr[Y_2 > 1.1] = 1-F_{Y_2}(1.1) = 7(1.1)^{-24}-6(1.1)^{-28} = .2946.$ ❐

Example 8.1-16 Distribution of the Median

Let X be uniformly distributed on $[0,1]$ and suppose a random sample of five observations is taken.

(a) Find the density and CDF for the median (3rd largest) observation.

(b) Find the mean and the variance of the median.

(c) Find the probability the median is less than 0.4.

Solution

Since X is uniform on $[0,1]$ we have $f(x)=1$ and $F(x)=x$ for $0 \le x \le 1$. Let Y_3 denote median (or 3rd smallest of 5 observations) from the standard uniform distribution. Then, with $k=3$ and $n=5$,

$$f_{Y_k}(y) = \frac{n!}{(n-k)!(k-1)!}F(y)^{k-1}f(y)[1-F(y)]^{n-k} = \frac{5!}{2!\,2!}(y^2)(1)(1-y)^2 = 30y^2(1-y)^2.$$

Thus $f_{Y_3}(y)=30y^2(1-y)^2;\ 0 \le y \le 1$.

This is a Beta distribution with $\alpha=\beta=3$, so that the mean and variance can be calculated from the standard formulas for the Beta in Chapter 6. Proceeding from scratch, however, the CDF is

$$F_{Y_3}(y) = \int_0^y 30t^2(1-t)^2\,dt = \int_0^y 30[t^2-2t^3+t^4]dt = 10y^3-15y^4+6y^5;\ 0 \le y \le 1.$$

Next, the expected value, calculated using the density function for Y_3 gives,

$$E[Y_3] = \int_0^1 30[y^3-2y^4+y^5]dy = \frac{1}{2}.$$

(No surprise there, since this is the expected median for observations from the standard uniform distribution.)

Similarly,

$$E[Y_3^2] = \int_0^1 30[y^4 - 2y^5 + y^6]dy = \frac{2}{7}.$$

Hence,

$$Var[Y_3] = \frac{2}{7} - \left(\frac{1}{2}\right)^2 = \frac{1}{28}.$$

Finally,

$$\Pr[Y_3 < 0.4] = F_{Y_3}(.4) = 10(.4)^3 - 15(.4)^4 + 6(.4)^5 = .3174.$$

❐

Exercise 8-24 Suppose that X is uniformly distributed on the interval $[0,10]$ and Y is uniformly distributed on the interval $[0,10]$ and that they are independent.

(a) Find the distribution function and the density function for $\min(X,Y)$.
(b) Find $\Pr[\min(X,Y) < 3]$.
(c) Find the expected value of the $\min(X,Y)$.
(d) Find the standard deviation of the $\min(X,Y)$.

Exercise 8-25 **(Dependence):** Suppose that X is uniformly distributed on the interval $[0,10]$ and suppose that $Y = 2X$. Note the Y is uniformly distributed on the interval $[0,20]$.

(a) Find the probability that the $\max(X,Y) > 7$.
(b) Find the probability that the $\min(X, 20 - Y) > 7$.

Exercise 8-26 Suppose that X is uniformly distributed on the interval $[0,10]$, Y is uniformly distributed on the interval $[0,10]$, and Z is uniformly distributed on the interval $[0,10]$ and that they are mutually independent.
(a) Find the expected value of the $\min(X,Y,Z)$.
(b) Find the standard deviation of the $\min(X,Y,Z)$.
(c) Make a conjecture about the expected value and standard deviation of the minimum and maximum of n independent random variables on the interval $[0,10]$.

Exercise 8-27 Suppose that X is uniformly distributed on the interval $[0,5]$, Y is uniformly distributed on the interval $[0,5]$, and Z is uniformly distributed on $[0,5]$. Suppose that the random variables are independent.
(a) Find the expected value of the $\max(X,Y,Z)$.
(b) What is the expected value of the maximum of n independent random variables that are uniformly distributed on $[0,5]$?
(c) Find $\Pr[\min(X,Y,Z) < 3]$.

Exercise 8-28 Suppose that the height of an American male is normally distributed with mean 5 feet 10 inches and standard deviation of 3 inches. How many men need you select at random so that you are at least 60% confident that the tallest male is 6'6" or taller?

Hint: Think of having a group of n people chosen randomly. The height of the ith person is X_i. Find n such that $\Pr[\max(X_1, X_2, ..., X_n) > 78] \geq .60$.

Exercise 8-29 Consider the following independent random variables: X has density function $f_X(x) = c \cdot x \cdot (10-x)$, Y is uniformly distributed on the interval $[0,10]$, and Z has exponential distribution with mean 5.

(a) Find the probability that the maximum of these three random variables is greater than six.

(b) Find the probability that the minimum of these three random variables is less than three.

(c) Find $E[\max(X,Y,Z)]$.

Exercise 8-30 Long-time friends Larry, Moe, and Curly bought a bottle of 1998 Dom Pérignon to be opened at the wake for the last friend to die. Assuming their future lifetimes in 1998 are independent and uniformly distributed between 0 and 30 years, find the expected age of the bottle of Dom Pérignon when it is opened.

Exercise 8-31 Suppose that X, Y, and Z are exponentially distributed with mean 3, 4, and 5 respectively. Further assume that they are independent. Find the expected value and variance of the $\max(X,Y,Z)$. *Hint*: Compute the distribution function first.

Exercise 8-32 Suppose that X is exponentially distributed with mean 5 and Y is exponentially distributed with mean 7 and that they are independent.

(a) Find the distribution function and the density function for $\min(X,Y)$.

(b) Find $\Pr[\min(X,Y) < 3]$.

(c) Find the expected value of the $\min(X,Y)$.

(d) Find the standard deviation of the $\min(X,Y)$.

(e) Find the distribution function and the density function for $\max(X,Y)$.

(f) Find $\Pr[\max(X,Y) < 3]$.

(g) Find the expected value of the $\max(X,Y)$.

(h) Find the standard deviation of the $\max(X,Y)$.

Exercise 8-33 **SOA EXAM P** **Sample Exam Questions** **#46**

A device that continuously measures and records seismic activity is placed in a remote region. The time, T, to failure of this device is exponentially distributed with mean 3 years. Since the device will not be monitored during its first two years of service, the time to discovery of its failure is $X = \max(T, 2)$.

Determine $E[X]$.

(A) $2 + \frac{1}{3}e^{-6}$ (B) $2 - 2e^{-2/3} + 5e^{-4/3}$ (C) 3 (D) $2 + 3e^{-2/3}$ (E) 5

Exercise 8-34 SOA EXAM P Sample Exam Questions #103

In a small metropolitan area, annual losses due to storm, fire, and theft are assumed to be independent, exponentially distributed random variables with respective means 1.0, 1.5, and 2.4.

Determine the probability that the maximum of these losses exceeds 3.

(A) 0.002 (B) 0.050 (C) 0.159 (D) 0.287 (E) 0.414

Exercise 8-35 SOA/CAS Course 1 2000 Sample Examination #35

Suppose the remaining lifetimes of a husband and wife are independent and uniformly distributed on the interval $[0,40]$. An insurance company offers two products to married couples:

One which pays when the husband dies, and
One which pays when both the husband and wife have died.

Calculate the covariance of the two payment times.

(A) 0.0 (B) 44.4 (C) 66.7 (D) 200.0 (E) 466.7

Exercise 8-36 SOA EXAM P Sample Exam Questions #66

A company agrees to accept the highest of four sealed bids on a property. The four bids are regarded as four independent random variables with common cumulative distribution function $F(x)=\frac{1}{2}(1+\sin \pi x)$ for $\frac{3}{2}\leq x\leq\frac{5}{2}$. Which of the following represents the expected value of the accepted bid?

(A) $\pi\int_{3/2}^{5/2} x\cos\pi x\, dx$

(B) $\frac{1}{16}\int_{3/2}^{5/2}(1+\sin\pi x)^4\, dx$

(C) $\frac{1}{16}\int_{3/2}^{5/2} x(1+\sin\pi x)^4\, dx$

(D) $\frac{1}{4}\pi\int_{3/2}^{5/2}\cos\pi x(1+\sin\pi x)^3\, dx$

(E) $\frac{1}{4}\pi\int_{3/2}^{5/2} x\cos\pi x(1+\sin\pi x)^3\, dx$

Exercise 8-37 CAS Exam 3 Fall 2006 #9

Claim sizes follow an exponential distribution with $\theta=200$. A random sample of size 10 has been drawn.

The probability density function of the order statistic Y_k with a sample size n is

$$\frac{n!}{(k-1)!(n-k)!}[F(y)]^{k-1}[1-F(y)]^{n-k}f(y).$$

Calculate the probability that the second smallest claim will be larger than 50.

(A) Less than 0.30
(B) At least 0.30, but less than 0.35
(C) At least 0.35, but less than 0.40
(D) At least 0.40, but less than 0.45
(E) At least 0.45

Exercise 8-38 CAS Exam 3, Segment 3L Fall 2008 #10

Let Y_1, Y_2, and Y_3 be the order statistics of a random sample of size 3 from a uniform distribution on $[0,1]$.

Calculate the probability that $Y_2 < \frac{1}{3}$.

(A) Less than 15%
(B) At least 15% but less than 20%
(C) At least 20% but less than 25%
(D) At least 25% but less than 30%
(E) At least 30%

Exercise 8-39 CAS Exam 3, Segment 3L Spring 2008 #7

You are given:

- A random sample of three claims
- The claim distribution is uniform over $50 < x < 700$

Calculate the probability that no more than two claims will exceed 300.

(A) Less than 40%
(B) At least 40% but less than 50%
(C) At least 50% but less than 60%
(D) At least 60% but less than 70%
(E) At least 70%

Exercise 8-40 CAS Exam 3 Spring 2005 #25

The probability density function of the kth order statistic of size n is:

$$\frac{n!}{(k-1)!(n-k)!}[F(y)]^{k-1}[1-F(y)]^{n-k} f(y).$$

Samples are selected from a uniform distribution on $[0,10]$.

Determine the expected value of the fourth order statistic for a sample of size five.

(A) Less than 6.5
(B) At least 6.5, but less than 7.0
(C) At least 7.0, but less than 7.5
(D) At least 7.5, but less than 8.0
(E) 8.0 or more

Exercise 8-41 CAS Exam 3 Fall 2005 #2

Let $Y_1, Y_2, \ldots, Y_5$ be the order statistics of a random sample of size 5 from a distribution having density function $f(x) = e^{-x}$ for $x \geq 0$, and $f(x) = 0$ elsewhere. Calculate the probability that $Y_5 > 1$.

(A) Less than 0.55
(B) At least 0.55, but less than 0.65
(C) At least 0.65, but less than 0.75
(D) At least 0.75, but less than 0.85
(E) At least 0.85

8.2 THE MOMENT-GENERATING FUNCTION METHOD

In Subsection 8.1.2, we showed how the CDF method can be used to find the distribution of a sum of two jointly distributed random variables. In this section we will consider extended sums of the form $S = \sum_{i=1}^{n} X_i$, but will restrict our attention to the case where the $(X_i)_{i=1}^{n}$ are independent members of the same distribution family. These results all make use of the basic properties of moment-generating functions first introduced in Section 5.6. We use these properties to derive the MGF of the sum, $M_S(t)$, and from there, to identify the distribution of S.

We summarize and review here a number of important results of this type, many of which have already been demonstrated in earlier sections.

Properties of the Moment Generating Function:

(1) $M_X^{(k)}(0) = E[X^k]$. The superscript (k) means the kth derivative.

(2) If $Y = a \cdot X + b$ for constants a and b, then $M_Y(t) = e^{bt} M_X(at)$.

(3) If $X_1, \ldots, X_n$ are independent random variables and $S = X_1 + \cdots + X_n$, then

$$M_S(t) = M_{X_1}(t) \cdot M_{X_2}(t) \cdots M_{X_n}(t)$$

(4) If $X_1, \ldots, X_n$ are independent random variables, all with the same distribution as X, then $M_S(t) = M_X(t)^n$.

Independent Sums of Specific Random Variables

Let $(X_i)_{i=1}^{n}$ be independent random variables and let $S = X_1 + X_2 + \cdots + X_n = \sum_{i=1}^{n} X_i$.

(1) If each X_i is ***binomial*** with n_i trials and with common probability of success p, then the sum S is binomial with $n_S = \sum_{i=1}^{n} n_i$ trials and probability of success p.

(2) If each X_i is ***negative binomial***, equaling the number of failures before the r_i^{th} success, all based on a common probability of success p, then the sum S is negative binomial, equaling the number of failures before the r_S^{th} success, where $r_S = \sum_{i=1}^{n} r_i$ and the probability of success is p.

(3) If each X_i is ***Poisson*** with parameter λ_i then the sum S is Poisson with parameter $\lambda_S = \sum_{i=1}^{n} \lambda_i$.

(4) If each X_i is ***Gamma*** with shape parameters α_i and common scale parameter β (that is, $X_i \sim \Gamma(\alpha_i, \beta)$) then the sum S is Gamma with shape parameter $\alpha_S = \sum_{i=1}^{n} \alpha_i$ and with scale parameter β (that is, $S \sim \Gamma(\alpha_S, \beta)$).

(5) If each X_i is ***Normal*** with mean μ_i and variance σ_i^2 ($X_i \sim N(\mu_i, \sigma_i^2)$), then the sum S is normally distributed with mean $\mu_S = \sum_{i=1}^{n} \mu_i$ and variance $\sigma_S^2 = \sum_{i=1}^{n} \sigma_i^2$.

Proof

(1) The binomial MGF for X_i is given by $M_{X_i}(t) = (q + pe^t)^{n_i}$. Then by the summation property of MGFs, $M_S(t) = \prod_{i=1}^{n} M_{X_i}(t) = \prod_{i=1}^{n} (q + pe^t)^{n_i} = (q + pe^t)^{n_S}$, which shows that S is binomial with n_s trials and probability of success p.

(2) The negative binomial MGF for X_i is given by

$$M_{X_i}(t) = \left(\frac{p}{1 - qe^t}\right)^{r_i}.$$

Thus,

$$M_S(t) = \prod_{i=1}^{n} M_{X_i}(t) = \prod_{i=1}^{n} \left(\frac{p}{1 - qe^t}\right)^{r_i} = \left(\frac{p}{1 - qe^t}\right)^{r_S},$$

which shows that S is negative binomial.

(3) The Poisson MGF for X_i is given by $M_{X_i}(t) = e^{\lambda_i(e^t - 1)}$. Thus,

$$M_S(t) = \prod_{i=1}^{n} M_{X_i}(t) = \prod_{i=1}^{n} e^{\lambda_i \cdot (e^t - 1)} = e^{\lambda_S \cdot (e^t - 1)},$$

which shows that S is Poisson with parameter λ_S.

Exercise 8-42 (4) Suppose each X_i is Gamma with shape parameters α_i and common scale parameter β. Assume $(X_i)_{i=1}^{n}$ are independent random variables and let $S = X_1 + X_2 + \cdots + X_n = \sum_{i=1}^{n} X_i$. Show that

$$M_S(t) = \left(\frac{1}{1 - \beta t}\right)^{\alpha_S}$$

for $t < \frac{1}{\beta}$, where

$$\alpha_S = \sum_{i=1}^{n} \alpha_i.$$

Exercise 8-43 (5) Suppose each X_i is Normal with mean μ_i and variance σ_i^2. Assume $(X_i)_{i=1}^n$ are independent random variables and let $S = \sum_{i=1}^{n} X_i$ denote the sum. Show that $M_S(t) = e^{\mu_S t + (1/2)\sigma_S^2 t^2}$ where $\mu_S = \sum_{i=1}^{n} \mu_i$ and $\sigma_S^2 = \sum_{i=1}^{n} \sigma_i^2$. This shows that $S \sim N(\mu_S, \sigma_S^2)$.

8.3 COVARIANCE FORMULAS

In this section we continue to work with sums of random variables, but now with the assumption of independence removed. In particular we focus on the ***covariance*** between sums of random variables.

The covariance of two jointly distributed random variables was introduced in Section 7.4. We summarize here the relevant definition and properties.

Covariance Definition and Properties

If X and Y are jointly distributed random variables then the ***covariance between X and Y*** is defined to be

$$Cov[X,Y] = E[(X-\mu_X)\cdot(Y-\mu_Y)].$$

Properties

(1) ***The covariance formula*** $Cov[X,Y] = E[X\cdot Y] - \mu_X \cdot \mu_Y$.

(2) $Cov[X,X] = Var[X]$.

(3) $Cov[aX,bY] = a\cdot b\cdot Cov[X,Y]$.

(4) $Cov[aX+bY,Z] = a\cdot Cov[X,Z] + b\cdot Cov[Y,Z]$.

Notes

As we saw in Chapter 7, the covariance formula in property (1) is usually the preferred method for calculating covariance. Property (2) follows immediately from the definitions of variance and covariance. Property (3) is a consequence of the definition and the fact that $E[a\cdot X] = a\cdot \mu_X$ and $E[b\cdot Y] = b\cdot \mu_Y$. Property (4) follows from the covariance definition as well:

$$\begin{aligned} Cov[aX+bY,Z] &= E[(aX+bY-\mu_{aX+bY})\cdot(Z-\mu_Z)] \\ &= E\big[((aX-a\mu_X)+(bY-b\mu_Y))(Z-\mu_Z)\big] \\ &= E[a(X-\mu_X)(Z-\mu_Z)+b(Y-\mu_Y)(Z-\mu_Z)] \\ &= a\,Cov[X,Z]+b\,Cov[Y,Z]. \end{aligned}$$

The key result for this section is the following procedure for dealing with the variance or the covariance of sums of random variables. The results are straightforward generalizations of the above four properties.

Variance and Covariance for Sums of Random Variables

(1) Suppose that $S = a_1X_1 + a_2X_2 + \cdots + a_mX_m = \sum_{i=1}^{m} a_iX_i$. Then

$$Var[S] = \sum_{i=1}^{m}\sum_{j=1}^{m} a_ia_jCov[X_i,X_j] = \sum_{i=1}^{m} a_i^2Var[X_i] + 2\sum_{i<j} a_ia_jCov[X_i,X_j].$$

(2) Suppose that $S = a_1X_1 + a_2X_2 + \cdots + a_mX_m = \sum_{i=1}^{m} a_iX_i$ and

$$T = b_1Y_1 + b_2Y_2 + \cdots + b_nY_n = \sum_{j=1}^{n} b_jY_j.$$

Then

$$Cov[S,T] = \sum_{i=1}^{m}\sum_{j=1}^{n} a_ib_jCov[X_i,Y_j].$$

Notes

(1) The easiest way to implement these formulas is to draw an $m\times m$ or $m\times n$ box, label the margins with the individual summands, fill in the individual cells, and then add the results to get the final answer.

(2) Students who have learned matrix multiplication may recognize these formulas as matrix products. For example, with S and T as above, let A be the $1\times m$ row vector $A=(a_1,a_2,\ldots,a_m)$ and let B be the $n\times 1$ column vector $B=(b_1,b_2,\ldots,b_n)^{\text{Transpose}}$. Finally, let W be the $m\times n$ matrix of covariances $\left(Cov[X_i,X_j]\right)\Big|_{i,j=1}^{m,n}$. Then the covariance of S and T is the matrix product of A, W, and B; that is, $Cov[S,T] = A\cdot W\cdot B$.

Example 8.3-1 Covariance of Two Sums

Let X and Y be random variables with $Var[X]=2$, $Var[Y]=3$, and $Cov[X,Y]=-1$. Let $U = 2X-Y$ and $V = -X+3Y$. Find $Cov[U,V]$.

Solution
We first set up the U,V table:

	U: $2X$	U: $-Y$
V: $-X$	$-2Var[X]$	$Cov[X,Y]$
V: $3Y$	$6Cov[X,Y]$	$-3Var[Y]$

Then,

$$Cov[U,V] = -2Var[X] - 3Var[Y] + 7Cov[X,Y]$$

$$= (-2)(2) + (-3)(3) + (7)(-1) = -20.$$ ❐

Exercise 8-44 As in Example 8.3-1, let X and Y be random variables with $Var[X]=2$, $Var[Y]=3$, and $Cov[X,Y]=-1$. Let $U=2X-Y$ and $V=-X+3Y$.
Find $Cov[U,V]=Cov[2X-Y,-X+3Y]$ using the definition of covariance.

Exercise 8-45 Suppose X and Y are random variables with $Var[X]=5$, $Var[Y]=2$, and $Cov[X,Y]=-1$. Find $Cov[2X-4Y, X+7Y]$.

8.3.1 COVARIANCE AND THE MULTINOMIAL DISTRIBUTION

In order to motivate the following results for the multinomial distribution we will first consider a simple example involving drawing balls out of an urn.

Example 8.3-2 A Multinomial Experiment

An urn contains 20% red balls, 30% white balls, and 50% blue balls. A game consists of a player drawing a single ball randomly from the urn, and receiving a penalty or reward dependent upon the color selected. The ball is then returned to the urn prior to the next player drawing. A total of 4 players participate. Let X be the number of times a red ball is selected, Y the number of times a white ball is selected, and Z the number of times a blue ball is selected.

(a) Calculate $\Pr[X=1, Y=2, Z=1]$.
(b) Calculate the expected value of X.
(c) Calculate the expected value of Y.
(d) Calculate the covariance of X and Y.

Solution
Since the plays are independent this is a multinomial distribution with 4 trials, where each trial has 3 possible outcomes (red, white or blue).

(a) Using the multinomial formula (see below)

$$\Pr[X=1,Y=2,Z=1] = \binom{4}{1\ \ 2\ \ 1}\cdot(.2)^1\cdot(.3)^2\cdot(.5)^1 = .108.$$

(b) If we think of red as a success, and any other color as a failure, then X, the number of red balls selected, is a binomial random variable with $n=4$ and $p=.2$. Thus, $E[X]=.8$.

(c) Similar reasoning shows that $E[Y]=(4)(.3)=1.2$.

(d) Calculating the covariance is harder, requiring the full joint probability table for X and Y. Since $n=4$ the joint probabilities are zero for $i+j>4$. For the others we have,

$$\begin{aligned} p_{ij} &= \Pr[X=i,Y=j,Z=4-i-j] \\ &= \binom{4}{i\ \ j\ \ 4-i-j}\cdot(.2)^i\cdot(.3)^j\cdot(.5)^{4-i-j};\, i+j\le 4. \end{aligned}$$

This results in the joint distribution table,

		Y-Values				
		0	1	2	3	4
X-Values	0	0.0625	0.1500	0.1350	0.0540	0.0081
	1	0.1000	0.1800	0.1080	0.0216	0
	2	0.0600	0.0720	0.0216	0	0
	3	0.0160	0.0096	0	0	0
	4	0.0016	0	0	0	0

Using this joint distribution we calculate $E[X\cdot Y] = \sum_{i=0}^{4}\sum_{j=0}^{4} i\cdot j\cdot p_{ij} = 0.72$, giving,

$$Cov[X,Y] = E[X\cdot Y]-E[X]\cdot E[Y] = 0.72-(.8)(1.2) = -0.24.$$ ❐

In Section 7.8 we presented properties for the general multinomial distribution (reproduced below). In this section we will give a proof for property (4), which provides a simple formula for the covariance we rather laboriously calculated in part (d) of Example 8.3-2.

Multinomial Random Variables

Consider a probability experiment with r distinct outcomes, and that $p_1, p_2, \ldots, p_r$ represent the probabilities of the respective outcomes, so that the p_i's are non-negative and sum to one.

Consider n independent repetitions of the experiment and let X_i represent the number of times outcome i occurs. Then,

(1) $X_1+X_2+\cdots+X_r=n.$

(2) Each X_i is binomial with parameters n and p_i.

(3) Given non-negative integers $k_i; i=1,\ldots,r$, with $k_1+\cdots+k_r=n$, then

$$\Pr[X_1=k_1, X_2=k_2,\ldots,X_r=k_r] = \binom{n}{k_1\ \ k_2\cdots k_r} p_1^{k_1}p_2^{k_2}\cdots p_r^{k_r}.$$

(4) For $i\ne j$, $Cov[X_i,X_j] = -np_ip_j$.

Proof of Property (4): Let *i* and *j* with $i \neq j$ be given. Let $X_i = U_1 + U_2 + \cdots + U_n$, where

$$U_k = \begin{cases} 1 & \text{if result of } k^{th} \text{ trial is outcome } i \\ 0 & \text{otherwise} \end{cases}.$$

In other words, U_k is the Bernoulli trial random variable associated with the kth repetition of the experiment and the ith outcome. Similarly, let $X_j = V_1 + V_2 + \cdots + V_n$, where V_k is the Bernoulli trial random variable associated with the kth trial and the jth outcome. We will calculate $Cov[X_i, X_j]$ using their representations as sums of Bernoulli trial random variables. Toward this end, note that since the trials are independent we have the covariances of the U's and the V's all equal to zero unless they represent the same trial. But, since only one of the outcomes *i* or *j* can occur, at least one of U_k and V_k (or both) must be zero, regardless of the outcome. Thus, $U_k \cdot V_k = 0$ and so $E[U_k \cdot V_k] = 0$. On the other hand, $E[U_k] = p_i$ and $E[V_k] = p_j$, so by the covariance formula we have

$$Cov[U_k, V_k] = E[U_k V_k] - E[U_k]E[V_k] = 0 - p_i p_j = -p_i p_j.$$

We can now use the box method for the covariance of the sums to write,

$Cov[X_i, X_j] =$

	V_1	V_2	⋯	V_n
U_1	$-p_i p_j$	0	⋯	0
U_2	0	$-p_i p_j$	⋯	0
⋮	⋯	⋯	⋱	⋯
U_n	0	0	⋯	$-p_i p_j$

$= -np_i p_j.$

Example 8.3-3 Using the Multinomial Covariance Formula

In Example 8.3-2, calculate the covariance of *X* and *Y* using property (4) of the multinomial distribution.

Solution

$$Cov[X, Y] = -(4)(.2)(.3) = -.24.$$ ❐

8.3.2 VARIANCE FORMULA FOR THE HYPER-GEOMETRIC DISTRIBUTION

Another piece of unfinished business involves the variance formula for the hyper-geometric distribution introduced in Section 4.5. To review the context, we have a box containing *B* objects of one type (type *B*) and *G* objects of another type (type *G*), for a total of $N = B+G$ objects. For simplicity's sake the letters *B* and *G* stand for both the number and the type. A random sample of size n is chosen, ***without*** replacement. This can be considered as a sequence of n Bernoulli trials, where the Bernoulli random variable corresponding to the kth trial is

$$U_k = \begin{cases} 1 & \text{if result of } k^{\text{th}} \text{ trial is } B \\ 0, & \text{if result of } k^{\text{th}} \text{ trial is } G \end{cases}.$$

Then the hypergeometric random variable W is the number of B's in the sample, and can be written as the sum of n Bernoulli trials, $W = U_1 + U_2 + \cdots + U_n$. The formulas for the mean and the variance of W, presented initially in Section 4.5, are,

Mean and Variance of the Hyper-geometric Random Variable

Suppose that W is a ***hyper-geometric random variable***. Then

(a) $\mu = E[W] = n \cdot \left(\frac{B}{B+G}\right) = n\frac{B}{N}$, and

(b) $\sigma^2 = Var[W] = n \cdot \left(\frac{B}{B+G}\right) \cdot \left(\frac{G}{B+G}\right) \cdot \left(\frac{B+G-n}{B+G-1}\right) = n\left(\frac{B}{N}\right)\left(\frac{G}{N}\right)\left(\frac{N-n}{N-1}\right)$.

Proof

(a) As demonstrated in Section 4.5, the U's are identically distributed Bernoulli trials with probability of success $\frac{B}{N}$.

Thus, $E[W] = E[U_1 + U_2 + \cdots + U_n] = n \cdot \frac{B}{N}$.

(b) First, we observe that, as Bernoulli trials, each of the U's has variance,

$$Var[U_i] = \left(\frac{B}{N}\right)\left(1 - \frac{B}{N}\right) = \frac{BG}{N^2}.$$

Although the U's are identically distributed, they are not independent since we are sampling without replacement. This means we need to calculate all of the covariances, $Cov[U_i, U_j]$. This is simplified by the observation that, just as the U's are identically distributed, so too do all of the pairs (U_i, U_j) (with $i \neq j$) possess identical joint distributions. Thus, it suffices to find $Cov[U_i, U_j]$ for the case $i = 1, j = 2$.

First, we work out the joint distribution of U_1 and U_2 using the basic multiplication principle. There are $N(N-1)$ outcomes for the first two draws, and, for example, the number of outcomes with 2 B's is $B(B-1)$. So $\Pr\{[U_1 = 1] \cap [U_2 = 1]\} = \frac{B(B-1)}{N(N-1)}$.

The complete joint table for U_1 and U_2 is:

	$U_2 = 0$	$U_2 = 1$
$U_2 = 0$	$\dfrac{G(G-1)}{N(N-1)}$	$\dfrac{BG}{N(N-1)}$
$U_2 = 1$	$\dfrac{BG}{N(N-1)}$	$\dfrac{B(B-1)}{N(N-1)}$

Thus, $E[U_1U_2] = \dfrac{B(B-1)}{N(N-1)}$ and hence,

$$\begin{aligned} Cov[U_1,U_2] &= \frac{B(B-1)}{N(N-1)} - \left(\frac{B}{N}\right)^2 \\ &= \left(\frac{B}{N}\right)\left(\frac{(B-1)}{(N-1)} - \frac{B}{N}\right) \\ &= -\left(\frac{B}{N}\right)\left(\frac{G}{N}\right)\left(\frac{1}{N-1}\right) = -\left(\frac{BG}{N^2}\right)\left(\frac{1}{N-1}\right). \end{aligned}$$

Now we look at the $n \times n$ box, which has n diagonal elements and $n(n-1)$ elements off the main diagonal to conclude,

$Var[W] =$

	U_1	U_2	$\cdots$	U_n
U_1	$\dfrac{BG}{N^2}$	$-\left(\dfrac{BG}{N^2}\right)\left(\dfrac{1}{N-1}\right)$	$\cdots$	$-\left(\dfrac{BG}{N^2}\right)\left(\dfrac{1}{N-1}\right)$
U_2	$-\left(\dfrac{BG}{N^2}\right)\left(\dfrac{1}{N-1}\right)$	$\dfrac{BG}{N^2}$	$\cdots$	$-\left(\dfrac{BG}{N^2}\right)\left(\dfrac{1}{N-1}\right)$
$\vdots$	$\cdots$	$\cdots$	$\ddots$	$\cdots$
U_n	$-\left(\dfrac{BG}{N^2}\right)\left(\dfrac{1}{N-1}\right)$	$-\left(\dfrac{BG}{N^2}\right)\left(\dfrac{1}{N-1}\right)$	$\cdots$	$\dfrac{BG}{N^2}$

$$\begin{aligned} &= n\frac{BG}{N^2} - n(n-1)\left(\frac{BG}{N^2}\right)\left(\frac{1}{N-1}\right) \\ &= \left(n\frac{BG}{N^2}\right)\left(1 - \frac{n-1}{N-1}\right) = n\left(\frac{B}{N}\right)\left(\frac{G}{N}\right)\left(\frac{N-n}{N-1}\right). \end{aligned}$$

8.4 THE CONDITIONING FORMULAS AND MIXTURES

We now return to an old and familiar problem, that of calculating the mean and the variance of a given random variable Y. The novelty here is that we are given partial information about Y that depends on a second random variable X. More formally, Y is *conditioned* on X. In Chapter 7, we discussed how to derive the conditional distribution of Y given X from the joint distribution, $f(x,y)$, of X and Y. Here, we will work backwards, deriving the mean and variance for the distribution of Y from the conditional distribution $Y \mid X$.

Suppose, for example, that an auto insurance company insures two classes of customers – good drivers and bad drivers. The random variable Y represents the annual claim amount for a customer, and the company wishes to calculate the expected claim amount, $E[Y]$, and the variance, $\text{Var}[Y]$, of the claim amount. Reasonably enough, the insurance claim amount should depend on the class of driver insured. So, we use the second random variable, X to signify the class of driver (good or bad). Then, the conditional random variable, $Y|X = \text{good}$, will have one probability distribution, while the conditional random variable, $Y|X = \text{bad}$, will likely have a different distribution. That is, the annual insurance claim is *conditioned* by class X.

The perspective here is quite similar to the Bayesian approach to finding probabilities, discussed in Section 2.6. However, the goal now is to find the overall expectation and variance of the claim amount by using the conditional distributions. This will give us two levels of distributions, the conditional distributions $Y \mid X$ of claims by class, and then the distribution of X, representing the classes themselves.

Expectation and Variance by Conditioning

Let X and Y be jointly distributed random variables. Then,

(1) $E[Y] = E_X\left[E_Y[Y|X]\right]$. This is called the ***double expectation formula***.

(2) $Var[Y] = E_X\left[Var_Y\left[Y|X\right]\right] + Var_X\left[E_Y\left[Y|X\right]\right]$.

These formulas, while quite useful, can be difficult to interpret and apply without practice. Before providing derivations we will illustrate their use with some examples. In reading through the examples, keep in mind that:

(1) $Y \mid X$ merely represents some random variable. Think of X as a known fixed value (like being a bad driver).

(2) Compute the inside quantities $E_Y\left[Y \mid X\right]$ and $\text{Var}_Y\left[Y \mid X\right]$ in the usual manner for computing mean and variance.

(3) Realize that these quantities $E_Y\left[Y \mid X\right]$ and $\text{Var}_Y\left[Y \mid X\right]$ are themselves random variables whose values depend on the given value of X. That is, these are both transformations of the random variable X.

(4) Therefore, we can compute the outside expectation and variance, which is calculated with respect to X.

(5) In many applications of these formulas, each value of X can be considered a group, such as the group of good drivers or the group of bad drivers. Formula (1) can be thought of as the weighted average of the group averages. The expression on the right in formula (2) is sometimes referred to as "the variance within groups" plus "the

variance between groups." This uses the language of ***analysis of variance*** (ANOVA), an important statistical tool.

Example 8.4-1 Expectation by Conditioning

A certain professor of our acquaintance works in Moon Township, PA and lives in Pittsburgh, PA, about a 25 mile commute. The professor randomly chooses from 3 different routes home in a futile attempt to evade rush hour traffic. The routes are identified by the name of a major bridge along the way. The professor has accumulated data over a lengthy period of time on the mean drive times of the three routes. Using the data summary that follows, calculate the overall expected drive time.

Route	Probability of Route	Expected Time of Route
Sewickley Bridge	0.2	55 minutes
Fort Pitt Bridge	0.4	50 minutes
Liberty Bridge	0.4	45 minutes

Solution

In the context of formula (1) we let X represent the route selected, and we let Y denote the drive time. Then X takes the three values: *Sewickley*, *Fort-Pitt*, and *Liberty*, with the probabilities shown. The drive times in the third column are conditioned upon the route chosen, and are the 3 values of $E_Y[Y \mid X]$. The calculation for the average drive time, $E[Y] = E_X\left[E_Y[Y \mid X]\right] = \sum_X \Pr[X] \cdot E_Y[Y \mid X]$, is easily found by summing the fourth column below:

X = Route	$\Pr[X]$	$E_Y[Y \mid X]$	$\Pr[X] \cdot E_Y[Y \mid X]$
Sewickley	0.2	55	$(0.2)(55) = 11$
Fort Pitt	0.4	50	$(0.4)(50) = 20$
Liberty	0.4	45	$(0.4)(45) = 18$
Total			49

This amounts to the same weighted average calculation used to determine the probability of an event by conditioning, as often displayed in a tree diagram. The only difference is that instead of conditional probabilities along the branches we have conditional expectations:

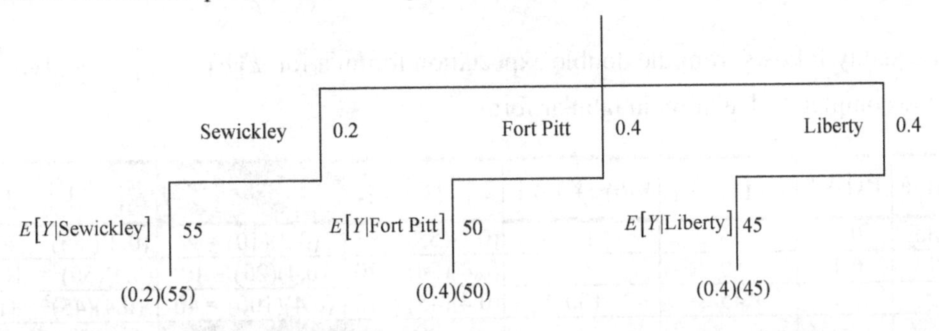

$$E[Y] = E[E[Y \mid X]] = E[Y \mid X = \text{Sewickley}] \cdot \Pr[X = \text{Sewickley}]$$
$$+ E[Y \mid X = \text{Fort Pitt}] \cdot \Pr[X = \text{Fort Pitt}]$$
$$+ E[Y \mid X = \text{Liberty}] \cdot \Pr[X = \text{Liberty}]$$
$$= (55)(0.2) + (50)(0.4) + (45)(0.4) = 49.$$

❐

Note

Strictly speaking, X is not a random variable since it is not numerical-valued. Its values are labels rather than numbers, but everything works out correctly. If this were an issue we could assign the numbers 1, 2, 3 to the routes, but nothing in the calculations would be affected.

Example 8.4-2 Variance by Conditioning

The professor of Example 8.4-1 has also calculated variances in the times of the three routes. The information is displayed in a new summary table:

X = Route	$\Pr[X]$	$E_Y[Y \mid X]$	$Var_Y[Y \mid X]$
Sewickley	0.2	55	10
Fort Pitt	0.4	50	25
Liberty	0.4	45	100

Using the conditional variances, calculate the overall variance in drive time.

Solution

We can play the same game as above, with the conditional variances, to calculate $E[Var_Y[Y \mid X]]$. But, as we see from formula (2) for the variance of Y, this is only half the story. We also need to calculate $\text{Var}_X[E_Y[Y \mid X]]$, which means the variance of the $E_Y[Y \mid X]$ column in the summary table. This will be done using the standard variance formula $\text{Var(whatever)} = E(\text{whatever}^2) - (E(\text{whatever}))^2$, which in this context means:

$$\text{Var}_X(E_Y[Y \mid X]) = E_X(E_Y[Y \mid X]^2) - (E_X[E_Y[Y \mid X]])^2$$
$$= E_X[E_Y[Y \mid X]^2] - (E[Y])^2.$$

The last equality follows from the double expectation formula for $E[Y] = E_X[E_Y[Y \mid X]]$. Here is the complete calculation in tabular form:

X = Route	$\Pr[X]$	$E_Y[Y \mid X]$	$Var_Y[Y \mid X]$	$E[E[Y \mid X]]$	$E[Var[Y \mid X]]$	$E[E[Y \mid X]^2]$
Sewickley	0.2	55	10	$(0.2)(55) = 11$	$(0.2)(10) = 2$	$(0.2)(55)^2 = 605$
Fort Pitt	0.4	50	25	$(0.4)(50) = 20$	$(0.4)(25) = 10$	$(0.4)(50)^2 = 1000$
Liberty	0.4	45	100	$(0.4)(45) = 18$	$(0.4)(100) = 40$	$(0.4)(45)^2 = 810$
				49	52	2415

$$Var\left[E[Y\mid X]\right] = E\left[E[Y\mid X]^2\right]-E[Y]^2 = 2415-(49)^2 = 14,$$

$$E\left[Var[Y\mid X]\right] = 52,$$

and

$$Var[Y] = E\left[Var[Y\mid X]\right]+Var\left[E[Y\mid X]\right] = 52+14 = 66. \quad \square$$

Note

Using the statistical language mentioned previously, the $Var[Y\mid X]$ column can be considered as the variance within routes, with $E\left[\text{Var}[Y\mid X]\right]=52$ being the (weighted) average variance within routes. Then, $\text{Var}\left[E[Y\mid X]\right]=14$ is the variance between routes. In this example most of the variance is within routes (52 out of a total of 66), with an especially high variance on the Liberty route. Since the mean route times differ by only a few minutes, the variance between routes is only 14.

An alternative approach to the route problem is to think of the total drive time Y as a ***mixture random variable***. This is a concept that was introduced in Section 5.4 in the context of mixed (part continuous and part discrete) random variables. From this perspective we construe Y as a mixture of three random variables, X_1, X_2, X_3, with weights 0.2, 0.4 and 0.4. Here, X_i represents the drive time for route i; $i=1,2,3$. This allows us to calculate the moments of Y as mixtures of the moments of X_i. That is,

$$E\left[Y^n\right]=0.2E\left[X_1^n\right]+0.4E\left[X_2^n\right]+0.4E\left[X_3^n\right].$$

Example 8.4-3 Mixing the Routes

You are given that the professor's drive time Y is a mixture random variable, using the data in the following table:

X = Drive Time	**Weight**	$E[X]$	$Var[X]$
X_1 = Sewickley	0.2	55	10
X_2 = Fort Pitt	0.4	50	25
X_3 = Liberty	0.4	45	100

(a) Calculate the mean and the variance of Y using the mixture formula for moments.
(b) Assume the drive times X_1, X_2, X_3 are normally distributed. Find the probability that the overall drive time Y is greater than 49 minutes.

Solution

For (a), we have

$$E[Y]=0.2E\left[X_1\right]+0.4E\left[X_2\right]+0.4E\left[X_3\right]=(.2)(55)+(.4)(50)+(.4)(45)=49.$$

This is the same calculation resulting from the double expectation formula.

Recall from Section 5.4 that we cannot directly mix the variances. Thus, we will need to first mix the second moments of X_1, X_2, X_3 to calculate the second moment of Y. Since $E\left[X_i^2\right] = Var[X_i] + E[X_i]^2$, we have,

$$\begin{aligned} E\left[Y^2\right] &= 0.2E\left[X_1^2\right] + 0.4E\left[X_2^2\right] + 0.4E\left[X_3^2\right] \\ &= (.2)\left(10 + 55^2\right) + (.4)\left(25 + 50^2\right) + (.4)\left(100 + 45^2\right) = 2467. \end{aligned}$$

Thus, $Var[Y] = E\left[Y^2\right] - E[Y]^2 = 2467 - 49^2 = 66.$

For (b) we make use of the fact that the CDF for Y is the mixture of the CDFs of X_1, X_2, X_3:

$$F_Y(49) = 0.2F_{X_1}(49) + 0.4F_{X_2}(49) + 0.4F_{X_3}(49).$$

Since we are now assuming X_1, X_2, X_3 are normal we can standardize to obtain,

$$\begin{aligned} F_Y(49) &= 0.2\Phi\left(\frac{49-55}{\sqrt{10}}\right) + 0.4\Phi\left(\frac{49-50}{\sqrt{25}}\right) + 0.4\Phi\left(\frac{49-45}{\sqrt{100}}\right) \\ &= (.2)(.0289) + (.4)(.4207) + (.4)(.6554) = .4362. \end{aligned}$$

Thus, $\Pr[Y > 49] = 1 - F_Y(49) = 1 - .4362 = .5638.$ ❐

Notes:

(1) Part (a) shows the average (mean) drive time is 49 minutes. Part (b) demonstrates that more often than not the drive time is above average. It often seems to the professor that ***all*** drive times are above average.[1]

(2) We emphasize that while we can mix the moments and the CDFs to obtain the moments and the CDF for Y, it is **NOT** the case that $Y = 0.2X_1 + 0.4X_2 + 0.4X_3$!!. For one thing, if that ***were*** the case then Y itself would be normal, so that the probability of Y being above average would be exactly 50%. Consequently, we see that the mixture Y is not normal even though the component random variables are.

Example 8.4-4 Mean and Variance by Conditioning

Let X and Y be jointly distributed random variables such that X has mean two and variance three. Suppose the conditional mean of Y, given $X = x$, is x and the conditional variance of Y, given $X = x$, is x^2. Calculate the mean and the variance of Y.

---

[1] This is an example of the so-called "Lake Wobegon effect." Lake Wobegon is Garrison Keillor's fictitious community where all children are above average. This example shows that a ***majority*** may be above average, but not all.

Solution

We are asked to find $E[Y]$ and $Var[Y]$. We are given $E[Y \mid X] = X$, $Var[Y \mid X] = X^2$, $E_X[X] = 2$, and $\text{Var}_X[X] = 3$. From the double expectation theorem,

$$E[Y] = E_X\left[E_Y[Y|X]\right] = E_X[X] = 2.$$

$$E_X\left[\text{Var}[Y \mid X]\right] = E_X[X^2] = 7.$$

Remember that $E[X^2] = \text{Var}[X] + \left(E[X]\right)^2 = 3 + 2^2$.

$$Var_X\left[E[Y|X]\right] = Var_X[X] = 3.$$

Finally,

$$Var[Y] = E_X\left[Var_Y[Y \mid X]\right] + Var_X\left[E_Y[Y \mid X]\right] = 7 + 3 = 10.$$

❐

Example 8.4-5 Random Parameter

Let X denote the amount of damage caused by a hurricane in millions. Suppose that X is exponentially distributed with mean Ω, where Ω is itself a random variable that is uniformly distributed on the interval $[5, 45]$. Calculate the mean and the standard deviation of X.

Solution

X is exponentially distributed with mean Ω, so we know that $E_X[X \mid \Omega] = \Omega$ and $Var_X[X \mid \Omega] = \Omega^2$. Since Ω is uniformly distributed on $[5, 45]$ we have $E[\Omega] = \frac{5+45}{2} = 25$ and $Var[\Omega] = \frac{(45-5)^2}{12} = \frac{400}{3}$. Thus,

$$E[X] = E_\Omega\left(E_X[X \mid \Omega]\right) = E_\Omega[\Omega] = 25.$$

Next,

$$Var_\Omega\left(E[X \mid \Omega]\right) = Var_\Omega(\Omega) = \frac{400}{3}.$$

Then,

$$E_\Omega\left[Var_X[X \mid \Omega]\right] = E_\Omega[\Omega^2] = Var_\Omega[\Omega] + E_\Omega[\Omega]^2 = \frac{2275}{3}.$$

And then,

$$Var[X] = E_\Omega\left(Var_X[X \mid \Omega]\right) + Var_\Omega\left(E_X[X \mid \Omega]\right) = \frac{2275}{3} + \frac{400}{3} = \frac{2675}{3} = 891.\overline{6}.$$

Finally, $\sigma_X = \sqrt{891.6} = 29.86$.

❐

Note:

(1) The hurricane loss in the above example can be thought of as being contingent on a risk factor Ω. The smaller values of Ω might correspond to homes in, say Laramie, Wyoming, where blizzards are common but hurricanes, not so much. Values of Ω close to the upper limit of 45 might correspond to communities like Galveston, Texas, with a heavy exposure to hurricane risk.

(2) The overall hurricane loss X may be thought of as a mixture random variable, where Ω is the mixing factor. In this case, instead of being a two or three point mixture, X is a continuous mixture as Ω varies continuously and uniformly from 5 to 45.

In general, for a continuous mixing variable the mixing weights are represented by the density function of the mixing variable.

Probabilities and The CDF of a Continuous Mixture

Let X be a random variable with a continuously varying parameter Λ with density function $f_\Lambda(\lambda)$. For a given $\Lambda = \lambda$ let X have a conditional CDF represented by $F(x \mid \Lambda = \lambda)$. Then,

$$F_X(x) = \int_{-\infty}^{\infty} F(x \mid \Lambda = \lambda) \cdot f_\Lambda(\lambda) d\lambda$$

Let E be an event in the domain of X. Then,

$$\Pr[E] = \int_{-\infty}^{\infty} \Pr[E \mid \Lambda = \lambda] \cdot f_\Lambda(\lambda) d\lambda$$

Example 8.4-6 A Continuous Mixture

Let X denote the amount of damage caused by a hurricane in millions. Suppose that X is exponentially distributed with mean $1/\Lambda$, where Λ is a random variable that is uniformly distributed on the interval $(0,10]$.

(a) Find a formula for the CDF, $F_X(x)$.

(b) Find the probability that X is greater than one.

Solution

Let λ be given, with $0 < \lambda \leq 10$. Then $F[x \mid \Lambda = \lambda] = 1 - e^{-\lambda x}$; $0 < x < \infty$.

Now, taking the overall mixture by integrating we have,

$$F_X(x) = \int_0^{10} \overbrace{\left(1 - e^{-\lambda x}\right)}^{F(x|\Lambda=\lambda)} \cdot \overbrace{\frac{1}{10}}^{f_\Lambda(\lambda)} d\lambda = \frac{1}{10}\left[\lambda - \left(\frac{e^{-\lambda x}}{-x}\right)\right]_{\lambda=0}^{10} = 1 - \left(\frac{1}{10x}\right)\left(1 - e^{-10x}\right).$$

Therefore, $\Pr[X > 1] = 1 - F_X(1) = \frac{1}{10}\left(1 - e^{-10}\right) = 0.099995.$ ❐

It is also possible for X to be discrete, conditioned on a continuous mixing variable. We can use the same ideas to calculate the probability function for X.

Example 8.4-7 A Poisson is Mixed with an Exponential

Let N denote the annual number of claims. Suppose that N has a Poisson distribution with mean Λ, where Λ is a random variable that is exponentially distributed with mean one. Find the probability function for N.

Solution

We are given that $f_\Lambda(\lambda)=e^{-\lambda};\ 0\le\lambda<\infty$. Also, $\Pr[N=n\,|\,\Lambda=\lambda]=e^{-\lambda}\cdot\dfrac{\lambda^n}{n!};\ n=0,1,2\ldots.$

Then,

$$\Pr[N=n]=\int_0^\infty \Pr[N=n\,|\,\Lambda=\lambda]\cdot f_\Lambda(\lambda)\,d\lambda=\int_0^\infty e^{-\lambda}\cdot\frac{\lambda^n}{n!}\cdot e^{-\lambda}\,d\lambda=\int_0^\infty\frac{\lambda^n}{n!}\cdot e^{-2\lambda}\,d\lambda=\frac{1}{n!}\int_0^\infty\lambda^n e^{-2\lambda}\,d\lambda$$

The last integral is of the gamma type. From Section 6.5.2, property (6) of the gamma distribution we have,

$$\int_0^\infty x^{\alpha-1}e^{-(1/\beta)x}\,dx \;=\; \beta^\alpha\cdot\Gamma(\alpha).$$

Thus, with $\alpha-1=n$ and $1/\beta=2$, we have $\alpha=n+1$ and $\beta=1/2$ and so,

$$\int_0^\infty\lambda^n e^{-2\lambda}\,d\lambda=\left(\frac{1}{2}\right)^{n+1}\Gamma(n+1)=\left(\frac{1}{2}\right)^{n+1}n!.$$

Therefore, $\Pr[N=n]=\dfrac{1}{n!}\displaystyle\int_0^\infty\lambda^n e^{-2\lambda}\,d\lambda=\dfrac{1}{n!}\cdot\left(\dfrac{1}{2}\right)^{n+1}\cdot n!=\left(\dfrac{1}{2}\right)^{n+1};\ n=0,1,2\ldots.$

This shows that the overall claim rate is a geometric random variable with $p=1/2$.[2] ❒

---

[2] It can be shown that in general the mixture of a Poisson with a Gamma results in a Negative Binomial. We omit the details.

Another common application of the conditioning formulas involves random sums. That is, situations where the number of terms N in the sum of random variables is itself a random variable.

Random Sums

Let $X_1, X_2, \ldots, X_N$ be a random sample from a population X with mean and variance given by μ_X and σ_X^2. Let the number of terms N be a random variable independent of X, and let μ_N and σ_N^2 denote the mean and variance of N. Let

$$S = X_1 + X_2 + \cdots + X_N$$

denote the random sum. Then,

(1) $E[S] = \mu_S = \mu_N \cdot \mu_X$, and

(2) $Var[S] = \sigma_S^2 = \sigma_N^2 \cdot \mu_X^2 + \mu_N \cdot \sigma_X^2$.

Proof

If the number of terms is given as $N = n$, we know that the sum S has exactly n terms. Then,

$$E_S[S \mid N = n] = E[X_1 + \cdots + X_n] = E[X_1] + \cdots + E[X_n] = n \cdot \mu_X$$

and

$$Var\left[S \mid N = n\right] = Var[X_1 + \cdots + X_n] = Var[X_1] + \cdots + Var[X_n] = n \cdot \sigma_X^2.$$

Next, we apply the double expectation theorem to account for the random number of terms N.

$$E[S] = E_N\left[E_S[S \mid N]\right] = E_N[N \cdot \mu_X] = \mu_X \cdot E_N[N] = \mu_N \cdot \mu_X.$$

$$\begin{aligned} Var[S] &= Var_N\left[E_S[S \mid N]\right] + E_N\left[Var_S[S \mid N]\right] \\ &= Var_N[N \cdot \mu_X] + E_N[N \cdot \sigma_X^2] \\ &= \mu_X{}^2 \cdot Var_N[N] + \sigma_X^2 \cdot E_N[N] = \sigma_N^2 \cdot \mu_X^2 + \mu_N \sigma_X^2. \end{aligned}$$

❐

Example 8.4-8 Random Sums

An insurance portfolio has a random number of claims given by N, where N is a Poisson random variable with mean of 100. The claim amount is X with mean of 50 and variance of 101. All the claim amounts are independent and the claim amounts are also independent of the number of claims. Estimate the 95th percentile of total claims.

Solution

The total claims random variable is the random sum $S = X_1 + X_2 + \cdots + X_N$. Since N is a Poisson random variable, the mean and the variance of N are both equal to 100. It follows from the above discussion that $\mu_S = \mu_N \cdot \mu_X = (100)(50) = 5{,}000$ and $\sigma_S^2 = \sigma_N^2 \cdot \mu_X^2 + \mu_N \cdot \sigma_X^2$ $= (100)(50^2) + (100)(101) = 260{,}100$. Then $\sigma_S = 510$.

Let $S_{.95}$ denote the 95th percentile of the sum. Making use of the Central Limit Theorem, we have,

$$.95 \;=\; \Pr\left[S \le S_{.95}\right] \;\approx\; \Pr\left[Z \le \frac{S_{.95} - \mu_S}{\sigma_S} = \frac{S_{.95} - 5000}{510} = 1.645\right].$$

Thus, $S_{.95} \;=\; \mu_S + z_{.95} \cdot \sigma_S \;=\; 5000 + (1.645)(510) \;=\; 5{,}838.95.$ ❐

Finally, we give derivations of the conditioning formulas. The exact details depend on whether the joint distribution of X and Y is discrete, continuous, or a mixture of the two. For example, in the driving route problems the route X is discrete and the drive time Y is continuous. For the proof below we will assume the joint distribution of X and Y is continuous on a domain R in the plane. This is the context of Section 7.5, and we will employ the notation and schematic diagram (reproduced below) from that section. The steps can be easily modified to relate to the other cases.

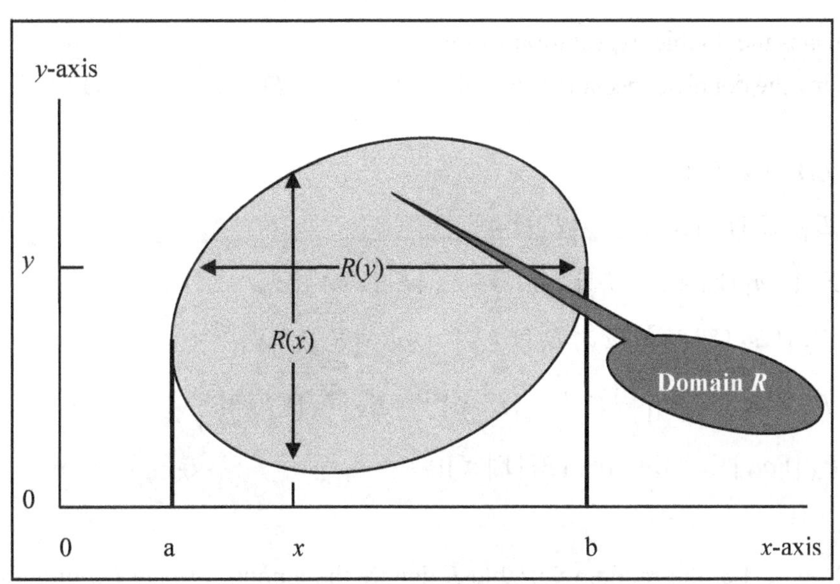

Proof of the conditioning formulas (joint continuous distribution case):

Let $f(x,y)$ be the joint density function for X and Y on the domain R. We make use of the fact that $f(y\,|\,X=x)=\frac{f(x,y)}{f_X(x)}$, or equivalently, $f(x,y)=f(y\,|\,X=x)\cdot f_X(x)$, for values of y along the vertical line segment $R(x)$. We describe the region R as consisting of the vertical line segments $R(x)$; $a \le x \le b$. Then,

$$
\begin{aligned}
E[Y] = \iint_R y\cdot f(x,y)\,dydx &= \int_a^b \int_{R(x)} y\cdot f(x,y)\,dydx \\
&= \int_a^b \int_{R(x)} y\cdot f(y\,|\,X=x)\cdot f_X(x)\,dy\,dx \\
&= \int_a^b f_X(x)\left[\int_{R(x)} y\cdot f(y\,|\,X=x)\,dy\right]dx \\
&= \int_a^b f_X(x)\,E_Y[Y\,|\,X=x]\,dx \\
&= E_X\left[E_Y[Y\,|\,X]\right].
\end{aligned}
$$

This establishes the double expectation formula.
Now, applying the double expectation formula to Y^2 gives, $E[Y^2]=E_X\left[E_Y[Y^2\,|\,X]\right]$.
Thus,

$$
\begin{aligned}
Var[Y] &= E[Y^2]-E[Y]^2 \\
&= E_X\left[E_Y[Y^2\,|\,X]\right]-E_X\left[E_Y[Y\,|\,X]\right]^2 \\
&= E_X\left(Var_Y[Y\,|\,X]+[E_Y[Y\,|\,X]^2\right)-E_X\left[E_Y[Y\,|\,X]\right]^2 \\
&= E_X\left[Var_Y[Y\,|\,X]\right]+E_X\left[E_Y[Y\,|\,X]^2\right]-E_X\left[E_Y[Y\,|\,X]\right]^2 \\
&= E_X\left[Var_Y[Y\,|\,X]\right]+\left\{E_X\left[E_Y[Y\,|\,X]^2\right]-E_X\left[E_Y[Y\,|\,X]\right]^2\right\} \\
&= E_X\left[Var_Y[Y\,|\,X]\right]+Var_X\left[E_Y[Y\,|\,X]\right].
\end{aligned}
$$

Exercise 8-46 Let the random variable B denote the amount of beer (in ounces) that you drink at a sporting event. This random variable depends on the sporting event that you attend, G. Consider the following table:

G	Pr(G)	Conditional *Beer* distribution, $B\,\|\,G$
Cubs at Wrigley	.4	$\sim N(\mu=50,\sigma^2=16)$
Steelers at Heinz	.3	$\sim Uniform[20,120]$
Red Sox at Fenway	.2	$\sim Exponential(\mu=65)$
Curling at Olympics	.1	24 ounces w/prob=1

Compute $E[B]$ and σ_B.

Exercise 8-47 Let X denote the amount of damage caused by a tornado. Suppose that X is uniformly distributed on the interval $[0,\theta]$, where θ is itself a random variable with exponential distribution and mean 5 million.

(a) Compute the expected damage caused by the tornado.
(b) Compute the standard deviation of the amount of damage caused by the tornado.
(c) What is the probability that at least 1 million dollars of damage is caused by the tornado?

Exercise 8-48 ***Compound Poisson***: Suppose that $X_1, X_2, \ldots$ is a random sample of independent losses. Let N denote the total (random) number of losses. Assume that N is independent of the X_i's and has Poisson distribution with mean μ_N. Show that for the total (aggregate) losses, $S = X_1 + X_2 + \cdots + X_N$ we have:

(a) $E[S] = E[X] \cdot E[N] = \mu_X \cdot \mu_N$, and
(b) $Var[S] = \mu_N \cdot E[X^2]$.

Exercise 8-49 Suppose that X is a random variable that denotes the amount of damage to your automobile. Suppose that $X \sim Uniform[5000, 20000]$. Let N denote the number of accidents that you have. Suppose that N is a Bernoulli random variable with probability of success p. That is, you have one accident with probability p and zero accidents with probability $q = 1 - p$. Find the expected value and variance for the total losses.

Exercise 8-50 SOA EXAM P Sample Exam Questions #54
An auto insurance company insures an automobile worth 15,000 for one year under a policy with a 1,000 deductible. During the policy year there is a 0.04 chance of partial damage to the car and a 0.02 chance of a total loss of the car. If there is partial damage to the car, the amount X of damage (in thousands) follows a distribution with density function

$$f(x) = \begin{cases} 0.5003\, e^{-x/2} & \text{for } 0 < x < 15 \\ 0 & \text{otherwise} \end{cases}.$$

What is the expected claim payment?

(A) 320 (B) 328 (C) 352 (D) 380 (E) 540

Exercise 8-51 Suppose that loss random variables X, Y, and Z are mutually independent and have the following probability distributions:

X	Pr(X)
0	.4
1	.3
2	.3

Y	Pr(Y)
0	.25
1	.35
3	.40

Z	Pr(Z)
0	.10
1	.15
2	.20
4	.55

Let $S = X+Y+Z$ represent the total loss amount.

(a) Complete the adjacent table for the probability distribution for $S = X+Y+Z$. *Hint*: This problem is quite tractable and we would draw a three stage tree to determine all the probabilities.

(b) Compute $E[S]$.

(c) Compute σ_S.

$S = X+Y+Z$	Pr[S]
0	.01
1	.0365
2	
3	
4	
5	
6	
7	.16975
8	.066
9	.066

Exercise 8-52 SOA/CAS Course 1 2000 Sample Examination #16
Micro Insurance Company issued insurance policies to 32 independent risks. For each policy, the probability of a claim is $\frac{1}{6}$. The benefit amount given that there is a claim has probability density function

$$f(y) = \begin{cases} 2(1-y) & 0<y<1 \\ 0 & \text{otherwise} \end{cases}.$$

Calculate the expected value of total benefits paid.

(A) $\frac{16}{9}$ (B) $\frac{8}{3}$ (C) $\frac{32}{9}$ (D) $\frac{16}{3}$ (E) $\frac{32}{3}$

Exercise 8-53 Compute the variance of the total benefits paid in Exercise 8-52.
Hint: The number of claims $N \sim \mathit{Binomial}\left(n=32, p=\frac{1}{6}\right)$.

Exercise 8-54 SOA EXAM P Sample Exam Questions #120

An insurance policy is written to cover a loss X, where X has density function

$$f(x) = \begin{cases} \frac{3}{8}x^2 & \text{for } 0 \le x \le 2 \\ 0 & \text{otherwise} \end{cases}$$

The time (in hours) to process a claim of size X, where $0 \le x \le 2$, is uniformly distributed on the interval from x to $2x$. Calculate the probability that a randomly chosen claim on this policy is processed in three hours or more.

(A) 0.17 (B) 0.25 (C) 0.32 (D) 0.58 (E) 0.83

Exercise 8-55 SOA EXAM P Sample Exam Questions #139

A driver and a passenger are in a car accident. Each of them independently has probability 0.3 of being hospitalized. When hospitalization occurs, the loss is uniformly distributed on [0,1]. When two hospitalizations occur, the losses are independent. Calculate the expected number of people in the car who were hospitalized, given that the total loss due to hospitalizations from the accident is less than 1.

(A) 0.510 (B) 0.534 (C) 0.600 (D) 0.628 (E) 0.800

Exercise 8-56

You are given:

i) Λ has a gamma distribution with a mean of 6 and a variance of 12,

ii) For a given $\Lambda = \lambda$, the number of losses N has a Poisson distribution with mean λ.

Calculate the (unconditional) probability that N is greater than zero.

8.5 POISSON PROCESSES REVISITED

In this section we return to our old friend the Poisson process, first introduced in Section 4.6, in order to cover two additional topics. The first concerns the situation when the rate parameter λ is allowed to be a variable rather than a constant. The second topic relates to the splitting of the occurrence events into several streams, each representing a different type of occurrence.

8.5.1 NON-HOMOGENOUS POISSON PROCESS

The Poisson processes studied in Chapter 4 can be generalized to handle situations where the rate parameter is a function, $\lambda(t)$, of time, t. These are termed ***non-homogenous Poisson processes***. A homogenous Poisson process is the special case of this where the rate parameter is constant.

When λ is constant we saw previously that the expected number of occurrences in the interval $(0,t)$ equals λt. We referred to this as the scalability property of the rate parameter λ. For the more general case, where the rate is not necessarily constant, we define $m(t)$ to be the ***expected number of occurrences*** over the interval $(0,t)$.

Let $N(t)$ be the ***actual number of occurrences*** over the interval $(0,t)$. In a non-homogenous Poisson process, $N(t)$ is a Poisson random variable with (constant) rate parameter $m(t)$.

To see why this is plausible, consider a partition of the interval $(0,t)$ into n subintervals of width $\Delta x = \frac{t}{n}$. We assume that Δx is sufficiently small so that $\lambda(t) \approx \lambda_i$, a constant on the interval $[x_i, x_i + \Delta x)$; $i = 0,1,\ldots,n-1$. Let X_i be the number of occurrences on the interval $[x_i, x_i + \Delta x)$. Then X_i can be approximated by a (homogeneous) Poisson random variable with rate parameter λ_i. Then, $E[X_i] \approx \lambda_i \Delta x$. Now,

$$N(t) = \sum_{i=0}^{n-1} X_i \text{ and } E[N(t)] = \sum_{i=0}^{n-1} E[X_i] \approx \sum_{i=0}^{n-1} \lambda_i \Delta x \xrightarrow[n\to\infty]{} \int_0^t \lambda(x)\,dx.$$

More generally, the expected number of claims over the time interval (s,t) is,

$$m(t) - m(s) = \int_s^t \lambda(x)dx.$$

Since $N(t)$ follows a Poisson distribution with mean $m(t)$, the probability function is,

$$\Pr[N(t) = k] = \frac{m(t)^k e^{-m(t)}}{k!};\ k = 0,1,\ldots$$

Further, the probability of k occurrences over the interval (s,t) is,

$$\Pr[N(t) = k] = \frac{[m(t) - m(s)]^k e^{-[m(t)-m(s)]}}{k!}.$$

It also follows that,

$$E[N(t)] = Var[N(t)] = m(t).$$

Example 8.5-1 Non-homogeneous Poisson process

You are given that the number of missed phone calls due to a customer service representative being on vacation for 7 days follows a non-homogenous Poisson process with a rate function $\lambda(t) = 2 + t$. Note that some of these missed calls may be repeat callers who are frustrated from not receiving callbacks. Compute the following:

(a) The expected number of missed calls during the first two days.
(b) The expected number of missed calls during the last two days.

Solutions:

The expected number of missed calls over the time interval (s,t) is,

$$\int_s^t (2+\tau)d\tau = 2\tau + \frac{\tau^2}{2}\Big|_s^t = \left(2t + \frac{t^2}{2}\right) - \left(2s + \frac{s^2}{2}\right).$$

(a) From time $s=0$ to $t=2$ the expected number of missed calls is $4+\frac{2^2}{2}-0=6$

(b) During the last two days, we set $s=5$ and $t=7$ to determine the expected number of missed calls to be 16.

Example 8.5-2 Non-homogeneous Poisson process

In a particular city with poor snow removal capabilities, the number of accidents over a certain two mile stretch of highway during a moderate three hour snow storm follows a non-homogeneous Poisson process with rate function $\lambda(t)=6t$, where t is the time in hours from the beginning of the snowstorm. Calculate,

(a) The probability of at least three accidents in the first hour.
(b) The probability of at least five accidents in the last hour.

Solution:

The expected number of accidents over the time interval (s,t) is,

$$m(t)-m(s) = \int_s^t 6\tau d\tau = \frac{6\tau^2}{2}\Big|_s^t = 3t^2 - 3s^2$$

(a) During the first hour, the expected number of claims is 3.
The probability of at least three accidents in the first hour equals,

$$1-e^{-m(1)} - m(1)e^{-m(1)} - \frac{m(1)^2 e^{-m(1)}}{2} = 1-e^{-3}(1+3+\frac{3^2}{2}) = 0.577$$

(b) During the last hour, the expected number of accidents is $m(3)-m(2)=15$.

The probability of at least five accidents during the last hour equals,

$$= 1-e^{-15}\left[1+15+\frac{15^2}{2}+\frac{15^3}{6}+\frac{15^4}{24}\right] = 0.99914.$$

Exercise 8-57 CAS Exam 3, Segment 3L Fall 2010 #11

You are given the following information:

- A Poisson process N has a rate function $\lambda(t) = 3t^2$
- You have observed 50 events by time $t = 2.1$.

Calculate $Var[N(3) | N(2.1) = 50]$

(A) Less than 10
(B) At least 10, but less than 20
(C) At least 20, but less than 30
(D) At least 30, but less than 40
(E) At least 40

Exercise 8-58 CAS Exam 3, Segment 3L Spring 2012 #9

Claims reported for a group of policies follow a non-homogeneous Poisson process with rate function:

$\lambda(t) = \dfrac{100}{(1+t)^3}$ where t is the time (in years) after January 1, 2011.

Calculate the expected number of claims reported after January 1, 2011, for this group of policies.

(A) Less than 45
(B) At least 45, but less than 55
(C) At least 55, but less than 65
(D) At least 65, but less than 75
(E) At least 75

Exercise 8-59 CAS Exam 3, Segment 3L Fall 2008 #1

The number of accidents on a highway from 3:00 p.m. to 7:00 p.m. follows a non-homogeneous Poisson process with rate function

$\lambda(t) = 4 - (t-2)^2$ where t is the number of hours since 3:00 p.m.

How many more accidents are expected from 4:00 p.m. to 5:00 p.m. than from 3:00 p.m. to 4:00 p.m.?

(A) Less than 0.75
(B) At least 0.75, but less than 1.25
(C) At least 1.25, but less than 1.75
(D) At least 1.75, but less than 2.25
(E) At least 2.25

Exercise 8-60 CAS Exam 3 Fall 2006 #28

Customers arrive to buy lemonade according to a Poisson distribution with $\lambda(t)$, where t is time in hours, as follows:

$$\lambda(t)=\begin{cases} 2+6t & 0\le t\le 3 \\ 20 & 3<t\le 4 \\ 36-4t & 4<t\le 8 \end{cases}$$

At 9:00 a.m., t is 0.

Calculate the number of customers expected to arrive between 10:00 a.m. and 2:00 p.m.

(A) Less than 63
(B) At least 63, but less than 65
(C) At least 65, but less than 67
(D) At least 67, but less than 69
(E) At least 69

Exercise 8-61 CAS Exam 3, Segment 3L Spring 2013 #9

You are given the following:

- An actuary takes a vacation where he will not have access to email for eight days.
- While he is away, emails arrive in the actuary's inbox following a non-homogeneous Poisson process where

$\lambda(t)=8t-t^2$ for $0\le t\le 8$. (t is in days)

Calculate the variance of the number of emails received by the actuary during this trip.

(A) Less than 60
(B) At least 60, but less than 70
(C) At least 70, but less than 80
(D) At least 80, but less than 90
(E) At least 90

Exercise 8-62 CAS Exam 3 Spring 2006 #33

While on vacation, an actuarial student sets out to photograph a Jackalope and a Snipe, two animals common to the local area. A tourist information booth informs the student that daily sightings of Jackalopes and Snipes follow independent Poisson processes with intensity parameters:

$$\lambda_J(t)=\frac{t^{1/3}}{5}$$

$$\lambda_S(t)=\frac{t^{1/2}}{10}$$

where $0\le t\le 24$ and t is the number of hours past midnight

If the student takes photographs between 1 p.m. and 5 p.m., calculate the probability that he will take at least 1 photograph of each animal.

(A) Less than 0.45
(B) At least 0.45, but less than 0.60
(C) At least 0.60, but less than 0.75
(D) At least 0.75, but less than 0.90
(E) At least 0.90

Exercise 8-63 CAS Exam 3 Fall 2005 #26

The number of reindeer injuries on December 24 follow a Poisson distribution with an intensity parameter:

$$\lambda(t) = \left(\frac{t}{12}\right)^{1/2} \quad 0 \le t \le 24 \text{ where } t \text{ is measured in hours}$$

Calculate the probability that no reindeer will be injured during the last hour of the day.

(A) Less than 30%
(B) At least 30%, but less than 40%
(C) At least 40%, but less than 50%
(D) At least 50%, but less than 60%
(E) At least 60%

8.5.2 THE POISSON THINNING PROCESS

The idea here is that we have a (homogeneous) Poisson process in which the occurrence event can be split up into several different types. For example, suppose claims are from a Poisson process with a mean rate of 7/day. Further assume that any given claim has a 60% probability to be from a male and a 40% probability to be from a female. Intuition suggests that the claims from males only will also be a Poisson process, with a mean rate of 7(0.6) = 4.2/day. Similarly, we'd guess that the claims from females will constitute a Poisson process with a rate of 7(0.4) = 2.8/day.

The following result confirms our intuition in this situation. This procedure is known as thinning (or splitting) the Poisson process and can summed up by the following general result.

Poisson Thinning Process

Let N be the Poisson random variable arising from a Poisson process with the associated phenomenon (event) occurring at the mean rate of λ per unit period. Assume the phenomenon can be classified into m types, occurring with probabilities, p_i; $i = 1,2,\ldots,m$ with $\sum_{i=1}^{m} p_i = 1$. Let N_i be the number of occurrences of Type i in a unit period. Then,

1. $N = \sum_{i=1}^{m} N_i$,
2. Each N_i is Poisson with mean rate of occurrence of λp_i per unit period[3],
3. $\{N_i : i = 1,\ldots,m\}$ are independent.

Proof:

Statement (1) is obvious. For simplicity in proving (2) and (3) we will consider the case of just two types, Type I and Type II, with probabilities p and $1-p$, respectively. We derive the joint distribution of N_1 and N_2 by conditioning on $N = n$. Now, $\Pr[N_1 = i, N_2 = j \mid N = n] = 0$ unless $i + j = n$, in which case the conditional distribution is binomial with,

$$\Pr[N_1 = i, N_2 = j \mid N = n] = \binom{n}{i} p^i (1-p)^j \,;\, i + j = n.$$

Thus,

$$\Pr[N_1 = i, N_2 = j] = \sum_{n=0}^{\infty} \Pr[N_1 = i, N_2 = j \mid N = n] \Pr[N = n] = \binom{n}{i} p^i (1-p)^j \cdot e^{-\lambda} \frac{\lambda^n}{n!},$$

where $i + j = n$. Now, writing $\lambda = \lambda p + \lambda(1-p)$, we have,

$$\Pr[N_1 = i, N_2 = j] = \frac{n!}{(i!)(j!)} p^i (1-p)^j \cdot e^{-\lambda p} \cdot e^{-\lambda(1-p)} \frac{\lambda^i \cdot \lambda^j}{n!}$$

$$= \left[e^{-\lambda p} \cdot \frac{(\lambda p)^i}{i!} \right] \cdot \left[e^{-\lambda(1-p)} \cdot \frac{(\lambda(1-p))^j}{j!} \right] = \Pr[N_1 = i] \cdot \Pr[N_2 = j],$$

where N_1 and N_2 are now seen to be independent Poisson random variables with mean rates λp and $\lambda(1-p)$, respectively.

The procedure is easily extended to m types using the multinomial distribution (See Exercise 8-64).

[3] For $i = 1,\ldots,n$ Occurrences of Type i phenomena constitute a Poisson process with mean rate λp_i.

Example 8.5-3 Poisson Thinning

Vehicles arrive at the Poisson Weight Loss Clinic according to a Poisson process at a rate of 8/hour. Assume that the number of passengers in these vehicles are independent, with the following probabilities:

Passengers in Vehicle	Probability
1	0.50
2	0.30
3	0.15
4	0.05

(a) Calculate the probability of 0 cars with 4 passengers in an hour.
(b) Calculate the probability of at least 3 cars with 2 passengers in an hour.

Solution:

(a) We first let N_i be the number of cars with i passengers in an hour and p_i the probability of i passengers in any given car. By the thinning process, each N_i will have a Poisson distribution with parameter $\lambda_i p_i$. For the case of 4 passengers, the parameter is $\lambda_4 = 8(0.05) = 0.40$. Thus, the probability of 0 cars with 4 passengers can be computed to be:

$$\Pr[N_4 = 0] = e^{-0.4} = 0.67\,.$$

(b) The number of cars with 2 passengers has a Poisson distribution with parameter $\lambda_2 = 8(0.30) = 2.40$ resulting in:

$$\Pr[N_2 \geq 3] = 1 - \Pr[N_2 = 0,1,2]$$

$$= \quad 1 - e^{-2.4} - 2.4e^{-2.4} - \frac{2.4^2 e^{-2.4}}{2!}$$

$$= \quad 1 - 0.0907 - 0.2177 - 0.2613$$

$$= \quad 0.4303$$

Example 8.5-4 Poisson Claims Thinning – Normal Approximation

Suppose claims are from a Poisson process with a rate of 82/day and that each claim will be in one of three disjoint categories with associated probabilities:

30%	Minor
50%	Moderate
20%	Severe

Use a normal approximation with correction to determine the probability of more than 110 severe claims in a week.

Solution:
From the thinning property of Poisson processes, the number of severe claims has a Poisson probability distribution with parameter $\lambda^* = 82(0.2)(7)/\text{week} = 114.8/\text{week}$. Also note that the expected value is $\lambda^* = 114.8$ with a standard deviation of $\sqrt{\lambda^*} = \sqrt{114.8} = 10.7$. If N is the random variable for the number of claims, then using a normal approximation gives:

$$P(N > 110) = 1 - P(N \leq 110) \approx 1 - \Phi\left(\frac{110.5 - 114.8}{10.7}\right) = 1 - \Phi(-.40) = \Phi(.40) = 0.656.$$

Example 8.5-5 Poisson Thinning with Normal Distribution

Job offers arrive for top actuarial science graduates according to a Poisson process at a rate of 1.5/month. The salaries of the offers have a normal distribution with mean \$60,000 and standard deviation of \$4,000. Two of these graduates are Louis and Joan. Louis and Joan consider acceptable salaries to be \$65,000 and \$61,000, respectively. Neither will accept an offer below these amounts. What is the expected time for each to receive a suitable employment offer?

Solution:
Let X_1 and X_2 represent the amounts of offers for Louis and Joan. For a given offer, the probability that it will be acceptable is:

Louis: $$\Pr[X_1 \geq 65{,}000] = \Pr\left[Z \geq \frac{65{,}000 - 60{,}000}{4000}\right] = 0.1056.$$

Joan: $$\Pr[X_2 \geq 61{,}000] = \Pr\left[Z \geq \frac{61{,}000 - 60{,}000}{4000}\right] = 0.4013.$$

Using the thinning property, the arrival rates for acceptable job offers for Louis and Joan are:

Louis: $\lambda_1 = (0.1056)(1.5)\ /\text{month} = 0.1584/\text{month}$

Joan: $\lambda_2 = (0.4013)(1.5)\ /\text{month} = 0.6020/\text{month}$

The time to the next occurrence is exponential, so that the expected time to obtain a suitable employment offer is

Louis: $$E[T_1] = \frac{1}{\lambda_1} = \frac{1}{0.1584} = 6.3 \text{ months}$$

Joan: $$E[T_2] = \frac{1}{\lambda_2} = \frac{1}{0.6020} = 1.7 \text{ months}$$

Exercise 8-64 Suppose claims are from a Poisson process with rate λ and that claims come from one of m groups where the probability of a claim being from group i is p_i. Demonstrate that the claims from each group i is a Poisson process with rate λp_i and that these Poisson processes are independent.

Exercise 8-65 CAS Exam 3, Segment 3L Spring 2009 #8
Bill receives mail at a Poisson rate of 10 items per day. The contents of the items are randomly distributed:

- 50% of the items are credit card applications.
- 30% of the items are catalogs.
- 20% of the items are letters from friends.

Bill has received 20 credit card applications in two days.
Calculate the probability that for those same two days, he receives at least 3 letters from friends and exactly 5 catalogs.

(A) A Less than 6%
(B) At least 6%, but less than 10%
(C) At least 10%, but less than 14%
(D) At least 14%, but less than 18%
(E) At least 18%

Exercise 8-66 CAS Exam 3, Segment 3L Spring 2009 #9
You are given the following information:

- Policyholder calls to a call center follow a homogenous Poisson process with $\lambda = 250$ per day.
- Policyholders may call for 3 reasons: Endorsement, Cancellation, or Payment.
- The distribution of calls is as follows:

Call Type	Percent of Calls
Endorsement	50%
Cancellation	10%
Payment	40%

Using the normal approximation with continuity correction, calculate the probability of receiving more than 156 calls in a day that are either endorsements or cancellations.
(A) Less than 27%
(B) At least 27%, but less than 29%
(C) At least 29%, but less than 31%
(D) At least 31%, but less than 33%
(E) At least 33%

Exercise 8-67 CAS Exam 3, Segment 3L Fall 2009 #11

You are given the following information:

- Claims follow a compound Poisson process.
- Claims occur at the rate of $\lambda = 10$ per day.
- Claim severity follows an exponential distribution with θ=15,000.
- A claim is considered a large loss if its severity is greater than 50,000. What is the probability that there are exactly 9 large losses in a 30-day period?

(A) Less than 5%
(B) At least 5%, but Jess than 7.5%
(C) At least 7.5%, but Jess than 10%
(D) At least 10%, but Jess than 12.5%
(E) At least 12.5%

Exercise 8-68 CAS Exam 3, Segment 3L Spring 2011 #10

You are given the following information about an insurance policy:

- Claim frequency follows a homogeneous Poisson process.
- The average number of claims reported each month is 10.
- Claim severities are independent and follow an exponential distribution with θ =10,000
- To monitor the impact of large individual claims on the policy aggregate losses, management receives a large loss report any time a claim occurs that is greater than 30,000.

Calculate the standard deviation of the waiting time (in months) until the second large loss report.

(A) Less than 1
(B) At least 1, but less than 3
(C) At least 3, but less than 6
(D) At least 6, but less than 12
(E) At least 12

Exercise 8-69 CAS Exam 3, Segment 3L Spring 2011 #11

For a collection of insured vehicles, windshield cracks are repaired at a Poisson rate of 150 per month.

Windshield crack repairs fall into two categories:

- 90% of the cracks are minor and cost \$100 to repair
- 10% of the cracks are major and cost \$1,100 to repair

Using the normal approximation, calculate the probability that total windshield crack repair cost in one month is more than \$40,000.

(A) Less than 0.9%
(B) At least 0.9%, but less than 1.0%
(C) At least 1.0%, but less than 1.1%
(D) At least 1.1%, but less than 1.2%
(E) At least 1.2%

Exercise 8-70 CAS Exam 3, Segment 3L Fall 2012 #9

You are given the following information:

- Claims are given by a Poisson Process with claims intensity $\lambda = 8$,
- Frequency and severity of claims are independent,
- Claim severity follows a discrete distribution that is given in the table below

Claim Amount Interval	Probability
Less than \$7,000	0.40
At least \$7,000 but less than \$20,000	0.50
At least \$20,000	0.10

What is the probability that by time 0.6 there will be at least two claims with severity less than \$7,000?

(A) Less than 0.2
(B) At least 0.2, but less than 0.4
(C) At least 0.4, but less than 0.6
(D) At least 0.6, but less than 0.8
(E) At least 0.8

8.6 CHAPTER 8 SAMPLE EXAMINATION

1. Let X be the random variable whose density function is given by $f_X(x) = 3x^2$; $0 \leq x \leq 1$. Find the density function $f_Y(y)$ of $Y = e^{2X}$.

2. Let X be exponentially distributed with mean 5 and let Y be exponentially distributed with mean 3. Let $Z = Min(X,Y)$.
   (a) Find $F_Z(z) = \Pr(Z \leq z)$.
   (b) How is $Z = Min(X,Y)$ distributed?
   (c) Find $E[Z]$.
   (d) Find $Var[Z]$.

3. Suppose that $X \sim N(\mu_X = 2, \sigma_X = 3)$ and $Y \sim N(\mu_Y = -1, \sigma_X = 5)$ are independent. Let $S = X + Y$ denote the sum.

(a) Find the moment generating function for the sum, $M_S(t)$.

(b) How is the sum distributed?

(c) Determine $\Pr(S < 4)$.

4. **SOA/CAS Course 1 2000 Sample Examination #15**

An insurance company issues insurance contracts to two classes of independent lives, as shown below.

Class	Probability of Death	Benefit Amount	Number in Class
A	0.01	200	500
B	0.05	100	300

The company wants to collect an amount, in total, equal to the 95th percentile of the distribution of total claims. The company will collect an amount from each life insured that is proportional to that life's expected claim. That is, the amount for life j with expected claim $E[X_j]$ would be $k \cdot E[X_j]$. Calculate k.

(A) 1.30 (B) 1.32 (C) 1.34 (D) 1.36 (E) 1.38

5. **SOA EXAM P Sample Exam Questions #86**

A city has just added 100 new female recruits to its police force. The city will provide a pension to each new hire who remains with the force until retirement. In addition, if the new hire is married at the time of her retirement, a second pension will be provided for her husband. A consulting actuary makes the following assumptions:

(i) Each new recruit has a 0.4 probability of remaining with the police force until retirement.

(ii) Given that new recruit reaches retirement with the police force, the probability that she is not married at the time of retirement is 0.25.

(iii) The number of pensions that the city will provide on behalf of each new hire is independent of the number of pensions it will provide on behalf of any other new hire.

Determine the probability that the city will provide at most 90 pensions to the 100 new hires and their husbands.

(A) 0.60 (B) 0.67 (C) 0.75 (D) 0.93 (E) 0.99

6. Suppose that $X \sim Exp(1)$ and $Y \sim Unif[0,2]$ are independent.
   (a) Find $\Pr[Max(X,Y) < .6]$.
   (b) Find $E[Max(X,Y)]$.

7. **SOA EXAM P Sample Exam Questions #135**
   The number of workplace injuries, N, occurring in a factory on any given day is Poisson distributed with mean λ. The parameter λ is a random variable that is determined by the level of activity in the factory, and is uniformly distributed on the interval $[0,3]$. Calculate $Var[N]$.
   (A) λ (B) 2λ (C) 0.75 (D) 1.50 (E) 2.25

8. **CAS Exam 3 Fall 2007 #12**
   Claim severity follows a single-parameter Pareto distribution (that is, Type I) with $\beta = 5{,}000$ and $\alpha = 1.2$.
   - $X_1, X_2, \ldots, X_5$ represent a random sample of claims.
   - $Y_1, Y_2, \ldots, Y_5$ are the order statistics associated with the random sample.

   Calculate the probability that Y_5 is greater than $25,000.

   (A) Less than 20%
   (B) At least 20% but less than 30%
   (C) At least 30% but less than 40%
   (D) At least 40% but less than 50%
   (E) At least 50%

9. **CAS Exam 3 Spring 2008 #8**
   Claim severity follows an exponential distribution with mean 1000. If two claims are sampled randomly, calculate the expected value of the larger claim.

   (A) Less than 1200
   (B) At least 1200, but less than 1400
   (C) At least 1400, but less than 1600
   (D) At least 1600, but less than 1800
   (E) At least 1800

10. **CAS Exam 3 Spring 2006 #8**

The joint probability distribution of the first and last order statistic, y_1 and y_n, from a sample of size n is

$$f(y_1, y_n) = n(n-1)\left[F(y_n) - F(y_1)\right]^{n-2} f(y_1) f(y_n).$$

Recall that $\int_a^b x e^{-\alpha x} dx = \left(\frac{x e^{-\alpha x}}{\alpha}\right)\Bigg|_a^b + \frac{1}{\alpha}\int_a^b e^{-\alpha x} dx$

For a sample of size two from the exponential distribution $f(x) = e^{-x}$, determine the expected value of the sample range.

(A) Less than 0.5
(B) At least 0.5, but less than 0.7
(C) At least 0.7, but less than 0.9
(D) At least 0.9, but less than 1.1
(E) At least 1.1

11. **CAS Exam 3, Segment 3L Fall 2013 #9**

You are given that claim counts follow a non-homogeneous Poisson process with $\lambda(t) = 30t^2 + t^3$. Calculate the probability of at least two claims between time 0.2 and 0.3.

(A) Less than 1%
(B) At least 1%, but less than 2%
(C) At least 2%, but less than 3%
(D) At least 3%, but less than 4%
(E) At least 5%

12. **CAS Exam 3, Segment 3L Fall 2013 #10**

You are given the following information:

- Cars arrive according to a Poisson Process at a rate of 20 cars per hour.
- 75% of cars are red and 25% of cars are blue.
- 28 red cars and 32 blue cars have arrived after three hours have passed.

Calculate the total expected number of red cars that will have arrived after eight hours have passed.

(A) Less than 80
(B) At least 80, but less than 100
(C) At least 100, but less than 120
(D) At least 120, but less than 140
(E) At least 140

13. **SOA EXAM P Sample Exam Questions #148**

The number of hurricanes that will hit a certain house in the next ten years is Poisson distributed with mean 4.

Each hurricane results in a loss that is exponentially distributed with mean 1000. Losses are mutually independent and independent of the number of hurricanes.

Calculate the variance of the total loss due to hurricanes hitting this house in the next ten years.

(A) 4,000,000 (B) 4,004,000 (C) 8,000,000 (D) 16,000,000 (E) 20,000,000

14. **SOA EXAM P Sample Exam Questions #149**

A motorist makes three driving errors, each independently resulting in an accident with probability 0.25.

Each accident results in a loss that is exponentially distributed with mean 0.80.

Losses are mutually independent and independent of the number of accidents.

The motorist's insurer reimburses 70% of each loss due to an accident.

Calculate the variance of the total unreimbursed loss the motorist experiences due to accidents resulting from these driving errors.

(A) 0.0432 (B) 0.0756 (C) 0.1782 (D) 0.2520 (E) 0.4116

15. **SOA EXAM P Sample Exam Questions #136**

A fair die is rolled repeatedly. Let X be the number of rolls needed to obtain a 5 and Y the number of rolls needed to obtain a 6. Use conditioning (the double expectation formula) to calculate $E\left[X \mid Y=2\right]$. **Note:** See also Exercise 4-41.

CHAPTER 9

SAMPLING DISTRIBUTIONS AND ESTIMATION

In Chapters 1-8 we covered many of the commonly occurring probability distributions and demonstrated how they are used to model a wide variety of real-world phenomena. We now turn our attention to how we can fully specify these models by estimating values for the parameters of the probability distributions.

The typical approach in mathematical modeling is to begin with a theory about which mathematical tools might best apply to the given situation. For example, to describe a rocket trajectory a logical place to begin would be with the differential equations studied in calculus and physics that govern the physical laws of motion. To fully specify the model we would also need to know the values of certain observations, such as the initial position and velocity of the rocket to supplement and complete the theory-based equations.

Similarly, in the case of a probability experiment we might have theoretical grounds for selecting a particular type of distribution. Virtually all of the distributions we encountered in earlier chapters had parameters as part of their descriptions and we would need to know the values of those parameters to complete the model.

To take one example out of many, suppose we theorize that the underlying phenomenon we wish to study appears to conform to the requirements for a Poisson process, as spelled out in Subsection 4.6.2. As we saw, the two main random variables in a Poisson process (the Poisson distribution for the number of occurrences and the exponential distribution for the time between occurrences) both depend on specifying the underlying ***rate*** parameter, λ. Consequently, the next step in the modeling process would be to determine a value for this parameter. We could simply guess or arbitrarily assign a number, but being scientists rather than soothsayers, we are inclined to rely instead on actual data. This data will arise from experimentation, which might be based on historical records or on observations of newly designed laboratory experiments. Either way, once we have relevant data we can *estimate* the true value of the parameter. For Poisson processes this procedure was amply illustrated by the examples and exercises given in Subsection 4.6.3, which are based on actual data sets.

One essential point here is that the *true* values of our model parameters can rarely be known with exactitude by mere mortals, and therefore, must be estimated using experimental, that is to say, *random* observations. Because they are random, our estimates are subject to all the vagaries of chance, and will depend on how and when our observations are made. Consequently, we have the somewhat paradoxical situation in which the parameter is fixed, but unknown, while the estimates are random, but observable.

The main theme of this chapter is to illustrate some of the standard parameter estimation procedures used in statistics; to study the distributions of these statistical estimators; and to introduce the concept of a ***confidence interval***, a device for quantifying the reliability, or accuracy, of our estimators.

9.1 THE SAMPLE MEAN AS AN ESTIMATOR

Our starting point is the idea of a random sample from a population random variable, using the terminology and structure introduced in Subsection 6.4.1. We summarize the main results from Chapter 6 below, including the Central Limit Theorem for sample means in Subsection 6.4.5.

Random Samples and the Sample Mean

Let X be a random variable with finite mean μ_X and variance σ_X^2. We refer to X as the ***population*** random variable.

Let $X_1, X_2, \ldots, X_n$ be independent, identically distributed observations of X, abbreviated as $(X_i)_{i=1}^n$ I.I.D. $\sim X$. Then $(X_i)_{i=1}^n$ is called a ***random sample*** of X.

Define the ***sample mean*** by $\overline{X} = \frac{1}{n} \cdot \sum_{i=1}^{n} X_i$.

Then,

(1) $\mu_{\overline{X}} = E[\overline{X}] = E[X] = \mu_X$.

(2) $\sigma_{\overline{X}}^2 = Var[\overline{X}] = \frac{\sigma_X^2}{n}$. Equivalently, $\sigma_{\overline{X}} = \frac{\sigma_X}{\sqrt{n}}$.

(3) The Central Limit Theorem: For large sample sizes n, the distribution of $\overline{X}$ is approximately normal with mean $\mu_{\overline{X}} = \mu_X$ and variance $\sigma_{\overline{X}}^2 = \frac{\sigma_X^2}{n}$. If the population X is itself normal, then $\overline{X}$ is also normal. That is, $\overline{X} \sim N\left(\mu_X, \frac{\sigma_X^2}{n}\right)$.

Property (1) expresses the most basic result in statistical estimation, namely that the sample mean $\overline{X}$ is an ***unbiased estimator*** of the population mean μ_X. The descriptor "unbiased" refers to the fact that the expected value of the sample mean equals the population mean. We will provide an in-depth discussion of the theoretical properties of estimators in Chapter 11, but for now suffice it to say that it is desirable, but not necessary, that estimators be unbiased.

The next issue concerns the accuracy of $\overline{X}$ as an estimator for μ_X. To address this in a meaningful way we need to know something about the sampling distribution of $\overline{X}$, and toward that end the Central Limit Theorem (property (3) above) plays a prominent role.

The usual way of conveying the accuracy, or the reliability, of an estimator for the population mean is through the use of a probability statement taking the form $1-\alpha = \Pr\left[\left|\overline{X}-\mu_X\right| \leq \varepsilon\right]$. The following terminology was introduced earlier in our discussion of the Law of Large Numbers in Subsection 6.4.2 and the interested reader may wish to refer back to that section. The quantity ε measures the absolute deviation of the sample mean from the population mean and is called the ***margin of error***. The probability $1-\alpha$ is called the ***confidence level*** and is equal to the likelihood that the sample mean and the population mean are within a specified margin of error.

Example 9.1-1 Determining the Margin of Error

Let X represent a score on a standardized test. Suppose we are given that X is normally distributed with an unknown mean and a known variance of 25. A random sample of 100 test scores is used to calculate $\overline{X}$ as an estimate of the population mean μ_X. Find the margin of error for a given confidence level of 95%.

Solution

We have $\alpha = .05 = 5\%$ and (property (2) of Random Samples above) $\sigma_{\overline{X}} = \frac{\sigma_X}{\sqrt{n}} = \sqrt{\frac{25}{100}} = 0.5$.

We must solve for ε in the equation,

$$1-\alpha = 0.95 = \Pr\left[\left|\overline{X}-\mu_X\right| \leq \varepsilon\right].$$

Since $\overline{X}$ is normal we can standardize and make use of the standard normal table.

$$1-\alpha = 0.95 = \Pr\left[\left|\overline{X}-\mu_X\right| \leq \varepsilon\right] = \Pr\left[\left|\frac{\overline{X}-\mu_X}{\sigma_{\overline{X}}}\right| \leq \frac{\varepsilon}{\sigma_{\overline{X}}} = \frac{\varepsilon}{0.5}\right] = \Pr\left[|Z| \leq \frac{\varepsilon}{0.5}\right].$$

Since the event $|Z| \leq \frac{\varepsilon}{0.5}$ occupies the central 95% of the standard normal distribution, it follows that $\frac{\varepsilon}{0.5}$ must be at the 97.5^{th} percentile of the standard normal. This leaves 2.5% in each tail of the distribution (see Figure 9.1.1).

FIGURE 9.1.1

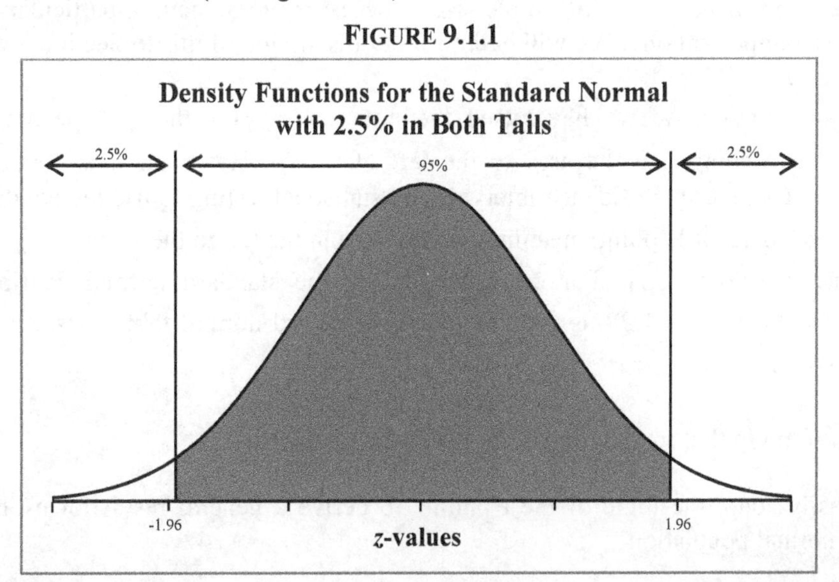

Using the Standard Normal Table, we have $.975 = \Phi(1.96) = \Pr[Z \le 1.96]$. Hence, $\frac{\varepsilon}{0.5} = 1.96$, so that $\varepsilon = (1.96)(0.5) = 0.98$.

We can conclude that there is a 95% probability that the sample mean and the population mean differ by no more than 0.98 on the standardized test score scale. ◻

Notes

(1) Suppose we calculate $\overline{X}$ based on our sample of size 100 and come up with the value $\overline{X} = 72.3$. Based on our margin of error of $\varepsilon = .98$ we can then be 95% certain that the population mean μ_X is within

$$\overline{X} \pm \varepsilon \;=\; 72.3 \pm .98 \;=\; (72.3 - .98, 72.3 + .98) \;=\; (71.32, 73.28).$$

This range is referred to as a 95% ***confidence interval*** for μ_X. We make the distinction between the ***point estimator*** $\overline{X}$ and the ***interval estimator*** $\overline{X} \pm \varepsilon$.

(2) People often make statements such as, "There is a 95% chance that μ_X is in the given confidence interval." But strictly speaking, this is not a correct statement about probability. Since μ_X is a fixed quantity, it is either within that interval or not within that interval. That is, the *probability* is either 0 or 1 that μ_X is inside the above interval (we just don't know which). This is the reason the term ***confidence level*** was introduced as a substitute for probability. Prospectively, that is, *before* the random sample is taken and $\overline{X}$ is calculated, it is perfectly correct to speak of the probability of the event $|\overline{X} - \mu_X| \le \varepsilon$, since we are making a statement about the distribution of the random variable $\overline{X}$. Retrospectively, with numerical values for $\overline{X}$ and the interval endpoints, it is proper to speak of the confidence level instead.

(3) The question of how large a sample size is necessary to achieve a particular margin of error is an important one. We will defer a fuller discussion of this to Section 9.7.

(4) Previously, we have used the notation z_p for the pth percentile of z, meaning that the area to the **left** of z is p. However, it is common in the context of confidence intervals and statistical testing to use the notation z_α for the $(1-\alpha)$ percentile point, meaning that the area in the tail to the **right** of z_α equals α. That is, $\alpha = \Pr[Z > z_\alpha]$. For example, if Z is the standard normal distribution and $\alpha = 10\%$, then $z_{.10} = 1.28$ is found from the standard normal table. Also, in this case, $z_{.90} = -1.28 = -z_{.10}$.

We will conform to this convention in the following chapters.

We can easily adapt the logic of the example to derive a general result for estimating the mean of a normal population.

Estimating the Mean of a Normal Population with known Variance

Let X be a normally distributed population random variable with a known variance σ_X^2. Let $\overline{X}$ be the sample mean from a random sample of size n. For a given confidence level $1-\alpha$ we have,

(1) The margin of error in estimating μ_X by $\overline{X}$ is given by $\varepsilon = z_{\alpha/2} \cdot \frac{\sigma_X}{\sqrt{n}}$, and

(2) $1-\alpha = \Pr\left[\left|\overline{X} - \mu_X\right| \le z_{\alpha/2} \cdot \frac{\sigma_X}{\sqrt{n}}\right]$.

We say that $\overline{X} \pm \varepsilon = \left(\overline{X} - z_{\alpha/2} \cdot \frac{\sigma_X}{\sqrt{n}},\ \overline{X} + z_{\alpha/2} \cdot \frac{\sigma_X}{\sqrt{n}}\right)$ constitutes a $1-\alpha$ symmetric ***confidence interval*** for estimating the population mean μ_X.

Proof

We have $1-\alpha = \Pr\left[\left|\overline{X} - \mu_X\right| \le \varepsilon\right] = \Pr\left[\left|\frac{\overline{X}-\mu_X}{\sigma_{\overline{X}}}\right| \le \frac{\varepsilon}{\sigma_{\overline{X}}}\right] = \Pr\left[|Z| \le \frac{\varepsilon}{\frac{\sigma_X}{\sqrt{n}}} = \frac{\varepsilon \cdot \sqrt{n}}{\sigma_X}\right]$.

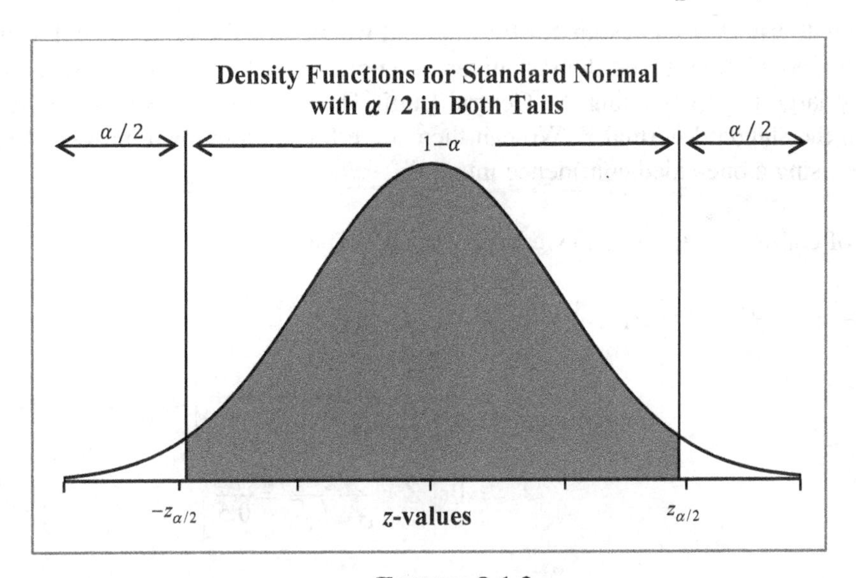

FIGURE 9.1.2

It follows that $\frac{\varepsilon \cdot \sqrt{n}}{\sigma_X} = z_{\alpha/2}$, and hence $\varepsilon = z_{\alpha/2} \cdot \frac{\sigma_X}{\sqrt{n}}$.

Notes

(1) Confidence intervals need not necessarily be symmetric about the estimator, although that is the most common approach to estimating the population mean with a sample mean. But any interval (A, B) whose endpoints are functions of a random sample and whose probability of containing the population mean is $1-\alpha$ can be considered a $1-\alpha$ confidence interval for μ_X. The symmetric confidence interval is the easiest to calculate,

and it is also generally the shortest possible confidence interval for a given α and ε, a desirable property.

(2) So far we have restricted our discussion to estimating the mean of a normal population with known variance. In this case the estimator $\overline{X}$ is also normally distributed and the probabilities for confidence intervals defined in terms of $\overline{X}$ are precise. If the population is not necessarily normal, but the sample size is *large* and the population variance is known, then the Central Limit Theorem allows us to conclude the confidence interval given above is approximately correct. A discussion of the accuracy of the Central Limit Theorem is given in Section 6.4.

Example 9.1-2 One-Sided Confidence Interval

Let X represent a score on a standardized test. We are given that X has a variance of 25. A random sample of 100 test scores is taken and the sample mean $\overline{X}$ is calculated to be 72.3. Find a number A for which we can be 95% confident that A constitutes a lower bound for the population mean, μ_X.

Solution
We continue to use $\overline{X}$ as an estimator for μ_X and we express the lower bound A in the form $\overline{X}-\varepsilon$. The population is no longer given as normal, but the sample size $n=100$ is sufficiently large to justify using the Central Limit Theorem to approximate the distribution of $\overline{X}$ with the standard normal Z. We can then solve for the unknown margin of error ε as before, but using a one-sided confidence interval.

The level of confidence is 95%, so we have $\alpha=.05$. Then,

$$\begin{aligned} 1-\alpha = .95 = \Pr\left[\mu_X > \overline{X}-\varepsilon\right] &= \Pr\left[\overline{X}-\mu_X < \varepsilon\right] \\ &= \Pr\left[\frac{\overline{X}-\mu_X}{\sigma_X/\sqrt{n}} < \frac{\varepsilon}{\sigma_X/\sqrt{n}}\right] \\ &\approx \Pr\left[Z < \frac{\varepsilon}{\sigma_X/\sqrt{n}} = \frac{\varepsilon}{0.5}\right]. \end{aligned}$$

In this case we now have $\frac{\varepsilon}{0.5} = z_{.05}$. Using the standard normal table we have $z_{.05}=1.645$, the 95th percentile of Z. Thus, $\varepsilon=(0.5)(1.645)=.8225$ and so we are 95% confident that μ_X exceeds $A = 72.3-.8225 \approx 71.48$. □

Exercise 9-1 Suppose that the weight, W, of a certain population of whales is known to be normally distributed with unknown mean and standard deviation of 500 pounds. The sample mean from a random sample of 16 whales was found to be $\overline{W}=2350$ pounds.

(a) Find the symmetric 90% confidence interval for the population mean μ_W.

(b) Find the weight B for which you can be 99% confident that B is an upper bound for the population mean μ_W. That is, $\Pr(\mu_W < B) \geq .99$.

Exercise 9-2 Suppose that the size of Wendy's single hamburger is normally distributed with mean 70 grams and standard deviation of 5 grams. If 4 hamburgers are selected at random, find the probability that the sample mean is less than 77 grams.

Exercise 9-3 A random variable X is known to be normally distributed with standard deviation $\sigma_X = 10$. A random sample of size n was taken and the 92% symmetric confidence interval was found to be $(8.50, 13.50)$.

(a) What was the sample mean $\overline{X}$?
(b) What was the sample size?

Exercise 9-4 A large population has mean 100 and standard deviation 16.

(a) What is the probability that the sample mean will be within ± 3 of the population mean if the sample size is $n = 100$?
(b) What is the probability that the sample mean will be within ± 3 of the population mean if the sample size is $n = 200$?
(c) What is the advantage of a larger sample size?

Exercise 9-5 A rather large (more than 30) random survey of Abby Township University students was conducted and their ages were ascertained. A 97% confidence interval for the average age was calculated to be $21.225 < \mu < 22.775$ years. The standard deviation of the population is 2.5 years.

(a) What was the sample mean?
(b) What was the exact size of the sample?

9.2 ESTIMATING THE POPULATION VARIANCE

So far, our discussion has been limited to situations where the population variance is known in advance. This assumption, while convenient, is hardly realistic. Consequently, we turn our attention now to using our random sample in order to devise a suitable estimate of the population variance σ_X^2. This estimate of the population variance will be called the sample variance (denoted S^2), and it will serve as a companion to the sample mean $\overline{X}$.

9.2.1 THE SAMPLE VARIANCE

Let X be a population random variable with mean μ_X and variance σ_X^2, and let $(X_i)_{i=1}^n$ be a random sample from X. Since variance is defined as the expected square deviation from the mean, it seems reasonable to base our estimator for σ_X^2 on the squares of the deviations,

$X_i - \overline{X}$, between the individual observations and the sample mean. If we treat all of the n observations as equally likely then a reasonable first guess for an estimator would be the average of the square deviations, that is, $\frac{1}{n}\sum_{i=1}^{n}(X_i-\overline{X})^2$.

If we could replace $\overline{X}$ with μ_X in this formula then the resulting quantity, $\frac{1}{n}\sum_{i=1}^{n}(X_i-\mu_X)^2$, would indeed be an unbiased estimator of σ_X^2. Since we do not know the value of μ_X, however, we have no choice but to estimate it using $\overline{X}$. It turns out though, in property (1) below, we can give a formula for the discrepancy introduced by using $\overline{X}$, and thereby account for it in designing an unbiased estimator for σ_X^2. Property (2) below shows the somewhat unexpected result that the sample mean is uncorrelated with each of the deviations from the sample mean.

Deviations from the Sample Mean

Let X be a population random variable with mean μ_X and variance σ_X^2, and let $(X_i)_{i=1}^n$ be a random sample from X.

(1) $\sum_{i=1}^{n}(X_i-\overline{X})^2 = \sum_{i=1}^{n}(X_i-\mu_X)^2 - n(\overline{X}-\mu_X)^2$.

(2) $Cov(\overline{X}, X_i-\overline{X}) = 0$ for each $i = 1, 2, \ldots, n$.

Proof

For (1), we have,

$$\begin{aligned}
\sum_{i=1}^{n}(X_i-\overline{X})^2 &= \sum_{i=1}^{n}\left[(X_i-\mu_X)-(\overline{X}-\mu_X)\right]^2 \\
&= \sum_{i=1}^{n}\left[(X_i-\mu_X)^2 - 2(X_i-\mu_X)(\overline{X}-\mu_X) + (\overline{X}-\mu_X)^2\right] \\
&= \sum_{i=1}^{n}(X_i-\mu_X)^2 - 2(\overline{X}-\mu_X)\sum_{i=1}^{n}(X_i-\mu_X) + \sum_{i=1}^{n}(\overline{X}-\mu_X)^2 \\
&= \sum_{i=1}^{n}(X_i-\mu_X)^2 - 2(\overline{X}-\mu_X)(n\overline{X}-n\mu_X) + n(\overline{X}-\mu_X)^2 \\
&= \sum_{i=1}^{n}(X_i-\mu_X)^2 - 2n(\overline{X}-\mu_X)^2 + n(\overline{X}-\mu_X)^2 \\
&= \sum_{i=1}^{n}(X_i-\mu_X)^2 - n(\overline{X}-\mu_X)^2.
\end{aligned}$$

Property (2) is easily established using the covariance techniques of Section 8.3. Since the X_i's are independent, $Cov(X_i, X_j) = 0$ for $i \neq j$ and, as always, $Cov(X_i, X_i) = Var[X_i] = \sigma_X^2$. We have for each i,

$$\begin{aligned} Cov(\bar{X}, X_i - \bar{X}) &= Cov(\bar{X}, X_i) - Var(\bar{X}) \\ &= \sum_{j=1}^{n} Cov\left(\frac{1}{n} X_j, X_i\right) - Var\left(\bar{X}\right) \\ &= \frac{1}{n} Var(X_i) - Var(\bar{X}) = \frac{\sigma_X^2}{n} - \frac{\sigma_X^2}{n} = 0. \end{aligned}$$

Since these covariances are zero, the sample mean and the sample deviations from the mean also have correlation of zero.

The Sample Variance

Let X be a population random variable with mean μ_X and variance σ_X^2, and let $(X_i)_{i=1}^n$ be a random sample from X. We define the ***sample variance*** by,

$$S^2 = \frac{1}{n-1} \sum_{i=1}^{n} (X_i - \bar{X})^2.$$

The sample variance S^2 is an ***unbiased*** estimator of σ_X^2. That is,

$$E[S^2] = \sigma_X^2 .$$

Proof

We begin with the identity in (1),

$$\sum_{i=1}^{n} (X_i - \bar{X})^2 = \sum_{i=1}^{n} (X_i - \mu_X)^2 - n(\bar{X} - \mu_X)^2 ,$$

and take the expected value of both sides.

Keep in mind that by definition, for any random variable X, $\sigma_X^2 = E\left[(X - \mu_X)^2\right]$.

Hence, $E\left[(X_i - \mu_X)^2\right] = \sigma_X^2$. Also, $Var[\bar{X}] = \frac{\sigma_X^2}{n}$, and so, $E\left[(\bar{X} - \mu_X)^2\right] = \frac{\sigma_X^2}{n}$.

It follows that,

$$\begin{aligned} E\left[\sum_{i=1}^{n} (X_i - \bar{X})^2\right] &= E\left[\sum_{i=1}^{n} (X_i - \mu_X)^2\right] - nE\left[(\bar{X} - \mu_X)^2\right] \\ &= n\sigma_X^2 - n\left(\frac{\sigma_X^2}{n}\right) = (n-1)\sigma_X^2 . \end{aligned}$$

Then, dividing both sides by $n-1$, we have $E[S^2] = \sigma_X^2$.

As advertised, we have devised an unbiased estimator S^2 for the population variance σ_X^2, which is based on a random sample of observations. At this juncture, however, without some knowledge about the population distribution we can say little about the actual distribution of S^2. If we make the further assumption that the population random variable X is normal, then it turns out we can fully describe the distribution of the sample variance. This is the objective of the next subsection.

Exercise 9-6 A random sample of size $n = 5$ from a population yielded the following data: $X_1 = 14.3$, $X_2 = 9.6$, $X_3 = 15.0$, $X_4 = 12.3$, and $X_5 = 10.6$. Find the unbiased estimates for the population mean and population variance.

9.2.2 SAMPLE VARIANCE FROM A NORMAL POPULATION

In this subsection we restrict our attention to sampling from a normal population. In this case we show below that a suitably standardized version of S^2 has a chi-square distribution, first introduced in Subsection 6.5.5. This result depends on the additional important and surprising fact that when drawing a random sample from a *normal* population, the sample mean $\overline{X}$ and the sample variance S^2 are *independent* random variables.

This result may be presaged by the fact that $\overline{X}$ and the deviations, $X_i - \overline{X}$, are uncorrelated, as shown in the preceding subsection. If the population X is normal, then the joint distribution of $\overline{X}$ and $X_i - \overline{X}$ is bivariate normal. Consequently, the fact that their covariance is zero implies that they are independent (please refer to Section 7.9). Since S^2 is a function of the deviations from $\overline{X}$, it seems plausible that $\overline{X}$ and S^2 would also be independent. However, the pair-wise independence of $\overline{X}$ and the individual deviations $X_i - \overline{X}$ is not sufficient to clinch the case. We will give an argument below based on the factorization of joint moment generating functions.

The following properties of S^2 are extremely important for statistical estimation, and they form the backbone for constructing the standard sampling distributions developed in the rest of this chapter. Before illustrating their use with a variety of examples, we first outline a proof in the remainder of this subsection.

Properties of the Sample Variance from a Normal Population

Let X be a ***normal*** population random variable with mean μ_X and variance σ_X^2, and let $(X_i)_{i=1}^n$ be a random sample from X.

(1) The sample mean $\overline{X}$ and the sample variance S^2 are independent random variables.

(2) The random variable $\dfrac{(n-1)S^2}{\sigma_X^2}$ is chi-square with $(n-1)$ degrees of freedom. That is, $\dfrac{(n-1)S^2}{\sigma_X^2} \sim \chi^2(n-1) = \Gamma\left(\frac{n-1}{2}, 2\right)$.

The key step in the demonstration of (1) uses property (2) of joint moment generating functions in Section 7.10. This states that two random variables are independent if and only if their joint moment generating function factors into the product of the individual moment generating functions.

Before proceeding we first draw upon some facts about higher dimensional versions of the bivariate normal distribution. The multivariate normal distribution can be thought of as a random vector $\mathbf{X} = (X_1, \dots, X_n)$, whose components (the marginal distributions) are themselves each normal. We summarize the required results next.

The Joint MGF of the Multivariate Normal Distribution

Let $\mathbf{X} = (X_1, \dots, X_n)$ be an n-dimensional vector with each $X_i \sim N(\mu_i, \sigma_i^2)$. For each pair (i, j) we denote $\sigma_{ij} = Cov(X_i, X_j)$; $1 \le i, j \le n$, where $\sigma_{ii} = \sigma_i^2 = Var[X_i]$. For a given n-tuple $(t_1, \dots, t_n)$ of real numbers, let,

$$T = t_1 X_1 + \cdots + t_n X_n = \sum_{i=1}^{n} t_i X_i .$$

Then T is a normal random variable with,

(1) $\mu_T = E[T] = \sum_{i=1}^{n} \mu_i t_i$ and $\sigma_T^2 = Var[T] = \sum_{i=1}^{n} \sum_{j=1}^{n} \sigma_{ij} t_i t_j$.

The joint moment generating function of $\mathbf{X} = (X_1, \dots, X_n)$ is given by,

(2) $M_{\mathbf{X}}(t_1, \dots, t_n) = M_T(1) = e^{\sum_{i=1}^{n} \mu_i t_i + \frac{1}{2} \sum_{i=1}^{n} \sum_{j=1}^{n} \sigma_{ij} t_i t_j}$.

The formulas for the mean and the variance of T are straightforward applications of previous results for the mean and the variance of linear combinations of random variables. In particular the general variance formula may be found in Section 8.3.

The joint moment generating function is presented in Section 7.10. Applying the definition here, we have

$$M_{\mathbf{X}}(t_1, \dots, t_n) = E[e^{t_1 X_1 + \cdots + t_n X_n}] = E[e^T].$$

By definition, $E[e^T] = M_T(1)$. Since T is normal we have $M_T(t) = e^{\mu_T t + (1/2)\sigma_T^2 t^2}$ (see property (6) for normal distributions in Section 6.3.3). It follows that

$$M_{\mathbf{X}}(t_1, \dots, t_n) = M_T(1) = e^{\sum_{i=1}^{n} \mu_i t_i + \frac{1}{2} \sum_{i=1}^{n} \sum_{j=1}^{n} \sigma_{ij} t_i t_j} .$$

Proof of the Properties of S^2

We turn now to a proof of properties (1) and (2) for the sample variance when the population X is normal.

First, note that $\bar{X}$ and each of the deviations $X_i - \bar{X}$ are linear functions of the independent identically distributed normal random variables $(X_i)_{i=1}^{n}$, and hence are themselves normal. We look at two multivariate normal distributions, $\mathbf{U} = (X_1-\bar{X},...,X_n-\bar{X})$, and its augmented cousin, $\mathbf{V} = (X_1-\bar{X},...,X_n-\bar{X},\bar{X})$. Both constitute multivariate normal random vectors, of dimensions n and $n+1$, respectively. We next derive their respective joint MGFs. Let $\mu_i = E\left[X_i - \bar{X}\right]$, and $\sigma_{ij} = Cov(X_i-\bar{X}, X_j-\bar{X});\ 1 \le i, j \le n.$

First, note that $\mu_i \;=\; E[X_i-\bar{X}] \;=\; \mu_X - \mu_X \;=\; 0$ for all $i = 1,\ldots,n.$

Then, using the above formula for the joint MGF, we have,

$$M_{\mathbf{U}}(t_1,t_2,\ldots,t_n) = e^{\frac{1}{2}\sum_{i=1}^{n}\sum_{j=1}^{n}\sigma_{ij}\, t_i\, t_j}.$$

We now consider the moment-generating function of $\mathbf{V}$. Again, $\mu_i = 0;\ i = 1,\ldots,n,$ and we let $\mu_{n+1} = E[\bar{X}] = \mu_X$. Let $T = \sum_{i=1}^{n} t_i\,(X_i-\bar{X}) + t_{n+1}\bar{X}$. Then $\mu_T = t_{n+1}\cdot\mu_X$ and,

$\sigma_T^2 =$

	$t_1(X_1-\bar{X})$	$t_2(X_2-\bar{X})$	$\cdots$	$t_n(X_n-\bar{X})$	$t_{n+1}\bar{X}$
$t_1(X_1-\bar{X})$	$t_1^2\sigma_1^2$	$t_1t_2\sigma_{12}$	$\cdots$	$t_1t_n\sigma_{1n}$	0
$t_2(X_2-\bar{X})$	$t_2t_1\sigma_{21}$	$t_2^2\sigma_2^2$	$\cdots$	$t_2t_n\sigma_{2n}$	0
$\vdots$	$\vdots$	$\vdots$	$\vdots$	$\vdots$	0
$t_n(X_n-\bar{X})$	$t_nt_1\sigma_{n1}$	$t_nt_2\sigma_{n2}$	$\cdots$	$t_n^2\sigma_n^2$	0
$t_{n+1}\bar{X}$	0	0	0	0	$t_{n+1}^2\dfrac{\sigma_X^2}{n}$

$$= \sum_{i=1}^{n}\sum_{j=1}^{n}\sigma_{ij}\,t_i\,t_j + t_{n+1}^2\frac{\sigma_X^2}{n}.$$

The zeros in the last row and column result from the fact that $Cov(\bar{X}, X_i - \bar{X}) = 0$ for each $i = 1,2,\ldots,n,$ and the entry in the lower right corner follows from $Var(\bar{X}) = \frac{\sigma_X^2}{n}$.
We can now write

$$\begin{aligned}
M_{\mathbf{V}}(t_1,t_2,\ldots,t_n,t_{n+1}) &= M_T(1) \\
&= e^{\mu_T+(1/2)\sigma_T^2} \\
&= e^{t_{n+1}\mu_X+\frac{1}{2}\left(\sum_{i=1}^{n}\sum_{j=1}^{n}\sigma_{ij}t_it_j+t_{n+1}^2(\sigma_X^2/n)\right)} \\
&= e^{\left(t_{n+1}\mu_X+\frac{1}{2}t_{n+1}^2\frac{\sigma_X^2}{n}\right)+\frac{1}{2}\sum_{i=1}^{n}\sum_{j=1}^{n}\sigma_{ij}t_it_j} \\
&= e^{t_{n+1}\mu_X+\frac{1}{2}t_{n+1}^2\frac{\sigma_X^2}{n}}\cdot e^{\frac{1}{2}\sum_{i=1}^{n}\sum_{j=1}^{n}\sigma_{ij}t_it_j} \\
&= M_{\overline{X}}(t_{n+1})\cdot M_{\mathbf{U}}(t_1,t_2,\ldots,t_n).
\end{aligned}$$

This provides the promised factorization of the joint MGF and shows that $\overline{X}$ and the n-dimensional multivariate $\mathbf{U}$ are independent. Since the sample variance S^2 is a function of $\mathbf{U}$, we have that $\overline{X}$ and S^2 are independent.

To prove property (2) of the sample variance we begin with the identity,

$$\sum_{i=1}^{n}(X_i-\overline{X})^2 = \sum_{i=1}^{n}(X_i-\mu_X)^2 - n(\overline{X}-\mu_X)^2$$

from Subsection 9.2.1. Now,

$$\begin{aligned}
\frac{(n-1)S^2}{\sigma_X^2} &= \frac{\sum_{i=1}^{n}(X_i-\overline{X})^2}{\sigma_X^2} \\
&= \frac{\sum_{i=1}^{n}(X_i-\mu_X)^2 - n(\overline{X}-\mu_X)^2}{\sigma_X^2} \\
&= \sum_{i=1}^{n}\left(\frac{X_i-\mu_X}{\sigma_X}\right)^2 - \left(\frac{\overline{X}-\mu_X}{\sigma_X/\sqrt{n}}\right)^2.
\end{aligned}$$

We let $U=\frac{(n-1)S^2}{\sigma_X^2}$, $V=\left(\frac{\overline{X}-\mu_X}{\sigma_X/\sqrt{n}}\right)^2$, and $W=\sum_{i=1}^{n}\left(\frac{X_i-\mu_X}{\sigma_X}\right)^2$ and then rewrite the above as $U+V=W$.

Since $\overline{X}\sim N\left(\mu_X,\frac{\sigma_X^2}{n}\right)$, we see that V is of the form Z^2, and hence has a $\chi^2(1)=\Gamma\left(\frac{1}{2},2\right)$ distribution. Therefore, $M_V(t)=\left(\frac{1}{1-2t}\right)^{1/2}$; $t<\frac{1}{2}$. (See Subsections 6.5.2 and 6.5.5). Similarly, since the $(X_i)_{i=1}^{n}$ are independent observations from X, we

have W is of the form $\sum_{i=1}^{n} Z_i^2 \sim \chi^2(n) = \Gamma\left(\frac{n}{2}, 2\right)$, and hence $M_W(t) = \left(\frac{1}{1-2t}\right)^{n/2}; t < \frac{1}{2}$.
Since $\overline{X}$ and S^2 are independent it follows that U and V are independent, so that, $M_{U+V}(t) = M_U(t) \cdot M_V(t) = M_W(t)$. Substituting for $M_V(t)$ and $M_W(t)$, we have,

$$M_U(t) \cdot \left(\frac{1}{1-2t}\right)^{1/2} = \left(\frac{1}{1-2t}\right)^{n/2}; t < \frac{1}{2}.$$

Thus,

$$M_U(t) = \left(\frac{1}{1-2t}\right)^{(n-1)/2} \text{ for } t < \frac{1}{2}.$$

It follows that

$$U = \frac{(n-1)S^2}{\sigma_X^2} \sim \Gamma\left(\frac{n-1}{2}, 2\right) = \chi^2(n-1).$$

9.2.3 CONFIDENCE INTERVAL FOR POPULATION VARIANCE

In general, we know that the sample variance is an unbiased estimator of the population variance. If we continue with the case of a normal population, then we can put our new-found knowledge of the distribution of the sample variance to work in order to assess its reliability as an estimator of the population variance.

Much as we used tables for the standard normal distribution $Z \sim N(0,1)$ to design confidence intervals for μ_X, we can make use of the chi-square tables (Appendix I) to derive confidence intervals for estimating σ_X^2. Consistent with our notation for z_α, we will let $\chi_\alpha^2(n)$ denote the $1-\alpha$ percentile of the chi-square with n degrees of freedom. That is, $\alpha = \Pr\left[\chi^2(n) \geq \chi_\alpha^2(n)\right]$. A typical table will list degrees of freedom n down the left-hand column, values of α along the top row, and values of $\chi_\alpha^2(n)$ in the cells of the table.

Example 9.2-1 Percentiles for the Sample Variance

A random sample of size 11 is taken from a normal population whose variance is 25.

(a) Find the 97.5^{th} percentile of the sample variance S^2.
(b) Find the 2.5^{th} percentile of the sample variance S^2.

Solution

We note that the standardized version of S^2 given by $\frac{(n-1)S^2}{\sigma_X^2} = \frac{10S^2}{25}$ is chi-square with $11-1=10$ degrees of freedom. The 97.5^{th} and the 2.5^{th} percentiles of the chi-square distribution mark the points with 2.5% of total area in the right-hand tail and the left-hand tail, respectively.

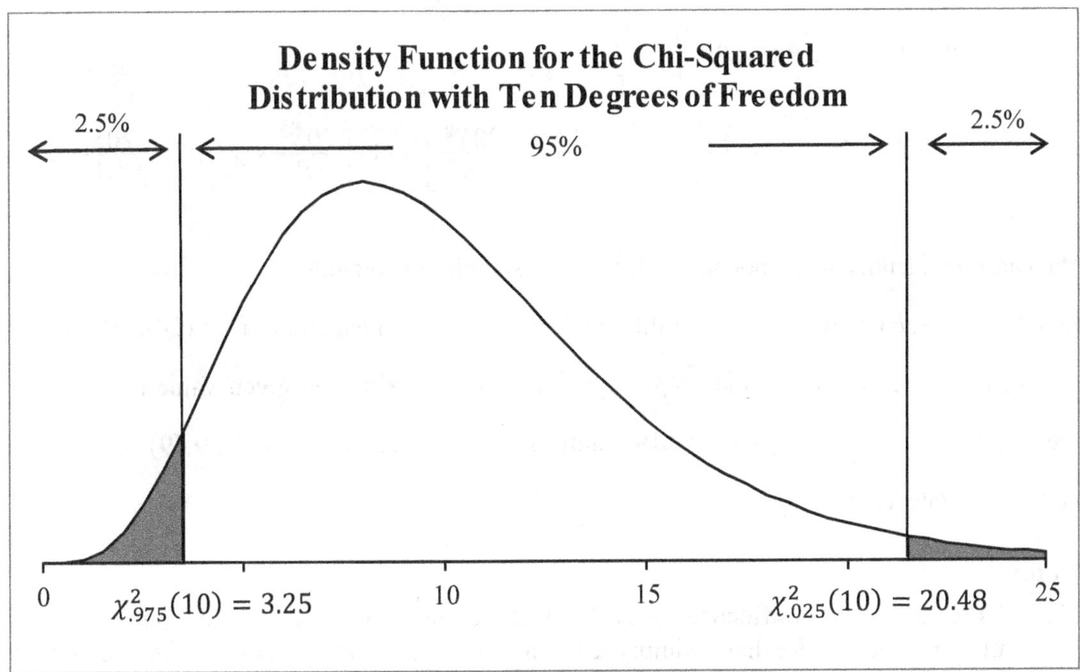

For (a), let A be the 97.5th percentile of S^2. We have, $.975 = \Pr[S^2 \le A] = \Pr\left[\frac{10S^2}{25} \le \frac{10A}{25}\right] = \Pr\left[\chi^2(10) \le \frac{10A}{25}\right]$. This means $\alpha = .025$ and $\frac{10A}{25} = \chi^2_{.025}(10)$. From the chi-square table we find $\chi^2_{.025}(10) = 20.48$. Thus, $A = \left(\frac{25}{10}\right) \cdot 20.48 = 51.21$.

For (b), let B be the 2.5th percentile of S^2. Now, we have, $.025 = \Pr[S^2 \le B] = \Pr\left[\frac{10S^2}{25} \le \frac{10B}{25}\right] = \Pr\left[\chi^2(10) \le \frac{10B}{25}\right]$. This is the left-hand tail with $\alpha = .975$ and $\frac{10B}{25} = \chi^2_{.975}(10)$.

From the chi-square table we find $\chi^2_{.975}(10) = 3.25$. Thus $B = \left(\frac{25}{10}\right) \cdot 3.25 = 8.12$. ❐

Example 9.2-2 Confidence Interval for σ_X^2

A random sample of size 21 is used to estimate the variance of a normal population X. The value of S^2 is calculated to be 16. Find values for A and B for which we can be 90% confident that $A < \sigma_X^2 < B$.

Solution

We use the fact that $\frac{(n-1)S^2}{\sigma_X^2} = \frac{20S^2}{\sigma_X^2} \sim \chi^2(20)$. Then,

$$.90 = \Pr\left[A < \sigma_X^2 < B\right] = \Pr\left[\frac{1}{B} < \frac{1}{\sigma_X^2} < \frac{1}{A}\right]$$

$$= \Pr\left[\frac{20S^2}{B} < \frac{20S^2}{\sigma_X^2} < \frac{20S^2}{A}\right] = \Pr\left[\frac{20S^2}{B} < \chi^2(20) < \frac{20S^2}{A}\right].$$

We can now identify these bounds with the corresponding percentiles for $\chi^2(20)$. To achieve an interval (A,B) with 90% probability, we can leave 5% in each tail of $\chi^2(20)$. The result is, $\frac{20S^2}{B} = \chi^2_{.95}(20) = 10.85$ and $\frac{20S^2}{A} = \chi^2_{.05}(20) = 31.41$. With the given value of $S^2 = 16$, we have, $A = \frac{20S^2}{\chi^2_{.05}(20)} = \frac{20 \cdot 16}{31.41} = 10.19$ and $B = 29.49$. Thus, (10.19, 29.49) is a 90% confidence interval for σ_X^2. ❐

Notes

(1) As we saw with confidence intervals for the population mean, the endpoints A and B are not unique. We have simply chosen a procedure that produces the confidence interval that results from splitting α equally between the two tails of the relevant chi-square distribution. Again, this will provide the most natural and efficient confidence interval for the population variance.

(2) When we computed confidence intervals for the population mean we used a standardized version of the *difference* $\overline{X} - \mu_X$ between the target parameter μ_X and the estimator $\overline{X}$. Here, we are using the *ratio* $\frac{S^2}{\sigma_X^2}$ between the estimator and the target. The point in either case is to have a random quantity with both estimator and target in the formula, but whose distribution does *not* depend on the parameter being estimated. Random variables with known distributions that can accomplish this are sometimes called ***pivots*** in statistical treatises.

We can follow the above procedure in general to derive the form of a $1-\alpha$ confidence interval for σ_X^2.

Confidence Interval for the Variance of a Normal Population

Let X be a *normal* population random variable with variance σ_X^2, and let $(X_i)_{i=1}^n$ be a random sample from X. Let S^2 be the sample variance. Then the interval,

$$\left(\frac{(n-1)\cdot S^2}{\chi^2_{\alpha/2}(n-1)}, \frac{(n-1)\cdot S^2}{\chi^2_{1-\alpha/2}(n-1)}\right)$$

constitutes a $1-\alpha$ confidence interval for estimating σ_X^2.

Proof

Proceeding as in the example, we let A and B be functions of S^2 such that,

$$1-\alpha = \Pr[A<\sigma_X^2<B] = \Pr\left[\frac{1}{B}<\frac{1}{\sigma_X^2}<\frac{1}{A}\right]$$

$$= \Pr\left[\frac{(n-1)S^2}{B}<\frac{(n-1)S^2}{\sigma_X^2}<\frac{(n-1)S^2}{A}\right]$$

$$= \Pr\left[\frac{(n-1)S^2}{B}<\chi^2(n-1)<\frac{(n-1)S^2}{A}\right].$$

We can then set $\frac{(n-1)\cdot S^2}{B}=\chi^2_{1-\alpha/2}(n-1)$ and $\frac{(n-1)\cdot S^2}{A}=\chi^2_{\alpha/2}(n-1)$. Then solving for A and B results in the required interval boundaries.

Note

If the situation requires a confidence interval for the standard deviation σ_X rather than the variance, then we can easily retrace our footsteps to find that,

$$\left(\sqrt{\frac{(n-1)\cdot S^2}{\chi^2_{\alpha/2}(n-1)}},\sqrt{\frac{(n-1)\cdot S^2}{\chi^2_{1-\alpha/2}(n-1)}}\right)$$

is the required interval.

Exercise 9-7 A random sample of size 13 is used to estimate the variance of a normal population X. The value of the sample variance is calculated to be 63. Find the 95% confidence intervals for σ_X^2 and σ_X.

Exercise 9-8 A random sample of size 10 was taken from a normal population with a standard deviation of 41.67. Find the 2.5th percentile of the sample standard deviation.

Exercise 9-9 Your scores from an afternoon of bowling were 128, 189, 156, 143, 117, and 152. Find the 90% confidence interval for the standard deviation of your bowling scores. Assume your bowling scores are approximately normally distributed.

Exercise 9-10 CAS Exam 3 Fall 2007 #10

You are given the following information from a random sample:

- X_i is from a normal distribution with mean = 1 and variance = 2.
- Sample size is 20
- S^2 is the unbiased sample variance

Calculate the critical value, c, which satisfies the equation $\Pr\left[S^2<c\right]=.95$.

(A) Less than 2.8
(B) At least 2.8, but less than 2.9
(C) At least 2.9, but less than 3.0
(D) At least 3.0, but less than 3.1
(E) At least 3.1

9.3 THE STUDENT *t*-DISTRIBUTION

We return to the problem of estimating the population mean with a sample mean, but now without the assumption that the population variance is known. Let X be a normal population and let $(X_i)_{i=1}^n$ be a random sample. The obvious approach is to estimate the unknown population variance using the sample variance S^2 developed in the preceding section.

Previously, we constructed confidence intervals for the mean using the pivotal random variable $\frac{\bar{X}-\mu_X}{\sigma_X/\sqrt{n}}$, whose distribution is standard normal. If we use $S=\sqrt{S^2}$ in place of σ_X then the pivotal random variable becomes $\frac{\bar{X}-\mu_X}{S/\sqrt{n}}$, which is no longer a normal distribution.

In this section we will discuss the family of distributions that arises in this way.

The Student *t*-Distribution

Let Z and V be ***independent*** random variables such that,

(1) Z has a standard normal distribution, and
(2) V has a chi-square distribution with n degrees of freedom.

Let $t(n)$ be a transformation of Z and V taking the form $t(n)=\frac{Z}{\sqrt{V/n}}$.

Then $t(n)$ is said to have the ***Student t-distribution with n degrees of freedom.***

The density function for $t(n)$ is given by,

$$f_T(t) = \frac{\Gamma\left(\frac{n+1}{2}\right)}{\sqrt{n\pi}\cdot\Gamma\left(\frac{n}{2}\right)}\cdot\left(\frac{1}{1+t^2/n}\right)^{(n+1)/2} \quad ; \ -\infty<t<\infty,$$

where Γ refers to the gamma function defined in Subsection 6.5.1.

The shape of the Student t density function is shown in Figure 9.3.1 for various degrees of freedom. The graphs for small degrees of freedom are shorter and squatter, with heavier tails. As the degrees of freedom increase the graphs approach the standard normal in shape, which is the tallest graph in the display.

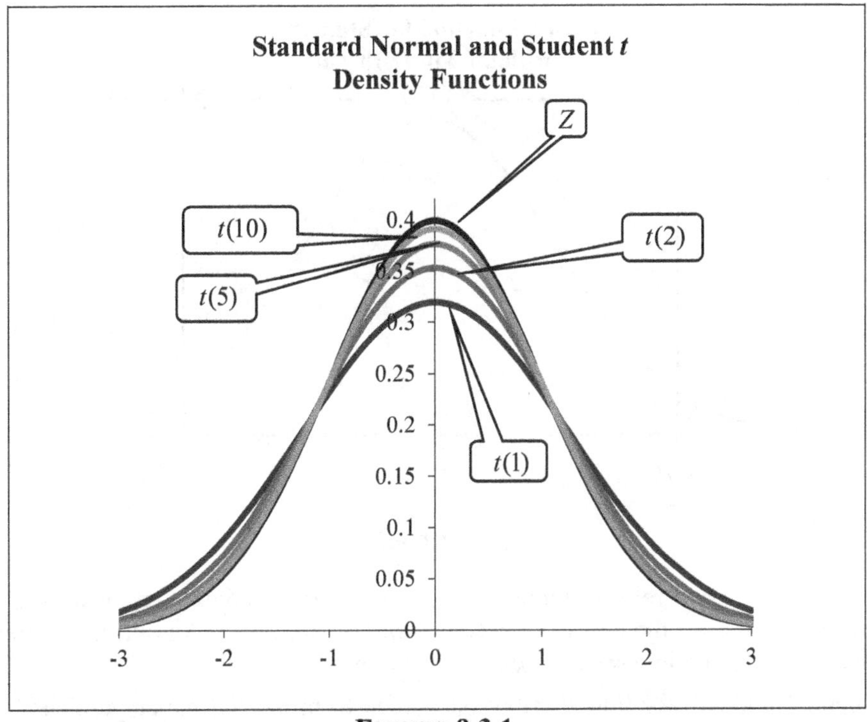

FIGURE 9.3.1

Notes

(1) Using basic calculus it is easily seen that for each n the density function for $t(n)$ is a symmetric bell-shaped curve centered at zero, as illustrated above.

(2) As n increases the corresponding distributions converge toward the standard normal.

(3) There is a standard table for the Student t-distributions in Appendix I. We will employ the usual notation $t_\alpha(n)$ for the $1-\alpha$ percentile, meaning $\alpha = \Pr[t(n) > t_\alpha(n)]$, where α is the area of the right-hand tail beginning at $t_\alpha(n)$. Tables vary in how they treat the entries and it is important to check carefully how the table is constructed. For example, many tables (including ours) assume a two-tail probability of the form $\alpha = \Pr[|t(n)| > t_{\alpha/2}(n)]$. This means the entry corresponding to α is $t_{\alpha/2}(n)$, with the area to the right of $t_{\alpha/2}(n)$ equal to $\alpha/2$. The extracted table used for both the actuarial statistics exam[1] and the VEE-Applied Statistics exam is of this type, and so we follow that convention with the table provided in Appendix I. The Microsoft Excel® function TINV also returns $t_{\alpha/2}(n)$ for input α. There is usually a schematic drawing, such as Figure 9.3.2, accompanying the table to clarify the issue.

[1] At the time of this writing, the CAS ST or S exam and, earlier, the statistics portion of CAS 3L.

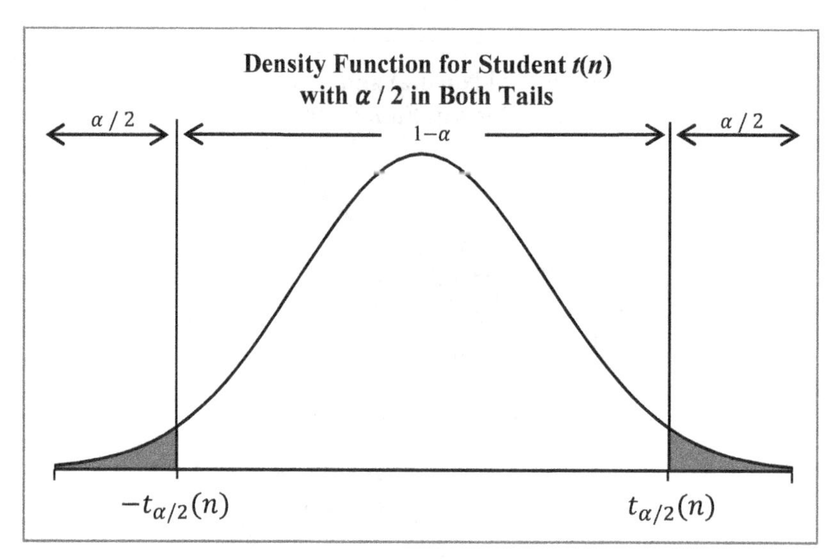

FIGURE 9.3.2

(4) Since statistical users rely on tables or technology for Student t applications, the rather imposing formula for the density function is rarely employed. An outline for deriving the density will be given in a series of exercises below.

(5) The name "Student" for this distribution comes from a 1908 paper published under the pseudonym *Student* by William Sealy Gosset[2]. This paper set forth properties of the t-distribution that Gosset had discovered in his statistical research while an employee of the *Guinness Brewery* in Dublin. Gosset was prevented by Guinness from publishing under his real name for fear of revealing proprietary information. In modern parlance you could say he leaked the material as an anonymous source.

[2] Student (March 1908). "The probable error of a mean." *Biometrika* 6 (1): 1–25

The main application of the Student t-distribution is in writing confidence intervals for the mean of a normal population.

Application of the Student t-Distribution

Let X be a normal population with $X \sim N(\mu_X, \sigma_X^2)$, with μ_X and σ_X^2 both unknown. Let $(X_i)_{i=1}^n$ be a random sample with sample mean $\overline{X}$ and sample variance S^2.

(1) The random variable $\dfrac{\overline{X} - \mu_X}{S / \sqrt{n}} \sim t(n-1)$.

(2) A symmetric $1-\alpha$ confidence interval for $\overline{X}$ is given by,

$$\left(\overline{X} - t_{\alpha/2}(n-1) \cdot \frac{S}{\sqrt{n}},\ \overline{X} + t_{\alpha/2}(n-1) \cdot \frac{S}{\sqrt{n}}\right)$$

Proof

By multiplying and dividing $\dfrac{\overline{X} - \mu_X}{\frac{S}{\sqrt{n}}}$ by σ_X and $\sqrt{n-1}$ we see that,

$$\frac{\overline{X} - \mu_X}{\frac{S}{\sqrt{n}}} = \frac{\overline{X} - \mu_X}{\frac{\sigma_X}{\sqrt{n}}} \cdot \frac{1}{\sqrt{\frac{(n-1)S^2}{\sigma_X^2}}} \cdot \frac{1}{\frac{1}{\sqrt{(n-1)}}} = \frac{Z}{\sqrt{\frac{U}{(n-1)}}} \sim t(n-1).$$

This follows since $\dfrac{\overline{X} - \mu_X}{\frac{\sigma_X}{\sqrt{n}}} = Z$, $\dfrac{(n-1)S^2}{\sigma_X^2} = U \sim \chi^2(n-1)$, and (not least) because $\overline{X}$ and S^2 are independent.

The confidence interval is now derived from the fact that $\dfrac{\overline{X} - \mu_X}{S / \sqrt{n}} \sim t(n-1)$:

$$1-\alpha = \Pr\left[\left|\overline{X} - \mu_X\right| \leq \varepsilon\right] = \Pr\left[\frac{\left|\overline{X} - \mu_X\right|}{S / \sqrt{n}} \leq \frac{\varepsilon}{S / \sqrt{n}}\right] = \Pr\left[\left|t(n-1)\right| \leq \frac{\varepsilon}{S / \sqrt{n}}\right].$$

Thus, $\frac{\varepsilon}{S/\sqrt{n}} = t_{\alpha/2}(n-1)$, and so

$$\underbrace{\varepsilon}_{\text{error}} = \underbrace{t_{\alpha/2}(n-1)}_{\substack{\text{value obtained}\\ \text{from a table}}} \frac{\overbrace{S}^{\text{sample st. dev.}}}{\sqrt{n}}.$$

This gives the confidence interval for μ_X in the form $\overline{X} \pm \varepsilon = (\overline{X} - \varepsilon, \overline{X} + \varepsilon)$.

As a practical matter confidence intervals based on the Student *t*-distribution are used for small samples when the population is assumed to be normal. Most tables extend out to no more than 30 degrees of freedom, at which point there is relatively little difference between $t_{\alpha/2}(n-1)$ and $z_{\alpha/2}$.

Example 9.3-1 Confidence Interval Using the Student *t*-distribution

Let X represent a score on a standardized test. We are given that X is normal with an unknown mean and variance. A random sample of 25 test scores is taken, resulting in values of $\overline{X} = 72.3$ and $S^2 = 16$. Calculate the 95% symmetric confidence interval for the population mean μ_X.

Solution

From the *t*-table we have $t_{\alpha/2}(n-1) = t_{.025}(24) = 2.064$, resulting in a margin of error $\varepsilon = t_{\alpha/2}(n-1)\frac{S}{\sqrt{n}} = 2.064 \cdot \frac{4}{5} = 1.65$, and a confidence interval of $72.3 \pm 1.65 = (70.65, 73.95)$. ❐

Notes

(1) If we had used $z_{.025} = 1.96$ instead of the *t*-distribution with value $t_{.025}(24) = 2.064$, then the resulting confidence interval would be (70.73, 73.87), an error of about 0.1%.

(2) The confidence interval based on the Student *t* is wider than the one calculated using the standard normal. This makes sense, because we are less confident using an estimate S in place of the true σ_X. This diminished confidence is reflected in a wider confidence interval.

It is the best practice to use the Student *t* in calculating confidence intervals for the mean of a normal population whenever the sample standard deviation is used to estimate σ_X. However, if the sample size is larger than the *t*-table accommodates then little harm is done by using *z* instead. It is also generally accepted to use the *z*-table even if the population is not normal, providing the sample size is large.

Exercise 9-11 A random variable Y is known to have a normal distribution. A random sample of size $n = 14$ resulted in values $\overline{Y} = -43.2$ and $S_Y = 17.9$. Find the 98% confidence interval for the population mean.

Exercise 9-12 A random sample of eleven obscure gas stations found the prices of a gallon of unleaded gasoline to be: 2.58 2.53 2.60 2.55 2.80 2.75 2.58 2.62 2.69 2.61 2.50

(a) What is the point estimate, $\overline{X}$, of the population mean price per gallon of gas?

(b) What is the point estimate, S_X, of the population standard deviation?

(c) Assuming a normal distribution, what is the 95% confidence interval estimate of the population mean price per gallon of gas?

Exercise 9-13 Derive the density function for the Student t-distribution using the following series of steps.

(a) Let $Y \sim \chi^2(n) = \Gamma\left(\frac{n}{2}, 2\right)$ and let $U = \sqrt{Y/n}$. Use the density function for Y (Subsection 6.5.2) and the transformation formula (Subsection 8.1.1) to show that

$$f_U(u) = \left(\frac{n}{2}\right)^{n/2} \cdot \frac{2}{\Gamma\left(\frac{n}{2}\right)} \cdot u^{n-1} \cdot e^{-\frac{nu^2}{2}}.$$

(b) Let Z be a standard normal random variable independent of Y, and let $T = \frac{Z}{U}$. Then, by definition, $T \sim t(n)$. Use the density function for the standard normal (Subsection 6.3.2) and the convolution formula for ratios (Subsection 8.1.3) in the form $f_T(t) = \int_0^\infty f_U(u) \cdot f_Z(tu) \cdot u \cdot du$ to show that,

$$f_T(t) = \left(\frac{n}{2}\right)^{n/2} \cdot \frac{1}{\Gamma\left(\frac{n}{2}\right)} \cdot \sqrt{\frac{2}{\pi}} \cdot \int_0^\infty u^n \cdot e^{-\frac{1}{2}(n+t^2)u^2} du.$$

(c) Use the substitution $v = u^2$ for the integral in (b) to reduce it to the non-constant portion of a gamma distribution. Evaluate this integral using the formula for the constant in a gamma distribution to finish the derivation.

9.4 THE *F*-DISTRIBUTION

Another important family of statistical distributions arises when we study the ratio of two independent chi-square random variables. This allows us to estimate the ratio of the variances of two independent populations. In the statistical procedures known as regression analysis and the analysis of variance, the issue is often whether or not two population variances are equal. Being able to construct a confidence interval for the ratio of the variances will help in resolving this question. The F-distribution is instrumental in making this determination.

We first give the formal definition of the F-distribution, and then provide some examples to illustrate how the distribution can be applied.

The *F*-Distribution

Let U and V be independent non-negative random variables with $U \sim \chi^2(m)$ and $V \sim \chi^2(n)$.

The random variable $F(m,n) = \frac{U/m}{V/n}$ is said to have ***the F-Distribution with degrees of freedom m and n.***

The density function for $F(m,n)$ is given by,

$$f_F(w) = \left(\frac{m}{n}\right)^{m/2} \cdot \frac{\Gamma\left(\frac{m+n}{2}\right)}{\Gamma\left(\frac{m}{2}\right)\Gamma\left(\frac{n}{2}\right)} \cdot \frac{w^{(m/2)-1}}{\left(1+\frac{m}{n}w\right)^{(m+n)/2}} \quad \text{for } w > 0$$

The outline for a derivation of the density function of $F(m,n)$ will be given in the exercises. Again, as with the chi-square and the Student t, the form of the density function plays a secondary role to the tables of percentiles.

Application of the F-Distribution

Let X and Y be independent normal populations with variances of σ_X^2 and σ_Y^2, respectively. Let $(X_i)_{i=1}^m$ be a random sample from X, and let $(Y_j)_{j=1}^n$ be a random sample from Y, with resulting sample variances S_X^2 and S_Y^2, respectively. Then,

$$\frac{S_X^2 / \sigma_X^2}{S_Y^2 / \sigma_Y^2} \sim F(m-1, n-1).$$

Proof

$$\frac{\frac{S_X^2}{\sigma_X^2}}{\frac{S_Y^2}{\sigma_Y^2}} = \frac{\frac{(m-1)S_X^2}{\sigma_X^2} \cdot \frac{1}{(m-1)}}{\frac{(n-1)S_Y^2}{\sigma_Y^2} \cdot \frac{1}{(n-1)}} = \frac{\frac{\chi^2(m-1)}{(m-1)}}{\frac{\chi^2(n-1)}{(n-1)}} \sim F(m-1, n-1).$$

Our table (Appendix I) of F percentiles is constructed with the degrees of freedom of the numerator (m) along the top and the degrees of freedom of the denominator (n) along the left-hand side. The numbers in the cells are the percentiles $F_\alpha(m,n)$ with $\alpha = .05$ and $\alpha = .01$ of $F(m,n)$. This is similar to the table provided for the actuarial statistics exam[3]. Consistent with our earlier notation, these are right-hand tail percentiles, meaning that $\alpha = \Pr\left[F(m,n) \geq F_\alpha(m,n)\right]$.

Properties of $F(m,n)$

(1) $F(n,m) = \dfrac{1}{F(m,n)}$.

(2) $[t(n)]^2 = F(1,n)$.

(3) $F_{1-\alpha}(m,n) = \dfrac{1}{F_\alpha(n,m)}$.

Proof

Property (1) follows immediately from the definition of the F-distribution. For property (2), recall the definition from the preceding section, $t(n) = \frac{Z}{\sqrt{V/n}}$, with Z the standard normal and V an independent chi-square random variable with n degrees of freedom. Then,

[3] Previously, CAS 3L and CAS ST or S at the time of this writing.

$$[t(n)]^2 = \frac{Z^2}{V/n} = \frac{\chi^2(1)}{\chi^2(n)/n} \sim F(1,n).$$

For property (3), we have

$$\begin{aligned} 1-\alpha &= \Pr[F(m,n) \geq F_{1-\alpha}(m,n)] \\ &= \Pr\left[\frac{1}{F(m,n)} \leq \frac{1}{F_{1-\alpha}(m,n)}\right] \\ &= \Pr\overbrace{\left[F(n,m) \leq \frac{1}{F_{1-\alpha}(m,n)}\right]}^{\text{by property (1)}} = 1-\Pr\left[F(n,m) > \frac{1}{F_{1-\alpha}(m,n)}\right]. \end{aligned}$$

This implies that $\alpha = \Pr\left[F(n,m) > \frac{1}{F_{1-\alpha}(m,n)}\right]$, which in turn shows that $F_\alpha(n,m) = \frac{1}{F_{1-\alpha}(m,n)}$. Equivalently, $F_{1-\alpha}(m,n) = \frac{1}{F_\alpha(n,m)}$.

Example 9.4-1 Confidence Interval using the *F*-Distribution

Let *X* and *Y* be independent normal populations with variances of σ_X^2 and σ_Y^2, respectively. Let $(X_i)_{i=1}^{21}$ be a random sample from *X*, and let $(Y_j)_{j=1}^{25}$ be a random sample from *Y*, with resulting sample variances $S_X^2 = 8$ and $S_Y^2 = 15$. Determine a 90% confidence interval for $\frac{\sigma_X^2}{\sigma_Y^2}$.

Solution

We need to determine values of *A* and *B* such that $.90 = \Pr[A < \sigma_X^2 / \sigma_Y^2 < B]$. We rearrange things in order to convert this into a question about the relevant *F*-distributions. Thus,

$$\begin{aligned} .90 = \Pr\left[A < \frac{\sigma_X^2}{\sigma_Y^2} < B\right] &= \Pr\left[\frac{1}{B} < \frac{\sigma_Y^2}{\sigma_X^2} < \frac{1}{A}\right] \\ &= \Pr\left[\frac{1}{B}\cdot\frac{S_X^2}{S_Y^2} < \frac{\sigma_Y^2}{\sigma_X^2}\cdot\frac{S_X^2}{S_Y^2} < \frac{1}{A}\cdot\frac{S_X^2}{S_Y^2}\right] \\ &= \Pr\left[\frac{1}{B}\cdot\frac{S_X^2}{S_Y^2} < \frac{S_X^2/\sigma_X^2}{S_Y^2/\sigma_Y^2} < \frac{1}{A}\cdot\frac{S_X^2}{S_Y^2}\right] \\ &= \Pr\left[\frac{1}{B}\cdot\frac{S_X^2}{S_Y^2} < F(m-1,n-1) < \frac{1}{A}\cdot\frac{S_X^2}{S_Y^2}\right] \\ &= \Pr\left[\frac{1}{B}\cdot\frac{S_X^2}{S_Y^2} < F(20,24) < \frac{1}{A}\cdot\frac{S_X^2}{S_Y^2}\right]. \end{aligned}$$

We now set $\frac{1}{B}\cdot\frac{S_X^2}{S_Y^2} = F_{.95}(20,24)$ and $\frac{1}{A}\cdot\frac{S_X^2}{S_Y^2} = F_{.05}(20,24)$. This results in

$$A = \frac{S_X^2}{S_Y^2}\cdot\frac{1}{F_{.05}(20,24)} = \left(\frac{8}{15}\right)\left(\frac{1}{2.02}\right) = .26$$

To calculate B we use property (3) to find,

$$B = \frac{S_X^2}{S_Y^2}\cdot\frac{1}{F_{.95}(20,24)} = \frac{S_X^2}{S_Y^2}\cdot F_{.05}(24,20) = \left(\frac{8}{15}\right)(2.08) = 1.11.$$

Therefore, the desired confidence interval is $(.26, 1.11)$. ❐

Note

Once again, the confidence interval limits calculated in the example are not uniquely determined by the random sample. We have elected to follow a procedure for calculating A and B that leaves equal probabilities of 5% in the two tails of the relevant F-distribution.

We follow the procedure of the example in order to write the general form for a confidence interval for the variance ratio. But once again, this is an instance where it is preferable to learn the method rather than memorize the formula.

Confidence Interval for the Ratio of Variances

Let X and Y be independent normal populations with variances of σ_X^2 and σ_Y^2, respectively. Let $(X_i)_{i=1}^m$ be a random sample from X, and let $(Y_j)_{j=1}^n$ be a random sample from Y, with resulting sample variances S_X^2 and S_Y^2, respectively. A $1-\alpha$ confidence interval for $\frac{\sigma_X^2}{\sigma_Y^2}$ is given by,

$$\left(\frac{S_X^2}{S_Y^2}\cdot\frac{1}{F_{\alpha/2}(m-1,n-1)}, \frac{S_X^2}{S_Y^2}\cdot F_{\alpha/2}(n-1,m-1)\right).$$

Proof

We determine values for A and B, where

$$1-\alpha = \Pr\left[A < \frac{\sigma_X^2}{\sigma_Y^2} < B\right]$$
$$= \Pr\left[\frac{1}{B} < \frac{\sigma_Y^2}{\sigma_X^2} < \frac{1}{A}\right]$$
$$= \Pr\left[\frac{1}{B}\cdot\frac{S_X^2}{S_Y^2} < \frac{\sigma_Y^2}{\sigma_X^2}\cdot\frac{S_X^2}{S_Y^2} < \frac{1}{A}\cdot\frac{S_X^2}{S_Y^2}\right] = \Pr\left[\frac{1}{B}\cdot\frac{S_X^2}{S_Y^2} < F(m-1,n-1) < \frac{1}{A}\cdot\frac{S_X^2}{S_Y^2}\right].$$

Set $\frac{1}{A}\cdot\frac{S_X^2}{S_Y^2} = F_{\alpha/2}(m-1,n-1)$, which implies $A = \frac{S_X^2}{S_Y^2}\cdot\frac{1}{F_{\alpha/2}(m-1,n-1)}$.

We set $\frac{1}{B}\cdot\frac{S_X^2}{S_Y^2} = F_{1-\alpha/2}(m-1,n-1) = \frac{1}{F_{\alpha/2}(n-1,m-1)}$, yielding $B = \frac{S_X^2}{S_Y^2}\cdot F_{\alpha/2}(n-1,m-1)$.

Exercise 9-14 A random sample of size eight from a normal population X had sample variance $S_X^2 = 81$. A random sample of size thirteen from a normal population Y had sample variance $S_Y^2 = 15$. Determine a 90% confidence interval for $\frac{\sigma_X^2}{\sigma_Y^2}$.

Exercise 9-15 A random sample of size fifteen (15) from a normal population X had sample standard deviation $S_X = 5.4$. A random sample of size thirteen (13) from a normal population Y had sample standard deviation $S_Y = 5.9$. Determine a 98% confidence interval for $\frac{\sigma_X}{\sigma_Y}$.

Exercise 9-16 Your scores from an afternoon of bowling were 128, 189, 156, 143, 117, and 152. Your sister's bowling scores were 204, 176, 189, 198, 218, 177, 190, and 203. Find the 90% confidence interval for $\frac{\sigma_{\text{You}}^2}{\sigma_{\text{Sister}}^2}$. Assume bowling scores are normally distributed.

The following exercise outlines a derivation for the density function of $F(m,n)$.

Exercise 9-17 Let U and V be independent random variables with $U \sim \chi^2(m)$ and $V \sim \chi^2(n)$.

(a) Use the transformation formula to show that $\frac{U}{m} \sim \Gamma\left(\frac{m}{2}, \frac{2}{m}\right)$ and $\frac{V}{n} \sim \Gamma\left(\frac{n}{2}, \frac{2}{n}\right)$.

(b) Let $S = \frac{U}{m}$ and $T = \frac{V}{n}$. Define $W = \frac{S}{T}$. Use the formula $f_W(w) = \int_1^\infty f_T(t) f_S(wt)\, t\, dt$ (Subsection 8.1.3) and the density function formula for gamma distributions to show that, $f_W(w) = c \cdot w^{(m/2)-1} . \int_0^\infty t^{\alpha-1} e^{-(1/\beta)t}\, dt$, where c is a constant (albeit a complicated one!), $\alpha = \frac{n+m}{2}$, and $\beta = \frac{2}{n+mw}$.

(c) Evaluate the integral in (b) by using the constant $\frac{1}{\beta^\alpha \Gamma(\alpha)}$ for a gamma distribution. Rearrange to match the formula for the density function of $F(m,n)$ given above.

9.5 ESTIMATING PROPORTIONS

In election polling at its most basic we ask a random sample of prospective voters to declare whether or not they intend to vote for candidate A (candidate A is usually the politician paying for the poll). For each person sampled, the answer to the polling question is either **yes** or **no** (for simplicity's sake we assume responses of "not sure" or "who cares?" are not counted, or preferably we ensure that respondents must choose either response **yes** or response **no** with no other options). We estimate the percentage of people favoring candidate A by calculating the proportion of **yes's** among the valid responses. Similar situations arise when we are estimating the proportion of life insurance applicants who are smokers, measuring free-throw shooting percentages in basketball, sampling for the proportion of defects in a manufacturing process, or sampling for the percentage of left-handed people taking actuarial exams.

The key element in common in these scenarios is that the population random variable X is a simple Bernoulli trial with probability of success equal to p, and the purpose of the sampling is to estimate the actual value of p. If the random sample is of size n then the number B of yes responses is a binomial random variable with parameters n and p. The random variable B/n is then often referred to as a ***binomial proportion***, although we prefer to think of it in more familiar terms as the sample mean, $\overline{X}$, arising from a random sample $(X_i)_{i=1}^n$ taken from the Bernoulli trial population X. In this case $(X_i)_{i=1}^n$ is a sequence of 0's and 1's, with the outcome 1 representing "yes" and the outcome 0 representing "no."

If X is a Bernoulli trial random variable, then we know (Section 4.2) that $\mu_X = p$ and $\sigma_X^2 = p(1-p)$. We also know from first principles (Section 9.1) that $E[\overline{X}] = \mu_X = p$, that $Var[\overline{X}] = \frac{\sigma_X^2}{n} = \frac{p(1-p)}{n}$, and that for sufficiently large sample sizes n, the sampling distribution of $\overline{X}$ is approximately normal with mean p and variance $\frac{p(1-p)}{n}$.

In what follows, we are using $\overline{X}$ as an unbiased estimator for the population mean, which in this case is the parameter p. Since we are estimating p, the near universal practice is to denote the estimator $\overline{X}$ by the symbol $\hat{p}$. We will conform to this practice, but the reader should keep in mind that, with a single exception, the underlying theory for $\hat{p}$ consists of nothing more than the properties of $\overline{X}$.

The exception referred to above has to do with the estimation of population variance σ_X^2. Rather than using the S^2 introduced in Section 9.2, we use the random variable $\hat{p}(1-\hat{p})$ to estimate $\sigma_X^2 = p(1-p)$. This seems, on the surface, to be a sensible estimator for $p(1-p)$, and it does have the advantage of being easy to calculate. As we note in the exercises, however, it is not an unbiased estimator of $\sigma_X^2 = p(1-p)$. Here is a summary of the results.

Confidence Interval for Estimating p

Let $(X_i)_{i=1}^n$ be a random sample taken from the Bernoulli trial population X. Let p be the probability of a success for the population X, and let $\hat{p} = \overline{X} = \frac{1}{n}\sum^{n} X_i$. Then,

(1) $\hat{p}$ is the proportion of successes in the sample.

(2) $\hat{p}$ is an unbiased estimator of p. That is $E[\hat{p}] = p$.

(3) For large n, $\dfrac{\hat{p}-p}{\sqrt{\frac{\hat{p}(1-\hat{p})}{n}}}$ is (approximately) a standard normal random variable.

(4) For large n, an (approximate) symmetric $1-\alpha$ confidence interval for p is given by $\hat{p} \pm \varepsilon$, where ε = margin of error $= z_{\alpha/2} \cdot \sqrt{\frac{\hat{p}(1-\hat{p})}{n}}$.

Proof

As indicated in the discussion above, we know that $\hat{p}$ is an unbiased estimator of p with $Var[\hat{p}] = \frac{\sigma_X^2}{n} = \frac{p(1-p)}{n}$. If n is sufficiently large, then by the Central Limit Theorem $\hat{p}$ is approximately normal with standard deviation $\sqrt{\frac{p(1-p)}{n}}$, which we approximate with $\sqrt{\frac{\hat{p}(1-\hat{p})}{n}}$. On that basis (3) follows, and we can write,

$$1-\alpha = \Pr[|\hat{p}-p| \le \varepsilon] = \Pr\left[\frac{|\hat{p}-p|}{\sqrt{\frac{\hat{p}(1-\hat{p})}{n}}} \le \frac{\varepsilon}{\sqrt{\frac{\hat{p}(1-\hat{p})}{n}}}\right] \approx \Pr\left[Z \le \frac{\varepsilon}{\sqrt{\frac{\hat{p}(1-\hat{p})}{n}}}\right].$$

It follows that $\dfrac{\varepsilon}{\sqrt{\frac{\hat{p}(1-\hat{p})}{n}}} = z_{\alpha/2}$, so that $\varepsilon = z_{\alpha/2} \cdot \sqrt{\frac{\hat{p}(1-\hat{p})}{n}}$.

Even the casual reader must sense the degree of looseness in the arguments presented. There is heavy use of the word "approximate" and not much guidance about how large the sample size needs to be. About all we can say in our defense is that "everyone does it." One common rule of thumb is that both the number of successes and the number of failures should be at least 10.

To illustrate, let's consider the simple experiment of tossing a fair coin 25 times in order to calculate an estimate of p, the probability of "heads," which we know in advance is 0.5. According to the result above based on the Central Limit Theorem, the margin of error in estimating p at the $1-\alpha$ level of confidence is given by $\varepsilon = z_{\alpha/2} \cdot \sqrt{\frac{\hat{p}(1-\hat{p})}{n}}$. We don't know in advance what value to use for $\hat{p}$. We do, however, know that the maximum possible value for $\hat{p}(1-\hat{p})$ occurs when $\hat{p} = 0.5$. Therefore, using a 95% confidence level we have $z_{.025} = 1.96$, and so an upper bound for the margin of error is given by

$$\varepsilon = 1.96 \cdot \sqrt{\frac{(0.50)(0.50)}{25}} \approx 0.20.$$

On the other hand, since $\hat{p}$ is a binomial proportion we can use the exact binomial probabilities to calculate the margin of error. Since $\hat{p}$ is discrete the confidence levels jump, but it can easily be shown from binomial tables that $\Pr[|\hat{p} - 0.5| \le .18] = .96$, which is fairly consistent with our approximated margin of error of 0.2 with confidence level 95%.

Exercise 9-18 Out of a total of 3000 adults of legal drinking age, 2070 responded YES to the question "Do you think beer costs too much at PNC Park?"

(a) What is the point estimate, $\hat{p}$, of the proportion of adults that feel beer costs too much?

(b) At 90% confidence, what is the margin of error?

(c) What is the symmetric 90% confidence interval for the true population proportion of people who believe that beer costs too much at PNC Park?

Exercise 9-19 Aaron was shooting his Koosh® basketball into the hoop above his door. He made 26 of 40 shoots. Find the 95% confidence interval for the true proportion of shots he makes over the long haul.

Exercise 9-20 The Wednesday October 29th, 2003 USA Today page 10A displayed poll results of support for the war in Iraq. Fifty seven (57) percent backed partial pullout. In small print, there were 1006 adults surveyed with a margin of error of ± 3 percentage points. How confident is USA Today in these results?

Exercise 9-21 Show that $E[\hat{p}(1-\hat{p})] = \left(\frac{n-1}{n}\right)p(1-p)$, which means that $\hat{p}(1-\hat{p})$ is not an unbiased estimator of $\sigma_X^2 = p(1-p)$. Hint: $E[\hat{p}(1-\hat{p})] = E[\hat{p}] - E[\hat{p}^2]$. Use the variance of $\hat{p}$ and the variance formula to find $E[\hat{p}^2]$ in terms of p.

9.6 ESTIMATING THE DIFFERENCE BETWEEN MEANS

A question that often arises in statistics is whether or not two populations have the same mean. A typical context might concern a standardized test administered to two different and independent populations, say Martians and Venusians. Do Martians tend to score significantly higher or significantly lower than Venusians, or are they about the same? The obvious approach toward answering such a question is to test separate random samples from each of the populations, and study the difference between the sample means.

If the sample sizes are sufficiently large the difference between sample means is approximately normal and, as we show below, we can use the methods of earlier sections to calculate appropriate confidence intervals. If the samples are small then we need to make several simplifying assumptions in order to derive meaningful results. We will begin with large samples and then turn to some partial results in the case of small samples in Subsection 9.6.2.

9.6.1 LARGE SAMPLES

We will begin with the understanding that the populations are normal and that their variances are known. These assumptions are somewhat restrictive but not completely unreasonable. We know that many populations that arise from a measurement (test scores, for example) are normal, or nearly so. It may also be that while the mean test score varies between populations, the variance remains fairly stable, so that historical data can be used in order to treat the variance as known. Whatever the case, if the sample sizes are sufficiently large (again, say 30 or more) then we can surmount these constraints.

The Difference of Means from Normal Populations, I

Let X and Y be independent normal populations with $X \sim N(\mu_X, \sigma_X^2)$ and $Y \sim N(\mu_Y, \sigma_Y^2)$. We assume that σ_X^2 and σ_Y^2 are known. Let $(X_i)_{i=1}^m$ be a random sample from X, and let $(Y_j)_{j=1}^n$ be a random sample from Y, with sample means $\overline{X}$ and $\overline{Y}$, respectively. Let $W = X - Y$ and let $\overline{W} = \overline{X} - \overline{Y}$. Then,

(1) $E[\overline{W}] = E[W] = \mu_X - \mu_Y$.

(2) $Var[\overline{W}] = Var[\overline{X}] + Var[\overline{Y}] = \dfrac{\sigma_X^2}{m} + \dfrac{\sigma_Y^2}{n}$.

(3) For a confidence level of $1-\alpha$, the margin of error, ε, in estimating $\mu_X - \mu_Y$ with $\overline{W} = \overline{X} - \overline{Y}$ is given by,

$$\varepsilon = z_{\alpha/2}\sqrt{\frac{\sigma_X^2}{m} + \frac{\sigma_Y^2}{n}}.$$

(4) $\overline{W} \pm \varepsilon$ is a symmetric $1-\alpha$ confidence interval for $\mu_X - \mu_Y$.

Notes

(1) Since X and Y are independent the sample means $\overline{X}$ and $\overline{Y}$ are also independent, so that the results for the expectation and variance of $\overline{W}$ follow immediately. The margin of error ε is also calculated exactly as shown in Section 9.1.

(2) If the sample sizes m and n are sufficiently large (30 or more, the usual rule-of-thumb), then we can drop the assumption of normality for the populations and rely on the Central Limit Theorem to show that $\overline{W}$ is approximately normal. In this case we may also forego the assumption that the population variances are known and replace them by the sample variances S_X^2 and S_Y^2.

(3) If the confidence interval calculated in property (4) does ***not*** contain zero, then it is reasonable to conclude that the difference between sample means is statistically significant. This means that at the stated level of confidence, $\mu_X \neq \mu_Y$.

Exercise 9-22 The 100 meter dash time, *X*, of students from Lake Hazel Middle School is known to be normally distributed with standard deviation $\sigma_X = 1.8$ seconds. The 100 meter dash time, *Y*, of students from Meridian High School is known to be normally distributed with standard deviation $\sigma_Y = 1.6$ seconds. A random sample of eighty Lake Hazel students resulted in a sample mean 100 meter time of $\overline{X} = 14.3$ seconds. A random sample of 50 Meridian students yielded a sample mean 100 meter time of $\overline{Y} = 13.8$ seconds.

(a) Find the 95% confidence interval for the difference in 100 meter time between Lake Hazel and Meridian students, $\mu_X - \mu_Y$.

(b) Is the difference in sample means between Lake Hazel and Meridian statistically significant?

9.6.2 SMALL SAMPLES

Matters are considerably more vexing when we are dealing with small samples, especially if we must estimate the population variances by sample variances. As we saw in Section 9.3 we can no longer dispense with the assumption of normal populations. Equally important, we must also account for the errors introduced in using sample variances in place of population variances by reverting to the Student *t*-distributions.

Even with these provisos it is still difficult to come up with a reliable confidence interval estimation of $\mu_X - \mu_Y$ without one additional assumption, namely, that the two populations have equal variances. Hence, we restrict our efforts here to that case only.

The Difference of Means from Normal Populations, II

Let *X* and *Y* be independent normal populations with means μ_X and μ_Y, and with equal (but unknown) population variances σ^2. Let $(X_i)_{i=1}^m$ be a random sample from *X*, and let $(Y_j)_{j=1}^n$ be a random sample from *Y*, with sample means $\overline{X}$ and $\overline{Y}$, and sample variances S_X^2 and S_Y^2, respectively. Let $W = X - Y$, $\overline{W} = \overline{X} - \overline{Y}$, and denote $\mu_W = E[W] = \mu_X - \mu_Y$. Define the ***pooled sample variance*** S_P^2 by,

$$S_P^2 = \frac{(m-1)S_X^2 + (n-1)S_Y^2}{(m+n-2)}.$$

Then,

(1) $E[S_P^2] = \sigma^2$.

(2) $\dfrac{(m+n-2)S_P^2}{\sigma^2} \sim \chi^2(m+n-2).$

(3) $\dfrac{\overline{W}-\mu_W}{S_P\sqrt{\frac{1}{m}+\frac{1}{n}}} \sim t(m+n-2).$

(4) For a confidence level of $1-\alpha$, the margin of error, ε, in estimating $\mu_X-\mu_Y$ with $\overline{W}=\overline{X}-\overline{Y}$ is given by,

$$\varepsilon = t_{\alpha/2}(m+n-2)\cdot S_P\cdot\sqrt{\frac{1}{m}+\frac{1}{n}}.$$

(5) $\overline{W}\pm\varepsilon$ is a symmetric $1-\alpha$ confidence interval for $\mu_X-\mu_Y$.

Proof

Since each of S_X^2 and S_Y^2 is an unbiased estimator of σ^2 it follows easily that

$$E[S_P^2] = \frac{(m-1)E[S_X^2]+(n-1)E[S_Y^2]}{(m+n-2)} = \frac{(m-1)\sigma^2+(n-1)\sigma^2}{(m+n-2)} = \sigma^2.$$

For (2) we have

$$\frac{(m+n-2)S_P^2}{\sigma^2} = \frac{(m-1)S_X^2}{\sigma^2}+\frac{(n-1)S_Y^2}{\sigma^2} \sim \chi^2(m-1)+\chi^2(n-1) = \chi^2(m+n-2).$$

We have used the fact that the two chi-square random variables in the decomposition of $\frac{(m+n-2)S_P^2}{\sigma^2}$ are independent, so that their sum is also a chi-square.

For (3) we must make use of the fact that $\overline{W}$ and S_P^2 are independent, which is related to the independence of the sample mean and the sample variance within the individual populations. Then,

$$\frac{\overline{W}-\mu_W}{S_P\sqrt{\frac{1}{m}+\frac{1}{n}}} = \frac{\dfrac{\overline{W}-\mu_W}{\sigma\sqrt{\frac{1}{m}+\frac{1}{n}}}}{\sqrt{\dfrac{(m+n-2)S_P^2}{\sigma^2(m+n-2)}}} \sim t(m+n-2).$$

We have used the fact that $Var[\overline{W}]=\sigma^2\left(\frac{1}{m}+\frac{1}{n}\right)$ to conclude that $\dfrac{\overline{W}-\mu_W}{\sigma\sqrt{\frac{1}{m}+\frac{1}{n}}} \sim N(0,1)$, and that $\sqrt{\dfrac{(m+n-2)S_P^2}{\sigma^2(m+n-2)}}$ is the square root of an independent chi-square random variable divided by its degrees of freedom. This shows the definition of a Student t with $m+n-2$ degrees of freedom is satisfied.

Finally, (4) and (5) follow from the calculation,

$$1-\alpha = \Pr\left[\left|\overline{W}-\mu_W\right| \le \varepsilon\right] = \Pr\left[\frac{\left|\overline{W}-\mu_W\right|}{S_P\sqrt{\frac{1}{m}+\frac{1}{n}}} \le \frac{\varepsilon}{S_P\sqrt{\frac{1}{m}+\frac{1}{n}}}\right]$$

$$= \Pr\left[\left|t(m+n-2)\right| \le \frac{\varepsilon}{S_P\sqrt{\frac{1}{m}+\frac{1}{n}}}\right].$$

Thus, $t_{\alpha/2}(m+n-2) = \dfrac{\varepsilon}{S_P\sqrt{\frac{1}{m}+\frac{1}{n}}}$ and so $\varepsilon = t_{\alpha/2}(m+n-2)\cdot S_P.\sqrt{\frac{1}{m}+\frac{1}{n}}.$

Exercise 9-23 Information from independent samples from two normal groups with unknown, but identical variances, is summarized below. Find the 98% confidence interval for $(\mu_1 - \mu_2)$.

Group 1	Group 2
$\overline{X}_1 = 104$	$\overline{X}_2 = 97$
$s_1 = 12$	$s_2 = 17$
$n_1 = 12$	$n_2 = 16$

Exercise 9-24 The assistant athletic director in charge of compliance at Moe College assumed that grade point averages (GPA) of students from athletic teams are all approximately normally distributed with essentially the same variances, but different means.

The sample GPAs from the women's basketball team are:

3.30, 3.95, 2.25, 3.80, 4.00, 2.80, 3.41, 3.79, 2.00, 3.61, 3.64, 2.85, and 3.33.

The sample GPAs from the men's ice hockey team are:

3.16, 2.24, 1.75, 2.50, 3.94, 2.00, 2.00, 2.93, 3.56, and 0.60.

Find the 90% confidence level for estimating the difference in the population mean GPAs of women's basketball and men's ice hockey.

9.7 ESTIMATING THE SAMPLE SIZE

In this section we will look more closely at the relationship between the confidence level, the margin of error, and the sample size, in the case of estimating a population mean. These connections are most easily seen in estimating the population mean with a symmetric two-sided confidence interval using the standard normal table.

Suppose first we are sampling from a normal population, as in Section 9.1. In that case we had the relationship $\varepsilon = z_{\alpha/2} \cdot \frac{\sigma_X}{\sqrt{n}}$, where $1-\alpha$ is the confidence level, ε is the margin of error, and n is the sample size. First, note that as the confidence level increases, the tails of Z shrink, meaning that $z_{\alpha/2}$ is increasing. It follows that, keeping n fixed, the margin of error ε increases. This leads to our first principle.

- For a fixed sample size, the higher the confidence level, the larger the margin of error.

Looking at the flip side of this,

- For a fixed sample size, the smaller the margin of error, the lower the confidence level.

In situations where we want to control both the margin of error and the confidence level our only choice is to manipulate the sample size.

Example 9.7-1 Choosing a Sample Size

A school district wants to estimate the mean on a standardized test without administering the test to the entire school population. They decide to give the test to a random sample of students, but are uncertain how reliable the results will be. They hire a statistical consultant to determine how large the sample size needs to be in order to have 90% confidence in the result with a margin of error of 2 points. From previous testing it seems reasonable to assume that the population is normal with a standard deviation of 10 points. What is the minimum required sample size?

Solution

From the standard normal table we have $z_{.05} = 1.645$. From the equation $\varepsilon = z_{\alpha/2} \cdot \frac{\sigma_X}{\sqrt{n}}$ we find that $n = \left(\frac{z_{\alpha/2} \cdot \sigma_X}{\varepsilon}\right)^2 \geq \left(\frac{(1.645)\cdot(10)}{2}\right)^2 = 67.7$. Rounding up to the next whole number for n, we would recommend a sample of 68 students to achieve the desired reliability. ❐

In political polling the common standard is a margin of error of 2.5 percentage points at the 95% confidence level.

Example 9.7-2 Choosing a Sample Size

A polling company wants to estimate what percentage of people will vote for candidate A.

(a) How many people should be sampled in order to achieve a margin of error of .025 at the 95% confidence level?
(b) Calculate the sample size if we allow the margin of error to be 5%.

Solution

(a) From Section 9.5 we have the formula $\varepsilon = z_{\alpha/2} \cdot \sqrt{\frac{\hat{p}(1-\hat{p})}{n}}$, which leads to $n = \frac{(z_{\alpha/2})^2 \, \hat{p}(1-\hat{p})}{\varepsilon^2}$. At the 95% confidence level we have $z_{.025} = 1.96$. Since we do not have a value yet for $\hat{p}$ the safest bet is to take the maximum possible value for $\hat{p}(1-\hat{p})$, which is $\left(\frac{1}{2}\right)\left(\frac{1}{2}\right)=\frac{1}{4}=0.25$. Then $n=\frac{(1.96)^2(.25)}{(.025)^2}=1536.6$. Thus, to assure achieving the desired reliability we would need a sample of 1537 voters.

(b) Since n is inversely proportional to the square of the margin of error, doubling the margin of error in (a) reduces the sample size by a factor of 4. Thus, $n=\frac{1536.6}{4}=384.1$, which rounds up to a sample size of 385. □

Exercise 9-25 Assume a population is normal with a standard deviation of 25. How large of a sample should be selected to provide a 98% confidence interval with a margin of error of ±4?

Exercise 9-26 A random variable Y is known to be normally distributed with standard deviation $\sigma_Y = 17$. How large a sample must be taken so that the width of the 95% symmetric confidence interval is smaller than 4?

Exercise 9-27 You are interested in determining the average starting salary of your profession. You want to determine a 99% confidence interval with a margin of error of at most ±$500. You trust that starting salaries are normally distributed with a known standard deviation of $4200. What is the smallest random sample of salaries you will need to make your estimate?

9.8 CHAPTER 9 SAMPLE EXAMINATION

1. A random sample of the number of hours that John Q. Student spent working on his homework per week is summarized below:

Sample Week 1	Sample Week 2	Sample Week 3	Sample Week 4	Sample Week 5
3 hours	5.5 hours	2 hours	30 minutes	4 hours

   (a) Calculate the sample mean and sample standard deviation of the aforementioned hours.

   (b) Assuming a normal distribution, calculate a symmetric 95% confidence interval for the true average number of hours per week that John spends working on his homework.

2. A 95% confidence interval for the mean of a normal population was reported to be 112 to 120. If the population standard deviation is 16.07, then

   (a) What was the sample mean?

   (b) What was the sample size used in this study?

3. **SOA/CAS Exam 110 November 1981 #4**

A random sample $X_1,\ldots,X_n$ of size n is selected from a normal population with mean μ and standard deviation 1. Later an additional independent observation X_{n+1} is obtained from the same population. What is the distribution of $(X_{n+1}-\mu)^2 + \sum_{i=1}^{n}(X_i-\overline{X})^2$, where $\overline{X} = \frac{1}{n}\sum_{i=1}^{n} X_i$?

(A) Chi-square with $n-1$ degrees of freedom
(B) Chi-square with n degrees of freedom
(C) Chi-square with $n+1$ degrees of freedom
(D) F-distribution with $n-1$ and 1 degrees of freedom
(E) F-distribution with 1 and $n-1$ degrees of freedom

4. A highway safety engineer used a radar gun placed in the walkway over Highway 40 and observed the following speeds of cars in the fast lane. Assume speeds are approximately normally distributed.

72	63	59	70
63	57	63	69

(a) Calculate the sample mean and sample standard deviation of the aforementioned speeds.

(b) Calculate the 98% confidence interval for the true average speed of cars that drive in the fast lane on Highway 40 under this walkway.

(c) Calculate the 90% confidence interval for the standard deviation of speeds driven on Highway 40.

5. A rather large (more than 30) random survey of Josieville University students was conducted and the ages of participants were recorded. A statistician calculated a 92% confidence interval for the average age under the assumption that student ages form a normal distribution. The confidence interval came out to be $(20.9, 22.3)$ in years. The standard deviation of the sample was 2.8 years.

(a) What was the sample mean?

(b) What was the exact size of the sample?

6. Let $X_1,\ldots,X_n$ be a random sample from a population random variable X. Which of the following are valid formulas for the (unbiased) sample variance S^2:

$$(1) \qquad S^2 = \frac{\sum_{i=1}^{n} X_i^2 - \frac{1}{n}\cdot\left(\sum_{i=1}^{n} X_i\right)^2}{n-1}$$

(2) $$S^2 = \frac{\sum_{i=1}^{n} X_i^2 - n \cdot \overline{X}^2}{n-1}$$

(3) $$S^2 = \frac{\sum_{i=1}^{n} X_i^2 - \left(\sum_{i=1}^{n} X_i \right)^2}{n-1}$$

(A) (2) only
(B) (1) and (2) only
(C) (1) and (3) only
(D) (2) and (3) only
(E) (2) and (3)

7. **CAS Exam 3 Fall 2005 #5**

The following information is based on a sample of 10 observations from a normal distribution:

$$\sum_{i=1}^{10} X_i = 110 \qquad \sum_{i=1}^{10} X_i^2 = 1{,}282$$

Which of the following are 95% confidence intervals for σ^2?

(1) $(0.0, 21.6)$
(2) $(3.8, 26.7)$
(3) $(4.3, \infty)$

(A) (2) only
(B) (1) and (2) only
(C) (1) and (3) only
(D) (2) and (3) only
(E) (1), (2), and (3)

8. **SOA/CAS Exam 110 November 1981 #6**

Let $T = \frac{k(X+Y)}{\sqrt{Z^2+W^2}}$, where X, Y, Z, and W are independent normal random variables with mean 0 and variance $\sigma^2 > 0$. For exactly one value of k, T has a t-distribution. If r denotes the degrees of freedom of that distribution, then the pair (k, r) is equal to:

(A) $(1,1)$ (B) $(1,2)$ (C) $(\sigma, 2)$ (D) $\left(\frac{\sqrt{2}}{2}, 1\right)$ (E) $\left(\frac{\sqrt{2}}{2}, 2\right)$

9. **SOA/CAS Exam 110 May 1982 #13**

Suppose $X_1,\ldots,X_6$ and $Y_1,\ldots,Y_9$ are independent, identically distributed normal random variables, each with mean zero and variance $\sigma^2 > 0$. What is the 95th percentile of $\dfrac{\sum_{i=1}^{6} X_i^2}{\sum_{i=1}^{9} Y_i^2}$?

(A) 2.25 (B) 2.31 (C) 3.37 (D) 5.06 (E) 5.90

10. **SOA/CAS Exam 110 November 1981 #44**

Suppose $X_1,\ldots,X_6$ and $Y_1,\ldots,Y_{10}$ are 16 independent normal random variables, all with unknown mean. Suppose each X_i has variance σ_X^2, and each Y_j has variance σ_Y^2. If S_X^2 and S_Y^2 are the usual unbiased estimators of σ_X^2 and σ_Y^2, respectively, then $\Pr[3S_Y^2 > S_X^2 \mid \sigma_Y^2 = 2\sigma_X^2 > 0]$ is closest to:

(A) .995
(B) .990
(C) .975
(D) .950
(E) Cannot be determined from the given information

11. **SOA/CAS Exam 110 May 1985 #41**

Let $X_1,\ldots,X_{10}$ be the values of a random sample from a normal distribution with unknown mean μ and unknown variance $\sigma^2 > 0$. Let $\overline{X}$ be the sample mean and let $S^2 = \frac{1}{9}\sum_{i=1}^{n}(X_i - \overline{X})^2$. Which of the following is a 95% confidence interval for μ?

(A) $\left(\overline{X} - 2.26\frac{S}{\sqrt{10}}, \overline{X} + 2.26\frac{S}{\sqrt{10}}\right)$

(B) $\left(\overline{X} - 2.26\frac{S}{\sqrt{9}}, \overline{X} + 2.26\frac{S}{\sqrt{9}}\right)$

(C) $\left(\overline{X} - 2.23\frac{S}{\sqrt{10}}, \overline{X} + 2.23\frac{S}{\sqrt{10}}\right)$

(D) $\left(\overline{X} - 2.23\frac{S}{\sqrt{9}}, \overline{X} + 2.23\frac{S}{\sqrt{9}}\right)$

(E) $\left(\overline{X} - 1.83\frac{S}{\sqrt{10}}, \overline{X} + 1.83\frac{S}{\sqrt{10}}\right)$

12. **SOA/CAS Exam 110 May 1990 #9**

A random sample of 9 observations from a normal population yields the observed statistics $\overline{X} = 5$ and $\frac{1}{8}\sum_{i=1}^{9}(X_i - \overline{X})^2 = 36$. What are the symmetric 95% confidence limits for the population mean?

(A) 5 ± 2
(B) 5 ± 4
(C) 5 ± 4.24
(D) 5 ± 4.62
(E) 5 ± 12

13. **SOA/CAS Exam 110 May 1990 #20**

If X is a random variable with mean μ and variance σ^2, then $\mu - 2\sigma$ is called the lower 2σ point of X. Suppose a random sample X_1, X_2, X_3, X_4 is drawn from a chi-square distribution with two degrees of freedom. What is the lower 2σ point of $X_1 + X_2 + X_3 + X_4$?

(A) 0
(B) 2
(C) 4
(D) 6
(E) 8

CHAPTER 10

HYPOTHESIS TESTING

Hypothesis testing constitutes an alternative framework for using data in the determination of model parameters. The cornerstone of this approach is the idea of a ***null*** hypothesis, denoted H_0. In general terms the null hypothesis posits a baseline position or belief. Most often in the statistical context this baseline position or belief is stated as a particular value for a distribution parameter. This value may be based on previous testing, received wisdom, or simply guesswork. Whatever the case, the null hypothesis represents the commonly accepted or status quo value of the parameter. The objective of the test is to determine on the basis of new data whether or not the null hypothesis should be accepted, or rejected in favor of an alternative hypothesis, which we denote by H_1.

In a typical test, the data are used to construct a so-called ***test statistic***. If the value of this statistic falls within a specified range called the ***rejection region*** (or, alternatively, the ***critical region***) then the null hypothesis is rejected in favor of the alternative. If the test statistic is *not* inside the rejection region then the null hypothesis *cannot* be rejected. Some statisticians quibble over whether or not this is the same as *accepting* the null hypothesis. We consider this a question of semantics over substance, and will treat "cannot reject" as equivalent to "accept."

10.1 HYPOTHESIS TESTING FRAMEWORK

Before providing examples of how this works in practice, we note that the ever present vicissitudes of statistical variation prevent us from accepting or rejecting the null hypothesis with absolute certainty. There will always be a real possibility of error in the process. Specifically, there are two types of errors we may commit. A ***Type I*** error occurs if we reject the null hypothesis even though it is true. A ***Type II*** error occurs if we accept the null hypothesis even though it is false. The following chart shows the four possible outcomes.

	H_0 is true	H_0 is false
Reject H_0	Type I Error	Good Call
Accept H_0	Good Call	Type II Error

In practical applications each of the two types of errors has negative consequences, and the goal in setting up the test is to minimize their likelihood of occurrence to the extent possible. As we shall see, reducing the chance of a Type I error increases the chance of a Type II error, and vice versa, so there are tradeoffs to be made. We first illustrate the basic ideas using standard medical testing terminology.

Example 10.1-1 Type I and Type II Errors

A medical test checks for the presence of a particular disease in the patient. The null hypothesis is that the patient is healthy. The alternative hypothesis is that the patient has this disease. If the test is in the rejection region then we say the test is ***positive***. If the test is not in the rejection region we say the test is ***negative***. Describe the Type I and Type II errors in this context, and list a possible consequence of each.

Solution

Since the null hypothesis is that the patient is healthy, a Type I error occurs if we mistakenly conclude the patient is sick. This will happen if the test is positive, and we refer to this outcome as a ***false positive***. Similarly, a Type II error occurs if we mistakenly conclude that the patient is well. This happens if the patient is sick, but the test is negative. We refer to this outcome as a ***false negative***. We summarize the possibilities and possible consequences in tabular form.

	H_0 **is true (Patient well)**	H_0 **is false (Patient sick)**
Reject H_0	1. **Type I Error** 2. False Positive 3. Unnecessary treatment with potential damaging side effects	1. Test is positive 2. Correctly conclude patient is sick
Accept H_0	1. Test is negative 2. Correctly conclude patient is healthy	1. **Type II Error** 2. False negative 3. Patient is denied early detection and proper treatment

Clearly, both types of errors result from a misdiagnosis with potentially serious consequences. In this example we did not have enough information about the test to actually calculate the probabilities of Type I and Type II errors. ❐

Exercise 10-1 Consider the null hypothesis, H_0 : God exists.

(a) State the appropriate alternate hypothesis, H_1.

(b) What is the Type I error?

(c) What is the Type II error?

In the following subsection we will present examples illustrating the relationships between the hypotheses and the two types of errors. But first we will establish some standard terminology and notation.

Hypothesis Testing – Simple Hypotheses

Let X be a population random variable with an unknown parameter θ. Consider a hypothesis test consisting of the following hypotheses:

$$\begin{aligned} H_0 &: \theta = \theta_0 \\ H_1 &: \theta = \theta_1 \end{aligned}, \text{ where } \theta_0 \neq \theta_1.$$

Let $X_1, \ldots, X_n$ be a random sample of size n from X and let $\hat{\theta}$ be a test statistic calculated from the random sample. Let C denote the ***rejection region***, consisting of the values of the test statistic $\hat{\theta}$ for which the null hypothesis will be rejected. Then,

(1) α denotes the ***probability of a Type I error***, meaning that,

$$\alpha = \Pr[\text{rejecting } H_0 | \ H_0 \text{ true}] = \Pr[\hat{\theta} \in C \,|\, \theta = \theta_0].$$

α is also referred to as the ***significance level*** of the test.

(2) β denotes the ***probability of a Type II error***, meaning that,

$$\beta = \Pr[\text{accepting } H_0 | \ H_1 \text{ true}] = \Pr[\hat{\theta} \notin C \,|\, \theta = \theta_1].$$

The complementary probability, $1-\beta$, is called the ***power*** of the test. This means,

(3) The ***power*** $= 1-\beta = \Pr[\text{rejecting } H_0 | \ H_1 \text{ true}] = \Pr[\hat{\theta} \in C \,|\, \theta = \theta_1]$.

10.1.1 FINDING THE LIKELIHOOD OF TYPE I AND TYPE II ERRORS

We begin with an example illustrating the regions where Type I and Type II errors may occur.

Example 10.1-2 Illustrating Type I and Type II Error Regions

Let X be a normal population random variable with unknown mean μ and a standard deviation of 5. A random sample of size 25 is taken. Consider the hypothesis test,

$$\begin{aligned} H_0 &: \ \mu = 5 \\ H_1 &: \ \mu = 8 \end{aligned}.$$

Let the sample mean $\overline{X}$ be the test statistic. Draw a schematic diagram of the distribution of $\overline{X}$ under both the null and the alternative hypotheses, and indicate the values of $\overline{X}$ corresponding to Type I and Type II errors.

Solution

Since the population is normal we know that $\overline{X}$ is normal with a standard deviation of $\frac{\sigma_X}{\sqrt{n}} = \frac{5}{\sqrt{25}} = 1$. Under the null hypothesis the distribution of $\overline{X}$ is centered at 5, and under the alternative hypothesis it is centered at 8. The diagram below shows these two distributions side by side, with the null hypothesis in light gray and the alternative in dark gray.

Since $\mu_0 = 5 < 8 = \mu_1$, large values of $\overline{X}$ would be evidence *against* the null hypothesis. Thus, a plausible rejection region C for the null hypothesis would consist of the right-hand tail of the distribution for $\overline{X}$ under the assumption that $\mu = 5$. This region takes the form $\overline{X} > K$, for some K greater than 5. This is shaded in light gray in the diagram and under the null hypothesis, values of $\overline{X} > K$ result in a Type I error.

Type II errors occur under the assumption that the alternative hypothesis $(\mu = 8)$ is true, and consist of those values of $\overline{X}$ in the complement of C. Therefore, the region where Type II errors occur consists of $\overline{X} \le K$, and is shaded in dark gray on the alternative hypothesis graph, centered at 8.

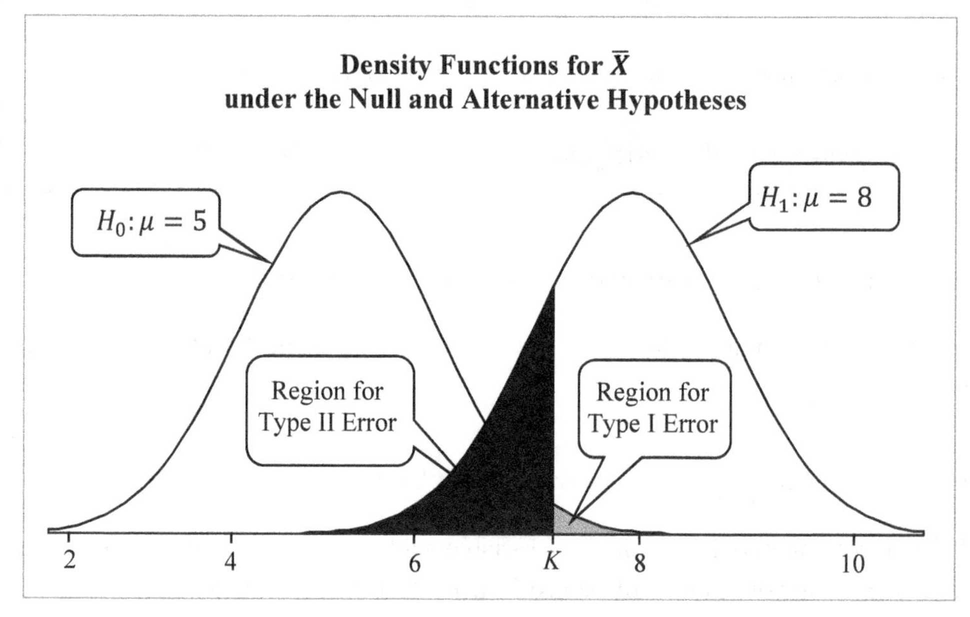

❐

Note

We will return to this example in Chapter 11. There, we will show how to determine the actual value for K, the number that demarcates the boundary of the rejection region C. This determination depends on the choices for α and β.

We next illustrate further the relationship between the two types of error and we calculate the likelihood of committing either a Type I or a Type II error. We present a scenario in which the hypotheses are simple and the test statistic has a familiar, easily calculated distribution.

Example 10.1-3 Calculating Type I and Type II Errors

A trick coin has been fabricated so that one side will come up in two-thirds of the tosses. No one knows or remembers whether that side is heads or tails. A statistical consultant is hired to help resolve the issue. She declares that p shall denote the probability of heads, and that she will conduct a hypothesis test as follows:

$$H_0: p = \frac{2}{3}$$

$$H_1: p = \frac{1}{3}$$

(a) The consultant decides to toss the coin 8 times and will reject the null hypothesis if the number of heads is less than or equal to r, where r is an integer between 0 and 8. Calculate the probability of a Type I error for each value of r.

(b) For each value of r between 0 and 8 calculate the probability of a Type II error.

(c) The consultant is informed that she must be more than 90% certain of her conclusion, meaning that both Type I and Type II errors must be less than 10%. She increases the number of tosses to 20 and will reject the null hypothesis if the number of heads is less than or equal to r. Find all values of r for which the probabilities of both Type I and Type II errors are less than 10%.

Solution

For parts (a) and (b) the test statistic is the random variable S equal to the number of heads in 8 tosses of the coin. Clearly, S is a binomial random variable with $n = 8$, and with probability of success p equal to either 2/3 or 1/3, depending on whether the null hypothesis or the alternative hypothesis is true. To facilitate writing the solutions we have calculated out the two relevant binomial distributions in the following tables. In the first table S denotes the binomial with $n = 8$ and $p = 2/3$. In the second table T is binomial with $n = 8$ and $p = 1/3$. Note that since failures in the first table correspond to successes in the second table, we have $S + T = 8$, so that $\Pr[S = k] = \Pr[T = 8 - k]$.

S binomial, $n = 8$, $p = 2/3$		
k	$\Pr[S = k]$	$\Pr[S \le k]$
0	0.0002	0.0002
1	0.0024	0.0026
2	0.0171	0.0197
3	0.0683	**0.0879**
4	0.1707	0.2586
5	0.2731	0.5318
6	0.2731	0.8049
7	0.1561	0.9610
8	0.0390	1.0000

T binomial, $n = 8$, $p = 1/3$		
k	$\Pr[T = k]$	$\Pr[T \le k]$
0	0.0390	0.0390
1	0.1561	0.1951
2	0.2731	0.4682
3	0.2731	**0.7414**
4	0.1707	0.9121
5	0.0683	0.9803
6	0.0171	0.9974
7	0.0024	0.9998
8	0.0002	1.0000

For part (a), the event of a Type I error corresponds to rejecting the null hypothesis, $H_0 : p = 2/3$, even though it is true. This in turn corresponds to the event $S \le r$ given that in fact $p = 2/3$. Thus, using conditional probability notation, we have,

$$\alpha = \Pr[\text{Type I error}] = \Pr\left[S \le r \mid p = \frac{2}{3}\right].$$

These are exactly the cumulative probabilities listed in the first table. For example, if we reject the null hypothesis when the number of heads in 8 tosses is less than or equal to 3 (see Figure 10.1.1), then the probability of a Type I error is 0.0879.

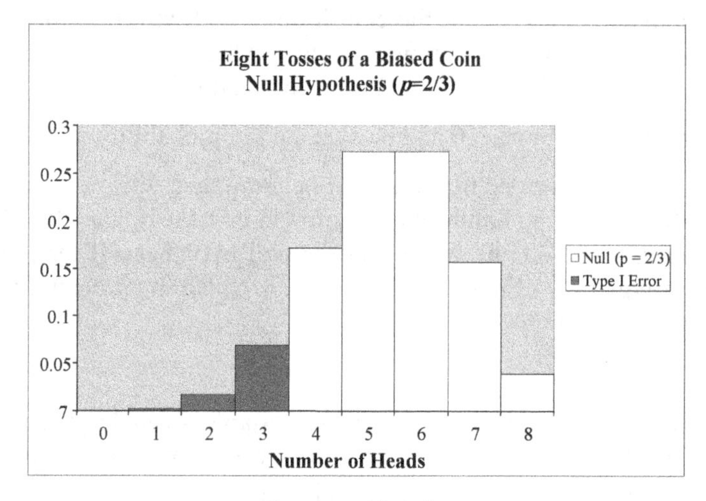

FIGURE 10.1.1

For part (b), the event of a Type II error corresponds to accepting the null hypothesis even though it is false. We now let T denote the number of heads in 8 tosses given that H_1 is true, that is, $p = 1/3$. Then, we will commit a Type II error if $T > r$ given that in fact $p = 1/3$. In this case we have,

$$\beta = \Pr[\text{Type II error}] = \Pr\left[T > r \mid p = \frac{1}{3}\right] = 1 - \Pr\left[T \le r \mid p = \frac{1}{3}\right].$$

Thus, for each r, the Type II error probabilities are the complements to the numbers in the cumulative probabilities of the second table above. For example again, if we reject the null hypothesis when the number of heads in 8 tosses is less than or equal to 3, then the probability of a Type II error is $1 - 0.7414 = 0.2586$. This region is shaded green in Figure 10.1.2.

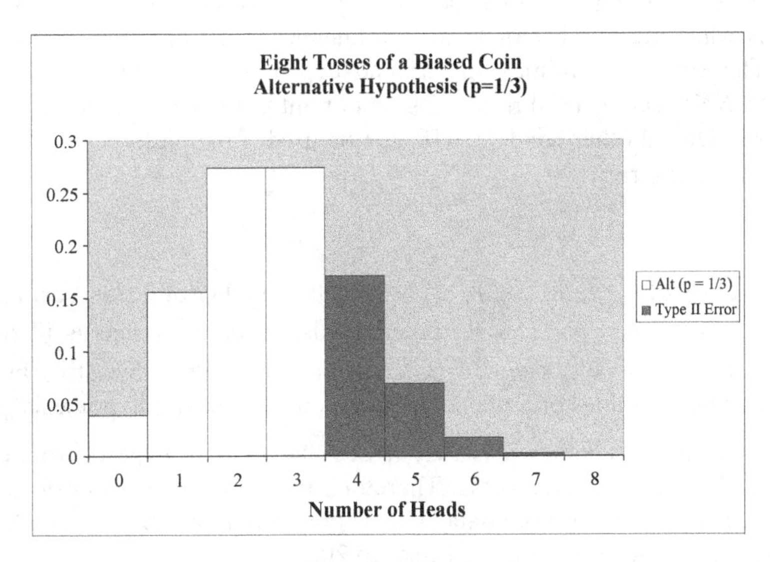

FIGURE 10.1.2

Here is a summary table showing the probabilities of Type I and Type II errors for each rejection region of the form $S \le r$.

r	Pr[Type I error]	Pr[Type II error]
0	0.0002	1 - 0.0390 = 0.9610
1	0.0026	1 - 0.1951 = 0.8049
2	0.0197	1 - 0.4682 = 0.5318
3	**0.0879**	**1 - 0.7414 = 0.2586**
4	0.2586	1 - 0.9121 = 0.0879
5	0.5318	1 - 0.9803 = 0.0197
6	0.8049	1 - 0.9974 = 0.0026
7	0.9610	1 - 0.9998 = 0.0002
8	1.0000	1 - 1.0000 = 0.0000

For part (c) the number of tosses is increased to $n = 20$. The only difference is that this time S is binomial with $n = 20$ and $p = 2/3$ or $p = 1/3$ depending on whether the null hypothesis or the alternative hypothesis is true. The table below shows an extract from these two binomial distributions.

r	Pr[Type I error] = $\Pr[S \le r \mid p = 2/3]$	$\Pr[S \le r \mid p = 1/3]$	Pr[Type II error] = $1 - \Pr[S \le r \mid p = 1/3]$
⋮	⋮	⋮	⋮
8	0.0130	0.8095	0.1905
9	**0.0376**	0.9081	**0.0919**
10	**0.0919**	0.9624	**0.0376**
11	0.1905	0.9870	0.0130
⋮	⋮	⋮	⋮

Thus, the consultant can reduce both Type I and Type II errors to under 10% by rejecting the null hypothesis when the number of heads is either less than or equal to 9 or less than or equal to 10. The error probabilities flip around depending on whether r is 9 or 10. The rejection region $S \leq 9$ can be used if the most important priority is to minimize the likelihood of a Type I error. On the other hand, $S \leq 10$ can be used if the preference is to minimize the likelihood of a Type II error. ❐

Notes

(1) Rejection regions of the form $S \leq r$, where S is the number of heads, seem sensible since the null hypothesis has $p = 2/3$. In other words, heads are twice as likely as tails, so smaller values of S would seem to be confirmation of the alternative hypothesis that $p = 1/3$, and larger values of S would seem to confirm the null hypothesis $p = 2/3$.

(2) From the two summary tables it is easy to see that smaller Type I errors correspond to larger Type II errors, and vice versa. Therefore, the only way to control for both Type I and Type II errors is to get more data. This is precisely how we approach the problem in part (c), by increasing the number of tosses to 20.

(3) There are lots of different ways to choose a rejection region if we are only interested in controlling the size of Type I errors. For example, in the case of 8 tosses, we have $\Pr[S = 2] = 0.0171$. Hence, the rejection region $S = 2$ leads to a Type I error probability of only 0.0171. The problem of course is that the rejection region $S = 2$ is a poor choice for controlling the likelihood of a Type II error. In fact, with the rejection region $S = 2$ we have $\Pr[\text{Type II error}] = \Pr[S \neq 2 \mid p = 1/3] = 1 - .2731 = .7269$. Furthermore, the rejection region $S = 2$ omits the outcomes $S = 0$ and $S = 1$, both of which would seem to be confirmation of the alternative hypothesis $p = 1/3$. In fact the rejection region $S \leq 2$ leads to an only slightly higher probability of 0.0197 for Type I errors and a significantly smaller probability of 0.5318 for Type II errors.

(4) In Chapter 11 we will look more deeply into the theory of hypothesis testing. We will then show definitively that in this problem, rejection regions of the form $S \leq r$ are in fact the most effective when controlling for both types of errors.

In Example 10.1-3 we were able to calculate the probabilities of both Type I and Type II errors because both the null and the alternative hypotheses were ***simple***. That is, each specified precisely one exact value of the population parameter, in this case p. In the standard examples to follow it is more typical to specify an exact value for the parameter in the null hypothesis, and then a complementary range of values for the alternative hypothesis. A hypothesis specifying a range of values for the parameter is called a ***composite hypothesis***.

For the remainder of this chapter we concentrate on standard applications of hypothesis tests and focus primarily on how to control the likelihood of Type I errors. We will investigate Type II errors more closely in Chapter 11.

Exercise 10-2 CAS Exam 3 Spring 2005 #24

Which of the following statements about hypothesis testing are true?

1. A Type I error occurs if H_0 is rejected when it is true.
2. A Type II error occurs if H_0 is rejected when it is true.
3. Type I errors are always worse than Type II errors.

(A) 1 only (B) 2 only (C) 3 only (D) 1 and 3 only (E) 2 and 3 only

10.1.2 THE SIGNIFICANCE LEVEL OF A TEST

When the alternative hypothesis is composite we can only calculate the probability of Type I errors. The only conclusion to be drawn from the experiment in this scenario is to either reject or not reject the null hypothesis, depending on if the test statistic is in the rejection region or not. In this type of test one usually specifies in advance what the allowable probability of a Type I error will be. This value is called the ***significance level*** of the test and it is equal to α, the probability that the test statistic is in the rejection region given that the null hypothesis holds.

Example 10.1-4 Testing the Mean in a Normal Distribution (Two Sided)

Let X be the score on a standardized test and assume X is a normally distributed population random variable whose standard deviation is 5. From past experience the population mean is thought to be 70. There is some concern that over time the population mean μ_X has drifted away from 70. In order to test for drift our statistical consultant is retained. She is instructed to be 95% certain of her conclusion. She proceeds to administer the test to a sample of 100 randomly chosen members of the population. The sample mean so obtained is 70.9. Set up and describe the results of a hypothesis test she can use for determining whether or not drift has occurred.

Solution

The null hypothesis is that the mean of X remains at 70, and the alternative hypothesis is that the mean has drifted away from 70. In symbols,

$$H_0: \ \mu_X = 70$$
$$H_1: \ \mu_X \neq 70$$

A reasonable choice for a test statistic is the sample mean $\overline{X}$. Since we are concerned about drift away from 70 in either direction, a logical choice of rejection region would be the event $|\overline{X} - 70| \geq K$, for some suitable $K > 0$. The right choice for K is determined by the mandate that the consultant be 95% certain of the conclusion. The consultant takes this to mean the significance level of the test is $\alpha = 5\%$. Since this is the same as the probability for a Type I error, we can write,

$$.05 = \Pr\left[|\overline{X} - 70| \geq K \mid H_0 \text{ true}\right] = \Pr\left[|\overline{X} - 70| \geq K \mid \mu_X = 70\right]$$

Since the mean of $\overline{X}$ is 70 and the standard deviation of $\overline{X}$ is $\frac{\sigma_X}{\sqrt{n}} = \frac{5}{\sqrt{100}} = 0.5$, we can standardize $\overline{X}$ to obtain,

$$.05 = \Pr\left[|\overline{X} - 70| \geq K \mid \mu_X = 70\right] = \Pr\left[\frac{|\overline{X} - 70|}{0.5} \geq \frac{K}{0.5}\right] = \Pr\left[|Z| \geq \frac{K}{0.5}\right].$$

As we saw in Chapter 9 (see, for instance, Example 9.1-1) this means the critical region in terms of the standard normal density function consists of the two tails, each with probability of 0.025. Thus,

$$\frac{K}{0.5} = z_{0.025} = 1.96, \quad \text{and} \quad K = (1.96)(0.5) = 0.98.$$

This means the rejection region consists of all values of $\overline{X}$ for which $|\overline{X} - 70| \geq 0.98$.

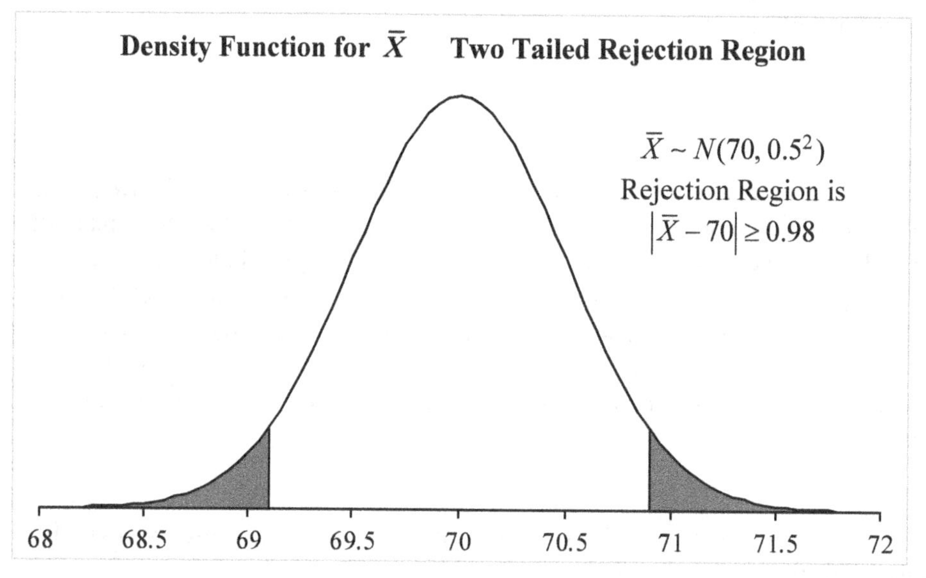

With a measured value of $\overline{X} = 70.9$, we have $|\overline{X} - 70| = |70.9 - 70| = 0.90 < K = 0.98$. Consequently, our value of $\overline{X}$ is **not** in the rejection region and the null hypothesis cannot be rejected. The consultant reports back that at the 95% confidence level (5% significance level) she **cannot** conclude that drift has occurred. ❒

Notes

(1) The null hypothesis specifies a single value of the parameter and hence it is a ***simple hypothesis***. The alternative hypothesis specifies a range of values, and is therefore a ***composite hypothesis***.

(2) The perceptive reader may detect a certain similarity between the calculations in the above example and the calculations for a confidence interval based on $\overline{X}$. Indeed, from Section 9.1 we have that the $1-\alpha$ confidence interval for μ_X is given as,

$$\overline{X} \pm \varepsilon = \left(\overline{X} - z_{\alpha/2} \cdot \frac{\sigma_X}{\sqrt{n}}, \overline{X} + z_{\alpha/2} \cdot \frac{\sigma_X}{\sqrt{n}} \right).$$

Hence, the 95% confidence interval centered at $\overline{X} = 70.9$ is $70.90 \pm 0.98 = (69.92, 71.88)$. Observe that this confidence interval **does** contain the null hypothesis value of $\mu_X = 70$. Below, we show that in general, rejecting the null hypothesis at the significance level of α is equivalent to having a $1-\alpha$ confidence interval that **misses** the null hypothesis value.

(3) Often in problems of this type it is not necessary to calculate the value of K and explicitly state the rejection region in terms of $\overline{X}$. One may state the equivalent rejection region in terms of the standard normal random variable Z that results from the standardized values of $\overline{X}$. The following example illustrates this.

Example 10.1-5 Testing the Mean in a Normal Distribution (One Sided)

Consider the same scenario as Example 10.1-4, except that this time the only concern is to test the assertion that the mean of test scores has actually **increased**. Using the same data and significance level, determine whether or not this is the case.

Solution

This time we state the hypothesis test as:

$$H_0: \ \mu_X = 70$$
$$H_1: \ \mu_X > 70$$

We will reject the null hypothesis in favor of the alternative providing $\overline{X}$ is sufficiently more than 70. That is, the rejection region is of the form $\overline{X} > 70 + K$, where, as in the previous example, K is determined by the significance level of 5%. When we standardize $\overline{X}$ this is equivalent to a rejection region in the 5% right-hand tail of Z. Since $z_{.05} = 1.645$, the rejection region consists of $z \geq 1.645$.

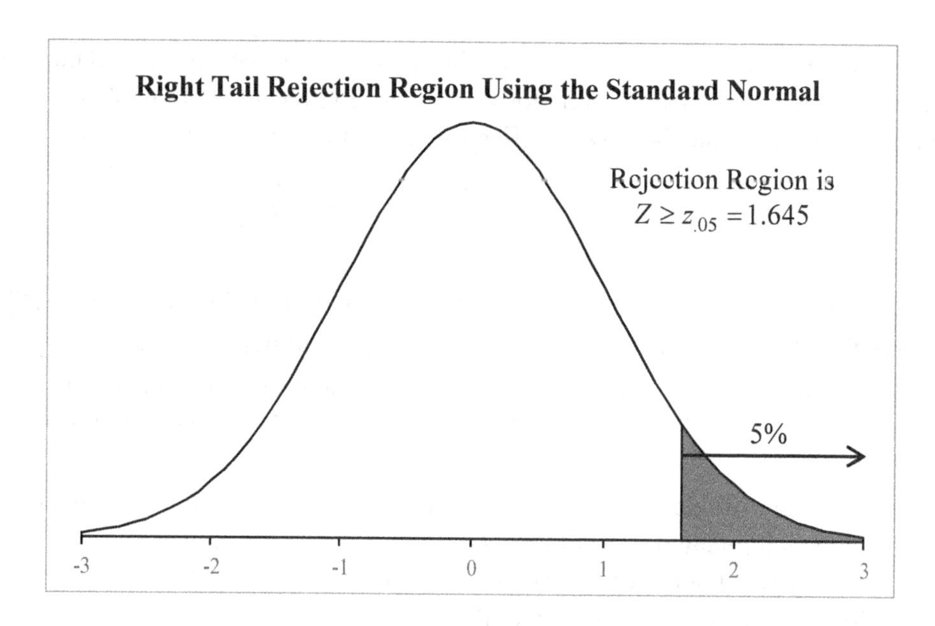

As in the previous example, under the null hypothesis the mean of $\overline{X}$ is 70, the standard deviation of $\overline{X}$ is 0.5, and the observed value of $\overline{X}$ is 70.9. Then $z = \frac{70.9 - 70}{0.5} = 1.80$. Since this is greater than 1.645, in this instance we conclude that at the 5% significance level we can reject the null hypothesis. That is, we conclude that mean test scores have in fact increased. ❐

Note
Once again we have a simple null hypothesis and a composite alternative. Since the alternative is one-sided some authors will write the null hypothesis as $H_0 : \mu_X \leq 70$, preferring to suggest that all possible values of the parameter are covered. However, even though the null hypothesis is stated as composite, the operative value is the boundary value, which in this case is $\mu_X = 70$. That is the value used in testing the null hypothesis.

The next example explores the duality relationship between hypothesis testing and confidence intervals for estimating the mean of a normal population. Although we won't attempt to explicitly state this connection in every possible application, we do want to point out that in most cases hypothesis testing and confidence intervals are two sides of the same coin. We illustrate the relationship using a normal population and a two-sided hypothesis test.

Example 10.1-6 Confidence Intervals and Hypothesis Testing

Let X be a normal random variable with known standard deviation σ_X and unknown mean μ_X. Let $\overline{X}$ be the sample mean from a random sample of size n. We wish to conduct the hypothesis test,

$$H_0 : \mu_X = \mu_0$$
$$H_1 : \mu_X \neq \mu_0$$

at the significance level α, using the test statistic $\overline{X}$. The critical region is of the form, $\left|\overline{X}-\mu_0\right| \geq K$, where K is chosen so that the probability of a Type I error is α. Show that the null hypothesis is rejected if and only if the $1-\alpha$ confidence interval for μ_X centered at $\overline{X}$ does **not** contain μ_0.

Solution

We first calculate the value of K. We have,

$$\alpha = \Pr\left[\left|\overline{X}-\mu_0\right| \geq K\right] = \Pr\left[\frac{\left|\overline{X}-\mu_0\right|}{\sigma_X/\sqrt{n}} \geq \frac{K}{\sigma_X/\sqrt{n}}\right] = \Pr\left[|Z| \geq \frac{K}{\sigma_X/\sqrt{n}}\right].$$

Hence, $\dfrac{K}{\sigma_X/\sqrt{n}} = z_{\alpha/2}$, so that $K = z_{\alpha/2}\dfrac{\sigma_X}{\sqrt{n}}$. This implies that the rejection region consists of all $\overline{X}$ such that $\left|\overline{X}-\mu_0\right| \geq K = z_{\alpha/2}\dfrac{\sigma_X}{\sqrt{n}}$. On the other hand, we note from Section 9.1 that the $1-\alpha$ confidence interval for μ_0 centered at $\overline{X}$ takes the form,

$$\left(\overline{X} - z_{\alpha/2}\cdot\frac{\sigma_X}{\sqrt{n}},\ \overline{X} + z_{\alpha/2}\cdot\frac{\sigma_X}{\sqrt{n}}\right).$$

Now, assume $\overline{X}$ is in the rejection region. Then $\left|\overline{X}-\mu_0\right| \geq z_{\alpha/2}\dfrac{\sigma_X}{\sqrt{n}}$ so that either $\mu_0 \leq \overline{X} - z_{\alpha/2}\cdot\dfrac{\sigma_X}{\sqrt{n}}$ or $\mu_0 \geq \overline{X} + z_{\alpha/2}\cdot\dfrac{\sigma_X}{\sqrt{n}}$. This shows μ_0 is **not** in the confidence interval

$$\left(\overline{X} - z_{\alpha/2}\cdot\frac{\sigma_X}{\sqrt{n}},\ \overline{X} + z_{\alpha/2}\cdot\frac{\sigma_X}{\sqrt{n}}\right).$$

Conversely, if μ_0 is **not** in the confidence interval then $\left|\overline{X}-\mu_0\right| \geq z_{\alpha/2}\dfrac{\sigma_X}{\sqrt{n}}$, so that $\overline{X}$ is in the rejection region. ❒

Exercise 10-3 CAS Exam 3 Fall 2007 #7

You are given the following information on a random sample:

- $Y = X_1 + \cdots + X_n$ where the sample size, n, is equal to 25 and the random variables are independent and identically distributed
- X_i has Poisson distribution with parameter λ
- $H_0 : \lambda = 0.1$
- $H_1 : \lambda < 0.1$
- The critical region to reject H_0 is $Y \leq 3$

Calculate the significance level of the test.

(A) Less than .50
(B) At least .50, but less than .60
(C) At least .60, but less than .70
(D) At least .70, but less than .80
(E) At least .80

10.1.3 THE *p*-VALUE OF A HYPOTHESIS TEST

Often, the result of a hypothesis test is conveyed by stating the so-called ***p-value*** of the test. The *p*-value is the probability that the test statistic is at least *as extreme as* the actual observed value, given that the null hypothesis is true. If α is the significance level of the test, then we reject the null hypothesis if and only if the *p*-value is smaller than α. The *p*-value conveys more information to the consumer of the test than simply stating whether or not the null hypothesis is rejected. For one thing, it quantifies how close we came to the boundary line between accepting or rejecting. Because of this, the *p*-value is often used as an alternative way of specifying the results of a hypothesis test.

To illustrate the idea, in Example 10.1-5 we tested the hypotheses,

$$\begin{aligned} H_0 &: \ \mu_X = 70 \\ H_1 &: \ \mu_X > 70 \end{aligned}$$

and came up with the test statistic value of $\overline{X} = 70.9$. The *p*-value is the probability that $\overline{X}$ could be as large as 70.9 (or more) under the assumption that the null hypothesis $(\mu_X = 70)$ is true. In symbols, we have, $p\text{-value} = \Pr\left[\overline{X} \geq 70.9 \mid \mu_X = 70\right]$. We will calculate this in the following example.

Similarly, in Example 10.1-4 we considered a two sided test under the same scenario, with a rejection region of the form $\left|\overline{X}-70\right| \geq K$. In this case, when we say "as extreme as" the observed value of $\overline{X} = 70.9$, we mean either $\overline{X} \geq 70.9 = 70{+}.9$ or $\overline{X} \leq 69.1 = 70{-}.9$. In other words in the two-sided case, $p\text{-value} = \Pr\left[\left|\overline{X}-\mu_X\right| \geq \left|70.9-70\right| = .9\right]$. This will be calculated as well in the next example.

Example 10.1-7 The *p*-values for Testing the Mean

We continue with the scenario of Examples 10.1-4 and 10.1-5 in which X is a normally distributed population random variable whose standard deviation is 5. A random sample of size 100 results in a sample mean $\overline{X}$ of 70.9.

(a) Consider the two-sided test, $\begin{aligned} H_0 &: \ \mu_X = 70 \\ H_1 &: \ \mu_X \neq 70 \end{aligned}$

conducted at the 5% level of significance. Calculate the resulting *p*-value and state whether or not the null hypothesis should be rejected.

(b) Consider the one-sided test, $H_0: \mu_X = 70$

$$H_1: \mu_X > 70$$

conducted at the 5% level of significance. Calculate the resulting p-value and state whether or not the null hypothesis should be rejected.

Solution

For part (a), since this is a two-sided test, a value of $\bar{X}$ at least as extreme as 70.9 means the event, $|\bar{X}-\mu_X| \geq |70.9-70| = .9$. We calculate this probability by standardizing $\bar{X}$ and using the standard normal table to determine probability. As in Examples 10.1-4 and 10.1-5, we have $\sigma_{\bar{X}} = \frac{\sigma_X}{\sqrt{n}} = \frac{5}{10} = 0.5$. Then,

$$\Pr\left[|\bar{X}-\mu_X| \geq |70.9-70| \mid \mu_X = 70\right] = \Pr\left[\frac{|\bar{X}-70|}{0.5} \geq \frac{|70.9-70|}{0.5}\right]$$

$$= \Pr[|Z| \geq 1.8] = 2[1-\Phi(1.8)] = 0.072.$$

Thus, the p-value for this result is 0.072. Since this is greater than the significance level of 0.05, we cannot reject the null hypothesis. Since the p-value is only slightly greater than 0.05 we can see that while the null hypothesis is not rejected, it is a fairly close call. This might prompt the investigator to seek more data to help resolve the issue.

Since part (b) involves only a one-sided test, the p-value is the probability of obtaining a value $\bar{X} \geq 70.9$. For this case we have that the p-value equals,

$$\Pr[\bar{X} \geq 70.9 \mid \mu_X = 70] = \Pr\left[Z \geq \frac{70.9-70}{0.5} = 1.8\right] = 1 - \Phi(1.8) = 0.036.$$

For this one-sided test we have the p-value of 0.036 is less than the significance level of 0.05, so we reject the null hypothesis. ❒

Exercise 10-4 A recent random survey found the following prices in cents per gallon of regular unleaded gasoline: 288.9, 286.9, 282.9, 282.9, 291.9 and 287.9. Assume prices follow a normal distribution with a population standard deviation of 3.5 cents. Is there sufficient evidence to claim that the average unleaded gas price is less than 288.9 cents? Give the p-value associated with this test.

We turn next to some of the standard applications of hypothesis testing, using the standard statistical distributions developed in Chapter 9 as our test statistics. We will approach these applications using a standard framework that we summarize in the following box.

General Procedure for Hypothesis Testing

Let X be a population random variable with an unknown parameter θ and let $X_1,\ldots,X_n$ be a random sample from X.

(1) Formulate the null and alternative hypotheses in terms of θ:

$H_0: \theta = \theta_0$ (stated as a simple hypothesis)
H_1: a composite hypothesis taking one of the forms,

(a) $\theta \neq \theta_0$ (two-tailed rejection region)
(b) $\theta > \theta_0$ (right-tailed rejection region)
(c) $\theta < \theta_0$ (left-tailed rejection region)

(2) Specify the significance level α.

(3) Specify a test statistic $\hat{\theta}$ based on the random sample $X_1,\ldots,X_n$. This test statistic should be formulated in such a way that its distribution follows one of the standard statistical random variables: z, t, χ^2, or F.

(4) Construct the ***rejection region*** using α, the distribution of $\hat{\theta}$, and the tail (tails) appropriate for the alternative hypothesis in (1).

(5) Evaluate the test statistic $\hat{\theta}$ using the observed values of the random sample $X_1,\ldots,X_n$ and determine whether or not it falls in the rejection region. Reject H_0 if $\hat{\theta}$ is in the rejection region and accept H_0 otherwise.

(6) If feasible, calculate the p-value of the test based on the calculated value of $\hat{\theta}$.

Note
In step (5) we are simply declaring a straight up-or-down decision on the null hypothesis based on where the test statistic falls. Calculating the p-value in step (6) supplements this information, and in some applications may be used as an alternative to stating a pre-ordained significance level α in step (2).

We will refer to the approach outlined in the box as the ***General Procedure***.

10.2 HYPOTHESIS TESTING FOR POPULATION MEANS

In this section we give a number of scenarios for testing hypotheses concerning the population mean. Let X be a population random variable with an unknown mean μ_X and let $X_1,\ldots,X_n$ be a random sample from X. In each of these scenarios the test statistic is based on the sample mean $\overline{X}$, but we will be using different tools depending on the distribution of X, the sample size n, and whether or not the population standard deviation σ_X is known or unknown.

The three different scenarios we study are,

(1) X is ***normal*** with a known standard deviation σ_X.

(2) The distribution of X is unknown but the sample size is sufficiently large so that we can conclude from the Central Limit Theorem that $\overline{X}$ is approximately normal.

(3) X is ***normal*** but the standard deviation σ_X is unknown and must be estimated using the sample standard deviation S.

In the first two scenarios we will be using the standard normal as the test statistic. In the third scenario we use a Student t test statistic. A typical rule-of-thumb is to use $n \geq 30$ for concluding that a sample is "sufficiently large" in the second scenario.

10.2.1 STANDARD NORMAL TESTS FOR THE POPULATION MEAN

The simplest case (scenario 1) occurs when the population X is assumed to be normal and the population standard deviation σ_X is given. This is exactly the situation we illustrated in Examples 10.1-4, 10.1-5, and 10.1-7. In this case $\overline{X}$ is normal with a standard deviation given by $\sigma_{\overline{X}} = \frac{\sigma_X}{\sqrt{n}}$. The ***test statistic***, Z, is given by,

$$Z = \frac{\overline{X} - \mu_X}{\sigma_{\overline{X}}} = \frac{\overline{X} - \mu_X}{\frac{\sigma_X}{\sqrt{n}}}$$

where μ_X is the value given in H_0.

Note

The quantity $\sigma_{\overline{X}} = \frac{\sigma_X}{\sqrt{n}}$ is sometimes referred to as the ***standard error*** of $\overline{X}$ instead of the ***standard deviation*** of $\overline{X}$.

We now repose the questions and give solutions following the outline in the General Procedure.

Example 10.2-1 Hypothesis Tests Using a Standard Normal Test Statistic, I

Let X be the score on a standardized test and assume X is a normally distributed population random variable whose standard deviation is 5. A random sample of size 100 is taken and the sample mean so obtained is 70.9.

(a) Consider the two-sided test, $H_0: \ \mu_X = 70$
$H_1: \ \mu_X \neq 70$

conducted at the 5% level of significance. Determine whether or not the null hypothesis should be rejected, and calculate the resulting p-value.

(b) Consider the one-sided test, $H_0: \ \mu_X = 70$
$H_1: \ \mu_X > 70$

conducted at the 5% level of significance. Determine whether or not the null hypothesis should be rejected, and calculate the resulting p-value.

Solution

The test statistic is given by Z and evaluates to $Z = \frac{\bar{X}-\mu_X}{\sigma_{\bar{X}}} = \frac{\bar{X}-\mu_X}{\frac{\sigma_X}{\sqrt{n}}} = \frac{70.9-70}{\frac{5}{\sqrt{100}}} = 1.8.$

For (a), we have a two-sided test for which the rejection region is given by $|Z| \geq z_{.025} = 1.96$. Since $Z = 1.8 < 1.96$ we cannot reject the null hypothesis. To calculate the p-value, we find that $\Pr[|Z| \geq 1.8] = 2 \cdot [1-\Phi(1.8)] = 2 \cdot [1-0.9641] = .0719$. Since this probability is <u>greater</u> than the significance level of 0.05, we do <u>not</u> reject the null hypothesis.

For (b) we have a rejection region given by $Z \geq z_{.05} = 1.645$. Since $Z = 1.8 > 1.645$, we reject the null hypothesis. This time the p-value is given by $\Pr[Z \geq 1.8] = 1-\Phi(1.8) = 1-0.9641$ $= .0359$. Because this is <u>less</u> than the significance level of 0.05 we <u>reject</u> the null hypothesis.

□

In the second scenario we make use of the Central Limit Theorem in order to treat $\bar{X}$ as at least approximately normal. In this scenario we will also be estimating the population standard deviation σ_X with the sample standard deviation S. As previously mentioned, many statistics textbooks consider sample sizes of at least 30 to be sufficiently large to use the Central Limit Theorem approximation for the distribution of $\bar{X}$. In this case approximating the population standard deviation with the sample standard deviation S is considered safe for most practical purposes.

Example 10.2-2 Hypothesis Tests Using a Standard Normal Test Statistic, II

A New York taxi cab company advertises an average waiting time for a cab outside of a popular Manhattan hotel of no more than 4 minutes. A hotel doorman, who is also an amateur statistician, suspects the advertised time of 4 minutes stretches the truth and that the waiting time is in fact significantly longer. He takes a random sample of 30 arrivals and determines that the sample mean is 5.75 minutes with a sample standard deviation of 5.85. Set up the appropriate hypothesis test and evaluate the cab company's claim according to the data at the 5% significance level. Determine the p-value for the observed data.

Solution
The doorman/statistician assumes that the sample mean will be approximately normal and that the sample standard deviation will be close to the population standard deviation. Let W be the waiting time random variable. Since the doorman believes the mean time is greater than 4, this is a one-sided right tail test. The hypotheses are:

$$H_0: \mu_W = 4$$
$$H_1: \mu_W > 4$$

Under the assumptions we use Z as the test statistic, with a rejection region of $Z > z_{.05} = 1.645$. From the data we have $Z \approx \frac{5.75-4}{5.85/\sqrt{30}} = 1.638 < 1.645$, so we cannot reject the null hypothesis. The cab company claim, while shaky, still stands after this test. The p-value, however, calculated as $\Pr[Z > 1.638] = 1 - \Phi(1.638) = 0.051$ is very close to the rejection value of 0.05. ❐

Example 10.2-3 The Doorman's Conundrum

The doorman of Example 10.2-2 wonders if his assumptions might be questionable, so he decides to consult an actuary of his acquaintance. From past experience she thinks it is reasonable to assume that the waiting time random variable has an exponential distribution. Thus, rather than using the Central Limit Theorem to approximate the distribution of $\overline{W}$, she recommends using the actual distribution based on an exponential population.

Reexamine the hypothesis test using the exact distribution of $\overline{W}$ and use this distribution to find the actual p-value of the data.

Solution
Under the assumption that W is exponential we can treat the wait for a cab as a Poisson process. From properties of the Gamma distribution and its relationship with Poisson processes described in Section 6.6, we ascertain that under the null hypothesis the sum, $\sum_{i=1}^{30} W_i$, of 30 independent observations, is a Gamma distribution with parameters $\alpha = 30$ and $\beta = 4$. That is, $\sum_{i=1}^{30} W_i \sim \Gamma(30,4)$. We will use a chi-square distribution as our test statistic. It follows from property (5) of the Gamma distribution (see Section 6.6) that,

$$\overline{W} = \frac{1}{30}\sum_{i=1}^{30} W_i \sim \Gamma(30, 4/30),$$

and therefore,

$$\frac{60}{4}\overline{W} = 15\overline{W} \sim \Gamma\left(30, 15\cdot\frac{4}{30}\right) = \Gamma(30,2) = \chi^2(60).$$

We can use a spreadsheet program or a χ^2 table to find that the 95th percentile of $\chi^2(60) = \chi^2_{.05}(60) = 79.082$. The null hypothesis is rejected for values of $\overline{W}$ such that $15 \cdot \overline{W} > \chi^2_{.05}(60) = 79.082$. Now, the observed value of the standardized test statistic is $(15)\cdot(5.75) = 86.25 > 79.082$, and consequently we are justified in rejecting the null hypothesis that $\mu_W = 4$.

Most tables don't list enough values of χ^2_α to permit accurate calculations of p-values. However, we can use the Microsoft Excel® function CHIDIST to find the p-value corresponding to 86.25 is .0148, well below the 5% threshold for rejecting the null hypothesis. ❐

Notes

(1) We could have used $\overline{W}$ itself as the test statistic, making use of the Excel functions GAMMADIST and GAMMAINV. We chose instead to scale $\overline{W}$ to be chi-square simply because the chi-square is one of our standard, tabulated statistical distributions.

(2) Since the assumed population distribution is exponential, a one parameter family, we did not need to make use of the sample standard deviation in constructing the test statistic.

(3) If we assume that the null hypothesis is true and that W is exponential then we can compare the distributions of $\overline{W}$ under the two different approaches. Under these assumptions we have that $\mu_W = \sigma_W = 4$. In the first approach, using the Central Limit Theorem, we have that $\overline{W}$ is ***approximately*** normal with a mean of 4 and a standard deviation of $\frac{4}{\sqrt{30}}$. In the second approach we have that $\overline{W}$ is ***exactly*** Gamma with parameters $\alpha = 30$ and $\beta = \frac{4}{30}$.

For purposes of comparison we plotted the two density functions on the same chart. The graphs are pretty close but exhibit enough of a gap to be readily apparent to the naked eye.

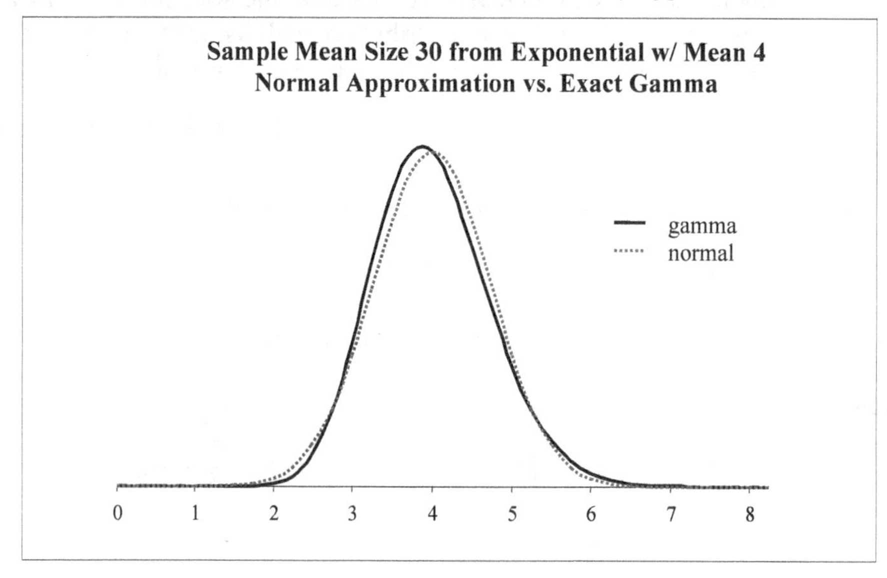

(4) An important point to the two doorman exercises is to illustrate the problematic nature of using a Z test when the population distribution and the standard deviation are unknown, especially when the sample size is borderline small according to the usual rule-of-thumb. The values we gave in Example 10.2-2 of 5.75 for the sample mean and 5.85 for the sample standard deviation arose from a simulation in Excel® of 30 values from an exponential distribution with an actual mean of 6. Repeated simulated samples of size 30 from the same exponential distribution of mean 6 showed that the two different approaches (Z test versus chi-square) resulted in different conclusions about rejecting the null hypothesis in approximately 15% of the trials.

10.2.2 STUDENT *t* TESTS FOR THE POPULATION MEAN

We now proceed to our third scenario in which we assume that the population X is normal but that the population standard deviation must be estimated using S, the sample standard deviation. In this case the test statistic is the Student t with degrees of freedom equal to $n-1$, where n is the sample size. This is in keeping with the result in Section 9.3 where it was shown that if X is normal then,

$$\frac{\overline{X}-\mu_X}{S/\sqrt{n}} \sim t(n-1).$$

This scenario is generally used for small sample sizes ($n<30$). For larger sample sizes we have some overlap with the second scenario in that the Central Limit Theorem shows once again that the test statistic is approximately normal. Indeed, most Student t tables are terminated or severely abridged at $n=30$. If, however, advanced calculators or statistical software is used, then the correct approach is to use the Student t for exact results regardless of sample size. Remember though, this scenario is predicated on the population being normal.

Example 10.2-4 Hypothesis Tests Using the Student *t*

Let X be the score on a standardized test and assume X is a normally distributed population random variable. A random sample of size 30 is taken and the sample mean so obtained is 70.9. Moreover, the sample standard deviation is calculated to be 5.2.

(a) Consider the two-sided test, $H_0: \ \mu_X = 70$

$$H_1: \ \mu_X \neq 70$$

conducted at the 5% level of significance. Determine whether or not the null hypothesis should be rejected and calculate the resulting p-value.

(b) Consider the one-sided test, $H_0: \ \mu_X = 70$

$$H_1: \ \mu_X > 70$$

conducted at the 5% level of significance. Determine whether or not the null hypothesis should be rejected and calculate the resulting p-value.

Solution
We calculate the Student t statistic as

$$t(29) = \frac{\overline{X} - \mu_X}{\frac{S}{\sqrt{n}}} = \frac{70.9 - 70}{\frac{5.2}{\sqrt{30}}} = .948.$$

For the two-tailed test in (a) the rejection region is given by $|t(29)| > t_{.025}(29) = 2.045$. Hence, we cannot reject the null hypothesis. We use the Excel function TDIST to find the p-value of 0.35, well above the .05 threshold for rejecting the null hypothesis.

In part (b) the rejection region is given by $t(29) > t_{.05}(29) = 1.699$. Since the t-statistic of .948 is smaller than 1.699, once again we do not reject the null hypothesis. The corresponding p-value is .1755 (half of the p-value in the two-tail case of part (a)). □

Note

If we had used the standard normal instead of the Student t then the rejection region in (a) would be $|z| > z_{.025} = 1.96$, and in (b) $z > z_{.05} = 1.645$. These are reasonably close to the corresponding Student t values and our conclusions would be the same.

Exercise 10-5 CAS Exam 3, Segment 3L Spring 2008 #6
You are given a random sample from a normal loss distribution:

- The sample mean is 42,000
- The sample standard deviation is 8,000
- There are 25 loss observations in the sample

Using a two-sided test with $H_0 : \mu = 45,000$ versus $H_1 : \mu \neq 45,000$, at which value of α would you reject the null hypothesis?

(A) Reject at $\alpha = .01$
(B) Do not reject at $\alpha = .01$, but reject at $\alpha = .02$
(C) Do not reject at $\alpha = .02$, but reject at $\alpha = .05$
(D) Do not reject at $\alpha = .05$, but reject at $\alpha = .1$
(E) Do not reject at $\alpha = .1$.

Exercise 10-6 CAS Exam 3 Fall 2006 #7
A random sample of 21 observations from a normal distribution yields the following results:

(i) $\overline{X} = 3.5$

(ii) $\frac{\sum_{i=1}^{21}(X_i - \overline{X})^2}{20} = 0.6156$

You are testing the following hypotheses:

$$H_0: \mu = 3$$
$$H_1: \mu \neq 3$$

Calculate the p-value for this test.

(A) Less than 0.002
(B) At least 0.002, but less than 0.004
(C) At least 0.004, but less than 0.006
(D) At least 0.006, but less than 0.008
(E) At least 0.008

Exercise 10-7 A survey claims that the average cost of a hotel room in Tulsa, Oklahoma, is less than \$49.21. In order to test this claim, a researcher selects a sample of 26 hotel rooms and finds the average cost equal to \$47.37 with a sample standard deviation of \$6.42. Assume the population random variable is normal. At $\alpha = 0.05$, comment on the claim.

Exercise 10-8 A survey reports that the average cost of a hotel room in Tulsa Oklahoma is less than \$49.21. In order to test this claim, a researcher selects a sample of 24 hotel rooms and finds the sample average cost equal to \$47.37 with a standard deviation of \$3.42. At $\alpha = 0.01$, comment on the claim.

Exercise 10-9 Assume the starting salary for diesel mechanics follows a normal distribution with a mean of \$52,000. Suppose that a random sample of nine recent graduates of the High Plains Technology Institute was taken, resulting in a sample average for their starting salary of \$55,000 with a sample standard deviation of \$4,000. Can one conclude at $\alpha = .01$ that High Plains students are more valued than the general starting diesel mechanic?

10.3 HYPOTHESIS TESTING FOR POPULATION VARIANCE

Assume that X is a normal population and that $X_1,\ldots,X_n$ is a random sample of size n. We now wish to construct a hypothesis testing protocol when the unknown parameter is population variance σ_X^2, or equivalently, the population standard deviation, σ_X. The obvious choice for the test statistic is a suitably standardized version of the sample variance $S^2 = \frac{1}{n-1}\sum_{i=1}^{n}(X_i - \bar{X})^2$. Following the results of Subsection 9.2.2, we have that,

$$\frac{(n-1)S^2}{\sigma_X^2} \sim \chi^2(n-1) = \Gamma\left(\frac{n-1}{2}, 2\right).$$

Example 10.3-1 Chi-Square Test for Population Variance

A random sample of size 21 is taken from a normal population and used to test the following hypotheses at the 5% level of significance. The sample variance is calculated to be 16. Determine whether or not the null hypothesis may be rejected.

(a) $H_0 : \sigma_X^2 = 10$

$H_1 : \sigma_X^2 \neq 10$

(b) $H_0 : \sigma_X^2 = 10$

$H_1 : \sigma_X^2 > 10$

Solution

We use the fact that $\frac{(n-1)S^2}{\sigma_X^2} = \frac{20S^2}{\sigma_X^2} \sim \chi^2(20)$. Then, using the data and assuming the null hypothesis to be true, we have $\chi^2(20) = \frac{(n-1)S^2}{\sigma_X^2} = \frac{(20)(16)}{10} = 32$.

In (a) we have a two-tailed test and the rejection region will consist of $\chi^2(20) < \chi^2_{.975}(20) = 9.59$, or $\chi^2(20) > \chi^2_{.025}(20) = 34.17$.

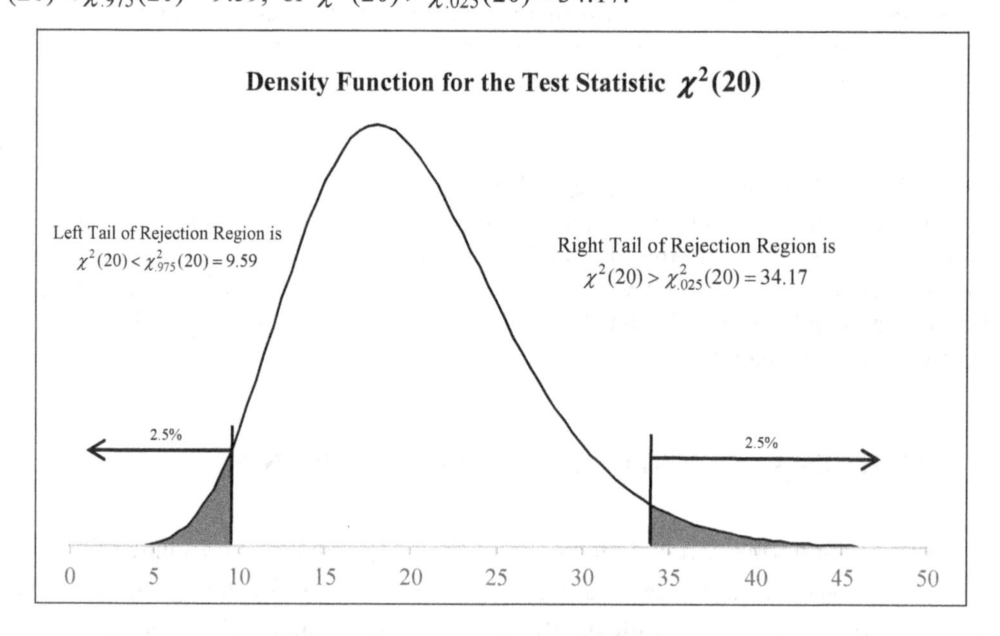

Therefore, with a test statistic of 32, the null hypothesis cannot be rejected.

In (b) we have a right-tailed test, so that the rejection region consists of

$$\chi^2(20) > \chi^2_{.05}(20) = 31.41.$$

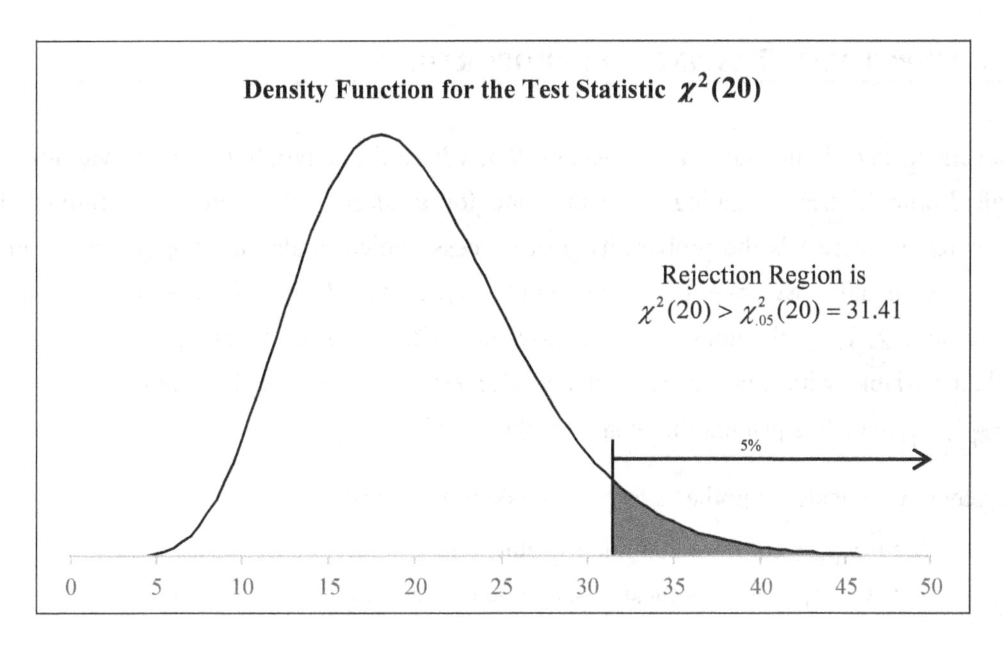

In this case we can reject the null hypothesis since the test statistic value of 32 exceeds 31.41.[1] ❐

Note

The hypotheses may be formulated in terms of the population standard deviation σ_X rather than the variance σ_X^2, but nothing essential changes in the solution procedure. For example, if the null hypothesis had been $H_0 : \sigma_X = \sqrt{10}$ and we were given that the sample standard deviation is 4, then the test statistic and rejection region would remain the same.

Exercise 10-10 CAS Exam 3 Fall 2005 #8

Claim sizes for a certain line of business are known to follow a normal distribution. A sample of claim sizes from this line of business are shown below:

Claim Number	Claim Size
1	3.3
2	5.4
3	7.1
4	8.9
5	23.5
6	29.8

For the hypothesis $\sigma^2 < 50$, in which of the following ranges does the p-value fall?

(A) $p \le 0.005$

(B) $0.005 < p \le 0.010$

(C) $0.010 < p \le 0.025$

(D) $0.025 < p \le 0.050$

(E) $p > 0.050$

---

[1] Note that the graphs are not drawn exactly to scale; the tails have been widened in order to better illustrate the rejection regions.

10.4 HYPOTHESIS TESTING FOR PROPORTIONS

The context here is the same as in Section 9.5, where the population random variable is a single Bernoulli trial X taking the value one for a success and zero for a failure. The parameter of interest is the probability p of success, which is also the population mean. In order to estimate p we take a random sample $X_1,\ldots,X_n$. This is a series of 0's and 1's, whose sum, $\sum_{i=1}^{n} X_i$, is the number of successes in n trials. This, of course, is just the binomial random variable with parameters p and n. Our estimator of p is the binomial proportion $\hat{p} = \frac{1}{n}\sum_{i=1}^{n} X_i$, which is just another name for the sample mean, $\overline{X}$.

We may now consider hypothesis tests for p taking the form:

$H_0 : p = p_0$ (stated as a simple hypothesis)

H_1 : a composite hypothesis taking one of the forms:

(a) $p \neq p_0$ (two-tailed rejection region)

(b) $p > p_0$ (right-tailed rejection region)

(c) $p < p_0$ (left-tailed rejection region)

Under the assumption that the sample is sufficiently large, we can invoke the Central Limit Theorem in order to treat the test statistic $\frac{\hat{p}-p_0}{\sigma_X/\sqrt{n}}$ as approximately normal.

Since X is a Bernoulli trial we have $\sigma_X = \sqrt{p(1-p)}$. Since we are assuming the null hypothesis to be true, the test statistic takes the form:

$$Z \approx \frac{\hat{p} - p_0}{\sqrt{\frac{p_0(1-p_0)}{n}}}.$$

Example 10.4-1 Testing a Poll Result

The campaign manager for a political candidate hears that a popular radio talk-show host has taken an informal poll of listeners that shows the manager's candidate trailing his opponent 45% to 55%. Strongly suspecting that the radio poll is biased against his candidate, he commissions a more scientific poll in order to show that his candidate is running significantly stronger than the radio poll suggests. His pollster canvases a random sample of 1250 likely voters with the result of this second poll showing that the race is dead even at 50% to 50%.

(a) Conduct a hypothesis test to determine whether or not the radio poll result should be rejected at the 5% significance level.

(b) Calculate the p-value of the test result.

Solution
Since the manager sets out to prove that his candidate is polling ***higher*** than 45% this is a one-sided right-tail test taking the form,

$$H_0 : p = .45$$
$$H_1 : p > .45$$

With the information supplied in the statement of the problem we have,

$$Z \approx \frac{\hat{p} - p_0}{\sqrt{\frac{p_0(1-p_0)}{n}}} = \frac{.5 - .45}{\sqrt{\frac{.45 \cdot (1-.45)}{1250}}} = 2.1251.$$

At the 5% significance level the rejection region is $Z > z_{.05} = 1.645$. Therefore, we can reject the null hypothesis that the candidate is favored by only 45% of likely voters.

The p-value of the second poll result (given the null hypothesis is true) is

$$\Pr[z > 2.1251] = .017.$$

❐

Exercise 10-11 CAS Exam 3 Spring 2005 #22
Drivers are classified as "good" or "bad." Results of the classification are assumed to be binomially distributed, with probability of being "good" equal to p. A sample consists of 100 drivers. Determine the critical value for testing the hypothesis $p < 0.5$ with significance level α of at most 0.05 using the normal approximation.

(A) Less than 52
(B) At least 52, but less than 54
(C) At least 54, but less than 56
(D) At least 56, but less than 58
(E) 58 or more

Exercise 10-12 Aaron was shooting his Koosh® basketball into the hoop above his door. He claims that he is an 80% shooter, $p = .80$. In a sample, Aaron made 25 of 36 shoots. Comment on his claim at the 5% significance level.

Exercise 10-13 Your college reports that at least 60% of all freshmen skip class on Friday mornings. Checking with teachers, a researcher found that only 43 of 80 freshmen skipped their Friday morning math class. Is there enough evidence to support the college report at $\alpha = 5\%$?

10.5 HYPOTHESIS TESTING FOR DIFFERENCES IN POPULATION MEANS

This section parallels the development of Section 9.6, covering confidence intervals for differences between means. The context consists of two independent populations for which we desire to compare population means. For large samples we use a Z test on the difference in sample means. This case encompasses testing for differences between population proportions. For small samples taken from normal populations the test statistic will be the Student t, as derived in Subsection 9.6.2.

10.5.1 LARGE SAMPLES

Here is a summary of the relevant results, using the machinery from Subsection 9.6.1.

Testing the Difference Between Population Means

Let X and Y be independent populations with means μ_X and μ_Y, and standard deviations σ_X and σ_Y, respectively. Let $(X_i)_{i=1}^{m}$ be a random sample from X, and let $(Y_j)_{j=1}^{n}$ be a random sample from Y with sample means $\overline{X}$ and $\overline{Y}$, and sample standard deviations S_X and S_Y, respectively. We assume that the sample sizes are greater than or equal to 30.

Let $W = X - Y$, with $\overline{W} = \overline{X} - \overline{Y}$, and let μ_{W_0} denote the null hypothesis value of $\mu_W = \mu_X - \mu_Y$. Then,

(1) $E[\overline{W}] = E[W] = \mu_X - \mu_Y = \mu_W$

(2) $Var[\overline{W}] = Var[\overline{X}] + Var[\overline{Y}] \approx \dfrac{S_X^2}{m} + \dfrac{S_Y^2}{n}$

(3) The *test statistic*, Z, used for determining whether to accept or reject the null hypothesis is given by:

$$\boxed{Z \approx \frac{\overline{W} - \mu_{W_0}}{\sqrt{\dfrac{S_X^2}{m} + \dfrac{S_Y^2}{n}}}}$$

Example 10.5-1 Testing the Difference in Means

A poll of 504 families in group A showed that they donated an average of \$2,400 to charities in the previous year, with a standard deviation of \$1,400. A similar poll of 348 families belonging to the group B (a population independent from group A) showed they donated an average of \$2,200 to charities, with a standard deviation of \$1,200.

(a) Determine whether or not there is a statistically significant difference in charitable giving between the two groups at the 1% level of significance.

(b) Calculate the p-value of the test.

Solution

Since the sample sizes are sufficiently large we can proceed with a Z test on the difference in sample means. Since the problem asks only whether or not there is a difference between the populations we set up a two-tailed test as follows:

$$H_0: \mu_W = \mu_X - \mu_Y = 0 \qquad H_1: \mu_W = \mu_X - \mu_Y \neq 0.$$

(a) The test statistic is $Z \approx \dfrac{\overline{W} - \mu_{W_0}}{\sqrt{\frac{S_X^2}{m} + \frac{S_Y^2}{n}}} = \dfrac{(2400-2200)-0}{\sqrt{\frac{1400^2}{504} + \frac{1200^2}{348}}} = 2.2323.$

The rejection region is given by $|Z| \geq z_{.005} = 2.5758$. Therefore, we <u>cannot</u> reject the null hypothesis that the two population means are the same at the 1% level of significance.

(b) The p-value equals $\Pr\left[|Z| \geq 2.2323\right] = 2 \cdot [1 - \Phi(2.2323)] = 2.56\%.$ ❐

The same approach may be used for differences in proportions, with the only adjustment coming in how the population standard deviations are estimated. In that regard we follow the procedure of Section 10.4, using estimates of the form $\sqrt{\frac{\hat{p}(1-\hat{p})}{n}}$. The most common problem of this type involves testing whether or not two different population proportions are equal (item (4) below).

Testing the Difference Between Population Proportions

Let X and Y be independent Bernoulli trial populations with means p_X and p_Y, respectively. Let $(X_i)_{i=1}^m$ be a random sample from X, and let $(Y_j)_{j=1}^n$ be a random sample from Y. Let $\hat{p}_X = \overline{X}$ and $\hat{p}_Y = \overline{Y}$ be the proportion of successes in the X sample and the Y sample, respectively. We assume that the sample sizes are greater than or equal to 30.

Let $(p_X - p_Y)_0$ denote the null hypothesis value of the difference in population means, and let $\overline{W} = \overline{X} - \overline{Y} = \hat{p}_X - \hat{p}_Y$. Then,

(1) $E[\overline{W}] = p_X - p_Y$.

(2) $Var[\overline{W}] \approx \dfrac{\hat{p}_X(1-\hat{p}_X)}{m} + \dfrac{\hat{p}_Y(1-\hat{p}_Y)}{n}$.

(3) The test statistic used for determining whether to accept or reject the null hypothesis is given by: $Z \approx \dfrac{\hat{p}_X - \hat{p}_Y - (p_X - p_Y)_0}{\sqrt{\frac{\hat{p}_X(1-\hat{p}_X)}{m} + \frac{\hat{p}_Y(1-\hat{p}_Y)}{n}}}$.

(4) If the null hypothesis is $H_0: p_X = p_Y$ (equivalently, $(p_X - p_Y)_0 = 0$) then the common value of p is estimated using a pooled proportion of successes given by:

$$\overline{p} = \frac{\sum_{i=1}^{m} X_i + \sum_{j=1}^{n} Y_j}{m+n} = \frac{m\hat{p}_X + n\hat{p}_Y}{m+n},$$

and the test statistic becomes, $Z \approx \dfrac{\hat{p}_X - \hat{p}_Y}{\sqrt{\overline{p}(1-\overline{p}) \cdot \left(\frac{1}{m} + \frac{1}{n}\right)}}$.

Example 10.5-2 Testing the Difference in Proportions

A researcher at a large university theorized, on the basis of his observations, that a larger proportion of women were smokers than of men. He conducted a randomized poll of the school's students to ascertain their true smoking habits. Of the 125 male respondents, 36 identified themselves as smokers. Of the 162 females polled, 58 were smokers.

(a) Test the researcher's hypothesis at the 5% level of significance.
(b) Find the p-value of the poll data.

Solution

(a) We let X denote the Bernoulli trial population random variable for males, with *success* signifying a smoker and *failure* signifying a non-smoker. Let Y be similarly defined for the population of females. The hypotheses are:

$$H_0 : p_X = p_Y, \text{ equivalently } H_0 : (p_X - p_Y)_0 = 0$$
$$H_1 : p_X < p_Y, \text{ equivalently } H_1 : p_X - p_Y < 0.$$

Since the researcher is testing the proposition that smoking is <u>more</u> prevalent among females, the alternative hypothesis is left-tailed with the way X and Y are defined. We have $\hat{p}_X = \frac{36}{125} = .29$ and $\hat{p}_Y = \frac{58}{162} = .36$. The pooled proportion of smokers under the null hypothesis is

$$\overline{p} = \frac{\sum_{i=1}^{m} X_i + \sum_{j=1}^{m} Y_j}{m+n} = \frac{36+58}{125+162} = .3275.$$

The Z-statistic is given by,

$$Z \approx \frac{\hat{p}_X - \hat{p}_Y}{\sqrt{\overline{p}(1-\overline{p}) \cdot \left(\frac{1}{m} + \frac{1}{n}\right)}} = \frac{\frac{36}{125} - \frac{58}{162}}{\sqrt{.3275(1-.3275) \cdot \left(\frac{1}{125} + \frac{1}{162}\right)}} = -1.253.$$

The left-tail rejection region is given by $Z < -z_{.05} = -1.645$. Therefore, the null hypothesis of equal proportions of smokers between males and females cannot be rejected.

(b) The p-value is $\Pr[Z < -1.253] = 1 - \Phi(1.253) = 10.5\%$. ❒

10.5.2 SMALL SAMPLES

As indicated in Chapter 9, testing for differences in means using small samples is considerably more complicated. We follow the development in Subsection 9.6.2, where we make the simplifying assumptions that,

(1) the two independent populations being tested are normal, and
(2) the two populations have the same population standard deviations.

The relevant conclusions from Chapter 9 follow:

The Difference of Means from Normal Populations II

Let X and Y be independent normal populations with means μ_X and μ_Y, and with equal (but unknown) population variances σ^2. Let $(X_i)_{i=1}^m$ be a random sample from X, and let $(Y_j)_{j=1}^n$ be a random sample from Y, with sample means $\overline{X}$ and $\overline{Y}$, and sample variances S_X^2 and S_Y^2. Let $W = X - Y$ with $\mu_W = \mu_X - \mu_Y$, and let $\overline{W} = \overline{X} - \overline{Y}$. Define the ***pooled sample variance*** by $S_P^2 = \dfrac{(m-1)S_X^2 + (n-1)S_Y^2}{(m+n-2)}$.

Then,

(1) $E[S_P^2] = \sigma^2$.

(2) $\dfrac{(m+n-2)S_P^2}{\sigma^2} \sim \chi^2(m+n-2)$.

(3) $\dfrac{\overline{W} - \mu_W}{S_P\sqrt{\frac{1}{m} + \frac{1}{n}}} \sim t(m+n-2)$.

The Student t statistic in (3) is used for testing hypotheses concerning the difference between the population means.

Example 10.5-3 Small Sample Differences in Population Mean

An administrator at the Central School District was interested in whether or not there was a measurable difference in mean SAT math scores between the AP calculus students at Ivanhoe High School versus those at Lancelot High School. When the records were searched, it turned out that a random sample of students in Ivanhoe's AP calculus class had a mean SAT score of 622 and a random sample of students in the Lancelot AP calculus class had a mean SAT score of 688. The administrator concluded that the difference of 66 points was indeed significant and real.

A statistical consultant reviewed the data and ascertained that the sample sizes were small, consisting of 12 students at Ivanhoe and 16 students at Lancelot. She also calculated the sample standard deviations, finding a value of 120 for Ivanhoe and 96 for Lancelot. She decided to test the administrator's ad hoc conclusion at the 5% level of significance under the assumption that the two relevant populations are normal and have the same standard deviation.

(a) Set up the pertinent hypothesis test and discuss the results in light of the administrator's conclusion.

(b) Calculate the p-value of the test, using statistical software.

Solution

(a) The administrator was interested in a "measurable difference" between the two schools, so we assume the difference could have gone either way. Hence, we use the two-tailed hypothesis test,

$$H_0: \ \mu_X = \mu_Y \qquad\qquad H_1: \ \mu_X \neq \mu_Y$$

where X represents the Ivanhoe population random variable and Y that of Lancelot. The consultant should use the Student t test, calculating a pooled variance of,

$$S_P^2 = \frac{(m-1)S_X^2+(n-1)S_Y^2}{(m+n-2)} = \frac{(12-1)\cdot 120^2+(16-1)\cdot 96^2}{(12+16-2)} = 11,409.23.$$

Thus, the pooled sample standard deviation is $S_P = 106.81$, and the test statistic is:

$$t(m+n-2) = t(26) = \frac{\overline{W}-\mu_W}{S_P \cdot \sqrt{\frac{1}{m}+\frac{1}{n}}} = \frac{(622-688)}{106.81 \cdot \sqrt{\frac{1}{12}+\frac{1}{16}}} = -1.618.$$

For a two-tailed test at the 5% level of significance the rejection region is,

$$|t(26)| \geq t_{.025}(26) = 2.0555.$$

Therefore, we cannot reject the null hypothesis that there is no difference in the sample means from the two schools.

(b) The p-value, calculated using the TDIST function in Excel® is given by,

$$\Pr\left[|t(26)| > 1.618\right] = 11.77\%.$$ ❐

A related situation involves two populations, but instead of being independent the two populations are matched. A typical situation might involve a "before and after" effect. For example, our school district administrator might be interested in the effect of an 8-week exam prep course on SAT math scores. Let X be the prior score and let Y be the score after the prep course for a random individual. Clearly X and Y are not independent, but if we assume that the scores have normal distributions then the pair (X, Y) has a bivariate normal distribution and the difference $W = Y - X$ is normal. Thus, we can apply a t test to a sample of differences to test the null hypothesis of zero difference versus the one-tailed alternative that the mean of W is positive.

Example 10.5-4 Matched Pairs Student *t* Test

Ten high school students were randomly chosen to participate in a study of the benefits of participating in a test preparation program. The first row in the following table represents their scores before the prep course, and the second row the scores after the prep course. Assume the scores constitute normal random variables, and use the t test to determine the p-value of the results in testing the null hypothesis of no change in score versus the alternative hypothesis that the scores tend to increase after the test prep course.

Student ID	1	2	3	4	5	6	7	8	9	10
X (Score before prep)	421	551	500	595	671	435	384	694	235	599
Y (Score after prep)	463	611	476	617	729	454	441	726	225	709

Solution:

Let $W = Y - X$ and take the differences in the matched pairs as a random sample $W_1,\ldots,W_{10}$ from W. That is, $W_1 = 463 - 421 = 42$, and so forth. Then $\overline{W} = 36.6$ and $S_W = 38.34$. This results in a t-value of $t(9) = \dfrac{\overline{W} - 0}{S/\sqrt{10}} = 3.0191$. Again, using the TDIST function in Excel®, we arrive at a one-tail p-value of 0.0072. This strongly suggests that there is a positive effect in using the exam prep c ourse.

Exercise 10-14 CAS Exam 3, Segment 3L Fall 2008 #11

You are given the following:

- $X_1,\ldots,X_{11}$ is a random sample of size 11 from a normal distribution with mean $= \mu_1$ and variance $= \sigma_1^2$, with sample mean 2, and the unbiased sample variance $= 4$.
- $Y_1,\ldots,Y_{13}$ is a random sample of size 13 from a normal distribution with mean $= \mu_2$ and variance $= \sigma_2^2$, with sample mean 1.5, and the unbiased sample variance $= 5$.
- $H_0 : \mu_1 = \mu_2$
- $H_1 : \mu_1 > \mu_2$
- The variance is unknown but is assumed to be the same between samples, (i.e., $\sigma_1^2 = \sigma_2^2$)

Calculate the test statistic to evaluate the null hypothesis and select the answer below that contains the observed significance level or p-value.

(A) Less than 0.5%
(B) At least 0.5%, but less than 1.0%
(C) At least 1.0%, but less than 2.5%
(D) At least 2.5%, but less than 5.0%
(E) At least 5.0%

Exercise 10-15 CAS Exam 3 Spring 2007 #29

You are given the information below:

- $X_1,\ldots,X_{11}$ is a random sample from a normal distribution and the sample mean is 10, the sample variance is 4, and the sample size is 11.
- $Y_1,\ldots,Y_{11}$ is a random sample from a normal distribution and the sample mean is 9, the sample variance is 12, and the sample size is 11.
- $H_0 : \mu_X = \mu_Y$
- $H_1 : \mu_Y > \mu_X$
- Assume that the underlying variance of the two normally distributed random variables are equal.
- Let t be the sample t-statistic used to test the difference of the means from two normally distributed random samples.
- Let T be the critical value at $\alpha = 0.05$.

Calculate the absolute value of the difference between the sample t-statistic and the critical value at $\alpha = 0.05$, $|t - T|$.

(A) Less than 0.75
(B) At least 0.75, but less than 0.85
(C) At least 0.85, but less than 0.95
(D) At least 0.95, but less than 1.05
(E) At least 1.05

Exercise 10-16 Lori claims that the female business students are much more intelligent than male business students. In fact, she believes female business students score at least 5 points higher on I.Q. tests than their male counterparts. A random sample was taken and the results are summarized below.

Female	Male
$\overline{X} = 104$	$\overline{Y} = 97$
$s_X^2 = 12$	$s_Y^2 = 17$
$n_X = 22$	$n_Y = 26$

Can one draw any conclusions regarding Lori's claim at the 90% confidence level? What assumptions are you making?

10.5.3 THE *F* TEST FOR EQUAL POPULATION VARIANCES

An important consideration in applying the Student t test for the difference in means from normal populations is whether or not the populations have equal variances. We can use the results of Section 9.4 in order to devise a statistical test for this assumption using the F-distribution for the ratio of two sample variances.

The applicable result from Chapter 9 is the following:

Application of the *F*-Distribution

Let X and Y be independent normal populations with variances of σ_X^2 and σ_Y^2, respectively. Let $(X_i)_{i=1}^m$ be a random sample from X, and let $(Y_j)_{j=1}^n$ be a random sample from Y, with resulting sample variances S_X^2 and S_Y^2, respectively. Then,

$$\frac{S_X^2 / \sigma_X^2}{S_Y^2 / \sigma_Y^2} \sim F(m-1, n-1).$$

Example 10.5-4 Testing the Hypothesis that Population Variances Are Equal

In Example 10.5-3, X represented test scores from Ivanhoe and Y represented test scores from Lancelot. Here is the relevant summary of the data:

	X	Y
Sample Size	12	16
Sample S.D.	120	96

Assuming independent, normal populations, use this information to test the following hypotheses at the 10% level of significance.

$$H_0 : \sigma_X = \sigma_Y$$
$$H_1 : \sigma_X \neq \sigma_Y .$$

Solution

Using the data summary we have,

$$F(m-1,n-1) = F(11,15) = \frac{S_X^2 / \sigma_X^2}{S_Y^2 / \sigma_Y^2} = \frac{120^2}{96^2} = 1.5625,$$

with the population variance terms canceling each other under the null hypothesis assumption.

Using an F table we find a rejection region consisting of,

$$F(11,15) < F_{.95}(11,15) = .3678,$$

and

$$F(11,15) > F_{.05}(11,15) = 2.5068 .$$

Since

$$F_{.95}(11,15) = .3678 < 1.5625 < 2.506 = F_{.05}(11,15),$$

we have no basis for rejecting the null hypothesis of equal variances. ❐

Exercise 10-17 CAS Exam 3 Spring 2007 #30

You are given the information below:

- $X_1,\ldots,X_8$ is a random sample, where X is normally distributed, the sample mean is 4, the sample variance, S_X^2, is 16.
- $Y_1,\ldots,Y_9$ is a random sample, where Y is normally distributed, the sample mean is 3, the sample variance, S_Y^2, is 9.
- $H_0 : \sigma_X^2 = \sigma_Y^2$
- $H_1 : \sigma_X^2 > \sigma_Y^2$
- Let F be the critical value for a one-sided test with significance level $\alpha = .05$.
- Let f be the test statistic calculated from the sample.
- The mean and variance for the two random samples are unknown.

Calculate the test statistic, f, and the critical value, F, then find the range below that contains the absolute value of the difference of the test statistic, f, and the critical value, F, $|f-F|$.

(A) Less than 0.6
(B) At least 0.6, but less than 1.2
(C) At least 1.2, but less than 1.8
(D) At least 1.8, but less than 2.4
(E) At least 2.4

Exercise 10-18 CAS Exam 3 Spring 2006 #7

The independent random variables X and Y are from separate normal distributions. Random samples from each distribution are shown below:

X	Y
1	1
5	4
9	5
10	5
17	15
36	

Using the sample data and a 5% significance level, which of the following are true statements regarding hypothesis testing?

I. Reject $H_0:\sigma_X^2=50$ in favor of $H_a:\sigma_X^2>50$
II. Reject $H_0:\sigma_Y^2=50$ in favor of $H_a:\sigma_Y^2<50$
III. Reject $H_0:\sigma_X^2=\sigma_Y^2$ in favor of $H_a:\sigma_X^2>\sigma_Y^2$

(A) I only (B) II only (C) I and II (D) I and III (E) II and III

10.6 CHI-SQUARE TESTS

There are many hypothesis tests that make use of the chi-square distribution. In this section we will illustrate several of the main applications of this technique. The context here generally involves a discrete distribution, either of a single random variable, or of a joint distribution of two variables. The data will be presented in the usual tabular form for discrete distributions, but using frequencies of occurrence rather than probabilities.

In these hypothesis tests there will be two versions of the frequency table. One version will consist of the *actual* data. The other table will consist of *expected frequencies*, calculated according to an assumption given by the null hypothesis. The null hypothesis is tested by constructing a statistic based on squared differences between the *actual* table frequencies and the *expected* table frequencies. This statistic, under reasonable restrictions, will be approximately chi-square in distribution, and the null hypothesis will be rejected if this statistic is sufficiently large.

With continuous random variables the data can be "discretized" by stating the frequencies for disjoint intervals of data. For example, data for a continuous age-at-death distribution can be presented as frequency of deaths within 5 year intervals.

We will now illustrate these ideas, applying them first to the concept of contingency tables.

10.6.1 CONTINGENCY TABLES

Suppose we have conducted a large national survey covering many variables and we want to analyze the relationship between two particular variables, let's say annual income levels and geographic location. The question we want to answer is whether or not these two variables are independent. The null hypothesis will be that they are independent, while the alternative hypothesis says the distribution of income is related to geographic location.

The vehicle for analyzing this question is the table exhibiting the joint frequency distribution between the two variables. Such a tabular display is called a ***contingency table***, or less formally, a ***crosstab***. In Excel it is called a ***pivot table***. Here is an example illustrating (make believe) survey data on 1,000 households, broken down by 4 income categories and 5 major U.S. cities.

TABLE 10.6.1

	Los Angeles	Denver	St. Louis	Pittsburgh	Boston	Row Total
Poverty	25	12	18	15	30	**100**
Lower Middle	55	25	55	65	100	**300**
Upper Middle	90	55	110	90	155	**500**
Wealthy	30	8	17	30	15	**100**
Column Total	**200**	**100**	**200**	**200**	**300**	**1000**

This table shows the *actual* frequency counts. The associated joint probability distribution is obtained by dividing each cell by the overall total of 1000. The null hypothesis is that, as a joint probability distribution, the two marginal random variables are independent.
Recall that the condition for independence is that, as a joint probability table, the contents of the cells on the inside (the joint probabilities) are the products of the associated probabilities on the margins. Under the null hypothesis of independence, we assume the marginal frequencies are given, and that the inside of the table is calculated accordingly. We demonstrate using Table 10.6.1.

Example 10.6-1 Calculating the Expected Frequencies

You are given the total row and column counts for the two variables, income-level and geographic location. Calculate the interior frequencies based on the assumption of independence between income-level and geographic location. The template is given by,

	Los Angeles	Denver	St. Louis	Pittsburgh	Boston	Row Total
Poverty						**100**
Lower Middle						**300**
Upper Middle						**500**
Wealthy						**100**
Column Total	**200**	**100**	**200**	**200**	**300**	**1000**

Solution

Let e_{11} denote the expected frequency in the upper-left cell corresponding to the frequency of Poverty in Los Angeles. As probabilities, we require for independence that,

$$\frac{e_{11}}{1000} = \Pr[\text{Poverty and Los Angeles}] = \Pr[\text{Poverty}] \cdot \Pr[\text{Los Angeles}] = \left(\frac{100}{1000}\right) \cdot \left(\frac{200}{1000}\right).$$

Thus, $e_{11} = \frac{(100) \cdot (200)}{1000} = 20$. The other cells are calculated similarly. The resulting table of expected frequencies is,

	Los Angeles	Denver	St. Louis	Pittsburgh	Boston	Row Total
Poverty	20	10	20	20	30	**100**
Lower Middle	60	30	60	60	90	**300**
Upper Middle	100	50	100	100	150	**500**
Wealthy	20	10	20	20	30	**100**
Column Total	**200**	**100**	**200**	**200**	**300**	**1000**

□

Note

For a shortcut, the cells in the first three rows and the first four columns can be calculated in this fashion, leaving the cells in the last row and last column to be filled in by simple addition/subtraction using the required row and column totals.

For the general situation we assume that X is the row variable and takes r values, and that Y is the column variable, taking c values. A commonly accepted notation is to use f_{ij} for the actual frequency, and e_{ij} for the expected frequency, in the cell corresponding to the ith row and the jth column. The ith row total is denoted $f_{i\bullet}$, and the jth column total is denoted $f_{\bullet j}$. We denote the total overall frequency by simply f (some authors use the double dot notation, $f_{\bullet\bullet}$).

Then the *actual* and *expected* frequency tables take the form,

	y_1	$\cdots$	y_c	
x_1	f_{11}		f_{1c}	$f_{1\bullet}$
$\vdots$				$\vdots$
x_r	f_{r1}		f_{rc}	$f_{r\bullet}$
	$f_{\bullet 1}$	$\cdots$	$f_{\bullet c}$	f

	y_1	$\cdots$	y_c	
x_1	e_{11}		e_{1c}	$f_{1\bullet}$
$\vdots$				$\vdots$
x_r	e_{r1}		e_{rc}	$f_{r\bullet}$
	$f_{\bullet 1}$	$\cdots$	$f_{\bullet c}$	f

Testing for Independence in an *r* by *c* Contingency Table

Let X and Y be discrete jointly distributed random variables with observed (***actual***) frequencies as given in the f_{ij} table above. Let e_{ij} be the frequencies resulting from the assumption that X and Y have independent marginal probability distributions.

(1) $e_{ij} = \dfrac{f_{i\bullet} \cdot f_{\bullet j}}{f}$; for $i = 1, \ldots, r$ and $j = 1, \ldots, c$.

Let the test statistic

$$T = \sum_{i=1}^{r} \sum_{j=1}^{c} \frac{(f_{ij} - e_{ij})^2}{e_{ij}}.$$

Then,

(2) $T \approx \chi^2((r-1)\cdot(c-1))$.

(3) The hypothesis of independence between X and Y can be rejected at the α level of significance providing $T \geq \chi^2_\alpha((r-1)\cdot(c-1))$.

❐

Notes

(1) As in Example 10.6-1, the formula for calculating e_{ij} follows from the fact that rescaled as probabilities the entries on the inside of the table are the products of the numbers in the corresponding marginal cells. Thus, $\frac{e_{ij}}{f} = \left(\frac{f_{i\bullet}}{f}\right) \cdot \left(\frac{f_{\bullet j}}{f}\right)$, and so $e_{ij} = \frac{f_{i\bullet} \cdot f_{\bullet j}}{f}$.

(2) The ***test statistic*** T represents the sum of suitably standardized square deviations between the actual frequencies and the expected frequencies under the assumption of independence. Thus, the null hypothesis of independence is rejected if T is sufficiently large.

(3) Since T is only approximately chi-square, there is the chance of gross error if certain conditions are not met. There is a rule of thumb that the frequencies in the individual cells should be sufficiently large. This is often expressed by requiring that each e_{ij} be

greater than or equal to 5. If this is not the case in the original data then categories of X or Y should be grouped to make sure it is the case.

(4) The expression $(r-1)\cdot(c-1)$ for the degrees of freedom can be rationalized as follows: In the table of actual frequencies suppose that the row and column sums (in other words, the marginal frequencies) must remain fixed, but that the inside frequencies f_{ij} are free to "wiggle." How many of them are free to wiggle before the row and column sum constraints kick in? A moment's thought will show that all the inside cells except for those in the last row and last column can be wiggled, and then the entries in the last row and column must be chosen to preserve the row and column sums. Thus, the number of "free" cells is $(r-1)\cdot(c-1)$, and the remaining cells are then constrained.

Example 10.6-2 Independence of Income and Geography

Use the data from the distribution of income by city in Example 10.6-1 to test whether or not the two variables of income and geographic location are independent at the 5% level of significance.

Solution

The actual and expected frequencies are given by,

Actual Frequencies

	Los Angeles	Denver	St. Louis	Pittsburgh	Boston	Row Total
Poverty	25	12	18	15	30	**100**
Lower Middle	55	25	55	65	100	**300**
Upper Middle	90	55	110	90	155	**500**
Wealthy	30	8	17	30	15	**100**
Column Total	**200**	**100**	**200**	**200**	**300**	**1000**

Expected Frequencies

	Los Angeles	Denver	St. Louis	Pittsburgh	Boston	Row Total
Poverty	20	10	20	20	30	**100**
Lower Middle	60	30	60	60	90	**300**
Upper Middle	100	50	100	100	150	**500**
Wealthy	20	10	20	20	30	**100**
Column Total	**200**	**100**	**200**	**200**	**300**	**1000**

Then, $T = \frac{(25-20)^2}{20} + \frac{(12-10)^2}{10} + \cdots + \frac{(15-30)^2}{30} = 28.31$. The degrees of freedom equals $(\underbrace{4}_{\text{wealth classes}} -1)\cdot(\underbrace{5}_{\text{cities}} -1) = 3\cdot 4 = 12$, and the rejection region is $T > \chi^2_{.05}(12) = 21.03$. We therefore reject the null hypothesis assumption of independence at the 5% level, and conclude that there *is* variation in income distribution depending on geographical area. ❐

Note

The p-value of the test can be calculated using CHIDIST in Excel and results in a value of 0.50% (which is, of course, less than the stipulated significance level of 5%).

Exercise 10-19 CAS Exam 3 Spring 2007 #19

A table has been created to test the hypothesis that the number of vehicles per policyholder is independent of the policyholder's rating classification. There are four possible rating classifications in the table crossed with the number of vehicles per policyholder. The number of vehicles per policyholder is limited to a value of 1, 2, 3, or 4.

Calculate the degrees of freedom to be used in a Chi-Square test of the hypothesis.

(A) 9 (B) 10 (C) 12 (D) 14 (E) 15

10.6.2 GOODNESS OF FIT TESTS

In this application of the chi-square the object is to determine whether or not observed frequencies from a single discrete random variable X conform to the expected frequencies under the null hypothesis assumption that X is a particular type of random variable.

For example, suppose that X takes the values $0, 1, 2, \ldots,$ and we wish to confirm that the actual observed frequencies for X arose from an underlying Poisson process. Then we would follow a two-step procedure. We first use the observations of X to estimate the mean rate of occurrence parameter, λ. Second, we calculate the expected frequencies using the Poisson probability formula and use them to calculate the chi-square statistic.

The degrees of freedom for the chi-square is then given by $m-r-1$, where m is the number of cells being compared, and r is the number of parameters being estimated.[2]

Example 10.6-3 The Prussian Cavalry Rides Again

In Example 4.6-2 we presented actual data for deaths in the Prussian cavalry caused by horse kicks over a 20 year period in the 19th century. The data is tabulated by corps-year for 10 corps over the 20 year timeframe, resulting in a total of 200 observed corps-years.

The actual frequencies are given as follows:

[2] One can also test the goodness-of-fit to a continuous distribution using the technique of this section. The approach would be to partition the observed data points into discrete intervals and to work with the frequencies in each interval. For example, observations from a continuous age-at-death random variable could be lumped together into 5 year intervals, say (0,5], (5,10], etc.

# of deaths by kicking	Corps-years
0	109
1	65
2	22
3	3
4	1
5 or more	0
Total:	200

(a) Estimate the Poisson parameter and calculate the resulting chi-square statistic.

(b) Determine whether or not to reject the null hypothesis that the data conforms to a Poisson distribution at the 5% level of significance.

Solution

The random variable X is the number of deaths in a single corps-year. The Poisson parameter λ is estimated by $\hat{\lambda}$, the sample mean based on the 200 observations. Using the data table, we obtain:

X	f	$f \cdot x$
0	109	0
1	65	65
2	22	44
3	3	9
4	1	4
5 or more	0	0
Total:	200	122

This results in $\hat{\lambda} = \frac{122}{200} = 0.61$. The expected frequencies are calculated using the resulting Poisson probabilities. If p_k is the Poisson probability of k deaths in a single corps-year, then the total number of years with k deaths is a binomial random variable with 200 trials and probability p_k. Thus, the expected number of corps-years, e_k, with k deaths is $200 \cdot p_k$. The result is,

k = number of deaths by horse-kicking X	$p_k = e^{-0.61} \frac{(0.61)^k}{k!}$	Predicted number of corps-years $e_k = 200 \cdot p_k$	*Actual number of corps-years* f_k
0	0.5434	108.7	*109*
1	0.3314	66.3	*65*
2	0.1011	20.2	*22*
3	0.0206	4.1	*3*
4	0.0031	0.6	*1*
5 or more	0.0004	0.1	*0*
Total:	1.0000	200.0	*200*

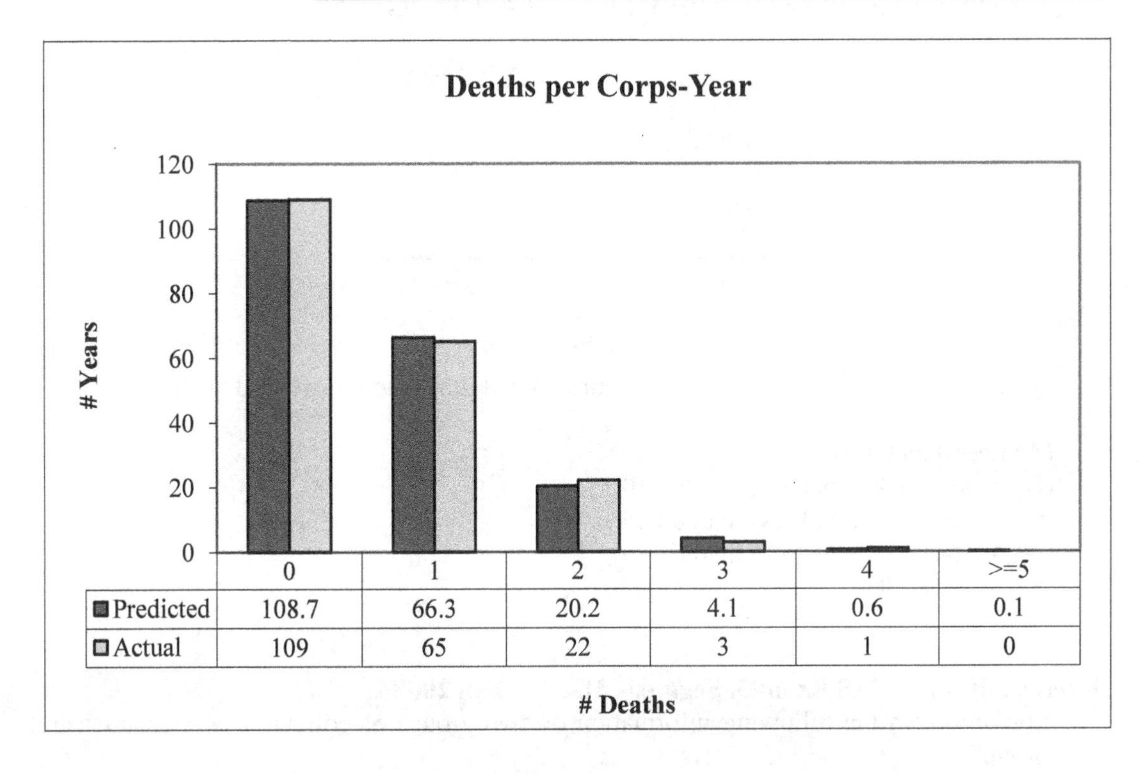

Since the cell frequencies for X equal to 4 and 5 or more are small, we combine them with X equal to 3. Then,

X	f	e
0	109	108.7
1	65	66.3
2	22	20.2
3 or more	4	4.8

and the chi-square statistic $T = \frac{(109-108.7)^2}{108.7} + \frac{(65-66.3)^2}{66.3} + \frac{(22-20.2)^2}{20.2} + \frac{(4-4.8)^2}{4.8} = 0.32.$

Since one parameter, λ, is estimated, the degrees of freedom equals

$$\underbrace{m}_{\text{cells}} - \underbrace{r}_{\text{parameters}} - 1 = 4 - 1 - 1 = 2.$$

The null hypothesis is rejected at the 5% level of significance for $T > \chi^2_{.05}(2) = 5.99$. Therefore, the null hypothesis that X is Poisson cannot be rejected. The resulting p-value of 85% can be verified using the function CHIDIST in Excel®. ❒

Exercise 10-20 CAS Exam 3 Spring 2007 #18

A company wants to determine whether sick days taken by its employees are randomly distributed throughout the five-day working week. A random sample of the sick days taken by a sample of 100 employees yielded the following data:

	Sick Days
Monday	32
Tuesday	18
Wednesday	18
Thursday	20
Friday	32
Total	120

Which of the following ranges contain the $p-$value that results from testing the hypothesis that sick days are randomly distributed throughout the five days listed above?

(A) Less than 0.005
(B) At least 0.005, but less than 0.010
(C) At least 0.010, but less than 0.025
(D) At least 0.025, but less than 0.050
(E) At least 0.050

Exercise 10-21 CAS Exam 3, Segment 3L Fall 2008 #9

You are given the following information on two groups of policyholders, Group 1 and Group 2:

- Group 1 and Group 2 claim counts follow the binomial distribution.
- A policyholder has either no claim or one claim during the policy period.
- A table summarizing the experience for one policy period is given below:

Policyholder Group	Policyholders with No Claims	Policyholders with One Claim
Group 1	40	10
Group 2	120	30

- H_0 : The probability of a policyholder having a claim is the same for the two groups.
- H_1 : The probability of a policyholder having a claim is not the same for the two groups.

Calculate the absolute value of the difference between the Chi-Square test statistic to test the null hypothesis and the critical region value at a 5% significance level.

(A) Less than 2.5
(B) At least 2.5, but less than 3.0
(C) At least 3.0, but less than 3.5
(D) At least 3.5, but less than 4.0
(E) At least 4.0

Exercise 10-22 CAS Exam 3 Fall 2007 #9

A claim department has operated under the following assumptions about expected automobile claims:

- 50% of the claims are for cars
- 20% of the claims are for motorcycles
- 15% of the claims are for vans
- 15% of the claims are for trucks

Using the following set of data, calculate the Chi-Square statistic that would be used in testing the claim department's assumption about expected claim counts.

	Claims
Cars	40
Motorcycles	24
Trucks	17
Vans	19
Total	100

(A) Less than 2
(B) At least 2, but less than 5
(C) At least 5, but less than 9
(D) At least 9, but less than 15
(E) At least 15

Exercise 10-23 CAS Exam 3 Fall 2006 #5

The following table displays the number of policyholders by territory by number of claims:

Number of Claims	Territory 1	Territory 2	Territory 3	Territory 4	Total
0	97	188	392	293	970
1	2	10	4	4	20
2	1	2	4	3	10
Total	100	200	400	300	1000

You are testing the hypothesis that the claim count distributions are the same in each territory using a Chi-Square goodness of fit test with significance level of 5%.

Calculate the absolute value of the difference between the test statistic and the critical value.

(A) Less than 0.30
(B) At least 0.30, but less than 0.40
(C) At least 0.40, but less than 0.50
(D) At least 0.50, but less than 0.60
(E) At least 0.60

10.7 CHAPTER 10 SAMPLE EXAMINATION

1. **CAS Exam 3 Spring 2005 #23**

   YG Insurance (YGI) estimates that 30% of its policyholders will file at least 1 claim each year. However, YGI also has reason to believe that policyholders who drive customized cars will file fewer claims than other policyholders.

   YGI has sampled 100 policyholders with customized cars and has determined the number of sample policyholders with at least 1 claim in the last year.

   YGI will charge a lower premium to owners of customized cars, if the sample results show that fewer than 30% of owners of customized cars file at least 1 claim each year. YGI requires that the sample results be subject to no more than 2% Type I error.

   Calculate the minimum allowable number of sampled policyholders with at least 1 claim in the last year.

   (A) Less than 16
   (B) At least 16, but less than 18
   (C) At least 18, but less than 20
   (D) At least 20, but less than 22
   (E) 22 or more

2. **SOA/CAS Exam 110 November 1981 #30**

   A random sample X_1, X_2, X_3 is taken from a normally distributed population with unknown mean μ and known variance σ^2. In order to test the null hypothesis $H_0 : \mu = 0$ against the alternative hypothesis $H_1 : \mu > 0$, it is decided to reject H_0 if and only if $X_1 + X_2 + X_3 > 20\sqrt{3}$. For what value of σ is the significance level of this test equal to 0.05?

   (A) 7.04 (B) 10.20 (C) 12.20 (D) 21.13 (E) 32.80

3. **CAS Exam 3, Segment 3L Spring 2008 #4**
One hundred insureds are assigned to one of three classes based on prior policy period claim experience, as follows:

Class	Number of Claims in Prior Policy Period	Number of Insureds in Class
A	No Claims	74
B	One Claim	16
C	Two or More Claims	10

Calculate the value of the chi square statistic that results from testing the hypothesis that claim frequency follows a Poisson distribution with mean 0.4.

(A) Less than 2
(B) At least 2, but less than 4
(C) At least 4, but less than 6
(D) At least 6, but less than 8
(E) At least 8

4. **SOA/CAS Exam 110 May 1982 #34**
Suppose $X_1,\ldots,X_{12}$ are independent random variables, each normally distributed with mean 0 and variance $\sigma^2 > 0$. The rejection region of size 0.025 for testing the hypothesis $H_0 : \sigma^2 = 10$ against the alternative hypothesis $H_1 : \sigma^2 < 10$ is given by:

(A) $\sum X_i^2 < 44.0$
(B) $\sum X_i^2 > 44.0$
(C) $\sum X_i^2 < 233.4$
(D) $\sum X_i^2 > 233.4$
(E) $\sum X_i^2 < 2567.1$

5. **SOA/CAS Exam 110 May 1982 #39**
Suppose that in 60 rolls of a die the outcomes 1,2,3,4,5, and 6 occur with frequencies $n_1, n_2, 14, 8, 10,$ and 8, respectively. Using the standard chi-square test, the hypothesis that the die is fair is rejected at the 1% level of significance if and only if n_1 is such that:
(A) $(n_1 - 10)^2 \geq 63.45$
(B) $(n_1 - 10)^2 \geq 71.75$
(C) $(n_1 - 10)^2 \geq 72.05$
(D) $(n_1 - 10)^2 \geq 125.9$
(E) $(n_1 - 10)^2 \geq 143.5$

6. **SOA/CAS Exam 110 May 1985 #8**

Let X_1, X_2, X_3, X_4 be a random sample from a normal distribution with unknown mean μ and unknown variance σ^2. The null hypothesis $H_0 : \mu = 10$ is to be tested against the alternative hypothesis $H_1 : \mu \neq 10$ at a significance level of 0.05 using the Student t statistic. If the resulting sample mean is $\bar{X} = 15.84$ and $S^2 = \frac{1}{3}\sum_{i=1}^{4}(X_i - \bar{X})^2 = 16$, then what are the critical t-value and decision reached?

(A) $t = 2.13$; reject H_0
(B) $t = 2.35$; do not reject H_0
(C) $t = 2.78$; reject H_0
(D) $t = 3.18$; do not reject H_0
(E) $t = 3.18$; reject H_0

7. **SOA/CAS Exam 110 May 1985 #33**

It is hypothesized that an experiment results in outcomes $K, L, M,$ and N with probabilities $\frac{1}{5}, \frac{3}{10}, \frac{1}{10},$ and $\frac{2}{5}$, respectively. Forty independent repetitions of the experiment have results as follows:

Outcome	Frequency
K	11
L	14
M	5
N	10

The chi-square goodness of fit statistic is used to test the above hypothesis. Let r be the observed value of the test statistic, and let s be the critical value corresponding to a significance level of .01. What are r and s?

(A) $r = \frac{95}{24}$ and $s = 13.28$

(B) $r = \frac{95}{24}$ and $s = 11.35$

(C) $r = \frac{28}{24}$ and $s = 0.30$

(D) $r = \frac{28}{24}$ and $s = 13.28$

(E) $r = \frac{28}{24}$ and $s = 11.35$

8. **SOA/CAS Exam 110 May 1990 #7**

Let $(X_1,Y_1),(X_2,Y_2),(X_3,Y_3)$ be a random sample of paired observations from distributions with means μ_X and μ_Y, respectively, and with positive variances. The null hypothesis $H_0 : \mu_X = \mu_Y$ is to be tested against the alternative $H_1 : \mu_X \neq \mu_Y$ by using the Student t statistic based on the difference scores $X_i - Y_i$. If the significance level of the test is 0.05, and the value of the test statistic is 4.10, what is the critical value of this test and what is the decision reached?

(A) 2.92, reject H_0
(B) 3.18, reject H_0
(C) 4.30, reject H_0
(D) 3.18, do not reject H_0
(E) 4.30, do not reject H_0

9. **SOA/CAS Exam 110 May 1990 #22**

Let $X_1,\ldots,X_{15}$ be a random sample from a normal distribution with mean μ and variance σ^2. Let $\bar{X} = \frac{1}{15}\sum_{i=1}^{15} X_i$ and $T = \sum_{i=1}^{15}(X_i - \bar{X})^2$. What is the appropriate critical (rejection) region for a test of $H_0 : \sigma^2 \leq 10$ versus $H_1 : \sigma^2 > 0$ at significance level equal to 0.05?

(A) Reject H_0 if and only if $T \geq 23.69$
(B) Reject H_0 if and only if $T \geq 25.00$
(C) Reject H_0 if and only if $T \geq 236.90$
(D) Reject H_0 if and only if $T \geq 250.00$
(E) Reject H_0 if and only if $T \geq 261.20$

10. **SOA/CAS Exam 110 May 1990 #40**

Let $X_1,\ldots,X_{25}$ and $Y_1,\ldots,Y_{25}$ be random samples from independent normal distributions with unknown means μ_X and μ_Y and common variance 100. The null hypothesis $H_0 : \mu_X = \mu_Y$ is to be tested against the alternative hypothesis $H_1 : \mu_X > \mu_Y$ using the statistic $W = \bar{X} - \bar{Y}$, where $\bar{X}$ and $\bar{Y}$ are the sample mean of the X and Y samples, respectively. What is the correct decision about H_0 if $W = 3.96$?

(A) Reject H_0 at the .025 level.
(B) Reject H_0 at the .05 level but not at the .025 level.
(C) Reject H_0 at the .075 level but not at the .05 level.
(D) Reject H_0 at the .10 level but not at the .075 level.
(E) Do not reject H_0 at the .10 level.

11. **SOA/CAS Exam 110 May 1990 #43**

Weekly repair costs for machines of a certain type have in the past averaged $1200 per machine. To determine whether this average μ_X has increased during a given week, the repair costs from a random sample of 10 machines are recorded, and the null hypothesis $H_0 : \mu_X = 1200$ is tested against the alternative $H_1 : \mu_X > 1200$. The sample mean was $1290 and the unbiased sample standard deviation was $102. Based on the appropriate Student t statistic for this test, what is the correct decision about the null hypothesis?

(A) Reject at the .01 level but not at the .005 level.
(B) Reject at the .025 level but not at the .01 level.
(C) Reject at the .05 level but not at the .025 level.
(D) Reject at the .10 level but not at the .05 level.
(E) Do not reject at the .10 level.

12. **CAS Exam ST Fall 2014 #16**

You are given the following information:

- Your company has developed a new training program for claims adjusters to increase their productivity.
- Assume that the number of claims closed follows a normal distribution, with equal variance before and after the training program.
- Observations for a given adjuster are not independent.
- H_0 : the number of claims closed is the same before and after training.
- H_1 : the number of claims closed is not the same before and after training.

Adjuster ID	1	2	3	4	5	6	7	8	9	10
# Claims Closed Before Training	7	6	7	8	4	7	5	7	8	2
# Claims Closed After Training	5	3	7	12	10	13	10	14	6	6

Calculate the minimum significance level at which you reject H_0.

(A) Less than 0.1%
(B) At least 0.1%, but less than 1.0%.
(C) At least 1.0%, but less than 2.0%.
(D) At least 2.0%, but less than 5.0%.
(E) At least 5.0%.

Note: The way H_1 is worded suggests a two-tail test, meaning that the training could possibly decrease productivity. Answer accordingly to be consistent with the exam question grading.

CHAPTER 11

THEORY OF ESTIMATION AND HYPOTHESIS TESTING

In this chapter we turn our attention to some of the general theory behind the statistical applications we have discussed in Chapters 9 and 10. Almost all of the discussion and examples in the preceding chapters have dealt with procedures for estimating the mean or the variance of a population. Here, we look at the estimation of parameters in general, not just the parameters of population mean and variance. We present some of the common properties of estimators that bear on their reliability, and we show some standard methods for developing parameter estimators.

We also look at more general hypothesis testing. In Chapter 10 the test statistic was usually a given, and we used it to find rejection regions (also called critical regions) that suited the context of the problem. Here, we will investigate methods for deriving the most appropriate test statistics and the best associated rejection regions as part of a general theory.

The starting point, as previously, is with a population random variable X, whose distribution will depend on a parameter, or possibly even several parameters. While these parameters may be the population mean or variance as in previous situations, they may also be different parameters entirely. As a simple example, our population X may have a gamma distribution, with parameters α and β, which, of course, are distinct from the mean and the variance of X. In this case, our goal would be to use a random sample from X to construct random variables that estimate α and β directly.

To capture the general nature of the discussion we will often use the symbol θ for a generic parameter, and write the density function of X as $f_X(x;\theta)$ in order to make explicit the dependence on θ. If we have multiple parameters, say k in number, then $\Theta = (\theta_1,\ldots,\theta_k)$ is a vector. Except for a few isolated cases, however, we restrict our attention to the case of a single parameter. In order to estimate θ we take a random sample $X_1,\ldots,X_n$ from the population X. Then we use some transformation $\hat{\theta} = u(X_1,\ldots,X_n)$ of the random observations as an estimator for θ.

The exact functional expression in terms of $X_1,\ldots,X_n$ for this estimator depends on the particularities of X and θ, but two important properties stand out. First, $\hat{\theta}$ is a ***random variable*** dependent on $X_1,\ldots,X_n$, and second, the expression $u(X_1,\ldots,X_n)$ for $\hat{\theta}$ is strictly a function of the observations and does not explicitly depend on the parameter θ being estimated. These two facts may seem self-evident, but it is nonetheless important to keep them in mind as we work through the theory of estimation in succeeding sections.

11.1 THE BIAS OF AN ESTIMATOR

In Chapter 9 we mentioned that an estimator is unbiased if its expected value is equal to the parameter being estimated. We now look more closely at the nature of bias (or its absence) as a general property of estimators.

Bias

Let X be a population random variable and let $f_X(x;\theta)$ be the density function for X. Let $\hat{\theta}$ be an estimator for the parameter θ based on a random sample of observations from X.

(1) We define the ***bias*** of $\hat{\theta}$ by $B(\hat{\theta}) = E[\hat{\theta}-\theta] = E[\hat{\theta}]-\theta$.

(2) If $E[\hat{\theta}]=\theta$, we say $\hat{\theta}$ is an ***unbiased estimator*** of θ, in which case $B(\hat{\theta})=0$.

We know from Chapter 9 that for any population with finite mean and variance, $\overline{X} = \frac{1}{n}\sum_{i=1}^{n} X_i$ is an unbiased estimator for μ_X and $S^2 = \frac{1}{n-1}\sum_{i=1}^{n}(X_i-\overline{X})^2$ is an unbiased estimator for σ_X^2. Here are some more examples of estimators, both biased and unbiased. Unless otherwise specified, we assume that our population X has finite mean and variance.

Example 11.1-1 Biased Estimator for Population Variance

Let $X_1,\ldots,X_n$ be a random sample from X and let $T^2 = \frac{1}{n}\sum_{i=1}^{n}(X_i-\overline{X})^2$ be an estimator for σ_X^2.

(a) Find a constant c such that cT^2 is an unbiased estimator for σ_X^2.

(b) Calculate the bias $B(T^2)$.

Solution

(a) We first note that $S^2 = \left(\frac{n}{n-1}\right)T^2$, where S^2 is the usual unbiased estimator of σ_X^2. Thus, the required c is $\frac{n}{n-1}$.

(b) We write $T^2 = \left(\frac{n-1}{n}\right)S^2$. Then,

$$B(T^2) = E[T^2]-\sigma_X^2 = E\left[\left(\frac{n-1}{n}\right)S^2\right]-\sigma_X^2 = \left(\frac{n-1}{n}\right)\sigma_X^2 - \sigma_X^2 = -\frac{1}{n}\sigma_X^2 .$$ ❐

Example 11.1-2 Uniform with Endpoint as Parameter

Let X be a uniform random variable on the interval $[0,\theta]$, where $\theta > 0$ is an unknown parameter. Let $X_1,\ldots,X_n$ be a random sample from X and let $U = Max(X_1,\ldots,X_n)$.

(a) Find a constant c such that cU is an unbiased estimator for θ.
(b) Calculate the bias $B(U)$.

Solution

(a) From Subsection 8.1.5 on Order Statistics we have that the density function for U is given by,

$$f_U(x) = nf_X(x)F_X(x)^{n-1} = n\left(\frac{1}{\theta}\right)\left(\frac{x}{\theta}\right)^{n-1} = \frac{n}{\theta^n}x^{n-1};\ 0 \le x \le \theta .$$

Thus, $E[U] = \int_0^\theta \left(x\frac{n}{\theta^n}x^{n-1}\right)dx = \frac{n}{\theta^n}\int_0^\theta x^n\,dx = \frac{n}{n+1}\theta$. It follows that $\frac{n+1}{n}U$ is an unbiased estimator of θ.

(b) To calculate the bias of U, we have

$$B(U) = E[U]-\theta = \frac{n}{n+1}\theta-\theta = -\frac{1}{n+1}\theta.$$

❐

Example 11.1-3 Estimating the Variance of a Bernoulli Trial

Let X be a Bernoulli trial population with parameter p representing the probability of a success. Let $X_1,\ldots,X_n$ be a random sample from X and let $Y = \overline{X}(1-\overline{X})$ be an estimator for $\sigma_X^2 = p(1-p)$.

(a) Find a constant c such that cY is an unbiased estimator for $\sigma_X^2 = p\,(1-p)$.
(b) Calculate the bias $B(Y)$.

Solution

(a) First, note that $E[\overline{X}] = \mu_X = p$ and, using the variance formula,

$$E[\overline{X}^2] = Var[\overline{X}]+E[\overline{X}]^2 = \frac{\sigma_X^2}{n}+p^2 = \frac{p(1-p)}{n}+p^2 .$$

Then,

$$\begin{aligned}E[Y] = E\left[\overline{X}(1-\overline{X})\right] &= E[\overline{X}]-E[\overline{X}^2] \\ &= p-\left[\frac{p(1-p)}{n}+p^2\right] \\ &= (p-p^2)-\frac{p(1-p)}{n} = p(1-p)-\frac{p(1-p)}{n} = \frac{n-1}{n}p(1-p).\end{aligned}$$

It follows that $\frac{n}{n-1}Y$ is an unbiased estimator for $p(1-p)$.

(b) To calculate the bias we have $B(Y) = \frac{n-1}{n}p(1-p)-p(1-p) = -\frac{1}{n}p(1-p)$.

❐

Note

In Chapters 9 and 10, in dealing with binomial proportions, $Y = \overline{X}(1-\overline{X})$ was the estimator of choice for $\sigma_X^2 = p(1-p)$ and was denoted $\hat{p}(1-\hat{p})$.

Example 11.1-4 The Square of an Unbiased Estimator

Let $\hat{\theta}$ be an unbiased estimator of the parameter θ. If $Var[\hat{\theta}] > 0$ then $\hat{\theta}^2$ is a ***biased*** estimator of θ^2.

Solution

This follows immediately from rearranging the variance formula for $\hat{\theta}$:

$$E[\hat{\theta}^2] = Var[\hat{\theta}] + E[\hat{\theta}]^2 > E[\hat{\theta}]^2 = \theta^2.$$ □

Note

In particular, this means that except for trivial situations, S, the square root of the unbiased variance estimator, S^2, is a ***biased*** estimator of σ_X.

In the first three examples above the bias formula turned out to be of the form $B[\hat{\theta}] = -\frac{1}{n}\theta$ or $B[\hat{\theta}] = -\frac{1}{n+1}\theta$. This means that as the sample size n increases the magnitude of the bias goes to zero. Estimators with this property are referred to as ***asymptotically*** unbiased.

Asymptotically Unbiased Estimators

Let $X_1, \ldots, X_n$ be a random sample from the population X and let $\hat{\theta}$ be an estimator of the parameter θ based on the random sample. We say $\hat{\theta}$ is an ***asymptotically unbiased*** estimator of θ if $\lim_{n\to\infty} B(\hat{\theta}) = 0$.

Exercise 11-1 CAS Exam 3 Fall 2007 #4

You are given the following information:

- $Y = X_1 + \cdots + X_n$ where $n = 100$
- $X_1, \ldots, X_n$ is a random sample from a Gamma distribution with $\alpha = 4$ and θ unknown

Calculate the value, c, to produce an unbiased estimate of θ from $c \cdot Y$.

(A) Less than .0035
(B) At least .0035, but less than .0045
(C) At least .0045, but less than .0055
(D) At least .0055, but less than .0065
(E) At least .0065

Exercise 11-2 CAS Exam 3 Fall 2006 #1

Claim sizes are described by an exponential distribution with parameter θ.

The probability density function of the order statistic Y_k with a sample size n is

$$\frac{n!}{(k-1)!(n-k)!}[F(y)]^{k-1}[1-F(y)]^{n-k}f(y).$$

For a sample of- size 5, determine the bias in using Y_3, the third order statistic, as an estimate of the median of the distribution.

(A) Less than 0.05θ
(B) At least 0.05θ, but less than 0.08θ
(C) At least 0.08θ, but less than 0.10θ
(D) At least 0.10θ, but less than 0.13θ
(E) At least 0.13θ

Exercise 11-3 Consider an exponential distribution with an unknown parameter β. To estimate β, you take a sample of size 20 and take the minimum of these 20 numbers as your approximation, $\hat{\beta}$. Calculate the bias of your estimate.

Exercise 11-4 Suppose that $W_1, W_2, \dots, W_n$ is a random sample from a population with density function $f(w) = \begin{cases} \dfrac{e^{\frac{-w}{\theta+4}}}{\theta+4} & \text{for } w > 0 \text{ and } \theta > -4 \\ 0 & \text{otherwise} \end{cases}$.

Find an unbiased estimator for θ.

11.2 BUILDING ESTIMATORS

At this point our portfolio of estimators is still pretty slim, consisting mainly of the sample mean and the sample variance, plus a few other examples from the previous section. Our aim now is to introduce some standard procedures for systematically constructing estimators out of random samples for a general parameter. We discuss two such methods in this section and will introduce a third (Bayesian estimation) at the end of the chapter.

11.2.1 METHOD OF MOMENTS

Let X be a population random variable and let $X_1, \dots, X_n$ be a random sample from X. The method of moments works well for multiple parameters, so this is a case where we can develop the procedure in general.

Thus, assume in general that the population has k parameters, $\theta_1, \theta_2, \ldots, \theta_k$. The first step is to evaluate the population moments $E[X^j]$; $j = 1, 2, \ldots, k$. In the continuous case this amounts to evaluating the k integrals $\int_{-\infty}^{\infty} x^j f_X(x)\,dx$. These integrals will be functions of the k parameters, $\theta_1, \theta_2, \ldots, \theta_k$. We will refer to the resulting formulas as the ***theoretical moments*** of X.

The next step is to calculate the sample moments, that is, the estimated moments resulting from the random sample values. We assume the sample values are all equally likely and calculate the jth sample moment as $\frac{1}{n}\sum_{i=1}^{n} X_i^j$.

The final step is to equate the theoretical moments with the sample moments and then algebraically solve simultaneously the k equations for $\theta_1, \theta_2, \ldots, \theta_k$ in terms of the sample moments. The resulting formulas provide the estimators $\hat{\theta}_1, \hat{\theta}_2, \ldots, \hat{\theta}_k$ for $\theta_1, \theta_2, \ldots, \theta_k$.

To illustrate the idea we will begin with the simplest situation, where there is a single parameter θ. In this case we only need the first moment, $\frac{1}{n}\sum_{i=1}^{n} X_i$, which of course is just the sample mean, $\overline{X}$. The theoretical first moment is $E[X]$, otherwise known as the population mean, μ_X. If the population parameter we seek to estimate is just μ_X, then the method of moments leads directly to the estimator $\overline{X}$ for μ_X. This is a reassuring, if less than startling, result.

To take a slightly more complicated example, we revisit the problem of estimating an unknown endpoint for a uniform distribution.

Example 11.2-1 Uniform on $[0,\theta]$ (Example 11.1-2 continued)

Let X be a uniform random variable on the interval $[0,\theta]$, where $\theta > 0$ is an unknown parameter. Let $X_1, \ldots, X_n$ be a random sample from X. Calculate the method of moments estimator for θ.

Solution
The theoretical first moment is the expected value of X, which is the midpoint of the interval, $\frac{\theta}{2}$. The sample first moment is the sample mean, $\overline{X}$. Equating the two, we have $\frac{\theta}{2} = \overline{X}$. Solving for θ, we calculate the estimator as $\hat{\theta} = 2\overline{X}$. □

Notes

(1) The estimator $\hat{\theta} = 2\overline{X}$ is easily seen to be unbiased.

(2) In Example 11.1-2 we used order statistics and showed that $\frac{n+1}{n} Max(X_1, \ldots, X_n)$ is also an unbiased estimator for θ. We will compare the properties of these two highly dissimilar estimators later in Section 11.3.

We look next at the method of moments applied to the case of two population parameters. The two most conspicuous candidates for population parameters are the population mean and variance, so we first consider that case in general.

Example 11.2-2 Moment Estimation for Mean and Variance

Let X be a population random variable and let $X_1,\ldots,X_n$ be a random sample from X. Let μ_X and σ_X^2 be the parameters for X whose values we wish to estimate. Use the method of moments to calculate estimators for μ_X and σ_X^2.

Solution

The theoretical first and second moments are $E[X] = \mu_X$ and, using the variance formula, $E[X^2] = \sigma_X^2 + \mu_X^2$. The first and second sample moments are $\overline{X}$ and $\frac{1}{n}\sum_{i=1}^{n} X_i^2$, respectively. Equating the theoretical and the sample moments leads to the two equations, $\mu_X = \overline{X}$, and $\sigma_X^2 + \mu_X^2 = \frac{1}{n}\sum_{i=1}^{n} X_i^2$. The solution for μ_X is obviously $\overline{X}$, and we can substitute that into the second equation to get, $\sigma_X^2 = \frac{1}{n}\sum_{i=1}^{n} X_i^2 - \overline{X}^2$. Thus, the general method of moments estimators for μ_X and σ_X^2 are $\overline{X}$ and $\frac{1}{n}\sum_{i=1}^{n} X_i^2 - \overline{X}^2$. ❐

Notes

(1) The method of moments estimator $\frac{1}{n}\sum_{i=1}^{n} X_i^2 - \overline{X}^2$ for σ_X^2 is identical to the biased estimator $T^2 = \frac{1}{n}\sum_{i=1}^{n}(X_i - \overline{X})^2$ introduced in Example 11.1-1. This can be seen by algebraically expanding T^2 as follows:

$$
\begin{aligned}
T^2 &= \frac{1}{n}\sum_{i=1}^{n}(X_i - \overline{X})^2 = \frac{1}{n}\sum_{i=1}^{n}(X_i^2 - 2\overline{X}X_i + \overline{X}^2) \\
&= \frac{1}{n}\left[\sum_{i=1}^{n} X_i^2 - 2\overline{X}\sum_{i=1}^{n} X_i + \sum_{i=1}^{n}\overline{X}^2\right] \\
&= \frac{1}{n}\left[\sum_{i=1}^{n} X_i^2 - 2\overline{X}\left(n\overline{X}\right) + n\overline{X}^2\right] = \frac{1}{n}\sum_{i=1}^{n} X_i^2 - \overline{X}^2.
\end{aligned}
$$

The formulation $\frac{1}{n}\sum_{i=1}^{n} X_i^2 - \overline{X}^2$ can be thought of as a sample variance shortcut formula for T^2.

(2) If X is normal, say $X \sim N(\mu_X, \sigma_X^2)$, then it follows immediately that the method of moments estimators for the parameters μ_X and σ_X^2 are $\bar{X}$ and T^2.

(3) In problems involving two parameters, instead of equating the theoretical and sample moments, it may be easier to equate the theoretical mean and variance to the method of moments mean and variance, that is, to the estimators $\bar{X}$ and T^2 derived above.

We illustrate the last point in the next example.

Example 11.2-3 Estimators for the Parameters of a Gamma Distribution

Let X be a population with a $\Gamma(\alpha, \beta)$ distribution and let $X_1, \ldots, X_n$ be a random sample from X. Derive the method of moments estimators for α and β.

Solution

In Subsection 6.5.2 we show that $\mu_X = \alpha\beta$ and $\sigma_X^2 = \alpha\beta^2$. We use note (3) above to derive estimators by equating these expressions to $\bar{X} = \frac{1}{n}\sum_{i=1}^{n} X_i$ and $T^2 = \frac{1}{n}\sum_{i=1}^{n}(X_i - \bar{X})^2$. Thus, we have, $\alpha\beta = \bar{X}$ and $\alpha\beta^2 = T^2$. These two equations are easily solved to yield the estimators, $\hat{\beta} = \frac{T^2}{\bar{X}}$ and $\hat{\alpha} = \frac{\bar{X}^2}{T^2}$. ❒

Exercise 11-5 CAS Exam 3, Segment 3L Fall 2006 #3

Claim sizes of 10 or greater are described by a single parameter Pareto distribution (Type I, Subsection 6.7.2), with parameter α. A sample of claim sizes is as follows:

10	12	14	18	21	25

Calculate the method of moments estimate α for this sample.

(A) Less than 2.0
(B) At least 2.0, but less than 2.1
(C) At least 2.1, but less than 2.2
(D) At least 2.2, but less than 2.3
(E) At least 2.3

Exercise 11-6 CAS Exam 3 Spring 2006 #1

Let $N = 0, 1, 2, \ldots$ be a random variable denoting the number of goals scored in a soccer game. Assume N has two parameters, r and β, where,

$$E[N] = r\beta \text{ and } Var[N] = r\beta(1+\beta).$$

A random sample of 20 games produced the following number of goals scored:

Goals Scored	Frequency
0	1
1	3
2	4
3	5
4	3
5	2
6	1
7	0
8	1

Use the sample data and the method of moments to estimate the parameter β.

(A) Less than 0.25
(B) At least 0.25, but less than 0.50
(C) At least 0.50, but less than 0.75
(D) At least 0.75, but less than 1.00
(E) At least 1.00

Note

The random variable N has a negative binomial distribution (see Section 4.4), where r is the number of successes and $p = \frac{1}{1+\beta}$ is the probability of success.

Exercise 11-7 CAS Exam 3, Segment 3L Fall 2007 #5

X is a two-parameter Pareto random variable (Type II, Subsection 6.7.2) with parameters α and β.

A random sample from this distribution produces the following four claims:

- $x_1 = 2,000$
- $x_2 = 17,000$
- $x_3 = 271,000$
- $x_4 = 10,000$

Find the method of moments estimate for α.

(A) Less than 2
(B) At least 2, but less than 3
(C) At least 3, but less than 4
(D) At least 4, but less than 5
(E) At least 5

Exercise 11-8 CAS Exam 3 Spring 2005 #19

Four losses are observed from a Gamma distribution. The observed losses are: 200, 300, 350, and 450. Find the method of moments estimate for α.

(A) 0.3 (B) 1.2 (C) 2.3 (D) 6.7 (E) 13.0

11.2.2 THE MAXIMUM LIKELIHOOD ESTIMATOR

Here is another method for estimating a parameter based on a random sample. The basic idea is to choose a value for the parameter that maximizes the probability of the observations that actually occurred in the random sample. This approach is most fruitful in the context of continuous population random variables with parameters that take values along a continuum of real numbers. But it also works for discrete random variables and parameters taking discrete values. Here is a simple example of the latter that illustrates the general reasoning.

Example 11.2-4 Maximum Likelihood Estimation

An urn contains 5 red balls and an unknown number θ of blue balls. A random sample of 3 balls is removed from the urn. The sample contains one blue ball and 2 red balls. Find the value of θ that maximizes the probability of this outcome.

Solution

Using a simple, intuitive approach we might reason as follows. The proportion of blue balls to red balls in the sample is 1:2. Consequently, the most likely ratio of θ to 5 would be 1:2. This leads to $\theta = 5/2$. Since θ must be a whole number, we would take either 2 or 3 as the most likely value for θ.

While this is the right general idea, we can sharpen the argument by treating θ as one of the parameters for the hypergeometric distribution introduced in Section 4.5. If X represents the number of blue balls in our sample of size 3 then, for a given θ,

$$\Pr[X=1] = \frac{\binom{\theta}{1}\binom{5}{2}}{\binom{\theta+5}{3}} = \frac{\theta \cdot 10}{\frac{(\theta+5)(\theta+4)(\theta+3)}{3!}} = \frac{60\theta}{(\theta+5)(\theta+4)(\theta+3)}; \ \theta = 0,1,2,\ldots,\infty.$$

The function of θ on the right is called the likelihood function, $L(\theta)$, based on the observation $X=1$. We can easily tabulate the values of $L(\theta)$.

θ	0	1	2	3	4	5	6	...
$L(\theta)$	0	.50	.57	.54	.48	.42	.36	...

Since $L(\theta)$ is decreasing for $\theta > 2$, we see that it is maximized for the parameter value $\theta = 2$. The conclusion is that $\hat{\theta} = 2$ is the maximum likelihood estimator (MLE) for θ. ❐

To develop this process in general let X be a population random variable with a parameter θ. Assuming X is continuous, we write the density function as $f_X(x;\theta)$ in order to emphasize the dependency on θ as well as x. Now let $X_1,\ldots,X_n$ be a random sample from X. We need to consider the multivariate joint distribution of the X_i's.

Since the observations are independent, the joint density function for the multivariate distribution $\mathbf{X}=(X_1,\ldots,X_n)$ can be written as the product of the individual density functions,

$$f_{\mathbf{X}}(x_1,x_2,\ldots,x_n;\theta) = \prod_{i=1}^{n} f_X(x_i;\theta).$$

Our perspective now is that the x's are held fixed as the observed values from the random sample, and so this expression is strictly a function of the parameter θ. For simplicity it is often just denoted $L(\theta)$, as in Example 11.2-4, and referred to as the ***likelihood function***. Since it is a joint density function it can be thought of as the continuous probability rate at the observed values $X_1 = x_1,\ldots,X_n = x_n$. The objective is to find the value of θ that maximizes this likelihood function. That value, if it exists, is the ***maximum likelihood estimate*** (MLE), $\hat{\theta}$, for θ.

Typically, locating the maximum for $L(\theta)$ is a calculus problem, solved by taking the derivative[1] of $L(\theta)$, setting it equal to 0, and solving for θ. Just as in calculus, there are important provisos. First, it is possible to get multiple so-called critical values of θ, and it is necessary to check which might be the absolute maximum. Second, it may also be necessary to check at the endpoints (if any) of the range of possible values for θ, since the maximum of $L(\theta)$ could occur there even though the derivative may not exist, or may exist but not be zero.

In implementing the maximum likelihood method there is one very important practical consideration. Since, in general, the likelihood function $L(\theta)$ described above is the product of n individual density functions, calculating its derivative could require multiple applications of the product rule for differentiation. This is not a minor matter and could serve as a serious impediment in applying the theory. Fortunately, there is a convenient "work-around," namely, using logarithmic differentiation rather than calculating the direct derivative.

Logarithmic differentiation means using the natural logarithm, $\ln L(\theta)$, in place of $L(\theta)$. This is feasible since the function $\ln(x)$ is monotonically increasing. Therefore, the value of θ that maximizes the function $\ln L(\theta)$ is exactly the same value of θ that maximizes $L(\theta)$. The real advantage though is that the logarithm converts the product into a sum, which is much easier to differentiate. The expression $\ln L(\theta)$ is called the ***log-likelihood function***.

---

[1] Technically this is a partial derivative with respect to θ and should be denoted $\frac{\partial L}{\partial \theta}$. However, since the x values are held fixed at the observed values throughout the calculations, and we never take derivatives with respect to the x's, we will keep things simple and use single variable derivative notation.

We now illustrate the maximum likelihood method with a number of examples. In the following, the use of a prime indicates differentiation with respect to the parameter.

Example 11.2-5 The MLE for the Poisson Rate Parameter

Consider a Poisson process with λ the mean rate of occurrence per unit period. Let X be the time between occurrences, so that X is an exponential random variable with density function given by $f_X(x;\lambda) = \lambda e^{-\lambda x}$; $\lambda, x > 0$ (see Subsection 6.2.2). A random sample $X_1,\ldots,X_n$ is taken from X, yielding the values $X_1 = x_1,\ldots,X_n = x_n$. Use these values to construct the maximum likelihood estimator for the parameter λ.

Solution
The likelihood function

$$L(\lambda) = \lambda^n \prod_{i=1}^{n} e^{-\lambda x_i} = \lambda^n e^{-\lambda(x_1+\cdots+x_n)};\ 0 < \lambda < \infty.$$

Then,

$$\ln L(\lambda) = n \ln \lambda - \lambda(x_1 + \cdots + x_n).$$

Next, we differentiate with respect to λ and set the result equal to 0. This gives, $[\ln L(\lambda)]' = \frac{n}{\lambda} - (x_1 + \cdots + x_n) = 0$. Solving for λ, we find that the MLE is given by $\hat{\lambda} = \frac{n}{x_1+\cdots+x_n} = \frac{1}{\overline{X}}$. ❐

Notes

(1) Since there is only one critical point in this case it is reasonably apparent that the calculation for $\hat{\lambda}$ is correct. To nail it down, though, we can use the second derivative test. We see that $[\ln L(\lambda)]'' = -n\lambda^{-2}$, so that $\ln L(\lambda)$ is concave down, indicating that $\hat{\lambda}$ is indeed the maximum point for $\ln L(\lambda)$.

(2) Since $\overline{X}$ is the mean waiting time for the random sample, $\overline{X}$ is an ***unbiased*** estimator for $\mu_X = \beta$, the reciprocal of the Poisson rate parameter λ. So it seems plausible that $\frac{1}{\overline{X}}$ is an estimator for λ. It is, however, a ***biased*** estimator for λ (see Exercise 11-14).

The next example shows how to construct an ***unbiased*** MLE for λ using the Poisson random variable. This also illustrates an application of the MLE method to a discrete distribution.

Example 11.2-6 The MLE for the Poisson Rate Parameter, Continued

Consider a Poisson process with $\lambda > 0$ the rate of occurrence per unit period. Let Y be the number of occurrences in a unit period, so that Y is a Poisson random variable with probability function given by $p_Y(n;\lambda) = e^{-\lambda}\frac{\lambda^n}{n!}$; $n = 0,1,\ldots$. A random sample $Y_1,\ldots,Y_m$ is taken from Y, yielding the values $Y_1 = n_1, \ldots, Y_m = n_m$. Use these values to construct the maximum likelihood estimator for the parameter λ.

Solution

Although the random variable Y is discrete, the likelihood function that arises from the Poisson probability function is a continuous function of λ given by,

$$L(\lambda) = \prod_{i=1}^{m} p_Y(n_i;\lambda) = \prod_{i=1}^{m} e^{-\lambda} \frac{\lambda^{n_i}}{n_i!} = e^{-\lambda m} \frac{\lambda^{n_1+\cdots+n_m}}{n_1!\cdots n_m!};\ \lambda > 0.$$

Then,

$$\ln L(\lambda) = -\lambda m + (n_1 + \cdots + n_m)\ln\lambda - \ln(n_1!\cdots n_m!).$$

Taking the derivative of $\ln L(\lambda)$ and setting it equal to 0 gives,

$$[\ln L(\lambda)]' = -m + \frac{(n_1+\cdots+n_m)}{\lambda} = 0.$$

Then, $\hat{\lambda} = \frac{n_1+\cdots+n_m}{m} = \overline{Y}$ is the resulting MLE for λ. ❐

Note

Since $\hat{\lambda}$ is the sample mean, it is an ***unbiased*** estimator for the population mean λ.

Example 11.2-7 Uniform on $[0,\theta]$ (Continuation)

Let X be a uniform random variable on the interval $[0,\theta]$, where $\theta > 0$ is an unknown parameter. Let $X_1, \ldots, X_n$ be a random sample from X. Calculate the maximum likelihood estimator for θ.

Solution

The density function for X is given by $f(x;\theta) = \frac{1}{\theta};\ 0 \le x \le \theta$. Let $x_1, \ldots, x_n$ be the observed values from the random sample. Let $u = Max(x_1, \ldots, x_n)$. Since for each i we have $x_i \le \theta$, it must be the case that $\theta \ge u$. Hence, the likelihood function is $L(\theta) = \left(\frac{1}{\theta}\right)^n;\ u \le \theta < \infty$. Now, we observe $L(\theta)$ is a strictly decreasing function on the interval $u \le \theta < \infty$. Therefore, the maximum of $L(\theta)$ is achieved at the left-hand endpoint u. The conclusion is that the MLE, $\hat{\theta} = U = Max(X_1, \ldots, X_n)$. ❐

Notes

(1) This is the estimator for θ we illustrated in Example 11.1-2, where we showed it was a biased estimator. On the other hand, the method of moments estimator (Example 11.2-1) is $2\overline{X}$, which is unbiased.

(2) The main point of this example is to illustrate a case where blind differentiation of $L(\theta)$, or $\ln L(\theta)$, leads to confusion, or worse (for this problem these derivatives are never zero).

Example 11.2-8 Estimating the Mean of a Normal Population

Let X be a normal population random variable with a known standard deviation σ. Let $X_1,\ldots,X_n$ be a random sample from X. Determine the resulting maximum likelihood estimator for the population mean μ.

Solution

The density function $f_X(x;\mu)$ for the normal distribution is given by,

$$f_X(x;\mu) = \frac{1}{\sigma\sqrt{2\pi}}e^{-\frac{1}{2}\left(\frac{x-\mu}{\sigma}\right)^2},$$

and so

$$L(\mu) = \prod_{i=1}^{n} f_X(x_i;\mu) = \left(\frac{1}{\sigma\sqrt{2\pi}}\right)^n e^{-\frac{1}{2}\sum_{i=1}^{n}\left(\frac{x_i-\mu}{\sigma}\right)^2}.$$

The log-likelihood function is,

$$\ln L(\mu) = \ln\left(\frac{1}{\sigma\sqrt{2\pi}}\right)^n - \frac{1}{2}\sum_{i=1}^{n}\left(\frac{x_i-\mu}{\sigma}\right)^2 = \ln\left(\frac{1}{\sigma\sqrt{2\pi}}\right)^n - \frac{1}{2\sigma^2}\sum_{i=1}^{n}(x_i-\mu)^2.$$

Differentiating with respect to the parameter μ yields,

$$(\ln L(\mu))' = \frac{1}{\sigma^2}\sum_{i=1}^{n}(x_i-\mu).$$

By setting the expression on the right equal to zero and solving for μ, we find the resulting estimator $\hat{\mu}$ is given by $\hat{\mu} = \frac{1}{n}\sum_{i=1}^{n} x_i$, which of course is identical to the sample mean $\overline{X}$. ❐

Example 11.2-9 CAS Exam ST Fall 2014

You are given a random sample of seven observations, 1.2, 1.9, 2.5, 2.7, 2.8, 3.3, and 3.6, from the following distribution:

$$f(x) = \frac{\theta^{-3}x^2e^{-\frac{x}{\theta}}}{2};\ x > 0.$$

Calculate the maximum likelihood estimate of θ.

Solution
We denote the observations as $x_1,\ldots,x_7$. The likelihood function is,

$$L(\theta)=\left(\theta^{-3}\right)^7\left(x_1\cdots x_7\right)^2 e^{-\frac{1}{\theta}(x_1+\cdots+x_7)}.$$

Then,

$$\ln L(\theta)=-21\ln\theta+\ln\left(x_1\cdots x_7\right)^2-\theta^{-1}\left(x_1+\cdots+x_7\right),\ \text{ so that,}$$

$$\ln L(\theta)'=-21\theta^{-1}+\theta^{-2}\left(x_1+\cdots+x_7\right)=0.\ \text{ This yields,}$$

$$\hat{\theta}=\frac{x_1+\cdots+x_7}{21}=\frac{18}{21}=\frac{6}{7}=0.8571.$$ ❐

Note
For a quick solution observe that the underlying distribution for X is a gamma, with parameters $\alpha=3$ and θ. Therefore, X can be thought of as the elapsed time in a Poisson process until the 3rd occurrence, where the mean time to an occurrence is θ. Thus, each x_i is the observed time to 3 occurrences, so that $x_1+\cdots+x_7=18$ is the elapsed time until the (3)(7) = 21st occurrence. It follows that the estimated mean time for a single occurrence is 18/21.

Maximum Likelihood Estimators possess some additional noteworthy theoretical features, which we mention now for completeness, although we will not be making further use of them here. First, they have an ***invariance*** property, meaning that if $\hat{\theta}$ is the MLE of θ, and $g(\theta)$ is a transformation, then $g(\hat{\theta})$ is the MLE for $g(\theta)$.

Secondly, MLEs have certain asymptotic properties. Recall from Section 11.1 that an estimator is asymptotically unbiased if the bias goes to zero as the sample size increases to infinity. The MLE is always asymptotically unbiased. It is also asymptotically normal, which means that as the sample size increases the MLE more closely approximates a normal distribution. Finally, in the next section we define an estimator as ***efficient*** if it has the minimum possible variance. In this sense, the MLE is also asymptotically ***efficient,*** that is, it approaches the minimum possible variance as the sample size goes to infinity.

Exercise 11-9 Let $X_1,\ldots,X_n$ be a random sample from an exponential distribution with mean β. Find the maximum likelihood estimate for β.

Exercise 11-10 Suppose that the life expectancy of a light-bulb is modeled by an exponential distribution. You randomly sampled 6 bulbs and found that they lasted 650 hours, 1123 hours, 67 hours, 1257 hours, 435 hours, and 1896 hours. Find the maximum likelihood estimate for the unknown parameter β.

Exercise 11-11 Suppose that $X_1,\ldots,X_n$ is a random sample from a geometric distribution with unknown parameter p. Determine the maximum likelihood estimate for p. *Hint*: Recall that $\Pr[X=k]=p\cdot(1-p)^k$ for $k=0,1,2,\ldots$.

Exercise 11-12 Suppose that $X_1,\ldots,X_n$ is a random sample from a distribution with density function $f(x)=\theta x^{\theta-1}$ for $0<x<1$ and $0<\theta<\infty$. Find the maximum likelihood estimate for θ.

Exercise 11-13 A random sample $X_1,X_2,\ldots,X_n$ is taken from a distribution with density function $f(x)=\begin{cases}(\theta+1)x^\theta & \text{for } 0<x<1 \\ 0 & \text{otherwise}\end{cases}$. What is the maximum likelihood estimator for θ?

Exercise 11-14 Let X be an exponential random variable with mean β and let $X_1,\ldots,X_n$ be a random sample from X. Then we know that $\overline{X}$ is an unbiased estimator for β, and from Example 11.2-5, that $\hat{\lambda}=\frac{1}{\overline{X}}$ is the MLE for the Poisson rate parameter, $\lambda=\frac{1}{\beta}$. The object of this exercise is to calculate the bias of $\hat{\lambda}$.

(a) Find the distribution of $\overline{X}$. **Hint:** Let $Y=\sum_{i=1}^{n}X_i$. What is the distribution type of Y (see Section 6.5)? What is the distribution type of $\frac{Y}{n}=\overline{X}$ (see Section 6.5 again)?

(b) Use the density function of $\overline{X}$ to calculate $E[\hat{\lambda}]=E\left[\frac{1}{\overline{X}}\right]=\int_0^\infty x^{-1}f_{\overline{X}}(x)\,dx$.

(c) Use the result of (b) to calculate $B(\hat{\lambda})$.

Exercise 11-15 CAS Exam 3 Spring 2005 #18

The following sample is taken from the distribution

$$f(x,\theta) = \left(\frac{1}{\theta}\right)e^{\frac{-x}{\theta}}$$

Observation	1	2	3	4	5	6	7
x	0.49	1.00	0.47	0.91	2.47	5.03	16.09

Determine the Maximum Likelihood Estimator of c, where $\Pr(X > c) = 0.75$.

(A) Less than 1.0
(B) At least 1.0, but less than 1.2
(C) At least 1.2, but less than 1.4
(D) At least 1.4, but less than 1.6
(E) At least 1.6

Exercise 11-16 CAS Exam 3 Spring 2006 #2

Annual claim counts follow a Negative Binomial distribution, with $p = \frac{1}{\beta+1}$. The following claim count observations are available:

Year	Claim Counts
2005	0
2006	3
2007	5

Assuming each year is independent, calculate the likelihood function of this sample.

(A) $\left(\frac{1}{\beta+1}\right)^{3r}\left(\frac{\beta}{\beta+1}\right)^{8}\frac{r^2(r+2)^2(r+4)}{3!5!}$

(B) $\left(\frac{1}{\beta+1}\right)^{3r}\left(\frac{\beta}{\beta+1}\right)^{8}\frac{r^2(r+2)^2(r+4)}{2!4!}$

(C) $\left(\frac{1}{\beta+1}\right)^{3r}\left(\frac{\beta}{\beta+1}\right)^{8}\frac{r^2(r+1)^2(r+2)^2(r+3)}{2!4!}$

(D) $\left(\frac{1}{\beta+1}\right)^{3r}\left(\frac{\beta}{\beta+1}\right)^{8}\frac{r^2(r+1)^2(r+2)^2(r+3)(r+4)}{2!4!}$

(E) $\left(\frac{1}{\beta+1}\right)^{3r}\left(\frac{\beta}{\beta+1}\right)^{8}\frac{r^2(r+1)^2(r+2)^2(r+3)(r+4)}{3!5!}$

Exercise 11-17 CAS Exam 3 Spring 2007 #10

Let $Y_1, Y_2, Y_3, Y_4, \ldots, Y_n$, represent a random sample from the following distribution with p.d.f.

$$f(x) = \begin{cases} e^{-x+\theta} & \text{where } \theta < x < \infty, -\infty < \theta < \infty \\ 0 & \text{elsewhere} \end{cases}.$$

Which one of the following is a maximum likelihood estimator for θ?

(A) $\sum_{i=1}^{n} Y_i$

(B) $\sum_{i=1}^{n} Y_i^2$

(C) $\prod_{i-1}^{n} Y_i$

(D) Minimum$[Y_1, Y_2, Y_3, Y_4, \ldots, Y_n]$

(E) Maximum$[Y_1, Y_2, Y_3, Y_4, \ldots, Y_n]$

Exercise 11-18 CAS Exam 3 Spring 2007 #11

The proportion of allotted time a student takes to complete an exam, x, is described by the following distribution:

$$f(x) = (\theta + 1)x^{\theta} \quad 0 \le x \le 1 \quad \text{and} \quad \theta > -1.$$

A random sample of five students produced the following observations:

Student	Proportion of Allotted Time
1	0.92
2	0.79
3	0.90
4	0.65
5	0.86

Using the sample data, calculate the maximum likelihood estimate of θ.

(A) Less than 0
(B) At least 0, but less than 1.0
(C) At least 1.0, but less than 2.0
(D) At least 2.0, but less than 3.0
(E) At least 3.0

Exercise 11-19 CAS Exam 3 Fall 2006 #2

Call center response times are described by the cumulative distribution function $F(x) = x^{\theta+1}$, where $0 \le x \le 1$ and $\theta > -1$.

A random sample of response times is as follows:

0.56 0.83 0.74 0.68 0.75

Calculate the maximum likelihood estimate of θ.

(A) Less than 1.4
(B) At least 1.4, but less than 1.6
(C) At least 1.6 but less than 1.8
(D) At least 1.8, but less than 2.0
(E) At least 2.0

Exercise 11-20 CAS Exam 3, Segment 3L Spring 2008 #3

You are given the following:

- A random sample of claim amounts:

  8,000 10,000 12,000 15,000

- Claim amounts follow an inverse exponential distribution, with parameter θ, whose density function is given by $f(x) = \dfrac{\theta e^{-\theta/x}}{x^2}$ for $x > 0$.

Calculate the maximum likelihood estimator for θ.

(A) Less than 9,000
(B) At least 9,000, but less than 10,000
(C) At least 10,000, but less than 11,000
(D) At least 11,000, but less than 12,000
(E) At least 12,000

Exercise 11-21 CAS Exam 3 Fall 2005 #4

When Mr. Jones visits his local race track, he places three independent bets. In his last 20 visits, he lost all of his bets 10 times, won one bet 7 times, and won two bets 3 times. He never won all three of his bets.

Calculate the maximum likelihood estimate of the probability that Mr. Jones wins an individual bet.

(A) $\frac{13}{60}$ (B) $\frac{4}{15}$ (C) $\frac{19}{60}$ (D) $\frac{11}{30}$ (E) $\frac{5}{12}$

Exercise 11-22 CAS Exam 3 Spring 2005 #20

Blue Sky Insurance Company insures a portfolio of 100 automobiles against physical damage. The annual number of claims follows a binomial distribution with $m = 100$.

For the last 5 years, the number of claims in each year has been:

Year 1	5
Year 2	4
Year 3	4
Year 4	9
Year 5	3

Two methods for estimating the variance in the annual claim count are:

Method 1: Unbiased Sample Variance
Method 2: Maximum Likelihood Estimation

Use each method to calculate an estimate of the variance. What is the difference between the two estimates?

(A) Less than 0.50
(B) At least 0.50, but less than 0.60
(C) At least 0.60, but less than 0.70
(D) At least 0.70, but less than 0.80
(E) 0.80 or more

Exercise 11-23 CAS Exam 3 Fall 2007 #6

Waiting times at a bank follow an exponential distribution with mean equal to θ. The first five people in line are observed to have the following waiting times: 10, 5, 21, 10, 7.

- $\hat{\theta}_A =$ Maximum Likelihood Estimator of θ
- $\hat{\theta}_B =$ Method of Moments Estimator of θ

Calculate $\hat{\theta}_A - \hat{\theta}_B$.

(A) Less than -0.6
(B) At least -0.6, but less than -0.2
(C) At least -0.2, but less than 0.2
(D) At least 0.2, but less than 0.6
(E) At least 0.6

Exercise 11-24 CAS Exam 3, Segment 3L Fall 2008 #4

You are given the following information:

- A random variable x has probability density function:

$$f(x;\theta) = \theta x^{\theta-1}, \quad \text{where} \quad 0 < x < 1 \text{ and } \theta > 0$$

- A random sample of five observations from this distribution is shown below:

0.25	0.50	0.40	0.80	0.65

Calculate the maximum likelihood estimator for θ.

(A) Less than 1.00
(B) At least 1.00, but less than 1.10
(C) At least 1.10, but less than 1.20
(D) At least 1.20, but less than 1.30
(E) At least 1.30

Exercise 11-25 CAS Exam 3, Segment 3L Fall 2008 #6

You are given the following:

- An insurance company provides a coverage which can result in only three loss amounts in the event that a claim is filed: \$0, \$500 or \$1000.
- The probability, p, of a loss being \$0 is the same as the probability of it being \$1000.
- The following 3 claims are observed:

\$0	\$0	\$1000

What is the maximum likelihood estimate of p?

(A) Less than 0.20
(B) At least 0.20, but less than 0.40
(C) At least 0.40, but less than 0.60
(D) At least 0.60, but less than 0.80
(E) At least 0.80

11.3 PROPERTIES OF ESTIMATORS

As we have seen, there are several ways to build estimators for parameters, and it would be useful to have some criteria for determining what makes one estimator better than another. One property we have already used extensively involves bias. Other things being equal, it is better to have an unbiased estimator, although, for large sample sizes an asymptotically unbiased estimator may work just as well.

In this section we introduce other criteria that bear on the quality of an estimator. We would wish that a proposed estimator be both accurate and reliable, and we develop mathematical tools that provide measures for these characteristics. In Section 11.3.1 we discuss ***consistency***, which, loosely speaking, bears on accuracy. In Section 11.3.2 we define ***efficiency***, a measure of variability, and hence reliability. This involves the variance of the estimator.

All things considered we would prefer an unbiased estimator with the minimum possible variance. In Section 11.4 we develop the concept of ***sufficiency*** for an estimator, and show how this is related to the construction of minimum variance unbiased estimators.

There is also an approach to estimator selection that is based on comparing the cost, or loss, of choosing one value for an estimator over another. This involves measuring the loss (for example, loss of accuracy) for all possible values of the parameter and choosing the estimator value that minimizes this loss over the parameter domain. This is most frequently done in Bayesian estimation (Section 11.6.2), where a probability distribution is assigned to the parameter domain. This leads to what is referred to as the ***mean square error*** criterion, still another measure of estimator quality.

11.3.1 CONSISTENT ESTIMATORS

Let $\hat{\theta}$ be an unbiased estimator for θ based on a random sample of size n. The quality of consistency for $\hat{\theta}$ refers to a form of convergence to θ as the sample size increases. This is closely related to the Chebyshev inequality and its application to the Law of Large Numbers in Subsection 6.4.2.

Consistency

Let X be a population random variable with parameter θ. Let $X_1,\ldots,X_n$ be a random sample of size n from X and let $\hat{\theta}$ be an unbiased estimator for θ based on the random sample. If, for any $\varepsilon > 0$,

$$\lim_{n\to\infty} \Pr\left[\left|\hat{\theta}-\theta\right| \le \varepsilon\right] = 1,$$

then $\hat{\theta}$ is said to be a ***consistent*** estimator of θ.

Notes

(1) Of course, for this definition to make sense we must be cognizant of the dependency of $\hat{\theta}$ on n, although for simplicity's sake we have not made this dependency explicit in the notation.

(2) In advanced mathematics courses the particular type of convergence in the definition of consistency is called ***convergence in probability***.

(3) Virtually all of the estimators discussed in this book satisfy the criterion of consistency. Also, sums, products and continuous transformations of consistent estimators are consistent as well for the new quantities being estimated.

The property of consistency says that for large n the absolute difference between $\hat{\theta}$ and its mean, θ, is small. Since variance refers to the square of this absolute difference, it is not surprising that consistency is in some sense related to the variance of $\hat{\theta}$ being small. In the next subsection we will more closely examine variance as a key measure of the performance of estimators. For now, though, we shall provide a sufficient condition for consistency in terms of the variance of $\hat{\theta}$.

Establishing Consistency

Let X be a population random variable with a parameter θ. Let $X_1,\ldots,X_n$ be a random sample of size n from X and let $\hat{\theta}$ be an unbiased estimator for θ based on the random sample.

(1) If $\lim_{n\to\infty} Var[\hat{\theta}]=0$, then $\hat{\theta}$ is a consistent estimator of θ.

(2) If X has a finite mean and variance, then $\overline{X}$ is a consistent estimator for μ_X.

Proof

This is based on the Chebyshev inequality applied to $\hat{\theta}$ and its mean, θ. With this notation, the Chebyshev inequality says,

$$\Pr\left[\left|\hat{\theta}-\theta\right|\le k\sigma_{\hat{\theta}}\right] \ge 1-\frac{1}{k^2}.$$

To show $\hat{\theta}$ is consistent, let $\varepsilon>0$ be given. Then to prove $\lim_{n\to\infty}\Pr\left[\left|\hat{\theta}-\theta\right|\le\varepsilon\right]=1$, we must show that for any given $\delta>0$ there exists a positive integer N such that for $n\ge N$ we have,

$$\Pr\left[\left|\hat{\theta}-\theta\right|\le\varepsilon\right] \ge 1-\delta.$$

First, choose k sufficiently large so that $\frac{1}{k^2}<\delta$. Since $\lim_{n\to\infty} Var[\hat{\theta}]=0$, we can choose a positive integer N such that for any $n\ge N$ we have $Var[\hat{\theta}]\le\left(\frac{\varepsilon}{k}\right)^2$. This is equivalent to $\varepsilon\ge k\sigma_{\hat{\theta}}$. Hence, using the Chebyshev inequality, we have,

$$\Pr\left[\left|\hat{\theta}-\theta\right|\le\varepsilon\right] \ge \Pr\left[\left|\hat{\theta}-\theta\right|\le k\sigma_{\hat{\theta}}\right] \ge 1-\frac{1}{k^2} \ge 1-\delta.$$

This establishes (1). Now (2) is an easy corollary since $Var[\overline{X}]=\frac{\sigma_X^2}{n}$, which converges to 0 as n goes to infinity.

11.3.2 EFFICIENT ESTIMATORS

Consistent estimators are reliable in the sense that they close in on the parameter as the sample size increases. Efficiency, on the other hand, refers to the reliability of an estimator based on random samples of a fixed size n. We continue to restrict our attention to ***unbiased*** estimators. The ***efficiency*** of an unbiased estimator refers to its variance. Given two unbiased estimators based on the same random sample, we would consider the one with the smaller variance to be more efficient than the other.

Relative Efficiency

Let X be a population random variable with a parameter θ. Let $X_1,\ldots,X_n$ be a random sample of size n from X and let $\hat{\theta}_1$ and $\hat{\theta}_2$ be unbiased estimators for θ based on the random sample. If

$$Var[\hat{\theta}_1] < Var[\hat{\theta}_2],$$

then we say $\hat{\theta}_1$ is ***relatively more efficient*** than $\hat{\theta}_2$. The relative efficiency of the estimators $\hat{\theta}_1$ and $\hat{\theta}_2$ is defined as $\dfrac{Var[\hat{\theta}_1]}{Var[\hat{\theta}_2]}$.

Example 11.3-1 Uniform on $[0,\theta]$ (Continuation)

Let X be a uniform random variable on the interval $[0,\theta]$, where $\theta > 0$ is an unknown parameter. Let $X_1,\ldots,X_n$ be a random sample from X and let $U = Max(X_1,\ldots,X_n)$.

In Example 11.1-2 we showed that $\hat{\theta}_1 = \frac{n+1}{n}U$ is an unbiased estimator for θ. In Example 11.2-1 we showed that $\hat{\theta}_2 = 2\overline{X}$ is also an unbiased estimator for θ. Calculate the relative efficiency, $\dfrac{Var[\hat{\theta}_1]}{Var[\hat{\theta}_2]}$.

Solution

We begin by calculating the variance of $U = Max(X_1,\ldots,X_n)$. From Example 11.1-2 we have $f_U(x) = \frac{n}{\theta^n}x^{n-1};\ 0 \le x \le \theta$. From that density function we calculated $E[U] = \frac{n}{n+1}\theta$. To get the variance we also need the second moment, calculated as

$$E[U^2] = \int_0^\theta x^2\left(\frac{n}{\theta^n}x^{n-1}\right)dx = \frac{n}{\theta^n}\int_0^\theta x^{n+1}\,dx = \frac{n}{n+2}\theta^2.$$

Thus, using the variance formula,

$$\begin{aligned}Var[U] &= \frac{n}{n+2}\theta^2 - \left(\frac{n}{n+1}\theta\right)^2 = \left[\frac{n}{n+2} - \frac{n^2}{(n+1)^2}\right]\theta^2 \\ &= n\left[\frac{1}{n+2} - \frac{n}{(n+1)^2}\right]\theta^2 \\ &= n\left[\frac{(n+1)^2 - n(n+2)}{(n+2)(n+1)^2}\right]\theta^2 = \frac{n}{(n+2)(n+1)^2}\theta^2.\end{aligned}$$

Since $\hat{\theta}_1 = \frac{n+1}{n}U$, we have $Var[\hat{\theta}_1] = \left(\frac{n+1}{n}\right)^2 Var[U] = \left(\frac{n+1}{n}\right)^2 \cdot \frac{n}{(n+2)(n+1)^2}\theta^2 = \frac{1}{n(n+2)}\theta^2.$

Calculating $Var[\hat{\theta}_2]$ is somewhat easier. For that, we have,

$$Var[\hat{\theta}_2] = Var[2\bar{X}] = 4 \cdot Var[\bar{X}] = 4 \cdot \frac{\sigma_X^2}{n} = \overbrace{\frac{4}{n} \cdot \frac{\theta^2}{12}}^{\text{Since } X \text{ is uniform on } [0,\theta]} = \frac{\theta^2}{3n}.$$

Finally, the relative efficiency is given by

$$\frac{Var[\hat{\theta}_1]}{Var[\hat{\theta}_2]} = \frac{\frac{1}{n(n+2)}\theta^2}{\frac{\theta^2}{3n}} = \frac{3}{n+2}.$$

❐

Notes

(1) For samples of size one the two estimators are identical. In all other cases the estimator based on the maximum of the observations is more efficient than the estimator based on the sample mean, and therefore more reliable in practice.

(2) Since $\hat{\theta}_1$ and $\hat{\theta}_2$ are unbiased estimators for θ, it follows that $\frac{\hat{\theta}_1}{2}$ and $\frac{\hat{\theta}_2}{2}$ are both unbiased estimators of the population mean $\frac{\theta}{2}$. Dividing each estimator by two doesn't affect their relative efficiency. This leads to the somewhat counter-intuitive result that <u>the sample mean is not always the most efficient estimator of the population mean</u>.

11.3.3 EFFICIENCY AND THE CRAMER-RAO INEQUALITY

The previous subsection discusses the relative efficiency of unbiased estimators by comparing their variances. This is fine as far as it goes, but it does raise further questions. For example, is there a ***most*** efficient estimator? That is, is there an unbiased estimator with the minimum possible variance? And if so, how will we know it when we see it? We use the abbreviation MVUE for minimum variance unbiased estimators.

It turns out that in many cases we can answer affirmatively based on the fact that for a given sample size there exists an absolute lower bound for the variance of any unbiased estimator. This lower bound is referred to as the ***Cramer-Rao Lower Bound*** (CRLB). If we have an estimator whose variance actually equals this lower bound then we know it is at least tied for being the most efficient possible unbiased estimator and is therefore an MVUE.

To begin addressing these issues we return to the context of maximum likelihood estimation discussed in Subsection 11.2.2. We will be delving into some of the deeper properties of the log-likelihood functions introduced in that section and logarithmic differentiation will play a major role here. To set the stage we summarize the main results we will need in the following derivations.

Logarithmic Differentiation

Let $L(\theta)$ be a differentiable function taking positive values. Then,

(1) $[\ln L(\theta)]' = \dfrac{L'(\theta)}{L(\theta)}$,

(2) $L'(\theta) = [\ln L(\theta)]' \cdot L(\theta)$.

Proof

(1) is just the chain rule applied to the composite function $\ln L(\theta)$, and (2) is just a re-write of (1) in multiplicative form.

Now, let X be a population random variable with a parameter θ and density function given by $f_X(x;\theta)$. We will be working only with populations that satisfy certain so-called ***regularity*** conditions. Basically, this means that the parameter θ cannot appear in the limits of integration for calculating expected values for X. Unfortunately, this eliminates one of our favorite examples, the uniform distribution on the interval $[0,\theta]$, but most populations will be ***regular*** in this sense. We will point out in the proof where the regularity condition is used.

Now, consider the transformation of X given by the odd looking formula,

$$Y = \frac{\partial[\ln f_X(X;\theta)]}{\partial\theta}.$$

For a given observation x of X this is just the derivative with respect to θ of the log-likelihood function based on our random sample of size one. This, in turn, is simply the expression we set equal to zero in order to find the MLE for θ. The main difference now is that we are using the partial-derivative notation, $\frac{\partial}{\partial\theta}$, rather than a simple prime to indicate differentiation with respect to θ. The reason is that we will be doing manipulations on both the random variable X and the associated parameter θ, so we will use the more precise, although somewhat cumbersome, notation of partial derivatives.

It turns out that the minimum possible variance of an unbiased estimator is closely related to the variance of the transformation random variable Y. So our first goal is to derive an expression for that variance.

The Mean and Variance of Y

Let X be a ***regular*** population random variable with a parameter θ and density function given by $f_X(x;\theta)$.

(3) $\frac{\partial}{\partial\theta}[f_X(x;\theta)] = \frac{\partial[\ln f_X(x;\theta)]}{\partial\theta} \cdot f_X(x;\theta).$

Let $Y = \frac{\partial[\ln f_X(X;\theta)]}{\partial\theta}$.

(4) $E[Y] = 0,$

(5) $Var[Y] = E[Y^2] = E\left[\left(\frac{\partial[\ln f_X(X;\theta)]}{\partial\theta}\right)^2\right] = -E\left[\frac{\partial^2[\ln f_X(X;\theta)]}{\partial\theta^2}\right]$

Proof

The expression in (3) is just the restatement of the logarithmic differentiation formula in (2) using partial derivative notation and applied to $L(\theta) = f_X(x;\theta)$.

For (4), we have, using the standard method for evaluating the expectation of a transformation of X,

$$E[Y] = \int_{-\infty}^{\infty} \frac{\partial[\ln f_X(x;\theta)]}{\partial\theta} \cdot f_X(x;\theta)\, dx.$$

To show this equals zero, we begin with the basic fact that the area under the density function $f_X(x;\theta)$ must equal one:

$$1 = \int_{-\infty}^{\infty} f_X(x;\theta)\, dx.$$

The next step is to differentiate both sides with respect to θ. For the right-hand side we must differentiate inside the integral sign, or, in other words, interchange the order of operation between integration with respect to x and differentiation with respect to θ. It is precisely here that we invoke the condition of ***regularity*** (see the note below) that justifies this step. The result is,

(6) $$0 = \int_{-\infty}^{\infty} \frac{\partial[f_X(x;\theta)]}{\partial\theta}\, dx = \overbrace{\int_{-\infty}^{\infty} \frac{\partial[\ln f_X(x;\theta)]}{\partial\theta} \cdot f_X(x;\theta)\, dx}^{\text{using formula (3)}} = E[Y].$$

Now that we have $E[Y] = 0$, we know from the variance formula that,

$$Var[Y] = E[Y^2] = E\left[\left(\frac{\partial[\ln f_X(X;\theta)]}{\partial\theta}\right)^2\right].$$

To show the last equality in (5) we begin with the expression (6) above for $E[Y]$:

$$0 = \int_{-\infty}^{\infty} \frac{\partial[f_X(x;\theta)]}{\partial\theta} dx = \int_{-\infty}^{\infty} \frac{\partial[\ln f_X(x;\theta)]}{\partial\theta} \cdot f_X(x;\theta)\, dx .$$

Differentiate again with respect to θ, once again interchanging integration and differentiation, to get,

$$\begin{aligned}
0 &= \int_{-\infty}^{\infty} \frac{\partial^2[f_X(x;\theta)]}{\partial\theta^2} dx \\
&= \int_{-\infty}^{\infty} \frac{\partial}{\partial\theta}\left\{ \frac{\partial[\ln f_X(x;\theta)]}{\partial\theta} \cdot f_X(x;\theta) \right\} dx \\
&= \int_{-\infty}^{\infty} \left\{ \overbrace{\underbrace{\frac{\partial^2[\ln f_X(x;\theta)]}{\partial\theta^2} \cdot f_X(x;\theta)}_{\text{derivative of the first times the second}} + \underbrace{\left(\frac{\partial[\ln f_X(x;\theta)]}{\partial\theta}\right) \cdot \left(\frac{\partial[\ln f_X(x;\theta)]}{\partial\theta} \cdot f_X(x;\theta)\right)}_{\text{the first times derivative of the second, using (3)}}}^{\text{product rule for differentiation}} \right\} dx \\
&= \int_{-\infty}^{\infty} \frac{\partial^2[\ln f_X(x;\theta)]}{\partial\theta^2} \cdot f_X(x;\theta)\, dx + \int_{-\infty}^{\infty} \left(\frac{\partial[\ln f_X(x;\theta)]}{\partial\theta}\right)^2 \cdot f_X(x;\theta)\, dx \\
&= E\left[\frac{\partial^2[\ln f_X(X;\theta)]}{\partial\theta^2}\right] + E\left[\left(\frac{\partial[\ln f_X(X;\theta)]}{\partial\theta}\right)^2\right].
\end{aligned}$$

Thus, we have

$$0 = E\left[\frac{\partial^2[\ln f_X(X;\theta)]}{\partial\theta^2}\right] + E\left[\left(\frac{\partial[\ln f_X(X;\theta)]}{\partial\theta}\right)^2\right],$$

which implies the last equality in (5).

Notes

(1) We used the generic limits of integration, from negative infinity to positive infinity but, as usual, the actual limits of integration will be determined by the domain upon which the density function $f_X(x;\theta)$ is positive.

(2) The ***regularity*** condition on $f_X(x;\theta)$ ensures that the actual limits of integration will not depend on the parameter θ. So long as that is the case, then taking the derivative with respect to θ inside the integral is comparable to the fact that the derivative of a sum is the sum of the derivatives. Since the integral is just a limit of sums, the validity of the operation carries over.

The expressions in (5) for the variance of Y are fundamental to the Cramer-Rao inequality, and are called the ***Fisher information*** for $f_X(x;\theta)$.

Fisher Information

Let X be a ***regular*** population random variable with a parameter θ and density function given by $f_X(x;\theta)$. Let $Y = \dfrac{\partial[\ln f_X(X;\theta)]}{\partial\theta}$.

The ***Fisher information*** for $f_X(x;\theta)$ is defined to be $I(\theta)$, where

$$I(\theta) = Var[Y] = E\left[\left(\frac{\partial[\ln f_X(X;\theta)]}{\partial\theta}\right)^2\right] = -E\left[\frac{\partial^2[\ln f_X(X;\theta)]}{\partial\theta^2}\right].$$

As we shall see, the Fisher information is inversely related to the lower bound for the variance of unbiased estimators. The above expressions for the Fisher information are somewhat formidable in appearance, but are often relatively easy to evaluate. Here is an example.

Example 11.3-2 Fisher Information for the Exponential

Let X be an exponential population random variable with density function $f_X(x;\theta) = \frac{1}{\theta}e^{-(x/\theta)}$; $0 \le x < \infty$ and $\theta > 0$.

(a) Calculate $I(\theta)$ using the expression $E\left[\left(\dfrac{\partial[\ln f_X(X;\theta)]}{\partial\theta}\right)^2\right]$.

(b) Calculate $I(\theta)$ using the expression $-E\left[\dfrac{\partial^2[\ln f_X(X;\theta)]}{\partial\theta^2}\right]$.

Solution

Begin on the inside with $f_X(X;\theta)$ and work your way out.

(a) First, $f_X(X;\theta) = \frac{1}{\theta}e^{-(X/\theta)}$. Then $\ln f_X(X;\theta) = -\ln\theta - \frac{X}{\theta} = -\ln\theta - X\theta^{-1}$. Next,

$$\frac{\partial[\ln f_X(X;\theta)]}{\partial\theta} = -\frac{1}{\theta} + X\theta^{-2} = -\frac{1}{\theta} + \frac{X}{\theta^2}.$$

The next step is to square this, giving,

$$\left(\frac{\partial[\ln f_X(X;\theta)]}{\partial\theta}\right)^2 = \frac{1}{\theta^2} - \frac{2X}{\theta^3} + \frac{X^2}{\theta^4}.$$

Finally, since X is exponential, we have $E[X] = \theta$ and

$$E[X^2] = Var[X] + E[X]^2 = \theta^2 + \theta^2 = 2\theta^2.$$

Thus,

$$I(\theta) = E\left[\left(\frac{\partial[\ln f_X(X;\theta)]}{\partial\theta}\right)^2\right] = \frac{1}{\theta^2} - \frac{2\theta}{\theta^3} + \frac{2\theta^2}{\theta^4} = \frac{1}{\theta^2}.$$

(b) From (a) we have

$$\frac{\partial[\ln f_X(X;\theta)]}{\partial\theta} = -\frac{1}{\theta} + X\theta^{-2} = -\theta^{-1} + X\theta^{-2}.$$

Taking the second derivative gives,

$$\frac{\partial^2[\ln f_X(X;\theta)]}{\partial\theta^2} = \theta^{-2} - 2X\theta^{-3} = \frac{1}{\theta^2} - \frac{2X}{\theta^3}.$$

Then,

$$I(\theta) = -E\left[\frac{\partial^2[\ln f_X(X;\theta)]}{\partial\theta^2}\right] = -E\left[\frac{1}{\theta^2} - \frac{2X}{\theta^3}\right] = -\left(\frac{1}{\theta^2} - \frac{2\theta}{\theta^3}\right) = \frac{1}{\theta^2}.$$ ❐

To derive the Cramer-Rao inequality we next expand the concept of Fisher information to random samples of arbitrary size n. We now let $X_1,\ldots,X_n$ be a random sample from X, where X is a population random variable with density function $f_X(x;\theta)$. For each observation X_i, let $Y_i = \frac{\partial[\ln f_X(X_i;\theta)]}{\partial\theta}$. From the above, we have for each i, $E[Y_i]=0$ and $Var[Y_i] = I(\theta)$.

We now consider the multivariate joint density function, $f_{\mathbf{X}}(x_1,x_2,\ldots,x_n;\theta)$, for $\mathbf{X} = (X_1,\ldots,X_n)$. We again construct a transformation based on the derivative with respect to θ of the log-likelihood function, this time letting,

$$V = \frac{\partial[\ln f_{\mathbf{X}}(X_1,X_2,\ldots,X_n;\theta)]}{\partial\theta}.$$

As in the one-variable case, we are interested in the variance of the random variable V.

The Mean and Variance of V

Let $X_1,\ldots,X_n$ be a random sample from X, where X is a population random variable with density function $f_X(x;\theta)$. For each observation X_i, let $Y_i = \frac{\partial[\ln f_X(X_i;\theta)]}{\partial\theta}$, and let $V = \frac{\partial[\ln f_{\mathbf{X}}(X_1,X_2,\ldots,X_n;\theta)]}{\partial\theta}$, where $f_{\mathbf{X}}(x_1,x_2,\ldots,x_n;\theta)$ is the multivariate joint distribution for $\mathbf{X} = (X_1,\ldots,X_n)$.

(1) $V = Y_1 + \cdots + Y_n$,
(2) $E[V]=0$, and,
(3) $Var[V] = nI(\theta) = nE\left[\left(\frac{\partial[\ln f_X(X;\theta)]}{\partial\theta}\right)^2\right] = -nE\left[\frac{\partial^2[\ln f_X(X;\theta)]}{\partial\theta^2}\right]$

Proof

Since the n observations $X_1,\ldots,X_n$ are independent, the joint distribution is, $f_{\mathbf{X}}(x_1,x_2,\ldots,x_n;\theta)=\prod_{i=1}^{n} f_X(x_i;\theta)$. Thus,

$$V = \frac{\partial\left[\ln f_{\mathbf{X}}(X_1,X_2,\ldots,X_n;\theta)\right]}{\partial\theta} = \frac{\partial\left[\ln \prod_{i=1}^{n} f_X(x_i;\theta)\right]}{\partial\theta}$$

$$= \frac{\partial\left[\sum_{i=1}^{n} \ln f_X(x_i;\theta)\right]}{\partial\theta} = Y_1+\cdots+Y_n.$$

This establishes (1), and (2) follows immediately since for each i, $E[Y_i]=0$. Since $Y_1,\ldots,Y_n$ are independent, (3) also follows from (1) and the fact that for each i, $Var[Y_i]=I(\theta)$.

Note

The quantity $nI(\theta)$ is sometimes referred to as the Fisher information for the sample of size n.

As we shall see next, the lower bound for the variance of unbiased estimators is given by $\frac{1}{Var[V]}$. Let $\hat{\theta}$ be any unbiased estimator for θ based on the random sample $X_1,\ldots,X_n$. This means that $\hat{\theta}$ is a transformation of the random variables $X_1,\ldots,X_n$, given by $\hat{\theta}=u(X_1,\ldots,X_n)$. To calculate the expected value of $\hat{\theta}$ requires an n-fold multiple integral using the multivariate density function $f_{\mathbf{X}}(x_1,x_2,\ldots,x_n;\theta)$. In the following we write this as,

$$E[\hat{\theta}] = \int_{-\infty}^{\infty}\cdots\int_{-\infty}^{\infty} u(x_1,\ldots,x_n)\cdot f_{\mathbf{X}}(x_1,x_2,\ldots,x_n;\theta)\,dx_1\cdots dx_n.$$

The Cramer-Rao Inequality

Let $X_1,\ldots,X_n$ be a random sample from X, where X is a population random variable with regular density function $f_X(x;\theta)$. Let $V=\frac{\partial\left[\ln f_{\mathbf{X}}(X_1,X_2,\ldots,X_n;\theta)\right]}{\partial\theta}$, where $f_{\mathbf{X}}(x_1,x_2,\ldots,x_n;\theta)$ is the multivariate joint distribution for $\mathbf{X}=(X_1,\ldots,X_n)$.

Let $\hat{\theta}=u(X_1,\ldots,X_n)$ be an unbiased estimator for θ. Then,

$$Var[\hat{\theta}] \geq \frac{1}{Var[V]} = \frac{1}{nI(\theta)} = \frac{1}{nE\left[\left(\frac{\partial\left[\ln f_X(X;\theta)\right]}{\partial\theta}\right)^2\right]} = -\frac{1}{nE\left[\frac{\partial^2\left[\ln f_X(X;\theta)\right]}{\partial\theta^2}\right]}.$$

The expressions for $\frac{1}{Var[V]}$ on the right are referred to as the ***Cramer-Rao Lower Bound*** (CRLB) for the variance of estimators based on $X_1,\ldots,X_n$. An estimator whose variance equals the CRLB is termed an ***efficient*** estimator.

Proof

Since $\hat{\theta} = u(X_1,\dots,X_n)$ is unbiased, we have,

$$\theta = E[\hat{\theta}] = \int_{-\infty}^{\infty}\cdots\int_{-\infty}^{\infty} u(x_1,\dots,x_n)\cdot f_{\mathbf{X}}(x_1,x_2,\dots,x_n;\theta)\,dx_1\cdots dx_n.$$

Once again, we differentiate both sides with respect to θ, employing the regularity property and logarithmic differentiation on the right hand side to obtain,

$$1 = \int_{-\infty}^{\infty}\cdots\int_{-\infty}^{\infty} u(x_1,\dots,x_n)\cdot\frac{\partial\left[\ln f_{\mathbf{X}}(x_1,x_2,\dots,x_n;\theta)\right]}{\partial\theta}\cdot f_{\mathbf{X}}(x_1,x_2,\dots,x_n;\theta)\,dx_1\cdots dx_n.$$

We now recognize the resulting multiple integral as the calculation for $E[\hat{\theta}\cdot V]$. Since $E[V]=0$, we have $Cov[\hat{\theta},V] = E[\hat{\theta}\cdot V] - E[\hat{\theta}]E[V] = E[\hat{\theta}\cdot V] = 1$.

Recall from Section 7.4 that the correlation between the two random variables $\hat{\theta}$ and V is given by $\rho = \frac{Cov[\hat{\theta},V]}{\sigma_{\hat{\theta}}\cdot\sigma_V}$, or equivalently, $Cov[\hat{\theta},V] = \rho\cdot\sigma_{\hat{\theta}}\cdot\sigma_V$. Recall, too, from Section 7.4 that $|\rho|\le 1$. Therefore, $1 = Cov[\hat{\theta},V] = \rho\cdot\sigma_{\hat{\theta}}\cdot\sigma_V \le \sigma_{\hat{\theta}}\cdot\sigma_V$. Squaring both sides gives, $1 \le Var[\hat{\theta}]\cdot Var[V]$, and so, $Var[\hat{\theta}] \ge \frac{1}{Var[V]}$. The remaining expressions for $\frac{1}{Var[V]}$ follow from the formulas in (3) for the variance of V.

Example 11.3-3 Efficient Estimator for the Exponential Parameter

Let X be an exponential population random variable with density function $f_X(x;\theta) = \frac{1}{\theta}e^{-(x/\theta)}$; $0 \le x < \infty$ and $\theta > 0$. Let $X_1,\dots,X_n$ be a random sample from X.

(a) Find the Cramer-Rao lower bound for the variance of unbiased estimators for θ.
(b) Determine whether or not $\overline{X}$ is an *efficient* estimator for θ.

Solution

In Example 11.3-2 we calculated the Fisher information $I(\theta) = \frac{1}{\theta^2}$. Therefore, the Cramer-Rao lower bound is given by $\frac{1}{nI(\theta)} = \frac{\theta^2}{n}$.

Since X is exponential we know that $\sigma_X^2 = \theta^2$. The estimator $\overline{X}$ is of course unbiased, and in general, $Var[\overline{X}] = \frac{\sigma_X^2}{n}$. But this shows that $Var[\overline{X}] = \frac{\theta^2}{n}$, exactly equal to the Cramer-Rao lower bound. We conclude that $\overline{X}$ must indeed be an efficient estimator for θ. ❐

Exercise 11-26 CAS Exam 3 Spring 2006 #4 (modified)

The random sample $x_1, x_2, \ldots, x_n$ is from a normal distribution with unknown mean and variance.

(a) Calculate the bias of $\sum_{i=1}^{n} \frac{x_i^2}{n}$ as an estimator of $E(X^2)$.

(b) Show that $T^2 = \sum_{i=1}^{n} \frac{(x_i - \bar{x})^2}{n}$ is a consistent estimator of the population variance.

Hint: Find the variance of T^2 and use (1) of **Establishing Consistency** in Subsection 11.3.1.

Exercise 11-27 SOA/CAS Exam 110 May 1985 #21

Let $X_1, \ldots, X_n$ be a random sample from a distribution with density function $f(x) = 3\theta x^2 e^{-\theta x^3}$ for $0 < x < \infty$, and 0, otherwise. What is the Cramer-Rao Lower Bound (CRLB) for the variance of unbiased estimators of θ?

(A) $\frac{3\theta}{n}$ (B) $\frac{2\theta}{n}$ (C) $\frac{3\theta^2}{n}$ (D) $\frac{2\theta^2}{n}$ (E) $\frac{\theta^2}{n}$

Exercise 11-28 Let X be a uniform random variable on the interval $[0, \theta]$, where $\theta > 0$ is an unknown parameter. Let $X_1, \ldots, X_n$ be a random sample from X and let $U = Max(X_1, \ldots, X_n)$. In Example 11.1-2 we showed that $\hat{\theta}_1 = \frac{n+1}{n} U$ is an unbiased estimator for θ. Determine whether or not $\hat{\theta}_1$ is a consistent estimator for θ.

Exercise 11-29 Let $B \sim Binomial(n, p)$ denote a binomially distributed random variable with n trials and probability of success p. Show that B/n is a consistent estimator for p.

Exercise 11-30 SOA/CAS Exam 110 May 1988 #5

Let $X_1, \ldots, X_n$ be a random sample from a population random variable with density function $f(x; \lambda) = \lambda e^{-\lambda x}$ for $x > 0$. What is the Cramer-Rao lower bound (CRLB) for the variance of an unbiased estimator of λ?

(A) $\frac{1}{n\lambda}$ (B) $\frac{1}{n\lambda^2}$ (C) $\frac{n}{\lambda^2}$ (D) $\frac{\lambda^2}{n-1}$ (E) $\frac{\lambda^2}{n}$

11.3.4 SUFFICIENT STATISTICS

In the previous subsection we showed how the ***Cramer-Rao Lower Bound*** (CRLB) can be used to check if a given unbiased estimator is efficient, that is, possesses the minimum possible variance in the class of unbiased estimators. In general a ***minimum variance unbiased estimator*** (abbreviated MVUE) may or may not have its variance equal to the CRLB. Moreover, checking the variance of an estimator against the CRLB presupposes we have such a candidate already in hand. In this subsection we discuss a procedure that – with certain restrictions – actually produces a MVUE.

The relevant concept involves the notion of ***sufficiency*** for a statistic. Roughly speaking, a sufficient statistic is one that extracts all the useful information that can be gained for estimating the parameter[2] from a given random sample. This is pretty loose language, and coming up with a mathematically precise definition that captures this idea is rather complicated.

Before proceeding with the definition let's consider a simple example that illustrates the general idea. Suppose we wish to estimate the probability p of heads in the flip of a possibly unbalanced coin. The common sense (and correct experimental) approach would be to flip the coin n times and count the number k of heads. The intuitively obvious estimator of p is k/n. Notice that the statistic k/n doesn't extract ***all*** of the information from the experiment. In particular, it ignores the exact sequence of heads and tails and just provides the executive summary (the number of heads). But would we gain any further useful information for estimating p from knowing the precise order of heads and tails? The answer is no, and it is in this sense that the statistic k/n is considered ***sufficient***. As we shall see, the statistic k in this example contains the same information as k/n and would work as well. Sufficient statistics are not necessarily unbiased, or minimum variance, estimators, but can often be used to construct an MVUE.

We will next provide the context for the definition of sufficiency, and then illustrate it with some examples.

As usual, let X denote a population with parameter θ and density function (or probability function in the discrete case) $f_X(x;\theta)$. Let $X_1,\ldots,X_n$ be a random sample of X and let

$$L(x_1,\ldots,x_n;\theta)=\prod_{i=1}^{n} f_X(x_i;\theta)$$

denote the joint density (probability) function of $X_1,\ldots,X_n$. This of course is just the likelihood function for the parameter θ for a given set of observations. Let T be a statistic based on $X_1,\ldots,X_n$, that is, $T=u(X_1,\ldots,X_n)$, where u is a real-valued function of the observations. Finally, let $g_T(t;\theta)$ be the density (probability) function for T.

[2] The theory in this subsection can be extended to multiple unknown parameters, but for simplicity's sake we will restrict our attention to the one unknown parameter case.

Sufficient Statistic

We define T to be a ***sufficient statistic*** if the conditional likelihood function for the random sample $X_1,\ldots,X_n$ given $T=t$ does not depend on θ. That is,

$$L(x_1,\ldots,x_n;\theta \mid T=t)=\frac{L(x_1,\ldots,x_n;\theta)}{g_T(t;\theta)}=h(x_1,\ldots,x_n),$$

with h a real-valued function of the observations $x_1,\ldots,x_n$ alone (not dependent on θ).

When we say that the conditional likelihood of the random sample for a given value of T does ***not*** depend on the parameter, we are saying there is no more useful information to be gained from knowing all the details of the random sample. We next consider a simple example to show how this definition works.

Example 11.3-4 Identifying a Sufficient Statistic

Let the population X be a single Bernoulli trial with parameter p, the probability of success. Let X_1 and X_2 be a random sample of size 2. Form the three statistics,

$$T_1=X_1,\ T_2=\frac{1}{2}(X_1+X_2) \text{ and } T_3=\frac{1}{3}(2X_1+X_2).$$

(a) Show T_1, T_2, T_3 are all unbiased estimators of p, with $Var[T_2]<Var[T_3]<Var[T_1]$.
(b) Use the definition to test the three estimators for sufficiency.

Solution
(a) Since X is a Bernoulli trial we know,

$$E[X]=E[X_1]=E[X_2]=p \text{ and } Var[X]=Var[X_1]=Var[X_2]=p(1-p).$$

Thus, $E[T_1]=E[X_1]=p$ and $Var[T_1]=Var[X_1]=p(1-p)$. Similarly,

$$E[T_2]=\frac{1}{2}(E[X_1]+E[X_2])=p \text{ and } Var[T_2]=\frac{1}{4}(Var[X_1]+Var[X_2])=\frac{1}{2}p(1-p),$$

and $E[T_3]=\frac{1}{3}(2E[X_1]+E[X_2])=p$ and $Var[T_3]=\frac{1}{9}(4Var[X_1]+Var[X_2])=\frac{5}{9}p(1-p)$.

Thus, part (a) follows.

(b) The probability function for X can be expressed as $f_X(x;p)=p^x(1-p)^{1-x}$; $x=0,1$. Let $X_1=x_1$ and $X_2=x_2$, where x_1 and x_2 take the values 0 or 1 only. Then,

$$L(x_1,x_2;p)=p^{x_1+x_2}(1-p)^{2-(x_1+x_2)};\ x_1,x_2=0,1.$$

Let $T_1=X_1=t; t=0,1$. Then $g_{T_1}(t;p)=p^t(1-p)^{1-t}$; $t=0,1$. In this case,

$L(x_1,x_2;p\,|\,T_1=t)=\dfrac{p^{t+x_2}(1-p)^{2-(t+x_2)}}{p^t(1-p)^{1-t}}=p^{x_2}(1-p)^{1-x_2}$. Since the expression on the right depends on p, T_1 is not a sufficient statistic.

Next, let $T_2=\frac{1}{2}(X_1+X_2)=t$. We note that T_2 is a binomial proportion, where t can take the values $0, \frac{1}{2}, 1$, with probabilities $(1-p)^2,\ 2p(1-p),\ p^2$, respectively.

Note that $g_{T_2}(t;p)=\binom{2}{2t}p^{2t}(1-p)^{2-2t}$; $t=0, \frac{1}{2}, 1$, and in this case,

$L(x_1,x_2;p\,|\,T_1=t)=\dfrac{p^{2t}(1-p)^{2-2t}}{\binom{2}{2t}p^{2t}(1-p)^{2-2t}}=\dfrac{1}{\binom{2}{2t}}$. Since the expression on the right does not depend on p it follows that T_2 is a sufficient statistic.

Finally, let $T_3=\frac{1}{3}(2X_1+X_2)=t$. Then t takes the values $0, \frac{1}{3}, \frac{2}{3}, 1$, with probabilities $(1-p)^2, (1-p)p,\ p(1-p),\ p^2$, respectively. Since there isn't a convenient functional form for $g_{T_3}(t;p)$ we will calculate $L(x_1,x_2;p\,|\,T_3=t)$ in tabular form.

x_1	x_2	$t=\frac{1}{3}(2x_1+x_2)$	$L(x_1,x_2;p)$	$g_{T_3}(t)$	$L(x_1,x_2;p\,\|\,T_3=t)=\frac{L(x_1,x_2;p)}{g_{T_3}(t)}$
0	0	0	$(1-p)^2$	$(1-p)^2$	1
0	1	$\frac{1}{3}$	$(1-p)p$	$(1-p)p$	1
1	0	$\frac{2}{3}$	$p(1-p)$	$p(1-p)$	1
1	1	1	p^2	p^2	1

Therefore, since $L(x_1,x_2;p\,|\,T_3=t)$ does not depend on p, it follows from the definition that T_3 is also a sufficient statistic.

Notes:

(1) For the non-sufficient statistic T_1 the conditional likelihood function $L(x_1, x_2; p \mid T_1 = t) = p^{x_2}(1-p)^{1-x_2}$ depends on p and is in fact maximized with $p = x_2$, showing that there is additional information to be gleaned from the random sample.

(2) It seems somewhat surprising that an inefficient estimator like T_3 turns out to be sufficient, but in this case the 4 values of T_3 are in one-to-one correspondence with the 4 unique outcomes of the random sample, and have the same probabilities. So T_3 trivially captures all the information in the sample.

(3) Sufficient statistics need not be unbiased. In fact any positive scalar multiples of T_2 and T_3, such as $X_1 + X_2$ and $2X_1 + X_2$, respectively, will also be sufficient statistics. More generally, invertible differentiable functions of sufficient statistics are also sufficient.

Example 11.3-5 A Sufficient Statistic for the Mean Waiting Time in a Poisson Process

The waiting time X is an exponential random variable with density function $f(x;\beta) = \frac{1}{\beta} e^{-\frac{x}{\beta}}$; $x > 0$, where the parameter β is the mean waiting time. Let $X_1, \ldots, X_n$ be a random sample from X, and let $S = \sum_{i=1}^{n} X_i$. Show that S is a sufficient statistic for β.

Solution:

The likelihood function $L(x_1,\ldots,x_n;\theta) = \frac{1}{\beta^n} e^{-\frac{1}{\beta}(x_1+\cdots+x_n)} = \frac{1}{\beta^n} e^{-\frac{1}{\beta}s}$, where $s = x_1 + \cdots + x_n$. We also know that S has a Gamma distribution with parameters n and β. Thus, $g_S(s;\beta) = \frac{1}{\beta^n \Gamma(n)} s^{n-1} e^{-\frac{1}{\beta}s}$. Then,

$$L(x_1,\ldots,x_n;\theta \mid S = s) = \frac{\frac{1}{\beta^n} e^{-\frac{1}{\beta}s}}{\frac{1}{\beta^n \Gamma(n)} s^{n-1} e^{-\frac{1}{\beta}s}} = \frac{\Gamma(n)}{s^{n-1}},$$

which does not depend on the parameter β.

Exercise 11-31 The population X has a Poisson distribution with unknown parameter λ. Let $X_1,\ldots,X_n$ be a random sample and let $S=\sum_{i=1}^{n} X_i$. Show that S is a sufficient statistic for λ.

Exercise 11-32 Let X have a Gamma distribution, with α known and β the unknown parameter , and let $X_1,\ldots,X_n$ be a random sample. Show that $S=\sum_{i=1}^{n} X_i$ is a sufficient statistic for β.

Exercise 11-33 Let X have a geometric distribution, with unknown parameter p. Let $X_1,\ldots,X_n$ be a random sample. Show that $S=\sum_{i=1}^{n} X_i$ is a sufficient statistic for p.

In the previous subsection we discussed the relative efficiency of unbiased estimators by comparing their variances. We also derived the ***Cramer-Rao Lower Bound*** (CRLB) formula for the lowest possible value the variance of an unbiased estimator can take. This is fine as far as it goes, but it does raise some important theoretical questions: Does there actually exist an MVUE for a given parameter? Can we construct a candidate for an MVUE in general? Is it possible that the variance of an MVUE is greater than the CRLB? In which case, how will we know an MVUE when we see one?

We shall see that sufficient statistics are important tools in deriving MVUEs. Our first step is to arrive at a simpler method for identifying sufficient statistics. The above definition can be hard to apply, especially in cases where the exact formula for the density (probability) function for the statistic (the denominator in the definition) is not available to us.

Notice the definition for T being a sufficient statistic for θ implies that,

$$L\left(x_1,\ldots,x_n;\theta\right)=g_T(t;\theta)h(x_1,\ldots,x_n),$$

where $g_T(t;\theta)$ is the density (probability) function for T. The ***factorization theorem*** states that instead of $g_T(t;\theta)$ in the factorization on the right-hand side, we can use any function $g(t;\theta)$.

Factorization Theorem for Sufficient Statistics

Let $X_1,\ldots,X_n$ be a random sample for the population X with unknown parameter θ. A statistic $T = u(X_1,\ldots,X_n)$ is sufficient if the likelihood function L can be written in the form,

$$L(x_1,\ldots,x_n;\theta) = g(t;\theta)h(x_1,\ldots,x_n),$$

where $u(x_1,\ldots,x_n) = t$, g and h are non-negative functions and h is not dependent on θ.

The factorization theorem is a huge aid in identifying sufficient statistics.

Example 11.3-6 Finding a Sufficient Statistic by Factoring *L*

Let X be a Bernoulli trial with probability of success p. Let $X_1,\ldots,X_n$ be a random sample and let $S = \sum_{i=1}^{n} X_i$. Show that S is a sufficient statistic for p.

Solution:

Let $X_i = x_i$; $i = 1,\ldots,n$, where each $x_i = 0$ or 1, and let $s = \sum_{i=1}^{n} x_i$. Then,

$$L(x_1,\ldots,x_n;\theta) = \prod_{i=1}^{n} p^{x_i}(1-p)^{1-x_i} = p^s(1-p)^{n-s}.$$

In this case, $g(s;p) = p^s(1-p)^{n-s}$ and $h(x_1,\ldots,x_n) = 1$.

Example 11.3-7 Finding a Sufficient Statistic by Factoring *L*

Let X be a population with a Weibull distribution, with density function $f_X(x) = \alpha\beta x^{\alpha-1}e^{-\beta x^{\alpha}}$; $0 \le x < \infty$. Assume α is known and β is the unknown parameter. Find a sufficient statistic for β.

Solution:

Let $X_1,\ldots,X_n$ be a random sample. Then,

$$L(x_1,\ldots,x_n;\beta) = \prod_{i=1}^{n} \alpha\beta x_i^{\alpha-1}e^{-\beta x_i^{\alpha}} = (\alpha\beta)^n (x_1\cdots x_n)^{\alpha-1} e^{-\beta\left(x_1^{\alpha}+\cdots x_n^{\alpha}\right)}.$$

Let $S = X_1^{\alpha} + \cdots + X_n^{\alpha}$. Then the above formula for L factors into,

$$L(x_1,\ldots,x_n;\beta)=\beta^n e^{-\beta\left(x_1^\alpha+\cdots x_n^\alpha\right)}\cdot\alpha^n\left(x_1\cdots x_n\right)^{\alpha-1}=\beta^n e^{-\beta s}\cdot\alpha^n\left(x_1\cdots x_n\right)^{\alpha-1},$$

where $s=x_1^\alpha+\cdots+x_n^\alpha$.

Then, $g(s;\beta)=\beta^n e^{-\beta s}$ and $h(x_1,\ldots,x_n)=\alpha^n\left(x_1\cdots x_n\right)^{\alpha-1}$, showing that *S* is sufficient.

Exercise 11-34 The population *X* is uniformly distributed on the interval $(0,\theta)$, where θ is an unknown parameter taking positive values. Let $X_1,\ldots,X_n$ be a random sample and let $T=Max\left(X_1,\ldots,X_n\right)$. Show that *T* is a sufficient statistic for θ.

Exercise 11-35 The population *X* is normally distributed, with μ_X the unknown parameter (assume σ_X^2 is known). Let $X_1,\ldots,X_n$ be a random sample and let $T=\overline{X}$. Show that *T* is a sufficient statistic for μ_X. Hint: Write $x_i-\mu_X=\left(x_i-\overline{X}\right)-\left(\mu_X-\overline{X}\right)$.

Exercise 11-36 The population *X* has density function $f_X(x;\theta)=\theta x^{\theta-1}$; $0\le x\le 1$, where θ is an unknown parameter. Let $X_1,\ldots,X_n$ be a random sample and let $T=\prod_{i=1}^{n}X_i$. Show that *T* is a sufficient statistic for θ.

Nearly every standard one-parameter population distribution we encounter, including the two preceding examples, falls into a category of distributions we call the ***exponential family***. Statistics *S* arising from exponential family distributions can be used to find MVUEs, not just for the parameter θ, but also for functions of θ, such as $1/\theta$ or $e^{-\theta}$.

The Exponential Family

Let $X_1,\ldots,X_n$ be a random sample for the population *X* with parameter θ. We say *X* is a member of the ***exponential family*** if the density (probability) function $f_X(x;\theta)$ can be written in the form,

$$f_X(x;\theta)=a(\theta)b(x)e^{-c(\theta)d(x)};$$

where *X* lives on a domain that does not depend on θ. If *X* belongs to the exponential family of distributions then,

(1) the statistic $S = \sum_{i=1}^{n} d(X_i)$ is sufficient for θ,

(2) if $\varphi = h(\theta)$, where h is a 1-1 differentiable function, and $T = v(S)$ is an unbiased estimator of φ, then T is a MVUE for φ.[3]

The result in (1) follows easily from the factorization theorem. It can be shown that the sufficient statistic S arising from a member of the exponential family is what is called a ***complete*** sufficient statistic. We'll use the term ***resultant statistic*** when we refer to the S in (1). We will not delve further into the details of this theory since it is quite technical and advanced. However we can easily make use of the results in the case of exponential families to illustrate the construction of MVUEs.

Example 11.3-8 Finding an MVUE in the Exponential Family

Let the population random variable X be exponential so that $f(x;\beta) = \frac{1}{\beta} e^{-\frac{1}{\beta}x}$; $x > 0$. Let $X_1, \ldots, X_n$ be a random sample from X and let $S = \sum_{i=1}^{n} X_i$.

(a) Show X is in the exponential family and can be factored in such a way that $d(x) = x$, so that S is the resultant sufficient statistic for β.

(b) Show that $T = \frac{1}{n} S$ is the resultant MVUE for β. Is this MVUE a CRLB estimator for β?

(c) Let $\lambda = \frac{1}{\beta}$ be the rate parameter for the underlying Poisson process. Use S to find the MVUE for λ. Is this MVUE a CRLB estimator for λ?

Solution:

(a) By inspection we have $a(\beta) = \frac{1}{\beta}$, $b(x) = 1$, $c(\beta) = \frac{1}{\beta}$, and $d(x) = x$. Hence, S is the resultant sufficient statistic.

(b) Since T is a function of S, and in fact equals $\overline{X}$, we have

$E[T] = \mu_X = \beta$ and $Var[T] = \frac{\sigma_X^2}{n} = \frac{\beta^2}{n}$.

Thus T is an MVUE. Example 11.3-3 demonstrated that the variance of T in fact attains the CRLB.

[3] The general form of this result is referred to as the Lehmann-Scheffé theorem. It can also be shown that the MVUE it produces is unique.

(c) Since $T = \frac{1}{n}S$ is an MVUE for β, it seems reasonable to try $\frac{1}{T} = \frac{n}{S}$ as an estimator for $\lambda = \frac{1}{\beta}$. However, it will turn out that this is a biased estimator. Nevertheless, we can try to find the right scalar multiple for $\frac{1}{S}$ to create an unbiased estimator for λ. Toward that end we will first calculate $E\left[\frac{1}{S}\right]$. For this purpose we note that $S = \sum_{i=1}^{n} X_i \sim \Gamma(n, \beta)$. Recall that,

$$f_S(s;\beta) = \frac{1}{\beta^n \Gamma(n)} s^{n-1} e^{-\frac{1}{\beta}s};\ s > 0, \text{ and } \int_0^\infty s^{n-1} e^{-\frac{1}{\beta}s}\, ds = \beta^n \Gamma(n).$$

Therefore,

$$E\left[\frac{1}{S}\right] = \frac{1}{\beta^n \Gamma(n)} \int_0^\infty \frac{1}{s} \cdot s^{n-1} e^{-\frac{1}{\beta}s}\, ds = \frac{1}{\beta^n \Gamma(n)} \int_0^\infty s^{n-2} e^{-\frac{1}{\beta}s}\, ds = \frac{\beta^{n-1}\Gamma(n-1)}{\beta^n \Gamma(n)} = \frac{1}{\beta(n-1)} = \frac{\lambda}{(n-1)},$$

providing n is at least 2.

It follows that $T = \frac{n-1}{S}$ is the unbiased estimator of λ. From property (2) of the Exponential Family we then know that T is an MVUE. We next check the variance of T to see if it attains the CRLB. It's easier to first calculate $Var\left[\frac{1}{S}\right]$. Toward that end,

$$E\left[\frac{1}{S^2}\right] = \frac{1}{\beta^n \Gamma(n)} \int_0^\infty \frac{1}{s^2} \cdot s^{n-1} e^{-\frac{1}{\beta}s}\, ds = \frac{1}{\beta^n \Gamma(n)} \int_0^\infty s^{n-3} e^{-\frac{1}{\beta}s}\, ds = \frac{\beta^{n-2}\Gamma(n-2)}{\beta^n \Gamma(n)} = \frac{\lambda^2}{(n-1)(n-2)},$$

providing n is at least 3. Thus,

$$Var[T] = (n-1)^2 Var\left[\frac{1}{S}\right] = (n-1)^2 \left\{ E\left[\frac{1}{S^2}\right] - E\left[\frac{1}{S}\right]^2 \right\}$$

$$= \lambda^2 (n-1)^2 \left[\frac{1}{(n-1)(n-2)} - \frac{1}{(n-1)^2} \right] = \frac{\lambda^2}{(n-2)},$$

providing n is at least 3.

Since Exercise 11-30 shows the CRLB for λ is $\frac{\lambda^2}{n}$, we now have an instance where the MVUE has variance strictly greater than the CRLB.

Example 11.3-9 Finding an MVUE in the Exponential Family

Let the population random variable X be Pareto Type I so that $f(x;\alpha,\beta)=\alpha\beta^{\alpha}x^{-\alpha-1};\, x>\beta$. Assume α is the unknown parameter, and $\beta>0$ is known, and let $X_1,\ldots,X_n$ be a random sample from X. Find an MVUE for $\frac{1}{\alpha}$.

Solution:
We first show X is in the exponential family by rewriting the density function as,

$$f(x;\alpha)=\alpha\beta^{\alpha}e^{\ln\left(x^{-\alpha-1}\right)}=\alpha\beta^{\alpha}\cdot e^{-(\alpha+1)\ln x};\; x>\beta.$$

Then $a(\alpha)=\alpha\beta^{\alpha}$, $b(x)=1$, $c(\alpha)=-(\alpha+1)$, and $d(x)=\ln x$. It follows that $S=\sum_{i=1}^{n}\ln X_i$ is the resultant sufficient statistic. We next calculate $E[\ln X]$ to see how we might manipulate S to get an unbiased estimate of $\frac{1}{\alpha}$. Using the density function for X, we see that,

$E[\ln X]=\alpha\beta^{\alpha}\int_{\beta}^{\infty}\ln x\cdot x^{-\alpha-1}\,dx$. We use integration by parts with $u=\ln x$ and $dv=x^{-\alpha-1}\,dx$.

Then, $du=x^{-1}dx$ and $v=\frac{x^{-\alpha}}{-\alpha}$ so that,

$E[\ln X]=\alpha\beta^{\alpha}\left[\frac{x^{-\alpha}\ln x}{-\alpha}-\int\frac{x^{-\alpha-1}}{-\alpha}dx\right]_{x=\beta}^{\infty}=\alpha\beta^{\alpha}\left[\frac{x^{-\alpha}\ln x}{-\alpha}-\frac{x^{-\alpha}}{\alpha^2}\right]_{x=\beta}^{\infty}=\ln\beta+\frac{1}{\alpha}$. Therefore,

$E[S]=nE[\ln X]=n\ln\beta+\frac{n}{\alpha}$. It follows that $T=\frac{S}{n}-\ln\beta$ is an unbiased estimator of $\frac{1}{\alpha}$.

Since T is a function of the sufficient statistic $S=\sum_{i=1}^{n}d(X_i)$, T is the MVUE for $\frac{1}{\alpha}$.

Exercise 11-37 The population X has density function $f_X(x;\theta)=\theta x^{\theta-1};\; 0\le x\le 1$, where θ is an unknown parameter. Let $X_1,\ldots,X_n$ be a random sample.

(a) Show X is a member of the exponential family with $d(x)=\ln x$.

(b) Is the resultant sufficient statistic S different than the sufficient statistic $T=\prod_{i=1}^{n}X_i$ identified in Exercise 11-36? Can you explain the difference?

(c) Find the MVUE based on S for $\frac{1}{\theta}$.

Exercise 11-38 The population X has a Weibull distribution, so that the density function is $f_X(x;\alpha,\theta) = \alpha\theta x^{\alpha-1} e^{-\theta x^\alpha}$; $x > 0$. Assume α is known and θ is the unknown parameter. Let $X_1,\ldots,X_n$ be a random sample. Show that X is in the exponential family and use the resultant sufficient statistic $S = \sum_{i=1}^{n} d(X_i)$ to find an MVUE for $\frac{1}{\theta}$.

11.4 HYPOTHESIS TESTING THEORY

We introduced framework and terminology for hypothesis testing in Chapter 10 and discussed many of the standard applications found in elementary statistics. We return now to the general theory of hypothesis testing. In previous applications we took a more or less ad hoc approach to choosing the test statistic and the rejection region. The choices depended on the particular problem and we used test statistics that seemed to make sense based on the context of the problem. Our choice of rejection region was dictated by the significance level α, which is the probability of a Type I error. We endeavored to choose a rejection region at the given significance level that would reduce the probability β of a Type II error. This is equivalent to increasing the power, $1-\beta$, of the test. A test statistic and rejection region that maximizes the power (and hence minimizes the probability of a Type II error) is called a ***most powerful*** test for the given significance level α.

We now show how to construct most powerful tests for a given pair of simple hypotheses. To set the stage we let X be a population random variable with an unknown parameter θ. We wish to resolve the following simple hypothesis test at the significance level α:

$$\begin{matrix} H_0: \ \theta = \theta_0 \\ H_1: \ \theta = \theta_1 \end{matrix}, \quad \text{where} \quad \theta_0 \neq \theta_1.$$

We proceed by accumulating data in the form of a random sample $X_1,\ldots,X_n$ of size n from X. We do not wish to prejudge which specific test statistic $\hat{\theta}$ will lead to the best, or most powerful, rejection region. Consequently, we will specify our rejection regions as subsets C of the n-dimensional sample space $\mathbf{R}^n$ arising from $\mathbf{X} = (X_1,\ldots,X_n)$, the multivariate distribution of the n independent observations of X. We calculate the probability of C, for a given value of the parameter, say $\theta = \theta_0$, by integrating the joint density function for $\mathbf{X}$ over C. This joint density function is just the likelihood function (introduced in Subsection 11.2.2) given by,

$$L(\theta) = f_{\mathbf{X}}(x_1, x_2, \ldots, x_n; \theta) = \prod_{i=1}^{n} f_X(x_i; \theta).$$

Hence, $\Pr[C \mid \theta = \theta_0] = \int\int\cdots\int_C f_{\mathbf{X}}(x_1, x_2, \ldots, x_n; \theta_0)\, dx_1\, dx_2 \cdots dx_n$. For simplicity we will abbreviate the integrand using the likelihood function notation, so that we write,

$$\Pr[C \mid \theta = \theta_0] = \int\int \cdots \int_C L(\theta_0)\, dx_1\, dx_2 \cdots dx_n .$$

The key question to resolve now is how we select the set of observations $\mathbf{X} = \mathbf{x} = (x_1, \ldots, x_n)$ that comprise the rejection region C. We have two criteria to guide us in this task. We want to make the probability of a Type I error less than or equal to the significance level α, while at the same time maximizing the power of the test, $1-\beta$, where β is the probability of a Type II error. In other words we want,

(1) The probability of a Type I error

$$\begin{aligned} &= \Pr[\text{rejecting } H_0 \mid H_0 \text{ true}] = \Pr[\mathbf{X} \in C \mid \theta = \theta_0] \\ &= \int\int \cdots \int_C L(\theta_0)\, dx_1\, dx_2 \cdots dx_n \le \alpha, \end{aligned}$$

and,

(2) The power

$$\begin{aligned} &= \Pr[\text{rejecting } H_0 \mid H_1 \text{ true}] = \Pr[\mathbf{X} \in C \mid \theta = \theta_1] \\ &= \int\int \cdots \int_C L(\theta_1)\, dx_1\, dx_2 \cdots dx_n \\ &= 1-\beta \end{aligned}$$

to be as large as possible.

This can be accomplished by having C consist of values of $\mathbf{X}$ for which $L(\theta_0)$ is small and $L(\theta_1)$ is large. It turns out theoretically that the best way to achieve this is to form the likelihood ratio $\frac{L(\theta_0)}{L(\theta_1)}$ and choose C to be the set of $\mathbf{X}$ values for which this likelihood ratio is small. The question of "how small" is settled by the choice of significance level, α. The main result, summarized below, is referred to as the ***Neyman-Pearson Lemma***.

Neyman-Pearson Lemma

Let X be a population random variable with an unknown parameter θ and consider the simple hypothesis test,

$$\begin{aligned} H_0 &: \theta = \theta_0 \\ H_1 &: \theta = \theta_1 \end{aligned}, \text{ where } \theta_0 \ne \theta_1 .$$

Let $\mathbf{X} = (X_1, \ldots, X_n)$ be a random sample of size n to be used to resolve the hypothesis test at the significance level α. Let C be a rejection region satisfying, for some $k > 0$,

(1) $\frac{L(\theta_0)}{L(\theta_1)} \le k$ for all values of $\mathbf{x} = (x_1, \ldots, x_n)$ in the rejection region C,

(2) $\dfrac{L(\theta_0)}{L(\theta_1)} > k$ for all values of $\mathbf{x} = (x_1, \ldots, x_n)$ in the complement C' of C,

(3) $\alpha = \Pr[C \mid \theta = \theta_0] = \iint \cdots \int_C L(\theta_0) dx_1\, dx_2 \cdots dx_n$.

Then the test statistic $\dfrac{L(\theta_0)}{L(\theta_1)}$, and the rejection region C comprised of all $\mathbf{x}$ for which $\dfrac{L(\theta_0)}{L(\theta_1)} \le k$, constitute a most powerful test at the significance level α.

Proof

Let $1-\beta$ be the power of the rejection region C, where by definition we have,

$$1-\beta = \Pr[C \mid \theta = \theta_1] = \iint \cdots \int_C L(\theta_1)\, dx_1\, dx_2 \cdots dx_n.$$

Let D denote another candidate for a rejection region in $\mathbf{R}^n$ at the significance level of α. This means that,

$$\alpha = \Pr[D \mid \theta = \theta_0] = \iint \cdots \int_D L(\theta_0)\, dx_1\, dx_2 \cdots dx_n.$$

We must show that the power of D, given by,

$$1-\beta' = \Pr[D \mid \theta = \theta_1] = \iint \cdots \int_D L(\theta_1) dx_1\, dx_2 \cdots dx_n,$$

is less than or equal to the power $1-\beta$ of C. To that end, we consider the Venn box diagram (see Subsection 2.2.4) of C and D and their complements:

	C	C'	
D	$C \cap D$	$C' \cap D$	α
D'	$C \cap D'$	$C' \cap D'$	
	α		

We have inserted α as the sum of the probabilities along both the first column and the first row, given that the null hypothesis is true. This follows since,

(4) $\alpha = \Pr[C \mid \theta = \theta_0] = \Pr[C \cap D \mid \theta = \theta_0] + \Pr[C \cap D' \mid \theta = \theta_0]$, and also,

(5) $\alpha = \Pr[D \mid \theta = \theta_0] = \Pr[C \cap D \mid \theta = \theta_0] + \Pr[C' \cap D \mid \theta = \theta_0]$.

Next, we observe the following three facts:

(6) $\Pr[C \cap D' \mid \theta = \theta_0] = \Pr[C' \cap D \mid \theta = \theta_0]$,

(7) on $C' \cap D$ we have $kL(\theta_1) < L(\theta_0)$,

(8) on $C \cap D'$ we have $\frac{1}{k} L(\theta_0) \le L(\theta_1)$.

The equality in (6) follows from subtracting equation (5) from equation (4). Fact (7) follows from (2) and the fact that $C' \cap D$ is a subset of C'. Similarly, (8) follows from (1) and the fact that $C \cap D'$ is a subset of C.

We now put this all together to show that D is a less powerful rejection region than C:

$$
\begin{aligned}
1-\beta' &= \Pr[D \mid \theta = \theta_1] \\
&= \iint \cdots \int_D L(\theta_1)\,dx_1\,dx_2 \cdots dx_n \\
&= \iint \cdots \int_{C \cap D} L(\theta_1)\,dx_1\,dx_2 \cdots dx_n + \iint \cdots \int_{C' \cap D} L(\theta_1)\,dx_1\,dx_2 \cdots dx_n \\
&= \iint \cdots \int_{C \cap D} L(\theta_1)\,dx_1\,dx_2 \cdots dx_n + \frac{1}{k} \iint \cdots \int_{C' \cap D} k \cdot L(\theta_1)\,dx_1\,dx_2 \cdots dx_n \\
&< \iint \cdots \int_{C \cap D} L(\theta_1)\,dx_1\,dx_2 \cdots dx_n + \frac{1}{k} \iint \cdots \overbrace{\int_{C' \cap D} L(\theta_0)\,dx_1}^{\text{this uses (7)}}\,dx_2 \cdots dx_n \\
&= \iint \cdots \int_{C \cap D} L(\theta_1)\,dx_1\,dx_2 \cdots dx_n + \frac{1}{k} \iint \cdots \overbrace{\int_{C \cap D'} L(\theta_0)\,dx_1\,dx_2 \cdots dx_n}^{\text{use (6) to replace } C' \cap D \text{ with } C \cap D'} \\
&= \iint \cdots \int_{C \cap D} L(\theta_1)\,dx_1\,dx_2 \cdots dx_n + \iint \cdots \int_{C \cap D'} \frac{1}{k} \cdot L(\theta_0)\,dx_1\,dx_2 \cdots dx_n \\
&\le \iint \cdots \int_{C \cap D} L(\theta_1)\,dx_1\,dx_2 \cdots dx_n + \iint \cdots \overbrace{\int_{C \cap D'} L(\theta_1)\,dx_1}^{\text{this uses (8)}}\,dx_2 \cdots dx_n \\
&= \iint \cdots \int_C L(\theta_1)\,dx_1\,dx_2 \cdots dx_n = \Pr[C \mid \theta = \theta_1] \\
&= 1 - \beta.
\end{aligned}
$$

We next provide some examples to show how the Neyman-Pearson lemma works in practice. The process begins with the likelihood ratio that defines C. As a rule, the likelihood ratio will be a cumbersome and inconvenient test statistic. Therefore, the strategy is to "work" this inequality until we arrive at a more convenient, or more recognizable, test statistic. At each stage the constant of the right-hand side keeps changing. The last step is to determine the value of the final constant by using the significance level α.

Example 11.4-1 Working the Likelihood Ratio

In Example 10.1-3 we tested the hypotheses,

$$
\begin{aligned}
H_0 &: \ p = \frac{2}{3} \\
H_1 &: \ p = \frac{1}{3}
\end{aligned}
$$

where p is the probability of heads when tossing an unbalanced coin. We now test the hypotheses by tossing the coin n times. Show that the most powerful rejection region for a given significance level is of the form $S \le r$, where S is the number of heads in n tosses.

Solution

To place this in the Neyman-Pearson context we let X be the Bernoulli trial random variable associated with a single toss of the coin, with heads being counted as a success. Then X is the population random variable and $\mathbf{X} = (X_1, \ldots, X_n)$ is an n-tuple of 0's and 1's, with $X_i = 1$ representing heads on the ith toss and $X_i = 0$ representing tails. The probability function for X can be written as,

$$f_X(x;p) = p^x \cdot (1-p)^{1-x}; \; x = 0, 1.$$

Let $S = \sum_{i=1}^{n} X_i$. Then S is the binomial random variable for the number of heads in the n tosses. Now, the likelihood function takes the form,

$$L(p) = f_{\mathbf{X}}(x_1, x_2, \ldots, x_n; p) = \prod_{i=1}^{n} f_X(x_i; p) = \prod_{i=1}^{n} p^{x_i} \cdot (1-p)^{1-x_i} = p^s \cdot (1-p)^{n-s}.$$

This implies that the likelihood ratio in this example is,

$$\frac{L\left(\frac{2}{3}\right)}{L\left(\frac{1}{3}\right)} = \frac{\left(\frac{2}{3}\right)^s \cdot \left(\frac{1}{3}\right)^{n-s}}{\left(\frac{1}{3}\right)^s \cdot \left(\frac{2}{3}\right)^{n-s}} = \frac{2^s}{2^{n-s}} = 2^{2s-n}.$$

From the Neyman-Pearson Lemma (1), the likelihood ratio rejection region takes the form,

$$2^{2s-n} \le k.$$

Using natural logarithms, this is seen to be equivalent to,

$$(2s - n)\ln 2 \le \ln k.$$

Since $\ln 2$ is positive, we can reduce this inequality to,

$$s \le \left(\frac{1}{2}\right) \cdot \left(\frac{\ln k}{\ln 2} + n\right).$$

The right-hand side is a constant, so if we let $r = \left(\frac{1}{2}\right) \cdot \left(\frac{\ln k}{\ln 2} + n\right)$ we see that the most powerful rejection region is indeed of the form $S \le r$. The exact value of r is determined by the given significance level α. The interested reader may revisit Example 10.1-3 for some specific examples of α and r values. ❐

Example 11.4-2 Working the Likelihood Ratio

Let X be a normal population random variable with unknown mean μ and known standard deviation σ. A random sample of size n is taken. Determine the most powerful rejection region at significance level α for testing the simple hypotheses,

$$\begin{matrix} H_0: \ \mu = \mu_0 \\ H_1: \ \mu = \mu_1 \end{matrix}, \text{ where } \mu_1 > \mu_0.$$

Solution

The density function for X is given by $f(x;\mu) = \frac{1}{\sigma\sqrt{2\pi}} e^{-\frac{1}{2}\left(\frac{x-\mu}{\sigma}\right)^2}$. Hence, for the random sample $\mathbf{X} = (X_1, \ldots, X_n)$, the likelihood function is,

$$L(\mu) = \prod_{i=1}^{n} f(x_i;\mu) = \left(\frac{1}{\sigma\sqrt{2\pi}}\right)^n e^{-\frac{1}{2}\sum_{i=1}^{n}\left(\frac{x_i-\mu}{\sigma}\right)^2}.$$

The likelihood ratio is given by,

$$\frac{L(\mu_0)}{L(\mu_1)} = \frac{\left(\frac{1}{\sigma\sqrt{2\pi}}\right)^n e^{-\frac{1}{2}\sum_{i=1}^{n}\left(\frac{x_i-\mu_0}{\sigma}\right)^2}}{\left(\frac{1}{\sigma\sqrt{2\pi}}\right)^n e^{-\frac{1}{2}\sum_{i=1}^{n}\left(\frac{x_i-\mu_1}{\sigma}\right)^2}} = e^{-\frac{1}{2}\sum_{i=1}^{n}\left(\frac{x_i-\mu_0}{\sigma}\right)^2 + \frac{1}{2}\sum_{i=1}^{n}\left(\frac{x_i-\mu_1}{\sigma}\right)^2}.$$

Therefore, the rejection region has the form $e^{-\frac{1}{2}\sum_{i=1}^{n}\left(\frac{x_i-\mu_0}{\sigma}\right)^2 + \frac{1}{2}\sum_{i=1}^{n}\left(\frac{x_i-\mu_1}{\sigma}\right)^2} \le k$. If we take natural logarithms and multiply through by $2\sigma^2$, this reduces to,

$$-\sum_{i=1}^{n}(x_i-\mu_0)^2 + \sum_{i=1}^{n}(x_i-\mu_1)^2 \le k',$$

where k' is the new constant on the right-hand side that results from this action. We next square out the expressions on the left to obtain,

$$\left(-\sum_{i=1}^{n}x_i^2 + \sum_{i=1}^{n}2\mu_0 x_i - \sum_{i=1}^{n}\mu_0^2\right) + \left(\sum_{i=1}^{n}x_i^2 - \sum_{i=1}^{n}2\mu_1 x_i + \sum_{i=1}^{n}\mu_1^2\right) \le k'.$$

After canceling and gathering together terms this becomes,

$$2(\mu_0-\mu_1)\sum_{i=1}^{n}x_i + n(\mu_1^2-\mu_0^2) \le k'.$$

Since $\mu_1 > \mu_0$, we have $\mu_0 - \mu_1$ is negative. Thus, dividing both sides by $\mu_0 - \mu_1$ reverses the direction of the inequality, so that we find,

$$\sum_{i=1}^{n} x_i \geq \frac{k' - n(\mu_1^2 - \mu_0^2)}{2(\mu_0 - \mu_1)} = k''.$$

Finally, dividing both sides by n leads to the rejection region $\overline{X} \geq \frac{k''}{n} = K$. Of course, a rejection region of the form $\overline{X} \geq K$ is exactly what we would have intuitively chosen in previous examples since a larger value of $\overline{X}$ is evidence in favor of the alternative hypothesis. ❐

Example 11.4-3 Controlling for Type I and Type II Errors

Let X be a normal population random variable with unknown mean μ and known standard deviation σ. A random sample of size n is taken. In Example 11.4-2 we determined that the most powerful rejection region for testing the simple hypotheses,

$$\begin{matrix} H_0: & \mu = \mu_0 \\ H_1: & \mu = \mu_1 \end{matrix}, \text{ where } \mu_1 > \mu_0,$$

is of the form $\overline{X} \geq K$.

(a) Given a significance level of α, find the corresponding value of K.
(b) Calculate the probability β of a Type II error for the most powerful test of part (a).
(c) If both α and β are given, find an expression for the smallest value of n that will make the probabilities of Type I and Type II errors less than or equal to α and β, respectively.

Solution

Since the rejection region is of the form $\overline{X} \geq K$, we must have $\alpha = \Pr[\overline{X} \geq K \mid \mu = \mu_0]$. Standardizing $\overline{X}$ leads to,

$$\alpha = \Pr\left[\frac{\overline{X} - \mu_0}{\sigma / \sqrt{n}} \geq \frac{K - \mu_0}{\sigma / \sqrt{n}}\right] = \Pr\left[Z \geq \frac{K - \mu_0}{\sigma / \sqrt{n}}\right].$$

It follows that $\frac{K - \mu_0}{\sigma/\sqrt{n}} = z_\alpha$ and so $K = \mu_0 + z_\alpha \frac{\sigma}{\sqrt{n}}$.

For (b), we have by definition that $\beta = \Pr[\overline{X} < K \mid \mu = \mu_1]$. If we use the value of K from (a) and standardize $\overline{X}$, we have,

$$\beta = \Pr\left[\bar{X} < K = \mu_0 + z_\alpha \frac{\sigma}{\sqrt{n}} \middle| \mu = \mu_1\right] = \Pr\left[Z < \frac{\mu_0 + z_\alpha \frac{\sigma}{\sqrt{n}} - \mu_1}{\frac{\sigma}{\sqrt{n}}}\right],$$

so that,

$$\beta = \Pr\left[Z < z_\alpha - (\mu_1 - \mu_0) \cdot \frac{\sqrt{n}}{\sigma}\right].$$

Assuming that $\beta < 0.5$, this means the z-values are in the left-hand tail, which makes $z_\alpha - (\mu_1 - \mu_0) \cdot \frac{\sqrt{n}}{\sigma}$ negative. Consequently, if we are looking up values in a standard normal table we would write,

$$\beta = \Pr\left[Z > (\mu_1 - \mu_0) \cdot \frac{\sqrt{n}}{\sigma} - z_\alpha\right] = 1 - \Phi\left((\mu_1 - \mu_0) \cdot \frac{\sqrt{n}}{\sigma} - z_\alpha\right).$$

For part (c), we note from the above relationship that $z_\beta = (\mu_1 - \mu_0) \cdot \frac{\sqrt{n}}{\sigma} - z_\alpha$. This means that $n = \left[\frac{(z_\alpha + z_\beta) \cdot \sigma}{(\mu_1 - \mu_0)}\right]^2$. ❒

Note

The rejection region $\bar{X} \geq K = \mu_0 + z_\alpha \frac{\sigma}{\sqrt{n}}$ requires that the alternative hypothesis value μ_1 be greater than μ_0, but beyond that it does ***not*** depend on the actual value of μ_1. Therefore, this rejection region is the same for all possible values of μ_1 in the composite alternative hypothesis test,

$$\begin{aligned} H_0 &: \ \mu = \mu_0 \\ H_1 &: \ \mu > \mu_0 \end{aligned}.$$

A rejection region for a given α that is most powerful regardless of the specific parameter value in the alternative hypothesis is called a ***uniformly most powerful*** rejection region.

Example 11.4-4 Controlling for Type II Errors

Let X be a normal population random variable with unknown mean μ and a standard deviation of 5. A random sample of size 25 is taken. Consider the hypothesis test,

$$\begin{aligned} H_0 &: \ \mu = 5 \\ H_1 &: \ \mu = 8. \end{aligned}$$

In Example 11.4-2 we determined that the rejection region for testing these simple hypotheses, is of the form $\bar{X} \geq K$.

(a) Given a significance level of $\alpha = 2.5\%$, find the corresponding value of K.

(b) Calculate the probability β of a Type II error for the most powerful test of part (a).

(c) Determine the minimum sample size and the most powerful rejection region for $\overline{X}$ that will provide an α of no more than 2.5% and a β of no more than 5%.

Solution

Given the null hypothesis of $\mu = 5$, we have the mean of $\overline{X}$ is 5. Also, the standard deviation of $\overline{X}$ is $\frac{\sigma}{\sqrt{n}} = \frac{5}{5} = 1$. Proceeding as in Example 11.4-3, we have,

$$0.025 = \Pr\left[\overline{X} \geq K \mid \mu = 5\right] = \Pr\left[Z \geq \frac{K-5}{1} = K - 5\right].$$

Thus,

$$K - 5 = 1.96 \quad \text{and} \quad K = 6.96.$$

Then,

$$\beta = \Pr[\overline{X} < 6.96 \mid \mu = 8] = \Pr\left[Z < \frac{6.96-8}{1} = -1.04\right]$$
$$= 1 - \Phi(1.04) = 1 - 0.8508 = 0.1492.$$

The situation is illustrated in Figure 11.4.1, where the light gray region in the right-hand tail of the light gray curve corresponds to Type I errors ($\overline{X} \geq 6.96$ given $\mu = 5$), and the dark gray region in the left-hand tail of the dark gray curve corresponds to Type II errors ($\overline{X} < 6.96$ given $\mu = 8$).

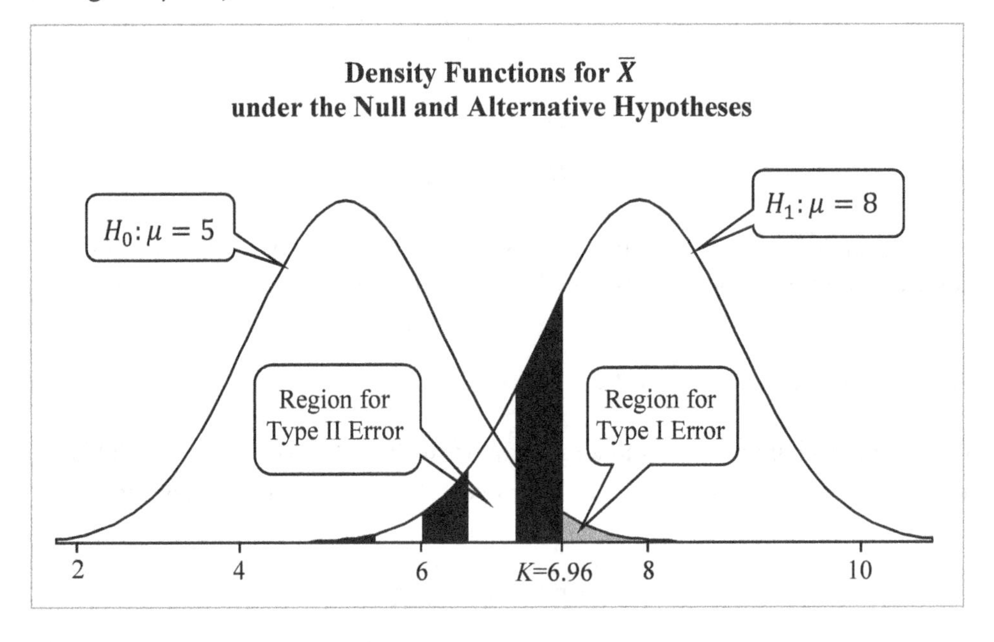

FIGURE 11.4.1

For part (c), we begin with,

$$\alpha = 0.025 = \Pr[\overline{X} \geq K \mid \mu = 5] = \Pr\left[Z \geq \frac{K-5}{\sigma/\sqrt{n}} = \frac{(K-5)\sqrt{n}}{\sigma} = \frac{(K-5)\sqrt{n}}{5}\right].$$

Hence,

$$(1) \qquad \frac{(K-5)\sqrt{n}}{5} = z_{0.025} = 1.96.$$

Next,

$$\beta = 0.05 = \Pr[\overline{X} < K \mid \mu = 8] = \Pr\left[Z < \frac{K-8}{\sigma/\sqrt{n}} = \frac{(K-8)\sqrt{n}}{\sigma} = \frac{(K-8)\sqrt{n}}{5}\right].$$

Since this event is in the left-hand tail of Z we have,

$$(2) \qquad \frac{(K-8)\sqrt{n}}{5} = -z_{0.05} = -1.645.$$

Divide (1) by (2) in order to eliminate n, and obtain,

$$\frac{K-5}{K-8} = \frac{1.96}{-1.645}.$$

Solving for K, we find $K = 6.63$, so the rejection region is $\overline{X} \geq 6.63$. Either equation (1) or equation (2) can now be solved for n to obtain $n = 36.1$. Rounding up, we find that the sample size of 37 is the smallest that will satisfy the required conditions. ❐

Note

We could have simply substituted into the general formulas developed in Example 11.4-3 to find these answers. However, as always, we believe it is far preferable for students to understand the concepts and master the procedures underlying the formulas. That applies especially in this case, since the formulas will vary depending upon whether we are conducting a two-sided, left-hand side or right-hand side hypothesis test.

Example 11.4-5 Type I and Type II Errors

Let X be a random variable with density function given by $f(x) = (1+\theta)x^{\theta}$; $0 \leq x \leq 1$. The hypotheses,

$$H_0: \ \theta = \frac{1}{2},$$
$$H_1: \ \theta = 2$$

are to be tested based on a single observation of X.

(a) Determine the form of the most powerful rejection region for this test.

(b) Determine the rejection region corresponding to $\alpha = \frac{1}{3}$.

(c) Find the probability β of a Type II error, and the power $1-\beta$ of this test when $\alpha = \frac{1}{3}$.

Solution

Since there is only one observation, we have the likelihood function, $L(\theta) = f(x;\theta) = (1+\theta)x^{\theta}$; $0 \le x \le 1$. Consequently, the likelihood ratio rejection region takes the form,

$$\frac{\left(1+\frac{1}{2}\right)x^{\frac{1}{2}}}{(1+2)x^2} = \frac{\left(\frac{3}{2}\right)x^{\frac{1}{2}}}{3x^2} = \frac{1}{2x^{\frac{3}{2}}} \le k.$$

This is equivalent to $x^{\frac{3}{2}} \ge \frac{1}{2k} = k'$, which reduces to a rejection region of the form $X \ge K$.

For (b), we have $\alpha = \Pr\left[X \ge K \,|\, \theta = \frac{1}{2}\right] = \int_K^1 \frac{3}{2}x^{\frac{1}{2}}\,dx = x^{\frac{3}{2}}\Big|_K^1 = 1 - K^{\frac{3}{2}}$. The result is that $K = (1-\alpha)^{\frac{2}{3}} = \left(\frac{2}{3}\right)^{\frac{2}{3}} = 0.7631$, and we reject the null hypothesis for observations $X \ge 0.7631$.

For (c), the power of the test is given by,

$$\Pr[X \ge (1-\alpha)^{\frac{2}{3}} \,|\, \theta = 2] = \int_{(1-\alpha)^{\frac{2}{3}}}^{1} 3x^2\,dx = 1-(1-\alpha)^2 = 1-\left(\frac{2}{3}\right)^2 = \frac{5}{9}.$$

The corresponding $\beta = \frac{4}{9}$. Figure 11.4-2 shows the rejection region in light gray under the null hypothesis density function $f\left(x;\frac{1}{2}\right) = \left(\frac{3}{2}\right)x^{\frac{1}{2}}$, and the Type II error region in dark gray under the alternative hypothesis density function $f(x;2) = 3x^2$. ❐

Note

The likelihood ratio method led to a most powerful rejection region of the form $X \ge K$, a choice that is not immediately apparent from the statement of the problem. A closer inspection of the two density function graphs reveals that larger values of X would favor the alternative hypothesis, since most of the area is concentrated to the right. Smaller values of X would tend to favor the null hypothesis, which has a relatively larger area concentrated to the left.

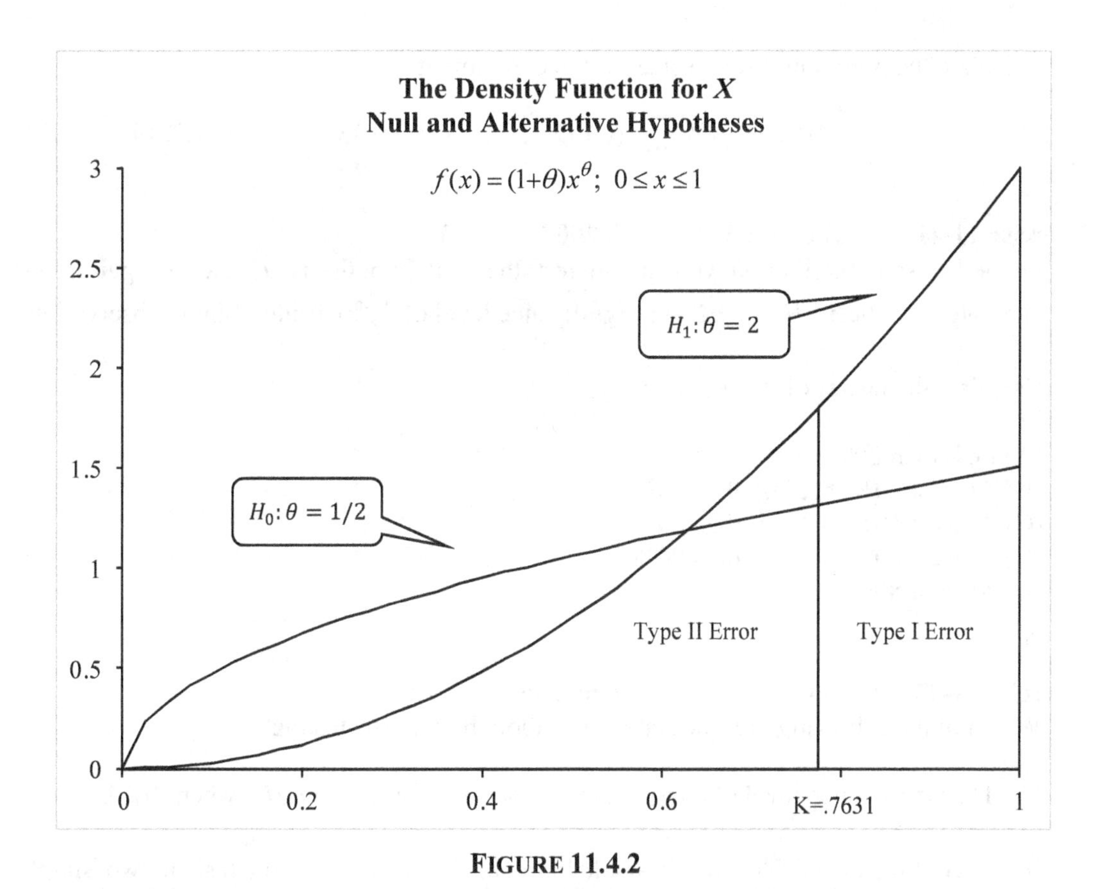

FIGURE 11.4.2

Exercise 11-39 How good of a free-throw shooter is Grant? Let p denote the probability of a made free-throw and assume each shot is independent of the others. You are testing the null hypothesis $H_0 : p = .8$ versus the alternative hypothesis $H_1 : p = .5$. Let $\alpha = .05$.

(a) Find the rejection region if Grant takes $n = 100$ shots.
(b) Find β, the Type II error.

Exercise 11-40 CAS Exam 3 Fall 2005 #7

You are given the following information about an experiment:

- The population is from a normal distribution
- Normal distribution values:

$\Phi(x)$	x
0.93	1.476
0.94	1.555
0.95	1.645
0.97	1.751

- $H_0 : \mu = 10$
- $H_1 : \mu = 11$
- $\sigma^2 = 1$
- The probability of a Type I error is 0.05.
- The probability of a Type II error is no more than 0.06.

Calculate the minimum sample size for the experiment.

(A) 10 (B) 11 (C) 12 (D) 13 (E) 14

Exercise 11-41 CAS Exam 3 Fall 2005 #3

For a Poisson distribution you are to test the null hypothesis $H_0 : \lambda = 1$ against the alternate hypothesis $H_1 : \lambda = 2$ at a significance level of 10% using a single observation.

Calculate the power of the test at $\lambda = 2$.

(A) Less than 20%
(B) At least 20%, but less than 40%
(C) At least 40%, but less than 60%
(D) At least 60%, but less than 80%
(E) At least 80%

Exercise 11-42 CAS Exam 3 Spring 2006 #5

Which of the following are true statements about hypothesis testing?

I. The significance level of a test is the probability of accepting H_0 when H_0 is true.

II. According to the Neyman-Pearson Lemma, the most powerful test of two simple hypotheses is determined by the ratio of the likelihood functions for the two hypotheses.

III. The critical region for a test cannot be determined using only a single observation.

(A) I only (B) II only (C) I and II (D) II and III (E) I, II and III

Exercise 11-43 CAS Exam 3 Fall 2007 #8

You have been given the following information to compare two hypotheses using the Neyman-Pearson lemma:

- X_i follows an exponential distribution where $f(x) = \lambda \exp(-\lambda x),\ x > 0$
- $H_0 : \lambda = 1$
- $H_1 : \lambda = 2$
- The numerator of the likelihood ratio test will hold the results for H_0 and the denominator will hold the results for H_1
- The test will be based on a sample of size 100 with $\overline{X} = 1.5$

Calculate the lower limit of the critical region value for the likelihood ratio test described above so that the significance level will be 5%.

(A) Less than 1.1
(B) At least 1.1, but less than 1.2
(C) At least 1.2, but less than 1.3
(D) At least 1.3, but less than 1.4
(E) At least 1.4

Exercise 11-44 CAS Exam 3 Fall 2006 #4

Claim counts, N, are described by a probability distribution that depends on θ, which can take on two values, θ_0 or θ_1. The possible probability distribution results are shown below:

n	0	1	2	3	4	5
$f(n;\theta_0)$	0.01	0.04	0.05	0.06	0.30	0.54
$f(n;\theta_1)$	0.04	0.04	0.09	0.05	0.40	0.38

The Neyman-Pearson Lemma and a single observation are used to test these hypotheses:

$$H_0 : \theta = \theta_0$$
$$H_1 : \theta = \theta_1$$

Which of the following points is in the critical region defined by a significance level of 5%?

(A) n is 0 or 1
(B) n is 2
(C) n is 3
(D) n is 4
(E) n is 5

Exercise 11-45 CAS Exam 3L Fall 2008 #8

You are given the following:

- A random variable, X, follows a normal distribution with unknown mean μ and known variance $\sigma^2 = 25$.
- $H_0 : \mu = 0$
- $H_1 : \mu = 1$
- The critical region is the best under the Neyman-Pearson lemma of size $\alpha = 0.05$.

What is the minimum sample size such that the power of the test (1 – the probability of a Type II error) is 0.95?

(A) 16 (B) 20 (C) 100 (D) 271 (E) 384

Exercise 11-46 SOA/CAS Exam 110 Fall 1981 #19

A random sample of size 5 is taken from a uniform distribution on the interval $[0,\theta]$. The null hypothesis $H_0: \theta = 1$ is to be tested against the alternative hypothesis $H_1: \theta = 2$, so that the significance level is exactly $\frac{1}{32}$. The test statistic is the maximum of the 5 observations, and the null hypothesis rejected if the test statistic is greater than c. Calculate c.

Exercise 11-47 SOA/CAS Exam 110 Fall 1986 #37

Let $\overline{X}$ be the mean of a random sample from a normal distribution with variance 9. The hypothesis $H_0: \mu = 100$ is rejected in favor of $H_0: \mu > 100$ if $\overline{X} > C$. If the significance level is to be .05, find the minimum sample size necessary to achieve a power of 0.5 when $\mu = 101$. The required sample size is the smallest integer larger than,

(A) $3(1.64)$ (B) $3(1.96)$ (C) $[3(1.64)]^2$ (D) $[3(1.96)]^2$ (E) $[3(1.64)]^{1/2}$

11.5 MORE GENERAL LIKELIHOOD RATIO TESTS

We next address the question of finding test statistics and rejection regions for tests with composite hypotheses. Once again, the issue can be approached by using the ratio of likelihood functions, similar to the setup in the Neyman-Pearson lemma. However, the likelihood functions must be modified to account for the possibility of multiple values of the parameter in at least one of the hypotheses.

We will continue using the notation and terminology of the Neyman-Pearson framework. In that regard, we let X denote a population random variable with a parameter θ. We assume X is continuous, and we write the density function as $f_X(x;\theta)$. Let $X_1,\ldots,X_n$ be a random sample from X, with the joint density function for $\mathbf{X} = (X_1,\ldots,X_n)$ given as the product $f_{\mathbf{X}}(x_1,x_2,\ldots,x_n;\theta) = \prod_{i=1}^{n} f_X(x_i;\theta)$. As previously, we assume the data, represented as $\mathbf{X} = \mathbf{x} = (x_1,\ldots,x_n)$, is given and fixed. We then revert to the likelihood function notation, designating L as a function of θ, so that,

$$L(\theta) \;=\; f_{\mathbf{X}}(x_1,x_2,\ldots,x_n;\theta) \;=\; \prod_{i=1}^{n} f_X(x_i;\theta).$$

We let Ω denote the set of all possible values of the parameter θ. Let ω be a subset of the parameter space Ω, and suppose the composite hypothesis test is given by,

$$H_0:\; \theta \in \omega$$
$$H_1:\; \theta \notin \omega.$$

As an example, if X is normal with mean θ, then $\Omega = (-\infty, \infty)$, and ω might be the subset $\omega = (-\infty, \theta_0)$, which results in a one-sided test for θ with the null hypothesis value of θ_0 (or any smaller value of θ). That is,

$$H_0: \ \mu \le \mu_0$$
$$H_1: \ \mu > \mu_0.$$

Also, ω might consist of just the single element θ_0, in which case we have the two-sided hypothesis test,

$$H_0: \ \mu = \mu_0$$
$$H_1: \ \mu \ne \mu_0.$$

In the case of simple hypotheses we used the likelihood ratio test statistic given by $\dfrac{L(\theta_0)}{L(\theta_1)}$.

The essence of the more general likelihood ratio test is to replace $L(\theta_0)$ with $L(\hat{\theta}_0)$, where $\hat{\theta}_0$ denotes the value of θ_0 that maximizes $L(\theta)$ over the set of values of θ contained in the null hypothesis range ω. This is just the maximum likelihood estimate for θ, restricted to the subset ω of parameter values.

Similarly, $L(\theta_1)$ is replaced by $L(\hat{\theta}_1)$, with $\hat{\theta}_1$ representing the maximum likelihood estimate of θ over the entire parameter space Ω. Just as in the Neyman-Pearson lemma, the rejection region is given by $\dfrac{L(\hat{\theta}_0)}{L(\hat{\theta}_1)} \le k$. As an example, consider the two-sided test for a normal population mean (see Section 10.2.1).

Example 11.5-1 The Likelihood Ratio Test for μ with X Normal

Let X be a normal random variable with known standard deviation σ and unknown mean μ. Let $\overline{X}$ be the sample mean from a random sample of size n. We wish to conduct the hypothesis test,

$$H_0: \ \mu = \mu_0$$
$$H_1: \ \mu \ne \mu_0$$

at the significance level α.

(a) Use the Likelihood ratio test to derive the test statistic $\overline{X}$ and a rejection region of the form, $|\overline{X} - \mu_0| \ge K$.

(b) Find the value of K in terms of α and n.

Solution

In this case ω consists of the single parameter value μ_0, so that the maximum likelihood estimator, denoted $\hat{\mu}_0$, is by default simply equal to μ_0. For the entire parameter space $\Omega = (-\infty, \infty)$ we have, from Example 11.2-8, that the MLE for μ is given by the sample mean $\bar{X}$. Hence, the likelihood ratio is given by $\frac{L(\hat{\mu}_0)}{L(\hat{\mu}_1)} = \frac{L(\mu_0)}{L(\bar{X})}$. Again, from Example 11.2-8 we have that the likelihood function is given by,

$$L(\mu) = \prod_{i=1}^{n} f_X(x_i;\mu) = \left(\frac{1}{\sigma\sqrt{2\pi}}\right)^n e^{-\frac{1}{2}\sum_{i=1}^{n}\left(\frac{x_i-\mu}{\sigma}\right)^2}.$$

Therefore, the likelihood ratio rejection region takes the form,

$$\frac{L(\mu_0)}{L(\bar{X})} = \frac{\left(\frac{1}{\sigma\sqrt{2\pi}}\right)^n e^{-\frac{1}{2}\sum_{i=1}^{n}\left(\frac{x_i-\mu_0}{\sigma}\right)^2}}{\left(\frac{1}{\sigma\sqrt{2\pi}}\right)^n e^{-\frac{1}{2}\sum_{i=1}^{n}\left(\frac{x_i-\bar{X}}{\sigma}\right)^2}} = e^{-\frac{1}{2\sigma^2}\left[\sum_{i=1}^{n}(x_i-\mu_0)^2-\sum_{i=1}^{n}(x_i-\bar{X})^2\right]} \le k.$$

The procedure now is similar to (but not the same as) that of Example 11.4-2. Taking natural logarithms gives,

$$-\frac{1}{2\sigma^2}\left[\sum_{i=1}^{n}(x_i-\mu_0)^2-\sum_{i=1}^{n}(x_i-\bar{X})^2\right] \le \ln k = k'.$$

Next, multiply through by $-\frac{1}{2}$, which reverses the direction of the inequality, to give,

$$\text{(1)} \qquad \frac{1}{\sigma^2}\left[\sum_{i=1}^{n}(x_i-\mu_0)^2-\sum_{i=1}^{n}(x_i-\bar{X})^2\right] \ge k''.$$

Now,

$$\frac{1}{\sigma^2}\left[\sum_{i=1}^{n}(x_i-\mu_0)^2-\sum_{i=1}^{n}(x_i-\bar{X})^2\right] = \frac{1}{\sigma^2}\left[\left(\sum_{i=1}^{n}x_i^2-2n\mu_0\bar{X}+n\mu_0^2\right)-\left(\sum_{i=1}^{n}x_i^2-2n\bar{X}^2+n\bar{X}^2\right)\right]$$

$$=\frac{n}{\sigma^2}\left(\bar{X}^2-2n\mu_0\bar{X}+\mu_0^2\right)=\frac{n}{\sigma^2}\left(\bar{X}-\mu_0\right)^2.$$

Therefore, substituting this back into the log-likelihood ratio test (1), gives,

$$\frac{n}{\sigma^2}(\bar{X}-\mu_0)^2 \ge k'', \text{ or } (\bar{X}-\mu_0)^2 \ge \frac{k''\sigma^2}{n} = k'''.$$

Finally, taking the square root of both sides yields the rejection region,

$$|\overline{X}-\mu_0| \geq \sqrt{k'''} = K.$$

This, of course, is the standard rejection region for the two-sided test for μ that we used previously. We find the exact value for K by standardizing $\overline{X}$ and using the standard normal table. Thus,

$$\alpha = \Pr\left[|\overline{X}-\mu_0| \geq K\right] = \Pr\left[|Z| = \frac{|\overline{X}-\mu_0|}{\sigma/\sqrt{n}} \geq \frac{K}{\sigma/\sqrt{n}}\right].$$

Therefore,

$$K = z_{\alpha/2}\frac{\sigma}{\sqrt{n}},$$

and the two-tail rejection region is given by

$$\overline{X} \leq \mu_0 - z_{\alpha/2}\frac{\sigma}{\sqrt{n}}, \text{ or } \overline{X} \geq \mu_0 + z_{\alpha/2}\frac{\sigma}{\sqrt{n}}.$$

❐

Likelihood ratio tests may also be constructed for discrete population random variables X, although in almost all cases the parameter space will be a continuum. This means that even though X is discrete, the likelihood function can be a standard continuous calculus function. Here is an example.

Example 11.5-2 Likelihood Ratio, Poisson (SOA Exam 110/CAS 2 – May 1983 #27)

Let X be a single observation from a distribution with probability function $f(x;\theta) = \frac{\theta^x e^{-\theta}}{x!}$; $x = 0,1,\ldots,\infty$, with $\theta > 0$. Which inequality determines the likelihood ratio rejection region for the test,

$$\begin{aligned} H_0 &: \theta = 2 \\ H_1 &: \theta \neq 2 \end{aligned}.$$

(A) $e^x \leq C$ (B) $\left(\frac{2e}{x}\right)^x \leq C$ (C) $\left(\frac{2}{x}\right)^x \geq C$ (D) $\left(\frac{2}{x}\right)^x \leq C$ (E) $\left(\frac{2e}{x}\right)^x \geq C$

Solution

Note that x takes only non-negative integers values and hence is discrete. The form of the probability function is that of a Poisson random variable X, with parameter θ.

The likelihood function for a single observation is just the probability function, $L(\theta) = \frac{\theta^x e^{-\theta}}{x!}$. Since the null-hypothesis is simple, we have $\hat{\theta}_0 = 2$ and, $L(\hat{\theta}_0) = L(2) = \frac{2^x e^{-2}}{x!}$.

The entire parameter space Ω for θ consists of the non-negative real numbers, and we showed in Example 11.2-6 that the MLE for θ is the sample mean $\overline{X}$. Thus, let $X = x$ be the random sample (of size one), so that $\overline{X} = x$, and,

$$L(\hat{\theta}_1) = L(\overline{X}) = L(x) = \frac{x^x e^{-x}}{x!}.$$

Then the likelihood ratio test takes the form,

$$\frac{L(\hat{\theta}_0)}{L(\hat{\theta}_1)} = \frac{\frac{2^x e^{-2}}{x!}}{\frac{x^x e^{-x}}{x!}} = \frac{2^x e^{-2}}{x^x e^{-x}} \le k, \text{ or } \frac{2^x}{x^x e^{-x}} = \left(\frac{2e}{x}\right)^x \le ke^2 = C, \text{ which gives answer (B).}$$ □

We note that these likelihood ratio tests can be thought of as goodness-of-fit tests in the sense that they use the data from a random sample to discriminate between one form of a distribution versus another. However, goodness-of-fit to a particular distribution family is likely to be more easily achieved by estimating the parameters and using the chi-squared methods of Section 10.6.[4]

We conclude this section with a comment on the asymptotic behavior of the likelihood ratio in general. Let Λ denote the value of the likelihood ratio, as defined above. Then ***Wilks' Theorem*** shows that under suitable conditions the expression $-2\ln\Lambda$ approaches a χ^2 distribution. The degrees of freedom is arrived at by taking the difference between the dimension of the parameter space of the denominator and the dimension of the parameter space of the numerator. The following example illustrates how this comes about using our likelihood ratio test for the mean of a normal distribution (see Example 11.5-1).

Example 11.5-3 Wilks' Theorem and the Test for μ with X Normal

Consider the hypothesis test of Example 11.5-1 for the mean of a normal distribution (σ known). Show that the expression $-2\ln\Lambda$ is a χ^2 random variable with one degree of freedom.

Solution

In Example 11.5-1 we show that,

$$\frac{L(\mu_0)}{L(\overline{X})} = e^{-\frac{1}{2\sigma^2}\left[\sum_{i=1}^{n}(x_i-\mu_0)^2 - \sum_{i=1}^{n}(x_i-\overline{X})^2\right]}.$$

Note that the dimension of the parameter space for the numerator is 0 since it consists of the single point μ_0. In the denominator the parameter space is the entire real line, of dimension 1. Thus, the predicted degrees of freedom is 1.

---

[4] There are other goodness-of-fit tests, such as the Kolmogorov-Smirnov, and the Anderson-Darling, which we do not cover in this text. Graphical goodness-of-fit techniques are discussed in Chapter 12.

As in Example 11.5-1, we have,

$$-2\ln\Lambda = -2\ln\frac{L(\mu_0)}{L(\bar{X})} = \frac{1}{\sigma^2}\left[\sum_{i=1}^{n}(x_i-\mu_0)^2 - \sum_{i=1}^{n}(x_i-\bar{X})^2\right]$$

$$= \frac{n}{\sigma^2}\left(\bar{X}^2 - 2n\mu_0\bar{X} + \mu_0^2\right) = \frac{n}{\sigma^2}\left(\bar{X}-\mu_0\right)^2 = \left(\frac{\bar{X}-\mu_0}{\sigma/\sqrt{n}}\right)^2.$$

Since under the null hypothesis, $\bar{X} \sim N\left(\mu_0, \frac{\sigma^2}{n}\right)$ we see that $-2\ln\Lambda$ takes the form $Z^2 \sim \chi^2(1)$. □

In the case of this example $-2\ln\Lambda$ actually equals a χ^2 random variable, so that the asymptotic convergence is automatically satisfied.

Exercise 11-48 CAS Exam 3L Spring 2008 #5

Let X be a random variable. X is normally distributed with σ=1.5 and either μ=1 or μ=5.

Consider the following hypotheses:

$H_0: X$ is normally distributed with $\mu = 1$ and $\sigma = 1.5$.
$H_1: X$ is normally distributed with $\mu = 5$ and $\sigma = 1.5$.

You perform hypothesis testing by observing one value of X and rejecting H_0 if this value exceeds k. If the probability of a Type I error is 2.5%, calculate the probability of a Type II error.

(A) Less than 10%
(B) At least 10%, but less than 20%
(C) At least 20%, but less than 30%
(D) At least 30%, but less than 40%
(E) At least 40%

Exercise 11-49 CAS Exam 3L Fall 2008 #7

You are given the distribution of the inter-event times between patients entering a waiting room is exponential with mean θ. You are testing the following hypotheses:

$$H_0: \theta = 10$$
$$H_1: \theta \neq 10$$

You are given the following five observations:

8 9 9 10 11

Calculate the Likelihood Ratio $L(H_0)/L(H_1)$.

(A) Less than 0.990
(B) At least 0.990, but less than 0.992
(C) At least 0.992, but less than 0.994
(D) At least 0.994, but less than 0.996
(E) At least 0.996

Exercise 11-50 SOA/CAS Exam 110 Fall 1981 #31

Suppose X is a random variable for which the following hypotheses are to be tested:

$H_0 : X$ is uniform on (0,1)

$H_1 : X$ has a normal distribution with mean 0 and variance 1.

What is the most powerful test with a significance level of $\alpha = .05$ based on one observation of X?

(A) Reject H_0 if $X \geq 1.64$

(B) Reject H_0 if $|X| \geq 1.96$

(C) Reject H_0 if $X \leq .95$

(D) Reject H_0 if $X \leq .05$ or $X \geq 1$

(E) Reject H_0 if $X \leq 0$ or $X \geq .95$

Exercise 11-51 SOA/CAS Exam 110 Fall 1982 #21

The null hypothesis states that a random sample of size n arises from a gamma distribution with $\alpha = 1$ and $\beta = 1$. The alternative hypothesis states that the random sample arises from a gamma distribution with $\alpha = 2$ and $\beta = 1$. What is the form of the most powerful rejection region?

(A) $\sum_{i=1}^{n} \ln X_i < C$

(B) $\sum_{i=1}^{n} \ln X_i > C$

(C) $\sum_{i=1}^{n} X_i < C$

(D) $\sum_{i=1}^{n} X_i > C$

(E) Does not exist.

Exercise 11-52 SOA/CAS Exam 110 Fall 1986 #18

A single observation is taken from a distribution having density function $f(x;\theta) = \frac{1}{\pi\left[1+(x-\theta)^2\right]}$ for $-\infty < x < \infty$. For testing $H_0 : \theta = 0$ against $H_1 : \theta \neq 0$ at the .05 significance level, the likelihood ratio test has what rejection region?

(A) $|X| > \tan(.025\pi)$

(B) $|X| > \tan^{-1}(.95\pi)$

(C) $|X| > \tan(.95\pi)$

(D) $|X| > \tan^{-1}(.475\pi)$

(E) $|X| > \tan(.475\pi)$

Note: $\int \frac{dx}{1+x^2} = \tan^{-1} x + C$.

Exercise 11-53 SOA/CAS Exam 110 Spring 1988 #11

Let $X_1, \ldots, X_n$ be a random sample from the Poisson distribution with probability function $f(x) = e^{-\lambda} \frac{\lambda^x}{x!}$ for $x = 0, 1, 2, \ldots$ where $\lambda > 0$. Let $\overline{X}$ be the sample mean. What is the form of the critical (rejection) region of the likelihood ratio test for testing $H_0 : \lambda = \lambda_0$ against $H_1 : \lambda \neq \lambda_0$?

(A) $\left(\frac{\lambda_0 e}{\overline{X}}\right)^{\overline{X}} < k$, for some $k > 0$

(B) $\overline{X} < \lambda_0 - k$ or $\overline{X} > \lambda_0 + k$ for some $k > 0$

(C) $\frac{e^{\overline{X}}}{\overline{X}} < k$, for some $k > 0$

(D) $\lambda_0^{\overline{X}} < k$, for some $k > 0$

(E) $\frac{\lambda_0}{\overline{X}} < k$, for some $k > 0$

11.6 BAYESIAN ESTIMATION

In this section we develop another procedure for estimating a population parameter. Previously, we constructed estimators as functions of the observations from a random sample. In both the method of moments of Subsection 11.2.1 and the maximum likelihood method of Subsection 11.2.2, we derived single point estimators of the parameter in this fashion. In Bayesian estimation we take a different perspective. Rather than viewing the parameter as a fixed, but unknown, quantity, we now treat the parameter as itself a random variable. We then assign a probability distribution to the possible values of the parameter. We usually start with a given distribution called the ***prior*** distribution. Next, we obtain data through a random sample of the population, and then use the resulting data to revise the probability distribution for the parameter. This new distribution is called the ***posterior*** distribution.

A point estimator can then be derived from this posterior distribution. For example, we could take the median, or the mean, of the posterior distribution as our actual estimator. The advantage of the Bayesian method is that by constructing an entire probability distribution we obtain a great deal more information about the parameter than simply a point estimate. For example we can use the distribution of the parameter to make interval estimates as well as point estimates. In Chapter 9 interval estimates were termed confidence intervals and were based on the distribution of the parameter estimator (a random variable dependent on the random sample). In the Bayesian context the interval estimate is based directly on the posterior distribution of the parameter, which is now treated as itself a random variable. These are called ***credible intervals***, to distinguish from the non-Bayesian confidence interval. We will discuss this further in Section 11.6.3.

Bayesian estimation can be confusing at first, especially since we are introducing a new random variable into the process, and we need to take into account various joint, marginal and conditional distributions. In order to simplify things as much as possible, we introduce a five-step plan, or recipe, for carrying out the procedure.

11.6.1 THE BAYESIAN RECIPE

Although the distributions involved can be either discrete or continuous (or even mixed), we will present the recipe using continuous density function notation.

Bayesian Estimation Recipe

Let X be a population random variable with the parameter θ and let Θ be the random variable for the parameter θ. Let $f_X(x;\theta)$ be the conditional density function of X for a given value $\Theta = \theta$, and let $\mathbf{X} = (X_1, \ldots, X_n)$ be a random sample from X.

(1) State $\pi(\theta)$, the density function of the ***prior distribution*** of Θ (usually given in the statement of a problem)[5].

(2) State the ***conditional density*** $f(x_1, \ldots, x_n \mid \Theta = \theta)$ of $\mathbf{X}$ for a given value of θ.

(3) Calculate the joint probability function of $\mathbf{X}$ and Θ,

$$f(x_1, \cdots, x_n; \theta) = f(x_1, \cdots, x_n \mid \theta)\pi(\theta) = (1)\cdot(2).$$

(4) Calculate the marginal density function of $\mathbf{X}$:

$$f(x_1, \cdots, x_n) = \int_\theta f(x_1, \cdots, x_n; \theta)\, d\theta\,.$$

(5) Calculate the conditional density $f(\theta \mid \mathbf{X})$ of Θ given $\mathbf{X}$, abbreviated as $\frac{(3)}{(4)}$.

This is called the ***posterior density*** of Θ.

Note

In step (3) we are just using the definition of a conditional density in Chapter 7 as "the joint divided by the marginal," rewritten in multiplicative form. In this case $\Theta = \theta$ is given and $f(x_1, \cdots, x_n \mid \theta) = \frac{f(x_1, \cdots, x_n; \theta)}{\pi(\theta)}$, where the prior distribution $\pi(\theta)$ is taken as the marginal density of Θ. The same principle applies in step (5), except now the observations $(x_1, \ldots, x_n)$ in the random sample $\mathbf{X}$ are taken as the givens.

[5] Deciding on a suitable prior distribution in real-world situations can be problematic.

Here is a typical illustration of how the recipe plays out.

Example 11.6-1 Bayesian Estimation

The claim size for a certain insurance policy is a Pareto random variable X with density function $f(x;\theta)=\frac{2\theta^2}{(x+\theta)^3};\ x>0$. The parameter θ is taken to be an observation from the random variable Θ whose prior distribution has the density function $\pi(\theta)=\frac{1}{\theta^2};\ 1\le\theta<\infty$. A claim of $X=3$ is observed.

(a) Find the density function for the posterior distribution of Θ.
(b) Find the cumulative distribution function for the posterior distribution of Θ.
(c) Calculate the posterior probability that $\Theta>2$.
(d) Find the expected value of the posterior distribution of Θ and compare this to the expected value of the prior distribution of Θ.

Solution
Implementing the 5 steps of the recipe, we have,

(1) The prior distribution of Θ is given by $\pi(\theta)=\frac{1}{\theta^2};\ 1\le\theta<\infty$.

(2) Since there is but one observation of X, we have the conditional density for X is given by

$$f(x;\theta)=f(x\,|\,\Theta=\theta)\ \overbrace{=f(x\,|\,\theta)}^{\text{shorthand abbreviation}}\ =\frac{2\theta^2}{(x+\theta)^3};\ x>0.$$

(3) The joint density function $f(x,\theta)$ of X and Θ is given by the product,

$$f(x,\theta)\ =\ f(x\,|\,\theta)\cdot\pi(\theta)\ =\ (1)\cdot(2)\ =\left[\frac{2\theta^2}{(x+\theta)^3}\right]\cdot\left[\frac{1}{\theta^2}\right]\ =\ \frac{2}{(x+\theta)^3};\ x>0,\ \theta\ge1.$$

(4) The marginal density for X is given by,

$$f_X(x)\ =\ \int_\theta\frac{2}{(x+\theta)^3}\,d\theta\ =\ \int_1^\infty 2(x+\theta)^{-3}\,d\theta\ =\ -(x+\theta)^{-2}\Big|_{\theta=1}^{\infty}\ =\ \frac{1}{(x+1)^2};\ x>0.$$

(5) Lastly, we write the conditional (posterior) density of Θ given X, which by definition is the joint density of X and Θ divided by the marginal for X. This is calculated as,

$$\frac{(3)}{(4)}\ =\ \frac{f(x,\theta)}{f_X(x)}\ =\ \frac{\frac{2}{(x+\theta)^3}}{\frac{1}{(x+1)^2}}\ =\ \frac{2(x+1)^2}{(x+\theta)^3}.$$

Since we are given that $X=3$, we now substitute in $x=3$, yielding the required posterior density function for Θ,

$$f(\theta \mid X=3) = \frac{2(3+1)^2}{(3+\theta)^3} = \frac{32}{(3+\theta)^3}; \theta \geq 1.$$

For (b), we denote the CDF for the posterior distribution of Θ as $G(\theta)$. Then,

$$G(\theta) = \int_1^\theta \frac{32}{(3+t)^3}\,dt = \int_1^\theta 32(3+t)^{-3}\,dt = -16(3+t)^{-2}\Big|_{t=1}^\theta = 1-16(3+\theta)^{-2}; \theta \geq 1.$$

For (c), we have $\Pr[\Theta > 2] = 1-G(2) = 16(3+2)^{-2} = \frac{16}{25}$.

For (d), we can calculate $E[\Theta \mid X=3]$ either from the density function in (a) or the CDF in (b). Since using the density function would require integration by parts, we use the CDF method to obtain,

$$E[\Theta \mid X=3] = 1+\int_1^\infty [1-G(\theta)]d\theta = 1+\int_1^\infty 16(3+\theta)^{-2}\,d\theta = 1-16(3+\theta)^{-1}\Big|_1^\infty = 5.$$

This can be compared to the prior expected value of Θ, which is calculated from the prior density function $\pi(\theta)=\frac{1}{\theta^2}$; $1 \leq \theta < \infty$. The result is,

$$\int_1^\infty \theta \cdot \frac{1}{\theta^2}\,d\theta = \int_1^\infty \frac{1}{\theta}\,d\theta = \ln\theta\Big|_1^\infty = \infty.$$

Technically, this means the expected value of the prior distribution of Θ does not exist. Thus, having a single observation with which to modify the distribution of Θ provides a significant improvement in the quality of our information. ❐

Notes

(1) There is a shortcut that can be used to combine steps (4) and (5) in problems such as this, where the numerical values of the random sample **X** are given. Those values can be substituted into the joint density function in step (3). Also, the marginal density for **X** calculated in step (4) reduces to a constant as well. This means essentially that the posterior density for Θ calculated as $\frac{(3)}{(4)}$, is just a constant times the joint density in (3). In the example, we have in step (3) that, $f(x,\theta)=\frac{2}{(x+\theta)^3}$. Thus, substituting $x=3$ shows that $f(\theta \mid X=3)=\frac{K}{(3+\theta)^3}$; $\theta \geq 1$. All that remains is to evaluate K, which we can easily do in the customary fashion. Since $f(\theta \mid X=3)$ is a probability density function we have,

$$1 = \int_1^\infty f(\theta \mid X=3)\,d\theta = \int_1^\infty K(3+\theta)^{-3}\,d\theta = K\frac{(3+\theta)^{-2}}{-2}\Big|_1^\infty = \frac{K}{32},$$

showing that $K = 32$ and $f(\theta | X{=}3) = \frac{32}{(3+\theta)^3}$; $\theta \geq 1$.

(2) In the example, X is a claim size, which is dependent on the value of the parameter θ. The insurer will want to know how the new claims experience (in this simple example, a single claim of size 3) will affect the expected claim size in the next period, using the posterior distribution of θ. Since the conditional distribution of X for a given θ is Pareto, type II (refer to Section 6.7.2), we know that $E[X | \theta] = \frac{\theta}{\alpha - 1}$. In this example $\alpha = 2$, so $E[X | \theta] = \theta$. We can use the double expectation formula to calculate the (posterior) expected value of X as, $E[X] = E[E[X | \theta]] = E[\theta] = 5$. In this fashion the insurance company will update it's claims expectation each period by incorporating the prior experience in the new (posterior) calculation for $E[X]$. This Bayesian updating process is an essential feature of what is referred to as credibility theory.

The logic of Bayesian estimation can be profitably applied in the context of hypothesis testing. Instead of simply accepting or rejecting the null hypothesis depending on the value of the test statistic and the level of significance, we can assign (posterior) probabilities to the two hypotheses based on the observations we obtain.

We illustrate this in our next example, which develops the Bayesian procedure for discrete distributions. This may seem eerily reminiscent of our tree diagram problems and the discussion of "cause and effect" found in Subsection 2.3.2 and Section 2.5, and a quick review of this material may add to your insights here.

Example 11.6-2 Bayesian Hypothesis Test (Continuation of Example 10.1-3)

A trick coin has been fabricated so that one side will come up in two-thirds of the tosses. However, no one knows whether that side is heads or tails. A statistical consultant is hired to help resolve the issue. She declares that p shall denote the probability of heads, and that she will conduct a hypothesis test as follows:

$$H_0: \ p = 2/3$$
$$H_1: \ p = 1/3.$$

The consultant decides to toss the coin 8 times and, depending on the number r of heads, will assign a probability distribution to the two hypotheses. Use Bayesian analysis to derive these probabilities for each value of $r = 0,1,\ldots,8$.

Solution

Let Θ be the Bernoulli trial random variable defined such that $\Theta = 0$ corresponds to the null hypothesis $H_0: p = 2/3$, and $\Theta = 1$ corresponds to the alternative hypothesis $H_1: p = 1/3$. Since we have no data yet, and therefore no reason to favor one hypothesis over the other, we take the prior distribution of Θ to be evenly split between H_0 and H_1. Schematically, following the Bayesian recipe we have,

Step (1): List Prior Distribution of Θ:

Event	Θ	Probability
$H_0 : p = 2/3$	0	0.5
$H_1 : p = 1/3$	1	0.5

Next, we let $\mathbf{X} = (X_1, X_2, \ldots, X_8)$ represent the outcomes of the experiment of conducting eight tosses of the coin. Each X_i is a Bernoulli trial, with heads representing a success. Since the probabilities depend only on the ***number*** of heads we will simplify the data using the binomial random variable *S*, the number of heads in eight tosses. Naturally, the probabilities of the outcomes $S = r;\ r = 0, 1, \ldots, 8$, are conditioned on whether $p = 2/3$ or $p = 1/3$.

The resulting joint and conditional distributions are most easily portrayed using the following tree diagram.

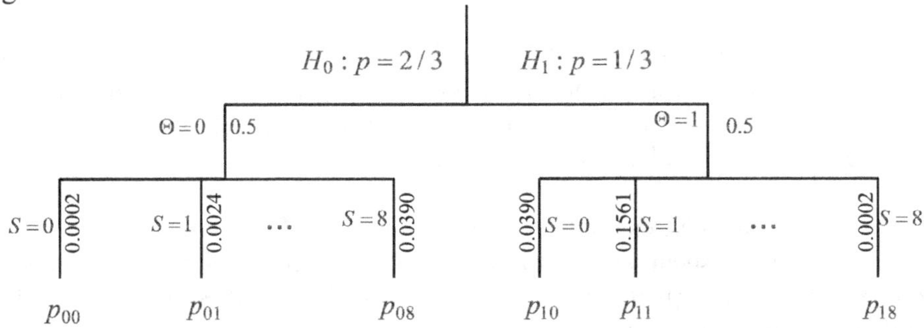

The top level of the tree displays the prior, 50-50, distribution of Θ. The next level of the tree shows the conditional distribution of *S*. It is abbreviated in order to prevent the diagram from becoming too cluttered. The distribution of *S* given $H_0 : p = 2/3$ $(\Theta = 0)$ is on the left-hand branches and the conditional distribution of *S* given $H_1 : p = 1/3$ $(\Theta = 1)$ is on the right-hand branches. The complete conditional distribution of S given Θ is contained in the following two tables.

Step (2): List the Conditional Distribution of *S* given Θ.

Null Hypothesis ($\Theta = 0$)	
k	$\Pr[S = k \mid p = 2/3]$
0	0.0002
1	0.0024
2	0.0171
3	0.0683
4	0.1707
5	0.2731
6	0.2731
7	0.1561
8	0.0390

Alt. Hypothesis ($\Theta = 1$)	
k	$\Pr[S = k \mid p = 1/3]$
0	0.0390
1	0.1561
2	0.2731
3	0.2731
4	0.1707
5	0.0683
6	0.0171
7	0.0024
8	0.0002

The probabilities p_{ij} at the bottom of the diagram, as usual in a tree diagram, represent the individual branch probabilities obtained by multiplying along the branch. The first subscript gives the value of Θ (0 or 1), and the second subscript represents the value k of S. So $p_{ij} = p[\Theta = i, S = j]$. Thus, for example,

$$\begin{aligned} p_{00} &= \Pr[\Theta=0]\Pr[S=0 \mid \Theta=0] \\ &= \Pr[\Theta=0]\Pr[S=0 \mid p=2/3] \\ &= (0.5)(0.0002) = 0.0001. \end{aligned}$$

Similarly,

$$\begin{aligned} p_{01} &= \Pr[\Theta=0]\Pr[S=1 \mid p=2/3] \\ &= (0.5)(0.0024) \\ &= 0.0012, \end{aligned}$$

and

$$\begin{aligned} p_{13} &= \Pr[\Theta=1]\Pr[S=3 \mid p=1/3] \\ &= (0.5)(0.2731) \\ &= 0.1366. \end{aligned}$$

Thus, the complete joint distribution table of S and Θ is the "product" of the top level of the tree (distribution of Θ) times the second level (conditional distribution of S given Θ). The complete table of p_{ij} follows.

Step (3): Calculate the Joint Distribution $(1)\cdot(2)$:

	S									
	0	**1**	**2**	**3**	**4**	**5**	**6**	**7**	**8**	
$\Theta=0$	0.0001	0.0012	0.0085	0.0341	0.0854	0.1366	0.1366	0.0780	0.0195	**0.5000**
$\Theta=1$	0.0195	0.0780	0.1366	0.1366	0.0854	0.0341	0.0085	0.0012	0.0001	**0.5000**
	0.0196	**0.0793**	**0.1451**	**0.1707**	**0.1707**	**0.1707**	**0.1451**	**0.0793**	**0.0196**	**1.0000**

Step (4): Calculate the Marginal Distribution for S:

The next step, (4), is to calculate the marginal probability function for S. These are the numbers along the bottom margin of the above joint distribution table. Mathematically, for each $k = 0, 1, \ldots, 8$, these are calculated as,

$$\Pr[S=k] = \Pr[\Theta=0]\Pr[S=k \mid p=2/3] + \Pr[\Theta=1]\Pr[S=k \mid p=1/3] = p_{0k} + p_{1k}.$$

The last step, (5), is to calculate the posterior (conditional) distribution of Θ given the observed value of S. Again, mathematically, this is identical to using Bayes formula and the results of the tree diagram.

Step (5): Calculate the Posterior Distribution of Θ given S:

$$\Pr[\Theta = 0 \mid S = k] = \frac{\Pr\left[(\Theta = 0) \cap (S = k)\right]}{\Pr[S = k]} = \frac{p_{0k}}{p_{0k} + p_{1k}},$$

and

$$\Pr[\Theta = 1 \mid S = k] = \frac{\Pr\left[(\Theta = 1) \cap (S = k)\right]}{\Pr[S = k]} = \frac{p_{1k}}{p_{0k} + p_{1k}}.$$

For example, if we observe 5 heads in 8 tosses of the coin, then the probability of the null hypothesis being true $(p = 2/3)$ is,

$\Pr[\Theta = 0 \mid S = 5] = \frac{p_{05}}{p_{05} + p_{15}} = \frac{0.1366}{0.1707} = 0.80$, and the probability of the alternative hypothesis being true $(p = \frac{1}{3})$ is $\Pr[\Theta = 1 \mid S = 5] = \frac{p_{15}}{p_{05} + p_{15}} = \frac{0.0341}{0.1707} = 0.20$. Five out of eight heads is evidence supporting $p = \frac{2}{3}$, and this is borne out by the posterior distribution of 80%-20% in favor of the null hypothesis.

A tabu lar listing of all the posterior distributions given $S = k$ is shown next.

	$S = k$								
	0	1	2	3	4	5	6	7	8
$\Pr[p = 2/3 \mid S = k]$	0.0039	0.0154	0.0588	0.2000	0.5000	0.8000	0.9412	0.9846	0.9961
$\Pr[p = 1/3 \mid S = k]$	0.9961	0.9846	0.9412	0.8000	0.5000	0.2000	0.0588	0.0154	0.0039

Note that small values of S give distributions weighted toward $p = 1/3$ and large values of S support $p = 2/3$. Not surprisingly, 4 heads out of 8 tosses leads to no change from the prior to the posterior in the distribution of Θ. ❐

11.6.2 THE LOSS FUNCTION AND MEAN SQUARE ERROR

So far we have emphasized the derivation of the entire posterior distribution of the parameter Θ based on a given set of observations. If a simple point estimate is desired then the one that is generally used is the expected value of the posterior distribution for Θ. Besides being intuitively appealing there is a logical rationale behind this choice as well.

In general terms, one might posit that for each specific choice of a point estimate θ_0 to represent Θ there is a cost, or loss, involved by choosing a single number as a surrogate for the full distribution. This cost is specified as a function, or transformation, of the random variable Θ and the candidate for point estimator, θ_0. It would then make sense to choose the estimator to be the number θ_0 that minimizes the expected value of this loss function.

To formalize this we need to select an appropriate loss function. The usual choice is the squared difference expressed by $L(\Theta,\theta_0)=(\Theta-\theta_0)^2$. The expected value of L is called the ***squared error loss***, or alternatively, the ***mean square error***. As we show below, the expected loss is minimized when θ_0 is taken to be the mean of the posterior distribution of Θ.

Minimum Squared Error Loss

Let Θ be a random variable with a mean of μ. For a given value θ_0 let the loss function L be the random variable given by $L=(\Theta-\theta_0)^2$. Then the value of θ_0 that minimizes the expected value of L is given by $\theta_0 = \mu$.

Proof

We add and subtract μ in the expression for L to obtain,

$$L = (\Theta-\theta_0)^2 = \left[(\Theta-\mu)+(\mu-\theta_0)\right]^2 = (\Theta-\mu)^2 + 2(\mu-\theta_0)(\Theta-\mu)+(\mu-\theta_0)^2.$$

We then take the expected value of both sides and observe that, since $E[\Theta]=\mu$, the middle term on the right becomes zero. The result is,

$$E[L] = E\left[(\Theta-\mu)^2\right]+(\mu-\theta_0)^2 = Var[\Theta]+(\mu-\theta_0)^2.$$

This shows that the minimum expected value for L is the variance of Θ, and this value is obtained when $\theta_0 = \mu$. The following example illustrates how this concept may appear in the formulation of a typical problem involving Bayesian estimation.

Example 11.6-3 Squared Error Loss (Exam 110 May 1982 #19)

Suppose one observation was taken of a random variable X which yielded the value 2. The density function for X is $g(x\,|\,\theta)=\frac{1}{\theta}$ for $0<x<\theta$, and $g(x)=0$ otherwise, and the prior distribution for the parameter θ is $h(\theta)=\frac{3}{\theta^4}$ for $\theta>1$, and $h(\theta)=0$ otherwise. If the loss function is squared error loss, then what is the Bayes estimate for θ?

Solution

The joint distribution of X and Θ has density function,

$$f(x,\theta) = \left(\frac{1}{\theta}\right)\cdot\left(\frac{3}{\theta^4}\right) = 3\,\theta^{-5};\, 1<\theta<\infty,\ 0<x<\theta.$$

Note that x appears in the joint density function only in the specification of the domain boundary condition, $0<x<\theta$. A diagram of the domain is shown in Figure 11.6-1, with the region corresponding to $1<\theta<\infty,\ 0<x<\theta$ consisting of the points above the black line.

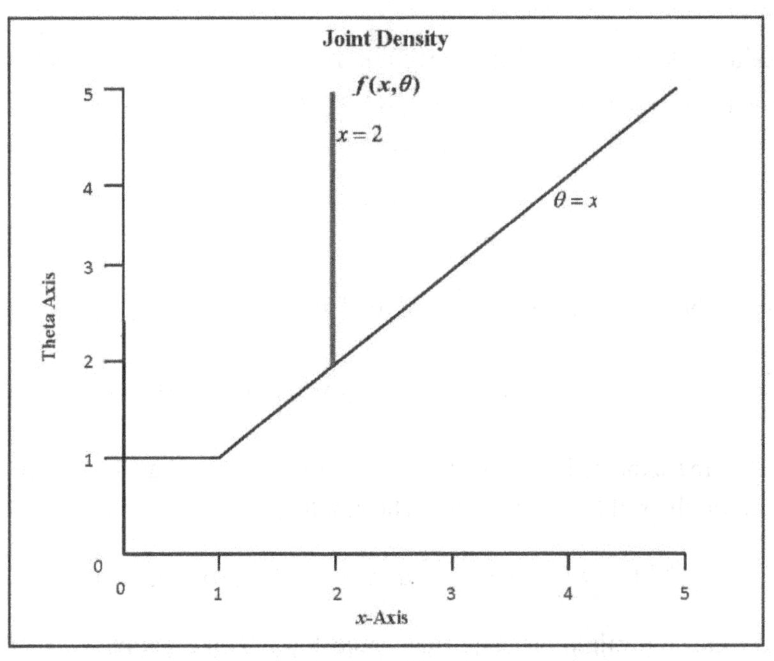

FIGURE 11.6.1

The posterior distribution of Θ lives on the vertical green line at $x=2$, where it has the density function $f(\theta\,|\,X=2)=c\,\theta^{-5};\, 2<\theta<\infty$. The value of c is found by integrating:

$$1 = c\int_2^\infty \theta^{-5}\,d\theta = c\left.\frac{\theta^{-4}}{-4}\right|_{\theta=2}^{\infty} = \frac{c}{64}.$$

Thus, $c=64$, and the Bayes estimate that minimizes the squared error loss is the expected value of the posterior distribution of Θ. Hence, the answer is found as,

$$E[\Theta\,|\,X=2] = 64\int_2^\infty \theta\cdot\theta^{-5}\,d\theta = 64\int_2^\infty \theta^{-4}\,d\theta = 64\left.\frac{\theta^{-3}}{-3}\right|_{\theta=2}^{\infty} = \frac{8}{3}.$$

□

Exercise 11-54 SOA/CAS Exam 110 #19 (expanded)

An auto insurance company insures drivers in four risk categories, with the following results:

Risk Category	Prob. of Accident	Proportion of Drivers
1	.06	.08
2	.03	.15
3	.02	.49
4	.04	.28

(a) Given an accident occurs, calculate the probability the driver is from category 1.
(b) Calculate the prior probability of an accident.
(c) Calculate the posterior probability of an accident (in the next period).

Exercise 11-55 SOA Course 4 Fall 2003 #14

Losses follow a Pareto with $f(x\,|\,\theta)=\frac{\theta}{(x+\theta)^2}$. Half of the policies have $\theta=1$ and half have $\theta=3$. A policy selected at random has first year losses of 5. Determine the posterior probability that losses for this policy in year two will exceed 8.

Exercise 11-56 SOA Course 4 Fall 2004 #33

In a portfolio of risks each policyholder has either 0 or 1 claim per year. The probability of a claim is q. The prior density for q is given by: $\pi(q)=\frac{q^3}{.07}$; $.6<q<.8$. A randomly selected policyholder has one claim in year 1 and zero claims in year 2. For this policyholder calculate the posterior probability that q is greater than .7.

Exercise 11-57 SOA Course 4 Fall 2004 #5

Claim Amount	Class I Probabilities	Class II Probabilities
250	.5	.7
2,500	.3	.2
60,000	.2	.1

Class I claims are twice as likely as Class II claims. A claim of 250 has been observed. Calculate the Bayesian estimate of the expected value of a second claim from the same policyholder.

Exercise 11-58 SOA Course 4 Fall 2002 #39

Class	# Insureds	# Claims=0	# Claims=1	# Claims=2	# Claims=3	# Claims=4
1	3000	1/3	1/3	1/3	0	0
2	2000	0	1/6	2/3	1/6	0
3	1000	0	0	1/6	2/3	1/6

A randomly selected insured has one claim in year 1. Determine the expected number of claims in year 2 for that insured.

Exercise 11-59 SOA Course 4 Fall 2002 #24

The amount of a claim X is uniform on the interval $[0,\theta]$. The prior density of θ is given by $\pi(\theta)=500\theta^{-2}$; $\theta>500$. Two claims, one of 400 and one of 600, are observed. Calculate the Bayesian estimate for the amount of the third claim.

Exercise 11-60 SOA Course 4 Fall 2001 #7

You are given the following information about 6 coins:

Coin #	Prob. of Heads
1 - 4	.50
5	.25
6	.75

A coin is selected at random and tossed four times, yielding the outcomes (HHTH). Determine the probability of heads on the 5th toss of that coin.

Exercise 11-61 SOA Course 4 Fall 2000 #28

You used to believe that claim sizes are Pareto with $\theta=10$ and $\alpha=1,2,3$ with equal likelihood. You then observe a claim of 20 for a randomly selected risk. What is your revised estimate for the probability that the next claim exceeds 30?

Exercise 11-62 SOA/CAS Exam 110 November 1981 #23

Given θ, the random variable X has a binomial distribution with $n=2$ and probability of success θ. If the prior density of θ is $f(\theta)=k$ for $\frac{1}{2}<\theta<1$, what is the Bayes' estimate of θ for squared error loss if $X=1$?

Exercise 11-63 CAS Exam 3L Spring 2008 #2

X is a Pareto random variable with the true parameter $\theta=5,000$. A computer simulation of n claims will produce the following estimator of θ:

$$\hat{\theta}_n = 5000 \times \frac{n}{n+1}, \text{ for } n = 1, 2, \ldots .$$

Which of the following statements are true about the estimator $\hat{\theta}_n$?

I. $\hat{\theta}_n$ is an unbiased estimator of θ.

II. $\hat{\theta}_n$ is an consistent estimator of θ.

III. The mean square error of $\hat{\theta}_{10}$ is more than 200,000.

(A) None are true
(B) I and II only
(C) I and III only
(D) II and III only
(E) I, II, and III

Exercise 11-64 CAS Exam 3L Fall 2008 #5

You are given the following information about a random variable X.

- X follows a Gamma Distribution
- $\alpha = 3$
- $\theta = 100$

Using the mean square error criterion, calculate the constant value, c, which is the best estimate of X.

(A) Less than 50
(B) At least 50, but less than 150
(C) At least 150, but less than 250
(D) At least 250, but less than 350
(E) At least 350

Exercise 11-65 CAS Exam 3 Fall 2005 #6

Claim sizes are uniformly distributed over the interval $[0, \theta]$. A sample of 10 claims, denoted $X_1, X_2, \ldots, X_{10}$, was observed and an estimate of θ was obtained as follows:

$$\hat{\theta} = Y = Max(X_1, X_2, \ldots, X_{10})$$

Recall that the probability density function for Y is:

$$f_Y(y) = \frac{10y^9}{\theta^{10}} \text{ for } 0 \le y \le \theta.$$

Calculate the mean square error $\hat{\theta}$ for $\theta = 100$.

(A) Less than 75
(B) At least 75, but less than 100
(C) At least 100, but less than 125
(D) At least 125, but less than 150
(E) At least 150

Exercise 11-66 Show that the ***mean square error*** of a point estimator $\hat{\mu}$ of a parameter μ, is equal to $Var[\hat{\mu}] + Bias^2$.

11.6.3 Conjugate Prior Distributions and Credible Intervals

We conclude our discussion of Bayesian estimation with some particular scenarios that arise frequently and merit special attention. In these scenarios the prior distribution and the posterior distribution for the parameter belong to the same distribution family. In these cases the new data results in a revision of the parameters within that family. Scenarios with this property are referred to as ***conjugate prior distributions***. We will also illustrate the calculation of interval estimates for the parameter, that is, ***credible intervals***, as they are called in Bayesian statistics.

In the following material we will employ the subscripts "old" and "new" to distinguish between the prior and posterior distributions for the population parameter. For example we will write $E_{\text{old}}[\Theta]$ and $E_{\text{new}}[\Theta]$ for the prior and posterior means of the parameter Θ.

We begin with the Gamma-Poisson case.

The Gamma-Poisson Conjugate Prior Scenario

Let the population random variable X have a Poisson distribution whose parameter λ is an observation from the random variable Λ, whose prior distribution belongs to the Gamma family with parameters α and β. Let $X_1,\ldots,X_n$ be a random sample from X, taking the values, $x_1,\ldots,x_n$, respectively.

(1) The resulting posterior distribution for Λ belongs to the Gamma family with updated parameters,

$$\alpha^* = \alpha + x_1 + \cdots + x_n \text{ and } \beta^* = \frac{\beta}{1+n\beta}.$$

(2) $$E_{\text{new}}[\Lambda] = \alpha^*\beta^* = \left(\frac{n\beta}{1+n\beta}\right)\overline{X} + \left(\frac{1}{1+n\beta}\right)E_{\text{old}}[\Lambda], \text{ where}$$

$$\overline{X} = \frac{1}{n}(X_1 + \cdots + X_n) \text{ and } E_{\text{old}}[\Lambda] = \alpha\beta.$$

Proof:

Following the Bayesian recipe, we have,

$\pi_{\text{old}}(\lambda) = \dfrac{1}{\beta^\alpha \Gamma(\alpha)}\lambda^{\alpha-1}e^{-\frac{1}{\beta}\lambda}$ and $f(x;\lambda) = e^{-\lambda}\dfrac{\lambda^x}{x!}$; $x = 0,1,2,\ldots$. Then the joint probability function is,

$$f(x_1,\ldots,x_n;\lambda) = \frac{1}{\beta^\alpha \Gamma(\alpha)}\lambda^{\alpha-1}e^{-\frac{1}{\beta}\lambda}\cdot\prod_{i=1}^{n} e^{-\lambda}\frac{\lambda^{x_i}}{x_i!} = C\lambda^{x_1+\cdots+x_n+\alpha-1}e^{-\left(\frac{1}{\beta}+n\right)\lambda}.$$

Since $\pi_{\text{new}}(\lambda)$ is proportional to $\lambda^{x_1+\cdots+x_n+\alpha-1}e^{-\left(\frac{1}{\beta}+n\right)\lambda}$, it follows that the posterior distribution for Λ is Gamma with parameters,

$\alpha^* = \alpha + x_1 + \cdots + x_n$ and $\frac{1}{\beta^*} = \frac{1}{\beta} + n$. Solving for β^* yields $\beta^* = \frac{\beta}{1+n\beta}$.

This shows (1). For (2) we note that,

$$E_{\text{new}}[\Lambda] = \alpha^*\beta^* = (x_1 + \cdots + x_n + \alpha)\left(\frac{\beta}{1+n\beta}\right)$$

$$= (x_1 + \cdots + x_n)\left(\frac{\beta}{1+n\beta}\right) + \left(\frac{\alpha\beta}{1+n\beta}\right) = \left(\frac{n\beta}{1+n\beta}\right)\overline{X} + \left(\frac{1}{1+n\beta}\right)E_{\text{old}}[\Lambda].$$

This result has an important insurance interpretation. Think of the population random variable X (for a given $\Lambda = \lambda$) as the claims frequency per period, where the rate parameter λ is a measure of the risk factor for a particular risk class. The risk factors Λ are assumed to be continuously distributed through the portfolio according to their prior distribution. Before the new data is taken into account the overall mean claims count is given by,

$$E[X] = E\left[E[X \mid \Lambda]\right] = E_{\text{old}}[\Lambda].$$

The random sample $X_1, \ldots, X_n$ represents new information about claim counts. Using the posterior distribution of Λ based on this sample we have a new expected value of X calculated as,

$$E[X] = E\left[E[X \mid \Lambda]\right] = E_{\text{new}}[\Lambda].$$

To distinguish the prior and the posterior expected values of X we can write $E_{\text{old}}[X]$ and $E_{\text{new}}[X]$, respectively. Now, formula (2) in the above box can be written as,

(2)′ $$E_{\text{new}}[X] = \alpha^*\beta^* = \left(\frac{n\beta}{1+n\beta}\right)\overline{X} + \left(\frac{1}{1+n\beta}\right)E_{\text{old}}[X].$$

This is a weighted average of $\overline{X}$ and $E_{\text{old}}[X]$. In the absence of any prior information we would take our new experience mean, $\overline{X}$, as our estimate of the overall expected claims rate. In the absence of any new experience we would take $E_{\text{old}}[X]$ as our expected claims weight.

As the experience count n increases, the reliability (or, in insurance parlance, the ***credibility***) of the new data increases, reflected in the fact that the coefficient of $\overline{X}$ converges to 1, while the coefficient of $E_{\text{old}}[\Lambda]$ goes to 0.

In the following example one may either use the formulas above, or work out the solution from the basic Bayesian recipe. Since the formulas are somewhat complicated and not particularly intuitive, we would encourage students to use the recipe, as demonstrated below.

Example 11.6-4: Gamma-Poisson Scenario (CAS ST Sample Problems)

For a large number of health insurance policies the annual numbers of claims follow the Poisson distribution with mean λ, which varies by policy. The prior probability density function of λ is

$$h(\lambda) = \frac{1}{2}\lambda^2 e^{-\lambda};\ \lambda > 0.$$

In Year I, four claims are reported for a policy.

Calculate the expected value of the posterior distribution of λ for this policy.

Solution:

a) The prior distribution is seen by inspection to be in the Gamma family with $\alpha = 3$ and $\beta = 1$. Let X denote the Poisson population. There is one observation, $X_1 = 4$. Therefore, the posterior density for λ is proportional to $\left(\lambda^2 e^{-\lambda}\right)\cdot\left(e^{-\lambda}\lambda^4\right) = \lambda^6 e^{-2\lambda}$, indicating a member of the Gamma family with $\alpha^* = 7$ and $\beta^* = \frac{1}{2}$.

   Therefore, $E_{\text{new}}[\Lambda] = \alpha^*\beta^* = 3.5$.

Example 11.6-5: Gamma-Poisson Scenario (CAS ST Sample Problem Modified)

For a large number of health insurance policies the annual numbers of claims follow the Poisson distribution with mean λ, which varies by policy. The prior probability density function of λ is

$$h(\lambda) = \frac{1}{2}\lambda^2 e^{-\lambda};\ \lambda > 0.$$

In Year I, four claims are reported for a policy.

a) Express the posterior mean as a weighted average of the sample mean and the prior distribution mean.
b) Use EXCEL or some other technological tool to find an interval estimate, or ***Bayesian credible interval***, for the posterior mean with 95% confidence.

Solution:

a) From the statement of the problem we have $E_{\text{old}}[\Lambda] = \alpha\beta = (3)(1) = 3$ and $\overline{X} = 4$. The previous example shows $E_{\text{new}}[\Lambda] = \alpha^*\beta^* = 3.5$. Therefore,

$$3.5 = 4a + 3b, \text{ where } a, b \geq 0 \text{ and } a + b = 1. \text{ Thus, } a = b = \frac{1}{2}.$$

b) There are many different intervals with probability 95% under the posterior distribution. We will take the interval from the 2.5th percentile to the 97.5th percentile. These values are difficult to find by hand, but easy using the EXCEL function GAMMA.INV with $\alpha^* = 7$ and $\beta^* = \frac{1}{2}$. This yields the 95% confidence interval of $(1.41, 6.53)$.

We next consider the Beta-Binomial case.

The Beta-Binomial Conjugate Prior Scenario

Let the population random variable X be a Bernoulli trial with probability of success p, where p is an observation from the random variable P, whose prior distribution belongs to the Beta family with parameters α and β. Let $X_1, \ldots, X_n$ be a random sample from X, taking the values, $x_1, \ldots, x_n$, (0's or 1's) respectively.

(1) The resulting posterior distribution for P belongs to the Beta family with updated parameters,

$$\alpha^* = \alpha + x_1 + \cdots + x_n \text{ and } \beta^* = \beta + n - (x_1 + \cdots + x_n).$$

(2) $E_{\text{new}}[\text{P}] = \dfrac{\alpha^*}{\alpha^* + \beta^*} = \left(\dfrac{n}{\alpha + \beta + n}\right)\overline{X} + \left(\dfrac{\alpha + \beta}{\alpha + \beta + n}\right)E_{\text{old}}[\text{P}]$, where

$$\overline{X} = \frac{1}{n}(X_1 + \cdots + X_n) \text{ and } E_{\text{old}}[\text{P}] = \frac{\alpha}{\alpha + \beta}.$$

Proof:

Again, following the Bayesian recipe, we have,

$\pi_{\text{old}}(\lambda) = \dfrac{\Gamma(\alpha + \beta)}{\Gamma(\alpha)\Gamma(\beta)} p^{\alpha - 1}(1 - p)^{\beta - 1}$ and $f(x; p) = p^x(1 - p)^{1 - x}$; $x = 0, 1$. Then the joint probability function is,

$$f(x_1, \cdots, x_n; p) = \frac{\Gamma(\alpha + \beta)}{\Gamma(\alpha)\Gamma(\beta)} p^{\alpha - 1}(1 - p)^{\beta - 1} \cdot \prod_{i=1}^{n} p^{x_i}(1 - p)^{1 - x_i} = Cp^{\alpha + x_1 + \cdots + x_n - 1}(1 - p)^{\beta + n - (x_1 + \cdots + x_n) - 1}$$

It follows that the posterior distribution for P is Beta with parameters,

$\alpha^* = \alpha + x_1 + \cdots + x_n$ and $\beta^* = \beta + n - (x_1 + \cdots + x_n)$. This shows (1). For (2) we note that,

$$E_{\text{new}}[\mathrm{P}]=\frac{\alpha^*}{\alpha^*+\beta^*}=\frac{\alpha+x_1+\cdots+x_n}{\alpha+x_1+\cdots+x_n+\beta+n-(x_1+\cdots+x_n)}=\frac{\alpha+x_1+\cdots+x_n}{\alpha+\beta+n}$$

$$=\left(\frac{n}{\alpha+\beta+n}\right)\left(\frac{x_1+\cdots+x_n}{n}\right)+\left(\frac{\alpha+\beta}{\alpha+\beta+n}\right)\left(\frac{\alpha}{\alpha+\beta}\right)$$

$$=\left(\frac{n}{\alpha+\beta+n}\right)\overline{X}+\left(\frac{\alpha+\beta}{\alpha+\beta+n}\right)E_{\text{old}}[\mathrm{P}].$$

Note: In some formulations of the Beta-Binomial case the population random variable is taken to be $S=\sum_{i=1}^{n}X_i$, which for a given $P=p$ is a binomial random variable. Then the random sample is $X_1,\ldots,X_n$, and the sum, $x_1+\cdots+x_n$ of the observations, is the number of successes in n trials. The results (1) and (2) are unchanged.

The insurance credibility application discussed after the Gamma-Poisson case applies here as well. We can think of X (given $P=p$) as the probability of a claim. Then $E[X]=E\big[E[X\,|\,P]\big]=E[P]$. Thus, (2) can be written as,

$$E_{\text{new}}[X]=\frac{\alpha^*}{\alpha^*+\beta^*}=\left(\frac{n}{\alpha+\beta+n}\right)\overline{X}+\left(\frac{\alpha+\beta}{\alpha+\beta+n}\right)E_{\text{old}}[X],$$

with the same interpretation as a weighted average of old and new information.

Example 11.6-6: Beta-Binomial Scenario (CAS ST Sample Problem Modified)

For a large number of workers' compensation policies the annual numbers of claims follow the binomial distribution with parameters $n=10$ and θ. The parameter θ, which varies by policy, follows the prior probability density function ,

$$h(\theta)=280\theta^3(1-\theta)^4;\ 0<\theta<1.$$

In Year I, four claims are reported for a policy.

a) Calculate the expected value of the posterior distribution of θ for this policy.
b) Express the posterior mean as a weighted average of the sample mean and the prior distribution mean.
c) Use EXCEL or some other technological tool to find an interval estimate, or ***Bayesian credible interval***, for the posterior mean with 95% confidence.

Solution:

As mentioned in the note above, the random sample has size $n=10$, and results in $S=\sum_{i=1}^{n} X_i = 4$ successes, that is, claims, in Year I. The probability of success is represented by the parameter θ. The prior density for θ is observed to be of the Beta type with $\alpha=4$ and $\beta=5$.

One may use the formulas of (1) and (2) to write down the posterior density and mean. Alternatively, we just follow the Bayesian recipe to write down the joint distribution of $X_1,\ldots,X_{10}$ and θ to recognize the form of the posterior (conditional) distribution for θ.

$$f(x_1,\cdots,x_{10};\theta)=C\theta^3(1-\theta)^4\cdot\prod_{i=1}^{10}\theta^{x_i}(1-\theta)^{1-x_i}=C\theta^{3+4}(1-\theta)^{4+10-4}=C\theta^7(1-\theta)^{10}.$$

We see that the posterior distribution for θ is in the Beta family with $\alpha^*=8$ and $\beta^*=11$.

(a) We have $E_{\text{new}}[P]=\dfrac{\alpha^*}{\alpha^*+\beta^*}=\dfrac{8}{19}=0.4211.$

(b) $\bar{X}=\dfrac{4}{10}$ and $E_{\text{old}}[P]=\dfrac{\alpha}{\alpha+\beta}=\dfrac{4}{9}.$ Thus, $\dfrac{8}{19}=\dfrac{4}{10}a+\dfrac{4}{9}b.$ Solving this simultaneously with $a+b=1$ shows that $a=\dfrac{10}{19}$ and $b=\dfrac{9}{19}.$ Then,

$$E_{\text{new}}[P]=\frac{8}{19}=\left(\frac{10}{19}\right)\bar{X}+\left(\frac{9}{19}\right)E_{\text{old}}[P].$$

(c) The EXCEL function BETA.INV is used to return percentiles of a Beta distribution. Using the posterior distribution for the parameter, where $\alpha^*=8$ and $\beta^*=11$, and the 2.5th and 97.5th percentiles, produces the 95% credible interval (0.2153, 0.6425).

We turn now to the Normal-Normal case.

The Normal-Normal Conjugate Prior Scenario

Let the population random variable X have a normal distribution with known variance σ_1^2 and mean θ, where θ is an observation from a normal random variable Θ with prior mean μ and variance σ_2^2. Let $X_1,\ldots,X_n$ be a random sample from X, taking the values, $x_1,\ldots,x_n$.

(1) The resulting posterior distribution for Θ belongs to the normal family with updated parameters,

$$\mu_{\text{new}} = \frac{\sigma_1^2\mu + \sigma_2^2\sum_{i=1}^{n} x_i}{\sigma_1^2 + n\sigma_2^2} \text{ and } \sigma_{\text{new}}^2 = \frac{\sigma_1^2\sigma_2^2}{\sigma_1^2 + n\sigma_2^2}.$$

(2) $$E_{\text{new}}[\Theta] = \mu_{\text{new}} = \left(\frac{n\sigma_2^2}{\sigma_1^2 + n\sigma_2^2}\right)\overline{X} + \left(\frac{\sigma_1^2}{\sigma_1^2 + n\sigma_2^2}\right)\mu, \text{ where}$$

$$\overline{X} = \frac{1}{n}(X_1 + \cdots + X_n).$$

Proof:
Using the density function formula for normal distributions we find the joint distribution of $X_1,\ldots,X_n$ and Θ takes the form,

$$f(x_1,\cdots x_n;\theta) = \frac{1}{\sqrt{2\pi\sigma_2^2}} e^{-\frac{1}{2\sigma_2^2}(\theta-\mu)^2} \cdot \prod_{i=1}^{n} \frac{1}{\sqrt{2\pi\sigma_1^2}} e^{-\frac{1}{2\sigma_1^2}(x_i-\theta)^2} = Ce^{-\frac{1}{2\sigma_2^2}(\theta-\mu)^2 - \frac{1}{2\sigma_1^2}\sum_{i=1}^{n}(x_i-\theta)^2}.$$

The expression in the exponent on the right is a quadratic function of θ, from which we can conclude that the posterior distribution for Θ is normal. However, identifying the new mean and variance requires some algebraic simplifications. Toward that end we will expand out the terms in the exponent and complete the square, as follows:

$$-\frac{1}{2\sigma_2^2}(\theta-\mu)^2 - \frac{1}{2\sigma_1^2}\sum_{i=1}^{n}(x_i-\theta)^2 = -\frac{1}{2\sigma_1^2\sigma_2^2}\left[\sigma_1^2(\theta-\mu)^2 + \sigma_2^2\sum_{i=1}^{n}(x_i-\theta)^2\right]$$

$$= -\frac{1}{2\sigma_1^2\sigma_2^2}\left[\left(\sigma_1^2 + n\sigma_2^2\right)\theta^2 - 2\left(\sigma_1^2\mu + \sigma_2^2\sum_{i=1}^{n}x_i\right)\theta + K\right] = -\frac{1}{2\sigma_1^2\sigma_2^2}\left[A\theta^2 - 2B\theta + K\right],$$

where $A = \sigma_1^2 + n\sigma_2^2$, $B = \sigma_1^2\mu + \sigma_2^2\sum_{i=1}^{n} x_i$, and K incorporates all the terms in the expansion not involving θ. By completing the square we see that the posterior density function for Θ has the form,

$$Ce^{-\frac{A}{2\sigma_1^2\sigma_2^2}\left[\left(\theta-\frac{B}{A}\right)^2+K'\right]} = C'e^{-\frac{A}{2\sigma_1^2\sigma_2^2}\left(\theta-\frac{B}{A}\right)^2}.$$

We can now identify the mean and the variance. We have,

$$\mu_{\text{new}} = \frac{B}{A} = \frac{\sigma_1^2\mu + \sigma_2^2\sum_{i=1}^{n} x_i}{\sigma_1^2 + n\sigma_2^2} \quad \text{and } \sigma_{\text{new}}^2 = \frac{\sigma_1^2\sigma_2^2}{A} = \frac{\sigma_1^2\sigma_2^2}{\sigma_1^2 + n\sigma_2^2}.$$

This completes the proof of (1). For (2), we can rewrite μ_{new} to see that,

$$\mu_{\text{new}} = \frac{\sigma_2^2}{\sigma_1^2 + n\sigma_2^2}\sum_{i=1}^{n} x_i + \frac{\sigma_1^2}{\sigma_1^2 + n\sigma_2^2}\mu = \frac{n\sigma_2^2}{\sigma_1^2 + n\sigma_2^2}\overline{X} + \frac{\sigma_1^2}{\sigma_1^2 + n\sigma_2^2}\mu.$$

Once again, we can interpret (2) in the insurance context as a weighted average of new and old information, this time concerning the expected claim amount, X. Since $E[X] = E\left[E[X \mid \Theta]\right] = E[\Theta]$, we can rewrite (2) in the form,

$$E_{\text{new}}[X] = \mu_{\text{new}} = \left(\frac{n\sigma_2^2}{\sigma_1^2 + n\sigma_2^2}\right)\overline{X} + \left(\frac{\sigma_1^2}{\sigma_1^2 + n\sigma_2^2}\right)E_{\text{old}}[X].$$

Example 11.6-7: Normal-Normal Scenario

Let X be an index that measures bias in a TV political newscast, with negative values indicating bias to the left and positive numbers bias to the right. Assume X is normal with a variance of 4 and an unknown mean θ that varies by news outlet. The prior distribution for θ is the standard normal. A random sample of the index for 8 news stories from a particular outlet results in the sample mean of negative one.

a) Find the posterior distribution for θ.
b) Find a 90% credible interval for θ.
c) Express the posterior mean for θ as a weighted average of the sample mean and the prior mean.
d) Based on the posterior distribution, estimate the probability that the outlet is biased to the left, that is, $\Pr[X < 0]$.

Solution:

(a) We use the formulas to calculate the posterior mean and variance. We have, $\sigma_1^2 = 4,\ \mu = 0,\ \sigma_2^2 = 1,\ n = 8$ and $\sum_{i-1}^{8} x_i = (8)(-1) = -8$. Then the posterior distribution is normal with,

$$\mu_{\text{new}} = \frac{\sigma_1^2 \mu + \sigma_2^2 \sum_{i=1}^{n} x_i}{\sigma_1^2 + n\sigma_2^2} = \frac{(4)(0) + (1)(-8)}{4 + (8)(1)} = -\frac{2}{3} \text{ and } \sigma_{\text{new}}^2 = \frac{\sigma_1^2 \sigma_2^2}{\sigma_1^2 + n\sigma_2^2} = \frac{(4)(1)}{4 + (8)(1)} = \frac{1}{3}.$$

(b) The 90% credible interval is given by $\mu_{\text{new}} \pm 1.645\sigma_{\text{new}} = (-1.616,\ 0.283)$.

(c) $$\mu_{\text{new}} = -\frac{2}{3} = \left(\frac{n\sigma_2^2}{\sigma_1^2 + n\sigma_2^2}\right)\overline{X} + \left(\frac{\sigma_1^2}{\sigma_1^2 + n\sigma_2^2}\right)\mu = \frac{8}{4+8}\overline{X} + \frac{4}{4+8}\mu = \left(\frac{2}{3}\right)(-1) + \left(\frac{1}{3}\right)(0).$$

(d) $$\Pr[X < 0] = \left[Z < \frac{0 - (-2/3)}{\sqrt{1/3}} = 1.15\right] = 0.8759.$$

Several of the following exercises are modified versions of SOA/CAS Sample Problems for Exam C.

Exercise 11-67 You are given:

(i) The number of claims incurred in a month by any insured has a Poisson distribution with mean λ.

(ii) The prior density function for λ is given by $\frac{100^6}{120}\lambda^5 e^{-100\lambda}$; $\lambda > 0$.

The experience is summarized below:

Month	Number of Insureds	Number of Claims
1	100	6
2	150	8
3	200	11
4	300	

Use the posterior distribution for λ to calculate the total number of expected claims for Month 4.

Exercise 11-68 You are given:

(i) The number of claims made by an individual insured in a year has a Poisson distribution with mean λ.

(ii) The prior distribution for λ is gamma with parameters $\alpha = 1$ and $\beta = 1.2$.

Three claims are observed in Year 1, and no claims are observed in Year 2.
Estimate the number of claims in Year 3.

Exercise 11-69 You are given:

(i) The conditional distribution of the number of claims per policyholder is Poisson with mean λ,

(ii) The variable λ has a gamma distribution with parameters α and β,

(iii) For policyholders with 1 claim in Year 1, the posterior mean for λ is 0.15.

(iv) For policyholders with an average of 2 claims per year in Year 1 and Year 2, the posterior mean for λ is 0.20.

Determine β.

Exercise 11-70 Professor Angst's Statistics final exam scores have a normal distribution X with mean Θ and standard deviation 8. The random variable Θ is also normal, with a prior mean of 65 and standard deviation of 6. A random sample of size 8 from this year's class had a sample mean of $\overline{X} = 60$.

(a) Find the posterior distribution for Θ and express its mean as a weighted average of $\overline{X}$ and the prior mean.

(b) Find a 90% credible interval for Θ.

(c) Using the posterior mean for Θ, estimate the probability that a score in this year's class is below 65.

Exercise 11-71 Let λ be an observation from a Gamma type random variable Λ with parameters α and β. For a given λ let X be an exponential random variable with density function $f(x;\lambda) = \lambda e^{-\lambda x}$; $x > 0$. Let $X_1,\ldots,X_n$ be random sample of X. Show that this is a conjugate prior scenario and derive the formulas for the posterior distribution of Λ.

11.7 CHAPTER 11 SAMPLE EXAMINATION

1. **CAS Exam 3 Fall 2005 #1**

The following sample was taken from a distribution with probability density function $f(x)=\theta x^{\theta-1}$, where $0<x<1$ and $\theta>0$.

0.21 0.43 0.56 0.67 0.72

Let R and S be the estimators of θ using the maximum likelihood method and method of moments, respectively. Calculate the value of $R-S$.

(A) Less than 0.3
(B) At least 0.3, but less than 0.4
(C) At least 0.4, but less than 0.5
(D) At least 0.5, but less than 0.6
(E) At least 0.6

2. **CAS Exam 3 Fall 2006 #6**

You are testing the hypothesis H_0 that the random variable X has a uniform distribution on $[0,10]$ against the alternative hypothesis H_1 that X has a uniform distribution on $[5,10]$. Using a single observation and a significance level 5%, calculate the probability of a Type II error.

(A) Less than 0.2
(B) At least 0.2, but less than 0.4
(C) At least 0.4, but less than 0.6
(D) At least 0.6, but less than 0.8
(E) At least 0.8

3. **CAS Exam 3 Spring 2005 #21**

An actuary obtains two independent, unbiased estimates, Y_1 and Y_2, for a certain parameter. The variance of Y_1 is four times that of Y_2. A new unbiased estimator of the form $k_1 * Y_1 + k_2 * Y_2$ is to be constructed. What is the value of k_1 that minimizes the variance of the new estimate?

(A) Less than 0.148
(B) At least 0.18, but less than 0.23
(C) At least 0.23, but less than 0.28
(D) At least 0.28, but less than 0.33
(E) 0.33 or more

4. **CAS Exam 3 Fall 2007 #11**

Given the following:

- A random sample, $X_1, \ldots, X_n$, where X_i is normally distributed with mean $= \mu_1$, variance $= \sigma_1^2$ and the sample size, n, is 12.
- A random sample, $Y_1, \ldots, Y_m$, where Y_i is normally distributed with mean $= \mu_2$, variance $= \sigma_2^2$ and the sample size, m, is 13.
- $H_0 : \sigma_2^2 = k\sigma_1^2$
- $H_1 : \sigma_2^2 > k\sigma_1^2$
- The test statistic is $\frac{S_2^2}{S_1^2}$.
- The critical value is 5.58 and $\alpha = 0.05$.

Calculate k, the multiplier that links σ_2^2 to σ_1^2 in the null hypothesis.

(A) Less than 1.2
(B) At least 1.2, but less than 1.4
(C) At least 1.4, but less than 1.6
(D) At least 1.6, but less than 1.8
(E) At least 1.8

5. **CAS Exam 3 Spring 2006 #3**

Mrs. Actuarial Gardener has used a global positioning system to lay out a perfect 20-meter by 20-meter gardening plot in her back yard.

Her husband, Mr. Actuarial Gardener, decides to estimate the area of the plot. He paces off a single side of the plot and records his estimate of its length. He repeats this experiment an additional 4 times along the same side. Each trial is independent and follows a Normal distribution with a mean of 20 meters and a standard deviation of 2 meters. He then averages his results and squares that number to estimate the total area of the plot.

Which of the following is a true statement regarding Mr. Gardener's method of estimating the area?

(A) On average, it will underestimate the true area by at least 1 square meter.
(B) On average, it will underestimate the true area by less than 1 square meter.
(C) On average, it is an unbiased method.
(D) On average, it will overestimate the true area by less than 1 square meter.
(E) On average, it will overestimate the true area by at least 1 square meter.

6. You are given:

(i) The annual number of claims for a policyholder follows a Poisson distribution with mean Λ,

(ii) The prior distribution of Λ is gamma with parameters $\alpha = 5$ and $\beta = \frac{1}{2}$.

An insured is selected at random and observed to have 5 claims during Year 1 and 3 claims during Year 2.

Determine the posterior mean of Λ.

7. **CAS Exam 3L Spring 2009 #17**

A random variable X follows a lognormal distribution. You are given a sample of size n and the following information:

$$\frac{\sum X_i}{n} = 1.8682 \text{ and } \frac{\sum X_i^2}{n} = 4.4817.$$

Use the method of moments to estimate the lognormal parameter σ.

(A) Less than 0.4
(B) At least 0.4, but less than 0.8
(C) At least 0.8, but less than 1.2
(D) At least 1.2, but less than 1.6
(E) At least 1.6

8. **CAS Exam 3L Spring 2009 #20**

You are given the following information about a random sample of 100 claims:

- Claim severity follows an exponential distribution with mean θ.
- $H_0 : \theta = 9{,}500$
- $H_1 : \theta = 12{,}500$
- Reject H_0 if the sample mean is greater than 11,000.

Calculate the probability of a Type II error.

(A) Less than 2.5%
(B) At least 2.5%, but less than 5.0%
(C) At least 5%, but less than 7.5%
(D) At least 7.5%, but less than 10.0%
(E) At least 10.0%

9. **CAS Exam 3 Spring 2009 #21**

You are given the following information:

- $X_1,\ldots,X_n$ is a random sample from a normal distribution with mean μ_X and standard deviation σ_X.
- $Y_1,\ldots,Y_n$ is a random sample from a normal distribution with mean μ_Y and standard deviation σ_Y.
- The unbiased sample variances S_X^2 and S_Y^2 are 4.0 and 1.3, respectively.
- $H_0:\sigma_X^2=\sigma_Y^2$
- $H_1:\sigma_X^2>\sigma_Y^2$
- Use of the F test results in rejection of the null hypothesis at a significance level of 5%.

Calculate the minimum possible value of the sample size, n.

(A) 7 (B) 8 (C) 9 (D) 10 (E) 11

10. **CAS Exam 3L Spring 2009 #22**

You are given the following information about a one-sided test that is being performed on a random sample taken from a normal distribution:

- The unbiased sample variance is 9.
- The sample size is 16.
- $H_0:\mu=80$
- $H_1:\mu>80$

Calculate the highest possible sample mean that would still result in the null hypothesis not being rejected at the 1% significance level.

(A) Less than 81.00
(B) At least 81.00, but less than 81.50
(C) At least 81.50, but less than 82.00
(D) At least 82.00, but less than 82.50
(E) At least 82.50

11. **CAS Exam 3L Spring 2009 #24**

You are given the following information:

- Claim severity follows a uniform distribution over the interval $(0,\alpha)$.
- A random sample of n claims is observed.
- Y_n denotes the largest claim amount in the sample and is used as an estimator for α.

Calculate the minimal sample size, n, necessary to ensure that the absolute value of the bias of the estimator Y_n is less than 5% of α.

(A) 19 (B) 20 (C) 21 (D) 22 (E) 23

12. The population X has a Poisson distribution with unknown parameter λ. Let $X_1,\ldots,X_n$ be a random sample and let $S = X_1 + \cdots + X_n$.

a) Find the CRLB for the variance of unbiased estimators of λ.
b) Let $\tau = \Pr[X = 0] = e^{-\lambda}$. Find the CRLB for the variance of unbiased estimators of τ.
c) Show X is in the exponential family and that S is the resultant sufficient statistic.
d) Find the resultant MVUE for λ. Compare the variance of this estimator to the CRLB found in part (a).
e) Find the value of a that makes a^S an MVUE for τ and compare its variance to the CRLB found in (b).

13. **CAS Exam ST Spring 2014 #25**

You are given the following information:

- Y follows a binomial distribution with parameters n and $p = \theta$.
- The prior distribution of θ follows a beta distribution with parameters $\alpha = 4$ and $\beta = 6$.
- A sample of size 15 is drawn from Y with $\sum_{i=1}^{15} y_i = 10$.

Calculate the estimate of θ for the posterior distribution.

14. **CAS Exam ST Spring 2014 #22**

You are given the following information for an insurance policy:

- Monthly claim frequencies follow a Poisson process with parameter λ.
- The prior distribution of λ follows a Gamma distribution with parameters $\alpha = 3$ and $\beta = 2$.
- In the first month a policy has 27 claims.

Calculate the posterior mean monthly claims for this policy.

15. **CAS Exam ST Spring 2014 #23**

You are given the following information:

- $X_1,\ldots,X_5$ are a random sample from a Poisson distribution with parameter λ, where λ follows a Gamma distribution with $\alpha = 2$ and β.
- The mean of this Poisson-Gamma conjugate pair can be represented as a weighted average of the maximum likelihood estimator for the mean and the mean of the prior distribution.
- Let W_{MLE} be the weight assigned to the maximum likelihood estimator.
- The maximum likelihood estimate for the mean is 1.2.
- The variance of the prior Gamma is 8.

Calculate W_{MLE}.

16. **CAS Exam ST Fall 2014 #5**

Calculate the Fisher information $I(q)$ based on a single observation from a Bernoulli trial random variable with probability of success q.

17. **CAS Exam ST Fall 2014 #7**

You are given the following information:

- $X_1,\ldots,X_7$ is a random sample of size 7 from an exponential distribution with mean θ.
- The sample mean is 400.

Calculate the MVUE (minimum variance unbiased estimator) of θ^2.

18. **CAS Exam ST Fall 2014 #22**

You are given:

i. For a general liability policy, the log of paid claims conditionally follows the normal distribution with mean μ, which varies by policyholder, and variance 1.

ii. The posterior distribution of μ follows the normal distribution with mean $\frac{n\overline{X}+2}{n+1}$ and variance $\frac{1}{n+1}$ where $\overline{X}$ denotes the sample mean and n denotes the sample size.

iii. The following sample of observed log of paid claims:

3.22 4.34 5.98 7.32 2.78

Calculate the upper bound of the symmetric 95% Bayesian confidence interval for μ.

19. **CAS Exam ST Fall 2014 #24**

For a large number of home insurance policies, the profit per policy is modeled using a Bayesian approach and a normal distribution with unknown mean μ and variance 10,000.

The prior mean expected profit per policy follows a normal distribution with mean 100 and variance 100. After the first year a sample of five policies returns a total profit of 5,000. Calculate the Bayesian posterior average profit per policy.

CHAPTER 12

A STATISTICS POTPOURRI

Chapter 12, as the name suggests, brings together a variety of statistical tools that we have not touched to any degree in previous chapters. We begin with the basic least-squares model for estimating a linear relationship between two variables. We begin in Section 12.1 with a simple non-parametric derivation of the main relationships in simple regression theory. We continue in Section 12.2 with the derivation of estimation procedures and model testing in the standard case of normally distributed error terms.

Section 12.3 introduces the topic of analysis-of-variance with a description of the one-factor ANOVA test for equality of means. In Section 12.4 we bring together a number of so-called non-parametric methods, which are applied to a variety of situations where we don't wish to assume that our data sets arise from a particular parameter model distribution.

Finally, in Section 12.5 we present a few data exploration tools that allow us to visually compare our data to potential theoretical models with graphs. In particular we describe both probability (*p-p*) charts and quantile (*q-q*) charts.

12.1 SIMPLE LINEAR REGRESSION: BASIC FORMULAS

In this section and the next we take up the question of estimating parameters in a multivariate distribution. This is a vast subject, and our intent is to provide only an introductory exploration of the widely used facet labeled ***simple linear regression***. Regression analysis in general refers to the idea of estimating the parameters of the conditional distribution of one random variable, Y, given values of additional random variables $X_1,\dots,X_n$. The random variable Y is considered a ***dependent*** variable whose value is determined, subject to random fluctuations, by the ***independent*** variables $X_1,\dots,X_n$.

The word ***simple*** in simple linear regression means we have just a single independent variable X. The appellation ***linear*** means we assume that the conditional mean of Y for a given X is a linear function of X. That is to say, $E[Y \mid X = x] = \alpha + \beta x$, where α and β are parameters that do not vary with changes in x. The general treatment for multivariate linear regression depends on more advanced tools of matrix algebra, so we will confine ourselves here to the simple linear regression case.

The word ***regression*** itself in this context appears to be unrelated to its usual meanings in ordinary discourse. It has historical roots going back to the 19th century scientist Francis Galton. Galton studied the relationship between the heights of fathers and their sons, observing that these two variables were indeed positively correlated. But the sons of extremely short or tall men tended to be less extreme in height, meaning that their heights ***regressed*** toward the mean through later generations. For whatever reason, the term ***regression line*** has stuck, and is universally used to describe the linear relationship between jointly distributed random variables.

In this section we take a straight-forward, data-driven approach in deriving the basic formulas of the simple linear regression model. In the following section we add some additional assumptions that will allow us to make use of the statistical estimation methods of earlier sections in order to derive confidence intervals and evaluate hypothesis tests.

12.1.1 THE LEAST SQUARES MODEL

We begin by assuming we have a two-variable data set consisting of N observations of pairs (X_i, Y_i); $i = 1, 2, \ldots, N$. As an aid in facilitating calculations, we display the information in tabular form:

i	X	Y
1	X_1	Y_1
2	X_2	Y_2
$\vdots$	$\vdots$	$\vdots$
N	X_N	Y_N

This data set could be regarded as a random sample from a joint distribution, and this will be our perspective in subsequent sections. For now, however, we will treat it as a self-contained entity. In this way we can employ the methods and notation of discrete joint distributions from Chapter 7. Thus, we begin by interpreting the data as a discrete joint distribution in which each of the pairs (X_i, Y_i) is assigned the probability $\frac{1}{N}$, with all other pairs $(X_i, Y_j)\,(i \neq j)$ having probability zero. The resulting joint distribution table, with the marginal random variables (X, Y) is,

	X_1	X_2	$\cdots$	X_N	
Y_1	$\frac{1}{N}$	0	$\cdots$	0	$\frac{1}{N}$
Y_2	0	$\frac{1}{N}$	$\cdots$	0	$\frac{1}{N}$
$\vdots$	$\vdots$	$\vdots$	$\vdots$	$\vdots$	$\vdots$
Y_N	0	0	$\cdots$	$\frac{1}{N}$	$\frac{1}{N}$
	$\frac{1}{N}$	$\frac{1}{N}$	$\cdots$	$\frac{1}{N}$	1

Here, X takes the values $X_1, \ldots, X_N$ and Y takes the values $Y_1, \ldots, Y_N$. It is common to depict the pairs in a two-dimensional graph referred to as a scatter diagram.

Example 12.1-1 Scatter Diagram for the Guinea Pig Experiment

The biology department has received a grant to study the relationship between the length and the weight of guinea pigs. In order to simplify the calculations as much as possible the consulting statistician recommends using standardized units, so that, for example, a negative weight means below average, and a positive weight means above average. Display the scatter diagram resulting from the given data.

Guinea Pig ID i	Weight X	Length Y
1	−2	0
2	0	−2
3	0	1
4	1	−1
5	3	1
6	7	4

Solution

We simply plot the pairs of points on a rectangular grid in the usual fashion.

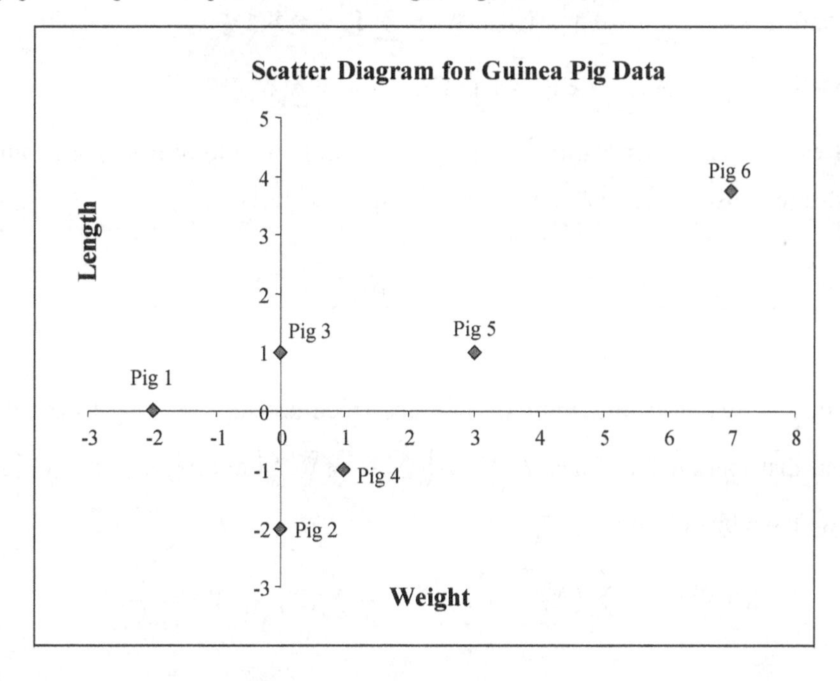

❒

Next, we use the discrete joint distribution of (X,Y) described above to derive the basic algebraic relationships used for linear regression calculations. We introduce the quantities S_{XX}, S_{YY}, and S_{XY}. These are sums of squares, or in the case of S_{XY} the sum of cross-products, and as we shall see, are closely related to $Var[X]$, $Var[Y]$, and $Cov[X,Y]$, respectively, for the joint distribution (X,Y). These quantities provide helpful shortcuts in expressing the regression formulas that follow. The box below summarizes the algebraic relationships involving these quantities. Rather than deriving these relationships from scratch, we use our previously developed formulas for variance, covariance and correlation (see Sections 7.1 and 7.4), applied to the discrete joint distribution (X,Y).

All sums in the following are from $i=1$ to N unless otherwise specified.

The Basic Quantities for Simple Linear Regression

Let the two-variable data set $(X_i,Y_i);\ i=1,2,\ldots,N$ be given, and let (X,Y) be the marginal random variables for the resulting discrete joint distribution with,

$$\Pr\left[(X_i,Y_i)\right] = \frac{1}{N} \text{ for } i=1,\ldots,N$$

and

$$\Pr\left[(X_i,Y_j)\right]=0 \text{ for } i,j=1,\ldots,N \text{ and } i\neq j.$$

(1) Define $\overline{X}=\frac{1}{N}\sum X_i$ and $\overline{Y}=\frac{1}{N}\sum Y_i$. Then $\overline{X}=E[X]$ and $\overline{Y}=E[Y]$.

(2) Define $S_{XX}=\sum(X_i-\overline{X})^2$. Then $S_{XX}=\sum X_i^2-N\overline{X}^2$.

(3) Define $S_{YY}=\sum(Y_i-\overline{Y})^2$. Then $S_{YY}=\sum Y_i^2-N\overline{Y}^2$.

(4) Define $S_{XY}=\sum(X_i-\overline{X})(Y_i-\overline{Y})$. Then $S_{XY}=\sum X_iY_i-N\overline{X}\overline{Y}$.

(5) Let ρ_{XY} be the correlation coefficient of X and Y resulting from their joint discrete distribution (X,Y) and define $R^2=\rho_{XY}^2$. Then,

$$R^2 = \frac{S_{XY}^2}{S_{XX}S_{YY}}.$$

Proof

Using the discrete joint distribution table described above, it is easy to see that the marginal distribution for X has $E[X]=\frac{1}{N}\sum_{i=1}^{n}X_i=\overline{X}$. Similarly, $E[Y]=\frac{1}{N}\sum_{i=1}^{n}Y_i=\overline{Y}$. Then, we have by definition,

$$Var[X] = \frac{1}{N}\sum_{i=1}^{n}(X_i-E[X])^2 = \frac{1}{N}\sum_{i=1}^{n}(X_i-\overline{X})^2 = \frac{1}{N}S_{XX}.$$

On the other hand, using the variance formula for X, we have,

$$Var[X] = E[X^2]-E[X]^2 = \frac{1}{N}\sum_{i=1}^{n}X_i^2-\overline{X}^2.$$

By equating these two expressions for $Var[X]$, we have,

$$\frac{1}{N}S_{XX} = \frac{1}{N}\sum X_i^2 - \bar{X}^2.$$

Multiplying through by N produces the formula in (2). The corresponding formula for S_{YY} follows similarly.

Now, by the definition of covariance,

$$\begin{aligned} Cov[X,Y] &= \frac{1}{N}\sum(X_i - E[X])(Y_i - E[Y]) \\ &= \frac{1}{N}\sum(X_i - \bar{X})(Y_i - \bar{Y}) = \frac{1}{N}S_{XY}. \end{aligned}$$

Using the covariance formula, we also have,

$$Cov[X,Y] = E[X \cdot Y] - E[X] \cdot E[Y] = \frac{1}{N}\sum X_i \cdot Y_i - \bar{X} \cdot \bar{Y}.$$

We equate the two expressions for $Cov[X,Y]$ and multiply through by N to obtain the shortcut formula for S_{XY} in (4).

Again, by definition (Section 7.4), $\rho_{XY} = \frac{Cov[X,Y]}{\sqrt{Var[X] \cdot Var[Y]}}$. Expressed in terms of S_{XX}, S_{YY}, and S_{XY} this becomes,

$$\rho_{XY} = \frac{Cov[X,Y]}{\sqrt{Var[X] \cdot Var[Y]}} = \frac{\frac{1}{N}S_{XY}}{\sqrt{\left(\frac{1}{N}S_{XX}\right)\left(\frac{1}{N}S_{YY}\right)}} = \frac{S_{XY}}{\sqrt{(S_{XX})(S_{YY})}}.$$

Therefore, $\rho_{XY}^2 = \frac{S_{XY}^2}{S_{XX} \cdot S_{YY}}$. ❒

Notes

(1) The main advantage of the alternative forms for S_{XX}, S_{YY}, and S_{XY} is that, like the variance and covariance shortcut formulas, they make computation easier.

(2) The definitions of $\bar{X}$ and $\bar{Y}$ are consistent with their previous usage for random samples from a population.

(3) From its definition as the square of the correlation, we see that R^2 is always between zero and one, and that it is a measure of the degree to which movements in X and Y tend to be related to one another. For example, if all the data points (X_i, Y_i) are on the same straight line, then $R^2 = 1$. As we will explain further below, R^2 is the most widely used quantity for evaluating the validity of the linear regression model.

We come now to the main idea, which is to model a linear relationship between the two variables X and Y. Our approach is to find the line $y = \alpha + \beta x$ that provides the best fit for the data points in the scatter diagram. This line will be the regression line. In this endeavor, the coefficients α and β are the unknowns. First, we need to define the term ***best fit*** according to some reasonable criterion, and it is the ***least squares error*** measure that is generally used.

To see how this works we need some additional notation. For each data pair (X_i, Y_i) we define $Y_i' = \alpha + \beta X_i$ as the corresponding y-coordinate on the (as yet, unknown) regression line. We can think of Y_i' as the ***predicted*** value of y for a given X_i. We then look at the error between the actual value Y_i and the predicted value Y_i'. This error is given by the difference $Y_i - Y_i'$, and we want to find the values α and β that will minimize the sum of the squared errors. This is what is meant by the ***least squares error*** criterion. One can imagine other approaches, such as minimizing the sum of the absolute differences of the errors. However absolute values are always problematic in calculus, and summing squared differences leads to the most tractable mathematical model.[1]

The following graph illustrates the errors using the scatter diagram for the guinea pigs in Example 12.1-1.

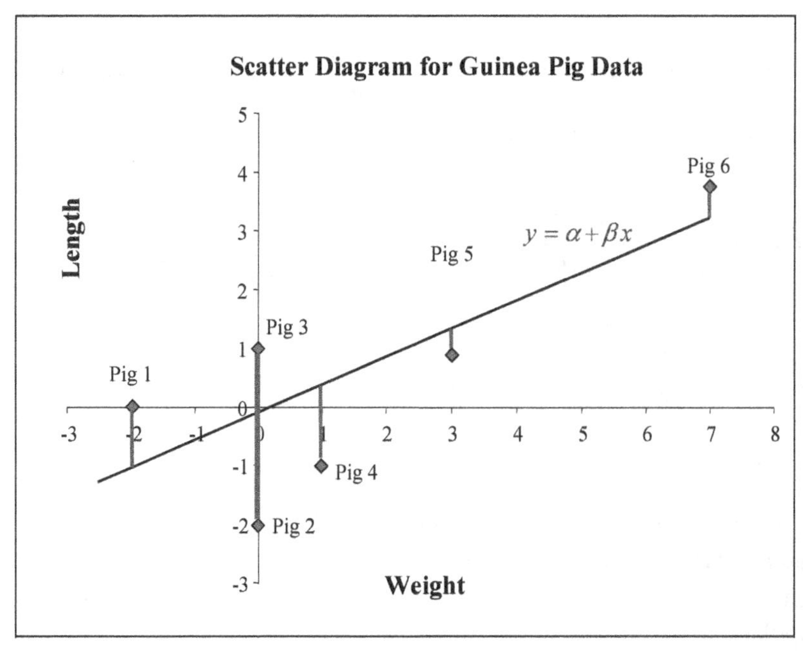

We have depicted the errors as the vertical green line segments on the graph.

Let $T = \sum_{i=1}^{N}(Y_i - Y_i')^2 = \sum_{i=1}^{N}(Y_i - \alpha - \beta X_i)^2$ be the sum of the squared errors. We next show how to solve for the values for α and β that minimize T.

---

[1] In the modern age of "big data" and super-fast computers the field of predictive modeling has renewed interest in algorithms that minimize absolute errors along with, or in place of, the traditional least squares errors.

Solution for the Least Squares Regression Line

Let the data set $(X_i,Y_i);\ i=1,2,\ldots,N$ be given and let $T=\sum_{i=1}^{N}(Y_i-\alpha-\beta X_i)^2$. The values of α and β that minimize S are given by,

(6) $\hat{\beta}=\dfrac{S_{XY}}{S_{XX}}$,

(7) $\hat{\alpha}=\overline{Y}-\hat{\beta}\overline{X}$.

The resulting line $y=\hat{\alpha}+\hat{\beta}x$ is called the ***least squares regression line.***

Let $\hat{Y}_i=\hat{\alpha}+\hat{\beta}X_i$, and define the ***error sum of squares*** (*ESS*) by,

(8) $ESS=\sum_{i=1}^{N}(Y_i-\hat{Y}_i)^2=\sum_{i=1}^{N}(Y_i-\hat{\alpha}-\hat{\beta}X_i)^2$. Then *ESS* is the minimum value of T

and,

$$ESS = S_{YY}\cdot(1-R^2).$$

Proof

For given values α and β let $T=\sum_{i=1}^{N}(Y_i-\alpha-\beta X_i)^2$. Using the joint distribution table for X and Y, we can rewrite this as,

$$T = \sum_{i=1}^{N}(Y_i-\alpha-\beta X_i)^2 = \sum_{i=1}^{N}\left[(Y_i-\beta X_i)-\alpha\right]^2 = \overbrace{N\cdot E\left[\left((Y-\beta X)-\alpha\right)^2\right]}^{\text{this uses the discrete joint distribution of }(X,Y)} .$$

We cast this in the context of the squared error loss function from the earlier section on Bayesian estimation (Section 11.6.2). Thus, let $\Theta = Y-\beta X$, and consider the loss function $L=(\Theta-\alpha)^2$. It follows from the result in Section 11.6 that the expected value of L is minimized when $\alpha = E[\Theta]=\overline{Y}-\beta\overline{X}$. This means that for any value of β, for the expression

$$T = \sum_{i=1}^{N}(Y_i-\alpha-\beta X_i)^2 = N\cdot E\left[\left((Y-\beta X)-\alpha\right)^2\right] = N\cdot E[L]$$

to be at a minimum we must have $\alpha=\overline{Y}-\beta\overline{X}$. We substitute this into the formula for T and then seek the value of β that minimizes,

$$\begin{aligned} T &= \sum_{i=1}^{N}(Y_i-\alpha-\beta X_i)^2 \\ &= \sum_{i=1}^{N}\left(Y_i-(\overline{Y}-\beta\overline{X})-\beta X_i\right)^2 = \sum_{i=1}^{N}\left[(Y_i-\overline{Y})-\beta(X_i-\overline{X})\right]^2. \end{aligned}$$

Squaring out the terms on the right gives,

$$\begin{aligned} T &= \sum(Y_i-\overline{Y})^2 - 2\beta\sum(Y_i-\overline{Y})(X_i-\overline{X}) + \beta^2\sum(X_i-\overline{X})^2 \\ &= S_{YY} - 2\beta S_{XY} + \beta^2 S_{XX} \end{aligned}$$

This is a simple quadratic expression in β and we can use calculus to find the value of β that minimizes T. We take the derivative of T with respect to β and set it equal to zero. Thus, $\frac{dT}{d\beta} = -2S_{XY} + 2\beta S_{XX} = 0$ results in the value $\hat{\beta} = \frac{S_{XY}}{S_{XX}}$. From the above, the corresponding value for α is given by $\hat{\alpha} = \overline{Y} - \hat{\beta}\overline{X}$.

The minimum value of T, denoted by ESS, is now given by,

$$\begin{aligned} ESS &= S_{YY} - 2\hat{\beta}S_{XY} + \hat{\beta}^2 S_{XX} \\ &= S_{YY} - 2\left(\frac{S_{XY}}{S_{XX}}\right)\cdot S_{XY} + \left(\frac{S_{XY}}{S_{XX}}\right)^2 \cdot S_{XX} \\ &= S_{YY} - 2\frac{S_{XY}^2}{S_{XX}} + \frac{S_{XY}^2}{S_{XX}} \\ &= S_{YY} - \frac{S_{XY}^2}{S_{XX}} = S_{YY}\cdot\left(1 - \frac{S_{XY}^2}{S_{XX}\cdot S_{YY}}\right) = S_{YY}\cdot(1-R^2). \end{aligned}$$

Notes

(1) The formula $ESS = S_{YY}\cdot(1-R^2)$ helps to explain the significance of R^2 in assessing the accuracy of the linear regression model. Clearly ESS is a measure of the degree to which the regression line fits the data. If ESS is small then R^2 must be close to one.

(2) The quantity S_{YY} can be thought of as the total variation of Y about its mean, and it constitutes an upper bound for ESS. Thus, if R^2 is zero then $\hat{\beta}$ is also zero, meaning that the regression line is horizontal, given by $y = \overline{Y}$, and $ESS = S_{YY}$.

(3) From formulas (5) and (6) we see that both $\hat{\beta}$ and R^2 are multiples of S_{XY}. It follows that $\hat{\beta}$ is zero if and only if R^2 is zero. Consequently, in either case we conclude the regression line is horizontal and the model implies no linear relationship between X and Y.

(4) The individual errors, $Y_i - \hat{Y}_i$ for $i = 1,\ldots,N$, are often referred to as ***residuals***.

Example 12.1-2 Regression for Guinea Pigs

Find the coefficients of the regression line and the value of R^2 for the guinea pig data given in Example 12.1-1.

Solution

The observational data is shown below in the first 3 columns. The ensuing calculations are laid out in tabular form.

#	X	Y	X^2	Y^2	XY
1	−2	0	4	0	0
2	0	−2	0	4	0
3	0	1	0	1	0
4	1	− 1	1	1	− 1
5	3	1	9	1	3
6	7	4	49	16	28
Totals	9	3	63	23	30

(1) $\overline{X} = \frac{9}{6} = 1.5$, and $\overline{Y} = \frac{3}{6} = 0.5$

(2) $S_{XX} = \sum X_i^2 - N\overline{X}^2 = 63 - (6)\cdot(1.5)^2 = 49.5$

(3) $S_{YY} = \sum Y_i^2 - N\overline{Y}^2 = 23 - (6)\cdot(0.5)^2 = 21.5$

(4) $S_{XY} = \sum X_i Y_i - N\overline{X}\overline{Y} = 30 - (6)\cdot(1.5)\cdot(0.5) = 25.5$

(5) $R^2 = \dfrac{S_{XY}^2}{S_{XX}S_{YY}} = \dfrac{(25.5)^2}{(49.5)\cdot(21.5)} = 0.6110$

(6) $\hat{\beta} = \dfrac{S_{XY}}{S_{XX}} = \dfrac{25.5}{49.5} = 0.5152$

(7) $\hat{\alpha} = \overline{Y} - \hat{\beta}\overline{X} = 0.5 - (0.5152)\cdot(1.5) = -0.2727$

The resulting regression line is $y = \hat{\alpha} + \hat{\beta}x = -0.2727 + 0.5152x$.

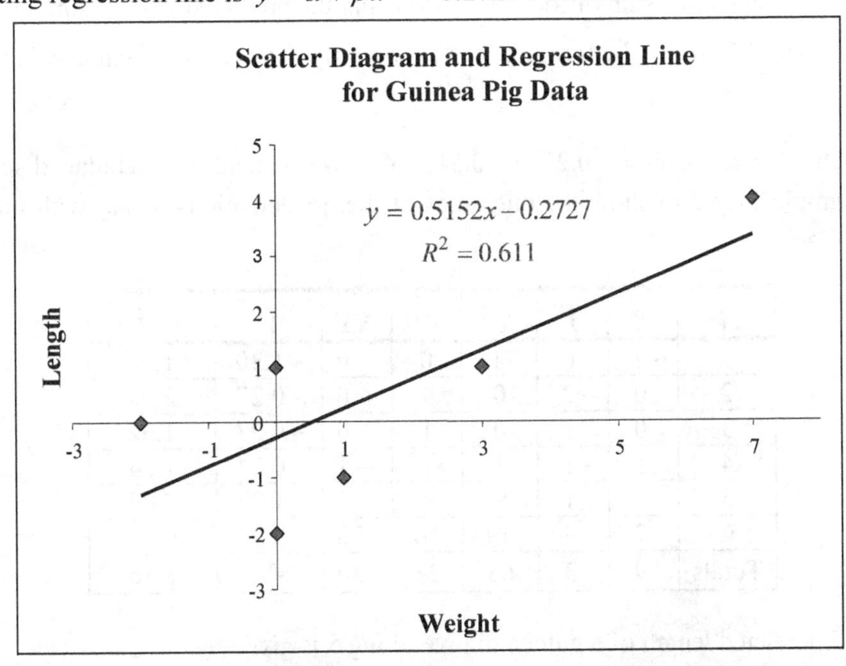

❐

Notes

(1) The graph indicates visually that the data points are pretty widely spread out around the regression line. This indicates that there is a linear trend between weight and length, but it is far less than perfect. Analytically, this is born out by the R^2 value of 0.6110, which indicates that about 61% of the variation in guinea pig length can be "explained" by their weight, leaving about 39% of the variation to other factors (for example, diet and lifestyle, as reflected in being cooped up in a cage all day), as well as random errors in the measurements.

(2) The slope, $\hat{\beta}$, is positive, indicating that the correlation between the variables is positive. A negative slope would indicate a negative correlation between the variables.

The regression line is often used to make predictions about Y for given values of X. The value of R^2, or *ESS*, is important in determining the accuracy of such predictions. This issue of accuracy will be explored further in the next section. We next examine the predictions determined by the guinea pig data.

Example 12.1-3 Predictions and Errors for Guinea Pigs

Use the results of Example 12.1-2 to make the following calculations.

(a) For each data pair (X_i, Y_i) calculate the corresponding predicted value $\hat{Y}_i$.

(b) A new guinea pig is discovered whose weight is 5. Use the regression line to predict the length of this new furry rodent.

(c) Calculate the error sum of squares (*ESS*) using the predicted values computed in (a).

(d) Calculate *ESS* using the formula in (8) and verify that this is the same as in (c).

Solution

(a) Each $\hat{Y}_i = \hat{\alpha} + \hat{\beta}X_i = -0.2727 + 0.5152X_i$. We extend the tabular display from Example 11.7-2 to show the outcomes of these calculations, along with the squared errors.

#	X	Y	X^2	Y^2	XY	$\hat{Y}$	$(Y-\hat{Y})^2$
1	−2	0	4	0	0	−1.30	1.70
2	0	−2	0	4	0	−0.27	2.98
3	0	1	0	1	0	−0.27	1.62
4	1	−1	1	1	−1	0.24	1.54
5	3	1	9	1	3	1.27	0.07
6	7	4	49	16	28	3.33	0.44
Totals	9	3	63	23	30	---	8.36

(b) The predicted length of a guinea pig weighing 5 is given by

$$\hat{\alpha} + \hat{\beta}(5) = -0.2727 + 0.5152(5) = 2.30.$$

(c) The calculation for *ESS* appears at the bottom of the last column, giving,

$$ESS = \sum_{i=1}^{6}(Y_i - \hat{Y}_i)^2 = 8.36.$$

(d) Using formula (8) and the previously calculated values of S_{YY} (formula (3)) and R^2 (formula (5)) we have $ESS = S_{YY} \cdot (1 - R^2) = 21.5 \cdot (1 - 0.6110) = 8.36.$ □

Using a spreadsheet program facilitates the calculations in problems such as this. The above results could also be compared to the built-in regression tools in Excel®. For larger data sets, such as those arising from real-world problems, a software package is indispensable. Students preparing for an actuarial exam, however, do not currently have this option, so learning the above framework is important, as many of the following exercises illustrate.

Exercise 12-1 You are given the following information about a simple linear regression model fit to 9 observations:

$$\sum_{i=1}^{9} X_i = 18, \ \sum_{i=1}^{9} Y_i = 90, \ \sum_{i=1}^{9} x_i^2 = 200, \ \sum_{i=1}^{9} y_i^2 = 1000,$$

and the sample correlation coefficient is $r = -.88$,

(a) Derive the regression equation $\hat{Y}_i = \hat{\alpha} + \hat{\beta} X_i$.
(b) Determine the predicted value of Y when $X = 6$.

Exercise 12-2 CAS Exam 3 Spring 2007 #31
The following table shows an independent variable, *x*, a dependent variable, *y*, and a least squares linear regression estimate, $\hat{y}$:

x	y	$\hat{y}$
1	10	9.60
2	12	11.3
3	11	13.0
4	15	17.7
5	17	16.4

Using the information from the table, calculate the coefficient of determination, R^2, of the regression line.

(A) Less than 0.75
(B) At least 0.75, but less than 0.80
(C) At least 0.80, but less than 0.85
(D) At least 0.85, but less than 0.90
(E) At least 0.90

Exercise 12-3 CAS Exam 3 Fall 2006 #8

Mr, Smith is buying a ring valued at $1,200 and wishes to insure it. The jeweler provides him with the following information to estimate the insurance premium, y_i, based on the value of the ring, x_i (in 000's):

$$y_i = \alpha + \beta x_i$$

$$\sum_{i=1}^{5} (x_i - \bar{x})^2 = 10$$

$$\sum_{i=1}^{5} (x_i - \bar{x})(y_i - \bar{y}) = 195$$

$$\bar{x} = 3$$

$$\bar{y} = 45$$

Using least squares regression, calculate the predicted premium for the ring.

(A) Less than 9.75
(B) At least 9.75, but less than 10.00
(C) At least 10.00, but less than 10.25
(D) At least 10.25, but less than 10.50
(E) At least 10.50

Exercise 12-4 CAS Exam 3 Spring 2006 #9

The following summary statistics are available with respect to a random sample of seven observations of the price of gasoline, *Y*, and the price of oil, *X*:

$$\sum x_i = 315$$

$$\sum x_i^2 = 14{,}875$$

$$\sum y_i = 12.8$$

$$\sum y_i^2 = 24.3$$

$$\sum x_i y_i = 599.5$$

Using the information and a linear regression model of the form $Y = \alpha + \beta X$, calculate the predicted price of gasoline if the price of oil reaches $75.

(A) Less than $2.85
(B) At least $2.85, but less than $2.90
(C) At least $2.90, but less than $2.95
(D) At least $2.95, but less than $3.00
(E) At least $3.00

Exercise 12-5 CAS Exam 3 Spring 2005 #27

Given the following information:

$$\sum X_i = 144$$

$$\sum Y_i = 1{,}742$$

$$\sum X_i^2 = 2{,}300$$

$$\sum Y_i^2 = 312{,}674$$

$$\sum X_i Y_i = 26{,}696$$

$$n = 12$$

Determine the least squares equation for the data.

(A) $\hat{Y}_i = -0.73 + 12.16X_i$
(B) $\hat{Y}_i = -8.81 + 12.16X_i$
(C) $\hat{Y}_i = 283.87 + 10.13X_i$
(D) $\hat{Y}_i = 10.13 + 12.16X_i$
(E) $\hat{Y}_i = 23.66 + 10.13X_i$

Exercise 12-6 CAS Exam 3 Fall 2005 #9

The following information is known about average claim sizes:

Year	Average Claim Size
1	\$1,020
2	\$1,120
3	\$1,130
4	\$1,210
5	\$1,280

Average claim sizes, Y, in a year X are modeled by: $Y = \alpha e^{\beta X}$

Using linear regression to estimate α and β, calculate the predicted average claim size in year 6. *Hint:* Apply the regression model to $\ln Y = \ln \alpha + \beta X$.

(A) Less than \$1,335
(B) At least \$1,335 but less than \$1,340
(C) At least \$1,340 but less than \$1,345
(D) At least \$1,345 but less than \$1,350
(E) At least \$1,350

Exercise 12-7 CAS Exam 3, Segment 3L Spring 2008 #9

You are modeling the price of gold, Y, as a function of the price of silver, X, using linear regression. You are given the following data points and summary statistics:

Price of Silver X	Price of Gold Y
\$4.67	\$343.80
\$5.97	\$416.25
\$6.39	\$427.75
\$9.04	\$530.00
\$13.01	\$639.75

$$\sum_{i=1}^{5} X_i = 39.08 \qquad \sum_{i=1}^{5} Y_i = 2.357.55 \qquad \hat{\beta} = 34.5$$

Calculate the residual (error) at $(6.39, 427.75)$.

(A) Less than -15
(B) At least -15, but less than -5
(C) At least -5, but less than 5
(D) At least 5, but less than 15
(E) At least 15

12.1.2 THE ANALYSIS OF VARIANCE PERSPECTIVE

We now have at hand the basic formulas for simple linear regression, and we have developed them with an eye toward the ease of performing calculations by hand. In most practical applications the computations will be handled by standard software, and the emphasis shifts to interpreting the output. With that thought in mind, we turn now to deriving some additional relationships in simple linear regression. The so-called ***analysis of variance*** (ANOVA) approach brings a different perspective that the student can carry over nicely to the general multivariate case when the time arises.

Before proceeding, it will prove convenient to bring together a number of relationships whose sole purpose is to aid in following the derivations in the ensuing narrative. Some of these are quite simple and have been utilized previously in one form or another.

Tools for Tomorrow

(9) $\sum_{i=1}^{N}(X_i - \bar{X}) = 0$ and $\sum_{i=1}^{N}(Y_i - \bar{Y}) = 0$

(10) $\sum_{i=1}^{N}(Y_i - \hat{Y}_i) = 0$

(11) $S_{XX} = \sum_{i=1}^{N}(X_i - \bar{X})\cdot X_i$ and $S_{YY} = \sum_{i=1}^{N}(Y_i - \bar{Y})\cdot Y_i$

(12) $S_{XY} = \sum_{i=1}^{N}(X_i - \overline{X})\cdot Y_i = \sum_{i=1}^{N}(Y_i - \overline{Y})\cdot X_i$

(13) $\sum_{i=1}^{N}(Y_i - \hat{Y}_i)\cdot X_i = 0$

(14) $\sum_{i=1}^{N}(Y_i - \hat{Y}_i)\cdot \hat{Y}_i = 0$

Proof

(9) $\sum_{i=1}^{N}(X_i - \overline{X}) = \sum_{i=1}^{N}X_i - \sum_{i=1}^{N}\overline{X} = N\cdot\overline{X} - N\cdot\overline{X} = 0$. The argument for Y is identical.

(10)
$$\sum_{i=1}^{N}(Y_i - \hat{Y}_i) = \sum_{i=1}^{N}(Y_i - \hat{\alpha} - \hat{\beta}X_i)$$
$$= \sum_{i=1}^{N}\left[Y_i - \left(\overbrace{\overline{Y} - \hat{\beta}\overline{X}}^{\text{formula (7)}}\right) - \hat{\beta}X_i\right]$$
$$= \sum_{i=1}^{N}\left[(Y_i - \overline{Y}) - \hat{\beta}(X_i - \overline{X})\right] = \overbrace{\sum_{i=1}^{N}(Y_i - \overline{Y}) - \hat{\beta}\sum_{i=1}^{N}(X_i - \overline{X})}^{\text{using formulas in (9)}} = 0$$

(11)
$$S_{XX} = \overbrace{\sum_{i=1}^{N}X_i^2 - N\overline{X}^2}^{\text{formula (2)}}$$
$$= \sum_{i=1}^{N}X_i^2 - N\overline{X}\left(\frac{1}{N}\sum_{i=1}^{N}X_i\right) = \sum_{i=1}^{N}(X_i^2 - \overline{X}\cdot X_i) = \sum_{i=1}^{N}(X_i - \overline{X})\cdot X_i$$

The calculation involving Y is identical.

(12) This calculation is almost identical to (11).

$$(13)\quad \sum_{i=1}^{N}(Y_i-\hat{Y}_i)\cdot X_i = \sum_{i=1}^{N}(Y_i-\hat{\alpha}-\hat{\beta}X_i)\cdot X_i = \sum_{i=1}^{N}\left[Y_i - \left(\overbrace{\overline{Y}-\hat{\beta}\overline{X}}^{\text{formula (7)}}\right) - \hat{\beta}X_i\right]\cdot X_i$$

$$= \sum_{i=1}^{N}\left[(Y_i-\overline{Y}) - \hat{\beta}(X_i-\overline{X})\right]\cdot X_i$$

$$= \overbrace{\sum_{i=1}^{N}(Y_i-\overline{Y})\cdot X_i - \hat{\beta}\sum_{i=1}^{N}(X_i-\overline{X})\cdot X_i}^{\text{using formulas in (11) and (12) next}}$$

$$= S_{XY} - \hat{\beta}\cdot S_{XX} = S_{XY} - \overbrace{\frac{S_{XY}}{S_{XX}}}^{\text{using formula (6)}}\cdot S_{XX} = 0$$

$$(14)\quad \sum_{i=1}^{N}(Y_i-\hat{Y}_i)\cdot \hat{Y}_i = \sum_{i=1}^{N}(Y_i-\hat{Y}_i)\cdot(\hat{\alpha}+\hat{\beta}X_i)$$

$$= \overbrace{\hat{\alpha}\sum_{i=1}^{N}(Y_i-\hat{Y}_i) + \hat{\beta}\sum_{i=1}^{N}(Y_i-\hat{Y}_i)\cdot X_i}^{\text{using formulas (10) and (13)}} = 0.$$

We return to the question of interpreting the strength of a linear relationship between the dependent variable Y and the independent variable X. Previously, we have used the formula (8) result $ESS = S_{YY}\cdot(1-R^2)$ to show that the error sum of squares (ESS) is a fraction of the variation in Y, as expressed by S_{YY}. Furthermore, that fraction is given by $1-R^2$, so that the smaller the total error, the closer R^2 is to one. Likewise, the larger ESS is (with an upper bound of S_{YY}), the closer R^2 is to zero.

We elaborate on this analysis by showing how the variation of Y from its mean $\overline{Y}$ can be decomposed into two parts. Simple algebra shows that,

$$Y_i - \overline{Y} = (Y_i-\hat{Y}_i) + (\hat{Y}_i-\overline{Y}_i).$$

This in turn shows that the deviation of Y from its mean is the sum of the prediction error plus the deviation from the regression line to the mean. Somewhat surprisingly, this decomposition continues to hold if we square each of the terms and sum over i.

Analysis of Variance in Simple Linear Regression

Define the ***Total Sum of Squares*** (TSS) in Y by $TSS = S_{YY} = \sum_{i=1}^{N}(Y_i-\overline{Y})^2$, and define the ***Regression Sum of Squares*** (RSS) by $RSS = \sum(\hat{Y}_i-\overline{Y})^2$. Then,

$$(15)\quad TSS = ESS + RSS.$$

$$(16)\quad R^2 = \frac{RSS}{TSS} = 1-\frac{ESS}{TSS}.$$

Proof

To establish (15) we begin with the above algebraic decomposition,

$$Y_i - \bar{Y} = (Y_i - \hat{Y}_i) + (\hat{Y}_i - \bar{Y}_i).$$

Then,

$$TSS = \sum_{i=1}^{N}(Y_i - \bar{Y})^2 = \sum_{i=1}^{N}\left[(Y_i - \hat{Y}_i) + (\hat{Y}_i - \bar{Y})\right]^2$$
$$= \sum_{i=1}^{N}(Y_i - \hat{Y}_i)^2 + 2\sum_{i=1}^{N}(Y_i - \hat{Y}_i)\cdot(\hat{Y}_i - \bar{Y}) + \sum_{i=1}^{N}(\hat{Y}_i - \bar{Y})^2.$$

Formulas (10) and (14) show that the cross-product term is zero:

$$2\sum_{i=1}^{N}(Y_i - \hat{Y}_i)\cdot(\hat{Y}_i - \bar{Y}) = \overbrace{2\sum_{i=1}^{N}(Y_i - \hat{Y}_i)\cdot\hat{Y}_i}^{\text{using formula (14)}} - 2\bar{Y}\overbrace{\sum_{i=1}^{N}(Y_i - \hat{Y}_i)}^{\text{using formula (10)}} = 0.$$

It follows that $TSS = \sum_{i=1}^{N}(Y_i - \bar{Y})^2 = \sum_{i=1}^{N}(Y_i - \hat{Y}_i)^2 + \sum_{i=1}^{N}(\hat{Y}_i - \bar{Y})^2 = ESS + RSS.$

To show (16) we rearrange terms to get $ESS = TSS - RSS$. Since TSS and S_{YY} are the same, we have,

$$ESS = TSS - RSS = S_{YY} - RSS = S_{YY}\cdot\left(1 - \frac{RSS}{S_{YY}}\right).$$

On the other hand, formula (8) shows that $ESS = S_{YY}\cdot(1 - R^2)$. Comparing the two formulas for *ESS* shows that $R^2 = \frac{RSS}{S_{YY}} = \frac{RSS}{TSS} = \frac{TSS-ESS}{TSS} = 1 - \frac{ESS}{TSS}$.

Note

As a practical matter it is rarely necessary to calculate *RSS* directly from its definition. Indeed in (15) and (16) we have two equations relating the four quantities, *TSS*, *ESS*, *RSS*, and R^2. Therefore, if we know any two of them we can calculate the remaining two. Typically, we would have already calculated *TSS* (in the form of S_{YY}) and R^2, which allows us to easily solve for *ESS* and *RSS*.

Exercise 12-8 You are given 5 observations from a simple linear regression model:

X	Y
6.8	0.8
7.0	1.2
7.1	0.9
7.2	0.9
7.4	1.5

(a) Determine $\hat{\alpha}$ and $\hat{\beta}$.

(b) Calculate the predicted value of Y when X equals 7.0.

(c) Calculate R^2, the coefficient of determination.

(d) Calculate the regression sum of squares, *RSS*.

(e) Calculate the error sum of squares, *ESS*.

Exercise 12-9 The least squares method is used to fit the following data to the model $Y_i = \alpha + \beta X_i$. The least squares line passes through the point $(3,3.3)$. Determine $\sum_{i=1}^{5} y_i$.

X_i	0	1	2	3	4
Y_i	.5	3.5	3	w	5

Exercise 12-10 (Graphing calculator or Spreadsheet recommended) In the field of actuarial science, it is believed that the starting salary of an actuarial analyst is related to the number of actuarial examinations that students have passed. Given the following data, use a simple linear least squares model to:

(a) Calculate the regression equation.

(b) Calculate the correlation coefficient.

(c) Predict the annual starting salary of a student with five examinations passed.

Student's name	Number of SOA/CAS examinations passed	Starting salary
Adam	3	53,021
Brenda	1	48,104
Colleen	3	50,004
Dennis	2	49,576
Eugene	1	46,500
Frederick	0	43,200
Gregory	4	56,299

Exercise 12-11 (Graphing calculator or Spreadsheet recommended) A Mathematics professor wants to know if faculty salary is correlated to the faculty weight. Consider the following:

Salary (in $1000), x	43	52	31	38	35	44
Weight (in pounds), y	225	320	115	175	155	240

(a) Create a Scatter Plot of the data with faculty salary as the independent variable.

(b) Calculate the correlation coefficient.

(c) Calculate the regression line. Graph the line on the scatter plot.

(d) If your business professor weighs 200 pounds, then predict her salary.

Exercise 12-12 (Graphing calculator or Spreadsheet recommended) It is conjectured that the number of children in a family is related to the amount of disposable income of the family. Use a simple linear regression model with the following data.

Number of children, x	0	0	1	2	2	3	5	9
Annual disposable income (in $1000), y	25	33	20	18	15	14	8	2

(a) Create a Scatter Plot of the data with the number of children as the independent variable.

(b) Calculate the correlation coefficient.

(c) Calculate the regression line. Graph the line on the scatter plot.

(d) If a family has four children, then predict their annual disposable income.

Exercise 12-13 (Graphing calculator or Spreadsheet recommended) A Mathematics professor wants to know if the composite ACT score is correlated to the final average in a basic statistics class. Use a simple linear regression model and the following data:

ACT score, x	20	26	32	15	21	30	23
Final Average in STAT, y	91%	74%	56%	98%	88%	63%	85%

(a) Create a Scatter Plot of the data with ACT scores as the independent variable.

(b) Calculate the correlation coefficient.

(c) Calculate the regression line. Graph the line on the scatter plot.

12.2 SIMPLE LINEAR REGRESSION: ESTIMATION OF PARAMETERS

In the previous section we treated a basic two-dimensional data set as a discrete joint distribution. From this bare-bones structure we derived formulas for the linear regression line using the least squares method. We now wish to treat those formulas as estimators in a statistical model so that we can measure the reliability of the estimates and compute confidence intervals. Accomplishing this requires that we impose more conditions on the underlying model.

We now assume that the values of X, $(X_1,\ldots,X_N)$, are fixed, non-random, observations, and we treat them throughout as given in advance. On the other hand, the values of Y will be considered as random variables conditioned on X. We give the precise relationships next in the following box.

The Normal Simple Linear Regression Model

Let $X_1,\ldots,X_N$ be given and assume that for each X_i; $i=1,\ldots,N$, there is a corresponding random variable Y_i such that,

(i) $Y_i = \alpha + \beta X_i + \varepsilon_i$ and

(ii) $\varepsilon_1,\ldots,\varepsilon_N$ are independent, identically distributed normal random variables, each with a mean of zero and a variance of σ^2.

The three quantities α, β, σ^2 are the ***parameters*** of the model and the random variables ε_i are referred to as the ***errors***.

Notes

(1) Since $E[\varepsilon_i]=0$, we have for each i that $E[Y_i \mid X_1,\ldots,X_n] = \alpha+\beta X_i$.

(2) Furthermore, the random variables Y_i are normal and independent. Each one is centered around its mean of $\alpha+\beta X_i$, and has a variance of σ^2.

(3) We treat the values $Y_1,\ldots,Y_N$ in the original data table as random observations from the model given by (i) and (ii).

These conditions really presume in advance a structured linear relationship between Y and X, with a predictable degree of statistical fluctuation. By predictable we mean that each of the errors ε_i between Y_i, and its linear predictor $\alpha+\beta X_i$, is normally distributed with a fixed variance and a mean of zero.

We show next that with the assumptions of a ***normal*** simple linear regression model in place, the $\hat{\alpha}$ and $\hat{\beta}$ derived in the previous section are normal random variables, and constitute unbiased estimators for the model parameters α and β. Furthermore, a suitably standardized version of *ESS* becomes a chi-squared random variable, independent of both $\hat{\alpha}$ and $\hat{\beta}$. We also show that *ESS* can be used to construct an unbiased estimator for σ^2, the variance of the error term. Among other things, this will allow us to build confidence intervals using student-*t* tables.

Estimators for the Normal Regression Model

Assume that the conditions of the normal simple linear regression model are satisfied, with errors $(\varepsilon_i)_{i=1}^N$ and $\sigma^2 = Var[\varepsilon_i]$. Using the notation and formulas from (6), (7) and (8) of the least squares model in Section 12.1, we let,

(6) $\hat{\beta} = \dfrac{S_{XY}}{S_{XX}}$,

(7) $\hat{\alpha} = \overline{Y} - \hat{\beta}\overline{X}$,

(8) $ESS = \sum_{i=1}^{N}(Y_i - \hat{\alpha} - \hat{\beta}X_i)^2 = \sum_{i=1}^{N}(Y_i - \hat{Y}_i)^2$, where $\hat{Y}_i = \hat{\alpha} + \hat{\beta}X_i$.

Then,

(17) $\hat{\beta} \sim N\left(\beta, \dfrac{\sigma^2}{S_{XX}}\right)$,

(18) $\hat{\alpha} \sim N\left(\alpha, \left[\dfrac{1}{N} + \dfrac{\overline{X}^2}{S_{XX}}\right] \cdot \sigma^2\right)$,

(19) $Cov(\hat{\alpha}, \hat{\beta}) = -\dfrac{\sigma^2 \overline{X}}{S_{XX}}$,

(20) $\dfrac{ESS}{\sigma^2} \sim \chi^2(N-2)$ and is independent of $\hat{\alpha}$ and $\hat{\beta}$,

(21) Define $S^2 = \dfrac{ESS}{N-2}$. Then S^2 is an unbiased estimator for σ^2 and is independent of $\hat{\alpha}$ and $\hat{\beta}$.

Proof

(17) $\hat{\beta} = \dfrac{S_{XY}}{S_{XX}} = \overbrace{\dfrac{1}{S_{XX}} \prod_{i=1}^{N}(X_i - \overline{X})Y_i}^{\text{formula (12)}}$. This shows that $\hat{\beta}$ is a linear combination of the normal random variables Y_i, and is therefore normal itself. Furthermore,

$$E[\hat{\beta}] = \frac{1}{S_{XX}}\sum_{i=1}^{N}(X_i-\overline{X})\cdot E[Y_i]$$

$$= \frac{1}{S_{XX}}\sum_{i=1}^{N}(X_i-X)\cdot E[\alpha+\beta X_i+\varepsilon_i]$$

$$= \frac{1}{S_{XX}}\left[\sum_{i=1}^{N}(X_i-\overline{X})\cdot(\alpha+\beta X_i)\right]$$

$$= \frac{1}{S_{XX}}\left[\alpha\sum_{i=1}^{N}\overbrace{(X_i-\overline{X})}^{\text{use formula (9)}} + \overbrace{\beta\sum_{i=1}^{N}(X_i-\overline{X})\cdot X_i}^{\text{use formula (11)}}\right] = \frac{1}{S_{XX}}(\beta S_{XX}) = \beta.$$

Since $Y_1,\ldots,Y_N$ are independent we can calculate the variance of $\hat{\beta}$ as,

$$Var[\hat{\beta}] = \frac{1}{S_{XX}^2}\sum_{i=1}^{N}(X_i-\overline{X})^2 Var[Y_i] = \frac{\sigma^2}{S_{XX}^2}S_{XX} = \frac{\sigma^2}{S_{XX}},$$

which completes the proof of (17).

(18) Since $\hat{\alpha}=\overline{Y}-\hat{\beta}\overline{X}$, we have that $\hat{\alpha}$ is also normal with

$$E[\hat{\alpha}] = E[\overline{Y}]-\overline{X}\cdot E[\hat{\beta}] = \frac{1}{N}\sum_{i=1}^{N}E[Y_i]-\beta\overline{X}$$

$$= \frac{1}{N}\sum_{i=1}^{N}[\alpha+\beta X_i]-\beta\overline{X} = \alpha.$$

Using the variance formula for sums we have,

$$Var[\hat{\alpha}] = Var[\overline{Y}-\hat{\beta}\overline{X}]$$

$$= Var[\overline{Y}]+\overline{X}^2\cdot Var[\hat{\beta}]-2\overline{X}\cdot Cov[\overline{Y},\hat{\beta}].$$

However,

$$Cov[\overline{Y},\hat{\beta}] = Cov\left[\frac{1}{N}\sum_{i=1}^{N}Y_i,\ \overbrace{\frac{1}{S_{XX}}\sum_{i=1}^{N}(X_i-\overline{X})Y_i}^{\text{formulas (6) and (12)}}\right]$$

$$\overbrace{= \frac{1}{NS_{XX}}\sum_{i=1}^{N}(X_i-\overline{X})\,Var[Y_i]}^{\text{independence of } Y_i\text{'s}} = \overbrace{\frac{\sigma^2}{NS_{XX}}\sum_{i=1}^{N}(X_i-\overline{X})}^{Var[Y_i]=Var[\varepsilon_i]=\sigma^2} = \overbrace{\frac{\sigma^2}{NS_{XX}}\cdot 0}^{\text{formula (9)}} = 0.$$

Therefore,

$$Var[\hat{\alpha}] = Var[\overline{Y}]+\overline{X}^2\cdot Var[\hat{\beta}] = \left[\frac{1}{N}+\frac{\overline{X}^2}{S_{XX}}\right]\sigma^2.$$

(19) $$Cov[\hat{\alpha},\hat{\beta}] = Cov[\overline{Y}-\hat{\beta}\overline{X},\hat{\beta}] = \overbrace{Cov[\overline{Y},\hat{\beta}]}^{\text{equals 0 from the calculation just above}} - \overline{X}\cdot Var[\hat{\beta}] = 0-\overline{X}\frac{\sigma^2}{S_{XX}} = -\frac{\sigma^2\overline{X}}{S_{XX}}.$$

(20) To establish this requires a decomposition of sums of squares. We begin with the random sample of normal errors, $\varepsilon_i = Y_i-\alpha-\beta X_i; i=1,\ldots,N$, and the sample mean $\overline{\varepsilon} = \overline{Y}-\alpha-\beta\overline{X}$. Then,

$$\begin{aligned}
\varepsilon_i = Y_i-\alpha-\beta X_i &= (Y_i-\hat{\alpha}-\hat{\beta}X_i)+(\hat{\alpha}-\alpha)+(\hat{\beta}-\beta)X_i \\
&= (Y_i-\hat{Y}_i)+\left[(\overline{Y}-\hat{\beta}\overline{X})-\alpha\right]+(\hat{\beta}-\beta)X_i \\
&= (Y_i-\hat{Y}_i)+\left[\overline{Y}-\alpha-\beta\overline{X}-(\hat{\beta}-\beta)\overline{X}\right]+(\hat{\beta}-\beta)X_i \\
&= (Y_i-\hat{Y}_i)+\left[\overline{\varepsilon}-(\hat{\beta}-\beta)\overline{X}\right]+(\hat{\beta}-\beta)X_i \\
&= (Y_i-\hat{Y}_i)+\overline{\varepsilon}+(\hat{\beta}-\beta)\cdot(X_i-\overline{X}).
\end{aligned}$$

We next square and sum both sides. Using formula tools as indicated, we find that all of the cross-product terms are zero. Thus,

$$\begin{aligned}
\sum\varepsilon_i^2 &= (Y_i-\hat{Y}_i)^2+N\cdot\overline{\varepsilon}^2+(\hat{\beta}-\beta)^2\cdot\sum(X_i-\overline{X})^2 \\
&\quad +2\left[\overline{\varepsilon}\cdot\overbrace{\sum(Y_i-\hat{Y}_i)}^{\text{using formula (10)}}+\overline{\varepsilon}\cdot(\hat{\beta}-\beta)\cdot\overbrace{\sum(X_i-\overline{X})}^{\text{using formula (9)}}\right. \\
&\quad \left.+(\hat{\beta}-\beta)\cdot\overbrace{\sum(Y_i-\hat{Y}_i)(X_i-\overline{X})}^{\text{using formulas (13) and (10)}}\right] \\
&= \sum(Y_i-\hat{Y}_i)^2+N\cdot\overline{\varepsilon}^2+(\hat{\beta}-\beta)^2\cdot\sum(X_i-\overline{X})^2.
\end{aligned}$$

Thus, $\sum\varepsilon_i^2 = ESS+N\cdot\overline{\varepsilon}^2+(\hat{\beta}-\beta)^2\cdot S_{XX}$. We next divide both sides by σ^2 and rewrite to find,

$$\sum\left(\frac{\varepsilon_i}{\sigma}\right)^2 = \frac{ESS}{\sigma^2}+\left(\frac{\overline{\varepsilon}}{\sigma/N}\right)^2+\left(\frac{\hat{\beta}-\beta}{\sigma/S_{XX}}\right)^2.$$

The left hand side is the sum of N squares of independent standard normal random variables, and hence $\chi^2(N)$ (please refer to Subsection 6.5.5). Similarly, $\overline{\varepsilon}\sim N(0,\sigma^2/N)$, and so the middle term on the right is $\chi^2(1)$. Using formula (17) we see that the third term on the right is also $\chi^2(1)$. Also, $\overline{\varepsilon}$ and $\hat{\beta}$ are independent. This is so because $\overline{\varepsilon}=\overline{Y}-\alpha-\beta\overline{X}$ and we showed above that $Cov[\overline{Y},\hat{\beta}]=0$, which is equivalent to independence for normal random variables.

Using a MGF argument quite similar to that of Subsection 9.2.2 (relating to the independence of the sample mean and the sample variance) it follows that $\frac{ESS}{\sigma^2}$ must be $\chi^2(N-2)$ and independent of both $\overline{Y}$ and $\hat{\beta}$. Since $\hat{\alpha} = \overline{Y} - \hat{\beta}\overline{X}$, it follows that ESS is also independent of $\hat{\alpha}$.

(21) Since by (20) we have $\frac{ESS}{\sigma^2} \sim \chi^2(N-2)$, it follows that $E\left[\frac{ESS}{\sigma^2}\right] = N-2$. Therefore, $E[S^2] = E\left[\frac{ESS}{N-2}\right] = \sigma^2$.

We next show how to construct confidence intervals in the context of the normal linear regression model, using S^2 as a replacement for σ^2. We use $1-\delta$ to denote the confidence level rather than the usual $1-\alpha$ to avoid any possible confusion with the model parameter α.

A General Confidence Interval Calculation

Let pairs $(X_i, Y_i);\ i = 1,\ldots,N$ satisfying the properties of a normal simple linear regression model be given, and let $\hat{\gamma}$ be a random variable satisfying,

(i) $\hat{\gamma} \sim N(\gamma, k\sigma^2)$, where k is a positive constant, and
(ii) $\hat{\gamma}$ and S^2 are independent.

Then, $\frac{\hat{\gamma}-\gamma}{\sqrt{kS^2}} = \frac{\hat{\gamma}-\gamma}{\sqrt{\frac{k \cdot ESS}{N-2}}} \sim t(N-2)$, a student t-distribution with $N-2$ degrees of freedom.
A symmetric two-sided $1-\delta$ confidence interval for γ is given by,

$$\gamma \pm e, \text{ where } e = t_{\delta/2}(N-2)\sqrt{kS^2} = t_{\delta/2}(N-2)\sqrt{\frac{k \cdot ESS}{N-2}}.$$

Note

The expression $\sqrt{kS^2} = \sqrt{\frac{k \cdot ESS}{N-2}}$ is often called the ***standard error*** of the random variable $\hat{\gamma}$.

Proof

By (i), we have that $\frac{\hat{\gamma}-\gamma}{\sqrt{k\sigma^2}} = Z$, a standard normal. Also,

$$\frac{\hat{\gamma}-\gamma}{\sqrt{kS^2}} = \frac{\hat{\gamma}-\gamma}{\sqrt{k\sigma^2}} \cdot \frac{1}{\sqrt{\frac{S^2}{\sigma^2}}} = \frac{\hat{\gamma}-\gamma}{\sqrt{k\sigma^2}} \cdot \frac{1}{\sqrt{\frac{ESS}{(N-2)\sigma^2}}} = \frac{Z}{\sqrt{\frac{\chi^2(N-2)}{N-2}}}.$$

Then (ii) shows that the last expression is by definition a student t-distribution with $N-2$ degrees of freedom. To find the confidence interval, we write,

$$1-\delta = \Pr\left[\left|\hat{\gamma}-\gamma\right| \le e\right] = \Pr\left[\frac{\left|\hat{\gamma}-\gamma\right|}{\sqrt{kS^2}} \le \frac{e}{\sqrt{kS^2}}\right] = \Pr\left[\left|t(N-2)\right| \le \frac{e}{\sqrt{kS^2}}\right].$$

It follows that $\frac{e}{\sqrt{kS^2}}$ is at the $1-\frac{\delta}{2}$ percentile of $t(N-2)$, that is,

$$\frac{e}{\sqrt{kS^2}} = t_{\delta/2}(N-2), \text{ so that } e = t_{\delta/2}(N-2)\sqrt{kS^2} = t_{\delta/2}(N-2)\sqrt{\frac{k \cdot ESS}{N-2}}.$$

Confidence Intervals for α and β

Let pairs $(X_i, Y_i);\ i = 1,\ldots,N$ satisfying the properties of a normal simple linear regression model be given. Let $S^2 = \frac{ESS}{N-2}$.

(22) A two-sided $1-\delta$ confidence interval for α is given by,

$$\hat{\alpha} \pm t_{\delta/2}(N-2)\sqrt{\left[\frac{1}{N}+\frac{\overline{X}^2}{S_{XX}}\right]S^2}.$$

(23) A two-sided $1-\delta$ confidence interval for β is given by,

$$\hat{\beta} \pm t_{\delta/2}(N-2)\sqrt{\frac{1}{S_{XX}}S^2}.$$

Proof

These results follow immediately from the formulas (17) and (16) for the distributions of $\hat{\alpha}$ and $\hat{\beta}$ together with the general confidence interval calculation given above.

Example 12.2-1 Guinea Pigs and the Normal Regression Model

Assume the lengths and weights of guinea pigs satisfy the properties of a normal linear regression model. Use the data and results of Examples 12.1-1, 12.1-2, and 12.1-3 on the weight and length of guinea pigs to calculate 95% confidence intervals for α and β. The data and summary table is repeated below.

#	X	Y	X^2	Y^2	XY	$\hat{Y}$	$(Y-\hat{Y})^2$
1	−2	0	4	0	0	−1.30	1.70
2	0	−2	0	4	0	−0.27	2.98
3	0	1	0	1	0	−0.27	1.62
4	1	−1	1	1	−1	0.24	1.54
5	3	1	9	1	3	1.27	0.07
6	7	4	49	16	28	3.33	0.44
Totals	9	3	63	23	30	---	8.36

Solution

From the solution to Example 12.1-2 we have $\overline{X}=1.5$, $S_{XX}=49.5$, $\hat{\alpha}=-0.2727$, and $\hat{\beta}=0.5152$. We also have from Example 12.1-3 that $ESS=8.36$. Hence $S^2=\frac{8.36}{4}=2.09$.

From tables we have $t_{.025}(4)=2.7764$. Thus, the confidence interval for α is,

$$\hat{\alpha}\pm t_{\delta/2}(N-2)\sqrt{\left[\frac{1}{N}+\frac{\overline{X}^2}{S_{XX}}\right]S^2} = -0.2727\pm 2.7764\sqrt{\left(\frac{1}{6}+\frac{1.5^2}{49.5}\right)(2.09)} = (-2.12,1.58).$$

The confidence interval for β is,

$$\hat{\beta}\pm t_{\delta/2}(N-2)\sqrt{\frac{1}{S_{XX}}S^2} = 0.5152\pm 2.7764\sqrt{\frac{1}{49.5}(2.09)} = (-0.06,1.09).$$ ❐

Note

The confidence interval for β includes negative as well as positive numbers. Consequently, at the 95% confidence level we cannot say for sure whether the regression has positive, negative or zero slope. Since β could be zero, R^2 could be zero as well. This means we cannot conclude that there is in fact a linear relationship between X and Y at the 95% level of confidence. Not surprisingly, the limited number of data points, plus the degree to which they are spread out around the regression line, makes our estimates highly unreliable.

An important application of the model is making predictions for the independent variable Y for a given value X_0. There are two types of prediction we can make. The first would be an estimate for the Y-value on the regression line corresponding to X_0. The corresponding Y value on the regression line is given by $E[Y\,|\,X=X_0]=\alpha+\beta X_0$. Since the coefficients of the regression line must be estimated, the best we can do is to approximate this using $\hat{Y}_0=\hat{\alpha}+\hat{\beta}X_0$. Thus, $\hat{Y}_0$ is an unbiased estimator for the point $\alpha+\beta X_0$ on the regression line corresponding to X_0.

The second type of prediction involves the notion of simulating an actual value for Y corresponding to X_0. This value is given in the model by,

$$Y_0 = \alpha+\beta X_0+\varepsilon_0.$$

Here, we are not really estimating a parameter, but instead, we are predicting the value of the random variable Y_0. In this case $\hat{Y}_0=\hat{\alpha}+\hat{\beta}X_0$ is an unbiased ***predictor*** for Y_0 in the sense that both random variables have the same expected value of $\alpha+\beta X_0$ (since the expected value of the error term is zero). In the following summary box we will refer to a ***prediction interval*** for Y_0 rather than a confidence interval. The prediction interval will always be wider (that is, allow more variability), than the corresponding confidence interval for a given X_0. This is because, in addition to estimating $\alpha+\beta X_0$ with $\hat{Y}_0=\hat{\alpha}+\hat{\beta}X_0$, we are also taking the variability of the error term ε_0 into account. We will illustrate this below with an example from the guinea pig data.

Confidence Intervals for Predictions

Let pairs $(X_i, Y_i);\ i = 1, \ldots, N$ satisfying the properties of a normal simple linear regression model be given. Let $S^2 = \frac{ESS}{N-2}$ and let X_0 be a given *X*-value. Let $\hat{Y}_0 = \hat{\alpha} + \hat{\beta} X_0$.

(24) A two-sided $1-\delta$ confidence interval for the *Y*-value $\alpha + \beta X_0$ on the regression line corresponding to X_0 is given by,

$$\hat{Y}_0 \pm t_{\delta/2}(N-2)\sqrt{\left[\frac{1}{N} + \frac{(X_0 - \bar{X})^2}{S_{XX}}\right] S^2}.$$

(25) A two-sided $1-\delta$ prediction interval for an actual *Y*-value corresponding to X_0 is given by,

$$\hat{Y}_0 \pm t_{\delta/2}(N-2)\sqrt{\left[1 + \frac{1}{N} + \frac{(X_0 - \bar{X})^2}{S_{XX}}\right] S^2}.$$

Proof

Since the estimators $\hat{\alpha}$ and $\hat{\beta}$ are normal and independent of S^2, it follows that $\hat{Y}_0 = \hat{\alpha} + \hat{\beta} X_0$ is also normal and independent of S^2. Also, $E[\hat{Y}_0] = \alpha + \beta X_0$, and, $Var[\hat{Y}_0] = Var[\hat{\alpha} + \hat{\beta} X_0] = Var[\hat{\alpha}] + X_0^2 \cdot Var[\hat{\beta}] + 2X_0 \cdot Cov[\hat{\alpha}, \hat{\beta}]$.

Using formulas (17), (18) and (19) we find that,

$$\begin{aligned} Var[\hat{Y}_0] &= \left\{\left[\frac{1}{N} + \frac{\bar{X}^2}{S_{XX}}\right] + \frac{X_0^2}{S_{XX}} - 2X_0 \frac{\bar{X}}{S_{XX}}\right\}\sigma^2 \\ &= \left[\frac{1}{N} + \frac{1}{S_{XX}}(X_0^2 - 2X_0 \cdot \bar{X} + \bar{X}^2)\right]\sigma^2 \\ &= \left[\frac{1}{N} + \frac{(X_0 - \bar{X})^2}{S_{XX}}\right]\sigma^2. \end{aligned}$$

The conclusion for (24) now follows from the general confidence interval calculation given above.

In the case of (25) we would simulate a value of $Y_0 = \alpha + \beta X_0 + \varepsilon_0$ using $\hat{Y}_0 + \varepsilon_0 = \hat{\alpha} + \hat{\beta} X_0 + \varepsilon_0$. Since we don't have an actual observation to use for the error term, we use its expected value (which is zero). However, in calculating the variance of the predictor we have,

$$Var[\hat{Y}_0 + \varepsilon_0] = Var[\hat{Y}_0] + Var[\varepsilon_0] = \left[\frac{1}{N} + \frac{(X_0 - \bar{X})^2}{S_{XX}}\right]\sigma^2 + \sigma^2$$

$$= \left[1 + \frac{1}{N} + \frac{(X_0 - \bar{X})^2}{S_{XX}}\right]\sigma^2.$$

Note

The magnitude of the variance in both cases depends on both N and the quantity $(X_0 - \bar{X})^2$. Not surprisingly, the larger N is, the smaller the variance. Also, the closer X_0 is to the mean of the existing data points, the more accurate the predictors. This is confirmation of the wisdom behind oft-repeated warnings against extrapolating from data.

Example 12.2-2 Predicting the Length of Guinea Pigs

Continuing with the guinea pig data, assume the normal linear regression model is used to predict the length of a guinea pig weighing 5.

(a) Find a 95% confidence interval for the Y-value on the regression line corresponding to an X-value of 5.

(b) You plan to use the regression model to simulate the length of a guinea pig weighing 5. Find a 95% prediction interval for the resulting predicted length.

Solution

From Examples 12.1-3 and 12.2-1, we have a predicted length of

$$\hat{\alpha} + \hat{\beta}(5) = -0.2727 + 0.5152(5) = 2.30,$$

with $\bar{X} = 1.5$, $S_{XX} = 49.5$, and $S^2 = 2.09$.

(a) The resulting confidence interval is,

$$\hat{Y}_0 \pm t_{\delta/2}(N-2)\sqrt{\left[\frac{1}{N} + \frac{(X_0 - \bar{X})^2}{S_{XX}}\right]S^2} = 2.30 \pm 2.7764\sqrt{\left[\frac{1}{6} + \frac{(5-1.5)^2}{49.5}\right]2.09}$$

$$= (-.28, 4.88).$$

(b) The resulting prediction interval is,

$$\hat{Y}_0 \pm t_{\delta/2}(N-2)\sqrt{\left[1+\frac{1}{N}+\frac{(X_0-\bar{X})^2}{S_{XX}}\right]S^2} = 2.30 \pm 2.7764\sqrt{\left[1+\frac{1}{6}+\frac{(5-1.5)^2}{49.5}\right]2.09}$$

$$= (-2.47, 7.07).$$

❐

Note

Practically speaking, the lengths of these confidence intervals suggest that any attempt at predicting the length of a guinea pig from its weight based on the available data would be essentially meaningless. How much of this is due to the small sample size and how much is due to the weak relationship between the two variables is difficult to say. In the real world the researchers would write a proposal for a bigger grant so they could expand the study and possibly take into account additional factors, such as the diet and exercise habits of the subjects.

The most common hypothesis test for the normal regression model is to determine at a given significance level whether or not there is any linear relationship between Y and X. This can be measured through the slope coefficient β of the regression line, where the hypothesis $\beta = 0$ signifies that no linear relationship exists. Since $\beta = 0$ if and only if $R^2 = 0$, this can be considered a hypothesis test for $R^2 = 0$ as well.

Hypothesis Test for $\beta = 0$

Let pairs $(X_i, Y_i);\ i = 1, \ldots, N$ satisfying the properties of a normal simple linear regression model be given. Consider the hypothesis test,

$$H_0 : \beta = 0$$
$$H_1 : \beta \neq 0$$

with a significance level of δ. The test statistic is $\dfrac{\hat{\beta}}{\sqrt{\dfrac{S^2}{S_{XX}}}} = \dfrac{\hat{\beta}}{\sqrt{\dfrac{ESS}{(N-2)S_{XX}}}}$, and the null hypothesis is rejected if,

$$\frac{|\hat{\beta}|}{\sqrt{\dfrac{S^2}{S_{XX}}}} = \frac{|\hat{\beta}|}{\sqrt{\dfrac{ESS}{(N-2)S_{XX}}}} \geq t_{\delta/2}(N-2).$$

Note

Since, from formula (17), $Var[\hat{\beta}] = \frac{\sigma^2}{S_{XX}}$, and under the null hypothesis the mean of $\hat{\beta}$ is zero, it follows from the General Confidence Interval Calculation that

$$\frac{\hat{\beta}}{\sqrt{\frac{S^2}{S_{XX}}}} = \frac{\hat{\beta}}{\sqrt{\frac{ESS}{(N-2)S_{XX}}}} \sim t(N-2),$$

a student-t random variable with $N-2$ degrees of freedom.

An alternative way of looking at this uses the F-distribution and ANOVA decomposition of Section 12.1-2, reproduced below.

Analysis of Variance in Simple Linear Regression

Define the ***Total Sum of Squares*** (TSS) in Y by $TSS = S_{YY} = \sum_{i=1}^{N}(Y_i - \overline{Y})^2$, and define the ***Regression Sum of Squares*** (RSS) by $RSS = \sum(\hat{Y}_i - \overline{Y})^2$. Then,

(15) $TSS = ESS + RSS.$

(16) $R^2 = \frac{RSS}{TSS} = 1 - \frac{ESS}{TSS}.$

For the null hypothesis $\beta = 0$, the test statistic used above is,

(26)
$$\frac{\hat{\beta}}{\sqrt{\frac{ESS}{(N-2)S_{XX}}}} \sim t(N-2),$$

Recall from Section 9.4 that the F-distribution is defined as the ratio of two chi-squares, each divided by its degrees of freedom. We also showed in the properties of the F-distribution that the square of a student-t with n degrees of freedom is an F-distribution with degrees of freedom 1 and n.

Thus, the square of the test statistic is,

(26a)
$$\frac{\hat{\beta}^2}{\frac{ESS}{(N-2)S_{XX}}} = \frac{\frac{\hat{\beta}^2 S_{XX}}{\sigma^2}}{\frac{ESS}{\sigma^2(N-2)}} \sim F(1, N-2).$$

Formula (20) shows that $\frac{ESS}{\sigma^2} \sim \chi^2(N-2)$, and so the bottom, $\frac{ESS}{\sigma^2(N-2)}$, is of the form $\frac{\chi^2(N-2)}{N-2}$. For the top we have, from Formula (16), that

$$RSS = R^2 \cdot TSS = \frac{S_{XY}^2}{S_{XX} \cdot S_{YY}} \cdot S_{YY} = \frac{S_{XY}^2}{S_{XX}} = \frac{S_{XY}^2}{S_{XX}^2} \cdot S_{XX} = \hat{\beta}^2 \cdot S_{XX}.$$

Hence, the top of (26a) takes the form $\frac{RSS}{\sigma^2} \sim \chi^2(1)$. Therefore, in testing the hypothesis $\beta = 0$, we can use the square of the t-statistic in (26a), which takes the form,

$$(27) \qquad \frac{RSS/1}{ESS/(N-2)} \sim F(1, N-2).$$

This leads to the equivalent formulation of the hypothesis test for $\beta = 0$ using the F-distribution.

Analysis of Variance Table for Testing $\beta = 0$

Let pairs (X_i, Y_i); $i = 1, \ldots, N$ satisfying the properties of a normal simple linear regression model be given. Consider the hypothesis test,

$$H_0 : \beta = 0$$
$$H_1 : \beta \neq 0$$

with a significance level of δ. The test statistic is $\frac{RSS/1}{ESS/(N-2)}$, and the null hypothesis is rejected if,

$$\frac{RSS/1}{ESS/(N-2)} \geq F_\delta(1, N-2).$$

Notes

(1) The smaller ESS is, the closer the data points conform to the regression line. The larger RSS is, the more the variation there is in the estimates $\hat{Y}_i$ about the horizontal line $y = \overline{Y}$, suggesting that the regression line is not horizontal. So a large value of the test statistic supports the premise of a non-zero value for β.

(2) When sums of squares like RSS or ESS are divided by their degrees the freedom, the resulting quantities are called ***mean squares***.

(3) In terms of the sum of squares decomposition of Formula (15) of the ***analysis of variance*** box above, we now have,

$$\frac{TSS}{\sigma^2} = \frac{ESS}{\sigma^2} + \frac{RSS}{\sigma^2} = \chi^2(N-2) + \chi^2(1) = \chi^2(N-1).$$

The calculations for the F-statistic for testing $\beta = 0$ are often organized in a so-called ANOVA table that looks like this:

Source	Notation	D of F	Mean Square	*F*-Statistic
Regression	RSS	1	$\frac{RSS}{1}$	$\frac{RSS/1}{ESS/(N-2)}$
Error	ESS	$N-2$	$\frac{ESS}{N-2}$	
Total	TSS	$N-1$		

Example 12.2-3 Guinea Pigs and the Hypothesis $\beta = 0$

Assume the lengths and weights of guinea pigs satisfy the properties of a normal linear regression model. Use the data and results of Examples 12.1-1, 12.1-2, and 12.1-3 to conduct the following hypothesis test at the 5% level of significance:

$$H_0 : \beta = 0$$
$$H_1 : \beta \neq 0$$

(a) Use a student-*t* test.
(b) Use an *F* test.

Solution
(a) We have, using results from previous calculations that,

$$\frac{\hat{\beta}}{\sqrt{\frac{ESS}{(N-2)S_{XX}}}} = \frac{.5152}{\sqrt{\frac{8.36}{(6-2)49.5}}} = 2.51 < 2.7764 = t_{.025}(4).$$

Hence, we cannot reject the null hypothesis.

(b) From previous results we have $TSS = S_{YY} = 21.5$ and $ESS = 8.36$. Then $RSS = TSS - ESS$ $= 21.5 - 8.36 = 13.14$. Using the template, we have,

Source	Notation	D of F	Mean Square	*F*-Statistic
Regression	$RSS = 13.14$	1	$\frac{RSS}{1} = 13.14$	$\frac{RSS/1}{ESS/4} = 6.28$
Error	$ESS = 8.36$	4	$\frac{ESS}{4} = 2.09$	
Total	$TSS = 21.50$	5		

From tables we find $\frac{RSS/1}{ESS/4} = 6.28 < 7.7086 = F_{.05}(1,4)$, and so we cannot reject the null hypothesis. ❒

Note

The *F*-statistic value of 6.28 in (b) is the square of the *t*-statistic from (a) of 2.51, and also $F_{.05}(1,4) = 7.7086 = 2.7764^2 = t_{.025}(4)^2$.

Most hypothesis tests of this type have a null hypothesis of $\beta = 0$, although other formulations are possible. The basic ideas are exactly the same, as the next example illustrates.

Example 12.2-4 A Non-zero Null Hypothesis (CAS ST Spring 2014 #20)

The model $y_i = \beta_0 + \beta_1 x_i + \varepsilon_i$ was fit using 6 observations. The estimated parameters are as follows:

- $\hat{\beta}_0 = 2.31$
- $\hat{\beta}_1 = 1.15$
- $\hat{\sigma}_{\beta_0} = 0.057$
- $\hat{\sigma}_{\beta_1} = 0.043$

The following hypothesis test is performed:

$$H_0: \beta_1 = 1$$
$$H_1: \beta_1 \neq 1$$

Determine the minimum significance level at which the null hypothesis would be rejected.

A. Less than 0.01
B. At least 0.01, but less than 0.02
C. At least 0.02, but less than 0.05
D. At least 0.05, but less than 0.10
E. At least 0.10

Solution

Under the null hypothesis $\hat{\beta}_1 \sim N\left(1, \sigma^2_{\hat{\beta}_1}\right)$, which implies that $\dfrac{\hat{\beta}_1 - 1}{\hat{\sigma}_{\beta_1}} \sim t(6-2)$. Thus, the test statistic is $\dfrac{1.15-1}{0.043} = 3.488$. From the t-table we have,

α	$t_{\alpha/2}(4)$
.02	3.747
.05	2.776

which implies answer (D).

Exercise 12-14 In a normal simple linear regression model you are given that the variance of the error term is 4. Four observations are taken, in which the X values are $X = -2, 1, 2, 3$. Calculate the variance in the estimate for the slope coefficient.

Exercise 12-15 In a normal simple linear regression model the t-statistic for testing the hypothesis that $\beta = 0$ is 2. Also, $\Sigma(X_i - \bar{X})(Y_i - \bar{Y}) = 4S$, where S^2 is the unbiased estimator for the variance in the error term. Calculate $\Sigma(X_i - \bar{X})^2$.

Exercise 12-16 A normal simple linear regression model has 20 observations. You are given,

$$\sum_{i=1}^{20} X_i = 100,$$

$$\sum_{i=1}^{20} Y_i = 120,$$

$$\sum_{i=1}^{20} X_i^2 = 2100,$$

$$\sum_{i=1}^{20} Y_i^2 = 7120,$$

$$\sum_{i=1}^{20} X_i Y_i = 2520.$$

Calculate the F-statistic for testing the hypothesis that $R^2 = 0$.

Exercise 12-17 Using the same data as in Exercise 12-16,

(a) Calculate the symmetric 90% confidence interval for the expected value of $\hat{Y}$ corresponding to $X = 10$.

(b) Calculate the symmetric 90% prediction interval for the value of $\hat{Y}$ corresponding to $X = 10$.

Exercise 12-18 Twenty-five applicants to the actuarial science program took both the SAT-M and the mathematics placement exam. The estimated correlation between the two scores came out to be 0.6. Calculate the F-statistic for testing the hypothesis that $R^2 = 0$. Determine the critical values at the 5% and the 1% level and state whether the null hypothesis can be rejected in each instance.

Exercise 12-19 The normal least squares model is used to fit $Y_i = \alpha + \beta X_i + \varepsilon_i$ to 25 data points. You determine $\hat{Y} = -3 + 2.9X$. The total sum of the squares is $TSS = 43$ and the regression sum of the squares is $RSS = 37$. Determine the value of the t statistic for testing the null hypothesis $H_0 : \beta = 0$ at the 5% level of significance.

12.3 COMPARING MEANS USING ANOVA

The analysis of variance framework developed in the previous section for simple linear regression has a broader range of statistical applications. In this section we illustrate the use of ANOVA for comparing means from different populations. Suppose we have a population that can be divided into a number of subpopulations and we wish to determine whether or not the subpopulations all have the same mean.

For example, we might administer a standardized test such as the SAT or ACT nationally. An academic researcher might ask if test scores are consistent across different time zones in the continental U.S. The subpopulations would consist of test scores from the Eastern, Central, Mountain and Pacific time zones. The null hypothesis is that mean scores are the same across time zones. In general this type of statistical test is referred to as ***one-factor analysis of variance***, the one factor in this example being the time zone.

The procedure, analogous to the regression ANOVA model, is to decompose the total square deviations from the mean for the overall sample into two components. The first component captures the square deviations from the mean ***within*** individual time zones. The second component is a measure of square deviations ***between*** individual time zone means and the overall mean. As in the regression model, under certain conditions of normality and independence, an appropriately scaled ratio of these two components has an F-distribution. If the (scaled) ratio of the square deviations ***between*** means over square deviations ***within*** means is large then this is evidence in favor of rejecting the null hypothesis that all the means are the same.

There is a standard terminology for ANOVA that might refer to the individual time zones as different ***treatment levels***. This makes sense if, for example, we are testing responses to a new drug, and our subpopulations consist of responses at different dosage, or treatment, levels. If there are just two treatment levels then the ANOVA procedure described here is equivalent to the t-test discussed in Section 10.5.2 for comparing two means.

Assume now we have k normal subpopulations, $X_1,\ldots,X_k$ all with the same (possibly unknown) variance σ^2 and respective means, $\mu_1,\ldots,\mu_k$. We then draw independent random samples, of sizes $n_1,\ldots,n_k$, respectively, from each of the k subpopulations. Let $X_{i1},\ldots,X_{in_i}$ denote the random sample from X_i, with sample mean $\overline{X}_i;\, i=1,\ldots,k$. Let $n = n_1 + \cdots + n_k$. Since the individual random samples are independent we can pool them together into an overall random sample of size n, whose sample mean we denote by simply $\overline{X}$.

The Square Deviation Definitions

Total Sum of Squares $= TSS = \sum_{i=1}^{k}\sum_{j=1}^{n_i}\left(X_{ij}-\overline{X}\right)^2$,

Error Sum of Squares $= ESS = \sum_{i=1}^{k}\sum_{j=1}^{n_i}\left(X_{ij}-\overline{X}_i\right)^2$,

Means Sum of Squares $= MSS = \sum_{i=1}^{k} n_i\left(\overline{X}_i-\overline{X}\right)^2$.

The quantity *ESS* measures the square deviations within treatments, and the quantity *MSS* measures the square deviations between treatments. We next summarize the relationships that lead to the analysis of variance hypothesis test for the one-factor case.

Single Factor ANOVA Essentials

(1) $TSS = ESS + MSS$,

(2) $\frac{TSS}{\sigma^2} \sim \chi^2(n-1)$,

(3) $\frac{ESS}{\sigma^2} \sim \chi^2(n-k)$,

(4) *ESS* and *MSS* are independent,

(5) $\frac{MSS}{\sigma^2} \sim \chi^2(k-1)$,

(6) $\frac{MSS/(k-1)}{ESS/(n-k)} \sim F(k-1,n-k)$

The null hypothesis $H_0 : \mu_1 = \mu_2 = \cdots = \mu_k$ is rejected at the critical level of α if $\frac{MSS/(k-1)}{ESS/(n-k)} = F(k-1,n-k) > F_\alpha(k-1,n-k)$.

Proofs:

(1) The demonstration here is quite similar to the simple linear regression case and uses the same methods.

$$TSS=\sum_{i=1}^{k}\sum_{j=1}^{n_i}\left(X_{ij}-\bar{X}\right)^2=\sum_{i=1}^{k}\left\{\sum_{j=1}^{n_i}\left[\left(X_{ij}-\bar{X}_i\right)+\left(\bar{X}_i-\bar{X}\right)\right]^2\right\}$$

$$=\sum_{i=1}^{k}\left\{\sum_{j=1}^{n_i}\left[\left(X_{ij}-\bar{X}_i\right)^2+2\left(X_{ij}-\bar{X}_i\right)\left(\bar{X}_i-\bar{X}\right)+\left(\bar{X}_i-\bar{X}\right)^2\right]\right\}$$

$$=\sum_{i=1}^{k}\left\{\sum_{j=1}^{n_i}\left(X_{ij}-\bar{X}_i\right)^2+2\left(\bar{X}_i-\bar{X}\right)\sum_{j=1}^{n_i}\left(X_{ij}-\bar{X}_i\right)+n_i\left(\bar{X}_i-\bar{X}\right)^2\right\}.$$

Note that the sum $\sum_{j=1}^{n_i}\left(X_{ij}-\bar{X}_i\right)$ in the middle is equal to zero (see formula (10) in Section 12.2.2). Thus,

$$TSS=\sum_{i=1}^{k}\sum_{j=1}^{n_i}\left(X_{ij}-\bar{X}_i\right)^2+\sum_{i=1}^{k}n_i\left(\bar{X}_i-\bar{X}\right)^2=ESS+MSS.$$

(2) The collection of all the random samples can be used to construct the usual unbiased sample variance estimator for σ^2, $S^2=\frac{1}{n-1}\sum_{i=1}^{k}\sum_{j=1}^{n_i}\left(X_{ij}-\bar{X}\right)^2=\frac{TSS}{n-1}$. Thus, following Subsection 9.2.2 (properties of S^2) we have that $\frac{(n-1)S^2}{\sigma^2}=\frac{TSS}{\sigma^2}\sim\chi^2(n-1)$.

(3) For each i, $S_i^2=\frac{1}{n_i-1}\sum_{j=1}^{n_i}\left(X_{ij}-\bar{X}_i\right)^2$ is the unbiased sample variance from the i-th random sample. Thus, $\frac{(n_i-1)S_i^2}{\sigma^2}=\frac{\sum_{j=1}^{n_i}\left(X_{ij}-\bar{X}_i\right)^2}{\sigma^2}\sim\chi^2(n_i-1)$. It follows that $\frac{ESS}{\sigma^2}$ is chi-squared with degrees of freedom equal to $\sum_{i=1}^{k}(n_i-1)=n-k$.

(4) The expression *ESS* depends only on the sample variances S_i^2, and the expression *MSS* depends only on the sample means $\bar{X}_i$ (the overall mean $\bar{X}=\frac{1}{n}\sum_{i=1}^{k}n_i\bar{X}_i$ is a linear

combination of the $\overline{X}_i$'s). Since sample means and sample variances are independent, we see that *ESS* and *MSS* are independent.

(5) Let $U = \frac{ESS}{\sigma^2}$, $V = \frac{MSS}{\sigma^2}$, $W = \frac{TSS}{\sigma^2}$.

Then, $W = U + V$, with U and V independent. Using moment-generating functions for the two chi-squared random variables W and U, we have,

$$M_W(t) = M_U(t) \cdot M_V(t).$$

It follows that,

$$M_V(t) = \frac{M_W(t)}{M_U(t)} = \frac{\left(\frac{1}{1-2t}\right)^{\frac{n-1}{2}}}{\left(\frac{1}{1-2t}\right)^{\frac{n-k}{2}}} = \left(\frac{1}{1-2t}\right)^{\frac{k-1}{2}}$$

This shows that,

$$V = \frac{MSS}{\sigma^2} \sim \chi^2(k-1).$$

(6) That $\frac{MSS/(k-1)}{ESS/(n-k)} \sim F(k-1, n-k)$ follows from (3), (4) and (5), and the definition of F.

As with simple linear regression, there is an ANOVA table that conveniently organizes the calculations for the F-test for $H_0: \mu_1 = \mu_2 = \cdots = \mu_k$.

Source	Notation	D of F	Mean Square	*F*-Statistic
Between Means	*MSS*	$k-1$	$\frac{MSS}{k-1}$	$\frac{MSS/(k-1)}{ESS/(n-k)}$
Within Means	*ESS*	$n-k$	$\frac{ESS}{n-k}$	
Total	*TSS*	$n-1$		

Example 12.3-1 ANOVA (Exam ST Spring 2014 #19)

You are given the following sales information from three car dealers:

Dealer	Number of Cars Sold	Average Sales Price ('000s)	Error Sum of Squares
A	6	23.8	6.83
B	4	25.3	2.75
C	5	27.4	37.20

You wish to test whether the average sales prices are the same among the dealers:

$H_0 : \mu_A = \mu_B = \mu_C$ (The mean sales price between the three dealers is the same)
H_1 : The mean sales price between the three dealers is not the same.

Create an ANOVA table and assume the total sum of squares is 81.6. Determine whether to reject the null hypothesis at either the 0.01 or the 0.05 levels.

Solution:
We are assuming (although the problem doesn't say this explicitly) that the underlying distributions are normal and independent. We have $n = 15$ and $k = 3$.

From the last column we see that $ESS = 6.83 + 2.75 + 37.20 = 46.78$. Also, $MSS = TSS - ESS = 81.6 - 46.78 = 34.82$. We can now complete the ANOVA table.

Sum of Squares	D of F	Mean Square	F - Statistic
$MSS = 34.82$	2	$\frac{34.82}{2} = 17.41$	$\frac{17.41}{3.90} = 4.47$
$ESS = 46.78$	12	$\frac{46.78}{12} = 3.90$	
$TSS = 81.60$	14		

From the F-tables we have $F_{.01}(2,12) = 6.93$ and $F_{.05}(2,12) = 3.88$. Therefore we can reject the null hypothesis at the 0.05 level, but not the 0.01 level.

Note that we did not use the average sales price by dealer in this solution. However, we could have calculated *MSS* directly using that information:

The overall $\overline{X} = \left(\frac{1}{15}\right)\left[(6)(23.8) + (4)(25.3) + (5)(27.4)\right] = 25.4$. Then,

$$MSS = (6)(23.8 - 25.4)^2 + (4)(25.3 - 25.4)^2 + (5)(27.4 - 25.4)^2 = 34.82.$$

Let's look more closely at the situation when there are just two different treatment levels. As mentioned above, the ANOVA procedure described here is equivalent to the t-test for differences between two means. The underlying idea is that in general (Section 9.4), $[t(n)]^2 = F(1, n)$. In the current context and notation, the t-test in Section 10.5.2 is as follows:

There are two normal populations X_1 and X_2 with independent random samples of sizes n_1 and n_2, with sample means $\overline{X}_1$ and $\overline{X}_2$. The sample variances are given by $S_i^2 = \frac{1}{n_i - 1}\sum_{j=1}^{n_i}\left(X_{ij} - \overline{X}_i\right)^2$; $i = 1,2$. The pooled variance in the t-test is given by $S_P^2 = \frac{(n_1 - 1)S_1^2 + (n_2 - 1)S_2^2}{n_1 + n_2 - 2} = \frac{ESS}{n_1 + n_2 - 2}$. Thus,

$$ESS = (n_1 + n_2 - 2)S_P^2.$$

If we assume the null hypothesis is true ($\mu_1 - \mu_2$) then the t-statistic is,

$$t(n_1 + n_2 - 2) = \frac{\bar{X}_1 - \bar{X}_2}{\sqrt{S_P^2\left(\frac{1}{n_1} + \frac{1}{n_2}\right)}}.$$

The overall mean is $\bar{X} = \frac{n_1\bar{X}_1 + n_2\bar{X}_2}{n_1 + n_2}$, and, $MSS = n_1\left(\bar{X}_1 - \bar{X}\right)^2 + n_2\left(\bar{X}_2 - \bar{X}\right)^2$. Now,

$\bar{X}_1 - \bar{X} = \bar{X}_1 - \frac{n_1\bar{X}_1 + n_2\bar{X}_2}{n_1 + n_2} = \bar{X}_1 - \frac{n_1}{n_1 + n_2}\bar{X}_1 - \frac{n_2}{n_1 + n_2}\bar{X}_2 = \frac{n_2\left(\bar{X}_1 - \bar{X}_2\right)}{n_1 + n_2}$. Similarly,

$\bar{X}_2 - \bar{X} = \frac{n_1\left(\bar{X}_2 - \bar{X}_1\right)}{n_1 + n_2}$. Substituting these expressions into MSS, we find,

$$MSS = n_1\left(\bar{X}_1 - \bar{X}\right)^2 + n_2\left(\bar{X}_2 - \bar{X}\right)^2 = n_1\left(\frac{n_2}{n_1 + n_2}\right)^2\left(\bar{X}_1 - \bar{X}_2\right)^2 + n_2\left(\frac{n_1}{n_1 + n_2}\right)^2\left(\bar{X}_2 - \bar{X}_1\right)^2$$

$$= \frac{n_1 n_2}{n_1 + n_2}\left(\bar{X}_1 - \bar{X}_2\right)^2 = \frac{\left(\bar{X}_1 - \bar{X}_2\right)^2}{\frac{1}{n_1} + \frac{1}{n_2}}.$$

Therefore,

$$F(1, n_1 + n_2 - 2) = \frac{MSS}{\frac{ESS}{n_1 + n_2 - 2}} = \frac{\left(\bar{X}_1 - \bar{X}_2\right)^2}{S_P^2\left(\frac{1}{n_1} + \frac{1}{n_2}\right)} = \left[t\left(n_1 + n_2 - 2\right)\right]^2.$$

Example 12.3-2 (Example 10.5-3 - Knight School Revisited)

An administrator at the Central School District was interested in whether or not there was a measurable difference in mean SAT math scores between the AP calculus students at Ivanhoe High School versus those at Lancelot High School. When the records were searched, it turned out that a random sample of students in Ivanhoe's AP calculus class had a mean SAT score of 622 and a random sample of students in the Lancelot AP calculus class had a mean SAT score of 688. The administrator concluded that the difference of 66 points was indeed significant and real.

A statistical consultant reviewed the data and ascertained that the sample sizes were small, consisting of 12 students at Ivanhoe and 16 students at Lancelot. She also calculated the sample standard deviations, finding a value of 120 for Ivanhoe and 96 for Lancelot. She decided to test the administrator's ad hoc conclusion at the 5% level of significance under the assumption that the two relevant populations are normal and have the same standard deviation.

(a) Use the ANOVA single factor method to test the null hypothesis that the two means are the same. Compare your results to the outcome using the t-test in Example 10.5-3.

(b) Calculate the p-value of the test, using statistical software for the relevant F-distribution.

Solution:

Let X_1 be the Ivanhoe population and X_2 the Lancelot population. The overall sample mean is given by,

$$\bar{X} = \frac{n_1\bar{X}_1 + n_2\bar{X}_2}{n_1 + n_2} = \frac{(12)(622)+(16)(688)}{12+16} = 659.71.$$

Next,

$$MSS = n_1\left(\bar{X}_1 - \bar{X}\right)^2 + n_2\left(\bar{X}_2 - \bar{X}\right)^2 = (12)(622-659.71)^2 + (16)(688-659.71)^2 = 29{,}869.71,$$

and,

$$ESS = (n_1 + n_2 - 2)S_P^2 = (n_1 - 1)S_1^2 + (n_2 - 1)S_2^2 = (11)(120)^2 + (15)(96)^2 = 296{,}640.$$

Sum of Squares	D of F	Mean Square	F - Statistic
$MSS = 29{,}869.71$	1	29,869.71	$\frac{29{,}869.71}{11{,}409.23} = 2.618$
$ESS = 296{,}640$	26	11,409.23	

In Example 10.5-3 we showed that the t-statistic was $t(26) = -1.618$, with a p-value of 11.77%, indicating that we cannot reject at the 5% level the null hypothesis that the two populations have the same means.

Using the ANOVA table we see that
$F(1,26) = 2.618 = (-1.618)^2 = [t(26)]^2 < 4.23 = F_{.05}(1,26)$. The p-value equals
$\Pr[F(1,26) > 2.618] = 11.77\%$ (same as in Example 10.5-3), using the Excel function F.DIST.RT.

Exercise 12-20 Exam CAS ST Spring 2014 #21

For $i = 1,2,3,4$ let X_i denote the speed of cars passing location *i*.

- Assume that each X_i is normal with mean μ_i and common variance σ^2
- $H_0 : \mu_1 = \mu_2 = \mu_3 = \mu_4$ (the means are all equal)
- H_1 : The means are not all equal
- x_{ij} is observation *j* at location *i*
- You have the following observations:

Location	x_{i1}	x_{i2}	x_{i3}	x_{i4}	x_{i5}	$\sum_{j=1}^{5} x_{ij}$	$\sum_{j=1}^{5} x_{ij}^2$
1	52.4	55.7	60.2	69.9	59.4	297.6	17,886.66
2	62.5	65.1	56.5	52.8	69.0	305.9	18,885.35
3	61.0	63.5	84.0	68.1	77.6	354.2	25,468.62
4	63.0	52.1	65.2	59.5	52.8	292.6	17,262.54
					Totals	1250.3	79,503.17

Calculate the *F*-statistic, including degreees of freedom, used to test H_0.

Exercise 12-21 (see Exercise 10-16)

Lori claims that the female business students are much more intelligent than male business students. A random sample was taken and the results are summarized below.

Female	Male
$\overline{X}_1 = 104$	$\overline{X}_2 = 97$
$s_1 = 12$	$s_2 = 17$
$n_1 = 22$	$n_2 = 26$

Calculate the ANOVA F – statistic and determine if the evidence is sufficient to reject the null hypothesis that the two population means are the same at the 10% level. Use statistical software to find the *p*-value of the test. State what assumptions you are making when conducting this test.

12.4 NON-PARAMETRIC TESTS

Most of the statistical tests we have discussed previously have certain assumptions about the underlying populations. For our strongest results in hypothesis testing we often required normal distributions for the populations under investigation.

There are a variety of alternative approaches that do not require the populations to conform to any particular distribution family. These are called non-parametric tests. In this section we bring together some of the more common non-parametric tests. Most of them require only that we are able to rank the data points according to size, say from smallest to largest. These tests can then be applied to ordinal data (meaning that the data points need not be numerical at all, but only possess an ordering relationship). Indeed, in our first examples below we require that the data only be divisible into two categories, such as "positive" and "negative".

12.4.1 SIGN TESTS

Sign tests are based on tracking how often random quantities are either positive or negative. The test statistic counts the number of positives (or negatives) and will have a binomial distribution. We will illustrate the use of a sign test in two different contexts. The first application is for testing a hypothesis concerning the median of a population.

Assume we have a random sample $X_1,\ldots,X_n$ from a continuous population random variable X, and we wish to test the null hypothesis that the median of X is m. The test consists of taking the differences between the observations and the supposed median m, and counting the number of positive differences. We assume the population to be continuous so that the probability of an observation being exactly equal to m is zero.

Let S be the number of positive differences. Then S is binomial with n trials. Under the null hypothesis positive and negative differences from m will be equally likely, so that the probability p of success (a positive difference) is ½. Too large or too small values of S constitute evidence against the null hypothesis. The rejection region, or the p-value of the test, will depend on how the alternative hypothesis is formulated.

Example 12.4-1 Exam ST Sample Question (modified) – Sign Test for the Median

For a health insurance policy, losses follow an unknown continuous distribution with unique median m. You are given the following random sample of ten losses:

7	6	8	12	2	3	13	1	14	15

Based on this sample and the **sign test** you are asked to test the null hypothesis $H_0: m = 5.5$.

a) Calculate the p-value of the test if the alternative hypothesis is $H_1: m > 5.5$.

b) Calculate the p-value of the test if the alternative hypothesis is $H_1: m \neq 5.5$.

Solution:
We form differences between observations and 5.5 and calculate S as the number of positive differences:

x_i	7	6	8	12	2	3	13	1	14	15
$x_i - 5.5$	1.5	0.5	2.5	6.5	−3.5	−2.5	7.5	−4.5	8.5	9.5
$\text{sign}[x_i - 5.5]$	+	+	+	+	−	−	+	−	+	+

We've tabulated the actual differences for future use. For our purposes here all we need is the number S of positives, which is 7 out of 10. For part (a) a large value of S is evidence that the median of X is greater than 5.5. The p-value is,

$$\Pr[S \ge 7] = \sum_{i=7}^{10} \binom{10}{i} \left(\frac{1}{2}\right)^{10} = 0.1719.$$

b) The p-value for a two-tailed test is $\Pr[S \ge 7] + \Pr[S \le 10 - 7 = 3] = (2)(0.1719) = 0.3438.$

Our second application of a sign test is for testing hypotheses concerning matched pairs of observations. Say we have a population X and we want to test the efficacy of a treatment on individuals in the population. For example, we might want to know if a certain program to boost SAT Math scores is effective. We could test individuals before and after "treatment," thereby generating a random sample of pairs (X_i, Y_i). Then X_i is the score before taking the test prep, and Y_i is the score after the test prep.

We performed just such a test in Example 10.5-4, under the assumption that we were sampling from normal populations. We will revisit that example several times in this section, using various nonparametric tests. For the sign test, the test statistic is the number of positive (or negative) differences. If the null hypothesis is that there is no change, then the differences $X_i - Y_i$ are equally likely to be positive or negative. We are assuming here that no other factors come into play except for the prep course.

Example 12.4-2 Matched Pairs Sign Test (see Example 10.5-4)

Ten high school students were randomly chosen to participate in a study of the benefits of participating in a test preparation program. The first row in the following table represents their scores before the prep course, and the second row the scores after the prep course. Use the sign test to determine the p-value of the results in testing the null hypothesis of no change in score versus the alternative hypothesis that the scores tend to increase after the test prep course.

421	551	500	595	671	435	384	694	235	599
463	611	476	617	729	454	441	726	225	709

Solution:
Let X be the prior score and let Y be the score after the prep course. Then,

x_i	421	551	500	595	671	435	384	694	235	599
y_i	463	611	476	617	729	454	441	726	225	709
$\text{sign}[x_i - y_i]$	−	−	+	−	−	−	−	−	+	−

Let S be the number of positive differences in $x_i - y_i$. A small value of S is evidence that $Y > X$, meaning that the prep course is effective in shifting the score distribution to the right. Since we see that $S = 2$, the p-value of the test is,

$$\Pr[S \le 2] = \sum_{i=0}^{2} \binom{10}{i} \left(\frac{1}{2}\right)^{10} = 0.0547.$$

By way of comparison, the result in Example 10.5-4, using a t-test, yielded a p-value of 0.0072.

12.4.2 RUNS TESTS

Suppose we have a series of Bernoulli trials with the outcomes S = "success" and F = "failure." Then, as we learned in Section 4.2 outcomes can be represented by words using the letters S and F. Suppose further that we have m "successes" and n "failures," so that outcomes are words of length $m+n$, with exactly m S's and n F's. Usually, as with the binomial distribution, we are only interested in the ***number*** of successes, not their exact placement within the word. Now, we will look more closely at how the S's and F's are distributed in the outcome.

Previously, we assumed the ***independence*** of the trials, in which case all words with a given number of S's and F's are equally likely, regardless of the distribution of the letters. Now we will concern ourselves about whether or not the word was generated by a truly random procedure, as would be the case with independent trials. For example, we might view with suspicion an outcome in which all of the S's preceded all of the Fs. Similarly, observing a word in which the S's and the F's alternated one-by-one all the way through might also raise doubts about the randomness of the process.

In order to pin this down we focus on the number of ***runs*** in the word. A ***run*** is a maximal (non-empty) sequence of S's or F's in the word. For example, the word,

$$\overbrace{SS}^{1}\underbrace{FF}_{1}\overbrace{SSS}^{2}\underbrace{FF}_{2}\overbrace{SS}^{3}\underbrace{FFFF}_{3}\overbrace{S}^{4}\underbrace{FF}_{4}$$

has 4 runs of S's and 4 runs of F's, for a total of 8 runs. Let R be the total number of runs. Then R acts as a test statistic for randomness in the generating of the word. The smallest possible value of R is 2 (assuming both m and n are positive) and an upper bound for R is $2\min(m,n)+1$. A value of R near either extreme is evidence of non-randomness. To use R as a test statistic we will need a formula for its probability function, and we focus on that next.

The Runs Formulas

Assume we are given a sequence consisting of exactly m S's and n F's. Let s be the number of runs of S's and let f be the number of runs of F's.

1. The number of ways to have s runs of S's is given by $\binom{m-1}{s-1}$.

2. The number of ways to have f runs of F's is given by $\binom{n-1}{f-1}$.

Proof:

The approach is taken from Section 1.3.2, where we discussed the number of ways of distributing balls into urns (equivalently, sampling with replacement, order immaterial). Here, each run's position can be considered an urn, so that s is the number of urns, and m is the number of balls (S's) to be distributed. We don't want any empty runs, so first we make sure each run has one ball. This leaves $m-s$ balls (S's) to be distributed. From Section 1.3.2, the number of distinct ways to do this is,

$$\binom{s+(m-s)-1}{m-s} = \binom{m-1}{m-s} = \binom{m-1}{(m-1)-(m-s)} = \binom{m-1}{s-1}.$$

The formula in (2) is derived in identical fashion.

We can now make use of the runs formulas above to calculate $\Pr[R=r]$ runs. First note that by the definition of runs, s and f can differ by at most one. In fact if r is even then s equals f; if r is odd then either $s=f+1$ or $f=s+1$, depending on which letter starts and ends the full word.

The Probability Function for R

Assume we are given a sequence consisting of exactly m S's and n F's, and that all possible arrangements are equally likely. Let R be the random variable for the total number of runs, and assume R takes the value r. Assume that m and n are positive, so that $r \geq 2$.

1. If r is *even* let $k = s = f = \frac{r}{2}$. Then $\Pr[R = r] = \dfrac{2\binom{m-1}{k-1}\binom{n-1}{k-1}}{\binom{m+n}{m}}$.

2. If r is *odd* let $k = \frac{r+1}{2}$. Then $\Pr[R = r] = \dfrac{\binom{m-1}{k-1}\binom{n-1}{k-2} + \binom{m-1}{k-2}\binom{n-1}{k-1}}{\binom{m+n}{m}}$

Proof:

The total number of distinct words with m S's and n F's is given by $\binom{m+n}{m}$.

If r is even then we use the multiplication principle to first choose the distribution of S's for s runs, and then the distribution of F's for f runs based on the runs formulas. In this case $k = r/2$, and the "2" in the formula comes about because we can either start the word with an S or an F.

In case r is odd, then either $s = f + 1$, and the word begins and ends with S, or $s = f - 1$, and the word begins and ends with F. The total number of such words is the sum of the two cases, reflected in the numerator of the expression in (2). The first term in the numerator corresponds to the case $s = f + 1$ and the second term corresponds to the case $s = f - 1$.

Under the assumption that all arrangements of m S's and n F's are equally likely, each individual word has probability $\dfrac{1}{\binom{m+n}{m}}$. The formulas in (1) and (2) follow.

Example 12.4-3 Run Count Test for Randomness

Professor Stuckinrut has designed a 20 question *True/False* history test. Believing that most unprepared students would expect an equal number of true and false questions, he cleverly decides to have just 7 *True* statements and 13 *False* statements. Here is his answer key:

T	F	F	T	F	F	T	F	F	T	F	F	T	F	F	T	F	F	T	F

.

Viscount Runs, a statistics consultant, is asked to determine the likelihood that this is a random distribution of 7 *T*'s and 13 *F*'s. How might he respond?

Solution:
The Viscount is inclined to use the eponymously named runs test and notes $R = 14$, a suspiciously large outcome given that the maximum possible number of runs is 15. The null hypothesis is that the sequence of 7 *T*'s and 13 *F*'s is randomly distributed. If the alternative hypothesis is that the pattern is too cyclical then we would reject the null hypothesis for large values of *R*. The *p*-value equals $\Pr[R = 14, 15]$. We have $m = 7$ and $n = 13$.

For $R = 14$, $k = 7$, and $\Pr[R = 14] = \dfrac{2\binom{7-1}{7-1}\binom{13-1}{7-1}}{\binom{7+13}{7}} = 0.0238.$

For $R = 15$, $k = 8$, and,

$$\Pr[R = 15] = \frac{\binom{7-1}{8-1}\binom{13-1}{8-2} + \binom{7-1}{8-2}\binom{13-1}{8-1}}{\binom{7+13}{7}} = \frac{\binom{6}{7}\binom{12}{6} + \binom{6}{6}\binom{12}{7}}{\binom{20}{7}} = 0.0102.$$

Note that in the last calculation the binomial coefficient $\binom{6}{7}$ is, by definition, zero, a reflection of the fact that with $r = 15$ and $m = 7$, the only possibility is that $s = 7$ and $f = 8$.

Since the *p*-value for $R = 14$ is $0.0238 + 0.0102 = 0.034$, Viscount Runs might advise the professor to consider changing the sequence to include one or two longer runs to appear more random.

We next consider how a runs test can be used to determine if two independent populations have the same distribution. Let *X* and *Y* be the independent populations. The null hypothesis is that *X* and *Y* have the same distribution, and the alternative is generally posed as a statement to the effect that one population is shifted (to the left or the right) of the other.

Assume we have a random sample of size *m* from *X* and a random sample of size *n* from *Y*. We then pool the $m + n$ observations from *X* and *Y* and write them in rank order from smallest to largest. Under the null hypothesis observations from *X* and *Y* should be randomly distributed. We next count the runs of *X*'s and *Y*'s in this pooled ordering. If one population is shifted relative to the other then the number of runs is likely to be lower than expected. The reason is that we are likely to see longer runs from the left-shifted population at the start of the listing, followed by longer runs from the right-shifted population toward the end of the listing.

Example 12.4-4 Run Count Test for the Equality of Two Populations CAS Sample Problem #2 for Exam ST

You are given the following samples of paid medical claims for two similar hospitals A, and B.

Hospital A	54	56	60	50	59	38	76	12		
Hospital B	62	63	57	58	65	61	45	39	87	92

You are asked to use the **run test** to test the hypothesis that the distributions of paid medical claims are identical for these two hospitals.

(a) Calculate the probability of observing the number of runs you find for the above samples.

(b) Find an appropriate rejection region at the 5% significance level. Can we conclude that the claim amounts from one hospital are significantly different than the other's at the 5% level?

Solution:
We have $m = 8$ and $n = 10$. The pooled rank ordering, along with the associated hospital, is given in the following table:

Amount	12	38	39	45	50	54	56	57	58	59	60	61	62	63	65	76	87	92
Rank	1	2	3	4	5	6	7	8	9	10	11	12	13	14	15	16	17	18
Hospital	A	A	B	B	A	A	A	B	B	A	A	B	B	B	B	A	B	B

(a) From the last line we see that $r = 8$, and so $k = 4$. Then,

$$\Pr[R=8] = \frac{2\binom{8-1}{4-1}\binom{10-1}{4-1}}{\binom{8+10}{8}} = \frac{2\binom{7}{3}\binom{9}{3}}{43,758} = 0.1344.$$

(b) The rejection region is of the form $R \le c$. Applying the formulas, we have,

$$\Pr[R=2] = \frac{2}{43,758},\ \Pr[R=3] = \frac{16}{43,758},\ \Pr[R=4] = \frac{126}{43,758},\ \Pr[R=5] = \frac{441}{43,758},$$

$$\Pr[R=6] = \frac{1512}{43,758},\ \text{and}\ \Pr[R=7] = \frac{3024}{43,758}.$$

Then, $\Pr[R \le 6] = \dfrac{2+16+126+441+1512}{43,758} = 0.0479$ and,

$$\Pr[R \le 7] = \frac{2+16+126+441+1512+3024}{43,758} = 0.1170.$$

Therefore, the desired rejection region is $R \le 6$. Since the observed $R = 8$ we cannot reject the null hypothesis that the two populations are the same.

Exercise 12-22 In Example 12.4-1 we had the following series of observed losses:

7	6	8	12	2	3	13	1	14	15

Assume that the population median is 7.5, and consider an observation to be a "success" if it is above the median, a "failure" elsewise. You are asked to use a run test (on the series of *S*'s and *F*'s) to determine if the sequence of losses is random, or if a trend is developing.

(a) Calculate the probability of the observed number of runs.

(b) Find the probability of a Type I error if the rejection region is $R \leq 3$.

12.4.3 SIGNED RANK TESTS

In Section 12.4.1 we discussed the sign test in relation to hypotheses involving the median. A legitimate criticism of the method is that it doesn't take into account the magnitudes of differences from the median, only whether they are positive or negative. Here we will present an alternative approach, called the ***Wilcoxon signed rank test***, which incorporates the magnitudes as well as the signs of differences. This expands on the idea of a "pooled ranking" employed in the previous application in Section 12.4.2.

Let m denote the hypothesized median of a population and let $X_1, \ldots, X_n$ be a random sample. As in the sign test, for each observation X_i we track the sign of $X_i - m$, denoted $\text{sign}[X_i - m]$, and also the ***rank***, from smallest to largest of the quantities $|X_i - m|$. Let $R_i = \text{Rank}\left[|X_i - m|\right];\ i = 1, \ldots, n$. In case of a tie we assign the average of the tied positions (see the example below). Thus,

$$\sum_{i=1}^{n} R_i = 1 + 2 + \cdots + n = \frac{n(n+1)}{2}.$$

The Wilcoxon statistic is,

$$W = \sum_{i=1}^{n} \text{sign}[X_i - m] \cdot R_i.$$

If m is the true median of the population then observations are equally likely to be positive or negative. Thus W should be close to zero. If ***all*** the observations were on one side of the median then $W = \pm\frac{n(n+1)}{2}$, so large (positive or negative) values of W would constitute evidence against m being the median.

Example 12.4-5 CAS Sample Problem #1 for Exam ST (modified)
Wilcoxon Test for the Median

For a health insurance policy losses follow an unknown continuous distribution with unique median m. You are given the following random sample of ten losses:

7	6	8	12	2	3	13	1	14	15

Based on this sample and the **Wilcoxon sign rank test** you are asked to test the null hypothesis $H_0 : m = 5.5$. Find the value of the Wilcoxon statistic.

Solution:
We build on the table constructed for Example 12.4-1.

x_i	7	6	8	12	2	3	13	1	14	15
$x_i - 5.5$	1.5	0.5	2.5	6.5	−3.5	−2.5	7.5	−4.5	8.5	9.5
$\text{sign}[x_i - 5.5]$	+	+	+	+	−	−	+	−	+	+
$R_i = \text{Rank}\left[\lvert X_i - 5.5\rvert\right]$	2	1	3.5	7	5	3.5	8	6	9	10
$\text{sign}[X_i - m]\cdot R_i$	+2	+1	+3.5	+7	−5	−3.5	+8	−6	+9	+10

$W = 2+1+3.5+7-5-3.5+8-6+9+10 = 26$.

Notes:

(1) The values 8 and 3 lead to a tie in 3rd and 4th place in the ranking of absolute differences. Therefore, we use 3.5 for those rankings.

(2) The statistic W is approximately normal (the bigger the data set the better) with $\mu_W = 0$ and, $\sigma_W^2 = \dfrac{n(n+1)(2n+1)}{6}$. See Exercises 12-24 and 12.6-10 for more details.

With $n = 10$, $\sigma_W = \sqrt{\dfrac{10\cdot 11\cdot 21}{6}} = 19.62$. If the alternative hypothesis is that $m > 5.5$ then the p-value of the test can be estimated as,

$$\Pr[W \geq 26] \approx 1 - \Phi\left(\frac{26-0}{19.62}\right) = 0.0926.$$

In Example 12.4-2 we used a sign test on matched pairs to test for a "before and after" effect. The Wilcoxon signed rank test can also be applied in that context. Suppose then that we have jointly distributed pairs (X,Y) and a random sample of observations $\{(X_k, Y_k)\}_{k=1}^{n}$. We form the differences $X_k - Y_k$ and again take the pooled ranks of the magnitudes of the n differences. Let T^+ denote the sum of the ranks of the ***positive*** differences and let T^- denote the rank sum of the ***negative*** differences. Under the null hypothesis we would expect these two sums to be about the same. If X is shifted to the left of Y then more of the differences will be negative and this will be reflected in a low value of T^+. Similarly, if X is shifted to the right of Y then we would expect to see a low value of T^-. The usual procedure is to let $T = \min\{T^+, T^-\}$ and to construct a rejection region of the form $T \le c$.[2] Critical values for T can be found in the Tables for CAS Exam ST and Exam S. We will illustrate this with the data from Example 12.4-2.

Example 12.4-6 Matched Pairs Wilcoxon Signed Rank Test

Ten high school students were randomly chosen to participate in a study of the benefits of participating in a test preparation program. The first row in the following table represents their scores before the prep course, and the second row the scores after the prep course. Use the Wilcoxon signed rank test to test at the 0.01 level of significance whether the prep course improved scores.

421	551	500	595	671	435	384	694	235	599
463	611	476	617	729	454	441	726	225	709

Solution:
Let X be the prior score and let Y be the score after the prep course. The null hypothesis is that there is no improvement, versus the alternative that $Y > X$. A low value of T^+ (equivalently, a high value of T^-) would be evidence against the null hypothesis. According to the tables, for $n = 10$ and $p = 0.01$ the rejection region for a one-sided test is $T \le 5$.

Here is the complete computation:

x_i	421	551	500	595	671	435	384	694	235	599
y_i	463	611	476	617	729	454	441	726	225	709
$x_i - y_i$	−42	−60	24	−22	−58	−19	−57	−32	10	−110
$\text{sign}[x_i - y_i]$	−	−	+	−	−	−	−	−	+	−
$\text{Rank}\left[\lvert x_i - y_i \rvert\right]$	6	9	4	3	8	2	7	5	1	10

We have $T^+ = 4 + 1 = 5 << 6 + 9 + 3 + 8 + 2 + 7 + 5 + 10 = 50 = T^-$, so that $T = 5$. Therefore, we can reject the null hypothesis of no change at the 0.01 level.

Exercise 12-23 Exam ST Spring 2014 #16

A random sample of size 8 is drawn from a population in order to test the null hypothesis that the median of the population is 2.3. You are given the following sample values:

1.8	0.2	1.5	2.5	2.7	0.4	2.4	3.0

Calculate the Wilcoxon (W) statistic.

[2] As any listener of sports talk radio knows, a low T can be worrisome.

Exercise 12-24 Let X be a continuous population with an unknown distribution assumed to be symmetric about its median. Assume that the null hypothesis is that the median is equal to m, and let $X_1,\dots,X_n$ be a random sample from X. Show that the Wilcoxon statistic W can be written in the form, $W = I_1 + 2I_2 + \cdots + nI_n$, where $\{I_k\}_{k=1}^n$ are independent and identically distributed, with I_k taking the values ± 1 with probabilities of ½ each.[3]

12.4.4 MANN-WHITNEY-WILCOXON *U* STATISTIC

We return to the issue of testing two independent populations for a significant difference in location. In Example 12.4-4 we used a run count test on the claim amounts arising from two different hospitals. We will show here how to conduct a Wilcoxon-type rank sum test. The statistic U that is most widely used is a variation of Wilcoxon's rank sum. The refinement, due to Mann and Whitney, is equivalent mathematically, but often easier to calculate in small samples. Critical values of the resulting U are found in tables, including those provided for CAS Exam ST and Exam S.

Assume we have independent populations X and Y, and random samples $X_1,\dots,X_m$ and $Y_1,\dots,Y_n$. The sample sizes may be different, but for definiteness we will assume that $m \le n$. We once again look at the pooled ranking from smallest to largest of the $m+n$ observations. We denote,

$$R_X = \sum_{i=1}^{m} \text{Rank}\left[X_i\right] \text{ and } R_Y = \sum_{j=1}^{n} \text{Rank}\left[Y_j\right],$$

where Rank refers to the position of the data point in the pooled ordering. We can assume that X and Y are continuous populations so that the chance of ties has probability zero. In practice, statisticians will average together tied ranks should they occur, as we did in Example 12.4-5.

As previously, if the null hypothesis is that the two populations are the same then R_X and R_Y will be close. If X is shifted to the left of Y then R_X will be small compared to R_Y, and of course the reverse is true if X is shifted to the right of Y.

The Mann-Whitney variation is to count the number of times, U, that an X value is smaller than a Y value in the two samples. Since there are mn pairs $\left(X_i, Y_j\right)$, U can vary from 0 to mn. A low value of U is evidence that X is shifted to the right of Y and vice versa for values of U close to mn. There is a formula that allows us to calculate U in terms of R_X.

The Mann-Whitney *U* Formula

$$U = mn + \frac{m(m+1)}{2} - R_X$$

[3] The coefficients in the sum for W are not equal, so the Central Limit Theorem doesn't strictly apply in the version we have been using. However, there is an extension of the theorem that is applicable here.

Proof of Formula:

We first order the n Y values from smallest to largest, denoting the resulting permutation of the Y's by,

$Y_{(1)} < Y_{(2)} < \cdots < Y_{(n)}$. Let U_j be the number of X values that are smaller than $Y_{(j)}$, and let R_j be the pooled rank of $Y_{(j)}$. Then the key point is that $U_j = R_j - j$. In words, there is a total of R_j data points in order up to and including $Y_{(j)}$. Exactly j of those data points are Y values, and the rest, $R_j - j$ in number, are the X values preceding $Y_{(j)}$. It follows that,

$$U = U_1 + \cdots + U_n = \sum_{j=1}^{n} \left(R_j - j\right) = R_Y - \frac{n(n+1)}{2}.$$

We can reverse the roles of X and Y and let V be the number of pairs $\left(X_i, Y_j\right)$ in which $Y_j < X_i$. Then, $U + V = mn$, and $V = R_X - \frac{m(m+1)}{2}$. Combining these two formulas, we have,

$$U = mn - V = mn - \left[R_X - \frac{m(m+1)}{2}\right] = mn + \frac{m(m+1)}{2} - R_X.$$

If the alternative hypothesis is that X is to the right of Y then rejection regions are of the form $U \le U_0$. If the alternative hypothesis is that X is to the left of Y then V will be small and the rejection region is $V \le U_0$. However, under the null hypothesis X_i is equally likely to be less than, or greater than, Y_j. Thus, by symmetry,

$$\Pr\left[U \le U_0\right] = \Pr\left[V \le U_0\right] = \Pr\left[mn - U \le U_0\right] = \Pr\left[U \ge mn - U_0\right].$$

The tables are organized under the assumption that $m \le n$, and m is designated as n_1 with n designated as n_2. Only small values of U are tabulated, so if U is large (X to the left of Y) we check to see if $mn - U$ is in the tabulated rejection region. We revisit the hospital claim example (Example 12.4-4) to show how this works.

Example 12.4-7 Mann-Whitney-Wilcoxon Test for the Equality of Two Populations

You are given the following samples of paid medical claims for two similar hospitals A, and B.

Hospital A	54	56	60	50	59	38	76	12		
Hospital B	62	63	57	58	65	61	45	39	87	92

You are asked to test the hypothesis that the distributions of paid medical claims are identical for these two hospitals using the Mann-Whitney-Wilcoxon U statistic.

(a) Calculate the quantities R_X, R_Y and U, and check the Mann-Whitney formula.

(b) Find an appropriate rejection region at the 5% significance level. Can we conclude that the claim amounts from one hospital differ significantly from the other's at the 5% level?

(c) Calculate the p-value for the observed U.

Solution:
We have $m=8$ and $n=10$. The pooled rank ordering, along with the associated hospital, is given in the following table:

Amount	12	38	39	45	50	54	56	57	58	59	60	61	62	63	65	76	87	92
Rank	1	2	3	4	5	6	7	8	9	10	11	12	13	14	15	16	17	18
Hospital	A	A	B	B	A	A	A	B	B	A	A	B	B	B	B	A	B	B

The rank sum R_X corresponds to hospital A. From the table,

$$R_X = 1+2+5+6+7+10+11+16 = 58 \text{ and}$$
$$R_Y = 3+4+8+9+12+13+14+15+17+18 = 113.$$

Note that $R_X + R_Y = 171 = 1+2+\cdots+18 = \frac{18\cdot 19}{2}$. The statistic U is calculated directly from the table by counting the number of A's preceding each of the B's. So,

$$U = 2+2+5+5+7+7+7+7+8+8 = 58.$$

Calculating U from the formula yields $U = (8)(10) + \frac{8\cdot 9}{2} - 58 = 58.$

From the tables with $n_1 = 8$ and $n_2 = 10$ we find that $\Pr[U \le 17] = 0.022$ and $\Pr[U \le 18] = 0.027$. Since this is a 2-tailed test a rejection region corresponding to a significance level of 0.05 would be

$$U \le 17 \text{ or } U \ge (8)(10) - 17 = 63.$$

The p-value of $U = 58$ is $\Pr[U \ge 58 \text{ or } U \le 80-58 = 22] = 2\Pr[U \le 22] = (2)(0.061) = 0.122.$ Either way, we cannot reject the null hypothesis at the 5% significance level.

12.4.5 RANK CORRELATION

In Sections 12.1 and 12.2 we developed the linear regression model for measuring the relationship between two paired random variables. An important indicator in this regard is the correlation coefficient. The general definition is found in Section 7.4, and its adaptation to paired data-sets is discussed in Section 12.1. An alternative approach, in the spirit of the present section, is to calculate the correlation using rank order rather than the actual data points. This is particularly useful when analyzing ordinal data, such as, for example, the rankings of football teams, or the order of finish in a horse race.

When we are working with data sets arising from samples, the calculation for correlation given in Section 12.1 is often referred to as the ***Pearson correlation coefficient***, which we will denote here as ρ_P. In the notation of Section 12.1, the formula is given by,

$$\rho_P = \frac{S_{XY}}{\sqrt{(S_{XX})(S_{YY})}} = \frac{\sum X_i Y_i - n\overline{X}\overline{Y}.}{\sqrt{\left[\sum X_i^2 - n\overline{X}^2\right]\left[\sum Y_i^2 - n\overline{Y}^2\right]}}$$

$$= \frac{\sum X_i Y_i - \frac{1}{n}\left(\sum X_i\right)\left(\sum Y_i\right)}{\sqrt{\left[\sum X_i^2 - \frac{1}{n}\left(\sum X_i\right)^2\right]\left[\sum Y_i^2 - \frac{1}{n}\left(\sum Y_i\right)^2\right]}}.$$

Suppose now we find the individual (not pooled) rank ordering of the X's and the Ys. We make the simplifying assumption that the samples are from continuous distributions, thereby avoiding the possibility of ties in the rankings.

We will let $R_i = \text{rank}\left[X_i\right]$ and $S_i = \text{rank}\left[Y_i\right]$; $i = 1,\ldots,n$. Note that $\{R_i\}_{i=1}^n$ and $\{S_i\}_{i=1}^n$ both consist of the integers. $i = 1,\ldots,n$, with just the order (possibly) permuted.

Replacing the data points in the above formula with the rankings generates an alternative coefficient called the ***Spearman rank correlation***, which we will denote by ρ_S. The resulting formula is,

$$\rho_S = \frac{\sum R_i S_i - \frac{1}{n}\left(\sum R_i\right)\left(\sum S_i\right)}{\sqrt{\left[\sum R_i^2 - \frac{1}{n}\left(\sum R_i\right)^2\right]\left[\sum S_i^2 - \frac{1}{n}\left(\sum S_i\right)^2\right]}}.$$

Since the rankings are just the integers from 1 to n, we can simplify the calculation for the Spearman correlation.

Spearman Rank Correlation Formula

Let $D_i = R_i - S_i$; $i = 1,\ldots,n$. Under the assumption of no ties in the rankings we have,

$$\rho_S = 1 - \frac{6\sum_{i=1}^{n} D_i^2}{n\left(n^2 - 1\right)}$$

Proof:

The demonstration rests heavily on the summation formulas,

$$\sum_{i=1}^{n} i = \frac{n(n+1)}{2} \text{ and } \sum_{i=1}^{n} i^2 = \frac{n(n+1)(2n+1)}{6}.$$

First,

$$\sum D_i^2 = \sum (R_i - S_i)^2 = \sum R_i^2 - 2\sum R_i S_i + \sum S_i^2 = -2\sum R_i S_i + 2\frac{n(n+1)(2n+1)}{6}.$$

Therefore,

$\sum R_i S_i = \frac{n(n+1)(2n+1)}{6} - \frac{1}{2}\sum D_i^2$. Then, substituting in the summation formulas,

$$\rho_S = \frac{\sum R_i S_i - \frac{1}{n}\left(\sum R_i\right)\left(\sum S_i\right)}{\sqrt{\left[\sum R_i^2 - \frac{1}{n}\left(\sum R_i\right)^2\right]\left[\sum S_i^2 - \frac{1}{n}\left(\sum S_i\right)^2\right]}} = \frac{\frac{n(n+1)(2n+1)}{6} - \frac{1}{2}\sum D_i^2 - \frac{1}{n}\left[\frac{n(n+1)}{2}\right]^2}{\frac{n(n+1)(2n+1)}{6} - \frac{1}{n}\left[\frac{n(n+1)}{2}\right]^2}$$

$$= \frac{\frac{n(n+1)(2n+1)}{6} - \frac{1}{n}\left[\frac{n(n+1)}{2}\right]^2 - \frac{1}{2}\sum D_i^2}{\frac{n(n+1)(2n+1)}{6} - \frac{1}{n}\left[\frac{n(n+1)}{2}\right]^2} = 1 - \frac{\frac{1}{2}\sum D_i^2}{\frac{n(n+1)(2n+1)}{6} - \frac{1}{n}\left[\frac{n(n+1)}{2}\right]^2}$$

$$= 1 - \frac{6\sum D_i^2}{2n(n+1)(2n+1) - 3n(n+1)^2} = 1 - \frac{6\sum D_i^2}{n(n+1)\left[(4n+2) - (3n+3)\right]}$$

$$= 1 - \frac{6\sum D_i^2}{n(n+1)(n-1)} = 1 - \frac{6\sum D_i^2}{n(n^2-1)}.$$

The critical values for hypothesis tests involving ρ_S are tabulated online and in statistics software packages. At the time of this writing, they are not included in the CAS ST Tables. The null hypothesis is that the two variables have a correlation coefficient of zero, and values of ρ_S close to ± 1 are evidence of a non-zero correlation.

We will illustrate the procedure for calculating the Spearman correlation coefficient using the data from Examples 12.4-2 and 6 on matched pairs of "before and after" tests. Since we are observing two tests from the same individuals taken at different times, we would expect a strong correlation if the test is designed to produce consistent results.

Example 12.4-8 Spearman Rank Correlation

In academic assessment a test is considered ***reliable*** if it produces consistent results[4]. Ten high school students were randomly chosen to participate in a study on the reliability of a certain aptitude test. They took the test twice, separated by a period of several months, generating the following data.

[4] In this application we are testing the reliability of the relative rankings in the two administrations of the test.

First Testing (x_i)	421	551	500	595	671	435	384	694	235	599
Second Testing (y_i)	463	611	476	617	729	454	441	726	225	709

(a) Calculate the Spearman rank correlation coefficient for the data and compare it to the Pearson correlation.

(b) The null hypothesis is that there is zero correlation, versus the alternative that the test is reliable and the results are positively correlated. Use the Spearman rank correlation to test at the 1% level of significance.

Solution:
We first calculate the Pearson correlation directly from the data using the formula,

$$\rho_P = \frac{S_{XY}}{\sqrt{(S_{XX})(S_{YY})}} = \frac{\sum X_i Y_i - \frac{1}{n}\left(\sum X_i\right)\left(\sum Y_i\right)}{\sqrt{\left[\sum X_i^2 - \frac{1}{n}\left(\sum X_i\right)^2\right]\left[\sum Y_i^2 - \frac{1}{n}\left(\sum Y_i\right)^2\right]}}.$$

This is a straightforward, if tedious, calculation by hand. It can be easily accomplished in Excel using the built-in CORREL function. The result is $\rho_P = 0.9766$, which indicates a high level of reliability on the test.

We next calculate the Spearman correlation. We could just apply the CORREL function in Excel to the ranks, but we will assume we are making the calculation by hand and therefore use the simpler Spearman formula.

x_i	421	551	500	595	671	435	384	694	235	599
Rank$[x_i]$	3	6	5	7	9	4	2	10	1	8
y_i	463	611	476	617	729	454	441	726	225	709
Rank$[y_i]$	4	6	5	7	10	3	2	9	1	8
D_i^2	1	0	0	0	1	1	0	1	0	0

The last row shows $\sum_{i=1}^{10} D_i^2 = 4$. The Spearman formula with $n = 10$ gives,

$$\rho_S = 1 - \frac{6\sum_{i=1}^{n} D_i^2}{n\left(n^2 - 1\right)} = 1 - \frac{(6)(4)}{(10)(99)} = 0.9758.$$

As to (b), there are many tables of critical values online, and they are not all in agreement. However, most of them provide a critical value of 0.745 at the 1% level of significance with $n = 10$. Therefore, we may reject the null hypothesis of no correlation at the 1% level.

We will conclude with another alternative correlation calculation that is used for non-parametric hypothesis tests on the relationship between two populations. This is the ***Kendall Tau Correlation Coefficient***. It is also based on ranks, but in this case we use the relative rankings between the ordered pairs of observations. As previously, we will assume for the sake of simplicity that we are sampling from continuous populations in order to avoid ranking ties.[5]

Consider any two different ordered pairs of data points. We say the pairs are ***concordant*** if the ranks of one of the pairs are strictly greater than the ranks of the other pair. Otherwise, the pairs are said to be ***discordant***. The absence of ties in rankings assures there will be no ambiguity in determining which is the case. Let C be the number of concordant cases and let D be the number of discordant cases. There are exactly $\binom{n}{2} = \frac{n(n-1)}{2}$ ways to choose two ordered pairs of data points, so $C + D = \frac{n(n-1)}{2}$.

Kendall Tau Correlation Formula

Let X and Y be continuous jointly distributed populations and let $\{(x_i, y_i)\}_{i=1}^{n}$ be n ordered pairs of observations. Let C be the number of concordant cases and let D be the number of discordant cases. We define the ***Kendall tau correlation*** as,

$$\rho_T = \frac{2(C-D)}{n(n-1)}.$$

Since $C + D = \frac{n(n-1)}{2}$, it follows that $-1 \le \rho_T \le 1$.

The easiest way to count C or D by hand is to list the ordered pairs of data according to the rank of x_i. Then all the concordant pairs (x_j, y_j) with $x_j > x_i$ and $y_j > y_i$ will be further down the list of ordered pairs. Thus, let C_i be the number of pairs further down the list from (x_i, y_i) with $y_j > y_i$. The total number of concordant pairs, C, is the sum of the C_i's.

We will demonstrate this with our testing data set.

Example 12.4-9 Calculating the Kendall Tau Correlation Coefficient

You are given the following 10 observations from the continuous joint distribution of X and Y. Calculate the Kendall Tau correlation for the data.

[5] There are many modifications discussed in the literature for taking into account tied rankings between data points.

421	551	500	595	671	435	384	694	235	599
463	611	476	617	729	454	441	726	225	709

Solution:
Since we are only concerned with the ordering of data points we could work directly with the data itself or the rankings of the data. Since the ranks are easier to work with we will use the results of Example 12.4-8:

x_i	421	551	500	595	671	435	384	694	235	599
$\text{Rank}[x_i]$	3	6	5	7	9	4	2	10	1	8
y_i	463	611	476	617	729	454	441	726	225	709
$\text{Rank}[y_i]$	4	6	5	7	10	3	2	9	1	8

We next reorder the pairs of ranks according to the x-rank:

$\text{Rank}[x_i]$	1	2	3	4	5	6	7	8	9	10
$\text{Rank}[y_i]$	1	2	4	3	5	6	7	8	10	9

Starting with the first pair, $(1,1)$, we count 9 subsequent pairs in which the y-value is greater than 1. Thus, $C_1 = 9$. Proceeding to the second pair, $(2,2)$, we count $C_2 = 8$. At the third pair, $(3,4)$, we have one inversion so that $C_3 = 6$. We continue in this fashion, so that we have,

$\text{Rank}[x_i]$	1	2	3	4	5	6	7	8	9	10
$\text{Rank}[y_i]$	1	2	4	3	5	6	7	8	10	9
# Concordant	9	8	6	6	5	4	3	2	0	0

Then, $C = 9+8+6+6+5+4+3+2 = 43$. Since $C + D = \frac{n(n-1)}{2} = 45$, we have $D = 2$ and,

$$\rho_T = \frac{2(43-2)}{10(9)} = \frac{41}{45} = 0.911.$$

12.5 GOODNESS OF FIT: SEVERAL EYEBALL TESTS

Goodness of fit is an important and recurring theme in statistical analysis. The basic idea is that we have a set of data and we wish to determine if it can be treated as a random sample from a particular probability distribution. For example, in Section 10.6 we discussed the chi-squared test for determining if a data set conforms to a theoretical probability distribution model. The method there was to compare how well the observed frequencies match up with expected frequencies derived from the model.

We return to the subject in this section, but instead of using a statistical test such as chi-squared, we simply rely on graphical displays. These are not intended to provide analytical results, just visual confirmation that we are, or are not, on the right track with our proposed theoretical model. This is in the spirit of data exploration, and we refer to these graphs somewhat facetiously as eye charts, or eyeball tests.

One basic chart is the frequency histogram; in Section 10.6 for example, the results of a chi-squared test in Example 10.6-3 were confirmed by a visual comparison between the actual, or observed, frequencies, and the predicted frequencies from a Poisson model. In that spirit we investigate here a few different statistical plots that allow comparisons between the data and the proposed model.

To get started we will need to identify each data point with an ***empirical*** percentile. ***Empirical*** in this context means a calculation based purely on the data without recourse to an underlying population model. There are a variety of different ways of accomplishing this. The method introduced in Section 3.3.5 will suffice here, although just below we will provide some theoretical justification for our choice.

Assume the data points are arranged in increasing order[6] with values $y_1 < \cdots < y_k < \cdots < y_n$. In Section 3.3.5 we defined the percentile of y_k to be $100 \cdot \frac{k}{n+1}$. That can be thought of as a characterization of how far through the data set y_k is positioned. For example, given $n = 9$ data points, the smallest data point is at the 10th percentile and the largest data point is at the 90th percentile.

Suppose now that our data results from a random sample of size n from a known continuous population X with density function $f(x)$ and CDF $F(x)$. Then the percentile corresponding to x (as discussed in Section 5.3.3) is just $100p$, where $p = F(x)$. Let Y_k be the k-th order statistic, that is, as defined in Section 8.1.5, the k-th value when the data is listed from smallest to largest. We show next that the ***expected*** percentile of Y_k is in fact equal to $100 \cdot \frac{k}{n+1}$, the empirical percentile defined in the preceding paragraph. It is this result that makes $100 \cdot \frac{k}{n+1}$ a reasonable choice for the empirical percentile of the k-th value y_k.

The Expected Percentile of the *k*-th Order Statistic

Let $\{Y_k\}_{k=1}^{n}$ be the order statistics resulting from a random sample of size n from a population random variable X with density function $f(x)$ and CDF $F(x)$. Then for each $k = 1, \ldots, n$ the expected percentile of Y_k is given by $100 \cdot \frac{k}{n+1}$.

[6] For simplicity, we again treat the data as a random sample from a continuous population so that the values are all distinct and there will be no ties in the subsequent ordering.

Proof:

The density function for Y_k is given in Section 8.1.5 as

$$f_{Y_k}(y) = \frac{n!}{(n-k)!(k-1)!} F(y)^{k-1} f(y)[1-F(y)]^{n-k}.$$

Also, the percentile of Y_k is defined as $F(Y_k)$. Thus, the expected value of this percentile is given by,

$$E[F(Y_k)] = \int_{-\infty}^{\infty} F(y) \cdot f_{Y_k}(y) \cdot dy = \frac{n!}{(n-k)!(k-1)!} \int_{-\infty}^{\infty} F(y) \cdot F(y)^{k-1} f(y)[1-F(y)]^{n-k} (y) \cdot dy$$

$$= \frac{n!}{(n-k)!(k-1)!} \int_{-\infty}^{\infty} F(y)^k \cdot [1-F(y)]^{n-k} \cdot f(y) \cdot dy.$$

This rather formidable looking integral can be simplified with the substitution $w = F(y)$. We recognize that $dw = F'(y)dy = f(y)dy$. Furthermore, since F is a CDF, the limits of integration for the substitution integral become 0 and 1, respectively. Thus,

$$E[F(Y_k)] = \frac{n!}{(n-k)!(k-1)!} \int_0^1 w^k \cdot (1-w)^{n-k} \cdot dw.$$

The resultant integral is of the Beta family type with $\alpha - 1 = k$ and $\beta - 1 = n - k$. Using the results of Section 6.6.1 we see that,

$$\int_0^1 w^k \cdot (1-w)^{n-k} \cdot dw = \frac{\Gamma(\alpha)\Gamma(\beta)}{\Gamma(\alpha+\beta)} = \frac{k!(n-k)!}{(n+1)!}.$$

Then,

$$E[F(Y_k)] = \frac{n!}{(n-k)!(k-1)!} \cdot \frac{k!(n-k)!}{(n+1)!} = \frac{k}{n+1}.$$

In words, the ***expected*** percentile of Y_k is $100 \cdot \frac{k}{n+1}$.

The graphs we describe here amount to comparisons of the empirical percentiles of a data set with the corresponding percentiles of the proposed model, or theoretical, distribution. The simplest graphical comparison is to plot the empirical CDF with the model CDF.

The graphical version of the empirical CDF that we use consists of the pairs $\left(y_k, \frac{k}{n+1}\right)$ joined by straight lines. The model CDF is based on the pairs $\left(y_k, F\left(y_k\right)\right)$.

Example 12.5-1 Empirical versus Model CDFs

Suppose we are given the 9 sorted data points below and wish to determine if they could arise as a random sample from an exponential distribution with mean 100.

5	15	20	45	80	110	150	160	190

Plot a graph of the empirical CDF with the model CDF.

Solution

The model CDF is given by $F(x) = 1 - e^{-x/100}$. The resulting graph is built from the following tabular calculations.

k	y_k	Empirical CDF $= \frac{k}{10}$	Model CDF $F(y_k) = 1 - e^{-y_k/100}$
1	5	0.1	0.05
2	15	0.2	0.14
3	20	0.3	0.18
4	45	0.4	0.36
5	80	0.5	0.55
6	110	0.6	0.67
7	150	0.7	0.78
8	160	0.8	0.80
9	190	0.9	0.85

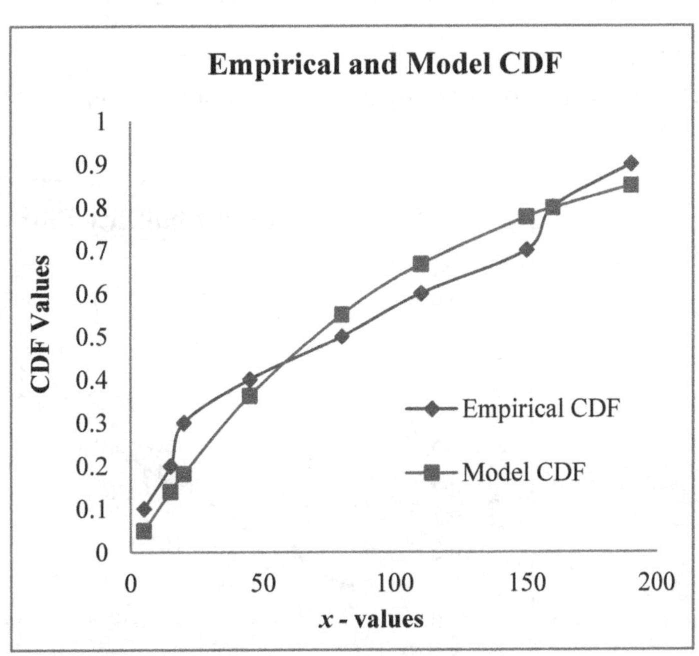

The graphs show a reasonably close relationship between the two, suggesting the theoretical exponential model is a decent starting point. But of course, any judgment is entirely subjective.

We next present two alternative graphical representations of the information in the CDF comparison chart. In the CDF chart the coordinates of the empirical CDF are given by

$\left(y_k, \frac{k}{n+1}\right)$; $k = 1,\ldots,n$. The coordinates of the model CDF are given by $\left(y_k, F(y_k)\right)$; $k = 1,\ldots,n$, where F is the CDF of the model population random variable X.

The ***probability plot***, or so-called ***p-p plot***, is constructed by matching the ordinates (that is, the second coordinates) of these two CDF graphs. Thus, the probability plot displays the points $\left(\frac{k}{n+1}, F(y_k)\right)$; $k = 1,\ldots,n$. If $\{Y_k\}_{k=1}^{n}$ is an ordered random sample from X then the above result states that we ***expect*** that $\frac{k}{n+1}$ equals $F(y_k)$, which would mean the plotted points are all on the line $y = x$, so this diagonal line is included in the probability plot for comparison purposes. Small deviations from the line can be explained by statistical fluctuations, but large systematic deviations would indicate the model is not correct.

Example 12.5-2 Probability Plot for Sample Data

Construct the probability plot for the scenario described in Example 12.5-1.

Solution

The graph is constructed by plotting the last two columns of the table in Example 12.5-1, as shown below.

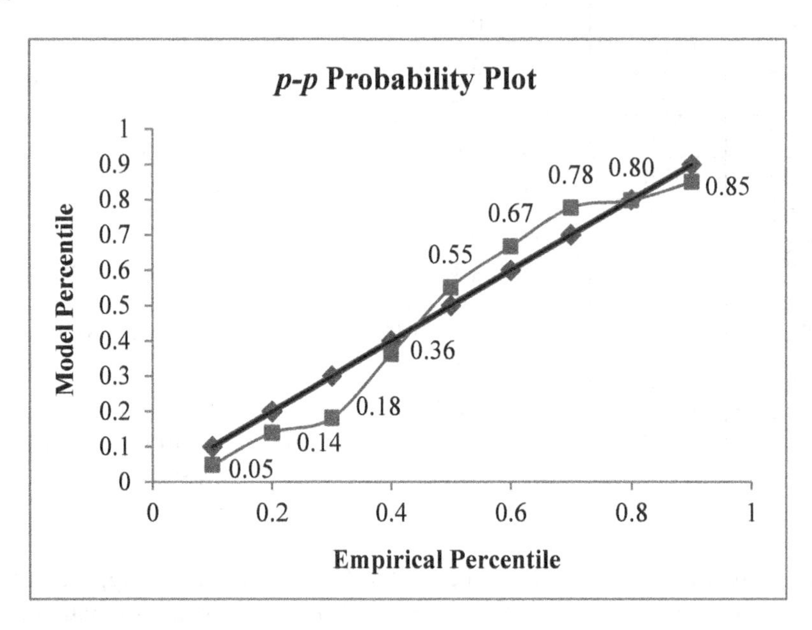

Note that the relative position of the plotted percentiles to the diagonal line is the same as the relative position of the empirical CDF to the model CDF in the CDF graph. The fit appears reasonably close, but the cyclical nature of the deviations suggests that the tails in the model are a little light (below the empirical probabilities) and the central portion is a little heavy (where the graph is above the line).

The second alternative plot to the CDF graph is called a ***quantile plot***, or a ***q-q plot***. In this version we use ***values*** (the abscissa, or first coordinates from the CDF graph) rather than the ***percentiles*** in the second coordinates. That is, we plot y_k versus the $100 \cdot \frac{k}{n+1}$ percentile of the model distribution for $k = 1, \ldots, n$. The $100 \cdot \frac{k}{n+1}$ percentile of the model is calculated as $x_k = F^{-1}\left(\frac{k}{n+1}\right)$. Again, if the data fits the model exactly then the plotted points will be on the line $y = x$.

Example 12.5-3 Quantile Plot for Sample Data

Construct the quantile plot for the scenario described in Example 12.5-1.

Solution

Since the model CDF is given by $F(x) = 1 - e^{-x/100}$, we have

$$x_k = F^{-1}\left(\frac{k}{n+1}\right) = -100 \ln\left(1 - \frac{k}{10}\right).$$

The table of values and the corresponding *q-q* plot are computed below.

k	y_k	$\frac{k}{10}$	$F^{-1}\left(\frac{k}{10}\right)$
1	5	0.1	10.54
2	15	0.2	22.31
3	20	0.3	35.67
4	45	0.4	51.08
5	80	0.5	69.31
6	110	0.6	91.63
7	150	0.7	120.40
8	160	0.8	160.94
9	190	0.9	230.26

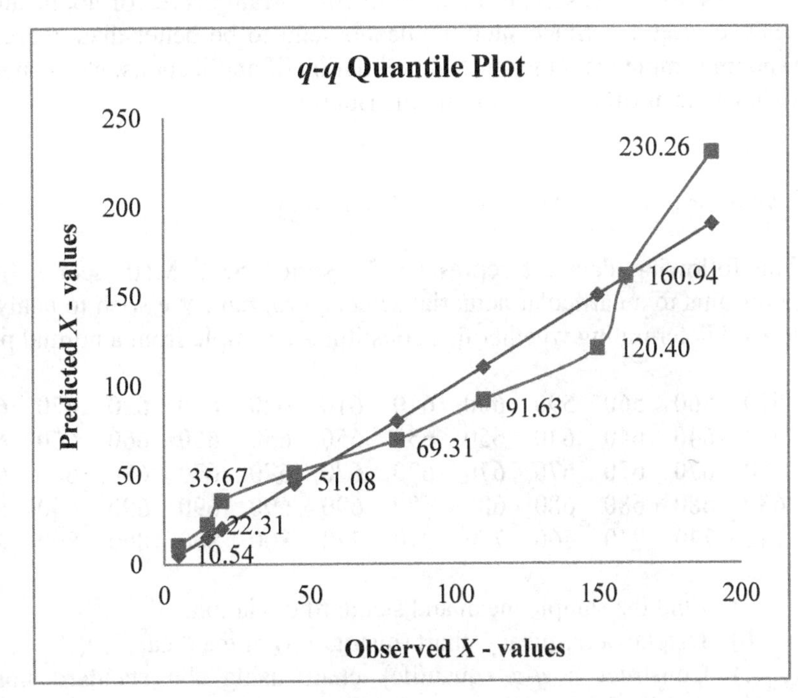

Note how the *q-q* plot reverses from the *p-p* plot the relative positions of the graph versus the diagonal line.

The most common application of the quantile plot is to the situation where the proposed model distribution is normal, that is $X \sim N\left(\mu,\sigma^2\right)$. It is common in this case to plot the sample values y_k against the corresponding percentiles z_k of the standard normal Z. If the $\{Y_k\}_{k=1}^{n}$ constitute an ordered sample from X then the plot will be nearly linear, but not necessarily on the diagonal $y = x$. The reason is, since $x_k = F^{-1}\left(\frac{k}{n+1}\right)$, we have,

$\frac{k}{n+1} = F(x_k) = \Pr[X \le x_k] = \Pr\left[Z \le \frac{x_k - \mu}{\sigma}\right]$, which implies that

$$z_k = \frac{1}{\sigma}x_k - \frac{\mu}{\sigma} \approx \frac{1}{\sigma}y_k - \frac{\mu}{\sigma}.$$

Therefore, a nearly linear quantile plot of the data against the percentiles of Z implies that the data conforms to a normal distribution. The parameters of X could also be estimated from the coefficients of the least-squares regression line. This allows for a comparison with the usual estimates of the sample mean and sample standard deviation.

Illustrations based on a small sample size, such as those above, tend to be inconclusive except in artificial situations. For a more realistic scenario let's look at some actual data. The next example is based on a set of 72 SAT-Math scores of applicants to a college actuarial science program. Since such applicants tend to be better than average at math this is not a random sample from the general population of applications. Nevertheless, we would expect it to have the attributes of a normal distribution.

Example 12.5-4 Sampling SAT-Math Scores

The following data set represents 72 sorted SAT Math scores from a sample of 2014 applicants to a particular actuarial science program. We wish to analyze the data with an eye toward determining whether this constitutes a sample from a normal population.

510	560	560	590	600	610	610	620	620	620	620	620	620	630	630
640	640	640	640	650	650	650	650	650	660	660	660	660	670	670
670	670	670	670	670	670	670	670	670	680	680	680	680	680	680
680	680	680	680	680	680	690	690	690	690	690	690	710	710	730
740	740	740	760	770	770	770	800	800	800	800	800			

a) Find the sample mean and standard deviation.
b) Display a frequency chart (histogram) of the data.
c) Construct a *q-q* (quantile) chart using the standard normal Z as the model distribution.
d) Construct the regression line for the chart and use the regression coefficients to estimate the mean and the standard deviation for the population.

Solution

All of the calculations were performed in Excel. The sample mean and sample standard deviation are 674.7 and 58.0 respectively. The following histogram is based on intervals of width twenty[7].

SAT-M	Frequency
500-520	1
530-550	0
560-580	2
590-610	4
620-640	12
650-670	20
680-700	18
710-730	3
740-760	4
770-790	3
800	5

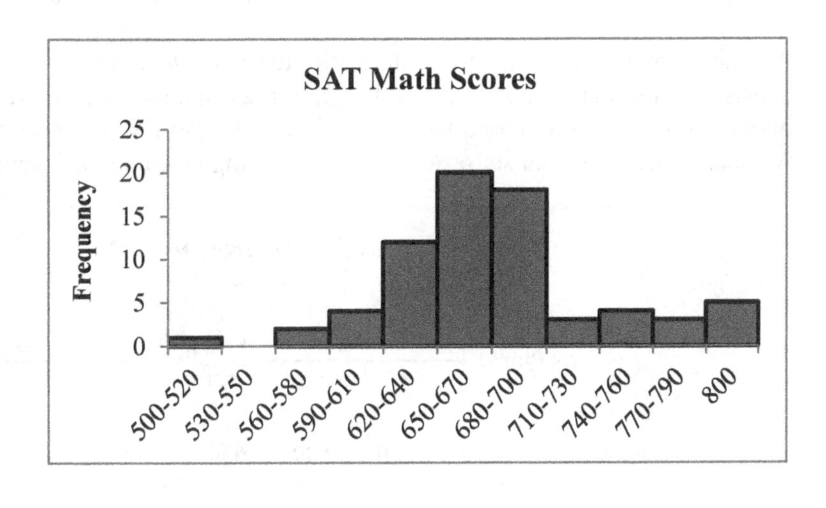

The histogram displays some of the characteristics of a normal distribution, but the right-hand tail is disproportionately heavy. This is mainly due to the five scores at 800, which distort the graph. For one thing, they are outliers in the sense that they are more than two standard deviations above the mean. More significantly, since SAT Math scores are capped at 800 these scores are "censored," meaning that if the test instrument allowed for it, the scores would be greater than or equal to 800, resulting in a more realistic right-hand tail.

The next graph shows the *q-q* plot using the standard normal percentiles of $\frac{100k}{73}; k = 1, \ldots, 72.$

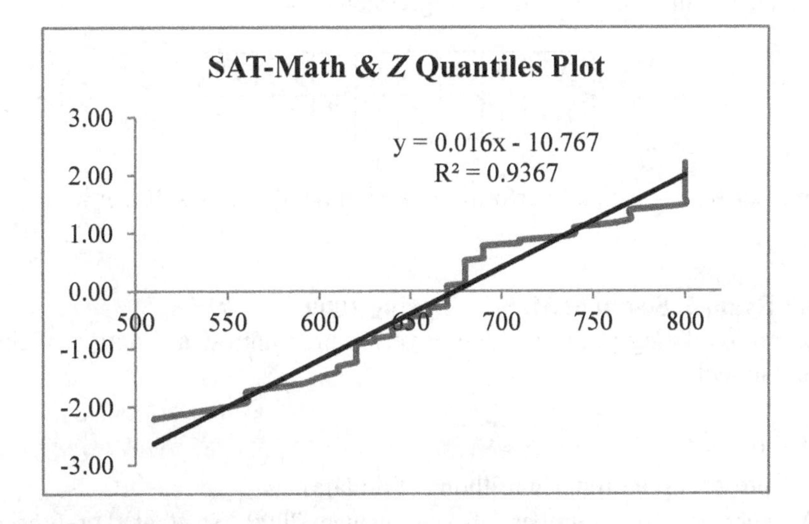

[7] This choice is arbitrary, and different class widths might result in significantly different shapes.

We've included the Excel calculations for the regression statistics. The R^2 value of 0.9367 suggests a relatively good fit with a normal distribution. Using the relationship $z_k = \frac{1}{\sigma}x_k - \frac{\mu}{\sigma}$ we can solve for $\sigma = \frac{1}{.016} = 62.5$ and $\mu = 10.767\sigma = 672.9$, which are not too far off from the sample mean and sample standard deviation in (a).

Another display that is used in data exploration is a ***box-and-whiskers*** chart. This generally consists of a graph of the 1st, 2nd and 3rd quartiles, sandwiched between the minimum and the maximum of the data. The quartiles are in a box, and the min and max are displayed as the whiskers. The SAT data set produces the following box-and-whiskers diagram.

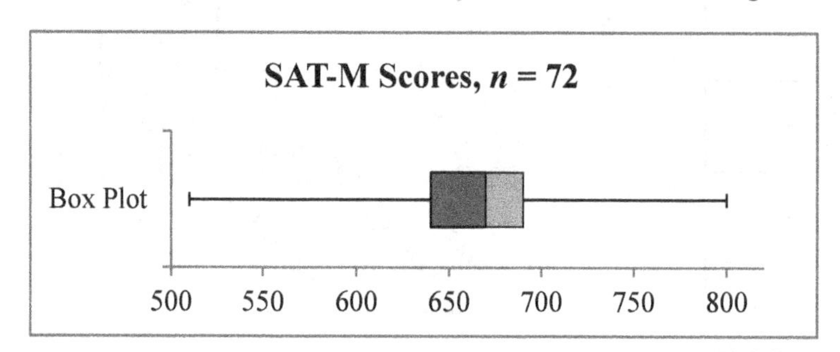

This diagram captures the concentration of scores at the higher end, as the 2nd quartile, or median, is closer to the 3rd quartile than the 1st. This slight skewing is not affected by the data points censored at 800 since the quartile calculations would not change even if the scores were allowed to extend past 800.

12.6 CHAPTER 12 SAMPLE EXAMINATION

1. Use the method of least squares to fit the model $Y_i = \alpha + \beta\sqrt{X_i} + \varepsilon_i$ to the following data. Determine the least squares regression equation.

X_i	0	0	1	4	4	9	25
Y_i	1	2	2	6	7	10	14

Hint: Let $W_i = \sqrt{X_i}$ and perform regression on $Y_i = \alpha + \beta W_i + \varepsilon_i$.

2. **CAS Exam 3, Segment 3L Spring 2009 #25**
You are modeling your company's yearly premium as a function of time using the linear model:

• $Y = \alpha + \beta X$
• Y represents premium (in millions of dollars).
• X represents the number of years since 2000 (so that 1 represents 2001, 2 represents 2002, etc.).

You use the following summary statistics to perform a least squares regression:

$$n = 6$$

$$\sum_{i=1}^{6} X_i = 15 \qquad \sum_{i=1}^{6} X_i^2 = 55$$

$$\sum_{i=1}^{6} Y_i = 468 \qquad \sum_{i=1}^{6} X_i Y_i = 1{,}363$$

Determine the first year that the estimated premium exceeds $150 million.

(A) 2007 (B) 2008 (C) 2009 (D) 2010 (E) 2011

3. **CAS Exam 3L Spring 2013 #25**

You are given the following data set with two variables, X and Y:

X	Y
10	22
13	20
20	6
15	18
5	10

Using the least squares method for a linear regression with Y as the dependent variable, calculate the absolute value of the residual where the x and y values are 12 and 18 respectively.

(A) Less than 1
(B) At least 1, but less than 2
(C) At least 2, but less than 3
(D) At least 3, but less than 4
(E) At least 4

4. **CAS Exam 3L Fall 2011 #25**

You are modeling your client's annual revenue as a function of time using the linear model:

- $y = \alpha + \beta x$
- y represents revenue (millions of dollars)
- x represents the number of years since the company started doing business

You have calculated the following statistics:

$$n=12,\ \sum x_i=78,\ \sum y_i=324,$$
$$\sum(x_i-\bar{x})^2=143,\ \sum(y_i-\bar{y})^2=1839,$$
$$\sum(y_i/x_i)=56,\ \sum(x_i-\bar{x})(y_i-\bar{y})=503.$$

Calculate the least squares estimate of β.

(A) Less than 1.00
(B) At least 1.00, but less than 3.00
(C) At least 3.00, but less than 5.00
(D) At least 5.00, but less than 7.00
(E) At least 7.00

5. Three types of brain food are each fed to random samples of statistics students in order to measure improvement on IQ scores. The increases are shown in the following table. Assume the improvements for each of the brain food types constitute normal distributions with means μ_1, μ_2, μ_3, respectively, and a common unknown standard deviation. Use the single factor ANOVA procedure and the following data on score improvements to test the null hypothesis that $\mu_1=\mu_2=\mu_3$ at the 5% level of significance.

Brain Food Type (i)	x_{i1}	x_{i2}	x_{i3}	x_{i4}
1	5	9	7	8
2	11	13	9	10
3	10	6	7	9

6. Fourteen students are selected at random from four different sections of statistics. The following table shows their scores on the final exam. Assume scores from individual sections constitute normal distributions with means $\mu_1, \mu_2, \mu_3, \mu_4$, respectively, and a common unknown standard deviation. Use the single factor ANOVA procedure to test the null hypothesis that $\mu_1=\mu_2=\mu_3=\mu_4$. Calculate the resulting F statistic and determine whether the null hypothesis can be rejected at the 5%, and the 1%, level of significance.

Section	Scores				
A	72	75	80	49	
B	73	94	54		
C	61	66			
D	86	114	93	94	94

7. The Pennsylvania Department of Motor Vehicles claims that the median wait time for driver license services is no more than 20 minutes. An independent contractor observed the following 20 wait times (in minutes). A sign test on the data is used to test the null hypothesis that the median is 20 versus the alternative hypothesis that the median is more than 20. Determine the p-value of the test.

9	20	24	31	34	13	35	31	9	31	22	11	16	28	31	23	28	22	21	23

8. Hoops Heaven is a college basketball wager consultant. Their non-parametric specialist was tasked with predicting the winner of the 2014 Pac-12 championship game between the Arizona Wildcats and the UCLA Bruins. Her procedure was to compare the margins of victory for both Arizona and UCLA against 17 common opponents over the course of the season. She then applied a sign test on the differences in margin of victory. Take as the null hypothesis that the two teams are even versus the alternative hypothesis that Arizona should be favored. Based on the following data, calculate the p-value of the test. The data consist of margins of victory, with a negative number meaning a loss.[8]

Opponent	Duke	WSU	at USC	ASU	CU	Utah	at Stan	at Cal	Ore	OSU
UCLA	-17	-18	10	15	18	14	-9	20	-4	5
Ariz	6	35	20	23	12	9	3	-2	2	22

at ASU	at Utah	at CU	Cal	Stan	at OSU	at Ore
15	14	13	12	17	-4	2
-3	4	27	28	13	5	-7

9. Miss Ormake is a basketball player who is considered to be a "streaky" shooter, meaning she has longer than expected runs of misses and makes for her shots. As a 40% shooter, she makes 10 out of 25 shots, in the following pattern. Here S means a success (made shot) and F means failure (missed shot). Use a runs test to calculate the p-value for the pattern under the assumption that a smaller than normal number of runs is evidence of a streaky shooter.

S	S	F	F	F	F	F	F	F	S	S	S	F	S	S	F	F	F	F	F	F	F	S	S	S

10. (See also Exercise 12-24.) Let $\{W_i\}_{i=1}^{n}$ be a set of n numbers, each of which is equally likely to be positive or negative. Let $\{I_i\}_{i=1}^{n}$ be n independent random variables, each taking the values +1 or −1 with equal probabilities. Then the random

[8] UCLA won.

variable $W = \sum_{i=1}^{n} \text{Rank}\left[|W_i|\right] \cdot I_i$ is the Wilcoxon sign rank statistic for $\{W_i\}_{i=1}^{n}$. Show that,

(a) $\mu_W = 0$.

(b) $\sigma_W^2 = \sum_{i=1}^{n} i^2 = \frac{n(n+1)(2n+1)}{6}$.

11. The non-parametric specialist at Hoops Heaven revisits the data displayed in Problem 8, above. She notices that the positive differences in point spreads seem larger than the negative differences (favoring Arizona). To capture the magnitude of the differences she calculates the Wilcoxon T statistic for the data.

    (a) Find the value of T she calculates.

    (b) Look up the critical value for a 5% level of significance in the CAS Exam ST (or Exam S) tables, assuming a one-tail test of the hypothesis that Arizona should be favored. Should the null hypothesis that the two teams are even be rejected?

12. Professor Azimuth has two small sections of statistics, one fully online and the other in a traditional classroom setting. He uses the final exam grades to determine if there is a significant difference between the student populations. Azimuth thinks there is a difference, but (despite his name) is uncertain of the direction. Here are the exam results:

Online Class	87	81	107	65	77	86		
On-ground Class	75	78	82	63	73	58	76	69

Calculate the Mann-Whitney-Wilcoxon U statistic and use the CAS Exam ST (or Exam S) tables to determine its approximate p-value (assuming a two-tail test).

13. A Hoops Heaven analyst is curious about the degree to which the pre-season college basketball rankings are correlated with the ending polls from the previous season. The following data shows the top ten college basketball teams in the 2014 USA Today pre-season poll, along with their final rankings in the Basketball Power Index poll from the previous 2013-14 season.

2014-15 Pre-Season Rank	Team	2013-14 Final BPI Rank
1	Kentucky	8
2	Arizona	1
3	Duke	13
4	Wisconsin	7
5	Kansas	6
6	North Carolina	29
7	Florida	2
8	Virginia	5
9	Louisville	3
10	Texas	51

(a) Calculate the Spearman rank correlation coefficient.
(b) Calculate the Kendall tau correlation coefficient.

14. **CAS Exam ST Spring 2014 #15**

You are given the following information:

Student	Number of Practice Exams	Score on Final Exam
1	8	9
2	6	8
3	5	7
4	4	6
5	3	5
6	2	10
7	1	4
8	0	3

Students are ranked in descending order for both the number of practice exams and the score on the final exam.

(a) Calculate Spearman's rank correlation coefficient between the number of practice exams and the score on the final exam.
(b) Calculate the Kendall tau correlation coefficient.

15. **CAS Exam ST Fall 2014 #17**

You are given the following daily temperatures from two different locations with only three observations from Location A and four from Location B:

Location A	45	57	62	
Location B	81	60	65	75

- Assume *m* is defined as the median of the data set.
- $H_0: m_A = m_{B.}$
- $H_1: m_A \neq m_{B.}$
- You perform a Mann-Whitney-Wilcoxon test using the exact calculation.

Determine the smallest significance level at which you reject H_0.

(A) Less than 1.0%
(B) At least 1.0%, but less than 2.5%
(C) At least 2.5%, but less than 5.0%
(D) At least 5.0%, but less than 10.0%
(E) At least 10.0%

16. **CAS Exam ST Fall 2014 #18 (modified)**

You are given the following information:

- From each of the classrooms X and Y, a random sample of students' test scores were selected for analysis.
- Five scores were selected from each classroom.
- Let $\{X_i\}_{i=1}^{5}$ be the sampled scores from classroom X and let $\{Y_i\}_{i=1}^{5}$ be the sampled scores from classroom Y.
- $H_0: m_X = m_Y$, where m_X and m_Y are the individual class medians.
- $H_1: m_X < m_Y$.
- The null hypothesis is rejected at the $\alpha = 0.05$ significance level.
- *W**U* is defined as the Mann-Whitney-Wilcoxon statistic.
- The following sample data is collected:

$X:$	86.5	85.2	70.8	96.2	90.6
$Y:$	65.2	99.4	82.3	92.2	89.8

Calculate the absolute value of the difference between the computed U and the critical value of the U implied by the significance level, using the normal approximation.

Hint: There is a normal approximation to U that uses $\mu_U = \dfrac{nm}{2}$ and

$\sigma_U^2 = \dfrac{nm(n+m+1)}{12}$, when n and m are both greater than 10, although for this problem $n = m = 5$.

(A) Less than 2
(B) At least 2, but less than 4
(C) At least 4, but less than 6
(D) At least 6, but less than 8
(E) At least 8

17 **CAS Exam ST Fall 2014 #19**

You are given the following information:

- A bowler claims to be equally skilled at bowling right-handed and left-handed.
- The bowler bowled 10 games with each hand. The scores are provided in the table below.

Game	*A*	*B*	*C*	*D*	*E*	*F*	*G*	*H*	*I*	*J*
Right-Handed	200	200	200	175	175	175	175	175	200	200
Left-Handed	175	175	175	175	175	175	175	175	175	175

- Assume m is defined as the median of the data set.
- $H_0: \ m_R = m_L$
- $H_1: \ m_R \neq m_L$

Calculate the smallest value for α for which the null hypothesis can be rejected using the Sign Test using the exact calculation.

(A) Less than 0.010
(B) At least 0.010, but less than 0.025
(C) At least 0.025, but less than 0.050
(D) At least 0.050, but less than 0.100
(E) At least 0.100

APPENDIX I

STATISTICAL TABLES

Entries represent the area under the standardized normal distribution from $-\infty$ to z, $\Pr[Z \le z] = \Phi(z)$. The value of z to the first decimal is given in the left column. The second decimal place is given in the top row.

NORMAL DISTRIBUTION TABLE

z	0.00	0.01	0.02	0.03	0.04	0.05	0.06	0.07	0.08	0.09
0.0	0.5000	0.5040	0.5080	0.5120	0.5160	0.5199	0.5239	0.5279	0.5319	0.5359
0.1	0.5398	0.5438	0.5478	0.5517	0.5557	0.5596	0.5636	0.5675	0.5714	0.5753
0.2	0.5793	0.5832	0.5871	0.5910	0.5948	0.5987	0.6026	0.6064	0.6103	0.6141
0.3	0.6179	0.6217	0.6255	0.6293	0.6331	0.6368	0.6406	0.6443	0.6480	0.6517
0.4	0.6554	0.6591	0.6628	0.6664	0.6700	0.6736	0.6772	0.6808	0.6844	0.6879
0.5	0.6915	0.6950	0.6985	0.7019	0.7054	0.7088	0.7123	0.7157	0.7190	0.7224
0.6	0.7257	0.7291	0.7324	0.7357	0.7389	0.7422	0.7454	0.7486	0.7517	0.7549
0.7	0.7580	0.7611	0.7642	0.7673	0.7704	0.7734	0.7764	0.7794	0.7823	0.7852
0.8	0.7881	0.7910	0.7939	0.7967	0.7995	0.8023	0.8051	0.8078	0.8106	0.8133
0.9	0.8159	0.8186	0.8212	0.8238	0.8264	0.8289	0.8315	0.8340	0.8365	0.8389
1.0	0.8413	0.8438	0.8461	0.8485	0.8508	0.8531	0.8554	0.8577	0.8599	0.8621
1.1	0.8643	0.8665	0.8686	0.8708	0.8729	0.8749	0.8770	0.8790	0.8810	0.8830
1.2	0.8849	0.8869	0.8888	0.8907	0.8925	0.8944	0.8962	0.8980	0.8997	0.9015
1.3	0.9032	0.9049	0.9066	0.9082	0.9099	0.9115	0.9131	0.9147	0.9162	0.9177
1.4	0.9192	0.9207	0.9222	0.9236	0.9251	0.9265	0.9279	0.9292	0.9306	0.9319
1.5	0.9332	0.9345	0.9357	0.9370	0.9382	0.9394	0.9406	0.9418	0.9429	0.9441
1.6	0.9452	0.9463	0.9474	0.9484	0.9495	0.9505	0.9515	0.9525	0.9535	0.9545
1.7	0.9554	0.9564	0.9573	0.9582	0.9591	0.9599	0.9608	0.9616	0.9625	0.9633
1.8	0.9641	0.9649	0.9656	0.9664	0.9671	0.9678	0.9686	0.9693	0.9699	0.9706
1.9	0.9713	0.9719	0.9726	0.9732	0.9738	0.9744	0.9750	0.9756	0.9761	0.9767
2.0	0.9772	0.9778	0.9783	0.9788	0.9793	0.9798	0.9803	0.9808	0.9812	0.9817
2.1	0.9821	0.9826	0.9830	0.9834	0.9838	0.9842	0.9846	0.9850	0.9854	0.9857
2.2	0.9861	0.9864	0.9868	0.9871	0.9875	0.9878	0.9881	0.9884	0.9887	0.9890
2.3	0.9893	0.9896	0.9898	0.9901	0.9904	0.9906	0.9909	0.9911	0.9913	0.9916
2.4	0.9918	0.9920	0.9922	0.9925	0.9927	0.9929	0.9931	0.9932	0.9934	0.9936
2.5	0.9938	0.9940	0.9941	0.9943	0.9945	0.9946	0.9948	0.9949	0.9951	0.9952
2.6	0.9953	0.9955	0.9956	0.9957	0.9959	0.9960	0.9961	0.9962	0.9963	0.9964
2.7	0.9965	0.9966	0.9967	0.9968	0.9969	0.9970	0.9971	0.9972	0.9973	0.9974
2.8	0.9974	0.9975	0.9976	0.9977	0.9977	0.9978	0.9979	0.9979	0.9980	0.9981
2.9	0.9981	0.9982	0.9982	0.9983	0.9984	0.9984	0.9985	0.9985	0.9986	0.9986
3.0	0.9987	0.9987	0.9987	0.9988	0.9988	0.9989	0.9989	0.9989	0.9990	0.9990
3.1	0.9990	0.9991	0.9991	0.9991	0.9992	0.9992	0.9992	0.9992	0.9993	0.9993
3.2	0.9993	0.9993	0.9994	0.9994	0.9994	0.9994	0.9994	0.9995	0.9995	0.9995
3.3	0.9995	0.9995	0.9995	0.9996	0.9996	0.9996	0.9996	0.9996	0.9996	0.9997
3.4	0.9997	0.9997	0.9997	0.9997	0.9997	0.9997	0.9997	0.9997	0.9997	0.9998
3.5	0.9998	0.9998	0.9998	0.9998	0.9998	0.9998	0.9998	0.9998	0.9998	0.9998
3.6	0.9998	0.9998	0.9999	0.9999	0.9999	0.9999	0.9999	0.9999	0.9999	0.9999
3.7	0.9999	0.9999	0.9999	0.9999	0.9999	0.9999	0.9999	0.9999	0.9999	0.9999
3.8	0.9999	0.9999	0.9999	0.9999	0.9999	0.9999	0.9999	0.9999	0.9999	0.9999
3.9	1.0000	1.0000	1.0000	1.0000	1.0000	1.0000	1.0000	1.0000	1.0000	1.0000

The table below gives the value χ_0^2 for $\Pr\left[\chi^2(n) \le \chi_0^2\right] = p$, where p is in the top row and the degrees of freedom n is in the left margin.

Percentiles Table of the Chi-Square Distribution

Degrees of Freedom	Value of p							
	0.005	**0.010**	**0.025**	**0.050**	**0.950**	**0.975**	**0.990**	**0.995**
1	0.00	0.00	0.00	0.00	3.84	5.02	6.63	7.88
2	0.01	0.02	0.05	0.10	5.99	7.38	9.21	10.60
3	0.07	0.11	0.22	0.35	7.81	9.35	11.34	12.84
4	0.21	0.30	0.48	0.71	9.49	11.14	13.28	14.86
5	0.41	0.55	0.83	1.15	11.07	12.83	15.09	16.75
6	0.68	0.87	1.24	1.64	12.59	14.45	16.81	18.55
7	0.99	1.24	1.69	2.17	14.07	16.01	18.48	20.28
8	1.34	1.65	2.18	2.73	15.51	17.53	20.09	21.95
9	1.73	2.09	2.70	3.33	16.92	19.02	21.67	23.59
10	2.16	2.56	3.25	3.94	18.31	20.48	23.21	25.19
11	2.60	3.05	3.82	4.57	19.68	21.92	24.72	26.76
12	3.07	3.57	4.40	5.23	21.03	23.34	26.22	28.30
13	3.57	4.11	5.01	5.89	22.36	24.74	27.69	29.82
14	4.07	4.66	5.63	6.57	23.68	26.12	29.14	31.32
15	4.60	5.23	6.26	7.26	25.00	27.49	30.58	32.80
16	5.14	5.81	6.91	7.96	26.30	28.85	32.00	34.27
17	5.70	6.41	7.56	8.67	27.59	30.19	33.41	35.72
18	6.26	7.01	8.23	9.39	28.87	31.53	34.81	37.16
19	6.84	7.63	8.91	10.12	30.14	32.85	36.19	38.58
20	7.43	8.26	9.59	10.85	31.41	34.17	37.57	40.00

Entries represent the percentile, $t_{\alpha/2}(n)$, where α (along top of table) is the two-tail area under the Student-t with n (down left side) degrees of freedom: $\alpha = \Pr\left[\left|t(n)\right| \le t_{\alpha/2}(n)\right]$.

Percentiles Table of the Student-*t* Distribution

n	0.1	0.05	0.02	0.01
1	6.314	12.706	31.821	63.657
2	2.920	4.303	6.965	9.925
3	2.353	3.182	4.541	5.841
4	2.132	2.776	3.747	4.604
5	2.015	2.571	3.365	4.032
6	1.943	2.447	3.143	3.707
7	1.895	2.365	2.998	3.499
8	1.860	2.306	2.896	3.355
9	1.833	2.262	2.821	3.250
10	1.812	2.228	2.764	3.169
11	1.796	2.201	2.718	3.106
12	1.782	2.179	2.681	3.055
13	1.771	2.160	2.650	3.012
14	1.761	2.145	2.624	2.977
15	1.753	2.131	2.602	2.947
16	1.746	2.120	2.583	2.921
17	1.740	2.110	2.567	2.898
18	1.734	2.101	2.552	2.878
19	1.729	2.093	2.539	2.861
20	1.725	2.086	2.528	2.845
21	1.721	2.080	2.518	2.831
22	1.717	2.074	2.508	2.819
23	1.714	2.069	2.500	2.807
24	1.711	2.064	2.492	2.797
25	1.708	2.060	2.485	2.787
26	1.706	2.056	2.479	2.779
27	1.703	2.052	2.473	2.771
28	1.701	2.048	2.467	2.763
29	1.699	2.045	2.462	2.756
30	1.697	2.042	2.457	2.750
40	1.684	2.021	2.423	2.704
60	1.671	2.000	2.390	2.660
120	1.658	1.980	2.358	2.617
∞	1.645	1.960	2.326	2.576

For each pair (m,n) the table below gives the two values $F_{.05}(m,n)$ and $F_{.01}(m,n)$, where $.05 = \Pr[f(m,n) \geq F_{.05}(m,n)]$ and $.01 = \Pr[F(m,n) \geq F_{.01}(m,n)]$. The value of $F_{.05}(m,n)$ is displayed in the shaded rows, and the value of $F_{.01}(m,n)$, is listed beneath in regular font. The degrees of freedom m are listed along the top and the degrees of freedom n are down the left and right margins

Percentiles of the *F*-Distribution

n \ *m*	1	2	3	4	5	6	7	8	9	10	11	12	14	16	20	24	30	40	50	75	100	200	500	∞	*n*
1	161	199	216	225	230	234	237	239	241	242	243	244	245	246	248	249	250	251	252	253	253	254	254	254	1
	4052	4999	5403	5625	5764	5859	5928	5981	6022	6056	6083	6106	6143	6170	6209	6235	6261	6287	6303	6324	6334	6350	6360	6366	
2	18.51	19.00	19.16	19.25	19.30	19.33	19.35	19.37	19.38	19.40	19.40	19.41	19.42	19.43	19.45	19.45	19.46	19.47	19.48	19.48	19.49	19.49	19.49	19.50	2
	98.50	99.00	99.17	99.25	99.30	99.33	99.36	99.37	99.39	99.40	99.41	99.42	99.43	99.44	99.45	99.46	99.47	99.47	99.48	99.49	99.49	99.49	99.50	99.50	
3	10.13	9.55	9.28	9.12	9.01	8.94	8.89	8.85	8.81	8.79	8.76	8.74	8.71	8.69	8.66	8.64	8.62	8.59	8.58	8.56	8.55	8.54	8.53	8.53	3
	34.12	30.82	29.46	28.71	28.24	27.91	27.67	27.49	27.35	27.23	27.13	27.05	26.92	26.83	26.69	26.60	26.50	26.41	26.35	26.28	26.24	26.18	26.15	26.13	
4	7.71	6.94	6.59	6.39	6.26	6.16	6.09	6.04	6.00	5.96	5.94	5.91	5.87	5.84	5.80	5.77	5.75	5.72	5.70	5.68	5.66	5.65	5.64	5.63	4
	21.20	18.00	16.69	15.98	15.52	15.21	14.98	14.80	14.66	14.55	14.45	14.37	14.25	14.15	14.02	13.93	13.84	13.75	13.69	13.61	13.58	13.52	13.49	13.46	
5	6.61	5.79	5.41	5.19	5.05	4.95	4.88	4.82	4.77	4.74	4.70	4.68	4.64	4.60	4.56	4.53	4.50	4.46	4.44	4.42	4.41	4.39	4.37	4.36	5
	16.26	13.27	12.06	11.39	10.97	10.67	10.46	10.29	10.16	10.05	9.96	9.89	9.77	9.68	9.55	9.47	9.38	9.29	9.24	9.17	9.13	9.08	9.04	9.02	
6	5.99	5.14	4.76	4.53	4.39	4.28	4.21	4.15	4.10	4.06	4.03	4.00	3.96	3.92	3.87	3.84	3.81	3.77	3.75	3.73	3.71	3.69	3.58	3.67	6
	13.75	10.92	9.78	9.15	8.75	8.47	8.26	8.10	7.98	7.87	7.79	7.72	7.60	7.52	7.40	7.31	7.23	7.14	7.09	7.02	6.99	6.93	6.90	6.88	
7	5.59	4.74	4.35	4.12	3.97	3.87	3.79	3.73	3.68	3.64	3.60	3.57	3.53	3.49	3.44	3.41	3.38	3.34	3.32	3.29	3.27	3.25	3.24	3.23	7
	12.25	9.55	8.45	7.85	7.46	7.19	6.99	6.84	6.72	6.62	6.54	6.47	6.36	6.28	6.16	6.07	5.99	5.91	5.86	5.79	5.75	5.70	5.67	5.65	
8	5.32	4.46	4.07	3.84	3.69	3.58	3.50	3.44	3.39	3.35	3.31	3.28	3.24	3.20	3.15	3.12	3.08	3.04	3.02	2.99	2.97	2.95	2.94	2.93	8
	11.26	8.65	7.59	7.01	6.63	6.37	6.18	6.03	5.91	5.81	5.73	5.67	5.56	5.48	5.36	5.28	5.20	5.12	5.07	5.00	4.96	4.91	4.88	4.86	
9	5.12	4.26	3.86	3.63	3.48	3.37	3.29	3.23	3.18	3.14	3.10	3.07	3.03	2.99	2.94	2.90	2.86	2.83	2.80	2.77	2.76	2.73	2.72	2.71	9
	10.56	8.02	6.99	6.42	6.06	5.80	5.61	5.47	5.35	5.26	5.18	5.11	5.01	4.92	4.81	4.73	4.65	4.57	4.52	4.45	4.41	4.36	4.33	4.31	
10	4.96	4.10	3.71	3.48	3.33	3.22	3.14	3.07	3.02	2.98	2.94	2.91	2.86	2.83	2.77	2.74	2.70	2.66	2.64	2.60	2.59	2.56	2.55	2.54	10
	10.04	7.56	6.55	5.99	5.64	5.39	5.20	5.06	4.94	4.85	4.77	4.71	4.60	4.52	4.41	4.33	4.25	4.17	4.12	4.05	4.01	3.96	3.93	3.91	
11	4.84	3.98	3.59	3.36	3.20	3.09	3.01	2.95	2.90	2.85	2.82	2.79	2.74	2.70	2.65	2.61	2.57	2.53	2.51	2.47	2.46	2.43	2.42	2.40	11
	9.65	7.21	6.22	5.67	5.32	5.07	4.89	4.74	4.63	4.54	4.46	4.40	4.29	4.21	4.10	4.02	3.94	3.86	3.81	3.74	3.71	3.66	3.62	3.60	
12	4.75	3.89	3.49	3.26	3.11	3.00	2.91	2.85	2.80	2.75	2.72	2.69	2.64	2.60	2.54	2.51	2.47	2.43	2.40	2.37	2.35	2.32	2.31	2.30	12
	9.33	6.93	5.95	5.41	5.06	4.82	4.64	4.50	4.39	4.30	4.22	4.16	4.05	3.97	3.86	3.78	3.70	3.62	3.57	3.50	3.47	3.41	3.38	3.36	
13	4.67	3.81	3.41	3.18	3.03	2.92	2.83	2.77	2.71	2.67	2.63	2.60	2.55	2.51	2.46	2.42	2.38	2.34	2.31	2.28	2.26	2.23	2.22	2.21	13
	9.07	6.70	5.74	5.21	4.86	4.62	4.44	4.30	4.19	4.10	4.02	3.96	3.86	3.78	3.66	3.59	3.51	3.43	3.38	3.31	3.27	3.22	3.19	3.17	

14	4.60	3.74	3.34	3.11	2.96	2.85	2.76	2.70	2.65	2.60	2.57	2.53	2.48	2.44	2.39	2.35	2.31	2.27	2.24	2.21	2.19	2.16	2.14	2.13	**14**
	8.86	6.51	5.56	5.04	4.69	4.46	4.28	4.14	4.03	3.94	3.86	3.80	3.70	3.62	3.51	3.43	3.35	3.27	3.22	3.15	3.11	3.06	3.03	3.00	
15	4.54	3.68	3.29	3.06	2.90	2.79	2.71	2.64	2.59	2.54	2.51	2.48	2.42	2.38	2.33	2.29	2.25	2.20	2.18	2.14	2.12	2.10	2.08	2.07	**15**
	8.68	6.36	5.42	4.89	4.56	4.32	4.14	4.00	3.89	3.80	3.73	3.67	3.56	3.49	3.37	3.29	3.21	3.13	3.08	3.01	2.98	2.92	2.89	2.87	
16	4.49	3.63	3.24	3.01	2.85	2.74	2.66	2.59	2.54	2.49	2.46	2.42	2.37	2.33	2.28	2.24	2.19	2.15	2.12	2.09	2.07	2.04	2.02	2.01	**16**
	8.53	6.23	5.29	4.77	4.44	4.20	4.03	3.89	3.78	3.69	3.62	3.55	3.45	3.37	3.26	3.18	3.10	3.02	2.97	2.90	2.86	2.81	2.78	2.75	
17	4.45	3.59	3.20	2.96	2.81	2.70	2.61	2.55	2.49	2.45	2.41	2.38	2.33	2.29	2.23	2.19	2.15	2.10	2.08	2.04	2.02	1.99	1.97	1.96	**17**
	8.40	6.11	5.18	4.67	4.34	4.10	3.93	3.79	3.68	3.59	3.52	3.46	3.35	3.27	3.16	3.08	3.00	2.92	2.87	2.80	2.76	2.71	2.68	2.65	
18	4.41	3.55	3.16	2.93	2.77	2.66	2.58	2.51	2.46	2.41	2.37	2.34	2.29	2.25	2.19	2.15	2.11	2.06	2.04	2.00	1.98	1.95	1.93	1.92	**18**
	8.29	6.01	5.09	4.58	4.25	4.01	3.84	3.71	3.60	3.51	3.43	3.37	3.27	3.19	3.08	3.00	2.92	2.84	2.78	2.71	2.68	2.62	2.59	2.57	
19	4.38	3.52	3.13	2.90	2.74	2.63	2.54	2.48	2.42	2.38	2.34	2.31	2.26	2.21	2.16	2.11	2.07	2.03	2.00	1.96	1.94	1.91	1.89	1.88	**19**
	8.18	5.93	5.01	4.50	4.17	3.94	3.77	3.63	3.52	3.43	3.36	3.30	3.19	3.12	3.00	2.92	2.84	2.76	2.71	2.64	2.60	2.55	2.51	2.49	
20	4.35	3.49	3.10	2.87	2.71	2.60	2.51	2.45	2.39	2.35	2.31	2.28	2.22	2.18	2.12	2.08	2.04	1.99	1.97	1.93	1.91	1.88	1.86	1.84	**20**
	8.10	5.85	4.94	4.43	4.10	3.87	3.70	3.56	3.46	3.37	3.29	3.23	3.13	3.05	2.94	2.86	2.78	2.69	2.64	2.57	2.54	2.48	2.44	2.42	
21	4.32	3.47	3.07	2.84	2.68	2.57	2.49	2.42	2.37	2.32	2.28	2.25	2.20	2.16	2.10	2.05	2.01	1.96	1.94	1.90	1.88	1.84	1.83	1.81	**21**
	8.02	5.78	4.87	4.37	4.04	3.81	3.64	3.51	3.40	3.31	3.24	3.17	3.07	2.99	2.88	2.80	2.72	2.64	2.58	2.51	2.48	2.42	2.38	2.36	
22	4.30	3.44	3.05	2.82	2.66	2.55	2.46	2.40	2.34	2.30	2.26	2.23	2.17	2.13	2.07	2.03	1.98	1.94	1.91	1.87	1.85	1.82	1.80	1.78	**22**
	7.95	5.72	4.82	4.31	3.99	3.76	3.59	3.45	3.35	3.26	3.18	3.12	3.02	2.94	2.83	2.75	2.67	2.58	2.53	2.46	2.42	2.36	2.33	2.31	
23	4.28	3.42	3.03	2.80	2.64	2.53	2.44	2.37	2.32	2.27	2.24	2.20	2.15	2.11	2.05	2.01	1.96	1.91	1.88	1.84	1.82	1.79	1.77	1.76	**23**
	7.88	5.66	4.76	4.26	3.94	3.71	3.54	3.41	3.30	3.21	3.14	3.07	2.97	2.89	2.78	2.70	2.62	2.54	2.48	2.41	2.37	2.32	2.28	2.26	
24	4.26	3.40	3.01	2.78	2.62	2.51	2.42	2.36	2.30	2.25	2.22	2.18	2.13	2.09	2.03	1.98	1.94	1.89	1.86	1.82	1.80	1.77	1.75	1.73	**24**
	7.82	5.61	4.72	4.22	3.90	3.67	3.50	3.36	3.26	3.17	3.09	3.03	2.93	2.85	2.74	2.66	2.58	2.49	2.44	2.37	2.33	2.27	2.24	2.21	
25	4.24	3.39	2.99	2.76	2.60	2.49	2.40	2.34	2.28	2.24	2.20	2.16	2.11	2.07	2.01	1.96	1.92	1.87	1.84	1.80	1.78	1.75	1.73	1.71	**25**
	7.77	5.57	4.68	4.18	3.85	3.63	3.46	3.32	3.22	3.13	3.06	2.99	2.89	2.81	2.70	2.62	2.54	2.45	2.40	2.33	2.29	2.23	2.19	2.17	
26	*4.23*	*3.37*	*2.98*	*2.74*	*2.59*	*2.47*	*2.39*	*2.32*	*2.27*	*2.22*	*2.18*	*2.15*	*2.09*	*2.05*	*1.99*	*1.95*	*1.90*	*1.85*	*1.82*	*1.78*	*1.76*	*1.73*	*1.71*	*1.69*	***26***
	7.72	5.53	4.64	4.14	3.82	3.59	3.42	3.29	3.18	3.09	3.02	2.96	2.86	2.78	2.66	2.58	2.50	2.42	2.36	2.29	2.25	2.19	2.16	2.13	

ANSWERS TO TEXTBOOK EXERCISES

CHAPTER 1 COMBINATORIAL PROBABILITY

Exercises

1-1	(c) 1/3					
1-2	(a) 6	(b) .4167				
1-3	(b) .5357	(c) .6429				
1-4	shoes – 48 outfits					
1-5	(a) 216	(b) 6	(c) .0278			
1-6	1,000,000,000					
1-7	137,540,000					
1-8	750					
1-9	7,200,000,000					
1-10	(a) 7!	(b) 13!	(c) 840	(d) 20	(e) 24	(f) Undefined
	(g) 10,626	(h) 1	(i) 120	(j) 10	(k) 35	(l) 1
1-11	(a) 17!	(b) 16!	(c) .0588			
1-12	.002976					
1-13	120					
1-14	11,880					
1-15	23					
1-16	4					
1-17	.1114					
1-18	.1183					
1-19	5					
1-20	(a) 7	(b) 13	(c) 13	(d) 10	(e) 1	(f) Undefined
	(g) 10	(h) 1	(i) 1	(j) 5867	(k) 35	(l) 1
1-21	220					
1-24	(a) 468	(b) 252	(c) 210			
1-25	(a) 3003	(b) 600				
1-26	825					

1-27 114,660

1-28 (a) 8550 (b) .00187

1-29 (a) .01997 (b) .2697

1-30 600

1-31 360,360

1-32 .08333

1-33 30

1-34 39,916,800

1-35 2.4×10^{34}

1-36 (a) 823,727,520 (b) 411,863,760

1-37 2.495×10^{11}

1-38 .02778

1-39 3360

1-40 560

1-41 4096

1-42 816

1-43 Ordered sample without replacement (3360)

1-45 .01961

1-46 (a) 21 (b) 6

1-47 (a) 34,320 (b) 1680

1-48 1001

1-49 ${}_{52}C_{13}$

1-50 2,598,960

1-51 166

1-52 ${}_{36}C_{21} \cdot {}_{45}C_{21} = 2.1 \times 10^{22}$

1-53 (a) $x^4 + 12x^3 + 54x^2 + 108x + 81$ (c) $64x^3 - 240x^2y + 300xy^2 - 125y^3$

1-57 15,504. The value of n is 19.

1-58 (a) $x^3 - 6x^2y + 12xy^2 - 8y^3 + 15x^2z - 60xyz + 60y^2z + 75xz^2 - 150yz^2 + 125z^3$

1-60 (a) .00077 (b) .01929 (c) .03086 (d) .03858
(e) .06173 (f) .15432 (g) .23148 (h) .46296

1-62 $-.4118$

Chapter 1 Sample Examination

1. 65,780
2. 7,990,000
3. 2,730
4. 1365
5. (a) ${}_{51}C_{17}$ (b) .30282
6. (a) 2,598,960 (b) 111,540
7. (a) 525 (b) .408
8. 18
9. 80,089,128
10. $32x^5 - 80x^4y + 80x^3y^2 - 40x^2y^3 + 10xy^4 - y^5$

12. .00995
13. .5152
14. .03985
15. .3846
16. C

CHAPTER 2 GENERAL RULES OF PROBABILITY

Exercises

2-1	(b) {3,5,7}	(c) {1,2,3,5,7,9}	(d) {1,9}
2-2	(a) {1,2}	(b) 2	(c) {water}
2-5	.97187		
2-6	.4444		
2-7	C		
2-14	4		
2-18	7		
2-19	B		
2-20	D		
2-21	D		
2-22	A		

2-23 D

2-24 D

2-25 B

2-26 (a) .5833 (b) .7143 (c) .3571

2-27 (a) .333 (b) .667

2-28 E

2-29 C

2-30 C

2-31 B

2-32 (a) .47575 (b) .8075

2-33 B

2-34 .3656

2-35 .1538

2-36 C

2-39 .2725

2-40 Dependent

2-42 Not necessarily

2-43 (c) .25 (d) Yes (e) .25 (f) No

2-44 .7 or .8

2-45 (b) .5 (c) Independent

2-46 (b) .6 (c) Independent

2-47 D

2-48 .0002

2-49 (a) .3226 (b) .5484 (c) .2258 (d) .2143 (e) .222 (f) No

2-50 (a) .1 (b) .6

2-51 B

2-52 .04977

2-53 (a) .4737 (b) .00253

2-54 E

2-55 A

2-56 (a) .07692 (b) .07843 (c) .07843

2-57 (b) .12941 (c) .2 (d) .07692 (e) .3882

2-58	(a) .058	(b) .32222	(c) .13793	(d) .497	(e) .9248		
2-59	D						
2-60	D						
2-61	D						
2-62	D						
2-63	B						

Chapter 2 Sample Examination

2.	.00144						
3.	(a) 17	(b) 65	(c) Dependent				
4.	(b) .28125	(c) .65625	(d) .26667	(e) Dependent			
5.	(a) .000181	(b) .999819	(c) .04	(d) .01370			
6.	(b) .125	(c) .4375	(d) 0	(e) 0			
7.	(b) .3	(c) .6	(d) .2	(e) .333			
8.	(b) .16667	(c) .8	(d) .6667				
9.	(a) .4	(b) .6	(c) .25	(d) .38462	(e) .8	(f) .625	(g) Dependent
10.	.15625						
11.	.234375						
12.	.875						
13.	.936						
14.	.8667						
15.	.333						
16.	.375						
17.	1						
18.	.67774						
19.	(a) .4	(b) .4259					
20.	C						
21.	C						
22.	B						
23.	B						
24.	D						
25.	D						
26.	D						

27. E

28. B

CHAPTER 3 DISCRETE RANDOM VARIABLES

Exercises

3-1 $\Pr(Y=0)=.159$ $\Pr(Y=1)=.477$ $\Pr(Y=2)=.318$ $\Pr(Y=2)=.045$

3-2 $\Pr(Z=i)=.1$ for $i=1,2,\cdots,10$

3-3 (a) .4737 (b) .1312 (c) $\Pr(S=s)=\left(\frac{20}{38}\right)^{s-1}\left(\frac{18}{38}\right)$ for $s=1,2,3,\cdots$

3-4 (a) Same as 3-3(c)

3-6 (a) $\Pr(A=20)=.41$

3-7 (a) $\Pr(H=h)={}_4C_h\cdot .3^{4-h}\cdot .7^h$ for $h=1,2,3,4$

3-9 .35

3-10 $E[W]=1.091$ $E[W^2]=1.636$

3-11 42 cents

3-12 48.64

3-13 2.11

3-14 D

3-15 0.7

3-16 C

3-17 (a) 60th (b) 44,456 (c) 48,250 (d) 68.48^{th}

3-18 (a) 25th to 75th

3-19 (a) 6.05 (b) 6.5 (c) 6.5 (d) 5.75 (e) 6.5

3-20 (b) 7 (c) all are 7 (d) 5 and 9

3-21 The ratio $m:n$ of girls to boys is 3.51:1.

3-22 (a) 166.14; 162; 162; 168 (b) 153 and 183 (c) 154; 162; 181
(d) 162 (e) 30 and 27

3-23 (a) 1.5 (b) .75

3-24 5.833

3-25 5.091; 2; 12; 2; 24; 7; 42.45

3-26 110.55

3-27 (b) 1.2051

3-28 (a) 79.67; 81; no mode; 75 (b) 48 and 102 (c) 94.2 (d) 15.43

3-29 A

3-30 50%

3-31 (a) 1.5625 (b) 72

3-32 Jen's orange.

3-34 33.7% and 25.6% respectively.

3-35 34.5%

3-36 (a) 4.0768 and .7871 (b) 4.26 and .7271

3-37 E

3-38 7.45

3-39 2900

3-40 (b) 1800 (c) 3333.33 (d) 2981.42^2

3-41 (a) .5 and .25 (b) 1 and .5 (c) $E[X+Y]=1.5$ (d) $1.25 \neq .75$

3-42 (a) Yes (b) 5.833

3-43 (a) 1 and .5 (b) .667 and .444 (c) 1.667 and .9444 (d) Yes

3-44 (b) .5 and .25 (c) 1.5 and .75

3-45 (a) 70.54 and 9.9 (b) 10.6 (c) 14.36%

3-46 (b) 1.5 and .75

3-47 (a) 3.5 (b) 1.025

3-48 (a) 1.25 (b) 44.72%

Chapter 3 Sample Examination

1. (b) 6 (c) 2 (d) 1.46
2. 1
3. (a) 6.5; 7; 9; 5 (b) 0 and 10 (c) 9 (d) $\sigma_C = 2.598$ (e) 39.97% (f) 1.73
4. (a) 25 (b) $\sigma_X = 50.74$ (d) -50 (e) 257,500 (f) 203%
5. D
6. (a) 6.13 (b) 2.4998
7. (a) .828 (b) .8

8.	(a) 19.47; 20; 20; 20.5	(b) 5; 1.23; 1.11	(c) 20.346	
9.	(a) 4.11	(b) 2.78	(c) 5	
10.	(a) 2592.59	(b) 472.13		
11.	(a) 2.71; 3; and .941	(b) 101.69		
12.	(a) 14.417; 14; 15 and 15	(b) 2	(c) $z = -2.845$	
13.	917.50 and 988.09^2			
14.	(a) mean is 12 and standard deviation is 29.5973	(b) 40	(c) 247%	
15.	115.48%			
16.	C			

CHAPTER 4 SOME DISCRETE DISTRIBUTIONS

Exercises

4-3	Option (b)			
4-4	$n(n+1)^2 / 4$			
4-5	0			
4-6	(a) .333	(b) 3.5	(c) 3.5	(d) 1.708
4-7	(a) .17	(b) 50.5	(c) None	(d) 833.25
4-8	(a) .444	(b) 4	(c) 64.55%	
4-9	(a) 7/9	(b) 10	(c) 5.164	
4-10	(a) 7/9	(b) 60	(c) 7.071	
4-12	(a) .381	(b) .4762		
4-14	$E[M]$ is 9.6; $Var[M] = 1.92$; and $\Pr = 44.17\%$			
4-15	.002; .931; 4; 3.2			
4-16	64.8%			
4-17	.027; .972; 2.7 and .27			
4-18	.1612; .7682; 3.6 and 2.16			
4-19	.06695; 5; .4148 and 1.9365			
4-20	A			
4-21	E			
4-22	E			
4-23	(a) .75	(b) .2503		

4-24 .428

4-26 D

4-27 E

4-28 .96; .25 and .375

4-29 .19; 9 and 9.49. Independence.

4-30 1.5

4-32 .64

4-33 No

4-34 1.41

4-35 $\Pr(Y = y) = q^{y-1}p$ for $y = 1, 2, 3, \cdots$

4-36 $1/p;\ \ q/p^2$

4-37 .0767

4-38 (c) 9.6 and 8.2365

4-39 B

4-40 E

4-41 D

4-43 rq/p and rq/p^2

4-44 (a) .1087 (b) .2352 (c) 5.571 and 15.92 (d) 8.571

4-45 (a) .00645 (b) 28 (c) 11.832

4-46 .1445

4-47 $\Pr(M = k) = {}_{r+k+1}C_k \cdot p^r \cdot q^k$ for $k = 0, 1, 2, \cdots$

4-48 D

4-49 B

4-50 .2092; 3.293 and 1.006

4-51 .0118

4-55 C

4-56 6.9

4-58 2

4-59 .1991

4-60 B

4-61 (a) .4755 and .4592

4-62 (a) .9288 and .9342

4-63 (b) 2 (c) 2.5

4-64 (a) 75 (b) $e^{-75}75^{70}/70!$

4-65 (a) .1247 (b) 5 (c) 2.236

4-67 The greatest integer in λ.

4-68 1.358×10^{-6}

4-69 D

4-70 .0996; 8 and 8

4-71 (a) 15 (b) .0008566

4-72 .00054

4-73 (b) No (c) .333

4-74 C

4-79 (a) .72 and .7416 (b) Poisson (c) .5894

4-80

	(1)	(2)	(3)	(4)	(5)	(6)
(a)	0.6000	0.6665	0.6941	0.7977	0.8665	0.7305
(b)	1.0000	0.9817	0.9900	0.9786	0.9437	0.9453
(c)	1.0000	0.9983	0.9996	0.9972	0.9822	0.9888

Chapter 4 Sample Examination

1. (a) .04365 (b) 1.1726
2. (a) .1667 (b) 5.58% (c) 3.35%
3. (a) .4167 (b) 88.5 (c) 47.9167
4. (a) .36 (b) 1.714 (c) 12.378
5. (a) 4 (b) 2 (c) .0183 (d) .762
6. −13,183.52
7. (a) .2240 (b) 1.17×10^{-10}
8. A
9. E
10. D
11. .117
12. B

13. C
14. A
15. E

CHAPTER 5 CALCULUS, PROBABILITY, AND CONTINUOUS DISTRIBUTIONS

Exercises

5-1	(b) $f(x) = 4e^{-4x}$ for $x \geq 0$		(c) .01798	(d) .01832	
5-2	(c) .124	(d) .992			
5-3	(a) 3	(c) .002355	(d) 1		
5-4	(a) 86	(b) 0	(c) .125		
5-5	A				
5-6	B				
5-7	C				
5-8	(a) 3.2	(b) 10.667	(d) .2539		
5-9	C				
5-10	.711				
5-11	.4267				
5-12	35.36%				
5-13	.2211				
5-14	(a) 2.0207	(b) 2.302			
5-15	(b) 3.333				
5-16	E				
5-17	.5				
5-18	.2929 and 1.7071				
5-19	(b) .667	(c) 1	(d) .7071		
	(e) .8832	(f) .366			
5-20	(a) 4	(c) .5333	(d) .57735		
	(e) .2929	(f) .333			
5-21	mean = 3.1416	median = 3.1416	modes = 1.571 and 4.712		
5-23	(a) .05 for $0 \leq w \leq 20$	(c) 10	(d) None	(e) 10	(f) 13.8

5-24 (a) 3.444; 3.6056 and 5

5-25 D

5-26 B

5-27 D

5-28 C

5-29 .05

5-30 (a) 80,000 and 17,321 (b) 80,000 and 16,966

5-31 D

5-32 C

5-33 B

5-34 C

5-35 C

5-36 C

5-37 9.9667

5-38 29.333

5-39 30

5-40 A

5-41 B

5-42 $\mu_x = .5$ $\sigma_x = .5$

5-43 $\mu_x = 3.8$ $\sigma_X^2 = 2.16$

5-44 (a) $.3 + .7e^t$ (b) .7 and .21

5-45 (a) $.5e^{-t} + .2e^{2t} + .3e^{5t}$ (b) 1.4 (c) $.5e^{-7t} + .2e^{2t} + .3e^{11t}$

5-46 (a) .5 and .45 (c) 0, 0, and 1

5-47 (a) 4 and .8944 (b) -25 and 6.2610

5-48 A

5-50 $MGF = \frac{3}{3-t}$ mean = .333 variance = .111

5-51 B

5-52 E

5-53 q/p and q/p^2

5-54 rq/p and rq/p^2

5-55 (a) 9.333 and 5.578 (b) .0397

5-56 (b) $e^{6e^{2t}+3t-6}$

5-58 3

Chapter 5 Sample Examination

1. (a) 3.6 and -2.4 (b) .2368 (c) .245 (d) .75
2. (a) .7769 (b) .231; 0 and .333 (c) .3499 (d) .222 (e) .333
3. (a) .844 (b) 0, 0, and 0 (c) .2028 (d) .2
4. 16.98%
5. .1667
6. C
7. 250; 0 and 250
8. C
9. 10.118; 6.667 and 0
10. (a) 12 (b) .0272 (c) .04 (d) .3333
11. (a) .551 (b) 0 (c) 1 (d) 1.6094 (e) 1
12. (a) 140 (b) 64
13. (a) $5/(5-t)$ (b) .2
14. (b) 2.5333
15. (a) 1.054 (b) .0384
16. B

CHAPTER 6 SOME CONTINUOUS DISTRIBUTIONS

Exercises

6-1 (b) 75 and 75 (c) 80.7 and 67.5 (d) .577

6-2 (b) 1.5 and 4.0833 (c) 1.5 and No Mode (d) .2857 (e) 0

6-3 (a) $f(t) = .2$ and $F(t) = t/5$ for $0 \le t \le 5$
(b) 2.5 hours (2:30P) and 1.4434 (c) .2 (d) .333

6-4 (a) .5774 (b) 1

6-5 mean = 10 standard deviation = 2.887

6-6 B

6-8 .5

6-9 (a) 5 (b) 3.466 (c) 0 (d) .2699 (e) .632

6-10 (b) .6321 (c) 519.86

6-12 (a) 10.099 (b) 0 (c) .09902 (d) .3905

6-13 10.986

6-14 2058

6-15 1944

6-16 −5,773,554

6-17 D

6-18 C

6-19 D

6-20 E

6-21 D

6-22 C

6-23 C

6-24 E

6-25 (a) .8375 (b) .1091 (c) .591 (d) .5346 (e) .0404

6-26 (a) 1.28 (b) 1.645 (c) 2.33 (d) -1.645 (e) .675
(f) does not exist

6-27 (a) .8 (b) 1.66

6-28 (a) .9772 (b) 81.1 inches

6-29 130.825

6-30 (a) 177.08 minutes (b) 252.08

6-31 (a) .1357 (b) 20.787

6-32 (a) .7764 (b) .6618 (c) 49.8

6-33 (a) −.333 (b) 53.56 and 78.64

6-34 (a) .5222 (b) 30.115

6-35 (a) A grade of 'C'

6-36 .0304

6-37 (a) 336.972 and 365,710.5 (b) .1647 (c) 763

6-38 (b) .3238

6-39 (a) 1.197 and .2487 (b) .4714 (c) .5724

6-40 .1215 is the approximation to .1144

6-41 (a) .0108 (b) .0179

6-42 .9484

6-43 B

6-44 (a) 75 and 208.33 (b) 75 and 2.78 (c) .1 and .849

6-45 (a) .7888 (b) .9876

6-46 D

6-47 C

6-48 B

6-49 C

6-50 (a) 135 and 9 (b) .9772

6-51 B

6-52 B

6-53 (a) 3.323 (b) 24 c) 1

6-54 1

6-55 30

6-56 (a) .1954 (b) .3679 (c) .6321 (d) .2325
(e) .265

6-57 (a) .1839 (b) 5

6-58 (a) 5 and 2 (b) $f(x)=\frac{1}{768}x^4e^{-\frac{1}{2}x}$; $x \geq 0$ (c) .1847
(d) −560

6-59 (a) 12 and 24 (b) 2688 (c) 3.393

6-60 (a) .857 (b) $f(x)=\frac{1}{6}t^3e^{-t}$; $t \geq 0$ (c) .143

6-61 (a) .5 (b).5188 (c) .5188

6-62 .05

6-63 (a) .0002525 (b) .1524

6-64 (a) 42 (b) .25 (c) .0625 (d) .138

6-65 (a) $F(x) = 3x^2 - 2x^3$; $0 \le x \le 1$ (b) .3 (c) .2236 (d) .5
(e) .5

6-66 (a) 19.69 (b) .4545 (c) .1953 (d) .582

6-67 $2.\overline{84}$

6-68 (a) 1 and 5 (b) 5 (d) \$242 (e) .1667 and .1409 (f) .00032

6-69 median = .26445 95th percentile = .5818

6-70 (a) .1004 (b) 2.5 (c) −2.5

6-71 (a) 5.6 (b) 5.9733 (c) 4.876 (d) .9415

6-72 (a) 5.77 (b) 1.28

6-73 (a) 1000 (b) .2963 (c) 519.84

6-74 (a) 4.5 (b) .7037 (c) 3.78

6-75 A

6-76 C

6-77 C

6-78 (a) .1353 (b) $2^{\frac{1}{3}} \times \Gamma\left(\frac{4}{3}\right)$

Chapter 6 Sample Examination

1. (a) 100; 100; No Mode and 100 (b) 200; 100; 3333.33 and 57.735
2. .3015
3. 481.88
4. 52,254,720
5. .75
6. 164.02
7. 478.72
8. .0064
9. D
10. 468.95
11. D
12. B
13. A

CHAPTER 7 MULTIVARIATE DISTRIBUTIONS

Exercises

7-1	(a) $e^{-3}3^x / x!$ for $x = 0,1,2,\cdots$		(b) 3		
7-2	7.45				
7-3	(a) .02778	(b) 1.253			
7-4	B				
7-5	(c) .5	(d) 3.91			
7-6	(b) 2.22	(c) .6173	(d) .5544	(e) 1.833	(f) .842
7-7	E				
7-8	C				
7-9	E				
7-10	.15				
7-12	.667				
7-13	3.2				
7-14	10				
7-15	C				
7-16	A				
7-17	.132				
7-18	$-21Cov[X,Y]$				
7-19	(a) 0	(b) $-.333$			
7-20	(a) .36	(b) .36	(c) $-.36$	(d) 0	
7-21	(a) Yes	(b) .75	(c) .25	(d) .125	
7-22	$-.248$				
7-23	$\rho = -1$				
7-24	E				
7-25	.0833				
7-26	(b) 0 and .4				
7-27	E				
7-28	D				
7-29	C				

7-30 B

7-31 D

7-32 E

7-33 Not a density function

7-34 .3

7-35 .6027

7-36 C

7-37 A

7-38 D

7-39 C

7-40 D

7-41 (c) .2691

7-42 (a) $f(x,y)=\frac{1}{x};\ 0<y<x<1$ (b) .5 (c) .5
(d) .25 (e) .8690

7-43 $(2.5-2y)/(4-3y)$

7-44 $y^2/3$

7-45 .866

7-46 D

7-47 E

7-48 .1125

7-49 (c) .6928

7-50 A

7-51 E

7-52 .333

7-53 (b) .5 (c) .5 (e) Dependent

7-54 D

7-55 C

7-56 B

7-57 A

7-58 C

7-59 D

7-60 C

7-61 2.75

7-62 .659

7-63 (a) .8 (b) .3085

7-64 D

7-65 E

7-66 E

Chapter 7 Sample Examination

1. 1/2

2. (b) 0 and .667 (c) .71875 (d) 0

3. (b) Both equal .5833 (c) Dependent (d) $(x+y)/(y+.5)$
(e) .5694 (f) 2.33 (g) $-.\overline{09}$ (h) 4.72

4. (a) .0303 (d) 1.727 and 2.182 (e) Dependent (g) 1.727
(h) 3.758 (i) $-.03108$ (j) .198 and .634 (k) $-.0110$ (l) 15.88

5. (a) 5 (b) 58

6. -6

7. (a) 27.333 (b) 2.768 (which is not possible, so the given information must be inconsistent)

8. D

9. C

10. D

11. C

12. C

13. B

14. C

15. B

16. B

17. C

18. A

19. $\frac{1}{800}\int_0^{60}\int_0^{60-x} e^{-x/40}e^{-y/20}\,dydx$ or $\int_0^{60}\int_0^{60-y} \frac{1}{800}e^{-x/40}e^{-y/20}\,dxdy$

20. C

CHAPTER 8 A PROBABILITY POTPOURRI

Exercises

8-2 (a) $1/y$ for $1 \le y \le e$ (b) 1.7183

8-3 (b) .88623 (c) .2146

8-4 (b) 1.341

8-5 E

8-6 (b) $1/4y^2$ (c) .4026 (d) .0381 and .1952 (e) No

8-7 .03125

8-8 A

8-9 E

8-10 E

8-11 A

8-12 $\lambda^2 xe^{-\lambda x}$

8-14 (b) 3

8-15 E

8-16 6; 8; 10; 6 and 8

8-17 8; 10; 10; 8 and 10

8-18 18.98; 12.06; -3.24; 22.03 and 4.48

8-19 8; 7; 5; 9 and 6

8-21 (b) 5.2034 and 3.939 (c) 5.1166 and 2.6746

8-22 (a) 12 (b) $6x^2 - 8x^3 + 3x^4$ (d) .3895 and .2582 (e) .40 and .20

8-24 (a) $(10-t)/50$ (b) .51 (c) 3.33 (d) 2.357

8-25 (a) .65 (b) 0

8-26 (a) 2.5 (b) 1.9365

8-27 (a) 3.75 (b) $5n/(n+1)$ (c) .936

8-28 241

8-29 (a) .2717 (b) .3012

8-30 22.5

8-31 (a) 7.4651 (b) 24.746

8-32 (b) .6425 (c) 2.917 (d) 2.917 (g) $9.08\overline{3}$ (h) 6.963

8-33 D

8-34	E		
8-35	C		
8-36	E		
8-37	B		
8-38	D		
8-39	E		
8-40	B		
8-41	E		
8-44	-20		
8-45	-56		
8-46	(a) 56.4	(b) 35.96	
8-47	(a) 2.5	(b) 3.323	(c) .5742
8-49	$12,500p$ and $175,000,000p - 156,250,000p^2$		
8-50	B		
8-51	(b) 5.2	c) 2.1	
8-52	A		
8-53	.79		
8-54	A		
8-55	B		
8-56	26/27		
8.57	B		
8.58	B		
8.59	D		
8.60	C		
8.61	D		
8.62	C		
8.63	A		
8.65	C		
8.66	C		
8.67	D		
8.68	B		

8.69 D

8.70 C

Chapter 8 Sample Examination

1. $3(\ln y)^2 / 8y$
2. (c) 1.875 (d) 3.516
3. (b) Normal random variable with mean 1 and variance 34 (c) .697
4. E
5. E
6. (a) .1354 (b) 1.43
7. E
8. E
9. C
10. D
11. B
12. C or D
13. C
14. B
15. 6.6

CHAPTER 9 SAMPLING DISTRIBUTIONS AND ESTIMATION

Exercises

9-1 (a) (2144.4,2555.6) (b) 2641.25

9-2 .9974

9-3 (a) 11 (b) 49

9-4 (a) .939 (b) .992

9-5 (a) 22 (b) 49

9-6 (a) 12.36 (b) 5.363

9-7 (a) (32.39,171.82) (b) (5.69,13.11)

9-8 22.82

9-9	(16.86,52.30)		
9-10	E		
9-11	(−55.9, −30.5)		
9-12	(a) 2.619	(b) .0924	(c) (2.557,2.681)
9-14	(1.86,19.28)		
9-15	(.455,1.785)		
9-16	(.882,3.882)		
9-18	(a) .69	(b) .01389	(c) (.6761, .7039)
9-19	(.575,.725)		
9-20	95%		
9-22	(a) (−.09,1.09)	(b) Not at the 95% level of confidence	
9-23	(−7.3, 21.3)		
9-24	(.240, 1.398)		
9-25	213		
9-26	278		
9-27	468		

Chapter 9 Sample Examination

1.	(a) 3 and 1.904	(b) (.64,5.36)	
2.	(a) 116	(b) 62	
3.	B		
4.	(a) 64.5 and 5.345	(b) (58.83,70.17)	(c) (3.77,9.6)
5.	(a) 21.6	(b) 49	
6.	B		
7.	E		
8.	B		
9.	A		
10.	B		
11.	A		
12.	D		
13.	A		

CHAPTER 10 HYPOTHESIS TESTING

Exercises

10-1 (a) God does not exist

(b) Rejecting God when He exists

(c) Believing God exists when He does not

10-2 A

10-3 D

10-4 .08

10-5 D

10-6 E

10-7 Do not reject at 5% level of significance

10-8 Reject the claim at 1% level of significance

10-9 No

10-10 D

10-11 E

10-12 Cannot reject his claim

10-13 Cannot reject the claim

10-14 E

10-15 E

10-16 Reject the null hypothesis and accept Lori's claim

10-17 C

10-18 A

10-19 A

10-20 E

10-21 D

10-22 B

10-23 E

Chapter 10 Sample Examination

1. D
2. C
3. D

4. A

5. A

6. D

7. B

8. E

9. C

10. D

11. B

12. E

CHAPTER 11 THEORY OF ESTIMATION AND HYPOTHESIS TESTING

Exercises

11-1 A

11-2 C

11-3 $-19\beta / 20$

11-4 $\overline{W} - 4$

11-5 E

11-6 A

11-7 C

11-8 E

11-9 $\overline{X}$

11-10 904.67

11-11 $1/(\overline{X}+1)$

11-12 $-n / \ln(x_1 \cdot x_2 \cdots x_n)$

11-13 $\{-n / \ln(x_1 \cdot x_2 \cdots x_n)\} - 1$

11-14 (b) $n\lambda / (n-1)$ (c) $\lambda / (n-1)$

11-15 B

11-16 E

11-17 D

11-18 E

11-19 D

11-20 C

11-21 A

11-22 D

11-23 C

11-24 E

11-25 C

11-26 (a) 0

11-27 E

11-28 Consistent

11-30 E

11-37 (b) $S = \sum_{i=1}^{n} \ln X_i = \ln T$ (c) $-\frac{S}{n}$

11-38 $\frac{S}{n}$

11-39 (a) If Grant make 73 or fewer shots (b) $\Pr(Z > 4.7)$

11-40 B

11-41 B

11-42 B

11-43 B

11-44 B

11-45 D

11-46 .9937

11-47 C

11-48 C

11-49 B

11-50 D

11-51 B

11-52 E

11-53 A

11-54 (a) .1584 (b) .0303 (c) .0352

11-55 .2126

11-56 .56

11-57 10,322

11-58 1.25

11-59 450

11-60- .5638

11-61 .1484

11-62 .6875

11-63 D

11-64 D

11-65 E

11-67 16.91

11-68 1.412

11-69 .01818

11-70 (a) $N(60.91, 6.545)$; $60.91 = (.8182)(60) + (.1818)(65)$

(b) (56.70, 65.12) (c) .9451

11-71 $\alpha^* = \alpha + n - 1,\ \beta^* = \dfrac{\beta}{1 + \beta\sum_{i=1}^{n} x_i}$

Chapter 11 Sample Examination

1. A
2. E
3. B
4. E
5. D
6. 3.25
7. B
8. E
9. E
10. C
11. B

12. (a) $\frac{\lambda}{n}$ (b) $-\frac{\tau^2 \ln \tau}{n} = \frac{\lambda e^{-2\lambda}}{n}$ (d) $\frac{S}{n}$ with Var $= \frac{\lambda}{n}$

(e) $a = \frac{n-1}{n}$ and $Var\left[a^S\right] = e^{-2\lambda}\left(e^{\lambda/n} - 1\right)$

13. 0.56

14. 20

15. .9091

16. $\frac{1}{q(1-q)}$

17. 140,000

18. 5.0735

19. 142.86

CHAPTER 12 A STATISTICS POTPOURRI

Exercises

12-1 (a) $\hat{Y} = 11.37 - .69 \cdot \hat{X}$ (b) 7.251

12-2 D

12-3 B

12-4 A

12-5 E

12-6 D

12-7 D

12-8 (a) -5.33 and .9 (b) .97 (c) .488 (d) .17 (e) .162

12-9 13.167

12-10 (a) $y = 2884.9x + 43{,}759$ (c) 58,184

12-11 (b) .998 (d) 39.99 thousand

12-12 (b) $-.9126$ (d) 13.2 thousand

12-13 (b) $-.986$

12-14 .2857

12-15 4

12-16 10.125

12-17 (a) (5.3,18.7) (b) $(-15,39)$

12-18 12.94 and 7.88

12-19 11.91

12-20 $F(3,16) = 2.9753$

12-21 .1123

12-22 (a) .2857 (b) .03968

12-23 -14

Chapter 12 Sample Examination

1. $Y = 1.08 + 265\sqrt{X}$
2. C or D
3. C
4. C
5. $F(2,9) = 4.445,\ F_{.05}(2,9) = 4.26$; reject null
6. $F(3,10) = 4.478,\ F_{.05}(3,10) = 3.71$; reject null,

   $F_{.01}(3,10) = 6.55$; do not reject null
7. .0577
8. .3145
9. .00884

11. $T = T^{-} = 50.5,\ T_{.05}(17) = 41$. Do not reject null
12. 9; .03
13. (a) .0788 (b) -.0222
14. (a) 0.6429 (b) 0.6429
15. E
16. D
17. D

INDEX